Human Visual Orientation

Human Visual Orientation

Ian P. Howard
York University, Toronto, Canada

1807 175 YEARS OF PUBLISHING 1982

JOHN WILEY & SONS
Chichester · New York · Brisbane · Toronto

British Library Cataloguing in Publication Data:

Howard, Ian P.
Human visual orientation.
1. Space perception
I. Title
152.1′42 QP491 80–41689

ISBN 0 471 27946 3

Photosetting by Thomson Press (India) Ltd., New Delhi.
Printed in the United States of America.

This book is dedicated to the memory
of my friend and colleague

W. B. Templeton

who died on August 31st, 1979

Acknowledgments

I wish to record my appreciation for the help I received from the following people in the preparation of this book. First and foremost my thanks are due to my wife Antonie, who patiently and thoroughly edited the whole text. My thanks are also due to Edward Keller, Joel Miller, Kenneth Money, Kenneth Nakayama, Christopher Tyler, and Hermann Schöne each of whom carefully read two or more chapters, and to Stuart Anstis, Fergus Campbell, Don Goodenough, Marc Jeannerod, Eileen Kowler, Janusz Kulikowski, Hiroshi Ono, Eric Sigman, Martin Steinbach, and Gerald Westheimer for making helpful comments on parts of the manuscript. Special thanks are also due to Mrs M. Adamson who typed the manuscript, to John Slavin who prepared many of the diagrams, and to my daughter Ruth who tracked down many of the references. I would also like to thank Dr Arthur Jampolsky and the governing body of the Smith-Kettlewell Institute of Visual Science in San Francisco for providing very pleasant office facilities during the year in which I completed the book.

Contents

Preface

This book was initially planned as a second edition of *Human Spatial Orientation*, which I wrote with W. B. Templeton in 1966. However, it soon became apparent that more work on this topic has been published since 1966 than had been published before that time. In order to make room for the new material, and at the same time make my task manageable, I decided to confine my attention to visual direction and orientation. Chapters on kinaesthesis and auditory localization have therefore been dropped. There is a chapter on the vestibular system, because this is required as a foundation for the chapter on vestibulo-ocular interactions. The chapter on geographical orientation has also been dropped, because this is not primarily a visual problem.

I have attempted to provide a fairly detailed coverage of the behavioural literature on the topics covered, and as much of the physiology as is required for the understanding of physiologically based theories of the bahaviour. I have provided references to major reviews on those topics which I treat only briefly, so that readers may pursue them in more detail if they wish.

Each time a technical term is mentioned for the first time in the text, it is printed in bold type and, where necessary accompanied by a definition.The page numbers where these definitions appear are also in bold type in the subject index. At various places throughout the book I suggest experiments which I think need to be done. These suggestions are printed in italics. If this book hastens the time when it will be out of date, I shall be well satisfied.

Preface

Chapter 1

Introduction

1.1 The problem

This book is about how human beings visually perceive the directions and orientations of objects. The perception of depth or distance is not discussed, and the perception of movement is discussed only insofar as it affects judgments about direction or orientation. Because I am a psychologist, my principal aim is to review the psychophysical and behavioural evidence bearing on these issues. But I also provide a fairly detailed account of physiological investigations in those areas where physiological procedures have been applied. Beyond Chapter 9, the physiological evidence becomes meagre or non-existent and very often the tasks are so little understood that it is not clear what type of physiological evidence would be relevant. I had intended to write a chapter on geographical orientation and the conception of spatial relationships, but I decided that these are not primarily visual problems, and, in any case, the book is long enough.

When people are asked to name the great unsolved problems of science, they usually mention cosmology, fundamental particles, and perhaps embryology—they rarely mention perception. Many people do not realize that perception is a problem; they perceive the world so effortlessly and continuously that they take the mechanism for granted. Perception is the most neglected of all the major problems of science, and this may be because it is the most difficult problem of them all. A good start has been made in understanding the sense organs, but hardly any progress has been made towards understanding the brain, which is where the perceptual mechanism resides. The problem lies beyond the 'art of the soluble' (Medawar, 1967), and frightens most scientists away. In recent years attempts have been made to simulate perceptual performance with computers, but even the most optimistic theorist will admit that very little progress has been made even towards defining the problem, let alone solving it (see Winston, 1975).

Any attempt to understand the mechanisms of perception involves a strange recursion; a perceiver, using his own perceptual mechanism, attempts to understand his own perceptual mechanism. The consequences of failing to recognize the significance of this recursion are not apparent in the study of sensory processes, but they are severe in the study of pattern recognition. It is so easy for an experimenter to forget that the way he perceives the stimuli which he presents to the subject may not be the way the subject perceives them. When a person recognizes an object, he relates it to a class of objects, other members of which are not in view, by rules about which he is usually unaware. I refer to such classes of objects as **stimulus domains**, and to the rules with which people compose descriptions (perceptions) as **descriptive domains**. But any object is describable (recognizable) in terms of a potentially infinite number of stimulus and descriptive domains. An experimenter perceives the stimuli he is using in his own way and may not know about the descriptive domain with which his subject is operating. He is therefore doomed not to understand how his subject performs. Difficulties of this type arise even in a simple task like asking a person to judge the sizes of objects at different distances. The experimenter may understand size to mean the size as measured by a ruler, but the subject may think that his task is to judge the relative projected sizes of the objects. It is no exaggeration to say that ambiguities about stimulus and descriptive domains plague most experiments in perception. Looked at another way, the problem of stimulus and descriptive domains is not a plague but the very problem that needs to be studied. This point was brought home to me as I was listening to a talk about a computer simulation that was designed to recognize 'natural scenes'. We were shown a scene of a house among trees and were told that the simulator was able to recognize successfully which was the house and which were the trees. But the device had done no such thing. It had merely extracted a region with sharp edges from regions with fuzzy edges and would have confused a matchbox among cotton wool with a house among trees. The speaker had unwittingly (I hope) used his own descriptive domain in interpreting the performance of the computer and thereby sidestepped the core of the problem with which he should have been concerned, which was to understand the structure of descriptive domains.

The first task in any enquiry into the mechanism of perception is to define carefully the stimulus and descriptive domains with which one is operating, and that is why, at the beginning of each chapter, I make an attempt to do this. In the early chapters, the stimulus domains consist of points, lines, and gratings in a frontal plane. In later chapters, more complex classes of stimuli are used and the problem of specifying them becomes all-important. This problem is discussed in more detail in Chapter 13.

In any experiment in perception it is also necessary to specify carefully the type of response which a subject is asked to make. Responses may be at any one of several levels of complexity, which I refer to as **response domains**. The simplest response domain is that of **detection**. A person visually detects a stimulus when he reliably reports that he can see it. A person **discriminates** one stimulus from another when he reliably detects a difference between them. If the stimuli being compared are identical when the difference in the attribute being discriminated is set at zero, we have what is called a **class A** discrimination (Brindley, 1970). An example of a class A experiment is provided by the task of deciding when there is a luminance difference between the two halves of a bipartite field. When no luminance difference exists, the boundary between the two halves is invisible, and one might say that the effective stimulus is no longer present. The observer is, in effect, being used like a null-reading instrument, and task ambiguity is reduced to a minimum. Class A experiments provide a powerful tool for investigating sensory functions, because one can argue that two identical sensations are caused by identical physiological events (see Boynton and Onley, 1962, for a critical discussion of this issue). However, in the study of perceptual mechanisms it is necessary to investigate how people judge differences between stimuli which differ in more than one respect. For instance, in studying the Müller–Lyer illusion, the subject is asked to set two lines to appear equal in length in the presence of distracting fins placed on the ends of the lines. At the point of subjective equality, the two stimuli being compared are still perceptibly distinct. This is a **class B** discrimination task. The probability of there being some ambiguity is very great in a class B task, and the interpretation of the results is much more difficult than in class A tasks.

In the experiments described in the first 10 chapters, class A and class B discrimination tasks were used. In Chapters 11 and 12, the experiments reviewed involved matching the position of a stimulus perceived in one sensory modality with a stimulus perceived in another modality. This type of task may also be regarded as a class B discrimination task. In Chapter 13, response domains at higher levels of complexity are described, which I refer to as **distinguishing between, identifying**, and **describing**. The definition of these response domains can be postponed until Chapter 13.

The two basic parameters of any human performance are **accuracy** and **precision**. Consider the task of setting a variable stimulus to equal a standard stimulus. The **point of objective equality** (P. O. E.) is the objective value of the standard and the **point of subjective equality** (P. S. E.) is the signed mean of a series of settings of the variable. Accuracy is indicated by the difference between the P. O. E. and the P. S. E., that is, the mean of the signed deviation scores with respect to the P. O. E. Where the P. O. E. is zero, accuracy is equal to the P. S. E. Accuracy is synonymous with **constant error** and **bias**. Precision is indicated by the mean of unsigned deviation scores with respect to the P. S. E. Related measures are the standard deviation, variance, and standard error of scores. Precision is synonymous with **sensitivity**, **variability**, and **variable error**. The term 'accuracy' is often used in the literature where 'precision' is meant; it is also often used inappropriately to signify the mean of unsigned deviation scores with respect to the P. O. E. The latter measure confounds accuracy and precision, and should be avoided. Its use has led to endless confusion. Accuracy and precision properly defined are independent, or orthogonal measures.

In assessing human performance on discrimination tasks it is useful to have some idea of the theoretical limit which it is possible for any detector, living or manufactured, to reach. A detector which performs at this theoretical limit is known as the **ideal observer** for the specified task. For instance, it is possible to calculate the performance of an ideal observer for the detection of light, given the quantal nature of light and the 'noise' within which the signal is presented. Once the ideal observer for a particular discrimination task is known, it provides a useful yardstick for assessing the performance of a human observer on that same task.

The concept of the ideal observer is applicable to tasks which involve the detection of a signal in the presence of noise, and which are limited by factors such as the resolving power of the detector. Usually, when the task is that of recognizing or describing an object, the observer has no difficulty in detecting the object, and the theory of the ideal observer is therefore not relevant. However, I suggest that there is a theoretical standard by which one may assess performance on these tasks. I call this standard the **ideal perceiver**. The ideal perceiver for the task of recognizing an object within a class of objects, or within a set of transformations of that object, is the complete specification of the set of rules which govern the composition of that class of objects, or of that set of transformations. For instance, the ideal perceiver for recognizing an object which rotates in three-dimensional space, and which is seen only as a projection in a plane, is one which embodies the mathematical theory of linear perspective for the class of objects being considered. *The ideal perceiver is the knowledge of a descriptive domain that is required to perform perfectly a prescribed task of recognition or description.* I have developed this theme in more detail in Howard (1975) and shall refer to it again in Chapter 13.

1.2 Direction and orientation defined

To specify the **direction** of an object it is first necessary to select a **reference point** O through which is drawn a **reference line**, or **fiducial line**. The direction of a point P on the object, from O, with respect to the fiducial line, is the signed angle between line PO and the fiducial line (see Figure 1.1). The angle is negative when PO is clockwise with respect to the fiducial line, but other sign conventions may be used, such as left–right, up–down or East–West. Some examples of the concept of direction as applied to vision are illustrated in Figure 1.2.

When we think of the **orientation** of an object, we do not usually think of its direction relative to some other object, but of its rotational position in space, that is, its attitude relative to some fixed axis of reference. To specify the orientation of an object it is first necessary to select a reference line, or axis. This may be an identifiable axis on another object, or an imaginary line joining two fixed reference points, such as the North and South poles. The two ends or directions along the reference line are distinctively labelled, as North–South, up–down,

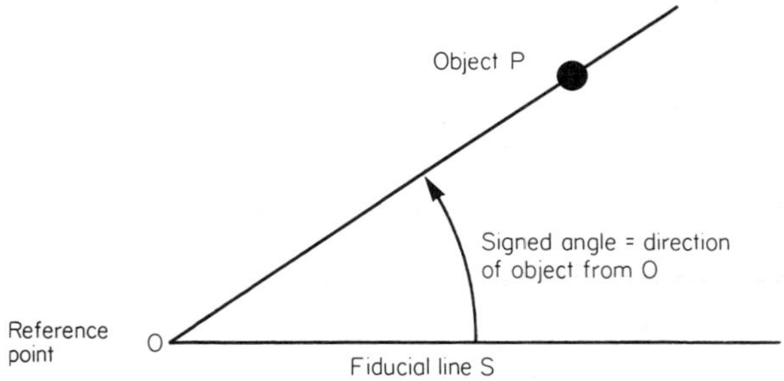

Figure 1.1 Illustrating the concept of direction

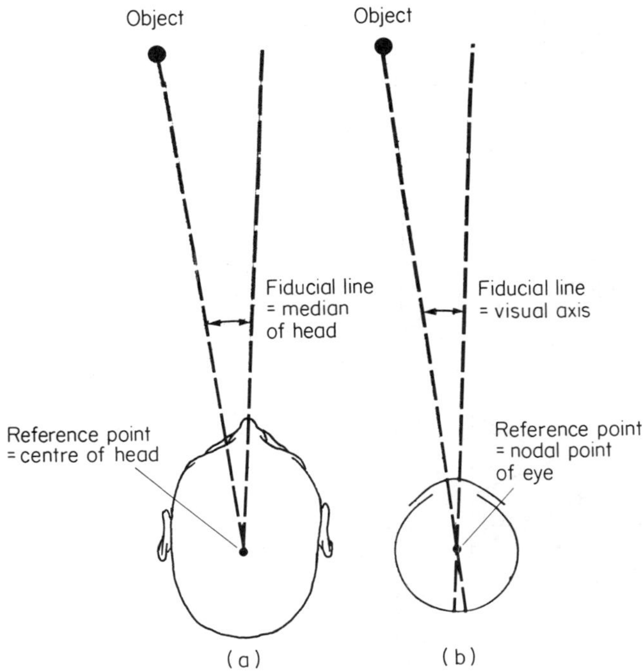

Figure 1.2 The concept of direction applied to the specification of (a) the headcentric direction of an object and (b) the oculocentric direction of an object

plus–minus, etc. A line, or axis, is then selected on the object whose orientation is to be specified (see Figure 1.3). This could be a man's body axis, the corner of a room, or the axis of a compass needle. If one end of the object differs from the other, the two ends are labelled. Such a labelled axis is a **polar axis**. If the object's axis and the reference axis are parallel and the poles of one axis point in a specified direction with respect to the poles of the other, the object has the same orientation as the reference axis. Otherwise, the orientation of the object is the angle through which the axis of the object is displaced with respect to the

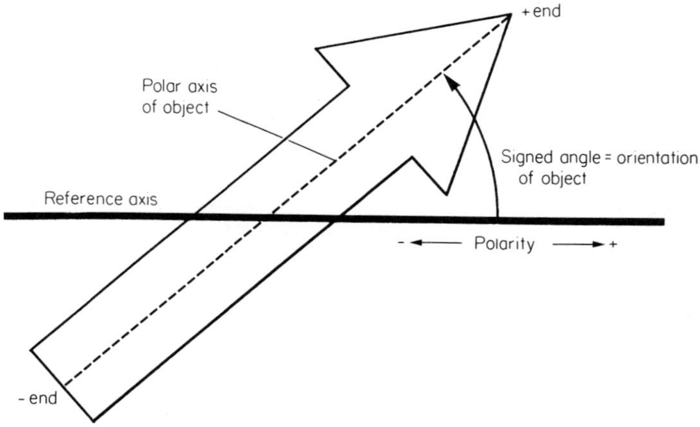

Figure 1.3 Illustrating the concept of orientation

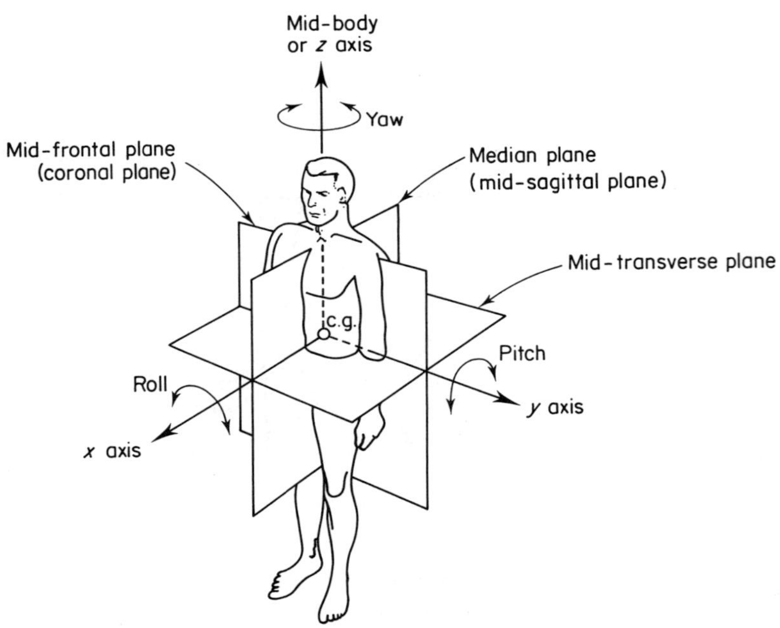

Figure 1.4 Principal planes and axes used for the specification of directions and rotations of the human body

reference axis. This angle may be signed by some suitable convention, such as calling a clockwise displacement positive.

These definitions specify the direction and orientation of an object in only two dimensions, that is, only with regard to the plane containing the object and the reference line. If the direction or orientation of an object is required in three-dimensional space, two orthogonal reference lines, or axes, must be specified.

In summary, when we talk about an object's direction we think of it as a point with a bearing and ignore its attitude, and when we talk about an object's orientation we ignore its direction and think only of its attitude.

The principal planes and axes of the human body are depicted in Figure 1.4. The mid-body axis, or ***z* axis**, passes along the length of the body through its centre of gravity. The ***x* axis** runs from back to front, and the ***y* axis** runs from side to side, both through the centre of gravity. The **median plane**, or mid-sagittal plane, is the plane of bilateral body symmetry containing the z and x axes (an approximation when the plane of symmetry does not contain the centre of gravity). A para-sagittal plane is any plane parallel to the median plane. The **mid-frontal**, or **coronal plane** is a plane at right angles to the median plane and containing the z and y axes. A frontal plane is any other plane parallel to the coronal plane. The **mid-transverse plane** is the plane at right angles to both the median and coronal planes and passing through the centre of gravity. A transverse plane is any other plane parallel to the mid-transverse plane.

A rotation of the head or body about the z axis is known as **yaw**, a rotation about the x axis is known as **roll**, and a rotation about the y axis is known as **pitch**. The orientations of the axes of the head or body are normally defined with respect to gravity. A roll of the head or body is referred to as **head tilt** or **body tilt** and a pitch of the head or body as **head slant** or **body slant**. In a similar way, a visual stimulus is said to be **tilted** when it is inclined to gravity within a frontal plane of an observer and to be **slanted** when it is inclined within a sagittal plane. A fuller treatment of the nomenclature of body axes and planes is provided in Hixson, *et al.* (1966).

1.3 A classification of judgments of direction and orientation

1.3.1 Orienting systems

The term 'orientation' is often used in an active sense to describe the process of restoring one thing into a given orientation with respect to another. Thus, an orienting device is one which restores itself to a given orientation or equilibrium position when disturbed. For instance, a magnetic compass or a hanging plumbline stores the kinetic energy which a disturbing force imparts to it and then releases the energy when allowed to return to its former state. An orienting device of this kind has no internal energy source of its own, and is therefore unable actively to resist the disturbing force. A device which is able to maintain a given orientation against disturbances can do so either because it is rigid or because it is a **servo-orienting system** and therefore capable of recruiting energy. Servos are superficially similar to rigid angular systems, but are quite distinct when considered part by part. The power which provides the restoring torque in a servo system is isolated from the power providing the disturbance. A control link is introduced between them, in such a way that a slight deviation from the equilibrium position caused by the disturbing force controls a valve (amplifier) which allows the restoring torque to operate and restore the balance. This is the

negative feedback loop of the system. Ideally, the system maintains absolute stability; in practice, there are time lags, which cause the system to oscillate, and power limits, which determine the range of effective stability. Stabilizers on ocean liners are servo-orienting devices. For a fuller treatment of servo theory, with special reference to biological systems, see Toates (1975).

The simplest behavioural response to disturbed orientation is the **righting reflex**. The most primitive righting reflexes occur in response to stimuli arising in one sensory modality. A good example is the reflex rotation of the eye about the visual axis that occurs when the head is tilted to the side. This torsional motion of the eye helps to keep the normally vertical meridian upright. Many invertebrates maintain their bodies in a given orientation relative to a particular stimulus gradient. Such responses are known as taxes. In **klinotaxis**, the animal possesses a single anterior light-sensitive region by means of which it successively compares the light intensity on either side of its body as it swings its head from side to side. Any asymmetry between successively sampled intensities on the eye spot induces an asymmetry in the swinging of the body in such a way that it comes to point away from the light source. A fly larva orients itself away from a light source in this way. In **tropotaxis**, the animal has a bilateral pair of sense organs which are capable of simultaneously comparing the light intensity on the two sides of the body. These two taxes are similar in that they both result in the animal aligning itself with the direction of the stimulus. Neither demands an image-forming eye. If two lights of equal intensity are put on at the same time, these mechanisms cause the animal to align itself with the line bisecting the angle between the two lights. In **telotaxis**, an animal with an image-forming eye can select one of several light sources as the effective stimulus towards which it orients its body. For further details on this topic see Fraenkel and Gunn (1940) and the thorough review provided in the recent book *Orientierung im Raum* by Schöne (1980).

Our point of departure is a human being, with an image-forming eye and vestibular, kinaesthetic, auditory and touch senses. The human postural mechanisms under the control of vestibular, kinaesthetic and touch senses are basically tropotaxic, and the human visual system is basically telotaxic, but the great variety and complexity of human orientation behaviour cannot profitably be described in terms of these simple mechanisms.

1.3.2 Types of directional judgments

There are three basic types of direction which are relevant to human perception. These are: proprioceptive direction, egocentric direction, and exocentric direction. Directional judgments may be made with respect to each of these directions.

Proprioceptive direction. This is the direction of one part of the body with respect to another part of the body. For instance, pointing the finger at the tip of the nose involves a proprioceptive directional judgment.

Egocentric direction. This is the direction of an external object with respect to a point in, or on the body of the observer. The reference axis may, or may not be fixed with respect to some part of the body of the observer. If it is so fixed, we have a pure egocentric direction. When the reference axis is part of an external reference system, such as the vertical, or a compass direction, I shall use the term **semi-egocentric direction**. For instance, when we judge whether an object is in the median plane of the head we make a purely egocentric judgment, and when we judge whether an object is to the North of us we make a semi-egocentric directional judgment, because the reference axis is no longer defined with respect to the observer. It is a characteristic of semi-egocentric directions that their value is not affected by which way the observer is facing with respect to the external stimulus. For instance, an object to the North of a person is still to the North when the person turns round.

There are three main types of egocentric direction: oculocentric, headcentric and bodycentric. The **oculocentric direction** of an object is objectively specified with respect to the centre of the eye and the visual axis. The **headcentric direction** of an object is specified with respect to the centre of the head and the median or mid-transverse plane of the head (Figure 1.4). The **bodycentric direction** of an object is specified with respect to some point in the spinal axis of the torso and a reference axis orthogonal to the spinal axis. Oculocentric direction is discussed in Chapter 2. Egocentric direction specified with respect to one or other of the planes of the head or body is discussed in Chapters 6 and 7.

Exocentric direction. This is the direction of an external object with respect to an external reference point and reference axis. An example is the compass direction of a town with respect to another town. Exocentric direction is often referred to as **allocentric direction**. The topic of exocentric direction is intimately connected with the subject of pattern vision on the one hand and the conception of geographical relationships on the other. Both these topics are beyond the scope of this book.

1.3.3 Types of orientational judgments

There are also the same three types of orientation.

Proprioceptive orientation. This is the orientation of one body part with respect to another body part. This topic crops up in Chapters 11 and 12, but it otherwise beyond the scope of this book.

Egocentric orientation. This is the orientation of an axis of an external object with respect to some axis in the body of the observer. Either axis can serve as the reference axis. For instance, setting a line to be parallel to the body axis is an egocentric orientation task, and so is standing upright, that is, setting the body to the vertical. If both axes are external to the body of the observer but one of them is anchored at one end to some point in the body, we have a **semi-egocentric orientation**. For instance, when a person judges whether a gun is pointing at him or whether an object is heading towards him on a collision course, he makes a semi-egocentric judgment because only a point on the reference axis is defined with respect to his body. It is a characteristic of semi-egocentric orientations that their value is not affected by a rotation of the observer. For instance, a gun still points to a person whichever way he is facing, and one cannot dodge an approaching missile by simply turning round. Semi-egocentric orientational judgments are discussed in Section 7.9. Egocentric orientation may be either oculocentric, headcentric or bodycentric. Oculocentric orientation is discussed in Chapters 3 and 4. Headcentric and bodycentric orientations are discussed in Chapters 10, 12, and 13.

Exocentric orientation. This is the orientation of an external object with respect to some external reference axis. When we judge the angle between two lines we make an exocentric judgment. Setting a line to the vertical is also an exocentric judgment, and will be discussed in Chapter 10. Pattern vision involves judging the relative exocentric orientations of parts of the stimulus, and this topic is dealt with in Chapter 13. Pattern recognition itself is dealt with only insofar as it is affected by the orientation of the stimulus to the observer (Chapter 14).

Directional or orientational judgments made in one sensory modality may interrelate with similar judgments made in another sensory modality. This issue of **intersensory spatial judgments** is dealt with in Chapter 11.

1.4 Spatial behaviour and spatial structures

Spatially coordinated behaviour is possible only because there are many constant features in the structure of the body and in the structure of the world. These constant features set a limit to chaos, and in that sense they are constraints; they are also constraints in the sense that, unless behaviour is conditioned by them, it will be non-adaptive. Spatially coordinated behaviour is conditioned by (1) the **internal constraints** of the body, that is, by the way in which the human body is constructed and normally develops, and by (2) **stimulus constraints** determinated by the nature of the physical world, that is, by environmental or ecological constraints.

1.4.1 Internal constraints

The most fundamental internal constraints are those provided by the mechanical structure of the neuromuscular system. Each joint has a limited direction and range of movement and every movement is controlled by contraction in particular muscles. This is the basic structural matrix of all operant behaviour. The second category of internal constraints consists of sensorimotor constraints, which may be reflexes, such as the postural reflexes, or learned sensorimotor coordinations.

1.4.2 Stimulus constraints

There are three types of stimulus constraint, which are listed in Table 1.1 and which will now be described.

Norms The first type of stimulus constraint consists of the stimulus norms that occur in many stimulus dimensions. For instance, the visual vertical is a norm in the dimension of visual tilt in the frontal plane. Tilt is a symmetrical stimulus dimension because an object may be tilted left or right of the vertical norm, which is a central, salient value of the stimulus dimension. Thus, a central value of a stimulus dimension is a *norm* and a sensory dimension in which a norm occurs is known as an **oppositional dimension** or scale. Other examples of norms are the straight ahead in the dimension of azimuth direction, eye level in the dimension of elevation, and stationary in the dimension of movement to the left or right. If an observer is exposed to stimuli on one side of the norm in an oppositional scale, he typically experiences a sensation in the opposite direction when subsequently exposed to the norm. The tilt aftereffect, to be discussed in Chapter 4, is an example of an aftereffect of this kind.

Frames of reference The second type of stimulus constraint consists of stimuli which maintain a constant value in some stimulus dimension. For instance, the North pole star maintains a constant geographical direction, and the walls of a room remain vertical. A constant stimulus of this sort provides what is known as a **frame of reference**, in terms of which other things present at the same time may be judged. If we enter a tilted room we still use the walls as our standard of reference in judging the verticality of other objects, including our own bodies. This can produce very compelling illusions, which will be discussed in Chapter 10. A common frame of reference illusion is the experience we have when we are sitting in a stationary train and the train next to it moves. That train is accepted as our frame of reference for motion and we falsely judge our own train to be moving. A frame of reference may be defined as *an attribute of a certain object which does not normally vary, and in terms of which variations of the same attribute in other objects perceived at more or less the same time may be judged.*

Stimulus covariances The final type of stimulus constraint consists of **stimulus covariances**. Changes in certain stimulus dimensions are normally accompanied by related changes in other stimulus dimensions. Such covariant changes in stimulation are, more than anything else, responsible for moulding behaviour. Stimulus covariances fall into three types, which are set out in Table 1.1.

The first type of stimulus covariance consists of incidental linkages between stimuli, which convey no useful information to the perceiver. An example is provided by the chromatic fringes produced by the lens of the eye. Although these fringes can be shown to exist optically, we are not aware of them. This suggests that we have a neural mechanism for cancelling out the perceptual effects of these colour fringes. Hay *et al.* (1963) found that artificially produced chromatic fringes are no longer perceived after about two days of continuous wearing of chromatically aberrant lenses. Perhaps the machanism responsible for adaptation to colour fringes is related to the contingent colour aftereffect discovered by McCollough (see Skowbo *et al.*, 1975). Here, too, a covariance between colour and contours generates the inverse function in the nervous system.

Table 1.1 Stimulus structures in the spatial domain (from Howard, 1978a)

		Stimulus covariances						
		Contamination (no information) covariance	Vantage-point dependent covariances				Vantage-point independent covariances	
			Exafferent: passive movement of observer or movement of object		Reafferent: motor–sensory covariance			
Stimulus symmetries or norms	Stimulus constants or frames of reference		Intrasensory	Intersensory	Self-observation reafference	Locomotory reafference	Taxonomic properties	Causally related properties
The visual vertical. Straight ahead	Familiar vertical objects. Large stationary objects	Eg., chromatic aberration of eye's lens. Contingent after-effects	Size–distance invariance. Shape of image & tilt of object	Sight & sound of bell. Seen size & felt size of an object	Sight of moving hand	Image motion due to eye movements. Feeling objects with mobile hand	Shapes & colours of flowers	Tipping of glass & pouring of liquid. Hitting of billiard ball & direction of travel

The two remaining classes of stimulus covariance convey useful information to the perceiver and are therefore not cancelled out, but rather coded in such a way that the information they contain is accurately extracted. In one class are those that depend on the observer's vantage-point, which are referred to as **vantage-point dependent covariances**. A person must be able to allow for the effects of changes in his vantage-point if he is to perceive the objective properties of objects. For instance, the shape of the retinal image of an object changes as the observer moves about. But a similar change in image shape can be produced by a movement of the object relative to the stationary observer or by a plastic deformation of the object. This type of ambiguity may be resolved in two ways. The first way involves covariant stimulation, the second involves a frame of reference. The distinction between the two should become clear in the following example.

When a person moves, there are velocity signals from the vestibular system,velocity and position signals from the joint receptors in the limbs, and, if the movement is self-generated, there are correlated motoneurone discharges. The function that relates these body motion signals to the resulting changes in the rentinal image is the **stimulus covariance function** for this situation. The term 'invariant' is often used to denote these constant relationships between stimuli, but it is not always consistently applied. A perceiver may be born with, or may acquire by experience, an internal representation of the stimulus covariance function, which I shall refer to as a **stimulus covariance schema**, or, simply, schema. A perceiver with a schema for the task we are considering 'knows' what image changes to expect as he moves in a certain way relative to a fixed object at a given distance from him. If the image changes differ from this standard, he will perceive the object as moving or deforming in such a way that will account for that difference. Needless to say, this is not normally a conscious process. This type of mechanism has been called a 'cancellation process' (Gregory, 1958), but the term 'cancellation' is a little misleading, for it suggests that perceived changes in the retinal image are cancelled by the body motion signals. They are not cancelled by the body motion signals but rather assessed with respect to those signals (MacKay, 1972). The image changes are still perceived, but they are perceived as due to body motion and not as due to object motion.

This servo-system model of perceptual constancies, in which the way one stimulus varies with another is compared with a covariance schema, has inspired most of the experiments on the effects of sensory discordance which are reviewed in Chapter 12. The idea is that by artificially changing the way in which two related stimuli covary we can study the way in which the system recalibrates its covariance schemata. This is a useful idea, but *in normal environments, covariance schemata might not be required for perceptual constancies*. Frames of reference often provide a simpler mechanism. Consider the example of body and object motion that we used before. When we move about in a normal room we assume that the room is not moving, and therefore any image motion may be ascribed to head or body motion without further ado.

Covariance schemata are studied by introducing a physical distortion or

discordance into the covariance function. For instance, one may have a subject view his arm through a prism which physically distorts the relationship between distal and proximal visual stimuli, and this has the effect of physically distorting the covariance function between the felt and seen position of the arm. In such studies, frames of reference should either be minimized by using highly abstract displays or their effects assessed by controls.

Studies of the effects of sensory discordance fall into three classes. In the first class, there are those studies concerned with the perceptual effects of discordance between stimuli in the same modality, or **intrasensory discordance**. Secondly, there are studies of the perceptual effects of **intersensory discordance**. Thirdly, there are studies of the perceptual effects of **motor–sensory discordance**. Examples of each type will occur at different places in the book.

Examples of **vantage-point independent covariances** are the relationship between the size of a animal and the pitch of the sound that it makes, and the relationship between the suit of a playing card and its colour. Relationships of this type between features, or properties, of objects are referred to in Table 1.1 as **taxonomic properties**. Other vantage-point independent covariances are listed under the heading of 'causally related properties'. Included here are such things as the causal relationship between the act of tipping a glass of water and the sight of pouring water. The possibility of studying taxonomic and causal perceptual schemata by the procedure of introducing discordance has been discussed in Howard (1975), but this topic is not relevant to the subject of this volume.

Chapter 2

Oculocentric Direction

2.1 The specification of oculocentric direction

The visually discriminable properties of the light that enters the eye are determined by four types of physical entity. These are: sources which emit light directly into the eye; surfaces which differentially absorb and reflect light; transparent media, such as mists, liquids, and glass, which refract or filter light; and opaque objects which occlude light. The **visual world** is the set of all such entities which have the potential to influence the visually discriminable features of the light entering the eye.

In the case of monocular viewing, it is convenient to define the location of the observer as the **nodal point** of the eye. In general, any system of lenses has two nodal points, which have the property that any incident ray of light directed towards the first (anterior) leaves the system as though from the second (posterior) with its direction unchanged. For most purposes, the two nodal points of the human eye may be considered to coincide at a point on the visual axis 17 mm in front of the retina. Thus, the nodal point of the eye is that location through which pass all those straight lines which join points in the object plane to their corresponding points in the image plane. Any such straight line is called a **visual line**, and the nodal point is the point of intersection of all visual lines. The visual line through the fixation point is the **visual axis**.

The total panorama of visible objects which a person can see by turning his body without changing his location constitutes his **visual surroundings**. For any particular position of the eyes, head, and body, an object, in order to be visible to a given eye, must occur within a three-dimensional cone-shaped segment of space which has its apex on the nodal point and is contained within the boundaries of the orbital ridges and nose or of whatever artificial aperture is placed before the eye. This segment of space is the **monocular field of view**, or visual field. The region of overlap between the two monocular fields of view is the **binocular field of view**. The term 'visual field' often refers to a projection of the three-dimensional field of view onto a hemisphere concentric with the eye or onto a plane at right angles to the line of sight. In perimetry, for instance, visual functions are plotted onto a **tangent screen**, which is a two-dimensional projection of the field of view from the nodal point of the eye. Each point in such a chart projects to one, and only one, point of the retina for a given position of the eye. For this reason, the term 'visual field' in ophthalmology is often used to refer to the retinal field, and the term 'object field' to the external field. The direction of a visual object is specified by its angle of eccentricity and its meridional angle. These two angles are defined with respect to a system of polar coordinates centred on the visual axis, as illustrated in Figure 2.1.

The **optical array**, or **distal stimulus**, is defined as *that set of sources, surfaces, and media within the field of view of an observer which determines the physical features of the light entering the pupil.* One may extend this definition to cover binocular vision by considering the light entering both pupils. In other words, the optical array is that array of sources, surfaces, and transparent media which could affect vision for a given position of the eyes, without consideration for the effects that the optics of the eye or the properties of the visual system may have in determining how things are imaged or perceived.

The retinal image of a given distal array is the **proximal visual stimulus**. This is essentially a similarity projection of the distal array onto a hemispherical surface. A similarity projection is one in which angles and relative distances are preserved. However, the local distribution of luminance over the image is not the same as would be produced by a perfect point-for-point projection of the distal array; each point in the object produces a blurred region of luminance in the

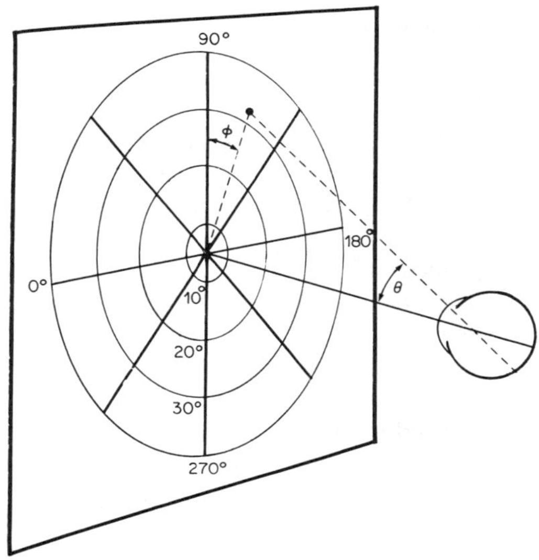

Figure 2.1 The oculocentric direction of a point P in the visual field is specified by its angle of eccentricity θ and its meridional angle ϕ with respect to polar coordinates lying in a frontal plane of the eye and centred on the fixation point

image. This topic will be discussed further in the next section. The proximal visual stimulus is the distribution of light formed on the retina. It thus takes account of the optical properties of the eye but not of the properties of the sensory transduction system.

After the light has been transduced into neural impulses, we can talk about the neural image or **afferent visual signal**. The properties of the transduction process and the anatomical features of the retina now impose themselves on the visual imputs. The determination of the detectability and discriminability of colours, shapes, movements, and other properties of the optical array, as a function of the position of the stimulus in the visual field, constitutes the branch of psychophysics known as **perimetry** or campimetry. Perimetry developed as a clinical test for the loss of visual functions due to damage in retina, visual pathways or visual cortex, but it is also a useful research tool (see Aulhorn and Harms, 1972). More detailed electrophysiological and psychophysical investigations of the retina and higher stages in the visual system are discussed in what follows.

There is inevitably a recursion in this set of definitions, because the distal stimulus was defined as the set of things which determine the physical properties of the light entering the eye. But we cannot know what the physical properties of light are until we have made observations with the aid of our sense organs, usually our eyes. In other words, what humans regard as physical objects are constructs derived from perceptual processing. The recursion in the above definitions is simply a recognition of the recursive nature of all attempts of a perceiving animal to understand its own perceptual mechanisms.

2.2 The specification of image quality

2.2.1 Introduction

Any system which forms a visual image of an array of objects is an **image-forming system** or **optical system**. The human eye, as far as the retina, is an optical system. Light energy is transduced into neural impulses in the retina and, when considering this and subsequent stages, I shall refer to the visual system as an **image-processing system**. This section is concerned with the spatial aspects of the image-forming properties of the human eye. The spatial quality of the image produced by any optical system may be specified by two basic features. The first is the extent to which the image is free from distortion and the second is the extent to which luminance contrast is preserved.

An image is distorted when it is not a similarity projection of the object. The image of an object lying at an angle to the frontal plane is necessarily distorted by perspective. But this type of distortion may be ignored if we consider only the distal array as it is projected in a frontal plane. I shall assume that the retinal image is a similarity projection of the visual scene (see Helmholtz, 1962, Vol. III, p. 154 for a discussion of this issue).

The normal measure of how well luminance contrast is preserved in the image produced by an optical system is known as the **resolving power** of the system. This is indicated by the number of black and white pairs of bars per degree of visual angle in the finest grating which is imaged as a grating rather than as an homogeneous patch of light. I shall now discuss the resolving power of the eye.

2.2.2 Resolution in the retinal image

In any optical system, the extent to which detail is preserved in the image depends upon the the extent to which light from each point in the object is brought to focus in one point in the image. This in turn depends on two main features of the optical system. The first is the amount of **light loss** and the other is the amount of **light dispersion**. The amount of light loss depends on the amount of absorption which occurs as light is transmitted through the optical components and through the media in which they are suspended. In addition to general light loss, there may be foreign objects in the system which are near enough to the image plane to cast shadows. Objects of this kind in a person's eye are called floaters. Some light is also lost because it is reflected out of the eye at the surfaces of the cornea and lens. The amount of light dispersion depends on the degree to which light is scattered in the optic media, the extent of internal reflections, diffraction through the aperture of the system, and the various refractive aberrations of the system, including spherical and chromatic aberrations and coma (see Davson, 1978, for a review of visual optics). The transmission of the ocular media is about 80% at 560 nm, although only about 2% of the total light evokes a detectable sensation (Van Meeteren, 1978). The complexity and diversity of these sources of image degradation complicate the task of specifying the quality of the optical image. However, the task is simplified by the fact that it is possible to summarize the effects of all the sources of optical degradation in one function, known as the **spatial modulation transfer function**. Two conditions must be met before such a function can be applied: the system must be **homogeneous**, that is, light entering from various directions must be affected in the same way, and the system must be **linear**, which I shall define later.

The spatial modulation transfer function is based on the idea of a **sinusoidal grating**, that is, a black and white grating like that shown in Figure 2.2, in which the luminance is modulated sinusoidally across the width of the grating. The **spatial frequency** of a sinusoidal grating is measured in cycles (black and white pairs of bars) per degree of visual angle. Such gratings have the interesting property that any visual display however complex may be synthesized by superimposing sets of sinusoidal gratings at different orientations with suitable

Figure 2.2 A spatial grating with a sinusoidally modulated luminance profile

amplitudes, phases, and wavelengths. These sets of gratings constitute the spatial **Fourier analysis** of the display. This fact makes spatial sine waves convenient components to work with. Another convenient property of spatial sine waves is that if a set of them is transmitted through an homogeneous linear system, the image consists of a set of sine waves with the same relative spatial frequencies. In fact, a **linear system** is defined as any system in which a set of sine-wave inputs emerges in the output as a set of sine waves with the same relative frequencies. A linear system may spatially magnify or minify the input, in which case all spatial frequencies are scaled up or down proportionally. A linear system may also displace, rotate or invert the image, because such a transformation does not affect spatial frequency. The modulation amplitudes of component spatial frequencies may also change in a linear system. The **modulation amplitude** is the difference in luminance between peaks and troughs of the sine wave (see Figure 4.5 for details). This difference is measured in an optical system by scanning the image with a photoelectric probe. If a sinusoidal grating of a given spatial frequency is transmitted through an optical system, its modulation amplitude is to some degree attenuated by the summed effects of loss and dispersion which have already been described. The **attenuation of modulation amplitude** produced by an optical system is expressed as the ratio of the amplitude modulation of the image grating to that of the object grating. This definition may be generalized to cover both optical systems and image-processing systems. In any practical system, the attenuation of modulation amplitude is complete for all spatial frequencies above a certain value. This simply means that the system cannot resolve gratings above a certain spatial frequency. The human visual system, the whole system that is, not merely the optics, is also insensitive to gratings below a certain spatial frequency. This is not an optical effect but is due to the inability of the nervous system to detect the gradual changes in luminance that are present in low-frequency gratings. The range of frequencies that is resolved by any system is its **spatial bandwidth** (or bandpass) at that mean luminance. The spatial bandwidth of the human visual system as a whole is illustrated in Figures 2.3 and 2.9. Gratings with frequencies outside the spatial bandwidth appear as homogeneous patches of light. The spatial modulation transfer function may now be defined.

The **spatial modulation transfer function** of an optical system is the function that relates the attenuation of modulation amplitude to the spatial frequency of a sinusoidal grating of constant mean luminance. In practice, the spatial modulation transfer function of an optical system is derived by using a photoelectric probe to measure the attenuation of amplitude modulation for sinusoidal gratings at each of several spatial frequencies within the spatial bandwidth of the system. The attenuation, or its reciprocal, **contrast transmission**, is then plotted as a function of spatial frequency, as in Figure 2.5. This function indicates how efficiently the system transmits spatial sine waves. However, a visual display of any complexity may be decomposed by Fourier analysis into sine waves. If the spatial modulation transfer function (MTF) of a linear system has been determined, it is possible to get an estimate of the quality of the image of any

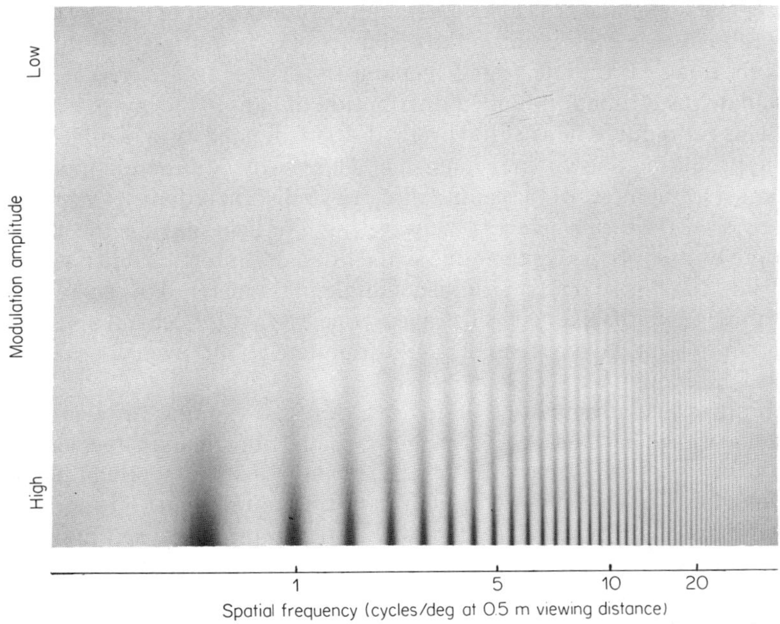

Figure 2.3 A sinusoidally modulated grating in which the spatial frequency increases from left to right and the luminance modulation amplitude increases from top to bottom. When the grating is viewed from a distance of about half a metre, the boundary of the region within which the grating can be detected gives an indication of the spatial bandwidth and the spatial modulation transfer function of the visual system. Compare with solid line in Figure 2.9 (Based on an unpublished photograph by F. W. Campbell and J. G. Robson)

display. A Fourier analysis of the display is first performed, then each sine-wave component is attenuated by an amount determined by the modulation transfer function. When the display is restored by Fourier synthesis, it defines the image produced by the system. The image is, in effect, the result of passing the visual display through a set of sine-wave luminance filters.

The concept of modulation transfer function may be usefully applied to any homogeneous and linear optical or image-processing system, but the method for determining the modulation amplitude of the output must be adapted to the type of output that the system produces. In optical systems, the output is a physically measured image. In the human visual system as a whole, the modulation amplitude of the output is indicated by the luminance contrast required for the detection of the grating at some specified criterion for detection. In general, we may define the spatial modulation transfer function of any system as a mapping of the output of the system onto a sinusoidal grating input of varying spatial frequency. An analogous transfer function may be derived in the time domain but is beyond the scope of this book.

An alternative measure of the optical quality of an image is known as the **point-spread function**. Even if the image is perfectly focused, diffraction of light by the pupil, dispersion in the optical media of the eye, and scatter of light by the surface of the retina cause each point in the distal array to be represented by a blurred disc in the image. If the effects of dispersion and scatter are ignored, it is possible to calculate the diffraction-limited distribution of light in the image of a point of light. This turns out to be a central bright disc surrounded by light and dark rings, with a profile as shown in Figure 2.4. This is the diffraction point-spread function. The diameter of the central disc, **Airy's disc**, in radians of visual angle is given by $\theta = 2.44(\lambda/a)$, where λ is the wavelength of light and a is the diameter of the pupil, both expressed in the same units. For a 2.3 mm pupil and a wavelength of 555 nm, the Airy's disc is 1′ of visual angle in diameter. This corresponds to about three times the diameter of a foveal cone, so that the central disc produced by a point of light, however small, could stimulate about seven cones if the light were bright enough.

A similar calculation can be done for a thin-line stimulus, in which case we have a **line-spread function**. A thin line in the spatial domain is analogous to a brief flash in the temporal domain. Both types of input are known as an **impulse**, or **delta function**. Impulses are used widely in testing communication systems, man-made and natural. The reason is that the Fourier transform of an impulse is a set of sine waves extending evenly across the whole frequency spectrum. The crests of the waves coincide at a place which defines the position of the impulse, and at all other places in the immediate neighbourhood crests and troughs cancel. When the eye is exposed to a thin line, it is as if a complete set of spatial sine waves were simultaneously injected into the visual system. The modulation transfer function of the eye determines how each component is attenuated and the line-spread function of the retinal image is the result of this interaction (convolution) between the Fourier transform of the impulse and the modulation transfer function. Another way of expressing this relationship is to say that *the line-spread function is the Fourier integral of the modulation transfer function.* This mathematical relationship forms the basis of many inferences about the performance both of the eye and of the visual system as a whole, as we shall see in later sections of this chapter.

It has been assumed up to now that the spatial modulation transfer function is

Figure 2.4 A diagrammatic representation of the luminance distribution across the image of a point source. About 84% of the light is contained in the central bright disc (Airy's disc)

the same for all orientations of the test gratings. If this is not so, then the system suffers from **astigmatism**.

There are no completely satisfactory methods for direct measurement of the spatial modulation transfer function of the human eye. If psychophysical procedures are used, the measurements reflect the behaviour of the whole visual system and not merely the properties of the optics of the eye. The light distribution may be measured in the image produced by the excised eye of a cadaver, but the results are affected by *post-mortem* changes. Another procedure is to form the image of a bright line in the eye and then allow the eye's optics to work in reverse to produce, by reflection, an image in space of this retinal image. The external image may be measured photoelectrically to give an indication of the optical quality of the image in the eye (Westheimer and Campbell, 1962; Krauskopf, 1962). The success of this method depends on the assumption that the retina is a perfect diffuser. Campbell and Gubisch (1966) were able to show experimentally that the retina is an almost perfect diffuser, and at the same time they improved the procedures for measuring the reflected image. Their measurements of the modulation transfer function of the human eye are shown in Figure 2.5. It can be seen that contrast is attenuated more and more as spatial frequency is increased, and that the rate of this attenuation is greater with a larger

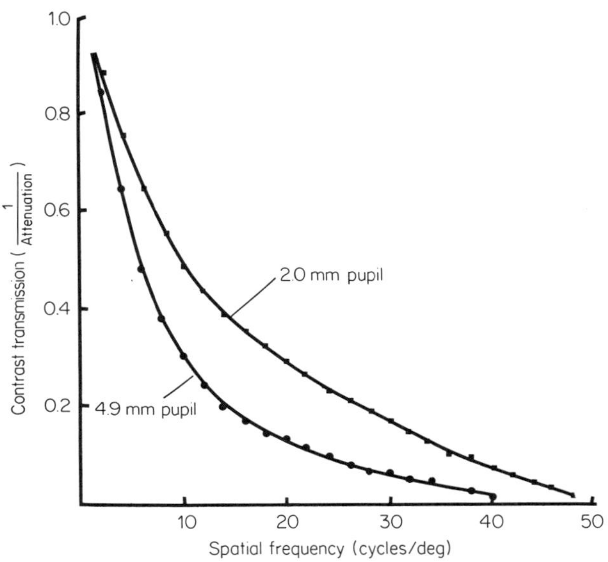

Figure 2.5 The modulation transfer function of a human eye for two sizes of pupil. The curves were obtained by Campbell and Gubisch (1966) from photoelectric measurements of the light reflected from the image lying on the fovea of a human eye. Note that contrast is attenuated only at the high-frequency end of the spatial-frequency spectrum. These curves are not directly comparable with that in Figure 2.9 because of the different ways in which spatial frequency is plotted. (Reproduced by permission of the *Journal of Physiology*)

pupil. The measured line-spread functions of the human eye for luminous lines imaged on the fovea at each of several sizes of pupil are shown in Figure 2.6. The dotted lines are the line-spread functions calculated from the degree of diffraction at each pupil diameter. The effects of diffraction diminish as pupil size increases. On the other hand, spherical aberration increases as pupil size increases and this widens the line-spread function. Thus, increasing the size of the pupil effects diffraction and spherical aberration in opposite ways. The best balance between the effects of diffraction and other aberrations is at a pupil

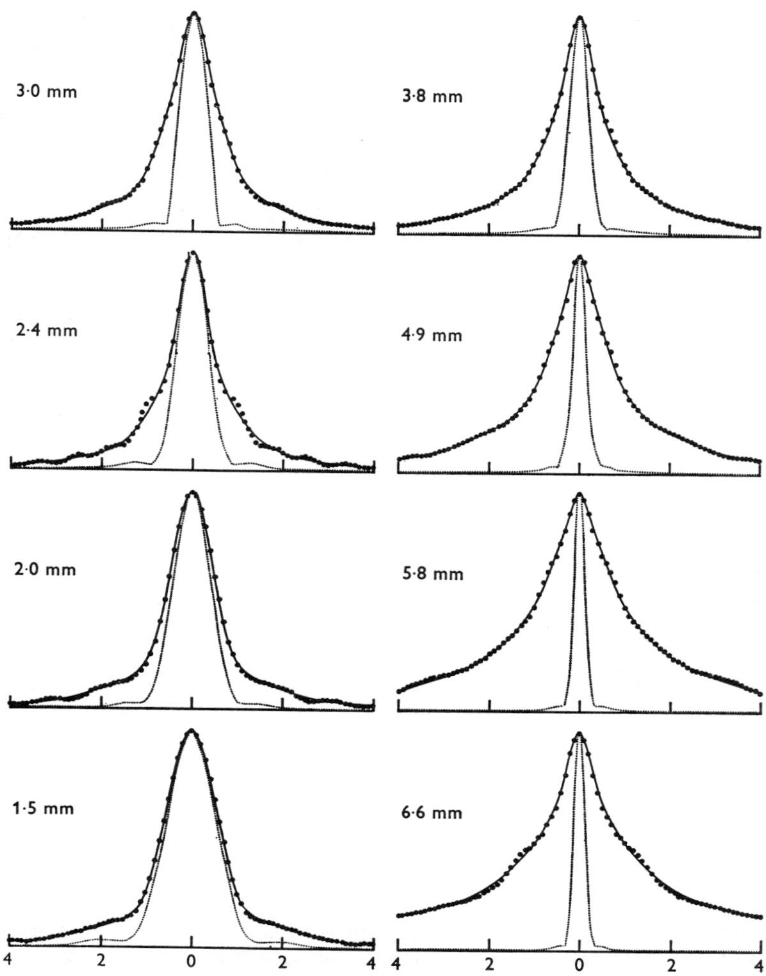

Figure 2.6 Optical line-spread functions of the human eye for several sizes of pupil. Each curve is the normalized distribution of illuminance in the foveal image of a thin line, measured as described in the text. The narrower curve in each case indicates the calculated diffraction image of the line at the given pupil diameter (From Campbell and Gubisch, 1966. Reproduced by permission of the *Journal of Physiology*)

diameter of 2.4 mm. (For line-spread functions in peripheral vision, see Jennings and Chapman, 1981). These measurements agree with psychophysical data, which I shall now describe.

Although it is not possible to measure the optical properties of the human eye psychophysically, it is possible to measure human visual acuity when optical factors have been eliminated, and in this way infer what contribution optical factors make to the visual acuity of the visual system. The refractive and diffractive processes of the eye may be bypassed by converging two beams of coherent light from neon–helium lasers onto the retina in such a way that they produce an interference pattern consisting of a sinusoidal luminance grating. It is a feature of such interference patterns that their image quality is independent of the diffraction and refractive properties of the eye. They are not immune to the effects of light scatter or reflection, but if due allowance is made for the effects of scatter and reflection, the capacity of a human observer to resolve sinusoidal interference gratings reflects the spatial acuity of the neural stages of image-processing. The observer's capacity to resolve ordinary gratings reflects the spatial acuity of the whole system. By comparing acuities obtained under these two conditions, it is possible to infer the contribution that the optical properties of the eye make to spatial acuity.

Arnulf and Dupuy (1960) and Westheimer (1960) were the first to apply such a procedure to measure the spatial modulation transfer function of the human visual system. However, the sources of coherent light then available were inadequate, and the experiment was repeated by Campbell and Green (1965) after the neon–helium laser had been invented. They psychophysically determined the threshold contrast for each of a set of laser-produced interference gratings on the retina at various spatial frequencies and, in each case, compared this with the threshold contrast for an equivalent grating presented in the ordinary way. The ratio between these two measures of contrast sensitivity indicates the contrast reduction due to the refractive and diffractive properties of the eye. In this way, Campbell and Green were able to show that the image formed by a well-focused eye, with a normal size pupil, is very good and that most of the loss of contrast in human vision is due to the neural processes in the retina and brain, rather than to optical factors.

Green (1970) extended these measurements into the retinal periphery and concluded that, for eccentricities greater than 5°, visual acuity is not limited at all by optical factors under optimal viewing conditions. Optical factors caused a reduction in acuity only for targets imaged within 5° of the fovea. He suggested that, for eccentricities greater than this, acuity is limited wholly by the properties of the retina. The most likely property is the density of the receptors and the extent to which they are linked into receptive fields (see Section 2.4.4). It can be shown that the angle subtended at the eye by one bar (half cycle) of the highest spatial-frequency grating that can be theoretically detected by a mosaic of independent receptors is equal to the mean spacing of receptors. Figure 2.7 shows that the acuity of the visual system for interference gratings reaches this limit for targets imaged within 3° of the fovea. It falls below this limit at greater eccentricities, presumably because of the increasing size of receptive fields. On

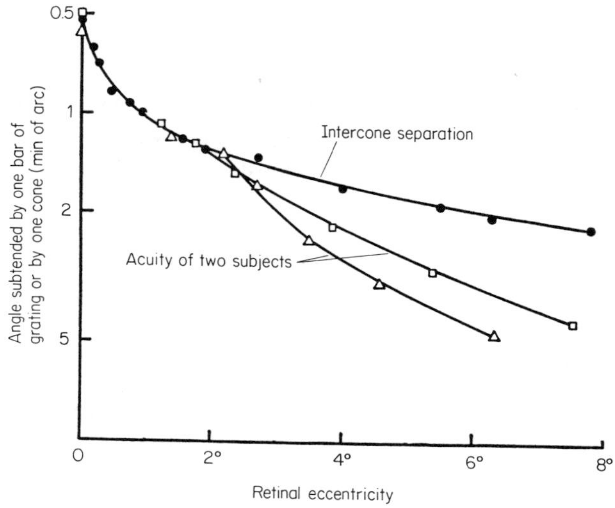

Figure 2.7 Visual acuity for interference gratings formed on the retina by converged laser beams, compared with mean intercone distance at various angles of eccentricity (From Green, 1970. Reproduced by permission of the *Journal of Physiology*)

the other hand, Leibowitz *et al.* (1972) found that movement sensitivity in the retinal periphery was higher after the optical quality of the image had been improved above its normal value. It looks as though movement detection is limited by optical factors and resolution by neural factors. This difference provides an explanation for the fact that the relative degradation of performance with eccentric viewing is less for motion detection than for resolution.

I have mentioned already that the spatial modulation transfer function is applicable only to homogeneous and linear systems. The human visual system is not homogeneous, because acuity falls rapidly as the image is moved away from the fovea. It is also not strictly a linear system, because it adapts to constant luminance contrasts. Furthermore, the appearance of a grating of one spatial frequency is affected by prior inspection of gratings of other spatial frequencies (see Figure 2.19). However, for regions near the fovea and for standard conditions of viewing, a function resembling the modulation transfer function may be derived for human vision which is at least useful for comparative purposes (see Cornsweet, 1970, for a fuller discussion of this issue). I shall use the term **contrast-sensitivity function** in preference to the term 'modulation transfer function' when talking about the visual system as a whole. This is meant to make it clear that no assumptions about homogeneity or linearity are being made. The contrast-sensitivity function is explained more fully in Section 2.3.3.

2.2.3 The directionality of retinal receptors

Stiles and Crawford (1933) concluded from psychophysical evidence that a ray of light stimulates a receptor more effectively when it has passed through the

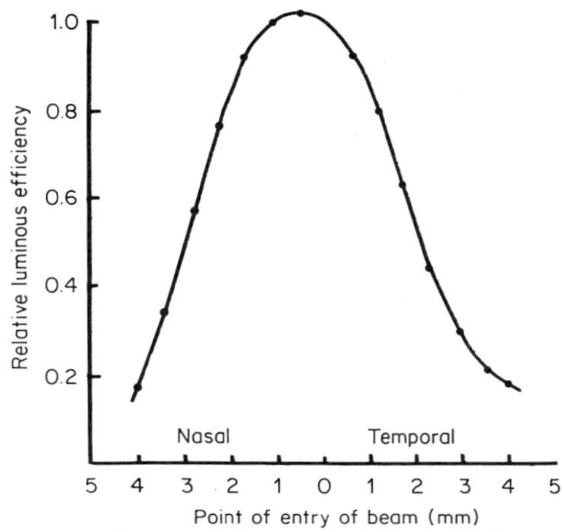

Figure 2.8 The relative luminous efficiency of a beam of light parallel to the optic axis and entering the pupil at various points along its horizontal diameter (From Stiles and Crawford, 1933)

centre of the pupil than when it has passed through the margin of the pupil (see Figure 2.8). This effect, which is known as the **Stiles–Crawford effect**, is due to the fact that the outer segments of the tube-shaped receptors are aligned with the centre of the exit pupil of the eye, so that light is most effectively trapped by the receptors when the beam approaches from this direction. In this respect, the receptors act like wave guides (O'Brien, 1946). Some alignment of receptors is present in the neonate and is presumably determined by the way in which the receptors are packed together. However, the fine alignment must be controlled by an active process because it has been found to recover after detachment of the retina, and this corrective recovery is confined to an area of localized disturbance (Campos *et al.*, 1978). Furthermore, Enoch *et al.* (1979) found that the magnitude of the Stiles–Crawford effect was reduced after an eye had been kept in the dark for several days, and returned to its normal value only after light was readmitted to the eye. In some manner yet to be determined, the receptors, at least the cones, are able to align themselves with the most probable direction of light which impinges on them.

2.3 The acuity of the human visual system

2.3.1 Introduction

Once the visual image has been transduced into neural impulses, the spatial information contained in it is analysed and transformed in many ways by a variety of parallel and sequential processes. The final output of the system, as far as an experimenter is concerned, consists of the judgments that a subject makes under specific stimulus conditions. The basic capacity of a subject to detect the spatial attributes of simple stimuli is known as **visual acuity**. However, the various spatial attributes of objects are analysed by at least partially distinct visual mechanisms, each of which imposes its own limits on the ability of the subject to process the kind of information with which it deals. For this reason, it is important to specify the type of visual acuity with which one is concerned. We can also talk about **temporal visual acuity**, which refers to the resolution of the intermittency of a flashing light. However, as I am concerned only with the spatial aspects of vision, I shall discuss only spatial acuities. There are three types of spatial acuity.

In the first type, which I shall call **luminance gradient detection**, the task is to detect a discontinuity of luminance. This task is spatial only in the sense that there is a spatial discontinuity in the stimulus, but the subject's task is to detect a change in luminance rather than any spatial attribute of the stimulus.

The second type of spatial acuity is resolution acuity. In one form this involves determining the highest spatial frequency for which a grating is distinguishable from an homogeneous display. Although this task involves a spatial attribute, namely spatial inhomogeneity, performance depends on the detectability of the luminance gradient between stimulus elements, and is therefore limited by many of the same factors that limit simple luminance gradient detection. We shall see that resolution acuity is best understood as a special case of contrast sensitivity, which may be defined as the luminance contrast required for the detection of a grating at each of several spatial frequencies over the spatial bandwidth of the visual system.

The final type of spatial acuity is acuity for spatial attributes which are more complex than mere spatial separation of points or lines. I shall call this type of acuity **pattern acuity**. The spatial attribute that the subject is required to detect may be that of a displacement break in a line, or **Vernier acuity**; an angular break in a line, or **chevron acuity**; a departure from the parallelity of two lines; the curvature of a line; or any other spatial feature. Westheimer (1976) has argued that pattern acuities are not limited in any simple way by the optical qualities of the eye, nor by the size of receptive fields. Spatial thresholds as measured in this type of task can be an order of magnitude smaller than resolution acuity. For instance, Vernier acuity has been reported to be as low as two or three seconds of arc, which is considerably smaller than the diameter of a retinal receptor. For this reason, Westheimer referred to pattern acuities as **hyperacuities**.

It has often been reported that luminance sensitivity, resolution acuity, and pattern acuity are all reduced when the test stimulus is surrounded by, or is next to other visual contours (Crawford, 1940; Westheimer, 1967; Lennie and MacLeod, 1973). I shall discuss this point more fully in Section 2.3.4.

2.3.2 Luminance gradient detection

Luminance gradient detection refers to the task of detecting a simple discontinuity of luminance in an otherwise homogeneous visual field. In a typical task, a line of light of a given length and luminance, on a background of a different luminance, is varied in thickness until the threshold for detectability of the line is found. In Section 2.2.2 it was shown that the image produced by a fine line is a line-spread function, which is essentially a gradient of luminance across several receptors. The greater the diffraction produced by the pupil, the greater will be the area over which the image is spread and the less steep will be the luminance gradient. A person's ability to detect the line at a given luminance will therefore depend on his ability to detect this luminance gradient. If the line is a bright line on a dark surround, there is no line so thin that it cannot be detected, because the luminance of the line may always be increased to compensate for the reduction in line width and guarantee a sufficiently high concentration of light quanta to produce a discriminable gradient. It is therefore meaningful to talk about the minimum resolvable thickness of such a line only if the contrast between line and surround is specified.

It has been estimated that a line of light seen against a dark background is detected at above chance level if two quanta of light are absorbed within an area of diameter D and within time T. The values of D and T vary according to the wavelength of the light and the retinal location stimulated. For green light (rod stimulation) at 7° from the fovea D is about 10′, and for red light (cone stimulation) D is about 2–4′ (Bouman, 1953). These diameters are only indirectly related to the diameters of the central areas of receptive fields (see Section 2.4.4).

The probability that the threshold number of quanta per unit time will fall in the right place clearly depends on the extent to which light is diffracted by the eye and on the length and width of the line. The longer and wider the line, the greater the likelihood that the required quantal events will occur (see Scholtes and Bouman, 1977, for a discussion of other factors). More quanta will be required if the line is set against an illuminated background, but in this case, what counts is not the number of quanta forming the image of the line, but the difference in mean density of quanta per unit time between the image of the line and the image of the background. The line may be a black line seen against a luminous background; in this case, it must be at least 0.5′ of visual angle wide to be detected, however high the background luminance (Hecht and Mintz, 1939). The visual target may also be the boundary between two unequally illuminated regions. In all cases, performance depends on the ability of the subject to detect a local luminance gradient, and the task may be described as that of detecting a luminance contrast between two regions. The subject is not required to respond to any spatial attribute of the stimulus.

2.3.3 Resolution acuity and contrast sensitivity

Resolution acuity is a measure of a person's ability to detect that two stimuli are spatially distinct. The simplest target that can be used consists of two small points of light of a given size and luminance on a dark background. The minimum spatial separation that is detected at a given level of probability is the **two-point resolution acuity**. The most commonly used target for measuring resolution acuity is a grating of alternating black and white lines. The luminance profile across the grating may be a square wave or it may be sinusoidal. It is reasonable to assume that performance on such tasks cannot exceed the physical limits set by the line-spread function of the optical system of the eye (Westheimer, 1976). The task is essentially that of detecting the local luminance gradients between the peaks and troughs of the grating (van der Wildt *et al.*, 1976). In this respect a spatial resolution task is similar to that of detecting a discontinuity of luminance. The crucial difference between the two tasks is that in a resolution task two luminance gradients run into one another, and the subject is required to distinguish the two peaks and not merely to detect the presence of a gradient. One consequence of this difference is that whereas a bright line, however thin it may be, can be detected if bright enough, a separation between two lines of less than a certain value is not detectable no matter how luminous the lines are. The theoretical limit to resolution acuity for any system is set by its point- or line-spread function. The central zone of these functions for the optics of the eye is about 1′ of visual angle wide for a 2.3 mm pupil at a wavelength of 555 nm. Resolution acuity for a centrally viewed target is also about 1′ of arc under the same conditions. Any practical system is also limited by the graininess of its detectors. In the fovea, this corresponds to the mean diameter of the cones. I have already mentioned that the narrowest bar of a grating that theoretically can be resolved is equal to the mean size of receptors. In the fovea, the cones are distributed densely enough to allow the system to operate near the limit of resolution set by the optics of the eye. In the periphery, where receptor density is lower and where several receptors converge their inputs, resolution acuity is limited by neural factors rather than by optical factors (Section 2.2.2).

Resolution acuity, as measured by a grating, depends on the subject's ability to detect a luminance contrast between the bars of a grating. **Grating acuity** is the highest spatial frequency for which a given luminance contrast can be detected. But the more general task, of which this is one instance, is that of detecting the luminance contrast of a grating over the whole range of spatial frequencies. Performance on this task is referred to as **visual contrast sensitivity**, and the function that relates contrast sensitivity to spatial frequency is the **contrast-sensitivity function**. I explained at the end of Section 2.2.2 that the contrast-sensitivity function gives some indication of the modulation transfer function of the visual system, that is, it provides a measure of performance over the whole range of spatial frequencies to which the visual system responds. It is a much more informative measure than resolution acuity, which measures visual performance at only one spatial frequency. It is possible for a person to have average sensitivity at one end of the spatial-frequency spectrum and subnormal sensitivity at the other end. For instance, Sekuler *et al.* (1980) found that, in the aged, contrast sensitivity for low spatial frequencies deteriorates more rapidly than sensitivity for high spatial frequencies. Only the contrast sensitivity function will reveal such a state of affairs. A recent review of this topic has been provided by Kelly (1977).

Visual contrast sensitivity may be measured by increasing the luminance modulation of a grating until the grating is detected. This process is repeated for several spatial frequencies of the grating and a contrast-sensitivity function plotted, as in Figure 2.9. The human visual system is most sensitive to a

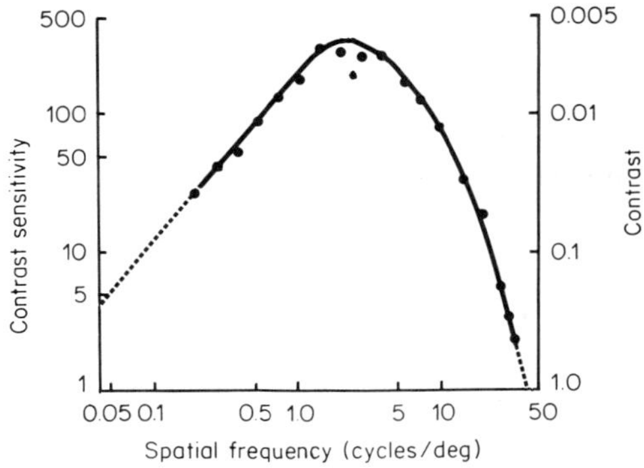

Figure 2.9 The human contrast-sensitivity function at photopic luminance levels. The low-frequency drop may be due to the reduced number of grating cycles rather than to spatial frequency (From Campbell and Robson, 1968)

sinusoidal grating with a spatial frequency of between 2 and 3 cycles/deg. The highest detectable spatial frequency is about 40–50 cycles/deg, at the fovea, and drops to about 9 cycles/deg, 10° into the retinal periphery (Sharpe and Tolhurst, 1973b). The sensitivity to a grating with a square luminance profile is determined largely by the frequency of the fundamental harmonic component; higher harmonics become important only for low-frequency square-wave gratings (Campbell and Robson, 1968). I shall comment on the significance of these facts in Section 2.4.6.

It is generally believed that diffraction and the size of the receptive fields in the retina are responsible for the high-frequency cut-off in the scotopic contrast-sensitivity function. The low-frequency cut-off is usually explained in terms of neural factors, but it could be due to the reduction in the number of bars in the stimulus area as spatial frequency is reduced (Hoekstra, *et al.*, 1974; Savoy and McCann, 1975). Howell and Hess (1978) found that the low-frequency fall-off was less steep when the grating always contained more than about 10 cycles. In assessing the influence of target size one must take account of the spatial components and luminance of the border (Estévez and Cavonius, 1976; McCann, *et al.*, 1978). When this was done, McCann and Hall (1980) found contrast sensitivity to be independent of spatial frequencies at low spatial frequencies.

Certain features of the contrast-sensitivity function may be explained in terms of the line-spread function of the visual system. This line-spread function may be determined psychophysically by measuring the effect of one line on the detectability threshold of another, parallel line at various separations (see Hines, 1976; H. R. Wilson *et al.*, 1979). A theoretical line-spread function may be derived by applying a Fourier transform to the contrast-sensitivity function (see Figures 2.9 and 2.10a). The line-spread functions derived in these two ways are very similar, and they also resemble the distribution of excitatory and inhibitory inputs across the receptive field of a ganglion cell as measured electrophysiologically (Figure 2.10b). Kelly (1977) concluded that the response characteristics of

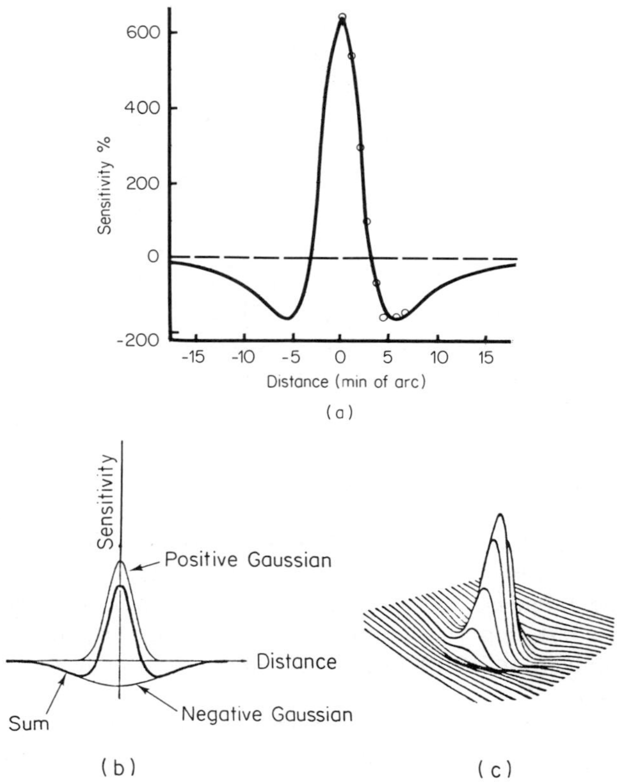

Figure 2.10 (a) The line-spread function of the human visual system derived from the Fourier integral of a contrast-sensitivity function resembling that of Figure 2.9. The data points are derived from an experiment by Hines (1976) on the effect of flanking lines at various separations on the visibility of a central line (From Kelly, 1977. Reproduced by permission of Taylor & Francis, Ltd.). (b) An idealized sensitivity profile across the receptive field of a ganglion cell, showing how the profile may be synthesized by summing two Gaussian distributions, a narrow and positive distribution representing the excitatory components of the cell's response and a broad and negative distribution representing the inhibitory components of the cell's response. (c) A hypothetical sensitivity profile of a ganglion cell across two dimensions. The 'volumes' of the excitatory and inhibitory regions are equal, which means that when the receptive field is evenly illuminated and responses are summed linearly, the firing rate of the ganglion cell is the same as it would be in the dark (Reprinted with permission from R. W. Rodieck, *Vision Research*, **5**. Copyright 1965, Pergamon Press, Ltd.)

the receptive fields of ganglion cells and their variation with eccentricity and other features of stimulation can probably account for the contrast sensitivity of human vision (see Section 2.4.4).

One of the other features of stimulation which must be taken into account is the temporal modulation of the display. Human sensitivity to spatial frequencies below about 1 cycle/deg is enhanced when the grating is moved or presented in short flashes (Robson, 1966; Kelly, 1974). This effect is illustrated in Figure 2.11. One suggested explanation for this effect is that lateral inhibition, which may be partly responsible for the visual system's insensitivity to low-frequency gratings, does not have time to develop when the grating is moved or flashed on and off. However, Tolhurst and Movshon (1975) have shown that temporal modulation does not change the spatial-frequency tuning of individual spatial-frequency selective cells in the visual cortex. The enhanced sensitivity to flickering low-frequency gratings is more likely to be due to the fact that there are cells in the visual system which are specifically sensitive to moving or flickering stimuli, and which at the same time are more sensitive to lower-frequency gratings than are cells most sensitive to stationary and steady stimuli. Wilson (1978b) has recently demonstrated the importance of temporal modulation in contrast sensitivity. He measured the line-spread function of the human visual system using a central line flanked by two lines of lower luminance at various separations. When the display was modulated temporally by a smooth change in luminance, the line-spread function had a narrower central excitatory zone and deeper but narrower flanking inhibitory zones compared with when the pattern was temporally modulated by a sudden change of luminance. He suggested that the smoothly modulated line-spread function resulted from the excitation of detectors tuned to static or slowly moving stimuli and that the suddenly modulated line-spread function resulted from the excitation of detectors tuned to flickering or rapidly moving stimuli. I shall have more to say about these two types of detector in Section 2.4.5.

Another factor which must be taken into account when relating the empirical and theoretical line-spread functions to the contrast-sensitivity function is

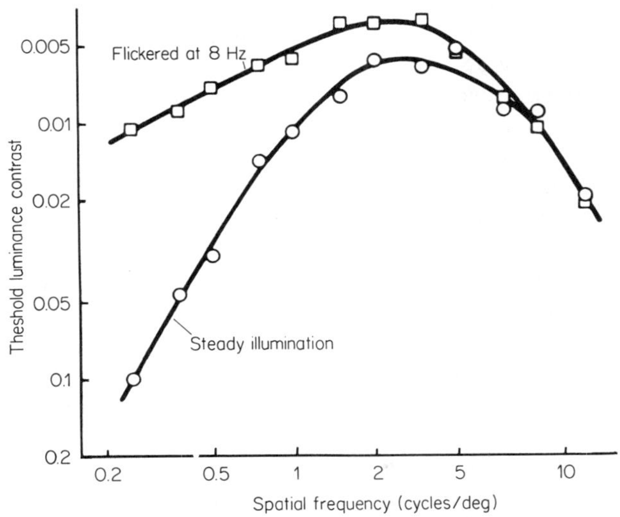

Figure 2.11 Showing how human contrast sensitivity for gratings of low spatial frequency is improved by flicker (From Kelly, 1974)

probablity summation. The displays used to measure the visual line-spread function contain a test line and one or more flanking lines. The several lines of such a display are more likely to be detected than one line, by any system of detectors, if the noise associated with each detector is independent of the noise associated with other detectors. This is because, when more detectors are stimulated, the signal in one or other of them is more likely to detectably exceed the noise level than when fewer detectors are stimulated. The effects of this factor can be calculated and allowed for. A function that has been used for this purpose is:

$$S = \left[\sum_{i=1}^{N} |S_i|^p \right]^{1/p} \qquad \text{(Quick, 1974)}$$

where N is the number of independent receptors responding to a pattern, S_i is the sensitivity of the ith detector, and P is the slope of the psychometric function in the neighbourhood of the point for 50% detection. In applying this function, a decision must be made about the spacing of the detectors over which probability summation is assumed to occur. Probability summation of luminance, in the manner just described, occurs over receptor units (receptors or converging groups of receptors) at the most peripheral level of processing. This is referred to as **incoherent**, or **probability summation**. But probability summation of other features of stimulation could also occur over detector systems for the relevant features at higher levels of processing. This is referred to as **coherent**, or **physiological summation**. Recent evidence suggests that probability summation is a more important factor in determining contrast thresholds than is physiological summation (Koenderink and van Doorn, 1980). I shall return to the subject of the line-spread function in Section 2.4.6.

It has been shown that contrast sensitivity is highest in the fovea and declines with increasing eccentricity. For spatial frequencies over 20 cycles/deg, contrast sensitivity decreases linearly with increasing eccentricity. For lower spatial frequencies, sensitivity declines only slightly until a certain eccentricity, which depends on spatial frequency, after which it decreases in the same manner as for higher spatial frequencies (Hilz and Cavonius, 1974).

Rijsdijk *et al.* (1980) measured the rate at which contrast sensitivity declined along various retinal meridia, using a small two-dimensional grating with sinusoidal orthogonal luminance profiles, displayed on a surround with the same mean luminance. Sensitivity declined least rapidly along horizontal meridia and most rapidly along vertical meridia. In other words, the loci of isocontrast sensitivity were elliptical, with the major axes horizontal.

Almost all studies of visual acuity in young infants have been concerned with resolution acuity, and have usually involved the use of a high-contrast square-wave grating of varying spatial frequency. Three types of response have been used to indicate that the child has detected the grating: the occurrence of optokinetic nystagmus, preferential looking (Held, *et al.*, 1979; Mayer and Dobson, 1980), and the visually evoked potential. This literature has been reviewed by Dobson and Teller (1978). There seems to be general agreement that resolution acuity improves during the first six months after birth, and recent evidence from the monkey reveals that adult sensitivity to spatial frequencies below 5 cycles/deg is reached some weeks before adult sensitivity to higher spatial frequencies (Boothe *et al.*, 1980). It seems that at least part of the increase in visual sensitivity with age is due to the growth of the eyeball. Olson and Freeman (1980) have produced evidence that the axial length of the cat's eyeball grows at a faster rate than the lateral dimensions of the retina during the first few weeks. As a result, a given array of receptors looks out on a region of the visual field that becomes progressively smaller with increasing age. The angular size of receptive fields therefore becomes smaller and spatial acuity more refined.

Objective measurements of refraction have revealed a high incidence of astigmatism in human infants in the first year of life. By about the age of 18 months the incidence declines to adult levels (Atkinson *et al.*, 1980).

It must be stressed that the visual contrast sensitivity to which I am referring, although measured at photopic luminance levels, involves threshold contrast levels. There is evidence that at higher contrasts the visual system has a uniform response over a wide range of spatial frequencies. For instance, gratings of identical high luminance contrast are perceived to have the same contrast even though varying widely in spatial frequency (Blakemore *et al.*, 1973; Georgeson and Sullivan, 1975). Furthermore, at suprathreshold levels, perceived contrast for both sine- and square-wave gratings is a linear function of stimulus contrast (Ginsburg *et al.* 1980). It seems that, in achieving a level response with high contrast inputs, the visual system increases the gain of those spatial frequencies most attenuated (blurred) by optical factors and by early stages of neural processing. To do this, the system requires a set of channels, each responsive to a limited range of spatial frequencies, and it needs to 'know' the degree of blur within each of these channels. Georgeson and Sullivan (1975) suggested that this could be achieved if the frequency-analysing system makes the 'assumption' that stimulation, averaged over a period of time, has a flat spectrum. Any systematic bias in the relative responses of different channels would signify loss of contrast due to optical factors or peripheral neural processes. The gains of the spatial-frequency channels could then be adjusted to restore a flat response in the system. The neural processes responsible for this improvement in visual performance are equivalent to the operation of a filter with characteristics which vary with the spatial frequency of the image (Snyder and Srinivasan,1979). The second law of thermodynamics imposes the constraint that an image can be deblurred only at the expense of increasing noise and that noise can be reduced only at the expense of increasing blur. Noise in the visual system contains high spatial frequencies, which a deblurring process will amplify. Only if characteristics of the signal or of the noise are predictable is it possible to overcome this limitation. Baker and Sullivan (1981) have shown that it is possible to deblur and at the same time improve the signal-to-noise ratio in an image, by a system which has no specific knowledge of stimuli, but which embodies only certain very general assumptions about them. They showed by computer simulation that a series of detectors with Gaussian spatial–sensitivity profiles, each sensitive to a particular spatial-frequency band, permits an easy implementation of such an optimal filter, which operates locally on each region of the image. Within each spatial-frequency channel of the system a thresholding operation is carried out which suppresses signals which are probably due to noise. The gain of each channel is then set to compensate for losses due to blur, and the image reconstituted. The mammalian visual system has the basic properties required for such operations, as we shall see in later sections of the chapter.

2.3.4 Pattern acuity

Pattern acuity is a measure of a person's ability to detect spatial features which are more complex than a simple non-uniformity of luminance or a spatial separation of stimulus elements. For instance, Vernier acuity is a pattern acuity because the subject must report when two lines no longer appear to be in alignment. He is not merely detecting an inhomogeneity of luminance, since the lines are clearly visible, nor is he detecting the presence of two lines, because the lines may be distinct even when aligned.

The distance between centres of adjacent foveal cones is about 2.0 μm, which is equivalent to a subtended angle of about 20″ of arc. A person can visually discriminate an offset in a line (Vernier acuity) when the offset subtends 2″ of arc at the eye. Thus, the spatial resolving power, or pattern acuity of the eye may be very much better than the 'grain' of the retinal mosaic. This is why Westheimer refers to this type of performance as hyperacuity, and it seems to contradict what was said in Section 2.4.1 about the resolving power of single receptors. A way out of this dilemma was suggested by Weymouth *et al.* (1923) on the basis of an idea first put forward by Hering. The idea is that the position of a line image is determined by averaging positional information from the receptors which lie along the length of the line. As we shall now see, the modern view is also based on the idea of averaging, but the limits and conditions under which it occurs have been specified.

According to the original averaging hypothesis, Vernier acuity should improve as the lines become longer. However, Sullivan *et al.* (1972) found that Vernier acuity is almost independent of the length of the lines. One task they used was to ask the subject to judge when two dots were out of vertical alignment. The minimum detectable offset, expressed as the angle of tilt of the imaginary line joining the two dots, was similar to the just discriminable tilt of an actual line extending between the dots. Sullivan *et al.* suggested that Vernier acuity measures the ability of the subject to detect the angle of the imaginary line joining the adjacent ends of the two lines. Westheimer and McKee (1977b) have confirmed these results. They varied the length of the lines in a Vernier task and also the vertical distance between the ends of the lines. Acuity improved with increasing vertical separation between the line ends, up to about 4′ of visual angle. At vertical separations of less than about 2′, offset acuity improved as line length was increased to about 5′ of visual angle, beyond which acuity remained constant. Thus, spatial information is averaged either over the region between the ends of the lines or over the length of the lines, but only over a distance of

4–5′ of visual angle. Andrews *et al.* (1973) also found that Vernier acuity, curvature acuity, and line-slope acuity improved with line length only over short distances.

Westheimer and McKee also used the two-point vertical alignment task, but, unlike Sullivan *et al.*, they found that performance was better with two points than with a line extending between them. They concluded that pattern acuity represents the limit of the ability to judge the relative positions of two distinguishable stimulus elements. Integration of spatial information over several receptors must be the basis for the high performance of the visual system, but the information is most efficiently integrated over quite small areas, at least in the fovea. These areas are about 5′ of visual angle in diameter and thus contain about 50 receptors. Furthermore, information is not integrated merely along lines of stimulated receptors, as the earlier theory suggested, but perhaps even more effectively over distinct regions of stimulation.

McKee and Westheimer (1978) reported that Vernier acuity improved with practice even when subjects were not given error feedback. This could have been due to an improvement in the sensitivity of whatever neural mechanisms are responsible for this ability, or to an increasing consistency of the subjects' use of the dimension of sensory stimulation which served as the basis of their judgments. McKee and Westheimer favoured the former explanation because in one case, when practice was confined to vertical and horizontal lines, improvement was also confined to those orientations. This argument is not conclusive, because the subject may simply have learned to use the salient main meridia of the retina consistently when judging the vertical and horizontal targets, and this criterion is not available for oblique targets.

Marshall and Talbot (1942) suggested that the constant motion of the retinal image produced by the fixation tremor of the eye improves visual acuity. It is well known that if the image motion is stopped, by artificially compensating for the motion of the eyes, then, after a short time, all objects fade from view. Eye movements are clearly needed for vision, but that is not to say that acuity is either improved or degraded by image motion in an image that has not faded. Tulunay Keesey (1960) found that visual acuity was not affected by image stabilization, before the image faded. Gilbert and Fender (1969) found that the contrast threshold of gratings was also unaffected by stabilization. Furthermore, Westheimer and McKee (1975, 1977a) found that imposed image motions at velocities up to 2.5°/sec had no effect on acuity. It has also been found that a person makes fewer microsaccades while attempting to detect fine detail (Winterson and Collewijn, 1976; Bridgeman and Palca, 1980). This would be a counterproductive strategy if microsaccades improved acuity. It seems therefore that Marshall and Talbot's dynamic theory of acuity is not correct and that small movements of the eyes have not evolved to improve acuity. The precision and accuracy of the comparison of distances between pairs of visual points up to 2° apart has been found to be, if anything, improved by image stabilization (Gilbert, 1969).

Pattern acuity, like other acuities, falls off rapidly as the image of the visual target is moved away from the fovea (see Figure 2.12). The basic reason for this drop in acuity is the decreasing density of retinal receptors with increasing eccentricity and the increasing extent to which receptors converge their outputs onto single bipolar and eventually onto single ganglion cells. These matters are discussed in Section 2.4.4. It is often claimed that there is a small region of isoacuity in the centre of the fovea. However, Millodot (1972) has shown that the area over which acuity is constant varies with the type of visual target used. Pattern acuities such as Vernier acuity have an isoacuity area of between 20 and 50′ of visual angle, whereas two-point resolution has an isoacuity area of only a few minutes of visual angle. This difference may be related to the different extents to which these types of acuity are affected by the density of receptors.

All three types of acuity deteriorate if there are other contours near the visual

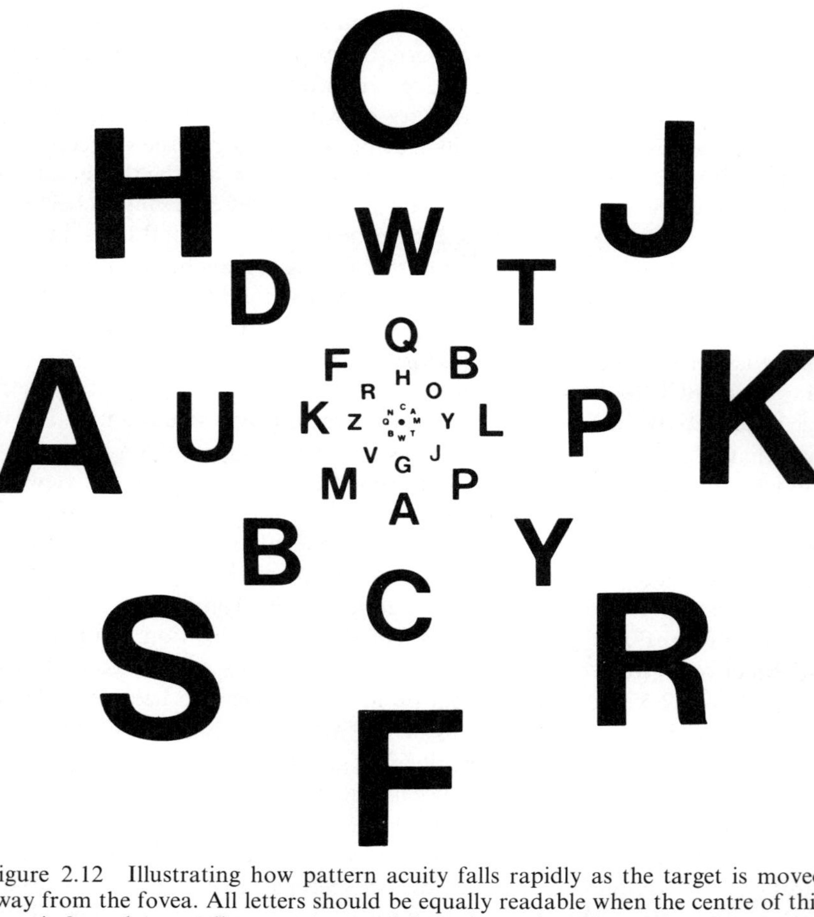

Figure 2.12 Illustrating how pattern acuity falls rapidly as the target is moved away from the fovea. All letters should be equally readable when the centre of this chart is fixated at any distance, since each letter is ten times larger than the smallest letter just readable by a person with average acuity (Reprinted with permission from S. M. Anstis, *Vision Research*, **14**. Copyright 1974, Pergamon Press, Ltd.)

target. This effect is particularly evident in amblyopic eyes and is referred to as **crowding**. Flom, Weymouth and Kahneman (1963), using a Landolt C target and distracting bars, found that the crowding effect was greatest when the distance between the bars and the target was such that the two objects could be just resolved and that the effect became negligible when the target and bars were about five times this distance apart. In peripheral vision the effect is stronger than in foveal vision even after allowing for the decreased acuity of the periphery (Jacobs, 1979). Crowding is not primarily due to optical factors, such as scattered light, because it persists when the visual target is presented to one eye and the distracting contours to the other (Flom, Heath and Takahashi, 1963). Nor is it due to the increase in total luminance. Under certain circumstances, adding luminance to the surround of a test patch decreases, rather than increases its threshold. For instance, adding a simple annular surround to a circular luminous disc of the same luminance has been found to decrease the luminance threshold of a test flash contained in the circular disc. But the addition of a thin, sharply contoured ring, which causes only a small change in luminance, always increases the threshold of a contained test flash (Lennie and MacLeod, 1973). One possible explanation of crowding is that it is due to the fact that the identity of a letter, especially in the retinal periphery, is determined by low spatial-frequency components, and that these are masked by the neighbouring contours.

2.4 Sensory coding in the retina

2.4.1 Types of coding

If any change in a distal stimulus evokes a change in the response of a single visual receptor, this can be due only to one of two things happening at the receptor: (1) a change in the rate at which light quanta strike the receptor, brought about by a change in the luminance of the stimulus; or (2) a change in the rate at which light of a given luminance is absorbed, brought about by a change in the wavelength of the light in relation to the spectral absorption characteristics of the receptor. A change in the direction from which the light strikes the receptor also changes the rate at which quanta are absorbed. This is the Stiles–Crawford effect that was described in Section 2.2.3. Once light is absorbed by a receptor, it initiates a **generator potential**. All visual generator potentials are essentially alike whether they are caused by a change in luminance or a change in the wavelength of the stimulus. The achromatic (luminance) and chromatic aspects of stimuli are distinguishable to the visual system only when the activities of several receptors with different spectral absorption characteristics are compared. Distinct spatial distributions of quanta within the boundary of a single receptor are without distinct effect on a receptor's output, which means that a receptor cannot spatially resolve stimuli. In this sense, the receptor is the ultimate unit of the spatial resolution system. However, as we saw in Section 2.3.4, pattern acuity is not limited in any simple way by the size of receptors.

From each local group of receptors, outputs converge through bipolar cells onto a single ganglion cell. Such a group of receptors is known as the **receptive field** of that ganglion cell (see Figure 2.13). A receptive field may be more responsive to a stimulus of a certain size or spatial periodicity than to stimuli of other sizes and periodicities, and in this sense a single receptive field has the capacity to spatially resolve stimuli. However, the location of a stimulus is not systematically coded in the discharge of the single ganglion cells. In other words, the single ganglion cell has a single local sign, and is in this respect like a single receptor. The output from each receptor branches to several bipolar cells, so that the receptive fields of neighbouring ganglion cells overlap. This means that, when two stimuli fall within the receptive field of a given ganglion cell, their spatial locations may be resolved only when the outputs from several ganglion cells are compared. Stimuli confined wholly to one receptor cannot be spatially resolved in any sense, but since effective stimulation of the retina is never confined to a single receptor, this limitation has little significance. Although a single receptor cannot resolve spatial features of the stimulus, it may resolve temporal features, such as flicker, because the ultimate unit of temporal resolution is not the receptor unit, but the nerve impulse. In this volume we are concerned only with spatial features of stimuli.

Identifiable neural spikes, or impulses, have not been detected in the receptors or bipolar cells, so that when I speak of nerve impulses in the eye, I refer to

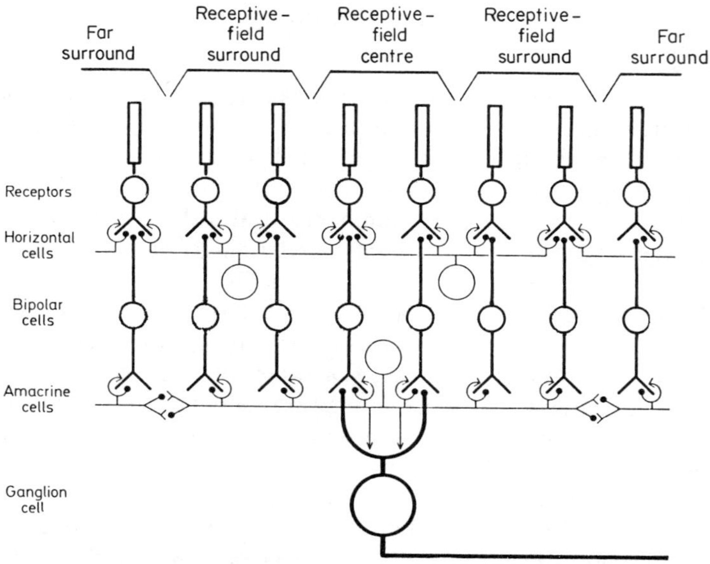

Figure 2.13 The arrangements of inputs to a ganglion cell from the different regions of its receptive field. It is assumed that horizontal cells are both excited and inhibited by receptors and that amacrine cells are excited by bipolar cells and inhibit bipolar cells and ganglion cells. Other synapses are one-way and excitatory (After Dowling and Boycott, 1966. Reproduced by permission of the Royal Society)

impulses recorded in ganglion cells. The nerve impulse and synaptic transmission may not be the only means by which information is conveyed from one cell to another in the central nervous system. For instance, it is possible that it is also transmitted by means of electrotonic coupling or by release of potassium ions. However, I shall assume that information is conveyed along axons of the optic nerve and tract wholly by means of nerve impulses. There are three main coding processes by which information is conveyed along the visual pathways, which I shall refer to as temporal, or impulse coding, cell-type coding, and topographic coding. Each of these coding processes will now be described.

Temporal, or **impulse coding**. All nerve impulses generated within a given neurone have essentially the same amplitude and conduction velocity, so that all information conveyed by the single axon must be coded in terms of the temporal features of the spike train. These features include:

1. The mean rate of impulses averaged over some moving period of time, with perhaps more weight given to the most recent impulses in this time period. Intensity attributes of stimuli, such as luminance and loudness, are coded in this way. Sensory adaptation and lateral inhibition allow temporal and spatial derivatives of stimulus intensity to be coded. The strength of the generator potential in a particular receptor, in a given state of adaptation, is uniquely determined by the stimulus energy which the receptor absorbs. No other changes in the proximal stimulus have any effect on the receptor's response. This is not true of the rate of discharge of nerve cells higher in the sensory chain. For instance, we shall see later that the frequency of impulses in the axon of a ganglion cell depends, not only on the stimulus energy absorbed by the receptors feeding into the ganglion cell, but also upon the positions of the stimuli within the cell's receptive field. Furthermore, the relative frequencies of firing over several ganglion cells are used for the central computation of many stimulus features other than luminance.
2. The duration of a burst of impulses.
3. The number of spikes in a burst of impulses. We shall see in Chapter 5 that these latter two characteristics of bursts of impulses are involved in the motor control of rapid eye movements.
4. Rhythms or temporal patterns occurring within a relatively short time period.
5. The timing, or phase, of bursts of impulses. Onset time is used, for instance, when the relative times of arrival of signals to the two ears are compared in auditory localization.
6. The frequency of bursts of nerve impulses allows us to perceive the intermittency of stimuli at frequencies below flicker fusion frequency. For a fuller discussion of temporal codes see Perkel and Bullock (1968).

Cell-type coding. A type of receptor or ganglion cell is defined in terms of the type and range of stimuli to which the cell normally responds. A cell's type is determined by three properties. In the first place, it is determined by how the mechanical structures of the sense organ in which the cell is embedded transmit or block the various forms of impinging stimulus energy. Thus, one reason why visual receptors respond to light and not to other stimuli is that the eye transmits light, while other forms of energy, such as vibrations and temperature fluctuations, are attenuated by the way the retina is maintained in the constant environment of the eyeball and by the way the eyeball is cushioned in the bony orbit of the skull. Secondly, the cell's type is determined by the physical and chemical composition of the cell itself. For example, there are three types of cone, each of which contains a specific pigment which determines the spectral sensitivity of the cell as a basis for colour vision. Thirdly, a cell's type is determined by the connections which it makes with other cells.

A stimulus feature the coding of which involves the type of receptor stimulated will be referred to as a **type feature**. For instance, colour is a type feature. For colour vision, as long as the three types of cone are intermixed, their positions on the sensory surface are not important – it is the type of pigment contained in the receptors that is crucial. Taste is also a type feature.

Topographic coding. The visual direction of an object is initially coded in terms of the relative spatial locations of excited receptors. No difference of type between receptors is necessary, but their spatial distribution on the receptor mosaic is crucial. This kind of sensory code will be referred to as a **topographic receptor code** and any stimulus feature which depends upon the integrity of this type of initial code will be referred to as a **topographic feature**.

The human topographic senses are vision, touch, pain, kinaesthesis, and audition. Of these, only vision is considered in this volume. It so happens that, at the receptor level, all spatial visual features are coded topographically, although not exclusively so because, as we shall see later, relative firing frequencies and type of receptor are also involved. On the other hand, non-spatial visual features, such as intensity, colour or flicker, are non-topographic. This correspondence between spatial features and topographic coding and non-spatial features and non-topographic coding is not logically necessary and, for instance, does not obtain in audition, where spatial location is initially coded non-topographically and frequency, a non-spatial feature, is coded, at least in part, topographically.

Note that in temporal coding the message is contained within the temporal characteristics of the impulse train; the identity of the neurones is not involved. That is why I refer to this type of code as an **impulse code**. The perception of luminance and loudness depends on impulse codes and so does the perception of flicker and the rhythm of a sound. In cell-type and topographic coding, the message is contained within the identity of the cells which are responding, within a given modality. In a topographic code, each cell has a unique value, so that it is the precise identity of the active cells that matters. In type coding, there are many of each type of cell, so that it is the class of cells to which each active cell belongs that matters. I shall refer to cell-type and topographic codes as **labelled-line codes**.

Impulse codes vary continuously over many values, and thus allow the single cell to respond differentially to different values of a stimulus dimension. Labelled-line codes have no significance for the single receptor. However, when the inputs of several receptors are related, these codes too vary more or less continuously over the stimulus continuum, and some of them have spatial-frequency bandwidths and modulation transfer functions analogous to the bandwidths and transfer functions of impulse codes (see Section 2.4.6).

Ultimately, the coding of any sensory quality depends on the internal connections which receptors make with other cells. We experience a visual sensation when the optic nerve is stimulated not because of the nature of the visual receptors but because of the central destination of the visual inputs. If the optic nerve were made to feed into the auditory cortex, then we would 'hear' light. These central coding processes will be discussed later.

So far, I have considered the coding processes discernible in the individual receptor units in a sense organ. I shall now consider how, when receptor units work in concert, they allow a much richer variety of features to be coded.

2.4.2 Channelled and distributed sensory codes

A stimulus feature or dimension will be said to be **channelled** at a specified level in the nervous system when a change in the stimulus dimension is accompanied by a characteristic change in the response of single neurones at that level. We can say that light intensity is channelled in the retina because it can be shown electrophysiologically that a change in light intensity is accompanied by a characteristic change in the firing rate of single ganglion cells, although the coding of intensity is complicated by adaptation and lateral inhibition, so that what is ultimately transmitted to the brain is some complex transform of intensity. Other stimulus features which are channelled in the retina are flicker, stimulus duration, and colour. Usually, different values of a stimulus dimension are represented in different populations of cells with overlapping response characteristics. The number of types of cell may be small, for instance there are only three types of cone responding to colour, or it may be very large, for instance position is coded into about one million ganglion cells each with its own 'local sign'. Note that the position code is very impoverished at this stage of processing, indicating only that such and such a group of cells in such and such a position has been stimulated. Single receptors or ganglion cells cannot provide unambiguous information about the position of even a point-like object, because the single receptor can never distinguish a point from an area of general illumination. The unambiguous coding of any stimulus feature requires the cooperative action of many neurones at several levels of processing. We shall see later that the size of a stimulus and the spatial periodicity of a grating also show signs of being channelled in the retina (Section 2.4.6). The direction of stimulus motion is also channelled in the retinae of lower mammals such as the rabbit, but this feature is not channelled in the higher mammals below the level of the visual cortex. There is some evidence that the mammalian retina contains channels which respond differentially to stimulus intermittency (Section 2.4.5). When a single neurone responds selectively to stimuli with certain features, it is called a **feature detector**. I suggest that a feature detector that occurs in the earliest stage of visual processing at which that feature is channelled be called an **initial feature detector** for that feature. A **secondary feature detector** for a given feature is one which receives inputs from neurones in which that feature is already channelled. A neurone may be an initial feature detector for one stimulus attribute and a secondary detector for other attributes. For example, neurones in the visual cortex which are initial detectors for orientation are secondary detectors for position and spatial frequency, because these latter features are already channelled in the neurones which feed into cortical cells.

There is a certain arbitrariness about which attributes of visual stimulation should be called features. Cells in the visual cortex change their response in characteristic ways to changes in any of several simple stimulus attributes. For instance, there are cells which respond best to stimuli in a particular orientation, moving at a certain speed and of a certain length. One could say that each of these features is channelled in the population of such neurones, or one could say that each neurone responds to a particular value of the **higher-order feature** 'orientation–speed–length', and the whole population of such neurones, which vary in their preferred values of this complex feature, may be regarded as a system for extracting this higher-order stimulus feature. I shall return to this theme in Section 3.3.8.

Stimulus features which are not channelled at a particular level will be said to be **distributed** at that level. For instance, a change in the orientation of a line in the frontal plane is not accompanied by any characteristic change in the responses of single neurones in the retina, and is therefore distributed at this level. The earliest stage at which we can say that orientation is channelled in mammals is in the visual cortex. At the retinal level, information regarding orientation is present only in the pattern of activity of several cells. Such a pattern of activity may be called a **distributed code** and any stimulus feature that is coded in this way may be called a **distributed feature** at that level of processing. All visual spatial features, except position and to some extent size and spatial frequency, are distributively coded in the retina. These features include the luminance polarity of an edge, orientation, curvature, length, and depth. Complex spatial features could be channelled by the receptors of a hypothetical eye if the stimulus domain were restricted to two-dimensional objects with a standard size, orientation, and luminance. For each shape that the system could detect, there would be receptors with an appropriately shaped mask on the receptor surface so that the cells would respond best to the shape that fitted the mask. Human vision does not operate in such a heavily constrained stimulus world, so that single receptors do not unambiguously code any feature of the world. In a varied stimulus world, it would be a disadvantage to have receptors with different and specific stimulus requirements in the same modality because most cells would then be inactive most of the time and the acuity of the system would suffer. It is much better to have a set of receptors of more or less the same kind with broad response characteristics and rely on central computation across the responses of a population of receptors for extracting information from the stimulus.

Minsky and Papert (1969) have shown that it is possible to specify the least number of independently acting simple detectors required to code a given stimulus feature, given that certain assumptions are satisfied. They called this

Figure 2.14 A set of detectors of determinate order cannot distinguish between these four shapes (From Minsky and Papert, 1969)

number the **order** of the stimulus feature. The order of any feature which is channelled at the level of the receptors is obviously 1. The feature of 'convexity' is order 3, because a decision about whether an edge is convex requires at least three simple detectors. Two detectors would confuse a straight edge with a curved edge. But this analysis depends on the assumption that the system knows that the stimulus is a continuous edge. In an ordinary visual world, three detectors could not distinguish between a convex edge and a large area of illumination, nor between a concave edge and two disconnected edges or regions. Thus, the concept of order is not easily applicable to vision.

To define any stimulus feature, the order of a feature, or the coding of a feature, it is always necessary to specify the total stimulus domain in which, *for that perceiver*, the feature is embedded (see Garner, 1962, for an expansion of this idea). This is one of the most neglected truths in the study of perception, learning, and intelligence, which I shall have occasion to stress again.

Minsky and Papert make a further important distinction. They point out that there are stimulus features which have no specifiable order. For instance, there are visual arrays for which a decision as to whether or not two regions are connected cannot be coded in a system of independently acting local detectors of finite order. An example of a set of such arrays is shown in Figure 2.14. Minsky and Papert call these unbounded features. They include 'connectedness' (the attribute of one thing being contained inside another) and parity (a feature related to there being an even or odd number of objects). Such features are important in figure–ground perception, and the proof of their existence provides a powerful argument against the belief that an adequate model of human perception can be constructed which takes only independent samples from local feature detectors. This class of device is known as a **perceptron**, an example of which is Pandemonium (Selfridge, 1956).

Up to now we have considered only features which are detected within a single sensory modality. Certain stimulus features require more than one sensory modality for their detection. For instance, for a decision as to whether retinal-image motion is produced by a motion of the seen object or of the observer's head, information from the eye and from those senses which indicate head motion is required. Such features are **intermodal**. Intermodal features are necessarily distributed at levels of processing below that at which the inputs from the sense organs concerned are combined.

2.4.3 Detecting many stimulus features in each small area of the retina

One could imagine a visual system in which each stimulus dimension had a distinct set of receptors devoted to it. Furthermore, each set of receptors could consist of subsets, each sensitive (tuned) to a particular range of values of that dimension. Such a system would have many types of receptor, so that it would take quite a large region of the retina to contain one of each kind, and only a fraction of the receptors would code any one stimulus feature. Acuity and sensitivity would be very low, and it might take time to recognize a small object with many features, because the image would have to be moved to several positions on the retina before its features could be fully sampled. To a very limited degree there is a segregation of receptors of different types in the human retina. In the fovea, there are only colour-coded cones connected to ganglion cells operating with high spatial acuity and low scotopic sensitivity, and in the periphery, there are mostly achromatic rods operating with low acuity and high scotopic sensitivity. High acuity is incompatible with high scotopic sensitivity, so that this division of labour allows each function to be more efficiently performed than would be possible if both were evenly distributed over all receptors. This increased efficiency is achieved at the price of requiring a saccadic eye movement system to bring the fovea to bear on the object of interest.

The other extreme would be to code all stimulus dimensions into each receptor and unscramble the components of the input later in the system. For example, many telephone conversations are transmitted along one cable and separated at the other end into distinct channels. Such a cable must be capable of transmitting a wide range of frequencies, that is, it must have a large bandwidth to accommodate the heavy information load. A typical neurone will not respond with a frequency higher than about 1000 impulses per second, which imposes a severe limit on the complexity of the message that it can transmit.

The solution that the eye adopts is to code all visual dimensions into as few receptors as possible in each small region of the retina. There are three basic devices for reducing the required number of receptors. The first device is to have several outputs from each group of receptors and partition them into distinct channels. For instance, several sets of outputs from red, green and blue sensitive cones converge locally in a characteristic way onto the opponent colour mechanisms in the retina and eventually form distinct opponent colour channels in the visual pathways. Other groupings of outputs from the same receptors form a system for coding the direction of motion of visual stimuli, either in the retina or in the visual cortex. There are yet other groupings and auxiliary neurones that serve to improve the relative border contrast of the input.

The second device for compressing the detection of multiple features into small retinal regions is to combine grouped inputs from one stage into other groupings at a later stage. For instance, the outputs of the colour and contrast groupings are converged in various ways onto ganglion cells to form the various kinds of receptive fields that are described in Section 2.4.4. As we shall see in Section 3.2, these fields are themselves converged in various overlapping groupings higher in the visual system. There is thus a compact set of parallel and hierarchical systems to serve each local region of the retina. I shall return to this subject in Section 3.3.6.

A third device is to have a small set of receptors broadly tuned to overlapping ranges of a given stimulus dimension and to compute the value of particular stimuli from the relative amounts of activity over this small set of receptors. I shall refer to sets of neurones with overlapping tuning characteristics as a **multichannel system**. For instance, there are three types of cone which are broadly tuned to overlapping regions of the spectrum, and this allows a smaller region of the retina to register the whole spectrum than would be the case if there were many narrowly-tuned cones.

At first glance, it might seem that a broadly-tuned system would not discriminate as many values of the stimulus dimension as would a finely-tuned system. However, as long as the different receptor types have overlapping ranges, the number of simple stimulus values that can be discriminated is not limited by the number of receptor types (Erickson, 1968). A limiting feature of a multichannel system of detectors is that different combinations of stimulus values produce indentical responses in the system. For instance, many combinations of wavelength produce the same colour sensation. This feature of multichannel systems is known as **metamerism**, and is more evident in broadly-tuned systems than in narrowly-tuned systems. Thus the human colour vision system cannot rival even an inexpensive spectrophotometer when it comes to analysing spectral components of complex mixtures of wavelengths.

Coding processes which involve only impulse codes are necessarily single-channel systems, because differences between neurones are not considered. On the other hand, all labelled-line codes are necessarily multichannel systems, because they involve different classes of neurone, each responding to a limited range of the total range to which the system responds. The range of stimulus values to which each detector responds is determined either by its tuning characteristics (type code) or by its position on the receptor surface (topographic code). The important point is that the response ranges ovelap in an ordered sequence which corresponds to the order of the stimulus continuum. The value of a particular stimulus along the stimulus continuum is indicated by the position of the mean (or mode) of the distribution of activity along the ordered set of detectors that is set up by the stimulus. The visual mechanisms for the detection of tilt and spatial frequency are multichannel systems, and so is the pitch-detecting system in the cochlea.

I cannot think of a dual-channel system that does not belong to a class of sensory coding mechanisms which I shall refer to as **opponent mechanisms**. In a dual-channel system, one class of detectors signals stimulus values on one side of a neutral point and the other signals magnitudes on the other side of the neutral point. The final sensation depends on the balance of activity, or **signed difference signal**, in each opponent pair of detectors. Not all opponent mechanisms involve two classes of detectors. For instance, each cupula of the vestibular canals is an opponent mechanism which signals the difference between hyperpolarization and depolarization of the hair cells contained within it. This system is a frequency-modulated system rather than a dual-channel system because only one class of cells is involved. Dual-channel opponent mechanisms may occur within the neural structures of the sense organ. For instance, the yellow–blue and red–green opponent colour mechanisms of the retina are each dual-channel mechanisms designed to signal the signed difference between complementary colour pairs. Dual-channel opponent mechanisms may also occur at a site in the brain where the inputs from sense organs on opposite sides of the head combine. For instance, the difference in time of arrival of a sound at the two ears is detected by an opponent mechanism in the olivary nucleus. Other examples are the push–pull combination of vestibular inputs in the vestibular nuclei that is described in Section 8.2.2, and the cortical mechanism for detection of binocular disparities. The general characteristic of all dual-channel opponent systems is that two initially distinct signals combine to generate a signed difference signal. When the difference signal is zero, the system is quiescent.

Multichannel mechanisms may be bandpass or cyclic mechanisms. In a **bandpass coding mechanism**, there is a lower and a higher threshold. For example, the set of cortical cell types, each of which is tuned to a specific range of spatial frequencies, constitutes a bandpass mechanism (see Section 2.4.6), because there are spatial frequencies which are too low and others which are too high to cause a response in the system. Note that a dual-channel opponent system is a bandpass system with an eliminated band at the neutral point.

A **cyclic coding mechanism** is one in which all possible values of the stimulus dimension cause a response in the system. Furthermore, the set of tuned channels has no 'ends', but forms a cyclic set. For example, the set of orientation detectors in the visual cortex is a cyclic system because a line in any orientation causes a response and because the channels overlap to form a cyclic set. I am not speaking here about the detection of changes in orientation or of rates of change in orientation (first and second derivatives). These stimulus attributes have bandpass characteristics, because they have lower and upper thresholds. The visual mechanism for the detection of the direction of a stimulus is cyclic, assuming that one is allowed to rotate the body so as to see all parts of the visual surroundings. The auditory localization of the source of a sound is also a cyclic mechanism.

Cyclic mechanisms are normally based on **closed** stimulus dimensions, that is, dimensions such as orientation, which are complete, or bounded. Bandpass systems are usually based on **open** stimulus dimensions, such as light intensity, which extend indefinitely in one or both directions. The sensation dimension of hue is peculiar; it has bandpass characteristics because the visual system responds to only a limited range of wavelengths, but it has cyclic characteristics because the two 'ends' of the multichannel response system interact to form the colour circle.

The processing of a given attribute may involve several stages, each involving a different type of mechanism. For example, in the coding of colour, the achromatic, or luminance component is coded by a single-channel system, the trichromatic component is coded by a three-channel system, and the opponent aspects are coded by a set of dual-channel opponent systems, and there are presumably other stages of processing yet to be discovered.

Finally, I shall consider the question of the **homogeneity of sensitivity** of multichannel systems over the range of stimulus values to which they are sensitive. The basic requirement of an homogeneous system is that the tuning functions of all the detectors are the same (except for differences in their positions along the stimulus continuum) or, if they vary, that the variation is random with respect to their positions along the continuum. This may be referred to as **channel homogeneity**. But even if this condition is met, some inhomogeneity of sensitivity (response frequency to standard input) is to be expected because of the finite number of detector types. Sensitivity to steady stimuli will tend to be highest at those points along the sensory continuum where the summed response is highest. On the other hand, sensitivity to changes in stimulation will tend to be highest where the difference in slope of the overlapping functions is greatest (see Section 3.3.8). In any multichannel system, even one in which the channels are homogeneous, there will therefore be fluctuations in absolute and differential sensitivity over the stimulus continuum.

In any bandpass system, there is an added source of inhomogeneity due to the fact that at each end of the stimulus range the number of overlapping tuned detectors declines to zero. If there are only a few types of detectors, each with a broad range of tuning, absolute sensitivity will tend to rise to a peak in the central region of the stimulus range, and the differential sensitivity will tend to show several fluctuations. For instance, the spectral luminosity curve is peaked smoothly in the centre of the visible spectrum (except at high luminance) and the spectral discrimination curve shows three peaks.

Cyclic mechanisms may be more or less homogeneous in sensitivity throughout their stimulus range or they may show major peaks in absolute and differential sensitivity. For instance, the visual orientation detection system is inhomogeneous in absolute sensitivity because the threshold for the detection of vertical and horizontal lines is lower than that for the detection of oblique lines, and in differential sensitivity because we are more sensitive to angular changes around the principal meridia than around other meridia (see Section 4.4). These inhomogeneities are probably due to channel inhomogeneities, such as higher sensitivities of channels tuned to the main meridia. This question will be discussed in Section 4.4.

2.4.4 Retinal receptive fields

The concept of receptive field was first used by Hartline (1940) to denote the retinal receptors that feed into one retinal ganglion cell. However, it is now applied quite generally to denote that region of the retina which, when stimulated in some more or less specific way, elicits a response in a specified neurone anywhere in the nervous system. This section is about the receptive fields of ganglion cells. The receptive fields of cells deeper in the visual system will be discussed in later sections.

One way to plot the receptive field of a ganglion cell is to record the electrical activity in the cell by means of a microelectrode while a spot of light is moved about the retina. This procedure reveals which receptors increase the spontaneous rate of firing of the ganglion cell and which reduce the spontaneous firing rate. By exploring with two spots of light simultaneously, it is possible to reveal regions which facilitate and regions which inhibit the response caused by stimulation of other regions. Stationary spots of light are not always the best stimuli for plotting receptive fields, because, depending on the animal being investigated, some ganglion cells respond only to stimuli which have a particular size, shape or direction of movement, and some respond only to a particular combination of such features. The most effective stimulus for a given neurone is known as its **trigger feature**. For instance, the trigger feature for certain cells in the frog's retina is a bug-sized, dark stimulus, moving against a stationary background (Maturana *et al.*, 1960). In mammals, the trigger features of ganglion cells are not so specific at this, this degree of specificity being reserved for cells at higher levels of the visual system.

Ganglion cells of mammals have two main types of receptive field, which were first described by Kuffler (1953). Both consist of two regions, a **centre**, which is roughly circular, and an annulus-shaped **surround**. In one type of receptive field, the discharge frequency of the ganglion cell increases from its resting frequency when the centre is stimulated by light, but this response is inhibited by the simultaneous stimulation of the surround. This is known as an **on-centre receptive field**. In the other type, the cell responds above its spontaneous rate only when the surround is stimulated, and this response is inhibited by stimulation of the centre. This is an **off-centre receptive field**. When both centre and surround regions are stimulated simultaneously, the on-centre type responds only at the onset of stimulation, whereas the off-centre type responds only at the offset of stimulation. For this reason, on-centre and off-centre receptive fields are known as B (bright-activated) and D (dark-activated) systems respectively (Levick, 1972; Robertson *et al.*, 1979). The receptors of the central region of a receptive field feed into bipolar cells which then converge their outputs onto a ganglion cell. Receptors of the outer region of the receptive field convey their inhibitory or excitatory influence on the centre indirectly through cells running laterally in the outer and inner plexiform layers of the retina (see Figure 2.13). Although *off* regions of a receptive field inhibit *on* regions, inhibition does not arise solely from *off* regions but has been found to arise from *on* regions when a stimulus is moved across such a region (Bishop *et al.*, 1973).

Consider the result of these excitatory and inhibitory influences on the response profile of an on-centre ganglion cell as a spot of light is moved in steps from the centre of the receptive field towards its edge. Sensitivity is highest in the centre and falls to zero at the edge of the central region, becoming negative as the spot encroaches into the inhibitory surround, until it finally rises to zero again at the outer edge of the receptive field. The sensitivity profile across the diameter of an on-centre receptive field is well fitted by superimposing two Gaussian normal functions, one narrow and representing the distribution of excitatory influences, and one broad and representing the inhibitory influences (Rodieck, 1965). The profile is often referred to as the **Mexican hat profile**, and it has also been referred to as a **DOG profile** (difference of Gaussians). It is probably no accident that the sensitivity profile of the receptive field of a ganglion cell resembles the profile of the line-spread function of the human visual system (see Figure 2.10). We shall see later that there has been some dispute about whether or not there is a disinhibitory zone outside the inhibitory zone.

According to one recent estimate there are about 33,000 receptive fields per degree of visual angle in the human fovea (Drasdo, 1977). This high density of receptive fields is achieved because they overlap. For example, a given receptor may be in the excitatory centre of several receptive fields (send inputs to several ganglion cells) and at the same time in the inhibitory surround of several other receptive fields. The net result is that neighbouring receptors mutually reinforce each other as far as certain ganglion cells are concerned, but mutually inhibit each other in terms of their effects on other ganglion cells. There is also evidence of recurrent inhibitory influences acting through horizontal cells, which control the transmission of signals from receptors to bipolar cells (Enroth-Cugell and Shapley, 1973). The effect of lateral and recurrent inhibition is to attenuate the output from areas of homogeneous stimulation and preserve the output from areas where there are stimulus gradients. Thus, the information-carrying capacity of the optic nerve is reserved for the most informative features of the stimulus, namely, lines, edges, and gradients (Barlow, 1961).

As the overall level of illumination falls, the inhibitory influences in the retina of the cat have been found to weaken and finally disappear (Barlow *et al.*, 1957). This change, predictably, results in increased sensitivity to light, but it also causes a reduction in contrast sensitivity (Enroth-Cugell and Robson, 1966; Bisti *et al.*, 1977). This dropping-out of inhibition is probably due to the high sensory threshold of the neurones that are responsible for lateral and recurrent inhibition. The process constitutes a form of automatic gain control and is responsible for the rapid phase of dark adaptation.

The size of a receptive field is usually specified in terms of its centre, although the size of this central region is not easy to determine in practice. One procedure which has been adopted is to define the radius of a receptive field as the distance from its centre to the point where its sensitivity is reduced by the exponential factor e(2.718). The size of receptive fields as determined electrophysiologically in the cat increases monotonically with increasing retinal eccentricity (see Figure 2.15a). Furthermore, it has been shown in the cat that the highest spatial frequency of a grating (dark bars per degree of visual angle) that can evoke a criterion response in a ganglion cell is inversely related to receptive field size (Figure 2.15b) and hence to stimulus eccentricity (Figure 2.15c). The increase in receptive field size with increasing eccentricity is accompanied by a proportional decrease in the density of ganglion cells (see Figure 2.16). If receptive fields did not overlap, this proportionality between field size and ganglion cell density would be inevitable. However, receptive fields overlap a great deal, and the following simple calculation shows that the proportionality between receptive field size and ganglion cell density implies that the number of receptive fields that overlap a given receptive field is constant across the retina. If the radius of a

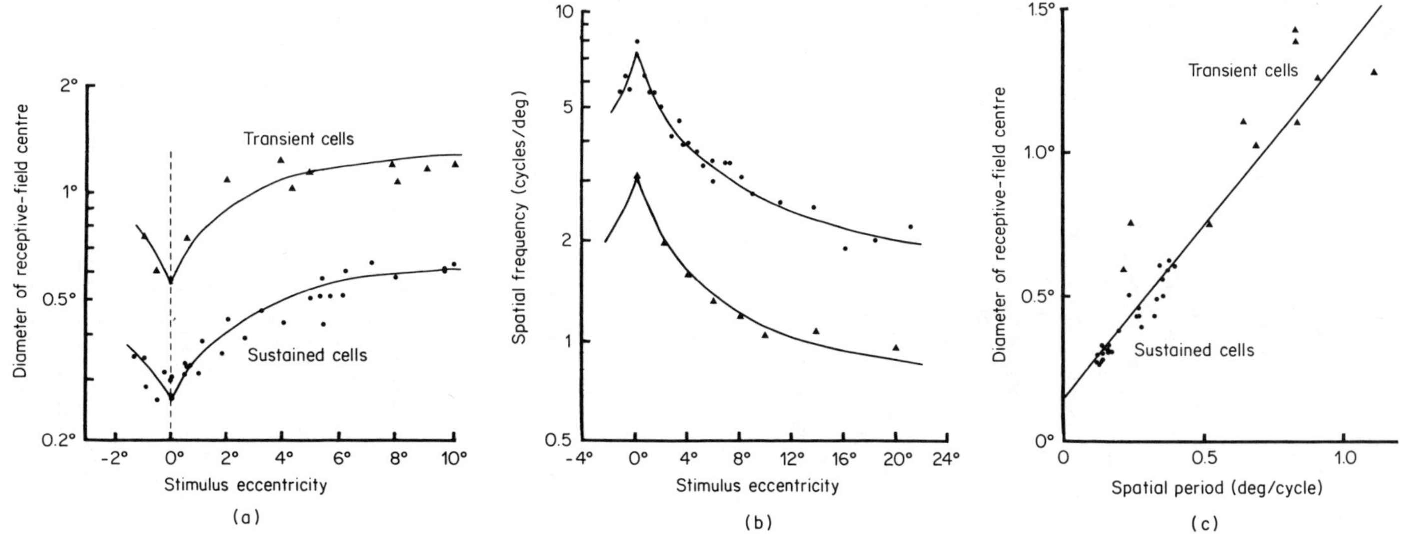

Figure 2.15 (a) Diameter of on-centres of ganglion cell receptive fields as a function of retinal eccentricity. The measurements were obtained by moving a bar stimulus along the vertical meridian of the retina while recording from ganglion cells in a single cat. Note that the linearity of relationships in these graphs and in the graphs shown in Figure 2.16 is obscured by the logarithmic scales on the ordinates. (b) The highest spatial frequency of a square-wave grating which evoked a criterion response in a ganglion cell as a function of the retinal eccentricity of the stimulus. Pooled data from four cats. All data in (a) and (b) were obtained from ganglion cells associated with briskly responding on-centre receptive fields of the cats' retinae. Two of the curves were obtained from cells showing a brisk sustained response and two from cells showing only a brisk transient response to the stimulus. (c) Showing that the diameter of receptive fields is linearly related to the reciprocal of the highest spatial frequency which evokes a response in a ganglion cell. The graph is based on data from (a) and (b) (From Cleland *et al.*, 1979. Copyright 1979 by the American Association for the Advancement of Science)

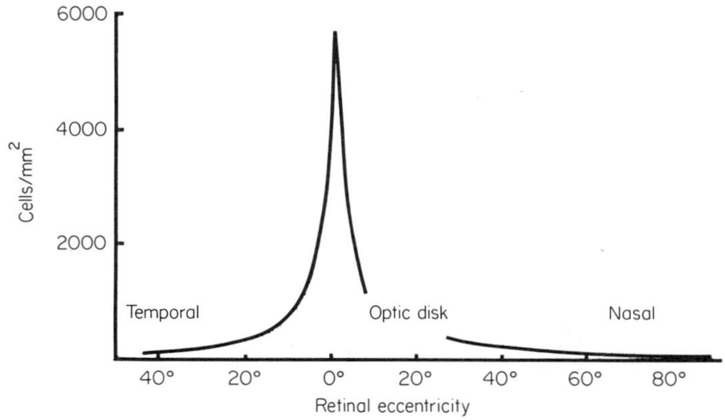

Figure 2.16 Ganglion cell density as a function of retinal eccentricity in the cat (From data provided in Stone, 1965)

receptive field is R, and the density of ganglion cells in that region is d, then the number of ganglion cells, k, that occur within the boundary of that receptive field is given by $k = \pi R^2 d$. As the area of the receptive field πR^2 is inversely proportional to d, k is a constant and gives an indication of the number of receptive fields that overlap a given receptive field (Fischer, 1973). A recent estimate, based on psychophysical data in the human, put k at 200 irrespective of retinal eccentricity. However, for receptive-field centres, the overlap number has been found not to be constant; it was about 32 at 10° eccentricity, falling to 13 at 70° (Ransom-Hogg and Spillmann, 1980). It looks as though receptive fields in the periphery of the retina, compared with those in the fovea, have centres which are smaller in proportion to their surrounds.

It is not possible to make electrophysiological measurements of the receptive fields of ganglion cells in humans. However, psychophysical procedures have been used to derive measures which are related to receptive-field size. One possibility is that receptive-field size is indicated by the size of the area within which Ricco's law holds, that is, by the area within which subthreshold stimulus intensities summate linearly. However, Scholtes and Bouman (1977) found that the Ricco diameter does not correlate with data on the density distribution of ganglion cells or receptors in the human retina, and concluded that the relation between spatial summation properties and receptive-field sizes is not straightforward. Another possibility is that receptive-field size is indicated by the diameter of point-spread functions inferred from contrast-sensitivity functions (Ronchi and Fidanzati, 1972). Westheimer (1965) developed a procedure which involves measuring the luminance increment threshold for small test spots flashed on a steadily illuminated disc. As the disc is increased in diameter, the threshold first rises because of the masking effect of the activity arising from the receptive-field centre. The disc diameter for which this threshold elevation is maximum gives an indication of the size of the centre of the receptive field. As the diameter of the disc is increased further, the threshold begins to fall and finally

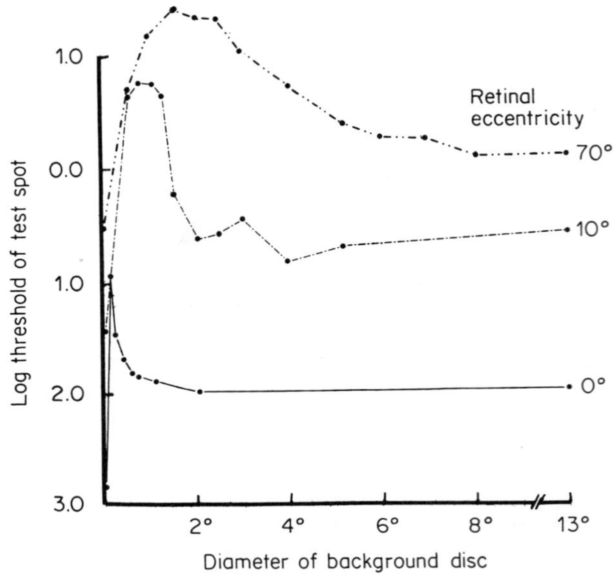

(a)

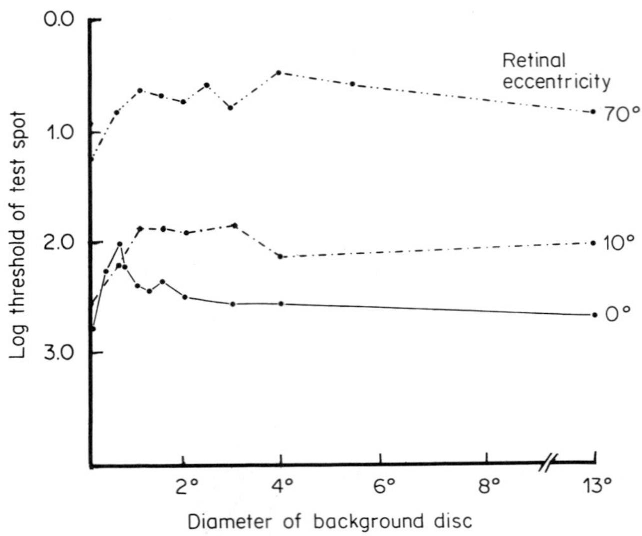

(b)

Figure 2.17 Showing how the threshold of a briefly exposed spot varies with the diameter of a steadily exposed background disc, at various retinal eccentricities. The curves in (a) were obtained when the disc and ambient background were illuminated at a photopic level (3 and 1.5 log trolands) and those in (b) when they were illuminated at a scotopic level (0.5 and 1.4 log trolands with red filter). These are a sample of the curves from one subject (Reprinted with permission from A. Ransom-Hogg and L. Spillmann, *Vision Research*, **20**. Copyright 1980, Pergamon Press, Ltd.)

levels off to a constant value. The fall in threshold is explained by assuming that the disc is impinging on the inhibitory surround of the receptive field, and it is assumed that the threshold becomes constant when the whole of the inhibitory surround is covered. The diameter of the disc when the threshold becomes constant indicates the total diameter of the receptive field. If the experiment is conducted on the dark-adapted eye, the surround inhibition becomes ineffective and consequently there is no fall in threshold with larger discs (Westheimer, 1965).

Ransom-Hogg and Spillmann (1980) applied Westheimer's procedure at various retinal eccentricities, at scotopic and photopic levels of illumination. A sample of their results for photopic levels is shown in Figure 2.17(a). The diameter of receptive-field centres, as indicated by the peaks of the curves, increased from less than 9′ at the fovea to 1° at 10° eccentricity and then more slowly to 1.75° at 70° eccentricity. The total diameter of receptive fields, as indicated by the points where the curves become level, increased from 1° at the fovea to about 6.5° at an eccentricity of 70°. It can be seen in Figure 2.17(b) that under conditions of dark adaptation the fall in threshold with larger discs is almost absent, reflecting the fact that surround inhibition has become largely ineffective. At the same time, the centre sizes show evidence of having expanded.

2.4.5 X and Y channels

We have seen that receptive fields may be classified according to their centre–surround organization. Enroth-Cugell and Robson (1966) showed that both on-centre and off-centre units may also be classified in terms of their spatial summation properties. Summation of excitatory and inhibitory inputs over some receptive fields was found to be approximately linear, and the ganglion cells into which receptive fields of this type feed were called **X cells**. Summation of inputs over other receptive fields was found to be non-linear, and their ganglion cells were called **Y cells**. A recent reveiw of X and Y visual channels has been provided by Lennie (1980a).

The response of an X cell to two equal but spatially separate stimuli at threshold luminances, within its receptive-field centre, is about twice the response from either stimulus alone. This is not true for a Y cell. Thus, only X cells summate inputs from their receptive-field centres in a linear fashion (Cleland and Enroth-Cugell, 1968). Note that these results concern spatial summation, not luminance increments, so that one cannot infer from them the relationship between stimulus intensity and impulse frequency. A difference in linearity of spatial summation in X and Y cells over the entire extent of their receptive fields is revealed when they are exposed to a sinusoidal grating which is alternating in luminance contrast. When such a stimulus is moved across the receptive field of an X cell, a position can be found at which the cell stops responding; this is known as the **null position**. Y cells respond for all positions of the stimulus within the receptive field; they have no null position (Hochstein and Shapley, 1976b; Hamasaki and Sutija, 1979). An X cell algebraically sums signals about luminance from the different regions of its receptive field, so that when a grating pattern is positioned to stimulate excitatory and inhibitory regions equally, the cell stops firing. Since Y cells do not sum signals in a linear fashion, no such null position can be found for them.

X and Y cells also respond in different ways to sinusoidal gratings which are moved over their receptive fields. As the grating moves, the response of an X cell is modulated by the rate at which bars of the grating traverse the receptive field, whereas the response of a Y cell is increased by the movement but shows only weak modulation, and even this modulation ceases at higher rates of transit. X and Y cells are distinguishable on the basis of spatial summation at both scotopic and photopic levels of illumination (Linsenmeier and Jakiela, 1979).

There is a difference in the organization of excitatory and inhibitory regions of the receptive fields of these two types of cell. The receptive field of an X cell is organized into clearly defined excitatory and inhibitory regions, with a strong antagonism between them. In the receptive field of a Y ganglion cell, the excitatory and inhibitory receptors are intermingled (Wright and Ikeda, 1974; Hammond, 1975; Robertson *et al.*, 1978; Bullier and Norton, 1979a). It has been suggested that the non-linearity of Y cells is due to several factors, including the non-linear behaviour of their bipolar cells, the geometry of ganglion cell dendrites, non-linear summation of signals within the ganglion cell, and the presence of rectifying subunits within the receptive field (Creutzfeldt *et al.*, 1970; Hochstein and Shapley, 1976a).

In the monkey, X cells are sensitive to colour opponency; their receptive-field centres contain only one type of cone and their surrounds also contain only one type of cone, but with a different spectral sensitivity (Ingling and Drum, 1973). Transient cells have the same types of cone in the centre as in the surround, although a few show evidence of some spectral imbalance in their centre–surround response (de Monasterio, 1978a). This insensitivity of the transient system to colour opponency may be responsible for the fact that if the colour of a patch is alternated, say between red and green, the change in colour is not seen if the rate of alternation is above about 10 Hz. The idea is that at these rates of alternation only the Y system is stimulated, and the change is seen as a change in luminance (Tolhurst, 1977). The Y system responds to the alternation of luminance but not to colour change, while the X system does not respond at all because of the intermittency of the stimulus.

The receptors of the fovea feed almost exclusively into X cells, and there are fewer X cells and more Y cells as one passes into the periphery of the retina (Ikeda and Wright, 1975b). The receptive fields of X cells are smaller than those of Y cells (see Figure 2.15a). Furthermore, the discharge of Y cells may be influenced by stimuli which fall well outside their receptive fields as ordinarily defined, whereas X cells are influenced little, if at all, by such stimuli (Cleland *et al.*, 1971). X cells have small-diameter axons which conduct more slowly than the larger axons of Y cells (Cleland *et al.*). However, the latency of X cells is shorter than that of Y cells, probably by an amount sufficient to outweigh the difference in conduction velocities (Ikeda and Wright, 1972b). Overall, there is no justification for calling one system 'fast' and the other system 'slow'.

Y cells respond particularly well to small stimuli travelling as fast as 200°/sec, whereas X cells respond best to stationary or slowly moving stimuli (Cleland *et al.*, 1971). The insensitivity of X cells to rapid movement seems to be due to the fact that the inhibition elicited by stimulation of the surround is sufficiently strong and has a sufficiently short latency to suppress the centre response of the receptive field. Surround inhibition in the receptive fields of Y cells is weaker and has a longer latency, so that it is ineffective when a stimulus moves rapidly across the receptive field but effective when the stimulus is stationary or moving slowly. Hamasaki and Cohen (1977) have shown that when the inhibitory regions of an X cell are masked, the on-centre of the field responds to small targets moving as fast as 1000°/sec.

In the cat, it has been found that X cells are most sensitive to spatial frequencies of about 0.5 cycle/deg and that Y cells are most sensitive to spatial frequencies below 0.2 cycle/deg (Derrington and Fuchs, 1979). X cells have been found to have a markedly greater contrast sensitivity than Y cells to gratings with spatial frequencies of 0.5 cycle/deg or more, but the sensitivities of the two types of cell have been found to be essentially the same for lower spatial frequencies (Lennie, 1980b). In spite of these differences in sensitivity at higher spatial frequencies, it has been reported that the highest spatial frequency to which X cells respond is only slightly higher than the highest spatial frequency to which Y cells respond (Lehmkuhle *et al.*, 1980).

Cleland, *et al.* (1971) suggested that X cells be called **sustained cells** because, in addition to their linear summation property, they maintain their response while stimulation lasts, and that Y cells be called **transient cells** because they respond best to stimulus onset and termination. This correlation between summation and duration-of-response properties, although widely accepted, must be qualified. In the first place, it holds only at photopic levels of illumination; when fully dark-adapted, both X cells and Y cells produce sustained responses, and at high scotopic luminance levels both produce largely transient responses (Jakiela *et al.*, 1976). Furthermore, more peripheral units, whether X or Y, tend to produce responses which are more transient, and since there is a greater number of Y cells than X cells in the retinal periphery, Y cells as a group will give responses which are more transient (Hochstein and Shapley, 1976a). The persistence of response in the two types of cell has been found to be much the same with low-contrast stimuli, and to become different only at higher-contrast levels (Lennie, 1980b). Finally, the temporal contrast sensitivities of the two types of cell have been found to be the same except at high temporal frequencies, where Y cells were found to have a slight advantage (Lennie, 1980b). Lennie argued that earlier evidence for a marked difference in temporal characteristics between X and Y cells was based on suprathreshold measurements, in contrast to his own, which were made at threshold levels. However, the former measurements are surely more relevant to understanding visual performance under normal conditions of viewing.

In the monkey, the axons of X cells synapse exclusively with cells having similar response properties in the parvocellular layers of the lateral geniculate nucleus (Dreher, *et al.*, 1976; Schiller and Malpeli, 1978). The axons of Y ganglion cells synapse with cells with matching properties in the magnocellular layers of the lateral geniculate nucleus, although some of them bifurcate before reaching the geniculate nucleus and send one branch to the superior colliculus (de Monasterio, 1978a). Post-geniculate fibres of the X system project exclusively to layers IVa, IVc and VI of area 17 in the monkey, and those of the Y system project perhaps entirely to layer IVc, although not to the same lamina within this layer to which cells of the X system project (Hubel and Wiesel, 1972). In the cat, Y cells project also to areas 18 and 19 (Le Vay and Ferster, 1977). It has been claimed that, at least in the cat, Y inputs project mainly to complex cells, particularly in area 18 (Stone and Dreher, 1973; Fukuda and Stone, 1974; Movshon *et al.*, 1978c). Ikeda and Wright (1975a) have questioned the basis for this conclusion, and report that simple cells and complex cells may both have either X or Y response properties in the cat (see also Singer *et al.*, 1975; Lee, Cleland and Creutzfeldt, 1977; Kulikowski *et al.*, 1979). Other evidence suggests that there are Y inputs to both simple and complex cells in the cat, but that X inputs feed only into simple cells and cells with non-oriented (concentric) receptive fields (Bullier and Henry, 1979c). I shall return to this question in Section 3.3.6, after I have described simple and complex cells.

The existence of separate X and Y channels in the visual system is supported by the results of studies on the effects of selective degeneration. For instance, it has been shown that if one eye of a kitten is kept closed during the first few months of the animal's life, the Y cells from that eye shrink in size and become less responsive, but that the smaller X cells are more immune to this treatment (Garey and Blakemore, 1977; Shapley and So, 1980). On the other hand, it has been claimed that X cells are more susceptible to the effects of certain neurotoxins (Vidyasagar, 1979) and to the effects of demyelinating diseases (King-Smith and Kulikowski, 1980). Furthermore, immobilization of one eye in the adult cat has been found to lead to a selective and substantial loss of binocularly driven X cells in the lateral geniculate nucleus (Brown and Salinger, 1975). Lesions of the visual cortex of the adult monkey have also been found to lead to a selective retrograde loss of X cells in the geniculate nucleus. Y cells survive cortical lesions probably because they are preserved by the still functioning collaterals which lead to the superior colliculus (Weller *et al.,* 1979). There is evidence that the larger Y cells recover faster than X cells after being damaged (Jacobson *et al.,* 1979).

More recently, a third type of ganglion cell has been recognized in the cat and called the W cell. W cells have very slowly-conducting axons (3–15 m/sec) but there is dispute about the structure of their receptive fields. Some respond to changes in luminance, and some only to slowly-moving stimuli in a given direction (Stone and Hoffmann, 1972; Levick and Cleland, 1974). W cells project directly to the superior colliculus (Hoffmann, 1973), but some also project to the lateral geniculate nucleus and visual cortex, particularly area 19 (Wilson and Stone, 1975; Toyama *et al.*, 1976). Cells with similar properties have been described in the monkey (Schiller and Malpeli, 1977), although most of the 5% of ganglion cells with unstructured receptive fields in the monkey have been reported to have properties more like X or Y cells than like W cells (de

Monasterio, 1978b). The functional significance of W cells remains obscure.

It has been widely believed that the X channel is designed for the detection of high spatial-frequency inputs which are stationary or move only slowly which, if true, would make it a system particularly suitable for pattern vision. According to this view, the Y channel is particularly suited for the detection of temporal features of stimuli, such as rapid movement and flicker. Those who hold it refer to the X channel as the **sustained system** and to the Y channel as the **transient system**. I shall briefly review some of the psychophysical evidence which points to distinctive functional roles for sustained and transient systems, but I must stress that there are two issues which have not yet been settled. The first issue is whether the X and Y channels, which are defined in terms of luminance summation within receptive fields, are coextensive with sustained and transient channels, which are defined in terms of the temporal persistence of responses. The second contentious issue is whether or not those cells which respond best to high spatial frequencies are the same cells which respond best to low temporal frequencies, and those cells which respond best to low spatial frequencies are the same cells which respond to high temporal frequencies.

It will be shown in Section 2.4.6 that the extent to which adaptation to one spatial frequency elevates the contrast threshold of other spatial frequencies reveals the presence of distinct spatial-frequency selective channels in the visual system. In the initial studies of this type, channels sensitive to spatial frequencies lower than 3 cycles/deg were not found. However, only stationary gratings, modulated at low temporal frequencies, were used. Channels most sensitive to low spatial frequencies form part of the low-resolution, movement-sensitive transient system and would be stimulated only if the adapting and test gratings were moved or flickered at high frequencies. In line with this suggestion is the fact that low spatial-frequency gratings, but not high spatial-frequency gratings, become more visible when moved or flickered (Robson, 1966; Van Nes *et al.*, 1967). When Tolhurst (1973) adapted the eye to low-frequency moving gratings, he found threshold elevation effects which were greatest at the frequency of the adaptation grating for spatial frequencies as low as 0.66 cycle/deg. These movement-sensitive, low-frequency responses were presumably due to activity in the transient channel.

Kulikowski and Tolhurst (1973) took sinusoidal gratings of various spatial frequencies and measured the contrast at which flicker could just be seen and the contrast at which the spatial structure was discriminable. The two thresholds varied independently as the spatial and temporal frequencies of the gratings were varied, suggesting that these thresholds are determined by two independent visual channels, one with the properties of the sustained channel and one with the properties of the transient channel. A similar difference of thresholds was found for flickering single lines (Tulunay-Keesey, 1972; King-Smith and Kulikowski, 1975b). However, application of a criterion-free procedure reveals that these differences between thresholds were exaggerated (Burbeck, 1981).

There is psychophysical evidence that the sustained and the transient channels inhibit each other (Georgeson, 1976a; von Grünau, 1978; Breitmeyer, 1978).

Breitmeyer and Ganz (1976) discussed the implications of this cross-channel inhibition for the study of visual masking. Kulikowski and MacCana (1980) found evidence of mutual inhibition between pattern and movement mechanisms only for gratings of low and medium spatial frequencies. They argued that it is only over this frequency range that pattern and movement are signalled by distinct channels, and that above about 20 cycles/deg pattern and movement are signalled by the same channel. Other psychophysical evidence for the existence of distinct transient and sustained channels has been presented by Harwerth and Levi (1978), Legge (1978), and Tolhurst (1975a, b).

Lennie (1980b) recently challenged the validity of the evidence for distinct sustained and transient channels in the visual system. He argued that mechanisms sensitive to high spatial frequencies have the same temporal properties as those sensitive to low spatial frequencies, and suggested that the apparent coupling between temporal properties and spatial properties can arise from (1) incidental movements of the retinal image due to eye tremor, (2) incidental changes in the retinal distribution of mechanisms which detect gratings of different spatial frequencies, and (3) unstable criteria adopted by subjects when judging spatial and temporal properties of stimuli under different conditions. He suggested that X and Y cells serve functions other than those associated with the detection pattern on the one hand and movement and flicker on the other. Most of the psychophysical evidence for a distinction between sustained and transient channels which Lennie challenged was based on measurements made at threshold contrast levels. Several pieces of evidence, which I have already cited, suggest that the two channels have similar properties at these levels and become markedly distinct only at high levels of contrast, levels at which pattern vision usually operates. Future efforts to distinguish between these two channels should be designed with this point in mind.

2.4.6 Spatial-frequency channels

Campbell and Robson (1968) suggested that the visual system behaves, not as a mechanism with a single broad-band spatial filter common to all receptors, but as a mechanism with a number of independent channels, each tuned to a particular band of spatial frequencies. In support of their suggestion they produced psychophysical evidence, which I shall describe later, and they cited physiological evidence of Enroth-Cugell and Robson (1966) that the contrast-sensitivity functions of individual ganglion cells of the cat are narrower than the overall contrast-sensitivity function. It seems therefore that spatial-frequency tuning begins in the retina. Campbell *et al.* (1969) later recorded from simple cells in the visual cortex of the cat and found that these cells are not only selectively responsive to the orientation of the stimulus, but also to its spatial frequency. Movshon *et al.* (1978c) determined the spatial-frequency characteristics of cells in areas 17 and 18 of the cat's visual cortex. The range of preferred spatial frequencies for cells in area 17 was 0.3–3 cycles/deg, and their spatial-frequency bandwidth (at half amplitude) varied from 0.7 to 3.2 cycles/deg. The most common bandwidth was approximately 1.3 octaves. Simple and complex cells (see Section 3.2) did not differ significantly in these respects. However, the preferred spatial frequencies of complex cells in area 18 were much lower than those of complex cells in area 17. Movshon *et al.* speculated that this difference might be due to a predominance of direct Y cell inputs to complex cells in area 18 of the cat's visual cortex.

Psychophysical evidence has accumulated which supports the idea that the visual system has separate spatial-frequency channels. I shall briefly review some of this evidence. A grating with a square luminance profile may be synthesized by superimposing a series of sine-wave gratings, one with the same frequency as the fundamental frequency of the square-wave grating and one at each higher odd harmonic of that frequency. The component gratings must be arranged in proper relative phase and their luminance contrasts must decrease in proportion to their increase in spatial frequency (see Figure 2.18).

The psychophysical evidence that Campbell and Robson (1968) produced in support of their original suggestion that the visual system has several spatial-frequency channels was, essentially, that a complex grating pattern, such as a

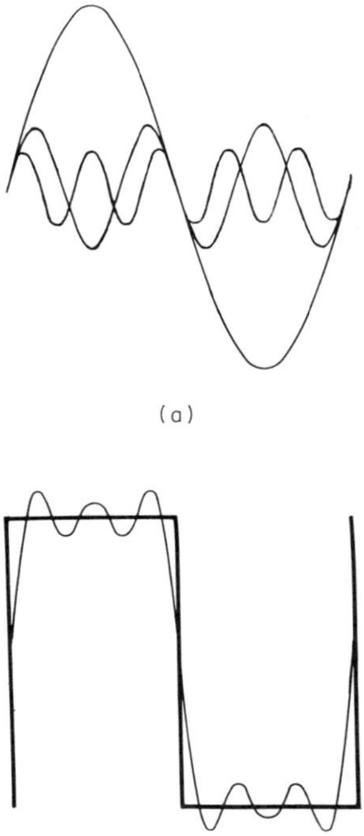

(a)

(b)

Figure 2.18 Illustrating how one period of a grating with a square-wave luminance profile may be built up by the superimposition of spatial sine waves whose frequencies are odd multiples of the fundamental frequency of the square wave arranged in proper phase and with decreasing contrast (amplitude). (a) shows the first three components, which add to produce the profile with the thin line shown in (b). As more components are added, the profile approaches a square wave

square-wave grating, becomes visible when the contrast of the fundamental component reaches the threshold of an isolated sine-wave grating with the same spatial frequency. However, as Furchner *et al.* (1977) have pointed out, this evidence is not conclusive. Consider a 5 cycle/deg square-wave grating. The lowest harmonic component is 15 cycles/deg, which has a contrast of only one-third that of the fundamental, and, in any case, the visual system is less sensitive to sine-wave gratings of 15 cycles/deg. It is not surprising that the visibility of a square-wave grating at threshold luminances is determined only by the fundamental component. This would be true whether or not the visual system had distinct spatial-frequency channels. The situation is different for low spatial-frequency square-wave gratings. For instance, for a square-wave grating of 0.25 cycle/deg, the lowest harmonic is 0.75 cycle/deg. Although this component has only one-third the contrast of the fundamental, the visual system is about three times more sensitive to it than to the fundamental (see Figure 2.3), so that the lowest harmonic is just as likely to affect the visibility of the square-wave grating as is the fundamental. By using complex spatial waveforms with two or more visible components, one can test for the presence of independent spatial-frequency channels. If the visibility of the complex pattern is no greater than the visibility of the most visible suprathreshold component when tested separately, this is evidence that the components are analysed separately by the visual system. In applying this test of channel independence, one must make allowance for the effects of probability summation (see end of Section 2.3.3).

Graham and Nachmias (1971) and Furchner *et al.* (1977) applied the above test for independence of spatial-frequency channels. They found that the visibility of a complex spatial waveform was no greater than that of its most visible component, unless its components were close together in spatial frequency. This supports the idea that there are independent spatial-frequency channels in the visual system which have partially overlapping tuning functions.

Furchner *et al.* also applied the following test of channel independence. One of two sine-wave gratings of different spatial frequencies was presented within one of two 1 sec intervals. The detection threshold for each grating was determined by asking the subject which interval contained the stimulus, and the threshold for discriminating between the two gratings was determined by asking the subject to say which of the two gratings had appeared. If the two gratings were processed by independent visual channels within non-overlapping regions of their tuning functions, the discrimination threshold should be similar to the detection thresholds, but if the two gratings were processed within a region where the tuning functions of two channels overlap, discrimination should require more contrast than is required for detection of the separate gratings. It was found that discrimination did not require more contrast than detection until the spatial frequencies of the two gratings were less than three octaves apart. The only evidence of a breakdown in independent processing beyond the limits of three octaves separation was that the contrast required to detect the fundamental component of a 0.25 cycle/deg square-wave grating was greater than that required to detect a sine-wave grating of that frequency, in spite of the fact that gratings of 0.25 and 0.75 cycle/deg were detected independently in other tests. Furchner *et al.* suggested that there may be some inhibitory interaction in this case, but another explanation may be provided by the phenomenon of the missing fundamental which I shall describe later.

Blakemore and Sutton (1969) showed that if the visual system becomes adapted to a grating of a given spatial frequency, the apparent spatial frequencies of gratings with similar frequencies are shifted in just the way that one would predict from the existence of distinct visual channels with overlapping spatial-frequency sensitivities (see Figure 2.19). In all studies on grating adaptation it is important to move the eyes to and fro across the adapting grating so as to avoid inducing an afterimage of the grating, which would produce an aftereffect in the space domain rather than in the spatial-frequency domain (see Jones and Tulunay-Keesey, 1975). Blakemore and Campbell (1969) found that the contrast threshold of a grating was increased above its normal value for some time after the subject was exposed for several minutes to a high-contrast grating in the same orientation (see Section 4.3.2). This elevated contrast threshold was as much as five times the normal value when the spatial frequencies of the inspection and test gratings were the same and fell off as the difference in spatial frequency between the two gratings was increased. They concluded that a channel maximally responsive to a given spatial frequency is not appreciably affected by a spatial frequency one octave higher or lower than its optimal frequency. K. K. De

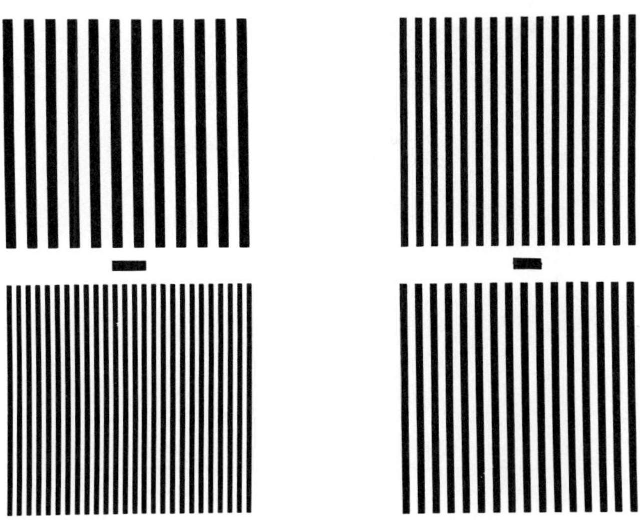

Figure 2.19 Spatial-frequency contrast effect. If the gaze is centred on the black rectangle on the left for about half a minute and is then transferred to the rectangle on the right, opposite apparent shifts in the spatial frequencies of the two identical gratings on the right will be noticed (From Blakemore and Sutton, 1969. Copyright 1969 by the American Association for the Advancement of Science)

Valois (1977a) extended the investigation of the aftereffect over a wider range of differences in spatial frequency between the test and inspection gratings. She confirmed that the threshold elevation of a test grating, after five minutes of exposure to the inspection grating, fell to zero when the spatial frequency of one grating was twice the spatial frequency of the other (one octave). Above two octaves, the contrast sensitivity to the test grating began to be enhanced (threshold reduced), and this enhancement reached a peak when the spatial frequencies of the gratings were separated by three octaves (see Figure 2.20). She explained the threshold elevation effect in terms of adaptation induced by exposure to the inspection grating within the one-octave region over which the tuning functions of neighbouring spatial-frequency channels overlap. She explained the threshold reduction effect by assuming that inhibition normally spreads over channels whose preferred spatial frequencies differ by several octaves and that adaptation to the inspection grating reduces the level of inhibition acting on the test grating. If this is so, one would expect that the threshold of a test grating presented simultaneously with a masking grating would be elevated by this inhibition when the two gratings are two octaves apart compared with when the test grating is presented alone. Tolhurst and Barfield (1978) found just such an effect. R. L. De Valois *et al.* (1978) conducted an equivalent electrophysiological experiment. They recorded from cortical cells in area 17 of the monkey and found that the normal response of a cell to a spatial grating was reduced when a second grating of four times the frequency was present, even though this second grating did not itself elicit a response in the cell.

Further support for the idea that spatial-frequency channels above one octave apart inhibit each other is provided by a phenomenon reported by Georgeson

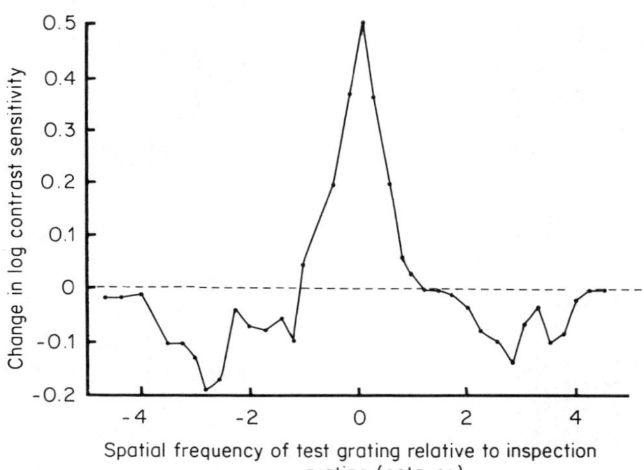

Figure 2.20 The luminance contrast threshold of a test grating as a function of the relative spatial frequency of an inspection grating. Mean data from three subjects (Reprinted with permission from K. K. De Valois, *Vision Research*, **17**. Copyright 1977, Pergamon Press, Ltd.)

(1976b). When a sinusoidal grating (preferably flickering in counterphase so as to avoid the formation of afterimages) has been inspected for a minute or two, and the gaze is then transferred to an homogeneous field, there appears an hallucinatory grating with a spatial frequency of about 1.5 octaves above that of the inspection grating. This hallucinatory grating may represent the rebound activity of spatial-frequency channels which were inhibited by the inspection grating.

Weisstein *et al.* (1977) found that gratings were masked by stimuli which shared orientational and spatial-frequency components with them even when the patterns were separated by over 4 of visual angle. They suggested that this implied the presence of processes of spatial-frequency analysis, which are independent of position, rather than processes of local adaptation in the space domain.

Inferences about the tuning width of spatial-frequency selective visual channels, based on adaptation effects, rely on the assumption that the perceptual changes result from changes in the distribution of activity over the population of spatial-frequency selective cells, and on the assumption that each cell suffers a uniform loss of sensitivity over its tuning range. Movshon and Lennie (1979) produced electrophysiological data which cast doubt on the second of these assumptions. They found that the frequency of firing of single cortical cells in the cat was reduced after they had been excited with a grating stimulus of a given spatial frequency, but that a cell's response to other spatial frequencies was hardly affected. They also found that the responses of cells in the LGN were not affected by grating adaptation, and concluded that frequency-selective, perhaps inhibitory, cortical mechanisms are responsible for adaptation to gratings. In spite of this complication, the physiological and psychophysical evidence which I have just reviewed leaves little room for doubt that the visual system contains channels specifically tuned to different spatial frequencies.

There has been some disagreement about the bandwidths of the various channels. Claims have been made that there are channels with very broad bandwidths which serve as line and edge detectors (Tolhurst, 1972; Kulikowski and King-Smith, 1973). Other claims have been made that there are channels with bandwidths as narrow as 0.25 octaves (Sachs *et al.*, 1971; Quick and Reichert, 1975). Other investigators have produced evidence that there are only channels with a medium bandwidth of about one octave (Stromeyer and Klein, 1975). More recent investigations have shown that studies supporting the existence of broad and narrow bandwidth channels failed to take probability summation (within the spatial-frequency domain or the space domain) properly into account, and it has been demonstrated that when this factor is allowed for, only medium-bandwidth channels, in the region of 1.5 octaves, are required to account for the results of experiments based on the subthreshold summation of gratings and aperiodic stimuli (King-Smith and Kulikowski, 1975a; Graham, 1977, 1980; Wilson, 1978a; Bergen *et al.*, 1979).

Albrecht *et al.* (1980) recently produced physiological data which they interpreted as evidence that cells in the visual cortex are tuned to the spatial frequency of correctly oriented sine-wave gratings rather than to the width of single bars with square-wave luminance profiles. They recorded the firing rates of single cortical cells as optimally oriented sine-wave gratings of various spatial frequencies or single bars of various widths were moved one at a time across the receptive fields of the cells. Almost all cells responded to the gratings only over a spatial-frequency range of about two octaves. They responded at a more or less constant rate to bars of all widths above a lower threshold, although never as well as to gratings at the optimal spatial frequency. The constant response to changes in bar width was explained by the fact that a bar of any width contains many spatial frequencies, so that for any given cell there were always components which fell within its tuning range. On the other hand, a bar has only part of its power within the frequency range of any one cell, which explains why bars did not excite any cell as well as a sine-wave grating with a frequency that matched the peak sensitivity of the cell. It is not clear from the results of this experiment whether the lack of tuning to bar width was due to the bar having a single spatial period or to its square luminance profile. Furthermore, it is not clear whether the bandpass tuning of cells to the sine-wave gratings was due to the gratings having an extended periodic structure or to their sine-wave luminance profile. *Further experiments are required to sort out these possibilities.*

There has also been a debate about whether the visual system carries out a Fourier analysis of the visual input. It has been suggested that such an analysis would simplify the process of pattern recognition, because a Fourier transform is independent of the spatial location of the pattern which produced it, and the relative harmonic content of the pattern is independent of the size and orientation of the stimulus (Blakemore and Campbell, 1969). According to this view, a pattern-recognition system based on Fourier analysis would have to store only the harmonic content of patterns. Schwartz (1980) has questioned this conclusion and argues that a Fourier analysis contributes nothing to the recognition of size- and orientation-invariant properties of the visual world.

Experimental support for the presence of some kind of Fourier analysis within the visual system is provided by the following evidence. Kelly and Magnuski (1975) and Kelly (1976) found that the contrast thresholds for circular and checkerboard patterns varied with spatial frequency in a way that was predicted from the two-dimensional Fourier components of such patterns (see also Carlson *et al.* 1977; Tyler and Chang, 1977; Ochs, 1979). Furthermore, Sullivan, Georgeson and Oatley (1972) found that the effects of adapting to bars which contain a broad spatial-frequency spectrum could be predicted by considering the Fourier components of the stimuli. On the other hand, Legge (1976) found that a Fourier-analysis model of visual processing did not account for the adaptation effects produced by exposure to a narrow bar of high luminance. Before considering other experimental evidence let us first look more closely at the properties required of any system if it is to carry out a Fourier analysis, and consider whether the visual system has these properties.

We shall see in the next chapter that the visual cortex contains cells which are selectively responsive to different spatial frequencies and orientations of the stimulus, which suggests that it is equipped to carry out a two-dimensional analysis of the visual input. But a true Fourier analysis requires many narrowly-tuned and independent spatial-frequency detectors for each location and each orientation in the visual field. At each location, sine and cosine spatial coefficients must be extracted. Furthermore, a true Fourier analysis requires that the tuning characteristics of these detectors be more or less the same at each position in the visual field; it requires an homogeneous retina. The visual system does not meet all these requirements. There are very few frequency-selective channels with distinct peak sensitivities for any one location of the visual field. As we shall see, one recent estimate indicates that there are only four. The most recent evidence also suggests that the frequency-selective channels are not narrowly tuned and that their tuning functions overlap, so that in this sense they are not fully independent. They also lack independence because of the mutual inhibition between them. Finally, the retina is far from homogeneous with regard to the spatial-frequency tuning of detectors at different retinal locations. This lack of homogeneity results from the fact that the mean size of receptive fields increases with retinal eccentricity.

Evidence that there are only four frequency-selective channels at each retinal location comes from an experiment by Wilson and Bergen (1979). They measured the line-spread function of the human visual system at each of several retinal eccentricities by recording the extent to which the threshold of a central line was affected by two flanking lines of lower luminance and at various separations. They also measured the threshold of a luminance distribution resembling the visual line-spread function (see Figure 2.10), formed from the difference of two Gaussian luminance distributions (which they named a DOG). Each stimulus pattern was temporally modulated, either smoothly or suddenly. These are the modulations which had been found by Wilson (1978b) to excite preferentially the sustained and transient visual channels respectively. After making allowance for the effects of probability summation, they obtained a good match between the results for the DOG pattern and the results predicted from the line-spread data by assuming that, at each location in the retina, there are only four spatial-frequency channels. Two of these channels had properties which suggested that they were in the sustained system, with peak sensitivities of 4.4 and 8.8 cycles/deg and a bandwidth at the fovea of one octave at 71% of the height of the tuning function. The other two channels had transient properties with peak sensitivities of 2 and 1 cycle/deg and a somewhat broader bandwidth than the first two channels. The bandwidth of all the channels increased with increasing retinal eccentricity. This work does not prove that there are only four spatial-frequency channels at each retinal location, but only that it is not necessary to have more than these four channels to account for the data which Wilson and Bergen obtained with the pattern they used. However, there are probably no more because, as I argued in Section 2.4.3, a multichannel system of coding which relies on a few broadly-tuned overlapping channels is a very economical way of packing a high-resolution system into each local area of the retina.

Such a small number of spatial-frequency channels at each location hardly qualifies as a basis for a proper Fourier analysis of the visual input.

Because of the spatial inhomogeneity of the retina, different ranges of spatial frequency are detected at different retinal locations. Wilson and Giese (1977) argued that the general use of extended periodic patterns has obscured the effects of spatial inhomogeneity, and that is why Wilson prefers to use the aperiodic DOG pattern. Wilson and Giese concluded that the two major factors involved in the scotopic visibility of gratings are the mean size of receptive fields as a function of retinal eccentricity and the variation in receptive field size at each retinal location, and that when both these factors are taken into account there is no need to assume that the visual system performs anything more than an extremely crude Fourier analysis. Crude though it may be, there is every indication that this quasi Fourier analysis of the visual input plays a very important role in visual processing, a role having to do with the preliminary processing of the visual input rather than with pattern recognition and memory (see Sections 2.3.3 and 2.4.7).

I shall now describe several characteristics of retinal organization that may be related to the spatial-frequency selectivity of the visual system. These fall under three headings: (1) variations in receptive-field size, (2) the number of excitatory and inhibitory zones within each receptive field, and (3) the spatial alignment of receptive fields. I shall consider each factor in turn.

Enroth-Cugell and Robson (1966) suggested that spatial tuning in the retina is a result of the varying sizes of the receptive fields of ganglion cells. The stimulus which is most effective in stimulating a given receptive field is one which just fits the excitatory region of the receptive field but which is not so large that it excites its inhibitory regions. A set of receptive fields of a given size will therefore respond most effectively to a spatial sine wave of a given mean spatial period (reciprocal of spatial frequency). They calculated that the range of diameters of receptive-field centres required to account for the range of spatial-frequency sensitivity which they had detected electrophysiologically was about 0.5–4.4° of visual angle, which is consistent with other determinations of receptive-field sizes (Wiesel, 1960). Movshon *et al.* (1978a) recorded from simple cells in the visual cortex of the cat and found that for most cells the width of the excitatory region of the cell's receptive field was the same as the width of a single bar of the grating to which the cell was most sensitive. All this evidence suggests that the width of excitatory regions of receptive fields of ganglion cells or of cortical cells has something to do with a cell's specificity to spatial frequencies, but, as we shall see, this is not the whole story.

Insofar as the mean size of receptive fields increases with increasing retinal eccentricity, it might be supposed that different parts of the retina are optimally tuned to different spatial frequencies; that the lower the spatial frequency, the more peripheral the position of maximum contrast sensitivity. The alternative view is that all spatial frequencies are detected optimally when the target is centred on the fovea. In recent studies on this issue, it was found that all spatial frequencies were detected optimally when the target was centred on the fovea, the only exception being that maximum visibility was reached outside the fovea at very low levels of luminance modulation. For a wide grating, of whatever spatial frequency, in a surround with equal mean luminance, contrast sensitivity was found to be determined entirely by a 2° wide strip of the grating closest to the fovea (Kroon and van der Wildt, 1980; Kroon *et al.,* 1980).

Sullivan, Georgeson and Oatley (1972) found that the human visual system is tuned not only to the width of each bar of a grating, but also to the spatial frequency of several bars. Kulikowski (1969) found that human contrast sensitivity improves as the number of cycles in a grating increases up to five, which again suggests that it is spatial frequency and not merely spatial period which is being detected. Ikeda and Wright (1972a) demonstrated electrophysiologically that there is a disinhibitory region surrounding the inhibitory annulus of receptive fields of ganglion cells of the X type and suggested that a receptive field has therefore five alternating excitatory and inhibitory regions across its diameter. Such a receptive field would be most responsive to a grating consisting of three bright bars and two dark bars. They found that the spatial period of the excitatory and inhibitory regions in a receptive field of a cat corresponds to the spatial period of the grating to which the cat is most sensitive. Further support for the existence of disinhibitory regions surrounding receptive fields was provided by Westheimer and Wiley (1970). They found that the detectability of a spot rose, fell, and rose again as a second spot was moved away about 1° of visual angle from the target spot. Thus, the visual point-spread function, which is what this experiment measured, shows an outer disinhibitory annulus, an inner inhibitory annulus, and an excitatory centre. Rentschler and Hilz (1976) and Beyerstein and Freeman (1977) found a similar disinhibitory region in psychophysically determined line-spread and point-spread functions. However, others have not found evidence of the disinhibitory zone after making allowance for the effects of probability summation, that is, for the greater probability of detecting simultaneously presented lines compared with a single line (Hines, 1976; Wilson, 1978a, b). (See Wilson, Phillips, Rentchler and Hilz (1979) for a discussion of this issue.) Furthermore, Wilson (1978b) produced evidence which suggests that the dependence of grating sensitivity on the number of bars of a grating is also due to spatial probability summation and to the variation in line-spread functions with retinal eccentricity. If this is so, there is no need to look for a specific neural mechanism to account for the fact that the visual system is more sensitive to several bars of a grating than to one.

Kelly (1975a) has shown that, irrespective of the size, composition, or degree of overlap of receptive fields, any regular spatial periodicity in their distribution in the retina will narrow the bandwidth of the response of any cortical cell into which a set of such receptive fields converges. Figure 2.21 shows the Fourier spectra, of the responses to a vertical grating, of horizontal rows of regularly spaced receptive fields with hypothetical spatial sensitivity profiles. Overlap between receptive fields has been arbitrarily set at zero. It can be seen that, as the number of regularly spaced receptive fields in a row increases, so does the narrowness of the effective bandwidth of the averaged response. Even a small departure from a random distribution of receptive fields is sufficient to generate an effective spatial-frequency filter. This may also be a reason why human

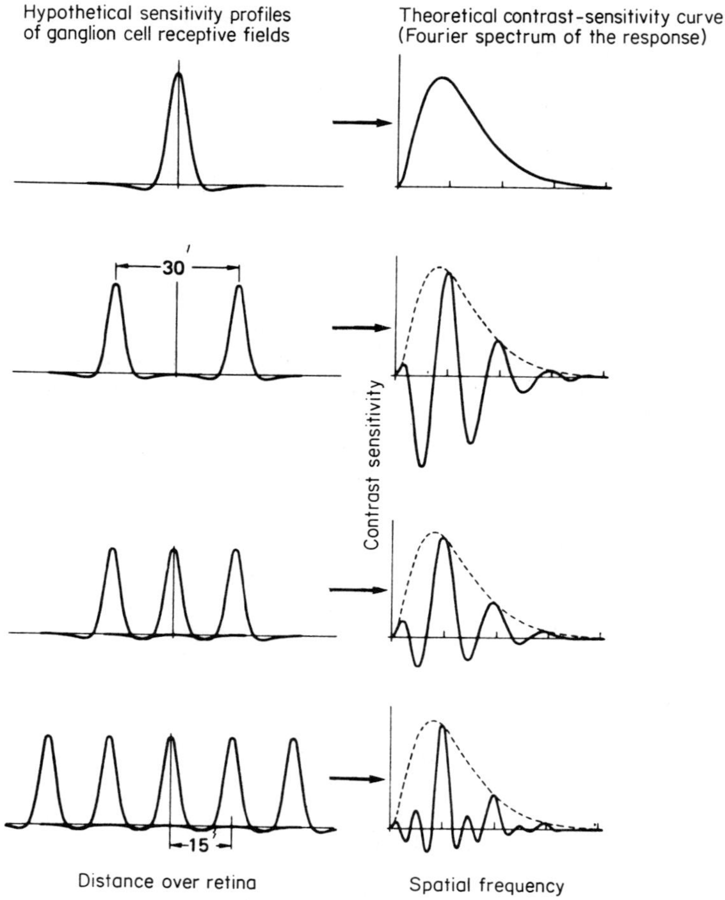

Figure 2.21 The curves on the left represent hypothetical line-spread functions of regularly spaced receptive fields in horizontal rows of from one to five fields. The curves on the right are the predicted contrast-sensitivity functions, for vertical gratings, of detectors which average the outputs of the rows of receptive fields. The contrast-sensitivity functions were derived from the corresponding line-spread functions by Fourier integration (Reprinted with permission from D. H. Kelly, *Vision Research*, **15**. Copyright 1975, Pergamon Press, Ltd.)

contrast sensitivity to a grating of a given spatial frequency improves as the number of bars is increased. Spatial-frequency selectivity in the visual system is probably also controlled by interactions among cells in the visual cortex.

I shall finish this section by discussing some evidence that the visual system is particularly sensitive to the Fourier components of edges even though it may not have low-level detectors especially sensitive to lines and edges.

If the fundamental component is removed from a square-wave grating with a spatial frequency of more than about 5 cycles/deg, the grating is no longer detectable, because all its harmonic components are too high in spatial frequency. If the fundamental is removed from a square-wave grating of 1 cycle/deg, the grating appears like a sine-wave grating of three times the spatial frequency of the fundamental. This is because, for a 1 cycle/deg grating, the only component visible, apart from the fundamental, is the lowest harmonic component (3 cycles/deg). If the second lowest harmonic component of a square-wave grating (the fifth harmonic) is made visible by increasing the luminance contrast of the grating or by selecting a grating with a lower spatial frequency, removing the fundamental does not change the square appearance of the grating (Campbell *et al.*, 1978). The rule seems to be that a grating with its fundamental spatial frequency removed still looks like a square-wave grating as long as at least two of its higher spatial components are above threshold. However, the higher spatial components must be in proper relative phase; if they are not, one sees both components in alternation (Atkinson and Campbell, 1974; Furchner and Ginsburg, 1978). The visual system is able to reconstruct the square-wave pattern from the partial harmonic components even when the fundamental component is missing. An analogous phenomenon has been known in audition for some time. In this case a complex tone seems to retain the same pitch when its fundamental is removed. Campbell *et al.* concluded from their observations that the visual system is designed to detect sharp edges by means of 'quasi Fourier analysis'.

2.4.7 Sensitivity to spatial phase

Of particular relevance to the subject of this book is the question of whether spatial frequency and spatial phase (position) are coded by the same or by distinct mechanisms. At first glance this may appear to be an empirical issue, requiring an experimental approach. However, at least part of the answer can be arrived at by a theoretical analysis, once the general properties of spatial-frequency channels are known. The basis of this analysis was provided by Gabor in 1946, and was developed with reference to temporal frequency and time rather than spatial frequency and position. Nevertheless, the conclusions, which I shall now summarize, apply to both domains.

If the frequency dimension of a signal is represented by the abscissa and the time dimension by the ordinate of Cartesian coordinates, we have what is known as a **time–frequency diagram**. Consider an impulse (delta function) which occurs at a well-defined time but has its energy distributed evenly over the whole frequency spectrum (it contains all Fourier components, all with the same amplitude). The impulse is represented in the time–frequency diagram as a horizontal line, as in Figure 2.22. An impulse is a useful test stimulus to apply to detection or communication systems because, if a system's response to a pulse is known, it is possible to infer the modulation transfer function of the system. Recall that the response to the impulse of any linear system is the Fourier integral of its modulation transfer function. Consider next a pure tone of infinite duration. This signal has a well-defined frequency but an undefined temporal epoch, and is therefore represented in the time–frequency diagram as a vertical

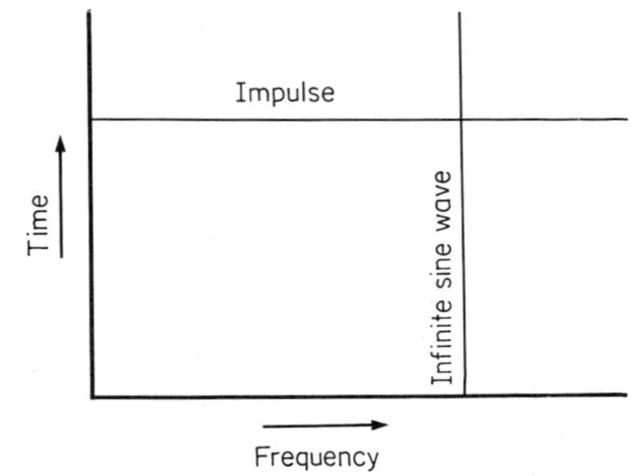

Figure 2.22 The time–frequency diagram showing the plot of the impulse (delta function) as a horizontal line and that of the infinite sine wave as a vertical line

line. Any finite signal may be regarded as intermediate between the pulse, which is determinate in time but not in frequency, and the infinite sine wave, which is determinate in frequency but indeterminate in time. A detector can be made so that it extracts time information from signals but disregards frequency or so that it detects frequency and disregards time. For instance, an ideal oscillograph with uniform response over the whole frequency range and a very short time constant is an instrument of the first type, and a bank of narrowly-tuned oscillators, each with a long time constant, may be regarded as an example of the second type of instrument. The same argument applies if we substitute spatial frequency for temporal frequency and position for time. Note that the ideal frequency detector has narrowly-tuned components with long time constants and the ideal time (or position) detector has broadly-tuned components with short time constants. Because of these opposed requirements, it is impossible to design an instrument which will extract both types of information with maximum efficiency. If both types of information are required from the same instrument, there must inevitably be a compromise, which is expressed by the fact that in any instrument:

$$\text{Time constant} \times \text{bandwidth} \geq 1/2.$$

A consequence of this relationship within the space domain is that, if we define the sensitivity of a detector to differences in position as Δs, and its sensitivity to spatial frequency as Δf, then the product of these uncertainties cannot be less than one-half, or:

$$\Delta s \times \Delta f \geq 1/2.$$

This reciprocity between these two uncertainties is essentially the same as that expressed in Heisenberg's principle of uncertainty, which states that it is impossible to know both the position and frequency characteristics of a fundamental particle at the same time.

The best compromise between the detection of position and that of spatial frequency is achieved when $\Delta s \times \Delta f$ is a minimum, which in the ideal case is 1/2. Gabor defined the characteristics of a detector for which this would be true. For a detector in the space domain, these requirements are that the sensitivity profile (profile of the line-spread function) of each component should be Gaussian, and that at each location there should be pairs of detectors in quadrature. Cortical cells with symmetrical (cosine) and antisymmetrical (sine) sensitivity functions are in quadrature, with sensitivity profiles phase-shifted by 90°. There is evidence that these types of all occur in pairs Pollen and Ronner (1981).

These two requirements may be summarized by the expression:

$$S(x) = e^{-x^2/2s^2} \begin{matrix} \cos \\ \text{or} \\ \sin \end{matrix} \; 2\pi F x \qquad \text{(Kulikowski, 1980)}$$

where $S(x)$ is the sensitivity profile across the x axis of a detector, s is the standard deviation of the sensitivity profile, F is the optimal spatial frequency to which the detector is tuned, and x refers to distance along the x axis. It is assumed that at each location of the detector surface there are pairs of such detectors, one member of each pair having a sine (antisymmetrical) sensitivity profile and the other member having a cosine (symmetrical) sensitivity profile. Different pairs of receptors are tuned to different regions of the spatial-frequency spectrum.

There is every indication that the sensitivity profiles of the receptive fields of visual spatial-frequency channels are approximately Gaussian, and we shall see in Section 3.2 that there are cells in the primary visual cortex some of which have receptive fields with symmetrical sensitivity profiles and some of which have receptive fields with asymmetrical receptive fields. It looks as though the spatial-frequency channels of the visual system are designed to achieve an optimal compromise between the detection of spatial frequency and the detection of position. This arrangement seems also to be ideally suited for optimally deblurring and reducing noise in the visual input, as I have shown in Section 2.3.2.

Studies which purport to demonstrate that spatial-frequency channels in the visual system are designed exclusively for the detection of spatial frequency and not at all for the detection of position (bars and edges) are misconstrued. Such studies (for example, that of De Valois *et al.*, 1979; Green, 1980) are based on the argument that cells in the visual cortex respond to the diagonal Fourier components of a square checkerboard pattern and not to the edges of the component squares, which are not represented in the Fourier transform of such a pattern. But the basis of this argument depends on the relationship between the size of the squares of the grating and the size of receptive fields. For a checkerboard in which the component squares are small in relation to the receptive fields of cortical cells, the diagonals are the only effective Fourier components, but they are also discriminable as bars running across the pattern. When the squares are large with respect to receptive-field size, the edges of the squares become effective local Fourier components and, at the same time, they are also detected as edges. In terms of the quasi Fourier analysis carried out by the visual system, a checkerboard pattern does not help to distinguish a Fourier model from a position-sensitive model of the visual system. On purely theoretical grounds, any attempt to make such a distinction is probably doomed.

I shall now review some studies on the detection of spatial phase by the human visual system.

Burr (1980) determined psychophysically the human threshold for the discrimination of differences in the relative phase of two superimposed sinusoidal gratings. The two gratings had spatial frequencies of f and $3\,f$ for various values of f between 0.33 and 10 cycles/deg. The contrast ratio was fixed at 3 to 1, but the luminance level was varied. The subject had to decide whether or not pairs of such displays had the same relative phase. The threshold for the discrimination of a difference in phase relative to a zero phase difference was found to be about 30° of phase angle for all spatial frequencies tested, as long as the two component sinusoids were clearly visible. Expressed in minutes of visual angle, the threshold varied between 5 and 0.3.

Westheimer (1978) conducted an experiment in which the subject had to detect the direction of a shift in phase when two sinusoidal gratings were presented in rapid succession. He found that sensitivity to changes in phase was about 10″ of visual angle at all spatial frequencies, rather than a constant phase angle. Burr argued that Westheimer's procedure measured motion sensitivity rather than phase sensitivity, and that this was why he obtained a constant absolute threshold.

Burr also found that sensitivity to a difference in phase was highest when one of the comparison displays had the same relative phase as components of a square wave. This is not surprising, because at this position the peaks and troughs of the component gratings are symmetrically disposed and a departure from symmetry should be easily detected.

Note that in Burr's experiment the contrast ratio was fixed at 3 to 1. There is evidence that phase discrimination is much reduced when the contrast ratio is less (Nachmias and Weber, 1975). The effects of differing frequency ratios on the sensitivity to differences in phase have yet to be determined. There is no doubt that humans are able to detect much finer differences in the offset of two lines than the phase sensitivity reported by Burr. This may partly be because the lines contain many spatial frequencies, but it is probably also due to the fact that Vernier offset taps the orientation-detection system (see Section 2.3.4).

A system which can detect the relative phases of Fourier components is obviously phase-sensitive as well as frequency-sensitive. However, the results of an experiment by Jones and Tulunay-Keesey (1980) were interpreted as indicating that channels sensitive to spatial frequency are distinct from those sensitive to spatial phase. In this experiment, an adapting grating was optically stabilized on the retina and flickered sinusoidally in counterphase so as to avoid local retinal adaptation. The elevations of thresholds of subsequently presented stabilized test gratings were found to be independent of the relative phases of adapting and test gratings. This suggests that phase information is extracted by a distinct system. However, the adapting grating in this experiment was 3 cycles/deg, and the results might have been different with gratings of lower spatial frequency, in fact Ross and Johnstone (1980) found that the thresholds of low-frequency compound gratings are affected by phase. Furthermore, since the adapting grating was flickered, the effect may have been confined to the transient channel of the visual system. Receptive fields of the transient channel

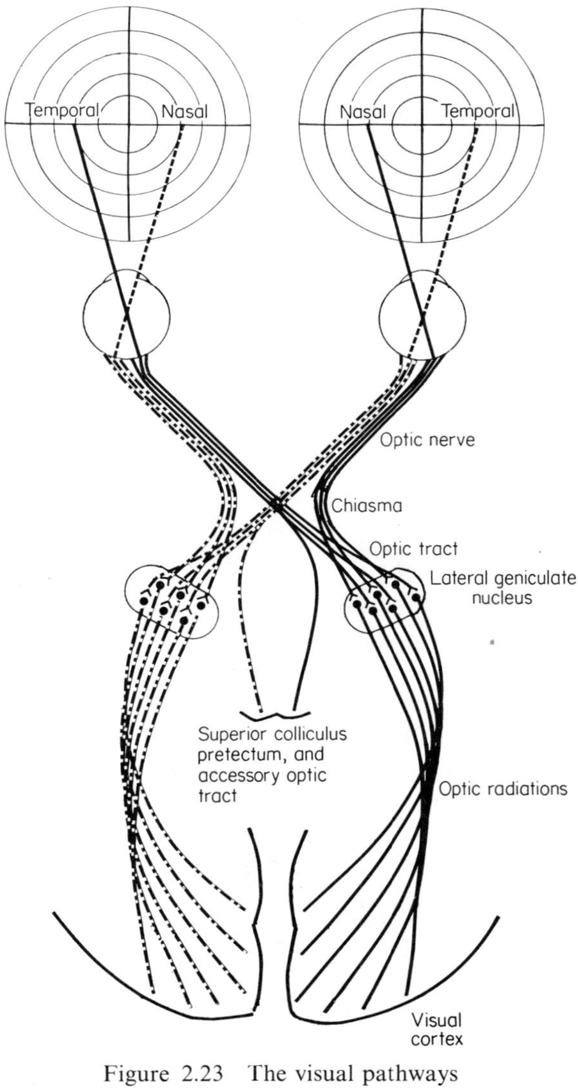

Figure 2.23 The visual pathways

are insensitive to changes in the spatial position of stimuli within their boundaries, and therefore operate as phase-insensitive spatial-frequency detectors. Receptive fields of the sustained system are sensitive to the position of the stimulus, and should therefore respond to spatial frequency and phase. Further experiments are required to elucidate the relation between spatial phase and frequency in visual channels, but, as I have already argued, spatial-frequency detectors with approximately a Gaussian sensitivity profile and sine and cosine types must be sensitive both to spatial frequency and to position, or phase.

2.5 The visual pathways

There are about 120 million rods and seven million cones in the human retina, which feed into about one million ganglion cells. The axons from ganglion cells leave the eye to form the **optic nerve**. Collateral branches of axons in the optic nerve pass to the superior colliculus in the midbrain (see Section 5.5), the **pretectum** (see Section 5.4.5) and through the **accessory optic tract** to the **inferior olive** (see Section 9.5.5). All these structures are subcortical. The optic nerve proceeds to the optic chiasma, where axons from the temporal half of the left eye join those from the nasal half of the right eye to form the left **optic tract**, which

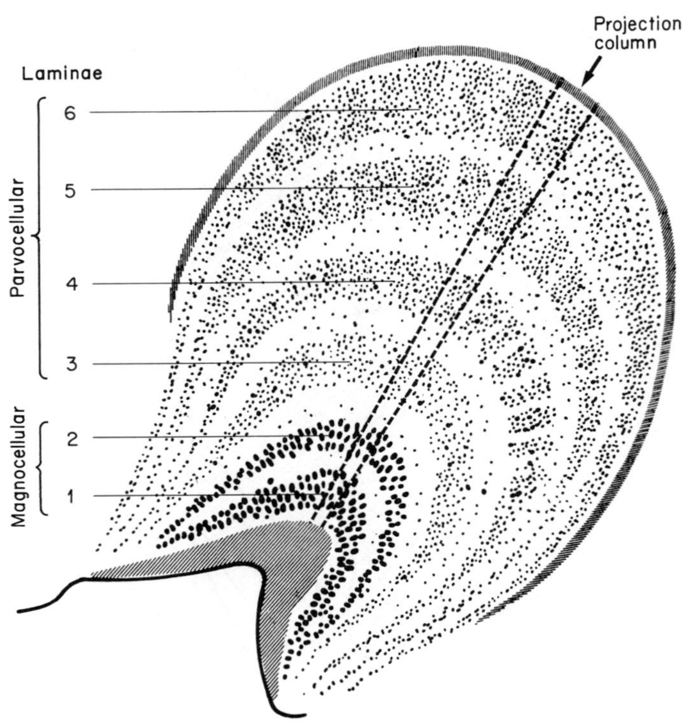

(a)

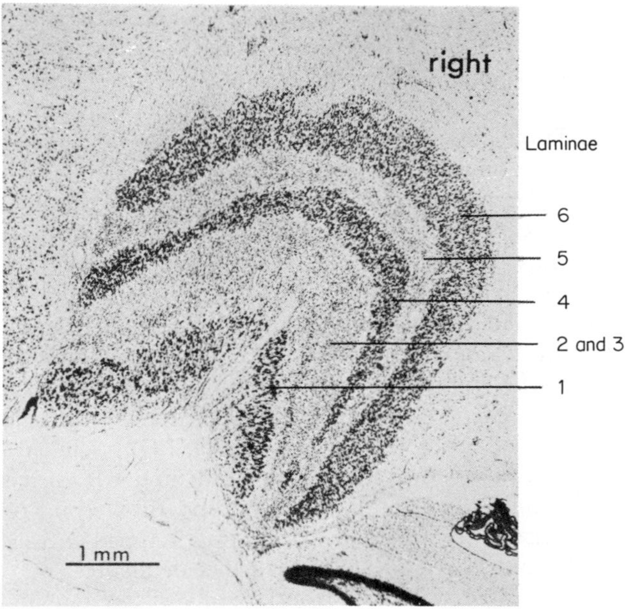

(b)

Figure 2.24 (a) Illustration of lamination and projection columns in a coronal section of the lateral geniculate nucleus of a monkey. A projection column is defined as containing 90% of cells having a single visual direction (From Szentágothai, 1973). (b) Coronal section through the right lateral geniculate nucleus of a monkey whose right eye had been removed soon after birth. The unstained laminae 2, 3, and 5 are those which have degenerated because of the removal of the ipsilateral eye. The stained laminae 1, 4, and 6 contain the surviving cells innervated from the intact eye (From Hubel *et al.*, 1977. Reproduced by permission of the The Royal Society)

projects through the **lateral geniculate nucleus**, or **LGN**, to the visual cortex in the left hemisphere (see Figure 2.23). Axons from the right half of each eye form the right optic tract, which projects to the right hemisphere. The net result of this reshuffling of axons, or **hemidecussation**, is that the left half of space is represented in the right hemisphere and the right visual hemifield in the left hemisphere. Thus, four distinct direct visual inputs may be distinguished: (1) geniculostriate, (2) collicular, (3) pretectal, (4) accessory.

It has been shown already that, because of the overlap of receptive fields, several ganglion cells receive inputs from the same area of visual space. Axons from ganglion cells which receive inputs from the same local region of one eye become grouped in the optic tract with axons receiving corresponding inputs from the other eye. The individual axons in such a local group differ in their colour coding and spatial-frequency characteristics, and some have X and some

Y characteristics. These various types of input from corresponding local regions of the two eyes converge into the same region of the LGN. In the primate, this region is a six-layered column of cells, more or less at right angles to the surface of the LGN (see Szentágothai, 1973). Within each column, the inputs from the two eyes, from the different colour systems, and from X and Y cells are kept separate in different layers, or **laminae** (see Figure 2.24).

Laminae 1, 4, and 6 receive inputs from the nasal half of the contralateral eye and laminae 2, 3, and 5 from the temporal half of the ipsilateral eye. The contralateral hemifield of visual space is thus represented in each lamina, with the six laminae in exact registration. The four dorsal laminae are known as the **parvocellular layers**, and it is these layers that receive inputs from the X ganglion cells. The two ventral laminae are known as the **magnocellular layers**, and these receive inputs from the Y ganglion cells. Incoming axons within each layer make synaptic contact with LGN relay cells, which in turn send axons along the optic radiations to the visual cortex (Hubel and Wiesel, 1972; Schiller and Malpeli, 1978). Each LGN column therefore receives all inputs from a small region of space which, within about 10 of the fovea, is approximately 0.5 of visual angle in diameter. An area this size in visual space is thus mapped into such a cylinder (see Freund, 1973, and Berkley, 1978, for further details).

Each LGN relay cell receives axons from one or very few ganglion cells. Thus, the functional characteristics of the retinal input and its receptive-field organization remain basically unaltered in the LGN. The only alteration found in the neural coding in the LGN of the cat is an attenuation of response frequency, particularly in Y channels (Bullier and Norton, 1979b). The elaborate assembly of cell types within the layers and columns of the LGN must therefore serve the purpose of providing for lateral inhibition, which is presumably responsible for this response attenuation. Several types of inhibition in the LGN have been described:

1. Interlaminar inhibition between cells in different laminae; for instance, inputs from one eye inhibit those from the other eye, which is perhaps part of the neural substratum for binocular rivalry. In the monkey, this interocular inhibition is apparently limited to the Y system (Rodieck and Dreher, 1979). Y inputs from an eye inhibit X inputs from the same eye (Singer and Bedworth, 1973). The significance of this fact will be discussed in Section 7.8.3.
2. Intralaminar inhibition between cells in the same lamina in neighbouring columns. This attenuates the input from regions of homogeneous luminance more than inputs from regions where there is a luminance gradient. It also increases the directional selectivity of motion detectors in lower mammals (Levick *et al.*, 1969).
3. Extrinsic inhibition (and excitation) from corticofugal neurones entering the LGN columns from areas 17, 18, and 19 of the visual cortex. These extrinsic inputs to the LGN presumably allow the visual inputs to be selectively controlled from above, but the significance of this control, which seems to be largely

confined to the X system, is still obscure (see W. Singer, 1977, and Tsumoto *et al.*, 1978, for a review of this topic).

Changes in potential in the LGN associated with eye movements have been recorded in the cat and monkey (Feldman and Cohen, 1968; Jeannerod and Putkonen, 1971; Magnin *et al.*, 1974). It is possible that these influences have something to do with the elevation of the visual threshold which is known to accompany rapid eye movements (see Section 7.8.3).

The afferent fibres from the LGN travel in the optic radiations to the visual cortex. In primates, it seems that the entire input from the LGN projects to layers IV and VI of the **primary visual cortex**, which occupies the occipital poles of the cerebral hemispheres. The primary visual cortex is also known as Brodmann's **area 17**, the **striate cortex**, and **visual area 1**, or simply V_1. In cats, other layers of area 17, as well as areas 18 and 19 and the lateral suprasylvian area, also receive direct visual inputs (Le Vay and Gilbert, 1976). This explains why extirpation of area 17 in the cat leaves area 18 responsive to visual stimulation (Doty, 1973). It would seem that the visual system is more hierarchically organized in primates than it is in cats. One may speculate that a hierarchical system is more suited to the extraction of more complex, or abstract features than is a parallel system. However, the fact that other visual areas receive all their visual inputs from area 17 in primates does not prove that information is handled in a hierarchical fashion. Information may well be parcelled out into discrete channels within these different visual areas, as Baizer *et al.* (1977) and Zeki (1978) have suggested.

2.6 The retinotopic organization of the visual cortex

The earliest indication that there is an orderly representation of retinal visual fields over the surface of the occipital cortex came from studies on the effects of local head injuries which produced localized visual defects. The First World War provided an abundance of such cases (Holmes, 1918). Early anatomical studies on the orderly projection of visual afferents from the LGN onto the cortex are reviewed in Polyak (1957). The first physiological study of retinotopic mapping was conducted by Talbot and Marshall (1941), who recorded evoked potentials from the exposed visual cortex of the monkey as stimuli were presented to various parts of the visual field. The subsequent electrophysiological studies of Daniel and Whitteridge (1961) and Cowey (1964) produced more refined maps of retinotopic organization in the monkey.

Visual cortical fields may also be explored in humans by recording evoked potentials through the skull with scalp electrodes. However, this procedure is complicated by the fact that the visual cortex is deeply fissured, with much of it lying in the calcarine fissure. In this fissure the cortical surface folds back on itself, so that the normally inside surface faces out towards the scalp. The polarity of the evoked potentials from this inverted region is thus reversed with respect to the polarity of potentials from other regions (Halliday and Michael, 1970). The convoluted nature of the visual cortex increases the difficulties of providing a prosthesis for the blind by electrical stimulation of the visual cortex. Nevertheless, some success has been achieved in generating recognizable patterns of phosphenes with stimulating electrodes chronically implanted in the visual cortex of human subjects. Subjects could discriminate phosphenes produced when the electrodes were 3 mm apart on the surface of the cortex (Brindley, 1973; Dobelle and Mladejovsky, 1974).

It has already been shown in Section 2.5 how the right hemifield of visual space projects to the left cortical hemisphere and the left hemifield to the right hemisphere. Receptive fields close to and overlapping the vertical meridian of each eye project to near the boundary between cortical areas 17 and 18. These boundary regions, and similar boundary regions between other areas of the visual cortex, are the only regions of the visual cortex to be connected directly by fibres in the corpus callosum. This suggests that the corpus callosum links cells whose receptive fields are close to, or straddle the vertical meridian of the visual field. These linkages presumably serve the same functions as intracortical fibres which link neighbouring cells in the rest of the visual cortex (Hubel and Wiesel, 1967). Recently it has been shown that, in the monkey, within a region of the visual field 0.5° on either side of the vertical meridian, ganglion cells which project to the ipsilateral cortex intermingle with cells which project to the contralateral cortex. This strip widens to 3° at the fovea (Bunt and Minckler, 1977). These anatomical facts provide an explanation of why we do not experience a discontinuity at the vertical meridian of the visual field, and they would also explain claims that patients with unilateral damage to the visual cortex typically preserve vision in the whole of the fovea—a phenomenon known as **macula sparing**.

The embryological mechanism that provides for the nerve cells in the optic nerve to make appropriate connections with cells in the retina and in the brain remains a mystery. It has been shown that optic nerve fibres in the chick embryo grow across the optic tectum in the chick's brain in search of their proper target cells, rather than binding to the first cells they encounter. The cells must possess a chemical label which allows them to recognize each other. Neurones in the visual system differentiate fully only after they have made the proper connections, but the topographic correspondences between cells are established even before they meet (see Kolata, 1974, and Gaze, 1970, for further discussion of this issue).

In animals, the distance apart of two points along the convoluted surface of the visual cortex can be determined for two corresponding points of light which are 1° apart in the visual field. This measure, the **magnification factor** M, indicates how many millimetres of cortex are devoted to each degree of visual field (Daniel and Whitteridge, 1961). In the monkey, and probably also in man, M decreases with increasing eccentricity from the centre of the visual field, but is similar for corresponding eccentricities on all meridia of the visual field. Cowey and Rolls (1974) estimated M for the human visual system. They used data from a study by Brindley and Lewin (1968) on the visual-field distribution of impressions of light (phosphenes) evoked by electrodes implanted at various locations in the visual cortex of a human subject. M was approximately 4 mm/deg at an eccentricity of 2° and declined monotonically to 0.5 mm/deg at an eccentricity of 25°. A more recent estimate puts M at 11.5 mm/deg at the human fovea (Drasdo, 1977). Hubel and Wiesel (1974b) have shown that a microelectrode must move a constant distance of between 2 and 3 mm over the surface of the visual cortex of the monkey before an entirely new region of the visual field is represented. It is reasonable to conclude from this that the same number of millimetres of visual cortex is devoted to each ganglion-cell receptive field. One recent estimate is that 0.88 mm of human cortex is devoted to one receptive-field diameter (Ransom-Hogg and Spillmann, 1980). It follows from the above relationships that the amount of cortex devoted to each degree of visual field is directly related to visual acuity and inversely related to the mean size of receptive fields (Rolls and Cowey, 1970; Rovamo and Virsu, 1979). (see Figure 2.25 and Section 2.4.4).

These facts may be briefly summarized by saying that ganglion-cell receptive fields, and hence areas of the retina which can just spatially resolve two stimuli, are represented by equal areas in the visual cortex. Because these spatial units are smaller and hence more numerous in the fovea than in the periphery of the retina, the fovea claims proportionately more of the cortical surface (its magnification factor is greater) than does the retinal periphery (see McIlwain, 1976).

Schwartz (1980) has drawn attention to the fact that the mapping of the retina

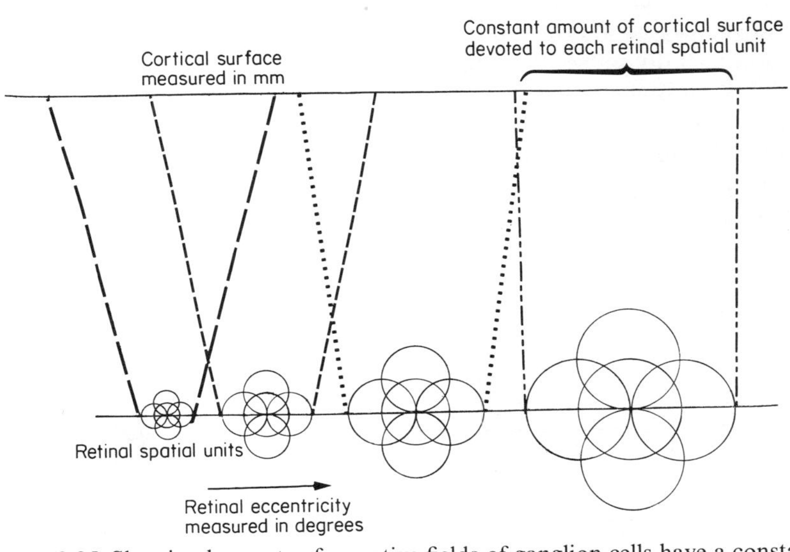

Figure 2.25 Showing how sets of receptive fields of ganglion cells have a constant overlap factor at different degrees of retinal eccentricity (see Section 2.4.4). Each set of overlapping fields is a hypothetical retinal spatial unit. All retinal spatial units have the same area of visual cortex devoted to them. The diagram also shows how the magnification factor (mm of cortical surface devoted to 1° of visual field) decreases with increasing eccentricity

onto the cortex may be described as a conformal logarithmic function of the form

$$w = \log(z + a)$$

where z is the retinal eccentricity and a is a constant. A function is conformal if its first derivative (in this case the magnification factor) is isotropic (independent of orientation) and if the sizes of local angles are preserved. Schwartz illustrates how, in a logarithmic mapping, the images of squares of different sizes centred on the retina are transformed into images of the same size and shape, but in different locations (see Figure 2.26). In a similar way, retinal images which are rotated with respect to each other are transformed into images which differ in spatial phase. In such a system, the recognition of size- and orientation-invariant features reduces to the computationally simpler process of deriving translation-invariant properties. Schwartz argues that the logarithmic retino-cortical mapping is part of a process for extracting size- and orientation-invariant properties of visual objects. One problem with this theory is that retinal images which differ in position are transformed into images which differ in size, and this would complicate the problem of deriving position-invariant features. The model of shape recognition which Schwartz proposes is essentially a template-matching model and adequate for the extraction of only very simple invariant features in a highly structured input. Furthermore, although Schwartz's model may be computationally sound for a computergraphics system, in which operations may be done directly on the geometry of a transformed display, it is not clear to me how it would apply to the visual cortex. The geometry of the cortical 'image' is not, in any obvious sense, represented in the activity of cortical cells, so that I do not see how this geometry can enter into spatial descriptions that the visual system constructs. It is safer to assume that the form of retinotopic mapping has to do with devoting equal central computational space to equally discriminable regions of oculocentric space. In addition, it has probably to do with bringing into juxtaposition cells which process similar features of oculocentric space so that local neural processes, such as facilitation and inhibition, may

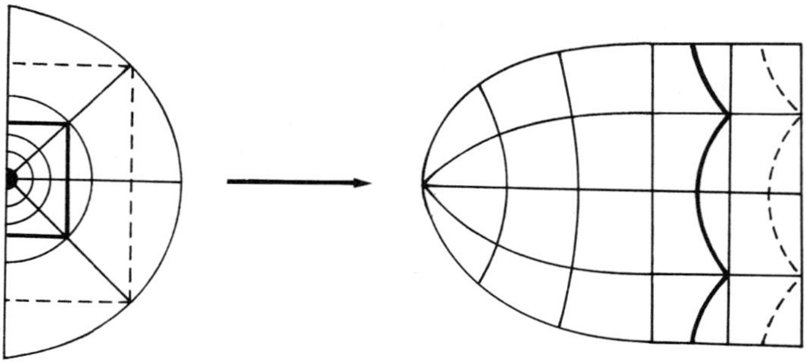

Figure 2.26 Showing how, in a logarithmic conformal mapping of the form $w = \log(z + a)$, concentric squares of different sizes are transformed into images which differ only in position (Reprinted with permission from E. L. Schwartz, *Vision Research*, **20**. Copyright 1980, Pergamon Press, Ltd.)

be economically achieved. We shall see in what follows that cells responsive to particular stimulus features are organized in distinct slabs of cortical tissue.

The mammalian visual cortex is a convoluted sheet of tissue about 1.5 mm thick consisting of six main **layers** of neural tissue designated layers I – VI, starting at the outer surface. Combined physiological and anatomical investigations in the monkey have revealed that, at least in that animal, X cells terminate in **stellate cells** in the deeper half of layer IVc and in layer IVa, and Y cells terminate in stellate cells in the superficial half of layer IVc, with both types of input sending collateral branches to layer VI (Hubel and Wiesel, 1977). The predominant projection from cells in layer IV seems to be to cells in layers II and III, which lie in the same column of tissue about 1 mm wide. Cells in layers II and III project to cells in layers V and VI, which in turn project back into layer IV (Gilbert and Wiesel, 1979).

Each neurone within the visual cortex responds only when an appropriate stimulus falls within a specific area of the retina. That area is defined as the **receptive field of the cortical cell**, and its retinal position is specified by the position of its centre. Receptive-field centres are represented retinotopically within each layer of area 17, although this retinotopic arrangement is perturbed by local random scattering. The receptive-field scatter is of the same order of magnitude as the size of the receptive fields at each location (Hubel and Wiesel, 1977). It must be assumed that the fine discrimination of locations that is revealed in human visual acuity depends upon the integration of information over local regions of the visual cortex.

Most cells in area 17 respond best when their receptive fields are stimulated by lines in a particular orientation. The optimal stimulus orientation for a given cell is known as its **preferred orientation**. The cells in each small column of cortical tissue have similar preferred orientations, and within each region of the visual cortex, columns containing cells with the same orientation preference are juxtaposed so as to form a sheet of neural tissue running at right angles to the cortical surface (see Figure 3.5). Different sheets lie parallel to each other in order of preferred orientation. The orientation preference rotates through its full range, 180°, as one traverses across the sheets through a distance of 0.5–1.0 mm. The details of this arrangement are given in the next chapter.

In the monkey, each local group of cells in layer IVc receives excitatory inputs from only one eye. These inputs are then relayed to cells in other layers in the same column of cortical tissue and, although most of these cells also receive inputs from the other eye through cells in layer IVc in neighbouring columns, they all respond most strongly to stimulation of the eye that is predominantly represented in their column. Cells which receive inputs from both eyes are known as **binocular cells** and the columns in which cells of a given eye dominance occur are known as **ocular dominance columns** (see Figures 2.27 and 2.28). Each binocular cell has two receptive fields, one in each eye. For any cell, these two receptive fields are practically identical with regard to their complexity, their orientation and direction preferences, and their oculocentric positions (Hubel and Wiesel, 1968). Columns with left eye ocular dominance form into a series of

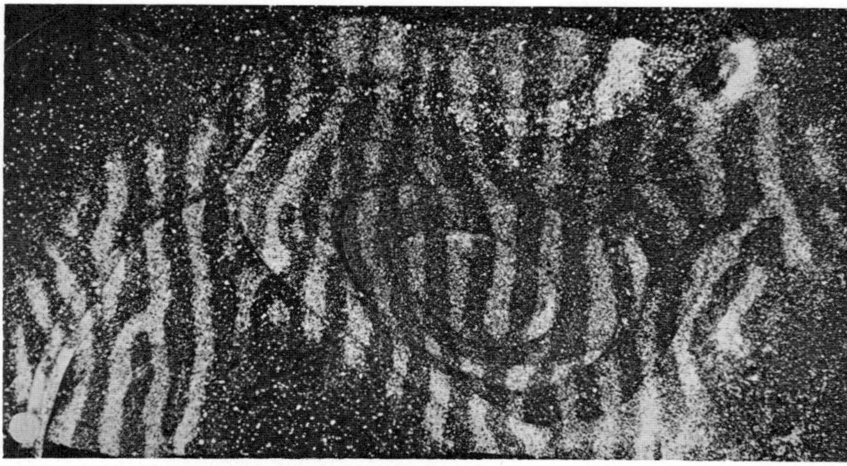

Figure 2.27 Alternating ocular dominance columns revealed in an autoradiograph of a tangential section of layer IVc of the monkey visual cortex. The ipsilateral eye was injected with tritiated proline 14 days before the radiographs were prepared, and the ocular dominance columns of this eye show up as bright bars in the autoradiograph. This plate was built up by superimposing concentric zones from several autoradiographs (From Hubel *et al.*, 1977. Reproduced by permission of The Royal Society)

sheets, each 0.25–0.5mm thick, interspersed with similar sheets with right eye dominance. Orientation sheets and ocular dominance sheets run roughly at right angles to each other. The whole visual hemifield is retinotopically mapped onto the ocular dominance sheets of the left eye, and again onto the ocular dominance sheets of the right eye. Along the sheets this representation is continuous, but across the sheets it is interrupted by the alternations between left and right eye sheets (see Hubel and Wiesel, 1977, for details of how these interrupted retinotopic representations are organized).

It has been shown that rearing cats and monkeys with an artificial strabismus induced by surgery, optical means, or immobilization of one eye, causes a reduction or absence of binocularly driven cortical cells (Blakemore, 1976; Bisti and Maffei, 1979; Bennett *et al.*, 1980; Crawford and Noorden, 1980). It is possible that a deficiency of binocularly driven cells is at least partly responsible for amblyopia in strabismic humans. Evidence reviewed in Section 12.5 suggests that the binocular orientation preferences of cortical cells may be modified in the presence of a small degree of cyclodisparity.

Ocular dominance sheets have been revealed anatomically in the monkey by tracing the areas of cell degeneration in the cortex produced by selective destruction of LGN cells arising from a given eye (Hubel and Wiesel, 1969). More recently, ocular dominance sheets, orientation sheets, and spatial frequency sheets have been revealed by the use of radioactive tracers (see Figure 2.27) (Silverman *et al.*, 1980). There are also direction-of-motion sheets (Payne *et al.*, 1981).

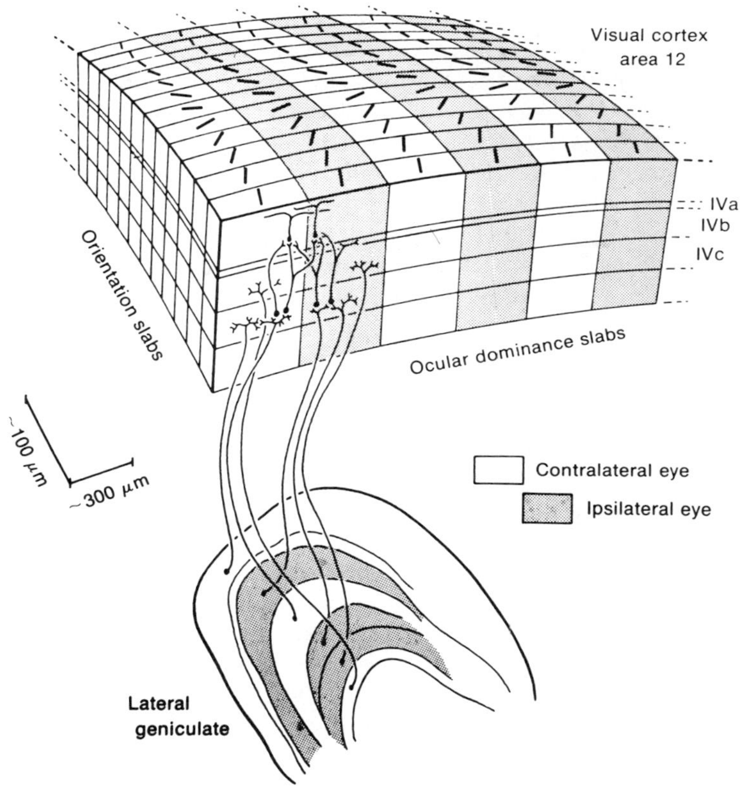

Figure 2.28 Diagrammatic representation of the columnar organization of the visual cortex of the macaque, as proposed by Hubel and Wiesel (1972). (Reproduced by permission of The Wistor Institute Press)

Another dimension of columnar organization is the degree of binocular disparity of the retinal receptive fields that feed into binocularly driven cortical cells. Barlow *et al.* (1966) showed that certain binocularly driven cortical cells respond selectively to stimuli which have a particular binocular disparity, that is, to stimuli which are normally produced by an object at a given distance from the plane of fixation. Blakemore (1970) found that cortical cells in the cat sensitive to a particular image disparity were, like orientation and ocular dominance cells, arranged in distinct columns of tissue, which he called constant depth columns. Of more importance to the subject of this volume is another type of columnar arrangement described by Blakemore. The binocularly driven cells in these columns were found to be driven by receptive fields in the contralateral eye which were all in the same region of the retina and by receptive fields in the ipsilateral eye which were scattered over several degrees. The cells in such a column therefore have a variety of preferred disparities, but they all respond to a stimulus which lies on a given oculocentric visual line of one eye. The column 'sees' along a tube of visual space that is lined up with an oculocentric visual direction of one eye. This topic is discussed further in Section 7.9.

The size of the region of visual cortex which contains a full 180° cycle of orientation preferences is roughly the same as the size of the region which contains a left-eye, right-eye ocular dominance set. Such regions are referred to as **hypercolumns**. They have a constant diameter of about 0.5–1.0 mm at all eccentricities, at least out to 15°. Regions of the cortex with completely non-overlapping receptive fields are about 2 mm apart; these regions easily contain all the cell types that are required to service that area of the visual field from which they derive their inputs (Hubel and Wiesel, 1974b). Neighbouring hypercolumns receive overlapping inputs, each from a constant number of overlapping retinal regions. Cells in neighbouring columns interact through inhibitory dendritic connections. Axons of cells in layer V, at least in the cat, have been found to spread 6–8 mm within area 17, which is equivalent to a distance of 15° of visual angle (Gilbert and Wiesel, 1979). The functional significance of these long axonal connections is still obscure. The organization of hypercolumns will be discussed again in Section 3.3.6, after I have dealt with the various types of cortical cells in more detail.

Axons which leave the primary visual cortex emerge from **pyramidal cells** situated in the inner and outer layers. Outputs from layers II and III project to each of a series of other visual areas which surround the striate cortex. These are known collectively as the **prestriate cortex** or **circumstriate belt** and, in primates, include at least three visual areas in Brodmann's area 18, known as V_2, V_3, and V_{3A}, and at least two visual areas in the posterior temporal lobe, called V_4 and the **motion area** (Zeki, 1974b; Wong-Riley, 1979). No doubt other visual areas remain to be discovered. The prestriate cortex in turn sends intracortical fibres to the inferotemporal cortex (Gross, 1973), the parietal lobe (Section 11.7.3), and the frontal lobe (Section 5.4.6).

The boundaries between one visual area and another are always comprised of cells which have receptive fields along either the vertical or horizontal meridian of the retina. These boundary regions are therefore recognizable by the sudden change in retinotopic representation that occurs there. One consequence of this juxtaposition of cells representing the main retinal meridia is that retinotopic representations of visual hemifields in succeeding visual areas are mirror images of each other (Cowey, 1979). Regions which represent the vertical meridia are the only regions which are connected by fibres running in the corpus callosum. In areas V_2, V_3, and V_{3A} the visual fields are mapped retinotopically, although not all regions of the visual field are necessarily represented in each area. Most of the cells in these three areas are orientation selective. In area V_4, only the central 20° of the visual hemifield is represented, which seems to be related to the fact that this area contains many colour-coded cells, and colour coding is confined to the central region of the retina. The so-called motion area in the posterior temporal lobe has multiple representations of the visual hemifields, but is nevertheless referred to as a single visual area because all the cells within it respond to similar features of the visual stimulus (Zeki, 1978). The picture that emerges is of a series of retinotopically coded visual areas, each of which is designed to handle different visual features. The system is hierarchical in the sense that each area receives its inputs from the striate cortex, but parallel in the sense that each area handles different visual features.

The retinotopic arrangement of cells within the various visual areas serves two purposes. In the first place, mutual inhibition occurs only between cells which are near neighbours. Inhibition is required only locally because one of its main functions seems to be to attenuate the response from regions of local homogeneous activity, thus, by comparison, accenting the response from regions with high gradients of activity. This ensures that what is conveyed to the next level of analysis is information regarding changes in stimulation. Secondly, pooling of spatial and other information is more often required over small regions than over larger regions, so that keeping spatially contiguous regions together economizes on the lengths of dendritic connections. The coding of visual location and other features depends ultimately on which axons connect to which, not on the spatial disposition of the pathways in the central nervous system. Once one gets beyond the retina, the notion of a topographic code ceases to have any significance; all features are coded in the central nervous system in terms of cells connections and the simultaneous and successive patterns of firing of the cells, not in terms of the spatial disposition of cells.

Outputs from pyramidal cells in layers V and VI of the primary visual cortex project to subcortical regions (Creutzfeldt, 1977), including the superior colliculus (see Section 5.5.2), the pulvinar, LGN and other areas of the thalamus (Gattas *et al.*, 1979), the caudate nucleus, and the cerebellum by way of pontine nuclei (Brodal, 1972a, b).

2.7 The effects of removal of visual cortex

2.7.1 Introduction

Many axons in the optic tract of mammals, including man, send branches to destinations other than the geniculostriate system. The superior colliculus, in the midbrain, is the principal recipient of extrageniculate visual inputs (see Section 5.5.2). One would expect those visual functions which depend on subcortical centres receiving direct visual inputs to survive the removal of the visual cortex (see Doty, 1973, for a review of the earlier literature). The evidence which I shall now review reveals that in monkeys certain visual functions do, indeed, survive cortical removal, although there is not full agreement about which functions survive. The evidence from humans which I shall also review is particularly controversial.

The visual cortex of a monkey may be ablated, either unilaterally or bilaterally, but care must be taken to ensure that all the relevant tissue is removed. Bilateral loss of the visual cortex in humans is very rare, and even when it is suspected, it is almost impossible to confirm that it is complete (Brindley *et al.*, 1969). Unilateral loss of visual cortex in humans is more common than bilateral loss. When the unilateral loss is more or less complete, it produces cortical blindness in the contralateral hemifield, or **cortical hemianopia**, although the region round the fovea is usually spared. When the loss is partial, it produces a local area of blindness, which is known as a **scotoma**. The loss of tissue in a region affected by a cortical lesion is not necessarily complete, and this complicates the task of interpreting residual visual functions in such a region. Perhaps the only procedure which guarantees complete unilateral loss is the operation known as **hemidecortication**, in which all of one cerebral hemisphere is removed.

Another problem in testing areas of cortical blindness is that light, which is supposed to stimulate only that part of the retina serving the defective region of the visual field, may be scattered in the eye and fall on other parts of the retina.

Finally, there is the problem of knowing whether deficits caused by cortical ablations are due simply to the loss of that tissue or whether they are due to a more widespread disturbance in the balance of excitatory and inhibitory activity in other centres. Sprague (1966) drew attention to this possibility by showing that the visual deficits caused by unilateral removal of striate cortex in the cat were less severe if the contralateral superior colliculus was also removed. The colliculus on one side is functionally associated with the visual cortex on the same side, so that it is the colliculus ipsilateral to the ablated half of the cortex which is probably responsible for compensating for cortical loss. But Sprague's finding indicates that this compensation can occur only if the ipsilateral colliculus is freed from inhibition from the contralateral colliculus. Sherman (1974) found that cats, after complete removal of the visual cortex, were effectively blind until the commissures connecting the two halves of the superior colliculus were cut, whereupon the animals recovered their ability to direct their movements visually. Goodale (1973) confirmed by electrophysiological procedures the presence of mutual inhibition between the two lobes of the cat's superior colliculus, and Zihl and von Cramon (1979) have revealed evidence of this mutual inhibition in humans by investigating interfield threshold-elevation effects in normal subjects and in patients with unilateral collicular damage. Thus, if one wishes to discover which visual functions survive removal of the visual cortex, one must also destroy the collicular commissures. However, nobody seems to have adopted this procedure in studies on the monkey.

2.7.2 Responses to luminance and movement in destriate visual fields

A monkey with complete bilateral ablation of the visual cortex appears to be blind on superficial examination. However, more careful testing reveals a variety of residual functions. The pupils still respond to changes in illumination (Magoun and Ranson, 1935), and the monkeys detect changes and differences in luminance (Klüver, 1941, 1942; Pasik *et al.*, 1959; Pasik and Pasik, 1971). When only area 17 is removed monkeys can also detect visual movement (Denny-Brown and Chambers, 1976).

Clinical studies have shown that human patients with hemianopia or scotomata, who are 'blind' when tested with conventional campimetric procedures, reveal residual visual functions when tested more carefully (Riddoch,

1917; Bender and Teuber, 1946; Bender and Krieger, 1951). For instance, Bender and Krieger found that patients could often detect a light seen against a dark background, and report when it was flickering and when it was moving. However, they were not able to determine the shape of a stimulus. In most of the 11 patients that Bender and Krieger studied the cause of the disorder was unknown, and in no case was it due to a complete loss of cortical tissue in the affected area. Several of the 11 patients recovered from the infirmity. Williams and Gassel (1962) also described patients with residual visual functions in parts of the visual field that were blind by conventional tests, but concluded that these functions remained as a result of tissue survival or were due to light scattering in the eye. Brindley *et al.* (1969) reviewed the clinical literature on cases of cortical blindness and concluded that the response of the pupils to changes in illumination is the only residual visual function that convincingly survives complete loss of visual cortex in man. The only other residual visual function which they found in two patients with bilateral loss of visual cortex was an ability to detect sudden changes in illumination, but they concluded that this ability was probably due to surviving cortical tissue.

Thus, a conservative way of interpreting this evidence that humans can detect luminance, flicker, and movement in hemianopic fields is that these residual functions depend on surviving cortical cells. Recent evidence does not support such a conservative view, as we shall see.

2.7.3 Eye movements into destriate visual fields

Destriate cats show optokinetic nystagmus in response to moving stimuli (Smith, 1937). The evidence on whether destriate monkeys show this response is conflicting (ter Braak and Van Vliet, 1963; Pasik and Pasik, 1964). Brindley *et al.* (1969) could not elicit optokinetic nystagmus in human decorticate patients, although ter Braak *et al.* (1971) observed it in a patient with what was claimed to be complete bilateral loss of striate cortex (area 17).

It has been reported that destriate monkeys are able to turn their heads and eyes towards a moving object (Denny-Brown and Chambers, 1958). Mohler and Wurtz (1977) were able to train monkeys to make saccades towards intensely illuminated visual targets in an hemianopic field as long as the superior colliculus was intact, and concluded that subcortical mechanisms are sufficient for the initiation of voluntary saccades.

Human patients with cortical hemianopia have been described who turned their gaze in the direction of a visual target presented in the hemianopic field even though they were unaware of any visual sensations (Pöppel *et al.*, 1973; Sanders *et al.*, 1974). It is possible that these patients relied on subcortical mechanisms, but the presence of surviving cortical tissue was not ruled out. However, Sharpe *et al.* (1979) found that patients in whom the whole of one hemisphere had been removed were able to make saccades towards visual objects in the hemianopic field even after the effects of light scattering had been allowed for. It looks as though saccadic eye movements which are controlled entirely subcortically occur in both monkeys and humans.

2.7.4 *Visually-guided reaching into destriate visual fields*

Humphrey and Weiskrantz (1967) found that a destriate monkey could reach towards small stationary objects. However, the monkey could do this only after it had searched for the object and then fixated the gaze upon it. This suggests that destriate monkeys cannot visually detect the oculocentric position of a visual target for purposes of reaching. Once the image is on the fovea, the task of reaching depends only on the ability to estimate the direction of gaze and the direction of the head on the shoulders, and these are not oculocentric skills. In a subsequent study on the same monkey, Humphrey (1974) found that the animal could avoid obstacles when moving about even though they were seen only by the peripheral retina. This ability may not have depended on non-striate tissue, because a *post-mortem* analysis revealed that a small portion of cortex serving the peripheral retina probably survived the ablation procedure.

Feinberg *et al.* (1978) repeated this experiment, taking greater care to ensure that all the visual cortex had been removed. They confirmed that decorticate monkeys are able to reach towards small objects, but they also noticed that the animals slowly scanned the field searching for the object, and reached for it only after it was fixated. So here again there is no evidence of oculocentric localization for purposes of reaching. Keating (1980) reported accurate reaching in monkeys which had had both the striate and prestriate cortex removed, although fixation was not controlled. *What is needed is an experiment in which the visual target is exposed briefly in the visual field, so that the animal does not have time to fixate it.* If a destriate monkey can reach for such an object, one could say that oculocentric visual localization survives decortication. As we shall see, such an experiment has been done only with human subjects, where it is less certain that all the visual cortex is absent.

Bender and Krieger (1951) reported that human patients with hemianopic cortical blindness could point towards the position of a light. It is not clear whether these patients first fixated the object. However, Weiskrantz *et al.* (1974) reported the same ability in an hemianopic patient who was unable to fixate on the peripheral targets to which he was pointing. Perenin and Jeannerod (1975) took the precaution of asking patients to point towards a light that was flashed briefly within their hemianopic visual field. Patients with post-geniculate lesions (which would cut off only cortical inputs) could point towards such targets, whereas patients with pre-geniculate lesions (which would cut off collicular as well as cortical inputs) could not. If light scatter had been the basis of the ability in the post-geniculate patients, this ability would have been present in the pre-geniculate patients also. These results suggest that humans are able to localize, oculocentrically, objects in hemianopic fields for purposes of guiding the hand. Perenin and Jeannerod (1978) reported this ability in patients who had been hemidecorticated as children, which suggests that it is not due to survival of cortical tissue.

If the capacity to reach towards a peripherally viewed visual target does survive after the visual cortex has been removed, the superior colliculus is the most likely subcortical structure that would serve this function. The superior colliculus, which is described in Section 5.5, is a midbrain structure that receives inputs from the eyes and from other sense organs and sends outputs to centres concerned with eye movements, head movements, and limb movements. There have been several reports that ablation of the superior colliculus abolishes the ability of hamsters and rats to orient themselves towards visual objects presented in the periphery of the visual field (Schneider, 1969; Goodale *et al.*, 1977). Keating (1974) found that monkeys could not reach accurately towards visual objects presented briefly in their peripheral visual fields after removal of the superior colliculus, although they had no difficulty with objects viewed centrally. Butter *et al.* (1978) confirmed these results with the monkey, taking greater care to control the direction of gaze of their animals. However, they found that performance of colliculectomized monkeys with peripherally placed targets was severely impaired only if the targets were exposed briefly; the monkeys had less difficulty with targets exposed for more than 1 sec.

It looks as though the colliculus is part of a mechanism for directing the animal's attention towards peripherally placed objects and guiding responses to these objects. It will be shown in Section 5.5.2 that the superior colliculus receives direct visual inputs from Y (transient) ganglion cells, mainly from the retinal periphery, and no input at all from X (sustained) cells. The fact that colliculectomized monkeys show a deficit in visually-guided reaching towards briefly exposed peripheral targets makes sense in terms of these anatomical facts.

2.7.5 *Pattern vision in destriate visual fields*

Cats have been found to be capable of discriminating between simple, large patterns after the removal of the whole of area 17 and most of area 18. On the other hand, pattern vision was absent or severely impaired after removal of suprasylvian gyri, including areas 19, 20, 21, and 7, leaving areas 17 and 18 intact (Sprague *et al.*, 1977). These suprasylvian regions receive visual inputs from the superior colliculus, the pretectum, the pulvinar, and parts of the LGN. Some of the pairs of test shapes were equated for total luminous flux, area and contour length and complexity, and were also varied in size and position from trial to trial. These precautions may not have been entirely adequate in preventing the animals from using extraneous cues, because the cats were free to move their eyes and may have detected differences in the rate of luminous flux as the images traversed the fovea. Destriate monkeys have also been shown to be capable of distinguishing between simple patterns equated for total luminance (Weiskrantz, 1963; Pasik, 1969). Dineen and Keating (1981) found that, after lengthy training, destriate monkeys showed transfer between patterns with different local features but similar overall shape. This suggests that they had true pattern vision.

Weiskrantz *et al.* (1974) reported that an hemianopic patient could distinguish between horizontal and vertical lines which were longer than 8° of visual angle, or between large isolated patterns such as O and X. The patient was not conscious of any visual sensations and felt that he was only guessing, when in fact he was performing almost perfectly. Before it is concluded from this behaviour that pattern vision is possible in a decorticate human, it is important that other explanations of this patient's performance are excluded. The most obvious explanation is that he was responding to the different degrees of image flicker as the eyes moved towards the patterns. Weiskrantz *et al.* claimed to have eliminated these sources of information by showing that the patient could still perform these tasks when the stimuli were exposed for only 150 msec, which is shorter than eye movement latency. Perenin and Jeannerod (1975) eliminated the effects of eye movements during stimulus exposure by flashing a bright pattern in the hemianopic field of patients with unilateral occipital lobectomy. When this flashed image was presented in the normal hemifield, it produced a visible afterimage which persisted for several seconds. However, the persisting after-image in the hemianopic field did not enable any of the patients to detect differences in the shapes of stimuli flashed in the defective hemifield, even though they could point to the positions of these stimuli. The occipital lobectomy was performed on these subjects when they were adult. In a subsequent study, it was reported that two adult patients who had been operated on when they were seven years old discriminated between patterns presented in the hemianopic field at well above the chance level of performance (Perenin,1978). These patients had presumably developed subcortical visual capacities which enabled them to discriminate patterns, although whether these capacities involved the discrimination of local differences in luminous flux or true pattern vision is not clear.

It seems safe to conclude that monkeys, and perhaps also humans, can detect sudden changes in luminance and make directed eye movements towards stimuli in destriate regions of the visual field. Reaching towards objects in destriate regions is certainly possible when the object is fixated and perhaps also when it is not fixated. The evidence for true pattern vision in destriate monkeys and humans is reasonable. In none of the studies reviewed here was the precaution taken of destroying the commissures of the superior colliculus. Perhaps, if this were done, a wider range of residual visual functions would be found. The superior colliculus is the subcortical centre which is most probably responsible for these residual visual abilities, and this question is discussed in Section 5.5.

Chapter 3

Orientation Detectors in the Visual System

3.1 Introduction

After the pioneering work of Hubel and Wiesel which began to appear in 1959 it became evident that orientation is one of the basic local attributes of visual stimuli, which, along with such attributes as position, colour, and motion, the mammalian visual system is specially equipped to detect. This chapter is about the physiology of the mechanisms responsible for the visual detection of local stimulus orientation. The basic stimulus domain to be considered consists of simple visual stimuli, such as lines, edges or gratings, in a frontal plane and in different orientations with respect to the normally vertical meridian of the retina of one eye. Other stimulus variables, such as retinal location, movement, and flicker, will also be introduced when these are relevant to specific issues.

The investigation of the neural mechanisms in the visual cortex thought to be responsible for the initial coding of the orientation of simple visual stimuli involves recording the activity of single cells, or groups of cells, by means of electrodes placed in or near the cells. For this reason, the physiological evidence reviewed in this chapter comes almost entirely from the cat and the monkey.

3.2 Basic features of cortical cells responsive to change in stimulus orientation

In the mammalian visual system, cells responsive to changes in stimulus orientation are not found to any significant extent at a level below the visual cortex. In the retina, orientation in the frontal plane is coded only distributively. Although cells have been found in the LGN of cats which respond best to stimuli in certain orientations (Daniels *et al.*, 1977; Levick and Thibos, 1980), these cells are not generally referred to as orientation detectors. Their specificity could be due to the fact that the receptive fields of such cells are non-circular or to a spatial alignment of receptors within receptive fields, or it could be due to corticofugal modulation of the otherwise isotropic responses of LGN cells.

Some cells in the primary visual cortices of cats and monkeys respond equally well to a line stimulus in any orientation and are described as **concentric** or **circular symmetric cells**. But most cells in the visual cortex respond best to a line or edge which has a particular orientation in the frontal plane; the stimulus can be a light–dark edge, a bar of light or a dark bar on a lighter background, and is most effective when moving. These cells are described as being **orientation selective**, and are often referred to as **orientation detectors**. They were first described by Hubel and Wiesel (1959) in the cat and by Hubel and Wiesel (1968) in the monkey. A recent review of this work is given in Hubel and Wiesel (1977).

Hubel and Wiesel found that in primates most, if not all, visual afferents from the LGN terminate in cells in layer IVc of area 17, and that all these cells are circular symmetric. Most of the cells in the other sublayers of layer IV and in layers near layer IV were found to be orientation selective and to require the stimulus to be precisely positioned. These cells were called **simple cells** and it was inferred that each of them receives its inputs from several circular symmetric cells in layer IVc. Hubel and Wiesel proposed the following scheme to account for their orientation selectivity.

Consider first a simple cell which responds to a bar of light. Such cells have been called **line detectors**, which, as we shall see later, may be misleading. The idea is that they receive inputs from several ganglion cells with on-centre receptive fields and that the circular receptive fields of these afferent fibres are arranged so that their on-centres form an elongated excitatory region and their off-surrounds form inhibitory flanks, as shown in Figure 3.1(c) and 3.2. This array of receptive fields of ganglion cells constitutes the receptive field of the simple cortical cell, and a line along the centre of the receptive field parallel to the boundaries separating excitatory and inhibitory regions in the **receptive-field axis**. According to Hubel and Wiesel's scheme, a slit of light which falls along the aligned on-centres stimulates the cortical cell maximally. This orientation of the stimulus is known as the cell's **preferred orientation**. As the stimulus line rotates from the preferred orientation, it impinges on more and more of the inhibitory flanks, causing a reduction in the rate of firing of the cortical cell. When the stimulus is at 90° to the preferred orientation, it lies across the central region and inhibitory flanks, and at this orientation inhibition is greatest, excitation least, and the cell is least likely to fire. The function that relates the rate of firing of the cell to the angle

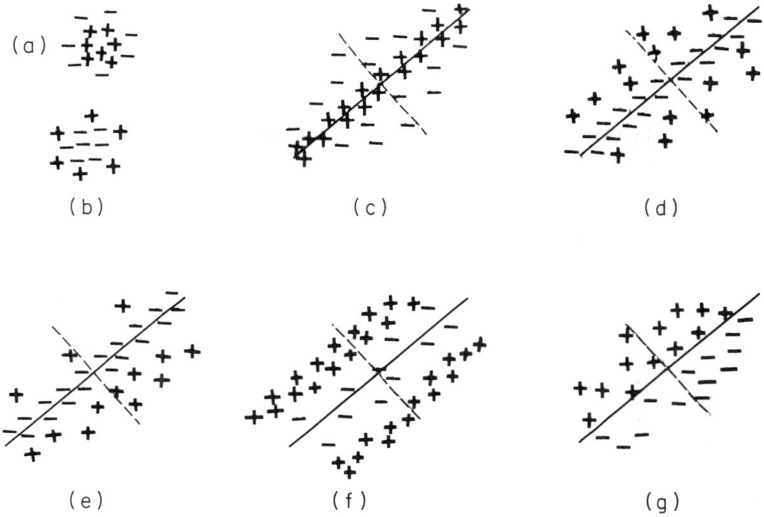

Figure 3.1 (a) An on-centre receptive field of a ganglion cell and (b) an off-centre receptive field of a ganglion cell. (c)–(g) Receptive fields of simple cells in the visual cortex of the cat as plotted by Hubel and Wiesel (1962). The receptive fields of the cortical cells are shown with preferred orientations of 45°, although cells with different orientation preferences were found. (Reproduced by permission of the *Journal of Physiology*)

of tilt of the stimulus is the **orientation-tuning function** of that cell. It is convenient to indicate the **orientation selectivity** of a cell as the width of one side of the tuning curve at the point where the cell's rate, or amplitude, of response is half the maximum amplitude (see Figure 3.3). In what follows, when the width of a tuning curve is referred to, it is understood that it is the width on one side of the mode at half amplitude. If the tuning curve is severely asymmetrical, the widths of both sides are specified. According to Hubel and Wiesel's scheme a cortical cell that responds to a dark bar on a brighter background receives inputs from a set of ganglion cells with off-centre receptive fields, as in Figure 3.1(d), (e), and (f). Hubel and Wiesel (1962) also described simple cells in the cat's visual cortex which have elongated receptive fields divided longitudinally into two halves. Such a cell is excited by illumination falling on one half of its receptive field and inhibited by illumination falling on the other half, as indicated in Figure 3.1(g). The optimal stimulus for a cell of this type is a dark–light border of the correct polarity lying along its longitudinal axis. These cells have been called **edge detectors** (Tolhurst, 1972), although this name implies a closer relationship between structure and function than is justified by the evidence. For reasons which I have already given in Section 2.4.7, so-called line detectors are best referred to as symmetrical or cosine detectors, and so-called edge detectors are best called antisymmetrical or sine detectors. The terms 'line detector' and 'edge detector' can be misleading because, for instance, a broad line can be

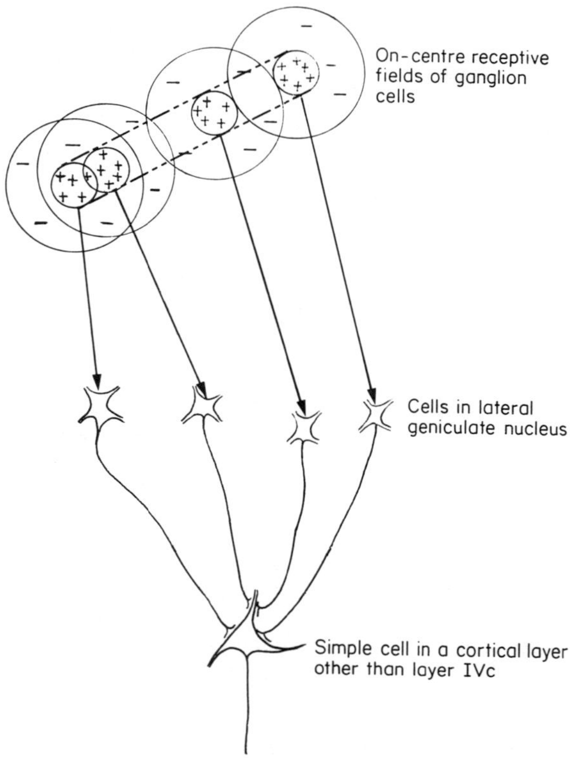

Figure 3.2 The scheme proposed by Hubel and Wiesel (1962, 1977) to account for the orientation selectivity of simple cells in the visual cortex. Recent evidence suggests that this scheme will have to be abandoned. (Reproduced by permission of the *Journal of Physiology*)

detected by a line detector or by two edge detectors of opposite polarity (King-Smith and Kulikowski, 1973).

Note that in this scheme the interactions between regions of excitation and inhibition which are thought to be responsible for the orientation tuning of a simple cell occur within the receptive field of each of the ganglion cells which contribute to the simple cell. Furthermore, the inhibition is thought to arise entirely from 'off' regions. This scheme is almost certainly too restrictive. Recent evidence confirms that a simple cell receives excitatory inputs exclusively either from on-centre or from off-centre LGN cells, but strongly suggests that inhibitory flanks of a cell are produced mainly by intracortical inhibition arising from other cortical cells with inputs from LGN cells having properties matching those of the excitatory centre (Heggelund, 1981a). In any case, other recent evidence suggests that we shall have to abandon the idea that cortical cells derive their orientation specificity from patterns of excitation and inhibition within or between channels which feed into them. This evidence, which

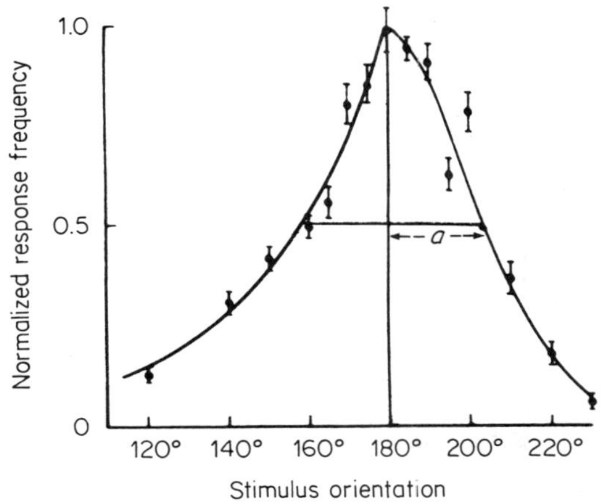

Figure 3.3 The orientation-tuning curve of a simple cell of the cat's visual cortex. Vertical bars indicate standard errors. The distance *a* is the width of the tuning curve on one side of the mode at half amplitude, which is the normal measure of a cell's orientation selectivity. Note that in this particular curve the widths of the two sides are not identical. (Reprinted with permission from P. Heggelund and K. Albus, *Vision Research*, **18**. Copyright 1978, Pergamon Press, Ltd.)

I shall review shortly, strongly suggests that the specificity of cortical cells derives mainly, if not entirely, from patterns of inhibition between cortical cells.

The orientation-tuning curve of a simple cell of a cat, based on averaged responses at 3° intervals of tilt, is shown in Figure 3.3. Note that the half width of this curve is about 22°, which is a typical value for the cat Orientation-tuning functions for simple cells in the monkey have been produced by Schiller *et al.* (1976a). Figure 3.4(a) is a histogram of the relative frequencies of cells with different tuning widths. There are no comparable data for humans, but Campbell and Maffei (1970), using evoked potentials from a man's scalp, found evidence of orientation-selective channels which gave completely distinct potentials when stimuli were separated by 15° of tilt.

Hubel and Wiesel (1965) described another class of orientation detectors which they called **complex cells**. Complex cells occur mainly in layers other than layer IV (Bullier and Henry, 1979a) and are found in areas 17 and 19 but most frequently in area 18. The tuning curves of complex cells are similar to those of simple cells although it can be seen in Figure 3.4 that orientation-tuning curves of complex cells are, on average, broader than those of simple cells. The receptive fields of complex cells are larger than those of simple cells and are not segregated into sharply defined excitatory and inhibitory regions when plotted by points of light. This means that their response is less dependent on the side-to-side position of an optimally oriented bar. Heggelund (1981b) has shown that each complex cell is activated by both on- and off-centre LGN cells and is inhibited by other cortical cells which also receive mixed inputs. This is in contrast to simple cells, which do not receive mixed inputs.

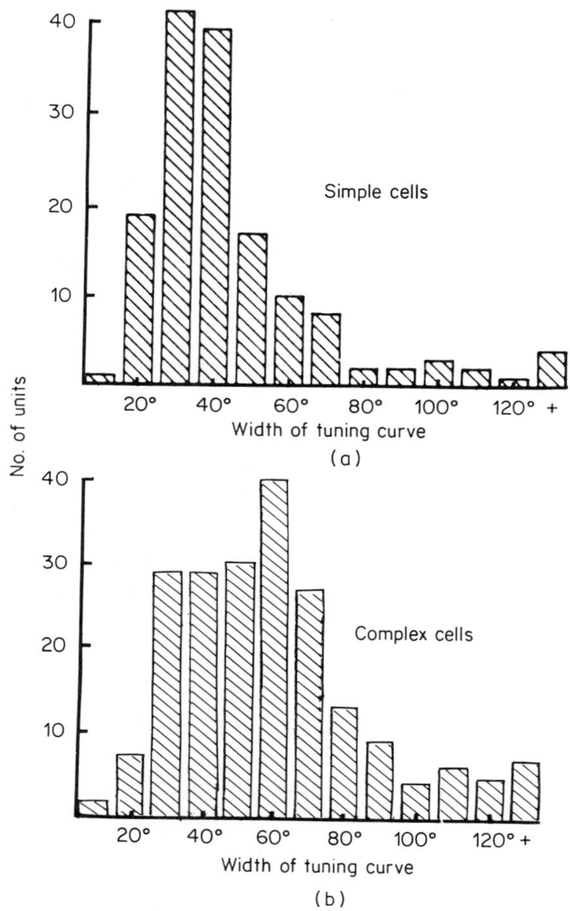

Figure 3.4 Distribution of tuning selectivity of (a) 151 simple cells and (b) 201 complex cells of the monkey's visual cortex. Selectivity is measured by the width of each cell's tuning curve, at maximum amplitude over $\sqrt{2}$ (From Schiller *et al.*, 1976a. Reproduced by permission of the *Journal of Neurophysiology*)

Hubel and Wiesel proposed that each complex cell derives its inputs from several simple cells with matching preferred orientations in the same column of cortical tissue, rather than directly from the LGN. However, it has been reported that some complex cells in the cat receive a direct input from Y afferents in the LGN (Stone and Dreher, 1973, see also Bullier and Henry, 1979b). Other evidence, which I shall review shortly, also argues against the hierarchical view, at least in subprimates.

Hubel and Wiesel also suggested that in the monkey, inputs from the two eyes, which are segregated into distinct alternating regions in layer 1Vc, converge on complex cells, not on simple cells as in the cat. More recent work with the monkey has revealed that a few simple cells and most complex cells are driven by inputs from both eyes (Schiller *et al.*, 1976a).

A third type of orientation-selective cortical cell described by Hubel and Wiesel (1962) is called the **hypercomplex cell**. The cells of this type were described as having larger receptive fields and were found mainly in the deep and superficial layers of areas 18 and 19, although cells with hypercomplex properties have been found in area 17 of the cat, though not inside layer IV (Gilbert, 1977). The main distinguishing feature of a hypercomplex cell is that it does not respond to slits or edges which extend further than a certain distance beyond one or both ends of the excitatory region of the receptive field. Testing the retinal receptive field of a hypercomplex cell reveals strong inhibitory end zones. A hypercomplex cell's selectivity to length is reduced, but not abolished, by the application of the drug bicuculline to the cortical layers in which the cell lies (Sillito and Versiani, 1977; Sillito, 1977). This drug abolishes intracortical inhibition because it antagonizes aminobutyric acid (GABA) which is the inhibitory neurotransmitter in the cortex. It looks as though inhibition between cortical cells is at least partly responsible for the length specificity of hypercomplex cells. Bicuculline did not entirely abolish length specificity, either because the dose was insufficient or because inhibition at some stage in the visual pathway and/or patterns of intracortical excitation must also be involved (Sillito and Versiani, 1977; Sillito, 1977). Hubel and Wiesel had suggested that the orientation specificity of hypercomplex cells results wholly from patterns of excitatory inputs from a few complex cells. But it is apparently patterns of intracortical inhibition rather than of excitation which are responsible, because the drug bicuculline abolishes the orientation specificity of hypercomplex cells, at least in the cat (Sillito and Versiani, 1977).

Hubel and Wiesel described hypercomplex cells as a distinct class of cortical cells, which receive their inputs solely from complex cells in the same cortical column, rather than directly from the retina. More recent evidence suggests that cortical cells fall into only two basic families, simple and complex, which derive their basic properties from differences in their inputs (Dreher, 1972), or, as more recent evidence suggests, from patterns of intracortical inhibition. These basic properties have to do with criteria which operate across the width of the receptive fields of the cells. In particular, the receptive fields of simple cells have clearly segregated zones of excitation and inhibition running parallel to their axis of preferred orientation, while the receptive fields of complex cells have intermixed zones of excitation and inhibition. Within each family, there are cells which show the hypercomplex property to a high degree and others which show it to a lesser degree or not at all. This criterion operates along the length of the receptive fields. Thus, a cell with the hypercomplex property has strong inhibitory end zones which cause the cell to stop firing when the stimulus exceeds a certain length. Cells with the hypercomplex property may be described as **end-stopped** and those lacking this property as **end-free**. The extent to which a cell's rate of firing is inhibited when the stimulus fully encroaches into the inhibitory end zones may be specified as the percentage inhibition. In the cat, there are only a few simple or complex cells with 100% end-stop inhibition and about half of both types of cell have end-stop inhibition in excess of 50% (Rose, 1977; Kato *et al.*,

1978). Within 5° of the fovea, cells of both the simple and complex families with an optimum stimulus length of less than 3° of visual angle are end-stopped to a significant extent, whereas cells with an optimal stimulus length of more than 3.5° are end-free (Kato *et al.*, 1978).

The view that the visual cortex of mammals is hierarchically organized has dominated the field since Hubel and Wiesel first proposed such a theory twenty years ago. According to this view, ever more complex visual features are extracted in the various layers and areas of the visual cortex, starting in layer IV of area 17, where simple cells are concentrated, through other layers of area 17 and through visual areas 18, 19, etc., where complex and hypercomplex cells are concentrated. In recent years several lines of evidence, which I shall now briefly review, have cast doubt on Hubel and Wiesel's simple hierarchical model, although most of this evidence was obtained from the cat and very little from the monkey or man (Rose, 1979; Bullier and Henry, 1979a, b).

According to the hierarchical model, simple cells should be mono- or disynaptically driven by stimulation of geniculocortical fibres, whereas complex cells should be di- or trisynaptically driven. This difference should show as a difference in mean latency of response to stimulation of geniculocortical fibres. Several investigators have reported that some, but not all, cells with complex properties in the cat respond with monosynaptic latencies (Singer *et al.*, 1975; Toyama *et al.*, 1977). It has been confirmed that complex cells do not occur in layer IV of the visual cortex (Gilbert, 1977), so that complex cells receiving a direct visual input must do so by geniculocortical fibres which terminate in other layers or on long dendrites of cells which have their cell bodies outside layer IV. Both types of terminations have been described (Garey and Powell, 1971; Leventhal, 1979). There is evidence that even complex cells with disynaptic latencies do not derive their inputs from simple cells. For instance, Sillito (1975) reported that the drug bicuculline, which eliminates intracortical inhibition, had little effect on the orientation selectivity of simple cells but almost abolished the selectivity of complex cells. This is difficult to explain in terms of Hubel and Wiesel's hierarchical model, and Stone *et al.* (1979), in a review of the evidence on this question, concluded that complex cells derive their inputs either directly from geniculocortical afferents or from cortical cells which do not themselves have any orientation specificity. In this scheme, the orientation specificity of most complex cells is derived from patterns of intracortical inhibition. Creutzfeldt *et al.* (1974) have suggested that all cortical cells, simple and complex, gain their orientation selectivity in this way. In support of this suggestion, Tsumoto *et al.* (1979) reported that an inhibitor of GABA synthesis plus bicuculline eliminated GABA mediated intracortical inhibition and completely destroyed the orientation selectivity of about a quarter of the cortical cells tested. Sillito *et al.* (1980) found that bicuculline in sufficient doses, completely eliminated the orientation selectivity of all the simple cells which they tested. Furthermore, it was reported in both studies that the excitatory receptive fields of cortical cells treated with bicuculline were round, and not elongated. This suggests that the elongated fields of such cells when functioning normally cannot be due to the alignment of excitatory on-centres of ganglion cell receptive fields in the manner suggested by Hubel and Wiesel. It looks as though the orientation selectivity of most, if not all, cortical cells arises from patterns of intracortical inhibition.

Further evidence that complex cells in the cat do not receive their inputs from simple cells is that Hammond and MacKay (1977) found that complex cells respond to a moving random texture, whereas simple cells do not. This would be an improbable state of affairs if simple cells fed into complex cells. In view of all this evidence it seems most likely that complex cells do not receive their only, or even their major inputs from simple cells, and that the two types of cell belong to at least partially distinct channels, presumably concerned with analysing different visual features.

The fact that simple cells receive a major input from X (sustained) cells suggests that the 'simple' channel is concerned with static or slowly moving patterns, and the evidence that complex cells receive inputs mainly from the Y (transient) system suggests that the 'complex' channel is concerned with the perception of flicker and motion. The case for this division of labour has been forcibly argued by Stone *et al.* (1979). Hoffmann *et al.* (1980) have confirmed that simple cells are more sensitive than complex cells to stationary patterns and complex cells are more sensitive than simple cells to moving patterns. However, they found that X cells, Y cells, and complex cells are all equally sensitive to stationary bars and equally sensitive to moving bars. It looks as though the differences between simple and complex cells are not due to difference between X and Y cells that feed into them. Hoffmann *et al.* concluded that the high sensitivity of simple cells to stationary stimuli and their low sensitivity to moving stimuli is not due to differences between X and Y inputs but to inhibitory inputs which they receive from complex cells.

Hubel and Wiesel's classification of cortical cells was based on a testing procedure in which points of light were used to probe the receptive field of a cortical cell. When probes of other shapes and sizes are used, other features of the cell's receptive field are revealed which do not always fit Hubel and Wiesel's simple picture (see Henry, Dreher and Bishop, 1974) and this led Henry (1977) to propose a more complex classification of cells in the visual cortex, which I shall describe briefly.

Henry (1977) has proposed that the terms 'simple' and 'complex' should be replaced by terms which do not suggest a hierarchical organization. He proposed that simple cells be called **S cells** and that complex cells with uniform receptive fields be called **C cells**. He suggested the distinct name **A cell** for complex cells with non-uniform receptive fields. Other types of cell that might be discovered would be given other letter designations. In this scheme, subtypes are designated by subscripts; thus a simple cell which has the hypercomplex property becomes an S_H cell.

Cortical cells of various types with similar orientation preferences are segregated into **sheets** of neural tissue lying perpendicular to the surface of the cortex. For historical reasons, these sheets are often referred to as **columns**. Sheets are packed in ordered sequence according to the orientation preference of the cells contained in them. As an electrode penetrates the cortex at right angles to its surface, the orientation preference of the cells which it meets remains more or less constant. As an electrode penetrates in a line parallel to the surface of the cortex, it may pass along a given sheet, in which case the orientation preference of the cells it meets will again be more or less constant, or it may pass across the sheets, in which case there will be a detectable change in the mean orientation preference of cells at least every 25 μm. These changes progress through an ordered sequence of orientations. Superimposed on this ordered sequence is some local variation, so that the ordered sequence is not a series of discrete steps, but a continuous gradation with local variability (Lee, Albus, Heggelund, Hume and Creutzfeldt, 1977). One complete 180° cycle of tilt occurs about every 1.0 mm of distance over the cortex, although every now and then, at unpredictable intervals, there is a large shift of orientation preference (Hubel and Wiesel, 1974a). Cells which lie above one another in a given column, or sheet, are connected exclusively by excitatory connections, whereas cells in one column are connected to cells in neighbouring columns only by inhibitory connections.

Hubel *et al.* (1977) were able to confirm the existence of orientation sheets by anatomical methods, They fed a monkey on radioactively labelled 2-deoxyglucose, the metabolic products of which accumulate in active neurones. The

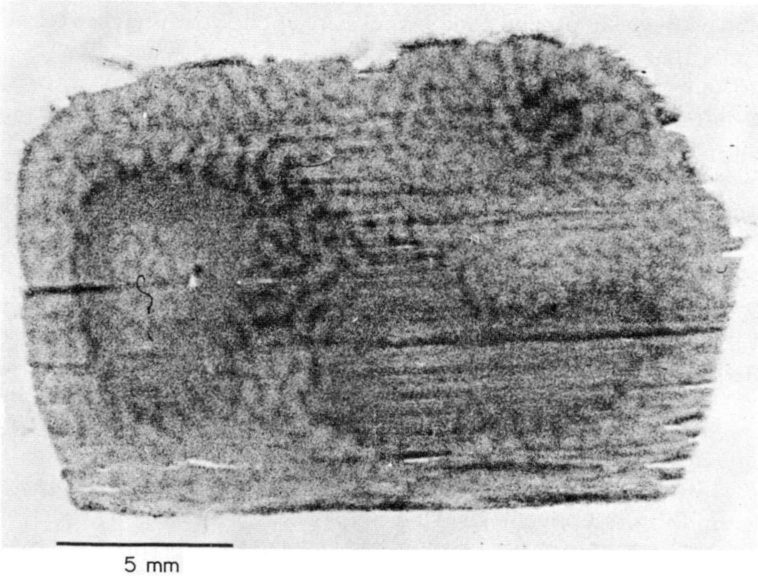

Figure 3.5 Autoradiograph of a tangential section through the dorsolateral surface of the visual cortex of a monkey. The dark meandering areas are the radioactively labelled sheets of cortical tissue containing cells which responded most vigorously to the vertical stripes to which the monkey had been exposed before it was killed. (Photograph, supplied by D. Hubel, is similar to one published by Hubel *et al.*, 1977)

animal was shown a display of moving vertical stripes for 45 minutes. The brain was then removed and thin sections of the visual cortex were pressed against X-ray film. In this way, photographs of orientation sheets such as the one shown in Figure 3.5 were produced. Such photographs are known as **autoradiographs.**

3.3 Response characteristics of cells in the visual cortex

In this section I shall present a list of response characteristics of orientation-selective cells in the visual cortex, which have been found to vary either as a function of the stimulus or as a function of the type of cell stimulated.

3.3.1 Linearity of spatial summation

Movshon *et al.* (1978a, b) investigated the **linearity of spatial summation** in simple and complex cells in the cat. This is a measure of the extent to which these cells sum the light-evoked activity across their receptive fields in a linear fashion. Linearity of a cell was tested by measuring the extent to which its firing rate kept pace with a regular series of optimally oriented light and dark bars moving across its receptive field, a procedure first used by Schiller *et al.* (1976b). The linearity of simple cells was indicated by the fact that their responses were modulated in synchrony with the stimulus over a wide range of spatial frequencies. Complex cells were revealed to be non-linear because their responses were not primarily governed by the frequency at which bars passed across their receptive fields. The way the two types of cell responded to luminance-modulated stationary gratings also indicated that simple cells have linear spatial summation and that complex cells do not. In the monkey, there is evidence that X ganglion cells, which are known to have linear receptive fields, feed predominantly into simple cortical cells, and that Y cells, which have non-linear receptive fields, feed predominantly into complex cells (Hubel and Wiesel, 1972). There is conflicting evidence on this point in the cat (see Section 2.4.5).

An indication of the line-spread function of a cortical cell may be obtained by measuring the luminance contrast sensitivity of the cell as a function of the position of an optimally oriented line. The line-spread function derived in this way may be called the **sensitivity profile** of a cell. A line-spread function of a cortical cell may also be obtained by measuring the contrast threshold of its response to a line in the centre of its receptive field while a second line, parallel to the first, is moved across its receptive field. I shall refer to a line-spread function derived in this way as a **subthreshold summation profile.**

Movshon *et al.* (1978a) obtained an indication of the sensitivity profile of a cortical cell by plotting the change in the frequency of its response while a fine optimally oriented line of fixed luminance contrast was moved in steps across its receptive field. The line could be either a bright line on a dark surround or a dark line on a bright surround. The functions for two simple cells are shown by the histograms in Figure 3.6. One of the cells has the usual excitatory centre and inhibitory flanks, but also outer facilitatory flanks which are referred to as

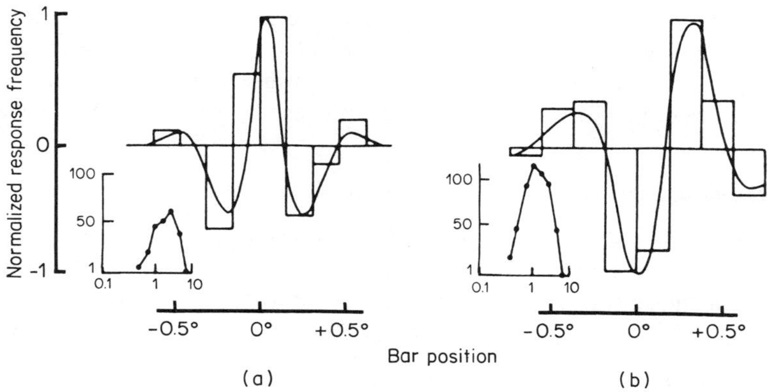

Figure 3.6 The histograms represent the sensitivity profiles of (a) a symmetrical and (b) an antisymmetrical simple cell in the visual cortex of the cat. Positive values indicate responses to bright bars and negative values indicate responses to dark bars. The continuous curves are the inverse Fourier transforms of the cells' spatial-frequency tuning functions, which are inset beside each diagram (abscissa, cycles/deg; ordinate, contrast sensitivity). (From Movshon *et al.*, 1978a. Reproduced by permission of the *Journal of Physiology*)

disinhibitory flanks because, when stimulated, they cancel the effect produced by the inhibitory flanks. This cell looks as though it is designed as a line detector. The other cell looks like an edge detector because it responds to light lines on one side of its receptive field and to dark lines on the other side. The solid curves in Figure 3.6 are the theoretical line-spread functions derived from the inverse Fourier integral of the spatial-frequency tuning curves (contrast-sensitivity functions) of each cell, as shown in the insets. (It was explained in Section 2.3.3 that the profile of the line-spread function of a linear system is the Fourier integral of its spatial-frequency tuning curve.) The close match between the sensitivity profile of a simple cell and the inverse integral of the spatial-frequency tuning curve indicates that simple cells integrate spatial information in a predominantly linear fashion.

Note that the sensitivity profile of the 'line detector' in Figure 3.6(a) is symmetrical, whereas the profile of the 'edge detector' in Figure 3.6(b) is asymmetrical. With regard to the centre of the receptive field, the symmetrical profile is approximately a cosine function and the asymmetrical profile approximately a sine function. For these reasons, so-called line detectors are called **symmetrical** or **cosine cells** and edge detectors are called **antisymmetrical** or **sine cells**. The theoretical significance of this terminology has been discussed in Section 2.4.7.

In a similar way, Movshon *et al.* (1978b) subsequently obtained sensitivity profiles of complex cells. One such profile is shown in Figure 3.7. It can be seen that the cell responded to light and to dark bars at all positions across its receptive field. In other words, the field of a complex cell is not segregated into excitatory and inhibitory regions. For a complex cell, the line-spread function derived from

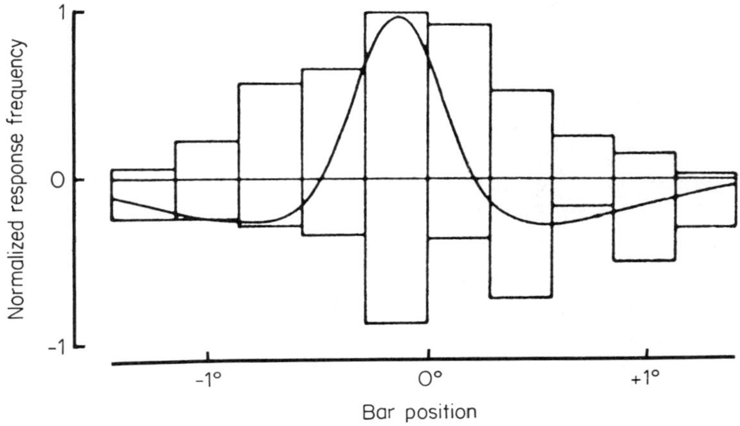

Figure 3.7 The histogram is the sensitivity profile of a symmetrical complex cell in the visual cortex of the cat. Positive values indicate responses to bright bars and negative values indicate responses to dark bars. The continuous curve is the inverse Fourier transform of the cell's spatial-frequency tuning function. (From Movshon *et al.*, 1978a. Reproduced by permission of the *Journal of Physiology*)

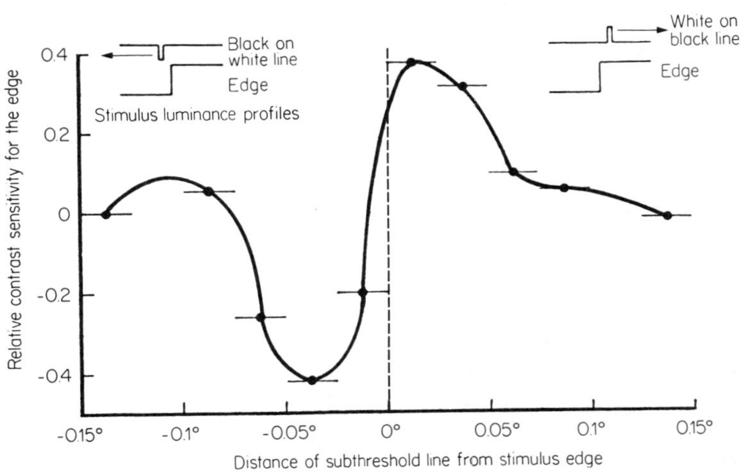

Figure 3.8 The sensitivity profile of a human 'edge detector'. This profile was determined by measuring the change in contrast sensitivity of an edge as a subthreshold line (white on black line on the excitatory side of the edge detector's receptive field and black on white line on the inhibitory side) was moved away from the edge in either direction. The bars on the graph indicate the width of the subthreshold line. (From Shapley and Tolhurst, 1973. Reproduced by permission of the *Journal of Physiology*)

the inverse Fourier integral of the spatial-frequency tuning curve does not conform to the sensitivity profile of the cell. This indicates that complex cells do not integrate spatial information in a linear fashion. However, it was found that the subthreshold summation profile of a complex cell did conform closely to the inverse Fourier integral of the spatial-frequency tuning curve. It was argued that the summation profile reveals the spatial characteristics of subunits within the receptive field of a complex cell rather than of the receptive field as a whole. By examining the interaction between such subunits as a function of the temporal interval between two lines, it was found that subunits interact in a facilitatory fashion and that facilitation outlasts the presentation of one line. This feature of complex cells makes them particularly suited for the detection of movement. Some indication of the sensitivity profiles and subthreshold summation profiles in humans may be obtained by psychophysical procedures. For instance, Shapley and Tolhurst (1973) measured the subthreshold summation profile of an edge detector using a central edge and a moving bar. The results are shown in Figure 3.8.

The tuning widths and preferred orientations of both simple and complex cells are not affected by a mean change in stimulus luminance. Changes in luminance merely alter the mean rate of firing of a cell by equal proportions at each orientation of the stimulus to which the cell is responsive (Bisti *et al.*, 1977).

3.3.2 Response variability of cortical cells

Cells in the visual cortex are also characterized by their **response variability**. By recording many times from the same cell for a given stimulus condition, Heggelund and Albus (1978b) determined the standard deviation of the cell's response, which they proposed as a measure of the cell's response variability. Response variability is a measure of the noise in the system, and is an important parameter of a cortical cell's function. The response variability of complex cells in the cat has been found to be, on average, only half that of simple cells (Heggelund and Albus, 1978b). This could be because complex cells are less dependent on the precise position of stimuli within their receptive fields, and hence less affected by minor perturbations of the stimulus.

It has been reported that most cortical cells have a distinct stimulus threshold and that response rate increases linearly with increasing contrast of a moving grating, until it levels off at high stimulus contrasts (Tolhurst *et al.*, 1981). These investigators also reported that the higher a cell's response rate the higher is its response variability and suggested that this relationship underlies the well-known Weber-Fechner law for contrast discrimination.

Response variability complicates the measurement of a cell's preferred orientation. Scobey and Gabor (1979) reviewed procedures for determining a cell's preferred orientation and concluded that the most reliable is one based on a determination of the preferred axis of movement. It has been claimed that cortical cells show large changes in preferred orientation over a period of hours (Donaldson and Nash, 1975). However, Hammond (1975) concluded that such changes are due to anaesthesia and variations in interocular alignment. Some cells have two preferred orientations, 90° apart (Hubel and Wiesel, 1965). These cells may be involved in the 'indirect effect' which is described in Section 4.6.1.

3.3.3 *Symmetry of orientation-tuning curves*

Cells in the visual cortex which are selectively responsive to the orientation of the stimulus may be characterized by the degree of symmetry of their tuning curves. This feature is specified by the **asymmetry coefficient**, which is the ratio of the partial widths of the tuning curve on either side of the mode at half the amplitude. Note that the tuning curve in Figure 3.2 is asymmetrical, so that the change in firing rate for each degree of change in the tilt of the stimulus is less on one side of the mode than on the other. It has been reported that about 60% of simple cells in the cat have asymmetrical tuning curves (Hammond and Andrews, 1978a). An asymmetrical tuning curve could result from a misalignment of receptive fields of LGN afferents that directly or indirectly feed into a given cortical cell, but it could also be due to intracortical inhibitory interactions.

3.3.4 *Tuning as a function of stimulus collinearity*

Cells in the visual cortex may be characterized by the degree of straightness of the line which stimulates them most effectively. Some cells in area 18 of the cat have been found to have a preferred orientation which differs according to which end of the cell's receptive field the stimulus occurs in (Hammond and Andrews, 1978b). This could be due to a kink or curve in the alignment of LGN receptive fields which feed into such cells. Such a cortical cell will clearly respond best to bent or curved lines occurring over the full extent of the cell's receptive field. Similar kinks near one end of a cortical cell's receptive field could account for the asymmetry of a cell's tuning curve.

3.3.5 *Effects of changes in stimulus length, width, and eccentricity*

The orientation-tuning curve of a simple cell becomes narrower as the length of the stimulus line is increased, up to about 4° of visual angle, after which it becomes slightly broader again. However, a cell's preferred orientation does not vary significantly with changes in the length of the stimulus (Henry, Bishop and Dreher, 1974). I have already explained that cells with the hypercomplex property do not respond to lines above a certain length.

If the stimulus is a bright bar on a dark surround or a dark bar on a light surround, the response of a cortical cell varies in a complex way as the bar becomes wider. A bar has two edges with opposite light–dark luminance profiles. About 50% of the simple cells of the cat respond to both edges and the other 50% respond to only one or the other of them. In the case of a cell that responds to both edges, the region of the cell's receptive field which produces a maximum response for one edge is not the same as the region which responds maximally to the other edge. The interval of time between the discharge induced by the leading edge of a moving slit and that induced by the lagging edge depends on the positions of these subregions and on the width of the slit and its speed of motion (Bishop *et al.*, 1971a, b).

Although the receptive-field size of simple cells increases with increasing retinal eccentricity, the mean tuning width of simple cells with receptive fields in the central region of the retina is much the same as the mean tuning width of simple cells with receptive fields in the retinal periphery. In the case of complex cells, there is some increase in tuning width with increasing retinal eccentricity of the cell's receptive field (Schiller *et al.*, 1976a).

3.3.6 Effects of stimulus motion, spatial frequency, and other stimulus features

Simple cells of the cat respond well to moving stimuli. The mean optimal stimulus velocity for simple cells is about 2.2°/sec along a path at right angles to the cell's axis of preferred orientation. This is their **preferred axis of motion**. Complex cells in the cat respond best to stimuli moving at about 19°/sec (Henry, 1977). Some of each type of cell respond only to movement one way along the preferred axis and others only to movement the other way. This defines their **preferred direction**. Some cells have a preferred axis of motion but no preferred direction of motion. The orientation tuning of cells in a given cortical column is not related to their preferred direction of motion, which suggests that these two features of a cell's response depend on distinct mechanisms (Schiller *et al.*, 1976a). The preferred velocity, axis, and direction of motion depend on the way in which subunits in the cell's receptive field respond to sequential stimulation. A coding process of this kind is called an **ordinal sequence code** (Henry, Bishop and Dreher, 1974). Directional selectivity has been found to depend on asymmetries in the cell's receptive field, on excitation spreading in the direction of preferred stimulus motion, and on inhibition spreading in the reverse direction (Michael, 1966; Emerson and Gerstein, 1977; Wyatt and Daw, 1975). In the cat, and probably in primates, some of these inhibitory interactions occur in the cortex. Patel and Sillito (1978) found that blocking cortical inhibition pharmacologically caused simple cells which normally responded only to slowly moving stimuli to respond to quickly moving stimuli.

Campbell *et al.* (1969) reported that cortical cells in the cat which are selectively responsive to the orientation of a grating are also selectively responsive to the spatial frequency of the grating. By recording evoked potentials from the human scalp, Campbell and Maffei (1970) found evidence of distinct channels selectively responsive to the spatial frequency of visual gratings. This linkage between orientation selectivity and spatial frequency prompted Blakemore and Campbell (1969) to suggest that orientation detectors may perform a Fourier analysis of the luminance profile of the retinal image along each of several different meridia (see Section 2.4.6).

Cells responsive to particular orientations are also selectively responsive to other stimulus dimensions such as colour (Poggio *et al.*, 1975), binocular disparity, and, of course, position.

In Section 2.4.2 I used the term **channelled system** to refer to a set of neurones in which the response of individual neurones is characteristically related to changes along a particular stimulus dimension. A stimulus feature which is mapped into the discharge of individual neurones is a **channelled feature** at that level of processing. Each cell in the visual cortex carries several channelled features. Considered individually, cortical cells cannot strictly be called orientation detectors or motion detectors, because their output is not uniquely determined by changes in any one stimulus feature. Presumably the various stimulus features to which cortical cells respond are portioned out into distinct cell populations or into distinctive signals at some higher level.

For each of the simple stimulus features of motion, binocular disparity, orientation, and spatial frequency there are sheets of cortical tissue ordered according to the stimulus preference of the cells within them. The periodicity of the various systems of sheets over the surface of the cortex is similar but their directions and phases vary. Note that the cells which constitute these various systems of cortical sheets are the same for each system; what varies is the stimulus preference of cells from one sheet to the next as each stimulus dimension cycles through its set of values. The whole system can be thought of as an array of neurones ordered in an n-dimensional feature matrix. I propose that such a system of multiple-feature detectors be called a **matricized system** and that the features be called **matricized features** at that level of processing. In a fully matricized system there are no correlations across cells between the preferred values of the various features which the system detects. In other words, the preferred value of a given feature to which a particular cell responds is not related to the preferred value of any of the other features to which it responds. It seems that the responses of cortical cells to orientation, direction of motion, position, and binocular disparity are independent, or fully matricized, or uncorrelated, at least within the major simple and complex cell categories. However, as we have just seen, preferred orientation is correlated with the preferred axis of motion. These may be referred to as **correlated matricized features** at this level of processing.

Certain cells in the visual cortex respond to stimulus features to which other cortical cells are insensitive. For instance, cells responding only to inputs from Y ganglion cells do not react to chromatic changes in the stimulus, whereas cells responding to X inputs do. I propose that a system in which there are distinct feature detectors be called a **partitioned system** and that the features be described as **partitioned features** at that level of processing.

Whether two features are correlated or independently processed depends on the two-dimensional distribution of excitatory and inhibitory regions within the receptive fields of cortical cells. The one-dimensional sensitivity profile is inadequate to specify the behaviour of a cortical cell. For instance, Daugman (1980) has shown that the independent tuning of spatial frequency and orientation requires receptive fields for which the two-dimensional Fourier transforms are the product of separable radial and angular functions. Cells which do not fulfil this requirement would change their preferred orientation as spatial frequency is changed.

This is as far as the analysis of visual processing has proceeded. What is needed now is a computational theory of how information may be extracted from matricized and partitioned feature-detecting systems. One must not fall into the trap of assuming that the processing of visual information progresses in one direction, from retina, through visual cortex, and beyond. At some stage the incoming information is confronted by attentional mechanisms, assumptions, and perceptual schemata. All the feature-detecting mechanisms that have been discovered operate at a very local level. For instance, each orientation detector operates most effectively over a line only a fraction of a degree long. There is no hint in the way they work of how global descriptive structures which underlie our ability to recognize and describe complex patterns arise. It may well be that global descriptions (not necessarily conscious) do not arise out of feature detectors; they may be imposed on visual inputs by a process akin to induction, by mechanisms operating relatively independently of the input. The role of feature detectors may be confined to that of supplying precise local information when called upon to do so by central global descriptive processes. I shall return to the question of descriptive processes in Sections 13.4 and 14.2.

3.3.7 Orientation tuning and head tilt

When the head of an animal is tilted, one would expect the preferred orientations of cortical cells with respect to a fixed external axis of reference to be tilted by the same amount. Horn *et al.* (1972) found that most orientation-specific cells in the cat's visual cortex behaved this way, but that about 5% of the cells changed their orientation preferences by between 10 and 15° as the animal was rotated to the side about its longitudinal body axis. These changes were in the direction of maintaining the orientation preferences of the cells upright with respect to gravity. The behaviour of these cells may be related to **orientation constancy**, that is, the ability of animals to make constant judgments about the orientation of stimuli as the head is tilted. The effect was not due to eye torsion, because the eyes of each cat were firmly attached to a metal ring, and, in any case, eye torsion would have affected all orientation detectors, not just a few. The effects were not eliminated by either bilateral labyrinthectomy or cervical transection of the spinal cord. Presumably if both these procedures had been carried out on the same cat the effects would not have occurred, but this was apparently not done. The cats in these experiments were anaesthetized, and so it is possible that the effects were due to an instability in the responses of cells due to the anaesthetic. However, cells which change their orientation preference with head tilt have also been reported in the alert cat by Denney and Adorjani (1972) and by Metzler and Spinelli (1979). On the other hand, Schwartzkroin (1972) could find no evidence of such cells in the alert cat.

Changes in the orientation preference of cells due to head tilt have not been reported in the monkey, and psychophysical measures of visual functions which are thought to reveal the preferred orientations of human orientation detectors have not revealed the presence of changes due to head tilt (see Sections 4.3 and 4.4).

3.3.8 The coding of orientation

Consider a hypothetical population of orientation detectors, all with the same sensitivity and selectivity and with preferred orientations spaced at equal intervals of 20°. Let the tuning curves of neighbouring units overlap in such a way that, for a stimulus at any one orientation, only two units are stimulated, as in Figure 3.9(a). With such a system, the orientation of a stimulus line could be indicated either by the **ratio** or by the signed **difference** in firing rates of adjacent units. There are two reasons why the firing rate of any one unit does not provide a basis for judging the angular position of the target. In the first place, except at the preferred orientation, there are always two stimulus orientations which produce the same firing rate in a given unit, and in the second place, the firing rate of a unit depends upon the luminance and position of the target as well as upon its orientation.

Whether we take the ratio of firing rates or the signed difference in firing rates of adjacent units does not materially affect the arguments that will be developed, but, because differences are easier to compute, these arguments will be expressed in terms of differences. The function that relates differences in firing rates of adjacent units to changes in the orientation of the stimulus has the same form as the tuning curves of the separate units. In the hypothetical case we are considering, the difference function is assumed to be sinusoidal, as can be seen in Figure 3.9(b). As a stimulus line rotates through a small angle, the rate of change of the difference in firing rates is given by the slope of this curve. The discriminability of a change in orientation of a line or of a difference in orientation between two lines will depend on the slope of this curve. Discriminability is highest at the point where the tuning curves of the adjacent units intersect, which in this case happens to be at the points of inflection of the curves. The curve in Figure 3.9(c) is the differential of the curve in Figure 3.9(b) and shows how discriminability should vary with the angle of orientation of the stimulus. Such a curve may be called the **orientation discrimination curve**. If the tuning curves, and hence the difference curves, are linear, rather than sinusoidal, the discriminability of differences in orientation is constant. In any case, we can take the slope of the difference function at the point of intersection of neighbouring tuning curves as one indication of the sensitivity of the system to differences in orientation. Anything that increases this slope should improve discriminability for orientation.

Discriminability should also be improved if the density of units is increased, with everything else held constant. Furthermore, the local variations in the orientation discrimination curve will be less when there is a greater density of

(a)

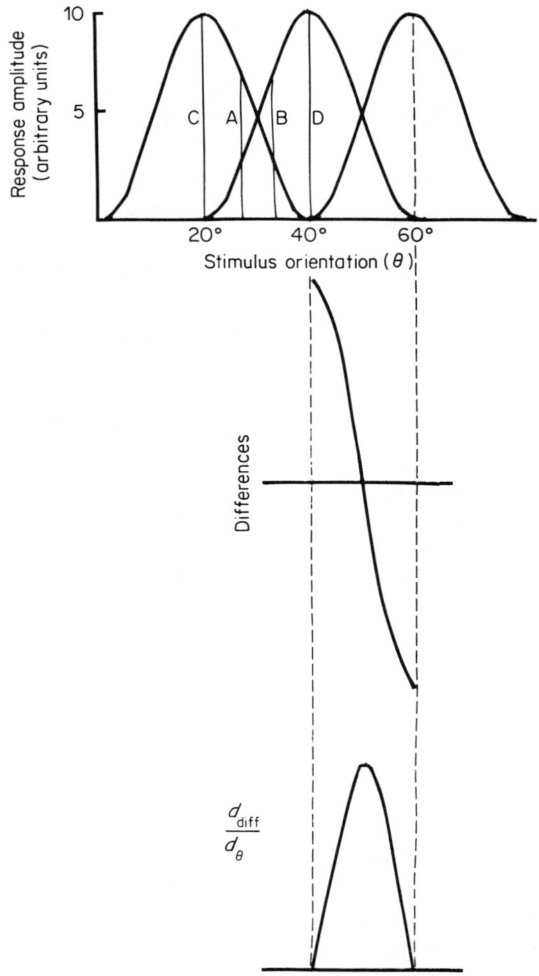

(b)

(c)

Figure 3.9 (a) Hypothetical tuning curves of three orientation detectors of the visual cortex. (b) The function of the signed difference in firing rate of detectors 2 and 3 as a stimulus line is rotated from 40 to 60°. (c) Orientation discrimination curve derived by differentiating the difference function.

units. Other factors that are likely to affect sensitivity to differences in orientation are the sensitivity of the orientation detectors, the width of their tuning curves, and the extent to which they inhibit each other. However, these factors are discussed in Section 4.4.6 and will therefore not be discussed further here.

The orientation of a single line is unambiguously coded in terms of the signed difference in firing rates of units whose tuning curves overlap at that point, given that the viewer knows that there is only one line present. However, consider two lines at the same retinal location but in different orientations, for instance the lines A and B in Figure 3.9(a). Let the orientations of these lines be symmetrically disposed about the point of intersection of the tuning curves. The 20° cell is stimulated by both lines, at three and seven arbitrary units of intensity as indicated on the ordinate. However, as far as this cell is concerned, this is equivalent to a single total stimulation of 10 units. Similarly, the 40° cell is also stimulated at a level of 10 units. But the two lines C and D would produce this same pattern of responses in these two cells. Furthermore, any pair of lines symmetrically disposed about the point of intersection of the curves would also produce this same pattern of responses, and so would a single line at the point of intersection of the tuning curves, if the luminance were suitably adjusted. This argument does not depend on the linearity of summation of stimulation produced by separate lines on the same cell. Pairs of lines that give identical responses can always be found, even if summation is not linear, because the luminance of the lines can always be adjusted until a matching response is found. If the tuning curves overlap more densely, the local range over which pairs of lines which produce identical responses can be found is reduced.

This is **metamerism**, and is analogous to colour metamerism. However, in colour metamerism, where different combinations of wavelengths produce identical responses, the process responsible is at the most peripheral level, so that the metamerism is obligatory. That is, there is no other information in the system that can be used to distinguish the equivalent sets of wavelengths. In orientation, the metamerism is not obligatory, because the relative orientations of two lines are encoded in terms of the distances which separate the ends of the lines as well as in terms of orientation. Furthermore, the orientations of intersecting lines are unlikely to be coded ambiguously, because such lines stretch across the receptive fields of several orientation detectors. If a stimulus line is shorter than the receptive field of an orientation detector, and if it is seen only briefly, so that the subject is prevented from moving the image onto other regions of the retina, then the metamerism should reveal itself as an uncertainty about the orientation of the line and about the number of lines present. Andrews (1967a) reported that short single flashed lines sometimes appear double or otherwise distorted. *I would predict that the confusion would be more marked with intersecting pairs of short lines.* The fact that two lines which intersect at an acute angle of less than about 3° are judged to be displaced towards each other may have something to do with metamerism in the orientation domain. This topic is discussed in Section 4.6.1.

The organization of cortical cells into sheets according to their preferred orientation can be understood as a mechanism for extracting the difference functions which I have described. Each cortical cell is responsive to many stimulus features so that its output is ambiguous with regard to any one of them. When differences between defined sheets of cells are extracted, the output of the difference detecting mechanism may well be unambiguous. There are different cortical sheets, with distinct or partially distinct periods and/or directions, for orientation, ocular dominance, movement, spatial frequency, etc. and the visual cortex may be understood as a mechanism for the extraction of distinct difference functions from a common matricized input (see Section 2.6).

3.4 The development of orientation detectors

Hubel and Wiesel (1963) recorded from cells in the visual cortex of two kittens which had been reared in the dark. One kitten was tested when eight days old, the other when 16 days old. Both were found to have a significant number of simple and complex cells which were selectively responsive to the orientation, and to the direction of motion, of moving bars, although these cells were rather sluggish in their response. Most of these orientation-selective cells were binocularly driven, as in the normal cat. When the kittens were retested after being kept in the dark for three months, the number of orientation-selective cells had declined and there were many responsive but non-selective cells and unresponsive cells. Futhermore, behavioural responses to visual stimuli were seriously impaired. Even with subsequent visual experience, these cats did not recover, physiologically or behaviourally (Wiesel and Hubel 1965). It was concluded that the stimulus selectivity of cells in the cat's visual cortex is largely innate, but that visual experience is required to maintain the normal performance of the cells beyond the first few weeks of life.

Pettigrew (1974) also found that many cortical cells in dark-reared kittens were binocularly driven and selectively responsive to the direction of stimulus motion, but found no cells which were selectively responsive to binocular disparity or orientation in such kittens. He argued that the cells which Hubel and Wiesel had found were probably not orientation selective, but were merely responding to the direction of motion of the stimulus slits. To prove that a cell is truly orientation selective, one must show that it is more responsive to a moving line that to a moving spot or that it is selectively responsive to stationary lines in various orientations.

More recent evidence supports the conclusions of Hubel and Wiesel. For instance, Blakemore and Van Sluyters (1975) found that about one-fifth of the cells which they sampled in dark-reared kittens, a few days old, were truly orientation selective, although broadly tuned by adult standards. After the kittens had been kept in the dark for three weeks, the number of orientation-selective cells had declined. The cells reached adult standards of tuning and sensitivity only if the kittens experienced high-contrast lines or edges in the appropriate orientation in the period between the third and fourth week. Blakemore and Van Sluyters suggested that Pettigrew failed to find orientation-selective cells because they tested dark-reared kittens older than five weeks, when the number of such cells had declined as a result of lack of stimulation. Sherk and Stryker (1976) also found orientation-selective cells in three-week-old, dark-reared kittens. In both these studies, care was taken to test with moving spots as well as moving lines, so as to avoid confusing movement selectivity with orientation selectivity. Buisseret and Imbert (1976) also confirmed Hubel and Wiesel's findings. They found that, at 12 days of age, 23% of cells in both normal and dark-reared kittens were selectively responsive to direction of stimulus motion and to stimulus orientation. The number of selective cells was the same in dark-reared and normal kittens up to the age of about three weeks; after this time, the selective cells began to be replaced by non-selective cells in the dark-reared kitten but increased in the normally-reared kitten (see also Frégnac and Imbert, 1978).

Wiesel and Hubel (1974a) found that in the monkey, too, the stimulus selectivity of many cortical cells is innate. Two macaque monkeys were tested at three and four weeks of age, after being reared in the dark. Their cortical cells were essentially like those in a normal adult monkey, although fewer of them were driven binocularly. This difference could be attributed to degeneration, because the pattern of ocular dominance in the cells of a normally-reared, two-day-old monkey was found to be normal. In this study, they were careful to test with stationary as well as with moving slits. Wiesel and Hubel were struck by the fact that there were fewer unresponsive or sluggish cells in the normal, two-day-old monkey than had been found in the normal, 12-day-old kitten. It seems that a newly born macaque is much more well developed than the newly born cat. This serves as a warning that one must not generalize about the relative contributions of innate and experiential factors to cortical development beyond the species being studied.

3.5 The modifiability of orientation detectors

The evidence reviewed in the previous sections suggests that the stimulus selectivity of many cells in the visual cortex is innately determined. In this section, it will be shown that visual experience also plays a role in determining the orientation selectivity and preferred orientation of cortical cells. There are several ways in which experience could have an effect.

(a) Experience could induce innately selective units to:
  1. become non-selective or totally unresponsive;
  2. become more selective;
  3. change their preferred orientation.

(b) Experience could induce non-selective units to become selective.

There is general agreement that cortical units lose their innate orientation selectivity or become unresponsive to stimulation of any kind if the animal is kept in the dark for the first few months of life. There is also little doubt that units with innately determined selectivity become more selectively tuned as a result of experience. The main debate is about whether, and under what circumstances, innately determined units radically change their stimulus preference. The first direct evidence that such radical changes may occur appeared in papers by Hirsch and Spinelli (1970, 1971). For the first three months of life kittens continuously wore masks which allowed one eye to see only three vertical black lines on a white ground and the other eye to see only three horizontal black lines. After three weeks, when recordings were taken from cells in the visual cortex, the only vertically-tuned units found were those which responded to vertical lines presented to the eye which had seen only vertical lines and the only horizontally-tuned units were those that responded to horizontal lines presented to the other eye. No units in either eye were found to be specifically tuned to oblique lines, although about half the cells were non-selective with regard to orientation. This preponderance of vertical and horizontal cells could have been due to the selective atrophy of cells which were innately tuned to other orientations, or to a change in the stimulus preference of other cells to vertical or horizontal. Hirsch and Spinelli's evidence did not settle this issue.

Stryker and Sherk (1975) at first failed to replicate these findings and suggested that Hirsch and Spinelli may have taken a biased sample of cells from too few locations in the visual cortex. However, Stryker and Sherk added a note to their paper describing a second experiment in which they replicated the findings of Hirsch and Spinelli (see also Stryker *et al.*, 1978). One factor which is likely to affect the outcome of such experiments is whether the animal is exposed in a striped drum or whether it is exposed to striped goggles worn on the head. The latter procedure should cause a tighter control over stimulus orientation, and therefore lead to the development of more cells tuned to a particular orientation. Gordon, Presson, Packwood and Scheer (1979) found that exposing kittens to horizontal lines in one eye and vertical lines in the other produced a greater preponderance of cells with corresponding preferences when the exposure was done with goggles compared with when a striped drum was used. Oddly, exposing both eyes to the same pattern produced no effect with either type of exposure.

In the same year that Hirsch and Spinelli made their report, Blakemore and Cooper (1970) reported an experiment in which kittens were reared for five months in an environment consisting entirely of vertical or horizontal stripes. After this experience, the cats behaved as if they were blind to lines in orientations other than the one they had experienced. More careful behavioural studies have revealed that cats reared in this way are not blind to non-experienced orientations but do show a permanent loss of visual acuity at those orientations (Muir and Mitchell, 1973, 1975; Blasdel *et al.*, 1977). When Blakemore and Cooper recorded from the visual cortices of their kittens, they found that 75% of the cells encountered responded only to the experienced orientation. The other cells were non-selective. The presence of such a high percentage of cells responsive to one orientation could have been due to the tuning of many non-selective or intially unresponsive cells to this orientation, coupled with the atrophy of cells originally tuned to other orientations. However, when the experimenters penetrated into different cortical columns, they did not notice any 'missing' columns that would correspond to the atrophied cells and concluded that cells originally tuned to other orientations had changed their preferred orientations. Pettigrew *et al.* (1973) also failed to find evidence of atrophied cells in cats raised in vertical environments, and agreed that there had been a true change in cortical organization and not merely a selective atrophy of cells. These experiments also showed that experience of oblique lines is required for the maintenance of obliquely-tuned cells; the experience of vertical or horizontal cells is not sufficient. However, the experiments do not prove that vertical and horizontal lines are the only stimuli that maintain vertically- and horizontally-tuned cells. In fact, Leventhal and Hirsch (1975) have shown that exposure to oblique lines is sufficient for the development, not only of obliquely-tuned cells, but also of vertically- and horizontally-tuned cells.

It looks as though cells tuned to vertical and horizontal orientations are organized in a different way to cells tuned to oblique orientations. Thus, Silverman *et al.* (1980) found the autoradiographic labelling in cortical layer IV produced by vertical and horizontal stimuli was lighter than that produced by oblique stimuli. These differences may be related to meridional differences in visual acuity (see Section 4.4). Removal of one eye has been found to produce a selective loss of obliquely-tuned cells (Frégnac, 1981).

Flood and Coleman (1979) recently obtained autoradiographs (see Figure 3.5) of cortical tissue from kittens raised in an environment of horizontal stripes. In these kittens, cortical columns were found to be more pronounced in the autoradiographs induced by exposure of the eye to horizontal stripes than in those induced by exposure to vertical stripes.

Cats exposed only to lines in a single orientation were found by Singer *et al.* (1981) to have wider and darker (more active) autoradiographic columns for that orientation. Columns serving the non-experienced orientation, especially in the foveal regions of area 17, were little in evidence outside layer IV.

A further demonstration that the stimulus selectivity of cells may change radically is provided by the experiments of Pettigrew and Freemen (1973) and Van Sluyters and Blakemore (1973). In these experiments, kittens were reared in an environment of white spots on a dark ground. At eight weeks of age, many more cortical units in these animals responded well to moving spots than was the case in normally-reared control animals.

Blakemore and Mitchell (1973) set out to determine just how much visual experience was required to bring about these dramatic and permanent effects. Inferring from earlier experiments that the peak period for inducing these changes was the twenty-eighth day, they exposed six dark-reared kittens to vertical stripes for totals of 1, 3, 6, 18, 27, and 33 hours around the twenty-eighth day. A control kitten was kept totally in the dark until the sixth week, when all the kittens were tested. The tests showed that even the kitten with only one hour of exposure to the vertical stripes had a preponderance of well-tuned cells responding preferentially to vertical stripes. The control kitten had only a few broadly-tuned cells; the rest were non-selective or unresponsive. Blakemore and Mitchell claimed that this latter finding with the control kitten conflicts with Hubel and Wiesel's observation that neurones in the inexperienced cat have an innate orientation selectivity. But Blakemore and Mitchell did not test their control kittens until they were six weeks old, by which time their innately-tuned cells may well have lost their selectivity. The main point, though, is that a period of just one hour is sufficient to induce radical and permanent changes in the visual cortex. The suggestion is that this experience must occur during a critical period of the animal's development, although Blakemore and Mitchell assumed, rather than proved this. Creutzfeldt and Heggelund (1975) exposed normally-reared adult cats to vertical stripes for one hour each day over a period of two weeks, during which time they were otherwise kept in the dark. They subsequently found that there were *fewer* cells responsive to vertical lines in these cats than in normal cats. These changes in the adult cat were probably only temporary, and look more like the sort of changes that could be produced by short-term adaptation or fatigue of orientation detectors (see Section 4.6).

Even though a very brief period of visual experience in the young cat triggers permanent changes in the selectivity of cortical neurones, it is reasonable to suppose that these changes take time to develop. Pettigrew and Garey (1974) have obtained evidence that these changes develop over a period of at least two days after the visual experience which has triggered them. Several dark-reared kittens were exposed to vertical lines for several hours at about the fourth week after birth. Immediately after this experience, it was found that there were *fewer* cells responsive to vertical lines than to lines in other orientations. The exposure to vertical lines had obviously suppressed the responsivity of vertical units, but this was only a temporary effect. Next day there was an increase in the number of cells which responded preferentially to vertical lines presented to the exposed eye, and on the second day most cells responded in this way. It looks as though a consolidation period is required for permanent changes in the orientation preferences of cortical cells.

Buisseret *et al.* (1978) paralysed the extraocular muscles of four dark-reared kittens with curare, and then, for several hours, passively moved the animals about the laboratory or allowed them to observe people moving about. These animals showed no evidence of cells with orientation selectivity when they were tested the next day. On the other hand, in animals which were not paralysed during exposure to the visual scene, the majority of cortical cells developed orientation selectivity. It was concluded that visual experience, coupled with self-produced ocular motility, is required for the development of stimulus selectivity. But this conclusion is unwarranted. In the first place, Pettigrew and Garey had obtained changes in the stimulus preferences of cells in animals with immobilized eyes. In the second place, the kittens with immobilized eyes in Buisseret and Imbert's study experienced a scene which was constantly moving with respect to their retinae, so that the animals probably experienced only smeared images. The kittens with mobile eyes would be able to track moving objects and thus stabilize their retinal images. It was probably this difference in retinal-image motion that caused the differences in stimulus preference.

All these experiments involved some kind of sensory deprivation. In a more recent study, Spinelli and Jensen (1979) have reported changes in the preferred orientation of cortical neurones as a consequence of quite brief periods of avoidance training in kittens which were otherwise reared in normal visual surroundings. In this experiment, kittens were trained to make a paw movement which turned off an electric shock and a pattern of vertical lines to one eye and turned on a 'safe' pattern of horizontal lines to the other eye. This training led to the development of a large number of cortical cells which had a preference for vertical stimuli in one eye and horizontal stimuli in the other. Cells with this characteristic are not found in normal animals. Many of these cells responded to vertical bars when tested with the eye that had been trained to avoid horizontal bars and to horizontal bars when tested with the eye that had been trained to avoid vertical bars. Paradoxically, the remaining cells of this type had the opposite characteristics. The significance of these findings remains puzzling and obscure, but we can expect many more experiments that explore the effects of visual training rather than of mere deprivation.

All these studies were done on cats, and one must be cautious about generalizing the findings. For instance, Mize and Murphy (1973) found that rabbits have a full complement of innate orientation-selective cells and that rearing with vertical lines has no effect. Blakemore (1974) has suggested that plasticity of orientation detectors is required in animals with binocular vision, because it is important for stereoscopic vision that the orientation preferences of corresponding inputs be the same. This would explain why the rabbit, which lacks stereoscopic vision, has no plasticity of its orientation detectors, and it would lead one to suppose that human beings have plastic orientation detectors at the time that the precise disparity mechanism underlying stereoscopic vision is developing. For further discussion of this point see Section 12.5.

The only comparable findings for human beings are from studies on people with a long history of severe optical astigmatism. Such people see lines in one orientation clearly and lines in another orientation much less clearly. If, when they are older, the optical astigmatism is corrected, it is found that they have a residual astigmatism along the same axis as the original astigmatism. This residual astigmatism shows as a reduction in grating acuity and in grating contrast sensitivity along the affected axis, and the evidence suggests that it is neuronal in origin (Freeman and Thibos, 1973, 1975a). But the whole question of astigmatism of neuronal origin will be reviewed in detail in Section 4.4.

Chapter 4

The Psychophysics of Oculocentric Orientation

4.1 Introduction

This chapter is about how people detect changes or discriminate differences in the orientation of simple visual stimuli, such as lines, edges, and gratings, presented in a frontal plane. The positions of the eyes, head, and body of the observer are held constant, so that the orientation of the stimulus may be specified oculocentrically, that is, with reference to retinal coordinates. Questions having to do with the orientation of simple stimuli with respect to the head and gravity are dealt with in Chapter 10, and broader questions about how people distinguish between and recognize the orientations of visual objects are dealt with in Chapter 13.

Basic psychophysical data concerning the discrimination of differences in the orientation of simple line stimuli are discussed in Sections 4.2 and 4.3. Many people have higher acuity for stimuli that are oriented vertically or horizontally than for stimuli at oblique angles. This is known as **meridional astigmatism**, and there has been a lot of interest in this subject during the last 10 years because of its relevance to recent physiological findings. This work is reviewed in Section 4.4. Meridional astigmatism is one form of **visual anisotropy**. There are also anisotropies in the apparent tilt of a line and this question is discussed in Section 4.5. Anisotropies in apparent length of lines, such as occur in the vertical-horizontal illusion, are beyond the scope of this book (see Thompson and Schiffman, 1974).

There is a great variety of visual tasks for which performance is superior when stimuli are horizontal or vertical rather than oblique. These tasks involve visual detection and acuity at one extreme, and identification and description of orientations at the other. There has been an unfortunate tendency to lump all visual effects of stimulus orientation under the same heading, 'oblique effects' (Appelle, 1972). This is very misleading, as Essock (1980) has pointed out. I shall deal with anisotropies of basic visual abilities, such as sensitivity and acuity, in this chapter, and with the effects of stimulus orientation on recognition of orientation and on shape perception in Chapters 13 and 14.

The study of tilt contrast has also been very popular recently, again because of its physiological relevance. It is discussed in Section 4.6. The reader is referred to Howard and Templeton (1966) for pre-1966 studies on these topics.

4.2 The discrimination of differences in the oculocentric orientation of lines

4.2.1 Introduction

When the oculocentric orientation of a line is changed, there is a corresponding change in the *orientation* of the retinal image of the line, which in turn results in a change in the pattern of excitation of cortical orientation detectors. At the same time, each point on the line, except the point of rotation, changes its oculocentric *location*, and this set of changes will also be detected by the visual system. Thus, any psychophysical measure of the discriminability of differences in the orientation of a line will reflect the acuity of the visual orientation-detection system as well as the acuity of the location-detection system. If the eyes are allowed to scan the stimulus line, the ability of the visual system to make use of information about the direction of an eye movement may also be involved.

Five procedures have been used to measure the acuity of the visual system to changes in the orientation of simple stimuli, but none of them is a class A, or identity-matching, procedure. Because of this, and because of the composite nature of the task, there is some doubt about whether the procedures measure the same thing. The five procedures will be described separately, with an indication in each case of the sensitivities that have been reported.

4.2.2 *Discrimination of departures from vertical*

The simplest procedure is to ask a subject to detect when a single luminous line on an otherwise homogeneous background is tilted from the vertical. In one recent determination, a human observer was able on 75% of trials to detect when a line 30′ of arc long was tilted from the vertical by as little as 25 of arc (Westheimer *et al.*, 1976). Similar values have been reported by Andrews *et al.* (1973). A change in line orientation of 25 of arc moves the position of the end of a short line only a few seconds of arc, which is less than the diameter of a single receptor.

There are no natural standards, or norms, at orientations other than vertical and horizontal. People can be taught to estimate other orientations verbally, but this is more of a recognition task than a discrimination task. Smith (1962) found average unsigned errors of about 10° in an experiment in which subjects were asked to estimate verbally the tilt of a line. This method is clearly too coarse for the determination of acuity to differences of tilt.

4.2.3 *Setting lines parallel*

The second task is that of setting a test line to be parallel with a comparison line or of indicating whether or not the two lines are parallel. The two lines may be presented in succession or at the same time. In all the studies reported, the two lines were presented in different locations and the comparison line was usually much longer than the test line. Performance in this task reflects the subject's uncertainty about the orientation of both the test line and the comparison line. Andrews (1967a) reported that a test line about 7 of visual angle long could be set parallel to a vertical or horizontal comparison line with a standard deviation of about 0.5° when the two lines were separated by about 1° of visual angle. The standard deviation rose to about 1° or more for other orientations of the comparison line. When the test line was flashed for 1 msec, and the method of constant stimuli used, the standard deviation rose to about 3° for a vertical comparison line. With briefly exposed lines, the subject is prevented from moving the image of the line by moving his eyes, so that the line excites only those orientation detectors that respond to a line in that location. Briefly exposed lines sometimes appeared to the subject to be as much as 30° away from their true orientation, and apparent shifts of 15° were commonly experienced. In spite of these apparent shifts, the standard deviation of responses was only a few degrees. I commented on the significance of these large apparent angular shifts in Section 3.3.8.

Andrews used the reciprocal of the variance of judgments as a measure of information (R. A. Fisher, 1966), and with this measure determined whether directional information from each eye combines without loss to give binocular orientation acuity. If the variances for the left eye, the right eye, and for both eyes are V_L, V_R, and V_{LR} respectively monocular information regarding orientation adds without loss if $(1/V_L)+(1/V_R)=1/V_{LR}$. His data suggested that there was indeed no loss of information about orientation.

4.2.4 *Discrimination of dot alignment*

In this procedure, first used by Solomon (1947), the subject adjusts the position of a small dot until it appears in alignment with the test line. Bouma and Andriessen (1968) reported that a dot could be set into alignment with a vertical or horizontal line 14 min of visual angle in length with a standard deviation of about 1° when the dot was 0.5° of visual angle away from the end of the line. At other orientations of the line, the standard deviation was between 2 and 3°. Like other investigators, they found that making the test line longer had little effect on orientation acuity, but the significance of this finding will be discussed later.

The measurement of Vernier acuity involves a task in which the subject detects when two lines are offset. For reasons which were given in Section 2.3.4, this task may be regarded as depending on the ability of the subject to detect that the orientation of the imaginary line joining the two neighbouring ends of the target lines are not in the same orientation as the target lines.

4.2.5 *Successive comparisons of line tilt*

Matin and Drivas (1979) claimed that none of the methods already mentioned is suitable for measuring acuity for changes in the orientation of gratings, and developed a method which is suitable for this purpose. In this method, the subject discriminates between the tilt of two gratings presented successively for 1 sec each to the same retinal area. Matin and Drivas used a signal-detection procedure and found that differences were more easily discriminated for lines near the vertical and horizontal meridia than for lines near oblique meridia. Their results cannot be compared with those from other studies, because they were not analysed in the same way. This method is not suitable for studying constant errors in perceived tilt. The presentation of the lines in the same location may predispose the subject to base his decision on differences in location of the ends of the lines, rather than on mechanisms for detecting differences in orientation. The subject may also be influenced by sensations of motion triggered by the succession of differently oriented lines (see Wenderoth and Tyler, 1979), although Matin and Drivas did not mention any such effect.

4.2.6 The ideal observer for oculocentric orientation

Andrews (1967b) described an ideal observer for the discrimination of line orientation, in terms of which he assessed the performance of the human observer. The retinal mosaic in Figure 4.1 was traced from a diagram in Polyak (1957). The centre of each cone was marked with a dot and it was determined that the dots were distributed in an approximately random fashion with a mean separation of 24″ of visual angle. The best possible estimate of the orientation of a stimulus line by an ideal observer with such an array of detectors was derived by fitting a line, by the method of least squares, to the centres of the set of retinal receptors which respond to the stimulus (see Figure 4.1). The effects of optical factors, duration of exposure, and luminance were ignored because these were assumed to be the same for the ideal and human observer, and were held constant in the actual experiments. It was assumed that the ideal observer (like the human observer) knows that the stimulus is a straight line.

The factor of interest was the length of the stimulus line. Andrews *et al.* (1973) wished to determine how efficiently human observers combine spatial information regarding the orientation of a line for lines of different lengths. The ideal observer is not subject to errors of combining spatial information from different parts of the retina. For instance, consider the task of judging the distance between two points. If two stimulus points are each located by the ideal observer with a certain error variance, the sum of these two variances should be the error variance of the estimate of the distance between the two points. The distance apart of the points should have no effect on the precision of the estimate of this distance. For the human observer, the precision of spatial judgments is known to

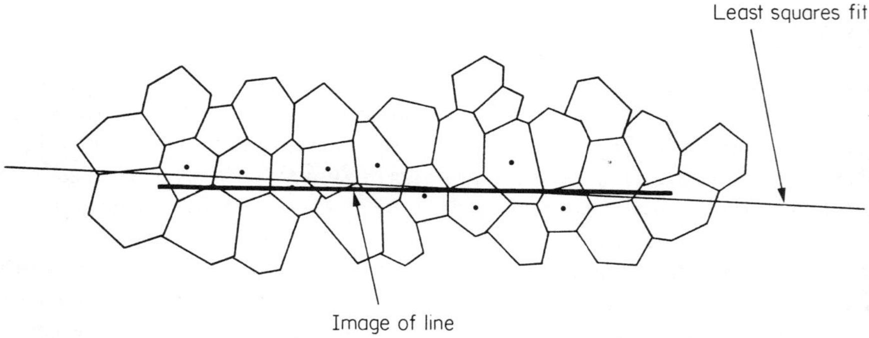

Figure 4.1 The retinal mosaic of the central fovea traced from Polyak (1957). The image of a line falling across the receptors is represented by the thick line. The thin line is the line of least squares fitted to the centres of the receptors lying on the line (Reprinted with permission from D. P. Andrews, *Vision Research*, 7, Copyright 1967, Pergamon Press, Ltd.)

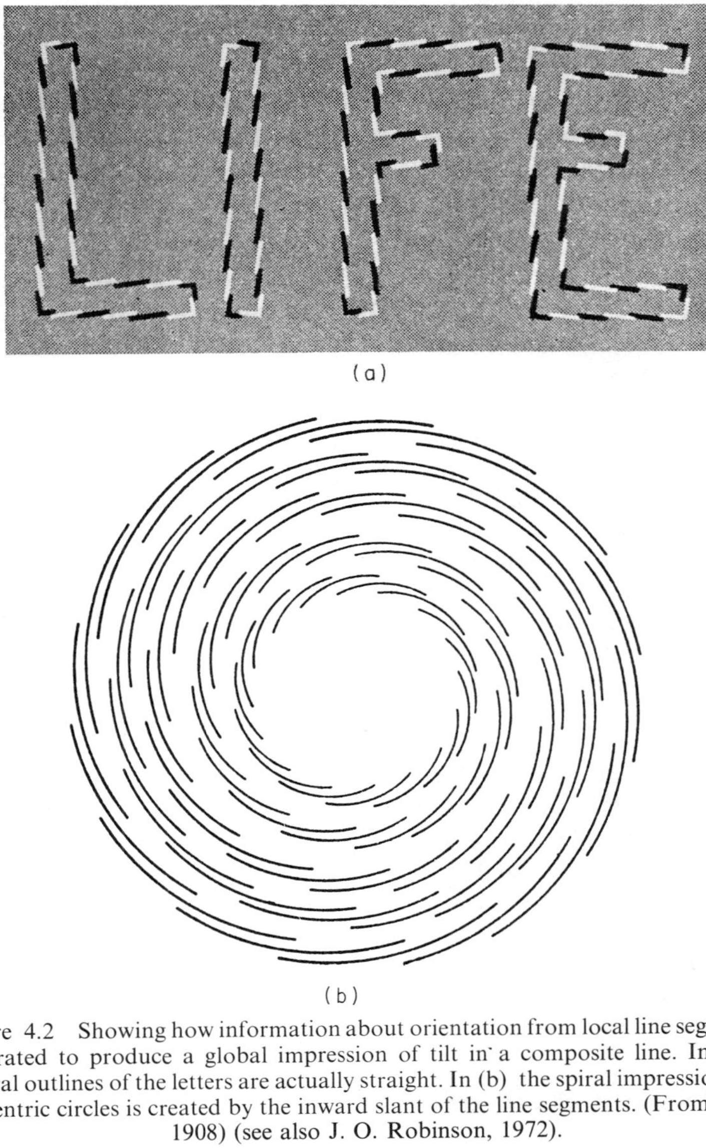

Figure 4.2 Showing how information about orientation from local line segments is integrated to produce a global impression of tilt in a composite line. In (a) the general outlines of the letters are actually straight. In (b) the spiral impression in the concentric circles is created by the inward slant of the line segments. (From Fraser, 1908) (see also J. O. Robinson, 1972).

be affected by the separation between parts of the stimulus. The ideal observer therefore provides a yardstick with which to assess the effects of the length of the stimulus line on the performance of the human observer. The relative performance of the ideal and the human observer is indicated by the ratio of the variance of the ideal observer to the variance of the human observer for the same task. This index contains certain constants related to mean cone size, etc., but the relative efficiency of human performance under different conditions is independent of the values of these constants. Andrews *et al.* measured variances for human observers in a task of setting a test line parallel to a long comparison line. This was repeated for test lines of from 3.5 to 60′ of visual angle long and for various separations between the two lines. The efficiency of the performance was constant and about 75% of the efficiency of ideal performance for lengths of line up to about 9′, after which the efficiency dropped and was only 25% for a line 18′ long. It was concluded that receptive fields of cortical orientation detectors are at least 9′ of visual angle long in the fovea. The drop in efficiency for longer lines was put down to losses incurred when the outputs of several orientation detectors are combined. Such losses could be due to metamerism within the orientation system (see Section 3.3.8), adaptation of orientation detectors to prolonged stimulation, and lateral inhibition between different detectors. These last two possibilities will be discussed in Sections 4.5 and 4.6.

When a person judges the orientation of a long line, he integrates the information provided by the local orientation detectors, each of which indicates the orientation of a small segment of the line. A visual indication of such an integrative process is provided by the illusions shown in Figure 4.2. The letters in Figure 4.2(a) are straight, but the inclined segments provide local slant information which, when integrated over the sides of the letters, falsely informs the visual system that the composite lines are tilted. In a similar way, the line segments in Figure 4.2(b) falsely inform the visual system that they form parts of a spiral, when in fact they are parts of concentric circles. Nakayama and Tyler (1976) measured the extent to which the tilt of small line segments disturbs the apparent tilt of a longer composite line. For tilts of the line segments up to between 5 and 15°, the apparent orientation of the composite line was shifted in the same direction as that of the component segments. For tilts of the line segments through larger angles, the apparent orientation of the composite line was shifted in the opposite direction, and this illusion peaked when the segments were tilted between 20 and 40°. It was suggested that the first effect is due to excitatory interactions between orientation detectors with small receptive fields, which respond to the line segments, and orientation detectors with larger receptive fields which respond to the orientation of sections of the composite line. The effect with the larger tilts of the segments was ascribed to inhibitory interactions between small-field and large-field orientation detectors.

4.3 Orientation-specific stimulus interactions

4.3.1 Introduction

Although it has not been possible to study human orientation detectors electrophysiologically, several psychophysical procedures have been proposed for indirectly determining the characteristics of these detectors. One must always be cautious in drawing parallels between electrophysiological and psychophysical data, especially when the two types of data are drawn from different species.

In each of these procedures, a measurement is made of some effect which an

induction line or grating at one orientation has on a **test** line or grating at another orientation. The effects that have been measured fall under four main headings: the **elevation of contrast thresholds**, the **reduction of contrast thresholds, shifts in spatial frequency**, and **tilt contrast**. I will be concerned mainly with threshold elevation and threshold reduction effects in this section. Tilt contrast is discussed in Section 4.6. Shifts in spatial frequency are mentioned briefly in Section 2.4.6.

All these effects show interocular transfer, that is, they occur when the inducing stimulus is presented to one eye and the test stimulus to the other eye (Gilinsky and Doherty, 1969; Ware and Mitchell, 1974b). Furthermore, Ware and Mitchell showed that stereoblind individuals, who lack binocularly driven cortical cells, do not show interocular transfer of these effects. These two facts strengthen the belief that the effects reflect properties of cortical neurones rather than of retinal neurones. They are a function of the relative oculocentric orientation of inducing and test stimuli, rather than of their relative egocentric orientation (Findlay and Parker, 1972; Mitchell and Blakemore, 1972). This fact is compatible with the belief that they reflect properties of cortical orientation detectors, because the orietation specificity of most, if not all, such detectors is oculocentrically fixed (see Section 3.3.7).

In order for these effects to reveal properties of orientation detectors, three types of quantitative comparison are required. First of all, we wish to know how each of the effects is related to the orientation of the test stimulus relative to that of the inducing stimulus. The idea is that this function indicates something about the tuning characteristics of orientation detectors, and about interactions between orientation detectors. Secondly, we wish to know whether the above function is affected by the orientation of the test stimulus, because this will indicate whether or not orientation detectors with different preferred orientations have the same characteristics. Finally, we wish to know whether the results obtained from the study of one effect are similar to those obtained from the study of another, because if they are, this strengthens the belief that the procedures are tapping the same fundamental visual mechanisms, and if they are not, it must be assumed that they are tapping different mechanisms.

It is not possible to make these comparisons unless some way is found of expressing the results obtained at different orientations and for different stimuli in an equivalent form. For this purpose, Blakemore and Nachmias (1971) introduced the **equivalent contrast transformation**. This is explained as follows, as it applies to the measurement of contrast threshold elevation at different orientations of the test grating. The contrast threshold of a test grating presented after a suprathreshold inspection grating in the same orientation is elevated in proportion to the contrast of the inspection grating. But the contrast threshold of a test grating presented after an inspection grating of fixed contrast is elevated by an amount which is inversely related to the angle between the test and inspection grating. Thus, one can adjust the contrast of a *vertical* inspection grating from a standard value so that it elevates the contrast threshold of a vertical test grating by the same amount that it is elevated by that same inspection grating at its standard contrast but at a given angle to the test grating. The contrast of the vertical inspection grating required for this matching threshold elevation is the **equivalent contrast**. The principle may be generalized as follows. The effect produced on a test stimulus at one orientation by an inspection stimulus of contrast C at another orientation can be equated with the effect produced by an inspection stimulus of contrast C_e when both test and inspection gratings are vertical. C_e is the equivalent contrast. The transformation may be applied to any simultaneous effect or aftereffect that viewing one grating may have on another grating, and of course this is the purpose of the equivalent contrast transformation. In the study of all these effects, several modes of stimulus presentation are possible according to whether the interacting stimuli are presented:

1. As lines or as gratings;
2. Simultaneously or successively;
3. Overlapping or adjacent;
4. Photopically or scotopically;
5. Stationary or moving;
6. At high spatial frequency or low spatial frequency;
7. With constant luminance or flickering.

The variety of stimulus conditions that has been used makes it difficult to compare the results of different experiments. A change in any one of the factors listed above can mean a change in the nature of the visual channel that is being stimulated. For instance, steadily illuminated stimuli are likely to stimulate the sustained system and flickering stimuli are likely to stimulate the transient system.

4.3.2 The threshold elevation effect

It is generally more difficult to detect a test stimulus when it is superimposed on another similar suprathreshold stimulus than when it is presented alone. The test stimulus is said to be **masked** by the other stimulus. The more similar the masking stimulus is to the test stimulus, the greater is the magnitude of the masking; that is, the higher the luminance contrast of the test stimulus must be in order for it to be detected. Campbell and Kulikowski (1966) optically superimposed a test grating of 10 cycles/deg on a masking grating of the same spatial frequency and phase. The test grating appeared on the masking grating for one second every alternate second. When it was not present, it was replaced by an homogeneous field of the same mean luminance as the grating.

The luminance contrast required for the detection of the test grating was determined as a function of the orientation of the masking grating when the test grating was vertical and also when it was inclined at 45°. This procedure measures the **simultaneous threshold elevation effect**. As the angle between the test and masking gratings increased, the amount of masking fell exponentially. With a vertical test grating, the amount of masking fell to half its maximum value when the angle between the gratings was 12°. With an oblique test grating, the masking fell to half when the angle between the gratings was 15°. I shall comment on these meridional differences in Section 4.4.2. Campbell and Kulikowski commented on the fact that these figures are in fair agreement with electrophysiological determinations of the half-width of the orientation-tuning curves of orientation detectors of the cat, and concluded that the angles of separation between induction and test gratings producing half-maximum masking are a convenient measure of angular selectivity of human orientation detectors. However, it will become apparent in what follows that the contrast threshold elevation effect may be the result of lateral inhibition between orientation detectors, and if this is so, it reflects the spread of lateral inhibition rather than the excitatory selectivity of the orientation detectors.

Blakemore and Nachmias (1971) measured the extent to which the contrast threshold of a vertical test grating was elevated by an inspection grating that preceded the test grating. This is the **successive threshold elevation effect**. They found that the aftereffect was reduced by a factor of two when the angle between the test and inspection gratings was increased from 0 to 6.75°. This is about half the value reported by Campbell and Kulikowski. However, when the Campbell and Kulikowski data were reworked by Blakemore and Nachmias in terms of the equivalent contrast transformation, the angle for half-maximum effect was reduced from 12 to 9.5°. Furthermore, Blakemore and Nachmias used lower mean luminances that did Campbell and Kulikowski, and inhibitory interactions between orientation detectors are probably weaker with lower luminances. Nevertheless, when Kulikowski (1972) made a direct comparison between simultaneous and successive contrast elevation with luminance equated across conditions, he found that simultaneous presentation still gave a narrower angular selectivity than successive presentation (see also Lovegrove, 1976). There thus seems to be a real difference in the magnitude of interactions between simultaneously and successively presented gratings. However, this may be due to the fact that superimposed gratings, but not successive gratings, produce a moiré pattern with a spatial frequency similar to that of the gratings. As far as I know, nobody has taken this factor into account. Furthermore, lines presented in rapid succession give rise to apparent movement, and there is evidence that apparent movement may be seen between two stimuli, even when one is masked by the other (Kolers, 1972). Wenderoth and Tyler (1979) eliminated apparent movement by using a target line and a masking grating. With these stimuli the simultaneous and successive threshold elevation effects had the same angular range. Thus the previously reported difference in range between the simultaneous and successive effects was probably not due to fundamental differences in neural processing.

Evidence reviewed in Section 3.3.6 suggests that the response of orientation detectors is a function of the spatial frequency of the stimulus as well as its orientation. One would therefore expect the threshold elevation effect (and other such effects) to depend on the similarity in spatial frequency of the test and inspection gratings. This has been shown to be the case, and the effect falls to zero when the two gratings differ by plus or minus one octave (Blakemore and Campbell, 1969; De Valois, 1977a). However, when the inspection and test gratings have the same spatial frequency, and when the data are rendered comparable by use of the equivalent contrast transformation, the magnitude of the contrast elevation is independent of the spatial frequency of the two gratings within the range 2.5–20 cycles/deg (Movshon and Blakemore, 1973).

The size of the successive threshold elevation effect also depends upon the duration of the inspection stimulus. Gilinsky (1968) showed that with exposures of a suprathreshold adapting grating of less than about 1 sec the aftereffect was facilitatory rather than inhibitory. Perhaps briefly exposed gratings produce a subthreshold afterimage which then summates with the test grating so as to reduce its threshold. We shall see in the next section that subthreshold gratings summate in this way. Blakemore *et al.* (1973) found that the magnitude of reduction in apparent contrast of a grating increased up to a maximum of about 0.8 log units after about 45 sec of exposure to an adapting grating. They also reported that longer adapting exposures, while not affecting the magnitude of the aftereffect, did lengthen the time that it took for the visual system to recover from the aftereffect. The rate of fading of the retinally stabilized image of a vertical test line has also been shown to be a function of the angular separation between the test line and a previously exposed inspection grating (Schmidt *et al.*, 1972).

4.3.3 The threshold reduction effect

If the luminance of a masking grating is subthreshold, then the contrast threshold for a superimposed identical test grating is decreased. This is the **contrast threshold reduction effect**, and is generally believed to result from the summation of excitation. Kulikowski *et al.* (1973) found that the simultaneous threshold reduction effect was reduced to half its maximum value when the test and induction gratings were only 3° of visual angle apart. The angular range of this effect is therefore much smaller than the range of the threshold elevation effect.

One possible explanation for the broader angular range of the threshold elevation effect is the presence of beats, or moire patterns, in superimposed gratings. These patterns are particularly pronounced in two superimposed gratings which are close in spatial frequency and orientation. With suprathreshold gratings, they may well produce a spurious widening of the range of orientations over which threshold elevation occurs. With subthreshold gratings, the moiré pattern, being of a lower spatial frequency than the gratings, would be very weak and ineffective.

Another possible explanation of the different angular ranges of the threshold elevation and threshold reduction effects is that the threshold reduction effect, which is a summation effect, depends upon the rather narrow angular selectivity of the excitatory response of single orientation detectors, and that threshold elevation effects reflect the broader spread of lateral inhibition between detectors. This interpretation is supported by the results of a study by Thomas and Shimamura (1975). In one experiment they showed that the detectability of the presence and orientation of a single line is predictable from psychophysical estimates of receptive field properties derived from studies of threshold summation. In a second experiment they showed that detectability of a pair of lines forming a cross at various angles to each other was lower than that predicted from the detectability of each line alone, but fitted the assumption that the two lines inhibit each other. Dealy and Tolhurst (1974) have produced evidence that orientation detectors can inhibit each other over a broader range of frequencies than that over which they summate. Blakemore and Tobin (1972), Benevento *et al.* (1972), and Fries *et al.* (1977) provided electrophysiological evidence for such inhibition.

Finally, the narrow angular range of the threshold elevation effect compared with the threshold reduction effect could be, in part, an artifact of the effects of probability summation between the distinct parts of the stimulus pattern.

Thomas and Gille (1979) estimated orientation tuning width of cortical cells by comparing the contrast required for line detection with the contrast required for orientation discrimination. They argued that the difference in orientation required to make the slope of the function relating the probability of detection to the probability of identification equal to unity sets the lower limit to the bandwidth of the orientation channel. Using this and other assumptions, they derived half-amplitude bandwidths of between 10 and 20°. This is in closer agreement with estimates from threshold elevation studies than with estimates from threshold reduction studies. One physiological measure of the orientation-tuning width of simple cells in the monkey has yielded a modal value of 30° (Schiller *et al.*, 1976a). This is broader than any of the psychophysical determinations. Perhaps the psychophysical procedures reflect the tuning width of the most narrowly tuned cells rather than of the modal tuning width of cells.

4.3.4 *After-effects in sustained and transient channels*

The gratings and lines used in the experiments described so far were stationary and continuous. It can therefore be assumed that they predominantly stimulated the sustained channel of the visual system. It may be that the angular widths of the inhibitory and excitatory functions are different in the transient system.

Sharpe and Tolhurst (1973a) used the equivalent contrast transformation to compare the threshold elevation effect using stationary continuous gratings with the threshold elevation effect using inspection and test gratings which moved back and forth or which were temporally modulated. They found, as did Blakemore and Nachmias, that the effect of a stationary and continuous inspection grating on the successive contrast threshold of a test grating was reduced to half the maximum value when the gratings were about 7° apart. However, when the gratings were moved back and forth every 1.5 sec or switched on and off at a frequency of 5 Hz, the after-effect fell to half value at angular separations of about 13°. But this only happened for low spatial-frequency gratings. The after-effects for high-frequency gratings were not affected by temporal modulation. This could be because low-frequency flickering gratings tap the transient system, and that this system has a wider spread of lateral inhibition than the sustained system which the steady low-frequency gratings and high-frequency gratings were tapping. *Nobody has determined whether simultaneous masking is similarly affected by temporal modulation of the gratings.*

Another piece of evidence that the width of inhibitory influences is greater for the transient channel than for the sustained channel is provided by the results of an experiment by Lovegrove and Over (1973). They found that the threshold of a grating in one colour was elevated more by an adapting grating of the same colour than by one of a different colour (see also Stromeyer *et al.*, 1980). However, this effect of colour was manifested only when the gratings were less than 15° apart in orientation. Beyond 15°, threshold elevation occurred, but was not colour specific. This suggests that the sustained channel has a narrower inhibitory range than the transient channel, if it is assumed that the chromatic components of stimuli are processed largely by the sustained channel and the achromatic components largely by the transient channel (Ingling and Drum, 1973).

4.4 Visual anisotropies

4.4.1 Introduction

It has been known since the nineteenth century that visual acuity is higher for targets which are aligned with the normally vertical or horizontal meridia of the retina than for targets aligned with oblique meridia. The effect is known as **meridional astigmatism**. The pre-1966 literature on this topic is reviewed in Howard and Templeton (1966, p. 30). Up to that time it had been shown that the effect is not due to eye movements, but it had not been finally decided whether it was due to optical or to neural factors. In 1966, Campbell, Kulikowski and Levinson conclusively settled the matter, at least as far as grating acuity is concerned. By converging two coherent light beams from a helium–neon laser onto the retina, they formed an interference pattern, consisting of a sinusoidal spatial grating, which could be varied in spatial frequency. An image formed in this way is essentially unaffected by the optical properties of the eye. The interference grating was set at a given luminance contrast, and the highest resolvable spatial frequency was determined for two subjects at each of several orientations of the grating. The same measurements were made using a normal sinusoidal grating. For both gratings, resolution was better at vertical and horizontal orientations than at oblique orientations, and the differences between orientations were similar for both gratings. This proves that optical factors are not responsible for meridional astigmatism. A similar experiment with similar results was reported at about the same time by Mitchell *et al.* (1967). Campbell *et al.* also found that meridional differences in acuity, as measured by spatial contrast sensitivity, increased as the spatial frequency of the grating increased from 3 to 35 cycles/deg (see Figure 4.3). Below about 5 cycles/deg, acuity was the same for all meridia tested. When the eye was artificially defocused, there was a uniform decrease in contrast sensitivity across these same spatial frequencies for a given orientation of the grating. That is, the decrease in contrast sensitivity due

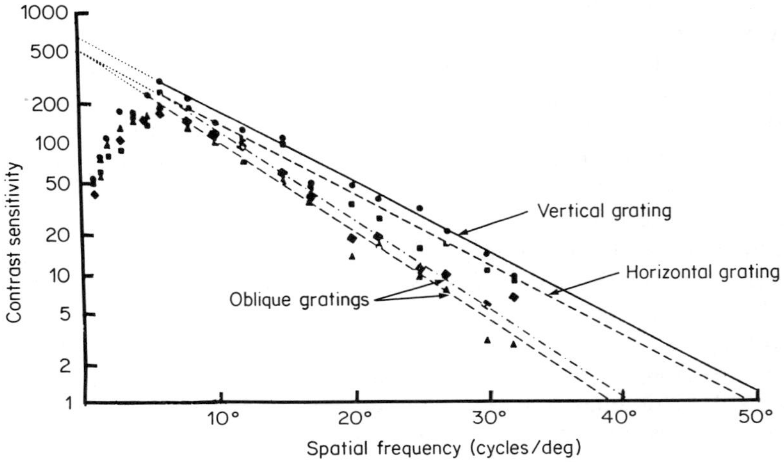

Figure 4.3 Showing how meridional differences in contrast sensitivity increase as the spatial frequency of the test grating is increased (Data from one subject from Campbell, *et al.*, 1966. Reproduced by permission of the *Journal of Physiology*)

to defocusing was not a function of spatial frequency, whereas meridional differences in acuity were affected by spatial frequency. This is further evidence that meridional astigmatism is not due to optical factors which defocus the image. Campbell *et al.* also argued that meridional astigmatism cannot be due to the hexagonal pattern into which foveal cones are packed, because if it were, acuity fluctuations would occur every 60°, not every 90°.

We may conclude that at least one type of visual anisotropy, namely meridional variations in grating acuity, is a property of the neural substratum of the visual system. Such anisotropies could be oculocentric, that is, tied to retinally defined coordinates, or they could be tied to gravitationally defined coordinates. By comparing the axes of meridional astigmatism when the head was vertical with those when the head was tilted 45° to one side of vertical, several independent investigators have shown that meridional astigmatism is oculocentric (Findlay and Parker, 1972; Lennie, 1974; Banks and Stolarz, 1975; Corwin *et al.*, 1977).

The evidence, which was reviewed in Section 3.5, that visual experience can modify the selectivity of orientation detectors in the visual cortex raises the possibility that meridional anisotropies are caused by a predominance of vertical and horizontal lines in the visual world. Before evidence for such a theory is reviewed, we must take a detailed look at the types of meridional anisotropy.

4.4.2 Anisotropies of grating acuity and contrast sensitivity

Grating acuity is the highest spatial frequency of a sinusoidal grating that can be detected as a grating for a given mean luminance level and luminance contrast between the lines of the grating. Figure 4.4 shows how grating acuity in human subjects varies with the orientation of the grating. Similar variations in grating acuity have been reported in the monkey (Bauer *et al.*, 1979; Boltz *et al.*, 1979). Grating acuity is essentially a measure of the ability to detect local luminance gradients. Thus, a measure which is closely related to grating acuity is **contrast sensitivity**, or its reciprocal, **contrast threshold**. The contrast threshold is the least luminance contrast required for the detection of a sinusoidal grating with a given spatial frequency and mean luminance level. The **luminance contrast** of a grating is defined as the difference between the maximum and minimum luminances of the grating divided by their sum (see Figure 4.5). Figure 4.3 shows how contrast sensitivity varies with spatial frequency at various orientations. It can be seen from the graph that meridional differences in contrast sensitivity do not occur for spatial frequencies of less than about 5 cycles/deg, which explains why Hirsch *et al.* (1974) failed to find them when using gratings of only 5 cycles/deg. This absence of meridional differences at low spatial frequencies suggests that meridional variations in grating acuity arise in the high spatial resolution, sustained channels of the visual system, and not in the low spatial resolution,

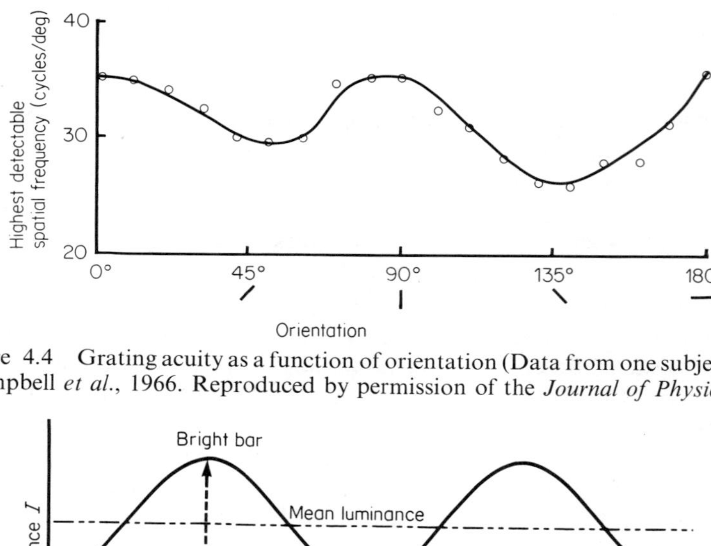

Figure 4.4 Grating acuity as a function of orientation (Data from one subject from Campbell *et al.*, 1966. Reproduced by permission of the *Journal of Physiology*)

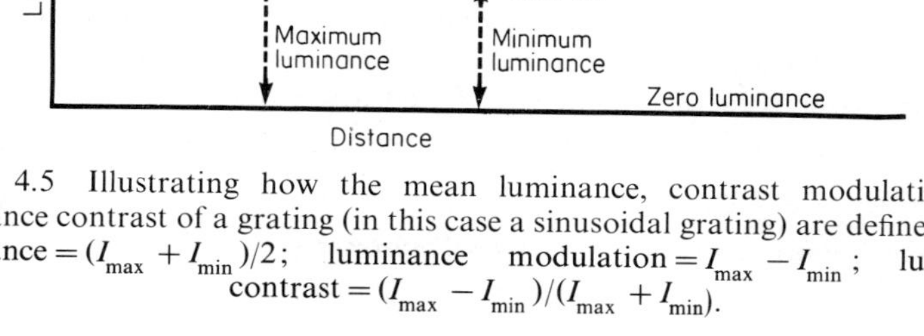

Figure 4.5 Illustrating how the mean luminance, contrast modulation, and luminance contrast of a grating (in this case a sinusoidal grating) are defined. Mean luminance $= (I_{max} + I_{min})/2$; luminance modulation $= I_{max} - I_{min}$; luminance contrast $= (I_{max} - I_{min})/(I_{max} + I_{min})$.

transient channels (see Section 2.4.5). Further support for this idea comes from a study by Berkley *et al.* (1975) in which they found that meridional variations in grating acuity are confined to the central area of the visual field within about 10° of the fovea, just where sustained, small-size receptive fields are concentrated. If meridional variations in grating acuity are confined to the sustained channel, they should disappear when a grating is flickered in counterphase at high frequencies, and, indeed, Camisa *et al.* (1977) found this to be the case.

Closely related to the contrast sensitivity of a grating is the apparent lightness contrast of a suprathreshold dark – light border, and this too is higher for main axes than for oblique axes (Kitterle, 1973). Furthermore, in complex patterns consisting of lines running in several directions, vertical and horizontal lines seem to stand out more clearly than oblique lines. Davidoff (1974) drew attention to this effect in the screen painting by François Morellet which is shown in Figure 4.6. Appelle (1976) found that figures embedded in distracting lines are

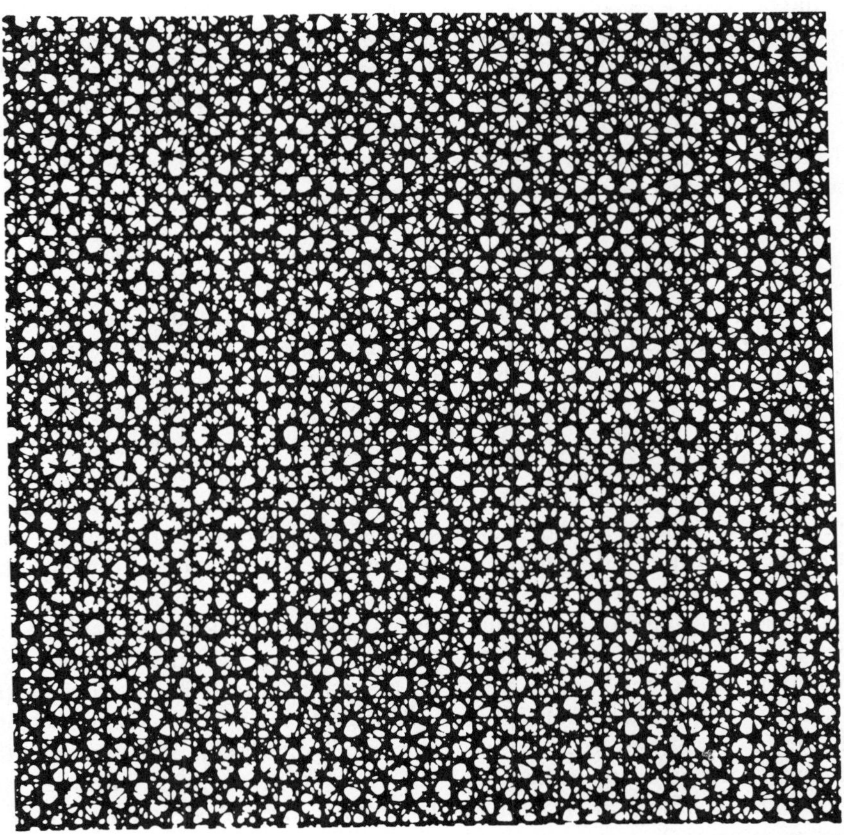

Figure 4.6 A screen painting by François Morellet showing how lines aligned with the normally vertical and horizontal retinal meridia are more visible than lines aligned with other retinal meridia (From Davidoff, 1974)

more easily seen when the lines composing the figures are oriented vertically or horizontally than when oriented obliquely.

4.4.3 Anisotropy of pattern acuity

The variability in setting a line parallel or perpendicular to another line is less for lines which fall on main axes than for those which fall on oblique axes (Andrews, 1967a, b; Bouma and Andriessen. 1968), and the variability in matching the orientations of successively presented lines is also less for lines which fall on main retinal axes (E. Matin and Drivas, 1979). Vernier acuity is essentially the ability to detect a change in the slope of the imaginary line joining the ends of two offset target lines (see Section 2.3.4), so that it is not surprising that Vernier acuity has also been shown to be better for vertical or horizontal lines (Ludvigh and McKinnon, 1967; McKee and Westheimer, 1978). The discrimination of the waviness of a line may be regarded as a periodic Vernier acuity, so that this too should be better when the line is vertical or horizontal. Tyler and Mitchell (1977) measured the just detectable amplitude of waviness of lines and found that the peak sensitivity occurred for lines with between 2 and 6 waves/deg of visual angle. It was also found that the meridional variation in sensitivity was the same for all spatial frequencies of waviness, which indicates that meridional anisotropy has effects across large retinal extents, even though requiring the presence of high spatial-frequency components in the stimulus line itself.

4.4.4 Anisotropy of luminance sensitivity

In those parts of the retina where acuity is high, scotopic luminance sensitivity is low, and *vice versa.* This inverse relationship between acuity and sensitivity is normally explained by saying that the small receptive fields and strong lateral inhibition of the fovea favour acuity but not sensitivity, whereas the large receptive fields and weak scotopic lateral inhibition of the periphery favour sensitivity but not acuity (Hallett, 1963). If meridional variations in spatial acuity are due to variations in the receptive-field sizes of cortical orientation detectors and/or to variations in the degree of lateral inhibition among them, then luminance sensitivity for lines should be inversely related to meridional variations in acuity.

Appelle (1973) found that the luminance threshold was lower for vertical and horizontal lines than for oblique lines, and this runs counter to the above prediction. However, Appelle measured luminance thresholds for lines in the periphery of the visual field, beyond five degrees from the fovea, just where meridional variations in acuity are very small or absent. Furthermore, the meridional astigmatism of the subjects was not determined. Although not required to report the orientation of the test lines, the subjects did not know in what orientation the test lines would appear, so that the meridional variations in threshold may have been due to meridional differences in the criterion for reporting a line. Ogilvie and Taylor (1958) and Vassilev and Penchev (1976) found that vertical or horizontal lines could be detected at lower luminances than could oblique lines. In these studies, the subjects had to say when they saw a line, not merely when they saw light. Taylor (1963) argued in the following way: when the stimulus is an oblique line, the visual system may detect small regions of luminance but is unable with precision to determine when these regions are collinear. The decision as to whether a line is seen is therefore withheld until more elements are visible. With a vertical line, the precision of collinearity judgments is higher, and therefore a line will be detected at lower luminances. Taylor thus sees luminance sensitivity for lines as a special case of spatial acuity. This argument is valid if the task is to detect the line as a line and if the subjects do not know where the line is, as in Taylor's earlier study with Ogilvie. Beyerstein and Freeman (1976) could find no evidence of meridional variations in luminance increment thresholds in normal subjects, although no measurements of anisotropies in acuity were made for these subjects. However, in subjects with a severe optical astigmatism, who showed a strong residual astigmatism after optical correction, increment thresholds were higher along the astigmatic axis than along other axes. But in this study also, the subjects did not know the orientation of the test line within the homogeneous luminous field until they detected it, and they reported only when they saw the line as a line. Therefore the task is more of an acuity than a detection task.

If an experiment were done in which subjects were always told the orientation of the luminous line, and asked to report the first impressions of *light* rather than the first impression of a *line,* then luminance thresholds for oblique lines might turn out to be lower than for vertical or horizontal lines. This would require that some light be visible from lines before the lines themselves are visible, and this requirement seems reasonable, because fixated lines and afterimages of lines fragment before they fade altogether (Wade, 1973a; Atkinson, 1972, 1973; Ellis, 1975). I decided to do such an experiment and, with the help of Peter Kaiser, measured the luminance thresholds of three subjects for the detection of light in a luminous line which subtended 0.5° of visual angle. On each trial the line was presented at one of several luminance values in one of two time periods and the subject was asked to say in which period it occurred. This is a criterion-free method for determining a threshold (Blackwell, 1952). This was done with a dark background (scotopically) and with a background at photopic levels of luminance. Two small luminous points indicated the orientation and position of the line to the subject at all times. This guarded against orientation-specific criteria effects. The results for both the scotopic and photopic conditions showed that vertical and horizontal lines had lower luminance thresholds than oblique lines. Meridional variations in grating acuity and chevron acuity were also determined. All significant meridional variations were consistent across tasks; for those meridia in which luminance sensitivity was highest, grating acuity and chevron acuity were also highest. This result argues against the idea that anisotropies in any of these visual abilities are due simply to variations in receptive-field sizes of cortical neurones or simply to variations in lateral inhibition between them.

4.4.5 *Anisotropy of flicker sensitivity*

P. J. Foley (1962) measured the frequency at which a flickering line of light fused into the impression of a steady light—the **critical fusion frequency,** or CFF. He found that the CFF was lower for oblique lines than for lines on principal axes. Kelly (1975b) used gratings flickering in counterphase, that is, gratings in which alternate stripes go from black to white as the other stripes go from white to black. The gratings had a constant mean luminance and extended over a central 10° visual field. He increased the luminance modulation until the subject detected the flicker. This procedure was repeated for temporal frequencies between 1 and 50 Hz, which is the bandwidth of human visual detectability of flicker for the mean luminance used. He found that the luminance modulation required for the detection of flicker was higher for oblique gratings than for vertical and horizontal gratings. Furthermore, the same degree of anisotropy was found over the whole temporal bandwidth. This latter finding strongly suggests that this type of meridional anisotropy is not confined to the sustained channel, even though, according to the evidence cited in Section 4.4.2, anisotropy of spatial acuity is.

Kelly went on to predict that flicker sensitivity should be isotropic for red and green gratings of equal luminance flickering in counterphase. In this connection he quoted a report by Hubel and Wiesel (1968) that, while orientation-selective neurones in the cortex of the monkey respond to the luminosity of the stimulus, most opponent-colour cells are insensitive to changes in orientation. Kelly's experimental results substantiated this prediction. *It would be interesting to know whether spatial acuity and grating acuity are also isotropic for steady chromatic gratings equated for luminance.*

4.4.6 *The physiological basis for visual anisotropies*

Human visual anisotropies of acuity and sensitivity are generally thought to be due to innate or experientially acquired differences between cortical units for which the preferred orientation is vertical or horizontal and units with preferences for oblique lines. Nevertheless, orientation bias in retinal ganglion cells cannot be ruled out (Levick and Thibos, 1980). Populations of cortical cells could differ with respect to any one or any combination of the following characteristics.

The number of each type There could be more cortical cells tuned to horizontal or vertical orientations than to oblique orientations (see Figure 4.7a). Hubel and Wiesel (1968) found no overrepresentation of cells with preferred horizontal or vertical orientations in the visual cortex of the monkey. However, only a few cells of their samples had receptive fields in the central area of the retina and, in humans at least, meridional astigmatism is confined to an area within 10° of the fovea (Mansfield, 1974). Pettigrew *et al.*, (1968) Albus (1975a, b) and Berman *et al.* (1981) found an excess of cells tuned to the vertical

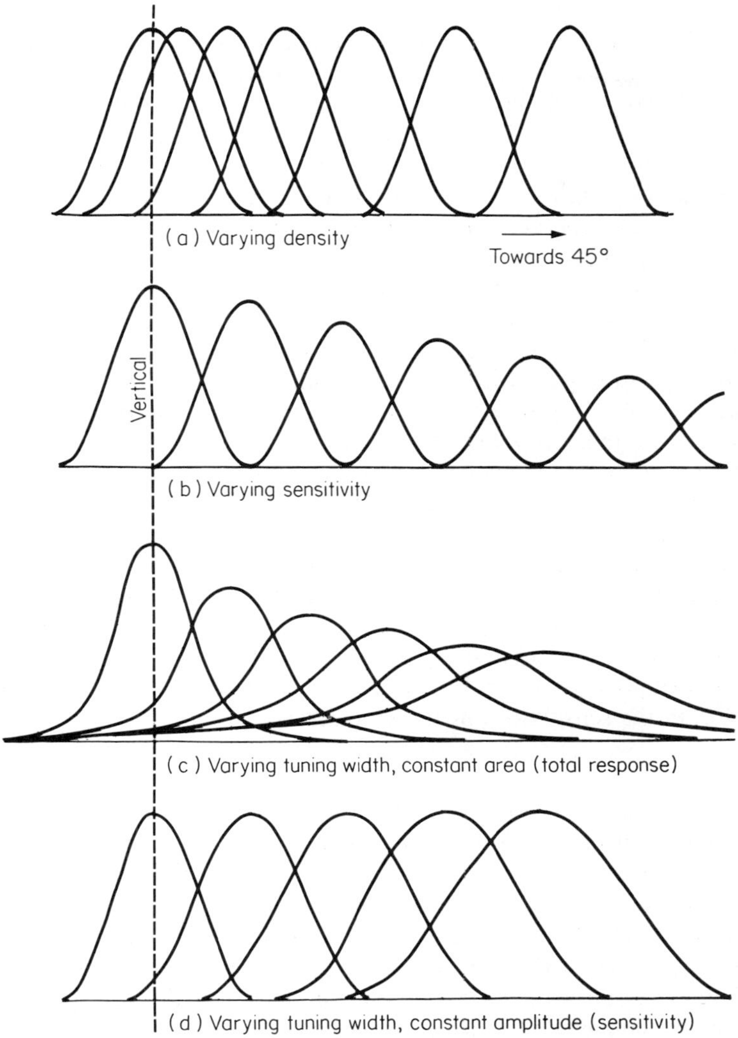

Figure 4.7 Hypothetical tuning curves of sets of orientation detectors showing how various characteristics may vary as a function of the preferred orientation of the detectors

or horizontal in the central region of the cat's visual cortex, and cats are known to show meridional astigmatism (Orban and Vandenbussche, 1979). Freeman and Pettigrew (1973) induced astigmatism in kittens by rearing them with cylindrical lenses before their eyes. When tested later, these kittens were found to have a preponderance of cortical cells selectively responsive to stimuli oriented along the axis of induced astigmatism.

Of more relevance to human meridional astigmatism is a study on the monkey by Mansfield (1974). He reported a considerable preponderance of 'vertical' and 'horizontal' cells with receptive fields in the centre of the retina and only a slight preponderance of such cells with peripheral receptive fields, and Mansfield and Ronner (1978) reported that the amplitude of evoked potentials arising from the synchronous discharge of many cortical cells was greater for vertical or horizontal stimuli than for oblique stimuli. On the other hand, Finlay *et al.* (1976) found no preponderance of cells tuned to vertical or horizontal stimuli in either foveal or peripheral cortical areas of the monkey. Some resolution of these different findings may be provided by a report by Bauer *et al.* (1981). They found a predominance of vertically-tuned cells in the upper layers of the monkey's visual cortex, but only within an area representing the central 1° of retina. There was a predominance of obliquely-tuned cells in other areas.

The only physiological evidence on humans is from a study by Maffei and Campbell (1970). They found that, for humans, the amplitude of the potential evoked on the scalp by a moving vertical or horizontal grating was greater than the potential evoked by a moving oblique grating. This difference was not found by recording from the electroretinogram, which reinforces the idea that the difference was cortical in origin. Differences in the amplitude of the human evoked potential have also been shown to be related to the incidence of meridional astigmatism (Freeman and Thibos, 1975b) and have been found to be tied to retinal meridia when the head is tilted (Frost and Kaminer, 1975). However, these differences in the evoked potential may not reflect differences in the relative numbers of orientation detectors – they could equally well be due to differences in the sensitivities of these detectors.

Suprathreshold sensitivity The suprathreshold sensitivity of a cortical cell is the rate of the cell's response to a stimulus in the preferred orientation of the cell, at a specified luminance above threshold. Let it be assumed that the orientation-tuning curves of orientation detectors have a standard shape, say a sinusoidal distribution, and that the resting discharge of the cells is the same for all the cells. Consider a population of such cells, over a 180° span of preferred orientations, and assume that all the cells have identical characteristics, except that, when stimulated by similar high-contrast lines, the cells responding to lines along the principal meridia fire at a higher maximum frequency than cells responding to lines along oblique meridia (Figure 4.7b). It may be assumed that the sensitivity of the system to a change in the orientation of a line is indicated by the rate of change of differences in firing rates of those units whose tuning curves overlap the angular position of the stimulus. The rate of change of these differences for a given change in stimulus orientation is greater for the cells near the principal meridia than for cells near oblique meridia. Contrast sensitivity and sensitivity for differences in orientation will therefore be higher where the suprathreshold sensitivity of orientation detectors is higher.

There is no direct evidence on how suprathreshold sensitivity varies with orientation. The amplitude differences of human evoked potentials may reflect differences in sensitivity.

Lateral inhibition between neighbouring cortical units The relationship between lateral inhibition and spatial acuity is obscure. There is no obvious reason why lateral inhibition between lines in neighbouring orientations should affect the visibility of single lines. It may be that the superior discrimination of differences in orientation in the region of vertical and horizontal meridia is due to greater lateral inhibition between cells with orientation preferences in these regions. Blakemore and Tobin (1972) electrophysiologically recorded the response of a cortical cell to an optimally oriented test bar when the visual field surrounding the test bar was stimulated by a grating in various orientations. The results, which are shown in Figure 4.8, demonstrate that the extent to which the cell was inhibited increased as the orientation of the surround grating approached that of the test bar. Thus, orientation detectors certainly inhibit each other, but we do not know whether those with preferred orientations in the neighbourhoods of the main meridia inhibit each other more than those with preferred orientations in other neighbourhoods.

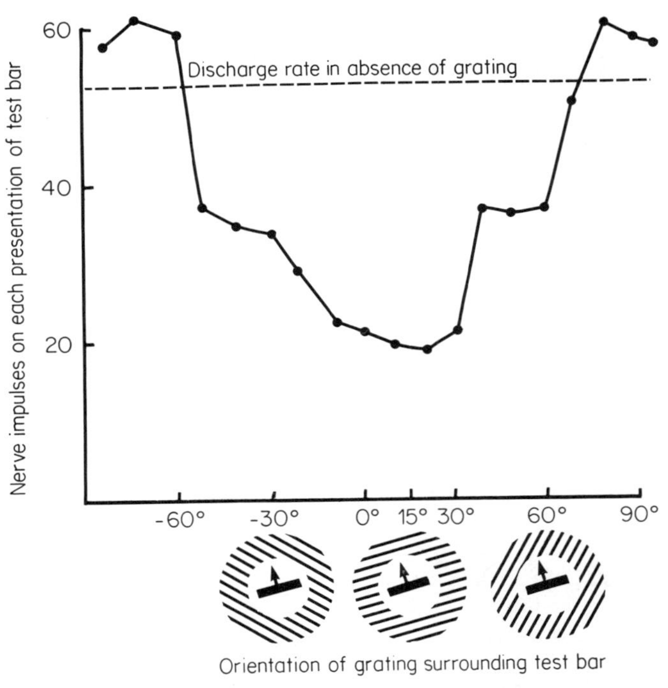

Figure 4.8 Showing how the response of a cortical cell to a bar moving across its receptive field at the optimum orientation is inhibited when the bar is surrounded by a grating in the same orientation (From Blakemore and Tobin, 1972)

Gilinsky and Mayo (1971) measured the extent to which a test grating is masked when presented immediately after an inspection grating. They found that, when the inspection grating was oriented near the main retinal meridia, the test grating had to be nearer in orientation to the inspection grating in order to be masked than it had when the inspection grating was oriented near oblique meridia. This suggests that there is more spread of inhibition in the orientation domain for oblique lines than for lines at the main meridia. However, the meridional differences they reported were very small, and it is probably the amount of lateral inhibition, rather than the degree of spread of inhibition, which is important for visual acuity. The only anatomical evidence on this issue is from a study by Colonnier (1964), in which he found more and longer dendritic fibres in cortical regions containing cells responsive to vertical or horizontal lines than in regions containing cells tuned to oblique lines.

It is difficult to see how differences in lateral inhibition between orientation detectors with neighbouring preferred orientations could account for superior acuity for vertical and horizontal gratings. What is required here is a greater

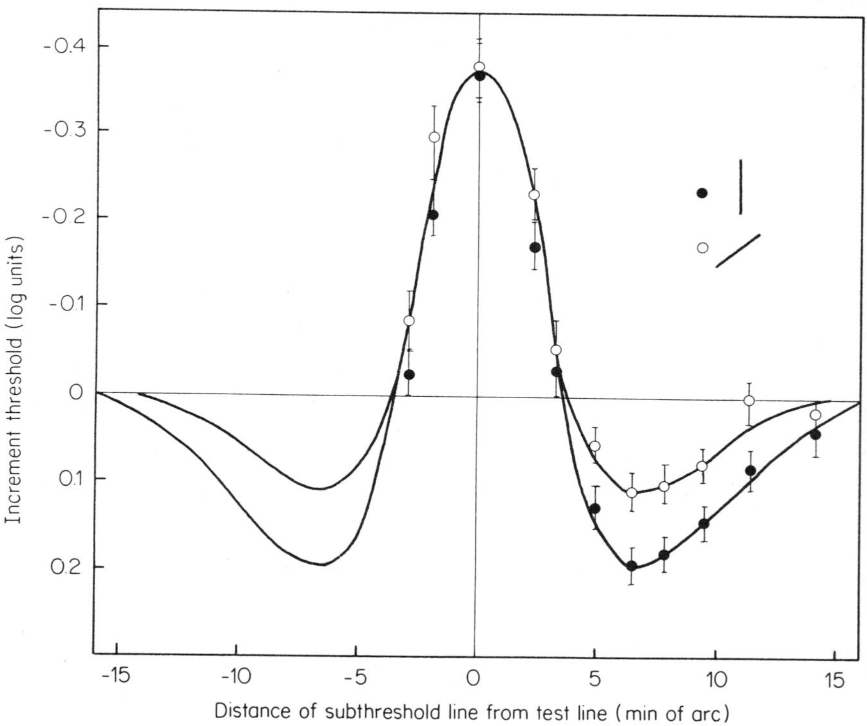

Figure 4.9 Each graph shows the extent to which the luminance threshold of a test line is affected by a subthreshold line with which it is parallel as a function of the distance between the lines. The top graph is for obliquely oriented lines and the bottom for vertically oriented lines (Reprinted with permission from I. Rentschler and A. Fiorentini, *Vision Research*, **14**, Copyright 1974, Pergamon Press, Ltd.)

lateral inhibition between units responding to lines in the same orientation when these lines are vertical or horizontal than when they are oblique. Just such a difference has been reported by Rentschler and Fiorentini (1974). They measured the luminance threshold for a test line when a subthreshold parallel induction line was located at various distances from it. When the distance between the two lines was less than about 2′ of visual angle, the threshold of the test line was lower by an amount that was independent of the orientation of the lines. When the induction line was 3–10′ from the test line, the threshold was elevated, showing the presence of lateral inhibition in the space domain, and this was greater at the main axes than at oblique axes (see Figure 4.9).

Response selectivity It has been suggested that regions within the population of orientation detectors in which the cells are narrowly tuned to orientation are better able to discriminate between one orientation and another (see, for instance, Rose and Blakemore, 1974). However, this would depend on whether the area under the tuning curves remains constant as tuning width varies, as in Figure 4.7(c), or whether the sensitivity of the units remains constant, as in Figure 4.7(d). In the former case, the curves intercept with steeper slopes where the tuning curves are narrow. Therefore, in such regions, the rate of change of the differences in firing rates will be higher, and hence orientation acuity will be higher. Note that the number of tuning curves overlapping any region of the orientation axis in Figure 4.7(c) is the same for all orientations.

Where the tuning curves change in selectivity with constant sensitivity (amplitude), as in Figure 4.7(d), the more narrowly tuned curves intercept at steeper angles, as in the former case, and this should result in higher orientation acuity. However, the number of curves overlapping a region of the orientation axis now increases as the width of the tuning curves increases, assuming of course that the density of units remains constant. This may be expected to slightly increase orientation acuity for the broadly-tuned units.

Mansfield (1974) found no difference in the breadth of tuning for different orientations in cortical cells of the monkey. Rose and Blakemore (1974) and Kennedy and Orban (1979) found that simple cells in the cat that responded best to vertical and horizontal lines had slightly narrower tuning curves than those responding best to oblique lines. However, Rose and Blakemore found that for complex cells the trend was the other way, so that overall there was no systematic difference.

The contrast sensitivity of a grating is reduced after inspection of a similar grating in another orientation. The function relating this masking effect to the degree of angular separation between inspection and test gratings has been suggested as a measure of the tuning characteristics of orientation detectors (see Section 4.3.2). Some investigators, using this measure, have concluded that tuning is finer for vertical and horizontal gratings than for oblique gratings. However, as the discussion in Sections 4.6.7 and 4.3.3 shows, differences in masking are more likely to be due to differences in lateral inhibition than to differences in width of tuning.

Related to the question of response selectivity is the question of the alignment of retinal receptive fields that feed into one cortical orientation detector. Orientation tuning would be finer for a cortical unit which receives inputs from precisely aligned receptive fields than for a unit which receives inputs from poorly aligned receptive fields.

If cells with vertical and horizontal preferred orientations receive inputs from receptive fields which are smaller than those which feed into cells with preferences for oblique lines, this would explain the fact that grating acuity is higher along the principal meridia. However, there is no reason why this factor should affect orientation acuity, as long as the stimulus line is longer than the receptive fields. The reason for this is that larger receptive fields are both wider and longer than small receptive fields of cortical orientation detectors, so that the angle through which an optimally oriented line must be rotated before it infringes into the inhibitory surround will be the same in both cases. There seem to be no physiological data on meridional variations in the alignment or sizes of receptive fields of cortical cells.

The only other evidence on the question of meridional variations in selectivity of orientation detectors is from a study by Foley (1962), in which he found that what he called the 'index of spatial summation' does not vary with the orientation of a test line. He concluded that neural summation does not vary with orientation. His index of summation involved measuring the ratio of the slope of the function relating critical frequency of flicker fusion (CFF) and stimulus area to the slope of the function relating CFF and luminance. If CFF varies with area in the same way that it varies with luminance, the index of summation is one, and summation of quanta within that area is complete. This index may have no bearing on the question of summation within the receptive field of a cortical orientation detector as it affects meridional acuity, because meridional variations in acuity seem to involve only the sustained visual channels, and these may be insensitive to high-frequency flicker, which was the basis of Foley's measurement.

The precision of calibration Even if orientation detectors for main meridia and oblique meridia were the same in all respects, the information provided by them may not be treated in the same way at some higher level. There is no doubt that the main meridia are the fiducial lines, or norms, for judgments of orientation. We can judge when a line is vertical or horizontal to within a fraction of a degree, and this is not true for any other angles (see Section 10.2.2). We precisely and accurately calibrate the norms and leave other orientations relatively uncalibrated. The reason is, of course, that much behaviour depends critically on the main visual axes, whereas nothing hangs on the precise perception of other angles.

McKee and Westheimer (1978) found that with long practice, without knowledge of results, Vernier acuity improved, but that the improvement was greater for oblique targets than for targets aligned with the main meridia. As a result, the acuity for main axes and oblique axes became equal. However, only one oblique angle was used, so that much of the improvement for targets at this angle may have been due to its improved internal calibration. The judgment about whether or not the two Vernier lines were aligned could then be supplemented by an estimate of whether the angle of target offset was to one or other side of this calibrated angle. In the case of vertically aligned targets, it is known that subjects use this criterion as a basis for their performance without training (see Section 2.3.4), which would explain why practice had relatively little effect with vertical targets.

The greater saliency of the main axes as norms could account for the greater sensitivity for angular changes at the main meridia. However, it is not clear how it could account for superior grating acuity or luminance sensitivity of lines at the main meridia. One possibility is that the saliency of main-line axes would affect the criterion for detection (β) rather than the sensitivity as such. In other words, the subject may be more willing to report the presence of vertical and horizontal lines than the presence of oblique lines, even though the 'signal' is just as strong for both. Weitzman *et al.* (1973) applied a signal-detection procedure to measure sensitivity (d') and criterion level (β) for a grating detection task. They found differences in β, and no differences in d', but they tested only vertical and horizontal lines—not oblique lines. *Another study is called for.*

No firm conclusion can be made about the physiological basis of anisotropies of visual acuity and sensitivity, but on balance, the most likely cause is a greater number of orientation-sensitive cells with preferred orientations at the main meridia, coupled with narrower tuning curves for these cells.

4.4.7 The effects of experience on visual anisotropies

Leehey *et al.* (1975) found that children as young as six weeks of age preferred to look at gratings aligned with vertical or horizontal retinal meridia than at gratings aligned with oblique meridia. Because this preference showed only for spatial frequencies which were at about the infant's threshold for resolution, the authors concluded that it was due to differential resolution of the gratings and not merely to the vertical and horizontal gratings being more attractive. A more recent study from the same laboratory has confirmed that children only a few weeks old have higher acuity for main axes, but the reported magnitude of the anisotropy was small at first and increased within the first six months (Gwiazda *et al.*, 1978). On the other hand, Teller *et al.* (1974) could find no evidence of meridional variations in grating acuity in infants between one and six months of age, and Mayer (1977) could find no evidence of anisotropies in children under eight years of age. She found that acuity for vertical and horizonal gratings increased from the age of eight to 30 years, but that acuity for oblique lines remained at about the same level.

Many people have an optical astigmatism which commonly affects the relative visibility of vertical and horizontal lines. It has been known for some time that an optical astigmatism which has been present from early life cannot be cured completely with corrective lenses. Mitchell *et al.* (1973) showed in 34 subjects with large optical astigmatism that the residual astigmatism after optical correction must have been neural in origin, because it was revealed with interference gratings formed on the retina by two beams of coherent light (see also Mitchell and Wilkinson, 1974). Such gratings are not affected by the optical properties of the eye (see Section 4.4.1). Furthermore, the axis of the residual astigmatism corresponded to the axis of the initial optical astigmatism, which suggests that the neural astigmatism in these subjects was induced by the optical astigmatism. On the other hand, Abadi (1974) could find no evidence of residual meridional astigmatism, as indicated by the extent of subthreshold masking, in highly astigmatic subjects.

Freeman and Thibos (1973) found that in patients with optically corrected astigmatism the evoked potential from the surface of the scalp over the visual cortex was smaller in amplitude when evoked by lines parallel to the axis of least resolution than when evoked by lines in other orientations. Abadi and King-Smith (1979) found that subjects with congenital nystagmic oscillation of the gaze had a lower sensitivity to gratings at right angles to the direction of the nystagmus when the effect of eye movements had been eliminated by exposing the gratings for short durations. In all these cases the residual neural astigmatism, like the optical astigmatism with which it was found to be associated, had only one axis, whereas the commonly reported anisotropies in acuity which we have been considering have two axes, so that, clearly, two-axis meridional astigmatism cannot be induced by optical astigmatism.

Annis and Frost (1973) reported that Cree Indians, who, it was claimed, live in an environment relatively devoid of vertical lines, show less meridional astigmatism than Canadians of European descent brought up in an urban environment. However, Timney and Muir (1976) found large differences in the magnitude of meridional astigmatism between urban Caucasians and urban Chinese, and suggested that the difference reported by Annis and Frost may have been a racial one and not due to environmental variations. Furthermore, Switkes *et al.* (1978) subjected photographs of indoor and outdoor urban and rural scenes to Fourier analysis and found that, for those spatial frequencies where meridional astigmatism is most marked, the rural environment had a *greater* relative preponderance of verticals and horizontals than had the urban scenes.

If experience can induce meridional astigmatism, then one might expect that experience may also abolish it. Evidence in support of this idea has been provided by Fiorentini *et al.* (1972), who found that meridional differences in contrast sensitivity were decreased in subjects who for five days had worn prisms which tilted the visual world by between 30 and 45°. In addition, the differences in the amplitude of cortical potentials evoked by vertical and oblique gratings which the subjects showed initially were reduced after five days of exposure to visual tilt. Furthermore, meridional astigmatism has been reported to be absent in people with unilateral or bilateral amblyopia due to uncorrected squint (Sireteanu and Singer, 1980). This suggests that the higher acuity for vertically oriented lines in people with normal vision may be an adaptation to the fact that disparity between vertical contours is the cue for stereopsis.

4.5 Tilt normalization

4.5.1 Introduction

The stimulus dimension of the visual orientation of a line in a frontal plane has two salient values—the vertical and the horizontal. I have already described how differences in the orientations of lines near these principal meridia are more easily discriminated than are differences between lines in other orientations. Thus, the perception of tilt is in this sense anisotropic, but I have not yet considered whether the subjective metric of visual tilt is anisotropic. The subjective metric of tilt is isotropic if all objectively equal changes in tilt are judged to be equal. Otherwise, the subjective metric is anisotropic. If the scale of judged tilt is displaced or inverted with respect to the objective scale, that is not an anisotropy, but rather a constant, or systematic, error. An anisotropy of the subjective metric is defined as *a non-linear (non-isometric) mapping of the objective scale of orientation onto the judgmental scale of orientation.* Although the evidence reviewed in Section 10.2.2 indicates that visual tilt is reasonable isotropic, the phenomenon known as tilt normalization suggests that there are permanent anisotropies in the mechanism responsible for the detection of visual tilt, and the phenomena of tilt adaptation and assimilation reveal that this mechanism may in addition become temporarily anisotropic. I shall now provide a geometrical definition of each of these effects, although it must be borne in mind that they are interrelated and difficult to distinguish in practice.

In **tilt normalization**, *a line near a principal visual meridian appears to rotate towards that meridian as the line is inspected.* Tilt normalization may be defined as occurring towards the normally vertical or horizontal retinal meridia or towards the gravitationally defined vertical and horizontal. These effects differ only when the head is tilted, and I shall assume for the time being that the head is upright. Inspection of a vertical or horizontal line will obviously not produce normalization, and if it is assumed that the vertical and horizontal norms are equally strong, one would expect a line at 45° to be also immune to the effects of normalization, because its tendency to normalize to the vertical would cancel its tendency to normalize to the horizontal. We shall see that in fact the two tendencies cancel at an angle of about 60° to the vertical.

In theory, tilt normalization may be localized or general. Localized normalization maximally affects the inspection line and affects lines in other orientations less or not at all. General normalization affects lines in all orientations equally. The phenomenon of tilt normalization resembles the phenomenon described in Section 10.3, in which tilted visual scenes appear to right themselves. Section 4.5.4 will be devoted towards deciding whether these two effects are in fact the same.

The vertical and horizontal are **natural norms** in the dimension of visual tilt. There are norms in other visual spatial dimensions; for instance, eye level is a norm in the dimension of elevation and the median plane is a norm in the dimension of azimuth. Generally speaking, a stimulus in a given sensory scale which is displaced with respect to a norm comes increasingly to resemble the norm as the stimulus is inspected. Thus, after a period of inspection, a curved line appears to get straighter (Gibson, 1933), a moving array appears to slow down, and two objects at different distances from the observer appear to become more equidistant (Howard and Templeton, 1964a). Any such apparent displacement of a stimulus attribute towards a norm is known as **normalization**.

Tilt contrast refers to *the apparent angular displacement of one line away from another line with which it forms an acute angle*. It may also be described as the apparent enlargement of the acute angle between two lines. If the lines are present at the same time, any such effect is known as **simultaneous tilt contrast**, and if they are presented one after the other, the effect is known as **successive tilt contrast**. Note that between any two lines which are not parallel there is always both an acute and an obtuse angle. Thus, if tilt contrast is defined as the apparent enlargement of acute angles, it must also involve the apparent diminution of obtuse angles. This point is explained more fully in Section 4.6.7.

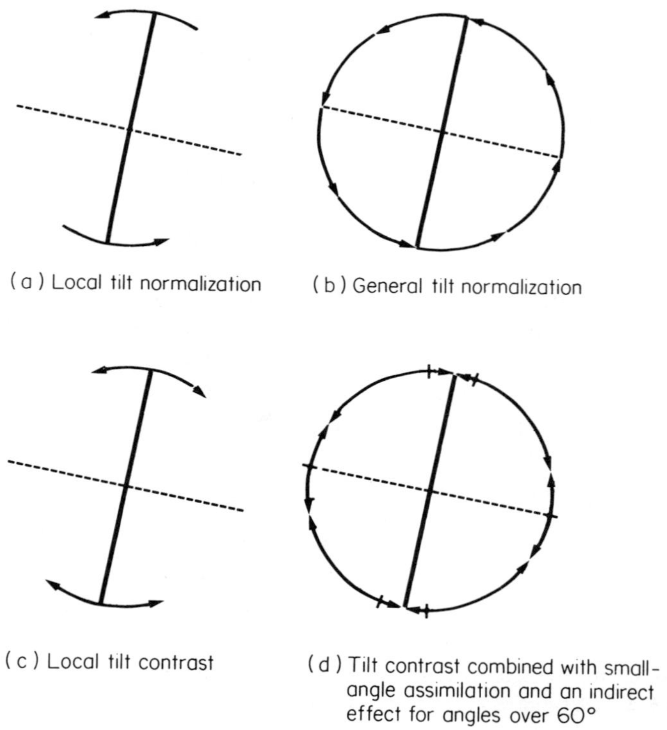

Figure 4.10 Illustrating various types of disturbance of the subjective metric of visual orientation produced by inspection of an off-vertical line. The bold line indicates the position of the hypothetical inspection line and the dashed line is the orthogonal orientation. The disturbance in each case is assumed to be limited to orientations spanned by arrows

Tilt assimilation is the opposite of tilt contrast. It is *the apparent angular displacement of one line towards another line with which it forms an acute angle.* We shall see later that tilt assimilation has been reported for angles of less than about 5° and for angles between about 60 and 90°. These two effects are usually referred to as the **small-angle assimilation effect** and the **indirect effect** respectively. Tilt assimilation has also been reported for lines which are well separated in space, but the cause of this effect remains obscure (Weintraub and Virsu, 1972). The following discussion is limited to cases where the two lines intersect.

As a way of summarizing what has just been written, consider the consequences of inspecting a line tilted at an angle of 10° to the vertical on the apparent orientation of that line and of lines in other orientations. If only local normalization occurs, the apparent orientation of only the inspection line and lines in neighbouring orientations will be affected, as in Figure 4.10(a). If normalization is generalized, the whole subjective metric of orientation will be rotated one way, as in Figure 4.10(b). If only tilt contrast occurs, the subjective metric will be displaced away from the inspection line, but a line at right angles to the inspection line will be unaffected, as in Figure 4.10(c). If small-angle assimilation and the indirect effect are superimposed on tilt contrast, a pattern of aftereffects like that shown in Figure 4.10(d) will result. Even more complex effects will be produced if normalization, contrast, and assimilation occur at the same time, and the task of measuring any one of them independently of the rest will require elaborate psychophysical procedures. We shall see in what follows that fully satisfactory procedures for isolating these various effects have not been applied.

4.5.2 Methods for measuring tilt normalization

The tilt aftereffect A vertical line seen after an inspection line tilted a few degrees with respect to the vertical appears to be tilted in the opposite direction with respect to the vertical for a period of several seconds. This phenomenon is known as the **tilt aftereffect**. The tilted line which induces such an aftereffect is called the **inspection line**, and the vertical line seen subsequently is known as the **test line**. It is not easy to measure the tilt normalization of a single line directly, because it is not easy to judge the orientation of a tilted line. The tilt aftereffect is measured by asking the subject to set the test line to the apparent vertical, and that is an easy judgment to make. That is why Gibson and Radner, and others since, have attempted to measure the tilt normalization of an inspection line indirectly, by measuring the tilt aftereffect. Gibson (1937a, b) and Gibson and Radner (1937) found the tilt aftereffect to have the following properties. It is limited to the location (but not the orientation) of the inspection figure, although it does transfer when the inspection line is presented to one eye and the test line to the other. The amount of aftereffect rises as the duration of the inspection period is increased, levelling off at a peak of 2° after between 45 and 90 sec of inspection. As the tilt of the inspection line is increased to between 5 and 20° from the vertical, the aftereffect increases to a maximum; for greater tilts of the inspection line, the aftereffect falls off, reaching zero when the inspection line is somewhere between 45 and 60° to the vertical. With an inspection line tilted more than the angle of zero effect, the whole scale shifts the other way, causing a vertical test line to be apparently shifted *towards* the position of the inspection line. Horizontal test lines are affected in a similar way by inspection lines oriented near the vertical. These reversed effects of inspection lines on test lines in neighbouring quadrants are known as **indirect effects**; they are not as large as the **direct effects**.

These are the basic features of the tilt aftereffect that were reported by Gibson and Radner and which have been substantially confirmed by other investigators, although, as we shall see, not all investigators have found the indirect effect.

Gibson and Radner thought that by measuring the tilt aftereffect they were exploring the process of tilt normalization of the inspection line to the nearest principal meridian. According to their theory of generalized normalization, vertical and horizontal lines and lines at an angle of somewhere between 45 and 60° should not normalize, and hence should not induce a tilt aftereffect. However, as we shall see in Section 4.6, tilt contrast operates between two lines irrespective of their relation to the principal meridia. For instance, it has been found that inspection of a vertical line affects the apparent orientation of a test line tilted 10° from the vertical (Templeton *et al.*, 1965). This effect cannot be due to normalization of the inspection line, since vertical lines do not normalize.

Thus, the tilt aftereffect and the indirect effect which Gibson and Radner used to measure tilt normalization were probably compound effects comprising components of normalization, contrast, and assimilation. Attempts to distinguish psychophysically between contrast and normalization will be reviewed in Section 4.5.3. The effect of normalization and that of the tilt aftereffect produced on a vertical test line by an off-vertical inspection line are in the same direction. Thus, the tilt aftereffect indicates the upper limit of tilt normalization, and one can at least be sure that if there is no tilt aftereffect there is no tilt normalization.

Scaling The most direct method for measuring tilt normalization is to ask the subject to estimate the tilt of a line while he inspects it. But estimates of the orientation of a line which is oriented away from the vertical and horizontal anchor points are too imprecise for the detection of a 2° shift, which is the approximate size of tilt normalization.

Parallel setting In this procedure, the subject is shown a series of briefly exposed test lines at well-spaced intervals of time while he is exposed to a tilted inspection line which is continuously in view. He is asked to select that test line which appears parallel to the inspection line. The test lines and the inspection line should be disposed at equal distances on either side of the point of fixation. It is assumed that the test lines are exposed for too brief a period for them to normalize. It is also assumed that the test lines are too far away from the inspection line to be affected by it; tilt normalization is reflected in the parallel-setting measure only to the extent that the normalization of the inspection line does not transfer to the test lines. Prentice and Beardslee (1950) used this method and reported about 2° of normalization to the vertical. Andrews (1967a) also used this procedure, except that his inspection line was much longer than the briefly exposed test line. In fact, Andrews assumed that the long continuously exposed line was perceived veridically, and that all constant errors arose from the misperception of the orientation of the briefly exposed test line. This explains why he reported the constant errors as an apparent repulsion of the briefly exposed line *away* from the nearest principal meridian, rather than as a normalization of the continuously exposed line *towards* the principal meridian. We shall see in what follows that this may be the best way to think of normalization. One problem with the procedure of measuring normalization by setting one line parallel to another is that the subject may rely on the cue of equality of distances between parts of the two lines, rather than on their orientation. Andrews attempted to circumvent this problem by making one line short. But short test lines are unreliable because, as Andrews himself showed, the apparent orientation of a short line is very variable.

Dot alignment Tilt normalization may be measured by asking the subject to align a dot with the perceived extension of a tilted inspection line. Bouma and Andriessen (1968) used this procedure and obtained maximum normalization effects for lines tilted about 15° to the nearest principal meridian. With lines shorter than about 1° of visual angle, normalization effects of up to 10° were found. Smaller effects were obtained with inspection lines longer than this. The reliability of this method depends on the assumption that the imaginary line joining the inspection line and the dot is not itself subject to normalization. This assumption is brought into question by the fact that these same authors found that a 'line' defined by two dots showed evidence of normalizing to the vertical.

Discrimination of angle size In this procedure, the subject is shown a display like the one in Figure 4.11 and is asked to adjust an arm of one of the two acute angles until the angles appear to be the same size (Lennie, 1971). This procedure does not measure tilt contrast, but only the extent to which tilt contrast varies as a function of the orientation of the angle in the frontal plane. Emerson *et al.* (1975)

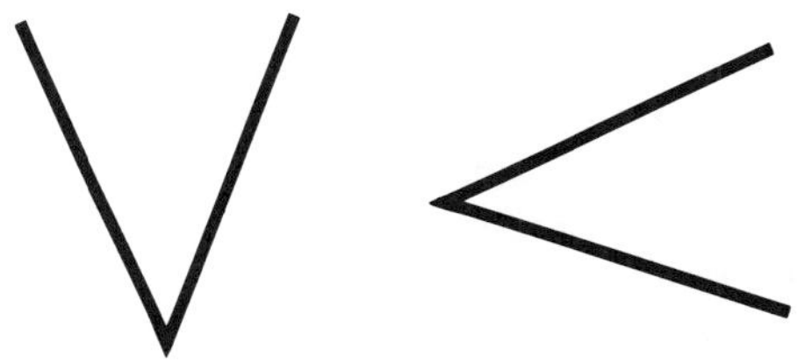

Figure 4.11 Arrangement used by Lennie (1971) to study constant errors in matching one angle to another as a function of the relative orientations of the two angles

noted that the constant errors reported by Bouma and Andriessen, using the dot-alignment procedure, were such that lines seemed to be nearer to the nearest main axis than they actually were. On the other hand, the constant errors reported by Lennie, using the angle-matching procedure, were in the opposite direction. Emerson *et al.* repeated these experiments and confirmed this discrepancy. It thus seems that the dot-alignment and angle-matching procedures do not measure the same thing. However, the directions of the constant errors in the dot-alignment task are the same as those which have been reported using the parallel-line test, so that the angle-matching procedure seems to be the odd man out.

Kinaesthetic matching In this procedure the subject sets a rod which he can feel but not see to match the orientation of a line, first when the line is exposed only briefly and then after it has been inspected for some time. Templeton (1972) used this method and reported a normalization effect of about 2° for an inspection tilt of 20°. This procedure depends on the unsubstantiated assumption that there are no constant errors in the felt position of the comparison rod.

It seem that no entirely satisfactory procedure has yet been devised for measuring tilt normalization.

4.5.3 Theories of tilt normalization

Gibson and Radner (1937) explained normalization in terms of adaptation processes in **oppositional** sensory dimensions. Examples of oppositional dimensions are 'up and down' in relation to eye level, 'movement right and movement left' in relation to the stationary state, and 'tilt right and tilt left' in relation to the vertical. In each case there is a neutral point in the scale. So-called **intensitive** sensory dimensions do not contain a neutral or symmetrical stimulus value; examples are weight and loudness. The point or axis of symmetry in an oppositional dimension is the norm; it is the point or axis from which the scale values increase in either direction. It is also the most frequent stimulus value to be experienced, and in many cases it is the stimulus value about which discrimination is most acute. Gibson and Radner suggested that inspection of an off-norm stimulus produces an alteration over the whole scale of correspondence between the physical values and the perceived values of the sensory dimension, although this shift in correspondence may be maximal in the region of the inspected value. This causes the subjective norm to be displaced away from the natural norm towards the value of the inspection stimulus. Such a shift persists for some time and manifests itself as an aftereffect when subsequent judgments of orientation are made. Gibson wrote, 'If a sensory process which has an opposite is made to persist by a constant application of its appropriate stimulus-conditions that quality will diminish in the direction of becoming neutral, and therewith the quality evoked by any stimulus for the dimension in question will be shifted temporarily towards the opposite or complementary quality' (Gibson, 1937b, p. 223).

I think that Gibson was correct in assuming that normalization is due to sensory adaptation. However, the idea of oppositional sensory systems is too general and obscures distinctions between opponent and multichannel systems, bandpass and cyclic systems, and homogeneous and inhomogeneous systems, all of which were described in Section 2.4.2.

On the basis of these distinctions, different predictions can be made about normalization and tilt contrast. In an opponent system, the effects of inspection of an off-centre stimulus would be, as Gibson and Radner suggested, an adaptation (response decrement) of activity on that side of the neutral point (norm) and hence an apparent shift of stimulus values over the whole scale in the opposite direction. In particular, stimulus values on the same side of the norm but more remote from the norm than the inspection stimulus will be apparently displaced *towards* the norm, and the inspection figure itself will be displaced towards the norm. Stimulus values on the opposite side of the norm will be displaced away from the norm. For instance, inspection of a yellow object causes that yellow and other yellows to appear greyer and grey and blues to appear bluer. In an homogeneous multichannel system, in which all the detectors have similar tuning functions evenly distributed over the stimulus range, the outcome will be different. The direction of the aftereffect will vary depending on which side of the inspection stimulus the test stimulus lies. If the test stimulus is between the inspection stimulus and the norm (assuming that the system has a norm) adaptation should cause it to be apparently displaced *towards* the norm, but if it is on the other side of the inspection stimulus it should be displaced *away* from the norm. There are other differences. In an opponent system an inspection stimulus should itself be displaced towards the norm, but in an homogeneous multichannel system there is no reason why inspection of a given stimulus should disturb the balance of response of the system with respect to that point. If an inspection stimulus in a multichannel scale does normalize, then one must look for some explanation other than adaptation or simple forms of interaction between detectors. One must assume that there are permanent inhomogeneities in the system. Furthermore, in an opponent system, inspection of the norm should have no effect, because opponent systems are quiescent in the presence of the norm. In an homogeneous multichannel system, inspection of any stimulus value should affect the system in much the same way as inspection of any other value of the stimulus dimension.

Note that a bandpass multichannel system which is homogeneous in the sense that the detectors differ only in the positions of their tuning functions may show a peak in overall sensitivity in the centre of the stimulus range, because of the greater number of overlapping tuning functions in the centre. If it is assumed that all detectors are adapted in proportion to the extent to which they are stimulated, then inspection of an off-centre stimulus in this kind of system would not change the relative firing rates of the various detectors, and the stimulus would therefore suffer no apparent shift, that is, it would not normalize.

Consider now what is to be expected in a non-homogeneous multichannel system in which detectors tuned to one region in the stimulus range (norm) have a higher sensitivity than those at other regions. Assume that adaptation reduces the firing rates of all optimally stimulated detectors by equal absolute amounts. Continued inspection of a stimulus with a value to one side of the region of highest sensitivity will adapt the more sensitive detectors, of which there are more on one side, proportionately less than it adapts the less sensitive detectors, of which there are more on the other side. This will cause the mode of excitation set up by the stimulus to be shifted towards the position of highest sensitivity. If the point of highest sensitivity is designated the norm of the scale, then the inspection stimulus will have normalized. Test stimuli with values similar to that of the inspection stimulus will apparently shift in the same direction as the inspection stimulus. Test stimuli with values more remote from the inspection stimulus on the side away from the norm may appear to be displaced away from the norm, because they will excite detectors not affected by the inspection stimulus. In this case one would say that the normalization is localized.

The presence of tilt normalization towards the vertical or horizontal suggests that orientation detectors for these meridia are more sensitive than those for other meridia, which would seem to be the most probable explanation for tilt normalization.

Evidence was presented in Section 4.4.2 that anisotropies of contrast sensitivity are confined to high spatial-frequency gratings. From this *it may be predicted that tilt normalization is confined to stimuli with high spatial-frequency components*. This prediction does not seem to have been tested.

We shall see in Section 4.6 that the extent of the apparent angular displacement of one line away from another line depends on the orientation of the two lines in the frontal plane. The most obvious conclusion is that this anisotropy of tilt contrast represents the effects of the normalization of each of the lines superimposed on the otherwise isotropic contrast effect.

It has been suggested that tilt normalization is due to a temporary state of adaptation (loss of sensitivity) of detectors for the principal meridia. If most natural scenes contained a preponderance of vertical and horizontal lines, then it would be reasonable to suppose that the detectors for these orientations are normally in a state of adaptation compared with detectors for other orientations. If a line is exposed in an orientation near a principal meridian, it will fall within a region in which adaptation, and hence excitation, is asymmetrical about the line. The outcome of this interaction will be a distribution of excitation that is skewed away from the nearest principal meridian. If we assume that judged tilt is a function of the mean of this distribution of excitation, the skewing of the distribution would cause the line to appear displaced *away* from the vertical. As the system recovers from its normally adapted state, the line will gradually appear to rotate *towards* the principal meridian. According to this view, it is the initial tilt of the line which is overestimated, and tilt normalization represents the recovery from this state.

There are two pieces of evidence that support the view that initially the tilt of a line is overestimated. Coren and Festinger (1967) measured curvature normalization by a method which involved estimating the height and width of the curved inspection line. They found that the curvature was initially overestimated and, during the course of the five-minute inspection period, came to be perceived more veridically. If curvature normalization represents a return to veridicality, then perhaps tilt normalization does so too. This idea is supported by the results of Templeton's experiment that was mentioned at the end of Section 4.5.2. He found that the tilt of an inspection line was initially overestimated, and returned towards a more veridical state after a period of inspection.

In spite of this supportive evidence, there are difficulties with the theory that tilt normalization is due to a temporary state of adaptation of vertical and horizontal orientation detectors. The fact that all or most of the normalization process is complete after the inspection line has been exposed for about 30 sec (Andrews, 1967a) reveals that this temporary state of adaptation persists for only this long. One would therefore expect that there would be no normalization of a line exposed after the subject has been allowed to sit in the dark for some time. As far as I know, nobody has reported that normalization is absent under these circumstances. The other difficulty with the theory is that many natural scenes do not have a preponderance of vertical and horizontal lines. Also, most horizontal objects are seen out of the frontal plane and project oblique images on the retina.

It seems most reasonable to conclude that tilt normalization is due to a permanently higher sensitivity of orientation detectors for the main visual meridia.

4.5.4 Tilt normalization and the visual frame effect

A tilted visual framework, such as a tilted room, may appear to be displaced towards the vertical through very large angles. This is the **visual frame effect** that is discussed in Section 10.3. The visual frame effect formally resembles tilt normalization, but may be distinguished from it on the following grounds.

1. The apparent shifts in orientation induced by a tilted frame may be 20° or more, whereas tilt normalization has a maximum value of only about 4° for all but very short lines.
2. The visual frame effect is very much more in evidence with frames which fall on the periphery of the retina than it is with frames presented in the central region of the visual field. The tilt aftereffect is only a little larger in the periphery than in the centre of the visual field (Muir and Over, 1970) and, insofar as the tilt aftereffect indicates the upper limit of tilt normalization, it may be concluded that normalization is also much the same in the two regions of the visual field.
3. A tilted visual frame affects all objects seen in the context of the frame, not only those which fall near the lines which constitute the frame. On the other hand, the tilt aftereffect is confined to within a few degrees of visual angle of the inspection line. The largest distance between inspection and test lines which has been found to produce some tilt aftereffect is 7° of visual angle (Morant and Mikaelian, 1960), but this was not confirmed by Muir and Over (1970), and most investigators have found that the effect is confined to within 1° of the inspection line. Because an absence of the tilt aftereffect indicates an absence of tilt

normalization, it must be assumed that tilt normalization is also confined to at least this extent.

4. I argue in Section 10.3 that, when a person views a tilted realistic scene which fills the entire field of view, the apparent vertical is displaced in the same direction, even for frame tilts of up to 75°, and possibly for frame tilts greater than this. On the other hand, tilt normalization reaches a maximum when the inspection line is between 10 and 20° from the vertical and reverses in sign when the inspection line is tilted at an angle of somewhere between 45 and 60°.

5. The magnitude of the visual frame effect has been reported to be a function of cognitive variables, such as the degree of realism in the tilted scene, the instructions given to the subject, and the attention of the subject. There are no reports that tilt normalization is affected by these variables.

I suspect that tilt normalization, like tilt contrast, is an oculocentric shift arising at an early stage in visual processing, probably in the primary visual cortex. The visual frame effect is probably not an oculocentric shift, but most likely depends on a higher level of processing involving the assessment of the orientation of objects relative to internal standards of vertical and horizontal.

4.6 Tilt contrast

4.6.1 Introduction

It has been known for a long time that the apparent orientation of a line is affected by another line in a different orientation. For instance, the two vertical lines of Figure 4.12 are apparently turned inwards at the bottom because of the influence of the background lines upon which they are superimposed. In Figure 4.13, the central discs of vertical lines are apparently tilted because of the influence of the surrounding lines.

Effects such as these are known as **tilt contrast**. Many well-known geometrical illusions, such as the Hering illusion and the Zollner illusion, depend, at least partly, on tilt contrast. The simplest stimulus array in which such an effect is produced consists of two lines meeting at an angle. One line is normally designated the **induction line** (line I) and the other the **test line** (line T). The angle between them is the **intersect angle**. Note that I use the term ‘induction line’ when this line is presented at the same time as the test line and the term ‘inspection line’ when it is presented for some time before the test line.

Most investigators agree that a test line is apparently rotated away from an induction line when the intersect angle is between 5 and 60°. And, of course, the induction line will be apparently rotated away from the test line. In other words, acute angles appear to be larger than they would if there were no tilt contrast. Jastrow (1892) reported that all acute angles are underestimated. However, his procedure involved asking subjects to reproduce angles by drawing. This is an unsatisfactory procedure, because, as Fisher (1969) pointed out, the reproduction must surely be subject to the same illusion as the pattern being reproduced. Heywood and Chessell (1977) also reported that acute angles are

Figure 4.12 The Zöllner illusion

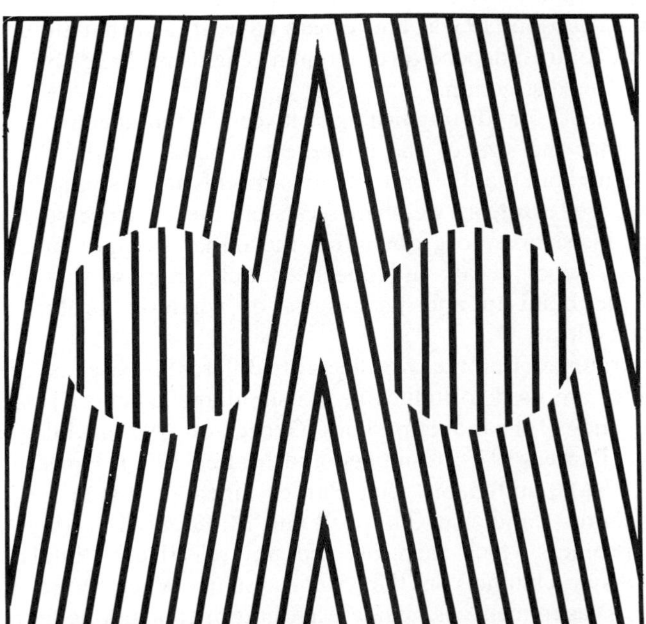

Figure 4.13 Tilt contrast. The lines in the two central discs are vertical, but appear tilted because of the contrasting tilt of the background lines

underestimated. They asked subjects to set dots to match the distance between the ends of the lines forming the angle. It is a poor strategy to attempt to measure an effect in the orientation domain by a judgment involving the position domain; changes in one are not necessarily related to changes in the other.

It will be assumed in what follows that angles between about 5 and 60° are *overestimated.* Several investigators have reported that angles of less than 5° are underestimated, which implies that the two lines are apparently rotated towards one another. Any apparent angular displacement of a line towards another line will be referred to as **tilt assimilation**. It is possible that tilt assimilation in small angles is due to metamerism in the orientation detection system (see Section 3.3.8). But whether or not there is tilt assimilation for small angles, it is generally agreed that tilt contrast is at a maximum when the angle between the induction and test lines is between 10 and 20°; the exact position of the maximum varies with the orientation of the angle in the frontal plane.

Several investigators have reported that acute angles of between 60 and 90° are also *underestimated.* This effect is known as the **indirect effect**. The theoretical significance of the small-angle assimilation effect and the indirect effect will be discussed in what follows.

Tilt contrast persists for a while after the induction figure has been removed. I shall refer to this as **successive tilt contrast** (see Figure 4.14). It is also referred to as the **tilt aftereffect**, but this term is often reserved for the case where the test line is vertical or horizontal, whereas the term successive tilt contrast is used in a more general sense. Gibson (1937a) and others have argued that simultaneous and successive tilt contrast depend on different mechanisms. I shall briefly review the evidence for this distinction.

One argument for distinguishing between simultaneous and successive contrast is that the size of the aftereffect depends on the duration of the inspection figure, but that the simultaneous effect is at full strength as soon as the lines are exposed. But, as we shall see later, the evidence suggests that the duration of inspection does not affect the initial strength of the aftereffect but only its rate of decay, and one need only suppose that the aftereffect represents the build-up and persistence of whatever process induces the simultaneous effect.

Another argument that the two effects are different is based on the claim that, when gratings are used to induce it, the simultaneous effect is specific to the spatial frequency of the induction lines (Georgeson, 1973), while the successive effect is not (Parker, 1972; Campbell and Maffei, 1971). However, Ware and Mitchell (1974a) pointed out that Parker failed to match his gratings for equivalent contrast and that Campbell and Maffei used gratings with square luminance profiles, which therefore did not selectively stimulate particular spatial-frequency channels in the visual system. When Ware and Mitchell repeated the measurements using sine-wave gratings and taking the precaution of holding the contrast of their gratings at a fixed increment above the contrast threshold, they found that the tilt aftereffect, like the simultaneous illusion and contrast elevation effects, is maximal when the induction and test gratings have the same spatial frequency.

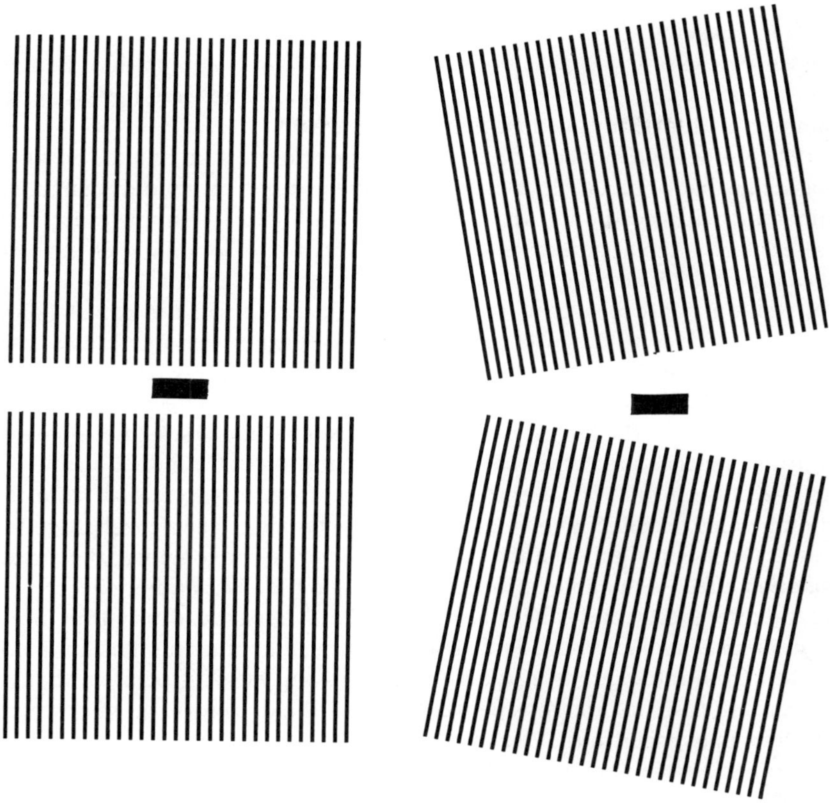

Figure 4.14 Successive tilt contrast. Inspect the small black rectangle on the right for about half a minute and then transfer the gaze to the rectangle on the left. Each test grating should appear to tilt in the opposite direction with respect to its corresponding inspection grating

The final argument that simultaneous and successive tilt contrast are fundamentally different is based on the claim that the successive effect shows the indirect effect (apparent angle contraction for intersect angles between 60 and 90°), whereas the simultaneous illusion does not. But, as we shall see later, indirect effects are not always found for the aftereffect and are often found for the simultaneous effect, so that the basis of this argument collapses. Magnussen and Kurtenbach (1980) have reported that a tilt aftereffect in one direction and a simultaneous effect in the other add linearly, which suggests a unitary mechanism. Since all three distinctions are unsupported by the evidence, I shall assume in what follows that the two effects are manifestations of the same neural processes.

One might also wonder whether the processes underlying tilt contrast are the same as those underlying the orientation-specific masking effects that were described in Section 4.3.2. Over *et al.* (1972) have argued that they are not, on the grounds that the two effects are affected in different ways by changes in the retinal location of the stimulus and that the indirect effects evident in tilt contrast are not evident in masking. Virsu and Taskinen (1975) have argued that the two processes are the same because both show with dichoptic stimulation. Perhaps tilt contrast is a composite phenomenon, and only one component has properties in common with orientation-specific masking.

In 1944 Kohler and Wallach described how prior inspection of a simple figure such as a line, bar or rectangle causes a subsequently seen neighbouring figure to be apparently displaced away from the inspection figure. A simple example of this aftereffect is the apparent repulsion of a line away from another line with which it is parallel. These displacement effects are also evident for some time after the inspection figure has been removed, and in this form are called **figural aftereffects**. Tilt contrast also perists after the inspection figure has been removed, which is the effect already referred to as successive tilt contrast. Therefore, tilt contrast is, in a formal sense, a figural aftereffect.

The magnitude of the figural aftereffect between parallel lines, like successive tilt contrast, reaches a maximum at a certain separation of the induction and test lines. The illusion falls off as the spacing is increased or decreased from this maximum value, an effect known as the **distance paradox**. The figural aftereffect may even reverse in sign when the two lines are very close together; this **figural assimilation** effect is analogous to the small-angle tilt assimilation effect.

Figural aftereffects, and other illusions involving apparent translation, occur in the *position domain* of visual coding. Tilt contrast and other tilt illusions could also result from interactions in the position domain, but it is possible that tilt illusions result from interactions between orientation detectors. In other words, tilt illusions may occur in the *orientation domain* of visual coding. The evidence bearing on this issue is reviewed in Section 4.6.7.

Even if it were known in which domain tilt contrast occurs, the precise nature of the neural processes responsible for the effect would still need to be specified. One possibility is that it is due to a reduction in the sensitivity (adaptation) of cells stimulated by the induction line and that this region of reduced sensitivity disturbs the distribution of excitation set up by the test line. The other possibility is that it is due to the effects of lateral inhibition between regions stimulated by neighbouring figures. This issue of adaptation *versus* inhibition as the cause of tilt contrast is reviewed in later sections.

I shall assume in what follows that it is the position of stimuli with respect to the retina, rather than with respect to gravity, which determines tilt contrast. Although there has been some dispute on this point (Day and Wade, 1969; Coltheart and Cooper, 1972; Wade and Day, 1973; Wallace and Moulden, 1973), it can at least be concluded that differences in relative retinal orientation are sufficient to induce simultaneous and successive tilt contrast.

4.6.2 Procedures for measuring tilt contrast

Types of display The simplest display for measuring tilt contrast is that shown in Figure 4.15(a). The single induction line forms an angle with a single

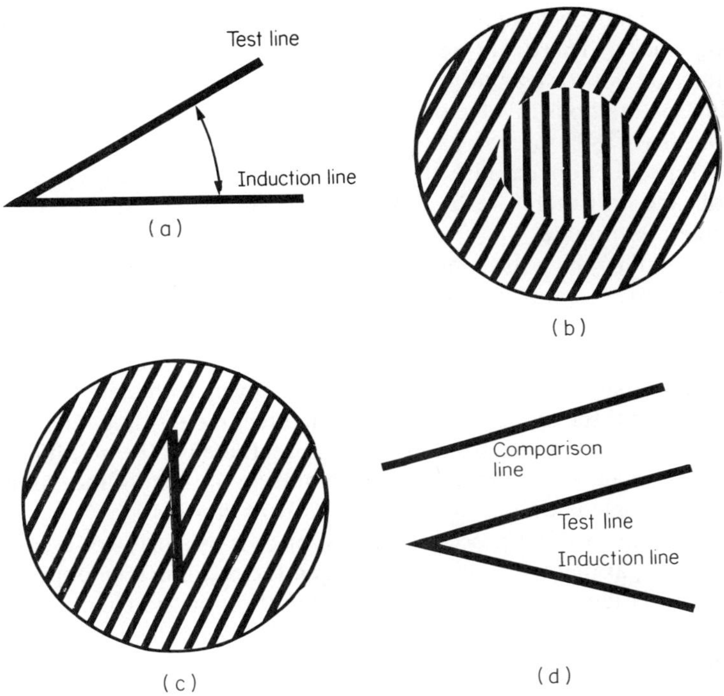

Figure 4.15 Stimulus displays for measuring tilt contrast. (a) Intersecting induction and test lines. (b) Disc of test lines with an annulus of induction lines. (c) Test line superimposed on induction lines. (d) Subject sets the comparison line to appear parallel to the test line in the presence of an induction line

test line. The two lines usually subtend about 1° at the eye. This display has the disadvantage that one end of the test line is nearer to the inspection line than the other end. A display which partly overcomes this problem is one in which a disc of test lines is surrounded by an annulus of induction lines, as in Figure 4.15(b). But even here the edge of the test grating is nearer than its centre to the induction grating. Tolhurst and Thompson (1975) were able to measure the difference in apparent tilt between the edge and centre of such a test grating. An S-shaped bending of the test lines in Figure 4.18 is, I think, apparent to casual inspection.

An even more satisfactory display from this point of view is one in which a test line is superimposed on a set of induction lines, as in Figure 4.15(c). However, in this display the test line does not have a continuous border when presented at the same time as the induction lines. In each type of display the induction and test lines may be presented simultaneously or in succession.

Let us suppose that a display consisting of a single induction line and a test line has been selected. Some means must now be found of measuring the apparent orientation of the test line when presented alone and again when presented with the induction line. The following methods have been used.

Angle estimation. The subject may be asked to give a direct verbal estimate of the intersect angle or he may be asked to reproduce it by drawing or arranging sticks. There are several problems with these methods. Verbal estimations are notoriously subject to anchoring effects and are not very precise. Reproduction methods involve the 'El Greco fallacy'. One cannot expect visual distortions (such as El Greco's astigmatism) to be revealed in the way the subject makes a reproduction because the reproduction will be subject to the same distortion and will therefore be drawn veridically by the subject in order to appear like the original.

Any method in which the subject judges the intersect angle indicates whether the angle is overestimated or underestimated and not whether the induction line has effected the apparent orientation of the test line. If the two lines affect each other equally, then the effect of the induction line on the test line is given by half the apparent change in the intersect angle. However, if the test line affects the induction line but the induction line does not affect the test line, there will be an apparent change in the intersect angle, although there will be no tilt contrast as measured by an apparent change in the orientation of the test line. An apparent change in an angle is not an ideal measure of tilt contrast.

Parallel setting. In this procedure, the subject is asked to set a comparison line to appear parallel with the test line when the test line is presented with an induction line (Figure 4.15d). It is essential that the induction line affects the test line but not the comparison line. This will probably be the case as long as the induction and test lines touch or overlap and the comparison line is at least 1° of visual angle away from the two other lines. This assumption is justified because evidence suggests that contrast effects, at least those in the central visual field, are confined to a region round the induction line subtending less than 1° (Campbell and Maffei, 1971; Tolhurst and Thompson, 1975; Weale, 1978). In applying this method, care must be taken to make control readings in the absence of the induction line because, as Carpenter and Blakemore (1973) found, subjects may show systematic errors in setting two lines parallel.

Vertical setting. In this method, the subject sets the test line to the apparent vertical or horizontal, with and without the presence of the induction line. The method is thus limited to cases where the test line is either vertical or horizontal.

Dot alignment. In this case, the subject sets a dot to be in apparent alignment with the test line. In one variant of this method, the subject is required to set a dot at the apparent point of intersection of two converging lines (Weintraub and Virsu, 1972). In another variant of the dot-alignment method, the subject adjusts the distance between two dots until it is apparently equal to the distance between the ends of two lines forming an acute angle (Heywood and Chessell, 1977).

It must not be supposed that results obtained by these various methods will necessarily agree. Wenderoth *et al.* (1979) measured tilt contrast by each method and found that the parallel-setting and vertical-setting methods produced similar results, which differed from those produced by the dot-alignment method. The parallel-setting and vertical-setting methods are similar in that they probably both involve averaging out any variations in apparent tilt along the length of the test line. Perhaps, when subjects use the dot-alignment method, they extrapolate the apparent direction of the test line more from the end near to the dot than from the other end.

I shall now describe some of the theories that have been advanced to explain tilt contrast.

4.6.3 Optical factors in tilt contrast

Chiang (1968) revised an earlier theory of why acute angles apparently expand. This theory is based on the fact that the images of lines are blurred by spherical, chromatic and other aberrations of the eye. Where two lines intersect at an acute angle, this blurring will tend to fill in the acute angle and thus shift the apparent point of contact of the two lines, making the angle appear more obtuse than it would otherwise. Coren (1969) produced evidence in support of this theory by showing that the illusion decreased when the optical aberration of the eye was reduced by use of an artificial pupil and a colour filter. However, the illusion did not decrease as much as predicted and it was concluded that optical aberrations only partly explain the acute angle illusion. In any case, it has been shown that acute angle illusions are still present when the region of intersection of two lines is omitted (Wallace, 1969; Oyama, 1975; Weale, 1978). The acute angle illusion and the tilt aftereffect have been found even in displays like that of Figure 4.16, in which the induction and test lines are 'subjective contours' (Smith and Over, 1976, 1977).

Further evidence against a purely optical explanation of tilt contrast is provided by the fact that successive tilt contrast is present when the induction line is presented to one eye and the test line to the other eye (Virsu and Taskinen, 1975). In other words, the effect shows **interocular transfer**. Campbell and Maffei

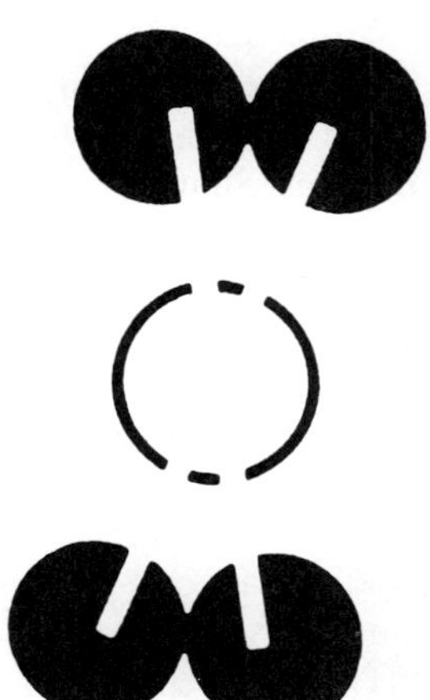

Figure 4.16 Showing the display used by Smith and Over (1977) to measure tilt contrast between subjective contours. (Reproduced by permission of Pion Ltd.)

(1971) found that the tilt aftereffect transferred 100% and argued that the mechanism underlying this effect differs from that underlying the threshold elevation effect, which shows only partial interocular transfer. However, Gibson (1937a) had found only about 70% interocular transfer of the tilt aftereffect. Mitchell and Ware (1974) confirmed that the tilt aftereffect transfers only partially, which casts doubt on Campbell and Maffei's conclusion that the effect differs from the threshold elevation effect. Mitchell and Ware demonstrated that interocular transfer of the tilt aftereffect and the threshold elevation effect is absent in stereoblind people, and increases with the degree of stereoacuity in people with normal stereovision (see also Ware and Mitchell, 1974b). This suggests that interocular transfer of the tilt aftereffect depends upon the number of binocularly driven cells in the visual cortex, which strongly supports the notion that successive tilt contrast is due to interactions between cortical neurones rather than between retinal neurones (although the site of the persisting image of the induction lines may be retinal rather than cortical). The idea of a cortical site for successive tilt contrast is strengthened by the fact that Tyler (1975) obtained a tilt aftereffect of normal size when the induction grating and test grating were stereoscopically synthesized as cyclopean images. In a cyclopean image of a grating there are no monocular images of lines, so that the presence of the cyclopean illusion provides a very powerful argument against any optical theory of the tilt aftereffect, in addition to strengthening the notion that the tilt aftereffect, and probably also successive tilt contrast, depends largely, if not entirely, on neurones in the visual cortex.

Further discussion will centre on the types of neural processes that may be involved.

4.6.4 *Tilt contrast and neural adaptation*

According to Köhler and Wallach (1944), who were among the first to describe figural aftereffects, protracted inspection of a visual stimulus leads to fatigue, or 'satiation', of a region of the visual cortex surrounding the location to which the image is projected. The distribution of cortical excitation set up by a neighbouring stimulus seen subsequently involves adapted tissue on one side and 'fresh' tissue on the other side. This causes the peak of excitation, and thus the apparent location, of the second stimulus to be displaced away from the first stimulus. According to this theory, the magnitude of the apparent repulsion decreases with increasing separation of induction and test lines. The theory depends on the notion of fatigue or adaptation within the position domain, rather than in the domain of orientation detectors. However, such a theory can account for many features of simultaneous and successive tilt contrast. Its physiological basis was understandably vague at the time that it was proposed, and arguments over these details need not concern us here. A modern version of the theory that successive tilt contrast is due to some kind of adaptation or response decrement in neural activity was proposed by Coltheart (1971). But this theory was based on the notion of adaptation within the orientation domain, rather than within the position domain. Coltheart suggested that continued inspection of a line in one orientation leads to adaptation of the orientation detectors which the line excites. This region of adaptation causes the response distribution over the population of orientation detectors set up by a test line in a neighbouring orientation to be skewed away from the position of the induction line. This process could explain simultaneous and successive tilt contrast when the lines are presented for long enough for adaptation to occur. The theory may also explain the apparent rotation of a test line towards the inspection line that several investigators have reported for intersect angles between 60 and 90°. This indirect effect could be due to the fact that complex cells may have two preferred orientations 90° apart (see Section 3.2). Thus, when two lines are at an angle of nearly 90°, each line could be 'repulsed' away from the 'complementary' excitation set up by the other line. There are facts which argue against Coltheart's general theory, but before these are discussed other theories must be described.

4.6.5 Tilt contrast and lateral inhibition in the position domain

Several investigators, including Ratliff (1965), Ganz (1966), and R. B. Howard (1971), have proposed that the apparent repulsion of one contour by another is due to lateral inhibition between neural elements in the retina and between neural elements in the visual cortex which carry information regarding the positions of contours in the visual field. Ganz proposed that the successive contour repulsion effects are due to lateral inhibition between the afterimage of the inspection figure and the image of the test figure. The general idea behind all these theories is that the distribution of neural activity set up by one contour is inhibited more on the side nearest the neighbouring contour than on the other side. The distribution of excitation for each contour is thus skewed away from its neighbour. If it is assumed that the mean or mode of the distribution of excitation carries positional information about a contour, then this skewing process will cause the two contours to appear displaced away from the position they would occupy if each were seen in isolation. Ganz reviewed physiological and psychophysical evidence that both the inhibition and apparent repulsion between two contours

Figure 4.17 The dashed line depicts the apparent shape of an acute angle that would be predicted from the attraction and repulsion of the two arms of the angle

are a decreasing linear function of the separation between the images of the two contours. Ganz suggested that the attraction of contours which are very close together is due either to summation of excitation outweighing the effects of inhibition or to small eye movements which carry an afterimage of the simultaneously present induction line or of the inspection line across the position of the test line. Ganz did not discuss angle illusions, but Figure 4.17 illustrates the apparent shape of an acute angle that his theory would predict. The repulsion should be largest somewhere near the apex of the angle and, if the arms are long enough, should fall towards zero away from the apex. In other words, the angle should appear to be bowed. With small angles, there should be angle assimilation. The theory does not predict the indirect effects for intersect angles between 60 and 90°. I shall now review the evidence bearing on the bowing of acute angles and on small-angle assimilation.

Several investigators have reported bowing of acute angles (for example, Schilder and Wechler, 1936; Chiang, 1968). Hotopf and Robertson (1975) devised the version of the Zöllner illusion shown in Figure 4.18(a), where bowing of the transversals is very evident. I also see a trace of the same bowing in the isolated transversal of Figure 4.18(b), especially when it is viewed peripherally. However, when Hotopf and Robertson, and Wenderoth *et al.* (1979) attempted to measure the bowing effect by asking subjects to set a line parallel to separate parts of a transversal, they were unable to find evidence of the effect. Perhaps people tend to estimate the inclination and parallelarity of lines by averaging along the whole length of the lines even when instructed not to. Wenderoth *et al.* (1978) did find psychophysical evidence that apparent angle expansion is greatest near the intersection of two lines by asking people to set a dot first in line with the abutting end, then in line with the free end of the test line shown in Figure 4.19. These measurements revealed that the vertical induction line apparently displaced the abutting end of the test line by 7° towards the normal, but displaced the free end by only 1°. When the test line was made short, the two apparent displacements were more alike, as one would expect from the theory that acute-angle expansion results from lateral inhibition in the position domain.

An apparent attraction between lines which are very close together has often been reported in the figural aftereffect literature. If the inhibitory processes invoked to explain the figural aftereffects also explain tilt illusions, then very small acute angles should appear to be contracted. Such an apparent contraction of small acute angles has been reported by several investigators. For instance, Oyama (1975) found evidence of contraction in angles of about 3° and less. Wallace and Moulden (1973) reported an inverse Zöllner illusion when the intercept angle was 5° or less.

A strong argument against the view that tilt contrast is due solely to interactions in the position domain can be based on the fact that tilt contrast is induced in a grating after inspection of another grating at a different angle. In this type of display, interactions in the position domain between the various lines should cancel out.

These then are the main features of the theory that tilt contrast is due to lateral

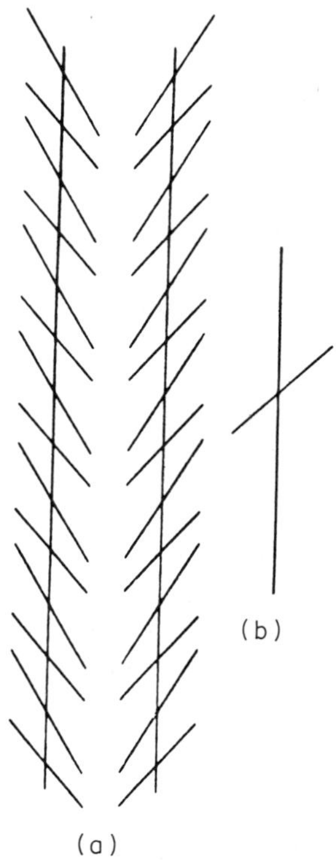

Figure 4.18 (a) A version of the Zöllner illusion in which the transversals appear to be bent. (From Hotopf and Robertson, 1975. Reproduced by permission of Psychonomic Society, Inc.) (b) A single transversal of the Zöllner illusion

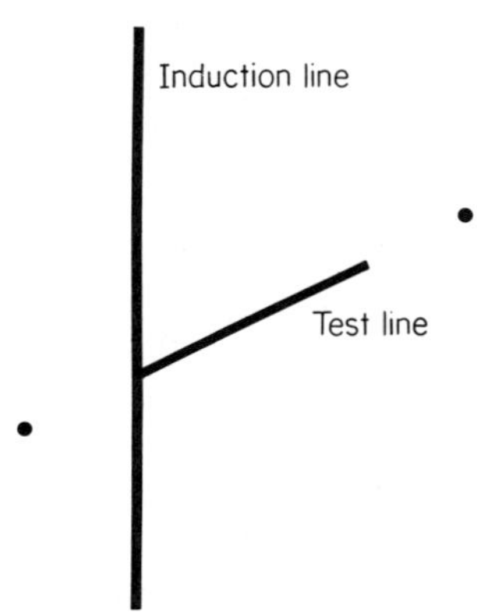

Figure 4.19 Stimulus display used by Wenderoth *et al.* (1978). The subject is required to align the dot with one or other end of the test line (see text)

inhibition in the position domain, and the evidence upon which the theory is based. But before the pros and cons of this theory can be assessed further, the outlines of two more must be presented.

4.6.6 Tilt contrast and interactions in the Fourier domain

In Section 2.4.6, I described how the visual system has separate channels sensitive to different spatial frequencies. The distance between the induction and test lines in a figural aftereffect display may be regarded as a low-frequency component of the spatial Fourier spectrum of the display, which is presented amongst the wide range of frequency components which define the lines themselves. The display shown in Figure 2.19 shows how the spatial frequency of a low-frequency display is apparently lowered after inspection of a high-frequency display. Perhaps figural aftereffects are due to a similar effect of the high spatial-frequency components of the display on the apparent value of the low-frequency component. This theory would explain the apparent attraction between a pair of parallel lines which are close together. If the distance between the lines is less than the width of the lines, then the spatial frequency of the distance component is higher than the fundamental frequency of the line-width component. The lines should therefore appear to get closer together.

There are no data bearing directly on this theory. However, it could be tested by *seeing whether the ratio of line width to interline distance is a crucial variable in the figural aftereffect*. One could also test to see whether *there is a figural aftereffect for a pair of neighbouring lines with sinusoidal luminance profiles*. If the theory is correct, there should be no effect with such a display because there is only one spatial frequency present.

It is interesting to note that the effect of inspection of one spatial frequency on another spatial frequency has been found to operate between gratings which are orthogonal to each other (Heeley, 1979). This effect presumably depends on cortical cells with dual preferred orientations, 90° apart; the same cells that are thought to be responsible for the indirect effect in tilt contrast.

Indirect evidence for the theory that figural aftereffects are related to interactions in the spatial-frequency domain comes from an experiment by Frome *et al.* (1979). They found that the effects of inspecting a sinusoidal grating on the apparent width of a single sharp-edged bar were more in accord with the notion of interactions between spatial-frequency channels than between channels sensitive merely to the widths of stimuli.

A similar, though more complex analysis in terms of interactions in the spatial-frequency domain could be applied to tilt contrast. However, there is evidence that argues against the appropriateness of such an analysis. De Valois (1977b) found that inspecting a bar affected the apparent width of subsequently presented bars of various widths only when the inspection and test bars had the same luminance polarity. For example, a dark bar on a bright surround did not affect the apparent width of a bright bar on a dark surround. On the other hand,

Magnussen and Kurtenbach (1979) found that tilt contrast is induced with equal magnitude whether or not the inspection and test figures have the same luminance polarity. Thus, tilt contrast and size contrast seem to depend on different mechanisms, and if, as the evidence produced by Frome *et al.* suggests, the size-contrast effect is related to interactions in the spatial-frequency domain, it must be concluded that tilt contrast does not operate within this domain.

4.6.7 *Tilt contrast and lateral inhibition in the orientation domain*

Soon after Hubel and Wiesel discovered orientation detectors in the visual cortex of the cat, several investigators, including Day (1962), Andrews (1967b), and Bouma and Andriessen (1970), suggested that tilt illusions result from lateral inhibition between cortical orientation detectors. Tilt contrast could result from inhibition in the orientation domain in a way analogous to that suggested in the position domain. The process was described by Carpenter and Blakemore (1973) with the aid of Figure 4.20. Each curve represents the hypothetical pattern of

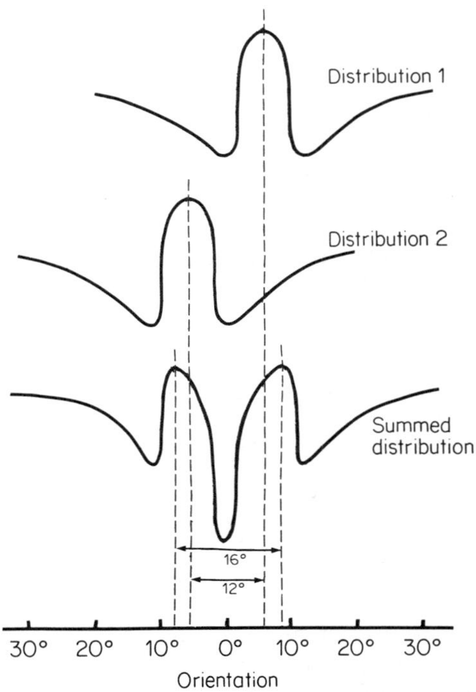

Figure 4.20 An illustration of the processes suggested by Carpenter and Blakemore (1973) as an explanation of tilt contrast. The curves represent hypothetical distributions of excitation across the population of cortical orientation detectors in response to a line at $+6^\circ$ (distribution 1), -6° (distribution 2), and two lines intersecting at an angle of 12° (summed distribution)

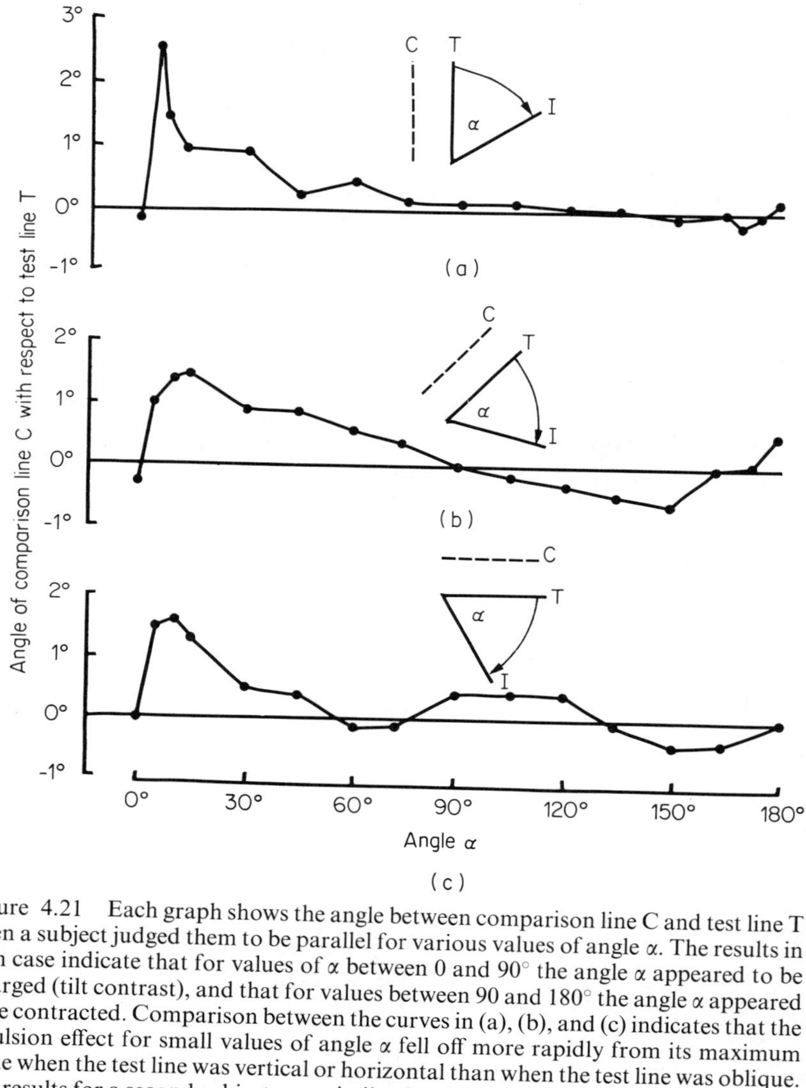

Figure 4.21 Each graph shows the angle between comparison line C and test line T when a subject judged them to be parallel for various values of angle α. The results in each case indicate that for values of α between 0 and 90° the angle α appeared to be enlarged (tilt contrast), and that for values between 90 and 180° the angle α appeared to be contracted. Comparison between the curves in (a), (b), and (c) indicates that the repulsion effect for small values of angle α fell off more rapidly from its maximum value when the test line was vertical or horizontal than when the test line was oblique. The results for a second subject were similar (From Carpenter and Blakemore, 1973)

activity across a population of orientation detectors resulting from the combined effects of both excitation and inhibition. Each detector in the population responds maximally to a line in one orientation, as indicated on the abscissa, and is inhibited by lines in neighbouring orientations. The two top curves represent the distribution of activity set up by each of two lines when presented in isolation, one tilted 6° to the left and one tilted 6° to the right. In each case a set of orientation detectors with preferred orientations similar to the stimulus is excited, and sets of orientation detectors with preferred orientations more remote from that of the stimulus are inhibited. The bottom curve represents the distribution of activity which results when the two lines are superimposed to form an acute angle. An inhibitory flank of each distribution overlaps one side of the excitatory region of the other distribution, resulting in a skewing of the two modes away from each other on the tilt axis. If we assume that the mode of any such distribution of activity across orientation detectors determines perceived tilt, then this process will result in an apparent enlargement of the angle between the two lines, from 12 to 16° in this hypothetical case.

The electrophysiological evidence for lateral inhibition between orientation detectors was reviewed in Section 4.5.4. This inhibition can be elicited in a cortical cell by stimuli which fall well outside the classical receptive field of an orientation detector, especially if the induction stimulus is moving. Failure of other investigators, such as Burns and Pritchard (1971), to obtain electrophysiological evidence of lateral inhibition in the orientation domain may have been due to their use of weak static patterns at 30° intersect angle, which is not the optimal angle for producing such effects.

Carpenter and Blakemore measured the tilt contrast illusion by the procedure of setting a comparison line parallel to a test line in the presence of an induction line. Induction and test lines each subtended 1° of visual angle in length and were presented at various angles to each other, with the test line at various angles in the frontal plane. The results for one subject are shown in Figure 4.21. It can be seen that the test line was apparently displaced away from the induction line by between 1 and 3° when the intersect angle was between 5 and 15°. There was little or no sign of an indirect effect for intersect angles between 60 and 90°, but there was some evidence of an attraction effect when the lines were over 90° apart. Carpenter and Blakemore explained this latter effect by pointing out that when the induction line is, for instance, at 170° with respect to the test line, as in Figure 4.22, the two lines have the same relative orientation as when the test line is in position T′. But line I causes an apparent anticlockwise rotation of T′, and therefore should cause an anticlockwise rotation of T. In the absence of other complicating factors, this should produce the apparent contraction of the obtuse angle that Carpenter and Blakemore found.

It can also be seen from Figure 4.21 that, as the angle between the two lines increased, the repulsion effect fell off more rapidly from its maximum value when the test line was vertical or horizontal than when the test line was at 45°. This finding reveals an anisotropy in the orientation domain. There are various ways in which such an anisotropy could be produced (Section 4.4.6), but the explanation favoured by Carpenter and Blakemore is that detectors maximally sensitive to the main meridia have narrower inhibitory flanks than those tuned to oblique meridia.

These investigators also presented evidence that the magnitude of these effects was not affected when the display was presented for 2 msec every 5 sec, rather than continuously. This result suggests that the processes responsible for tilt contrast have a short time constant. Carpenter and Blakemore also showed that a second induction line presented near the first induction line reduces rather than

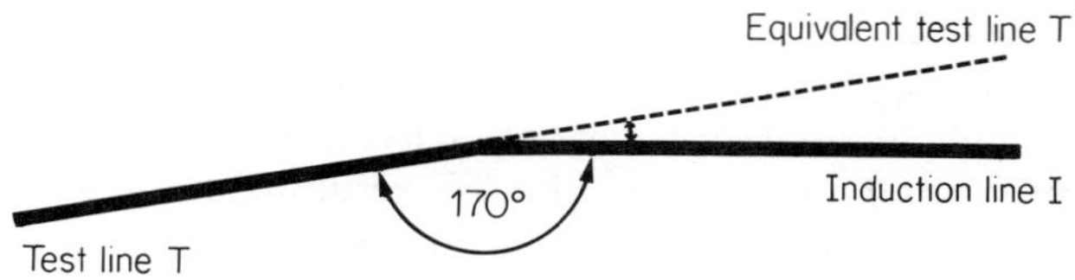

Figure 4.22 Showing how an angle of 170° between a test line and induction line is equivalent to an angle of 10° between the induction line and an extension of the test line

enhances the apparent displacement of the test line. This suggests that the second induction line disinhibits the effect of the first induction line.

The effects that moving stimuli may have on tilt contrast have some bearing on the question of whether tilt contrast is caused by interactions in the orientation domain. Some orientation detectors have preferred directions of stimulus motion in addition to preferred orientations. One would therefore expect a larger tilt aftereffect when the inspection line and test line move in the same direction, compared with when they move in opposite directions. Over and Broerse (1972) put this idea to experimental test. The inspection figure and the test figure were both square-wave gratings moving horizontally within the confines of the same fixed circular frame, which each of them filled. The inspection grating was tilted at various angles with respect to the vertical and the subjects had to set the subsequently exposed test grating to the apparent vertical. The apparent tilt of the test grating was unaffected by whether the two gratings moved in the same direction or whether they moved in opposite directions. However, Over and Broerse did not require their subjects to fixate a stationary object, so that the subjects would almost certainly have pursued the moving stimuli and therefore nullified any retinal motion. *This experiment should be repeated with controlled fixation.*

The tilt aftereffect is invariant over the colour of the inspection and test gratings as long as they are both the same colour and luminance is held constant (Lovegrove and Over, 1973). A grating in which the lines differ in chromatic contrast but not in luminance apparently induces successive tilt contrast (Elsner, 1978). However, it has been claimed that successive tilt contrast is reduced if the inspection and test gratings are not the same colour, which suggests that some orientation detectors are colour specific, a suggestion that has received some physiological support (Gouras, 1972). Lovegrove and Over (1973) found that this colour specificity of successive tilt contrast was absent when the inspection and test stimuli were presented dichoptically, and concluded that colour-specific orientation detectors are monocularly rather than binocularly driven. Wade (1980) found no effects of colour differences on simultaneous tilt contrast, a finding difficult to reconcile with previous findings.

This completes the description of the theories of tilt contrast which I wish to consider. There are other theories which I have not described, for instance those that have to do with apparent depth in displays of lines intersecting at acute angles (Gregory, 1963, 1965). But these theories are more relevant to complex displays than to those being considered here. I shall now proceed to discuss the relative merits of the theories outlined above.

4.6.8 The relative merits of theories of tilt contrast

The possible roles of adaptation and cortical inhibition in tilt contrast would be easier to assess if the time constants of the various processes were known. Sekuler and Littlejohn (1974) obtained a full-strength successive tilt contrast when the induction grating was exposed for only 18 msec. However, the subject was allowed to see the 18 msec grating followed by the test line as many times as he wished while he adjusted the test line, and perhaps these multiple exposures summated. In such experiments it is also important to distinguish between the initial strength of an aftereffect and the rate at which it decays. For instance, the magnitude of the figural aftereffect has been reported to be independent of the duration of the inspection figure, even though the effect is more prolonged if the inspection figure is exposed for longer (Ikeda and Obonai, 1953; Oyama, 1953). O'Toole (1979), in a study which overcame these problems, found that successive tilt contrast developed after a 10 msec exposure of a low spatial-frequency grating (2.7 cycles/deg), but required a full second of exposure with a high spatial-frequency grating (10.9 cycles/deg). He suggested that inhibitory processes occur in the transient (low spatial-frequency) system and in the sustained (high spatial-frequency) system with different time constants.

E. Matin (1974b) has shown that the apparent orientation of a test line is affected by an induction line which is presented up to 200 msec *after* the test line has been removed from view. She called this effect **backward tilt induction**. Inhibitory influences from the induction line presumably begin to affect the processing of tilt information from the test line before this processing is complete. It is not clear which physiological process is responsible for this retroactive tilt contrast, but adaptation is ruled out.

An argument that lateral inhibition, rather than adaptation, is responsible for tilt contrast is based on Carpenter and Blakemore's report that a second induction line reduces the effectiveness of an induction line. One would expect, on the basis of adaptation, that the effects of two induction lines would summate rather than cancel. Carpenter and Blakemore argued that the disinhibitory effect of a second induction line supports the notion that tilt contrast is due to recurrent inhibition between orientation detectors.

The effects of adding additional induction lines are not so easily explained. For instance, the Zöllner illusion (see Figure 4.12) is *increased* in magnitude as more background lines are added (Wallace and Crampin, 1969). Weale (1978) recently showed that the magnitude of the Zöllner illusion did not begin to decline until the spacing of the background lines was less than three or four times the threshold spacing for resolution of the lines. Thus, the lines do not disinhibit each other at spacings greater than this. In the Zöllner illusion, the induction lines are parallel, and each one intersects the test line, whereas in the display used by

Carpenter and Blakemore, the induction lines were not parallel and one induction line was placed between the test line and the other induction line. *More experiments are needed to unravel the possible contributions of these factors to summation or cancellation of the effects of several induction lines.*

If one had to decide between inhibition and adaptation as the sole cause of tilt illusions, inhibition would be the best choice. But such a choice need not be made, because there is no reason why both processes should not contribute to tilt illusions.

Even if lateral inhibition is the main cause of tilt contrast, there is still the issue of the relative contributions of inhibition in the position domain and inhibition in the orientation domain. Carpenter and Blakemore (1973) argued that if tilt contrast is due to inhibition in the position domain it should be limited to the vertex of the figure, that is, acute angles should appear bowed. They claimed that such bowing does not occur, but I have already shown that bowing of acute angles may be observed under certain circumstances. In any case, it is not clear to me why lateral inhibition of orientation detectors should not also be limited to the vertex of an angle. After all, most simple orientation detectors respond maximally to a line in a given retinal location and lateral inhibition between them falls off as the lines are separated, as Carpenter and Blakemore showed in their own experiments. I suggest that the presence or absence of bowing does not distinguish between inhibition in the position domain and inhibition in the orientation domain.

A more convincing argument that tilt contrast is due to interactions in the orientation domain can be based on the fact that tilt contrast is anisotropic. Tilt contrast is anisotropic because the function that relates the magnitude of the effect to the intercept angle between the induction and test lines is itself a function of the orientation of the test line in the visual field. There have been several reports, going back to Helmholtz, that the apparent size of an angle varies as a function of the orientation of the angle in the frontal plane (Onley and Volkmann, 1958; Weene and Held, 1966; Fisher, 1969; Maclean and Stacey, 1971). The illusions which depend on intersecting lines, such as the Zöllner illusion, have also been found to vary in magnitude with changes in the orientation of the figure (Weintraub and Krantz, 1971; Wallace and Crampin, 1969; Wallace and Moulden, 1973; Oyama, 1975; White, 1975).

Evidence already reviewed suggests that orientation detectors for vertical and horizontal stimuli interact over a narrower range in the orientation domain than do detectors for oblique stimuli, either because the vertical and horizontal detectors are more narrowly tuned, or because they have a narrower range of lateral inhibition. If interactions in the orientation domain cause tilt contrast, one would therefore expect the intersect angle producing the largest effect to be smaller for vertically and horizontally oriented angles than for obliquely oriented angles. Wallace and Moulden (1973) found that the intersect angle producing the largest Zöllner illusion was 15° when the induction lines were vertical and 20° when the induction lines were inclined at 30° to the vertical. White (1975) confirmed this finding and reported that the illusion also has a narrow intercept angle for the maximum effect when the induction lines are horizontal.

Campbell and Maffei (1971) found successive tilt contrast only when the test line was vertical or horizontal. This extreme form of anisotropy is not predicted on any theory. However, Mitchell and Muir (1976) found oblique test lines were affected by successive tilt contrast, and O'Toole and Wenderoth (1977) found that oblique lines were affected by simultaneous tilt contrast. In both these studies the function relating the magnitude of the effect to the intersect angle was much the same for oblique test lines as it was for vertical and horizontal test lines, even to the extent of showing the indirect effect (assimilation for intersect angles between 60 and 90°). This evidence is at the other extreme, in that no significant anisotropy is revealed. These conflicting results complicate the task of assessing the role of lateral inhibition in tilt contrast.

No definite conclusion about the cause of tilt contrast can be made. Perhaps all the effects mentioned contribute to tilt contrast in greater or lesser degree, depending on the conditions of stimulation.

Chapter 5

Eye Movements I Basic Structures and Physiology

5.1 Introduction

The aim of Chapters 5 and 6 is to present a general review of the subject of eye movements, especially those aspects that are relevant to the topics discussed in Chapters 7, 8, and 9. For a more detailed treatment of the subject of eye movements, the reader is referred to the following books: Bach-y-Rita *et al.* (1971), Baker and Berthoz (1977), Brooks and Bajandas (1977), Carpenter (1977), Dichgans and Bizzi (1972), Lennerstrand and Bach-y-Rita (1975), Monty and Senders (1976) and Senders *et al.* (1978).

The primary function of eye movements, and the one for which they first evolved, is to prevent the retinal image from moving as the animal moves about, or as the object which the animal is observing moves. Vision is seriously impaired by large movements of the retinal image, a fact which may easily be verified by observing the appearance of the stationary background as one allows the gaze to follow one's finger while it is moving from side to side. The eyes must be held reasonably steady with respect to the object of regard for at least as long as it takes to recognize the object. If the head and the object of regard are stationary, holding the eyes steady presents no problem. But animals must be able to see while their heads are in motion, and very often while the object of interest is moving. Two kinds of eye-stabilizing reflex are therefore required, one to compensate for movements of the head and the other to compensate for movements of the visual object. All animals with mobile eyes possess these two basic reflexes, and in lower animals eye movements probably serve no other purpose (Walls, 1962).

The eye movements that compensate for head movements are driven by stimuli arising in the utricles and semicircular canals of the vestibular system. For this reason they are known as **vestibulo-ocular reflexes**. For instance, as an upright person is rotated to the left about a vertical axis, stimuli arise in the horizontal canals of the vestibular system which cause a rotation of the eyes to the right, that is, in a direction which helps to stabilize the retinal image. After the eyes have rotated a certain distance, they quickly return to begin another compensatory movement. During this return movement, the image is smeared across the retina, and vision is impaired. Therefore, the sooner this movement can be accomplished the better, and indeed this return sweep is very rapid, reaching a velocity of 800°/sec or more. Such a rapid eye movement is called a **saccade**, whatever stimulus provokes it, and the saccadic phase of the vestibulo-ocular reflex is known as the **quick phase**. The slower compensatory phase is, not surprisingly, called the **slow phase**. Any rhythmic motion of the eyes is called **nystagmus**. Slow and quick phases usually alternate, but **pendular nystagmus** consists only of alternating slow phases. A nystagmic movement of the eyes tends to occur about a constant mean position of gaze. This mean position will be referred to as the **deviation component** of the nystagmus. If the mean position of gaze is the primary position of the eyes, the deviation component is zero. But this is not usually the case. As we shall see later, under certain circumstances the mean deviation of the eyes itself undergoes a slow oscillation, in which case we have a rapid nystagmic oscillation superimposed on a slower oscillation of the eyes. Under other circumstances, there is a steady deviation of the eyes induced by vestibular stimulation without nystagmus.

The nystagmus induced by vestibular stimulation is known as **vestibular nystagmus**. Vestibular nystagmus occurs within the plane in which the head rotates. Thus, rotation of the head about the body axis induces **horizontal nystagmus**, rotation of the head in the median plane induces **vertical nystagmus,** and rotation of the head in the frontal plane induces nystagmus about the visual axis—a response known as **torsional nystagmus**. A persisting torsional **deviation** of the eyes of up to about 8° is induced when the head is held in a tilted position, and reflex elevation and depression of gaze are induced by inclining the head forwards and backwards, respectively (Ebenholtz and Shebilske, 1975). These eye deviations are induced by stimulation of the otolith organs in the inner ear (see Section 9.2.8). Nystagmus and eye deviations are also induced by stimulation of joint receptors in the neck. These **cervico-ocular** reflexes are revealed in pure form by moving a person's torso relative to his stationary head. Horizontal eye movements are induced by rotating the torso about the vertical axis of the upright body, and vertical eye movements by inclining the torso forwards or backwards. However, eye torsion is apparently not induced by stimulation of neck receptors (Section 9.2.4).

The reflex eye movements that compensate for movements of the visual scene relative to the animal are driven by the motion of the retinal image. If the movement of the scene continues in one direction, slow pursuit movements alternate with rapid saccadic return movements. This nystagmic sequence is known as **optokinetic nystagmus**. If the scene oscillates sinusoidally back and forth, the eyes also swing back and forth, but because of the time delays between the visual input and muscular response, there is a phase lag between the stimulus motion and the eye motion. A similar, but much smaller phase lag occurs in the response of the eyes to vestibular stimulation resulting from a sinusoidal motion of the head. This difference in phase lag between optokinetic and vestibular nystagmus is clearly illustrated by the following observation. Rotate the head rapidly from side to side and observe the effect on the clarity of a page of print. Now swing the page of print from side to side at about the same rate, while keeping the head still, and note that vision is more severely degraded than it was when the head was moved. The phase lag of the eyes to a regular oscillation of a visual target is reduced as the system begins to anticipate the motion.

All these compensatory reflexes are very primitive and are controlled almost entirely by subcortical centres. They are very difficult to inhibit consciously and occur in animals that have no fovea. Their prime purpose, even in animals with foveate eyes, is not to maintain the image of an object on the fovea, but to stabilize the image as a whole. Animals that have a fovea, that is, an area of the retina within which vision is particularly acute, have the added requirement of maintaining the image of an object of interest on this central part of the retina. Such animals have voluntary, or **foveal pursuit eye movements**. For these movements, the velocity-mismatch signal that drives involuntary nystagmus is supplemented by signals indicating the perceived headcentric velocity of the moving object and by signals indicating the retinal eccentricity of the pursued object. These foveal pursuit movements are designed to stabilize the image of the object of interest, rather than that of the whole visual scene. They therefore require the intervention of high-level decision processes at a cortical level.

The possession of a fovea imposes a further demand on eye movements. Visual acuity falls off rapidly with increasing eccentricity from the fovea. Only in about one ten-thousandth part of the visual field is the acuity good enough for reading ordinary print. The eyes must be able to bring the image of an object of potential interest rapidly and accurately into the centre of the visual field and then rapidly fixate on its various parts. These voluntary changes of fixation are rapid, or saccadic, for three reasons. (1) The image is blurred during eye movements, so that it would be a disadvantage to spend longer than necessary in moving the eyes. (2) If the eyes moved slowly over a stationary scene, the resultant image motion would automatically trigger an optokinetic eye movement in the opposite direction, thus trapping the eyes in their original position. A saccadic eye movement is so quick that it occurs within the reaction time of the optokinetic response. In this way the optokinetic response is not invoked. (3) A mechanism for saccadic eye movements, in the form of the quick phase of nystagmus, had already evolved by the time voluntary changes of fixation were required, so that the neural commands for fixation control could simply be funnelled into the existing mechanism.

Voluntary changes of fixation are related to the nature of the visual object and to the interests of the observer. For this reason, high-level processes involving neural activity in many parts of the cerebral cortex are associated with voluntary changes in fixation.

The most recently evolved types of eye movement are those which involve the simultaneous coordinated movements of both eyes. In animals, like fish or lizards, in which the eyes are on the sides of the head, there is little binocular overlap of the two visual fields, and the image in one eye does not resemble that in the other. In such animals the fibres in the optic nerves decussate completely so that the inputs from the two eyes project to distinct parts of the nervous system, and coordinated movements of the two eyes are not required. In animals, like man, with **frontal vision**, the two monocular visual fields overlap. In this binocular region, the images in the two eyes are essentially the same, except for disparities due to depth parallax. Stereopsis, or the detection of these small disparities for the purpose of perceiving relative distances, requires that the corresponding inputs from the two eyes be brought together in the visual cortex. It is for this reason that the nerve fibres from the nasal half of each retina cross over, or decussate, to pair with corresponding fibres from the temporal half of the opposite retina (Figure 2.22). The types of eye movement just described are listed in Table 5.1.

Table 5.1 A functional classification of eye movements

Functions	Specific types
Stabilization of the retinal image	Utricular and proprioceptive reflexes Vestibular nystagmus—slow phase Optokinetic nystagmus—slow phase
Repositioning of the eyes	Quick phase of nystagmus
Maintaining a particular image on the fovea (foveate eyes only)	Foveal pursuit movements Maintenance of steady fixation
Changes of fixation (foveate eyes only)	Reflex and voluntary saccadic changes of fixation
Movements serving stereopsis	Coordinated version and vergence

Table 5.2 Terminology for eye movements

Monocular movements	Binocular movements	
Ductions	Versions	Vergences
Adduction (towards nose)	Dextroversion	Convergence
Abduction (away from nose)	Levoversion	Divergence
Elevation	Supraversion	Right (left) supravergence
Depression	Infraversion	Right (left) infravergence
Incycloduction	Dextrocycloversion	Incyclovergence
Excycloduction	Levocycloversion	Excyclovergence

When the movements of only one eye are being considered, they are called **ductions**. Coordinated movements of the two eyes in the same direction are called **versions**, and coordinated movements in opposite directions are called **vergences**. The terminology for the various types of duction, version, and vergence is listed in Table 5.2 (see also Section 5.3.5).

When the movements of the two eyes are not properly coordinated, the person is unable to converge the two visual axes onto the object of interest, and is said to suffer from squint, the technical term for which is **strabismus**. For details about different types of strabismus, see Burian and Noorden (1974).

In a person with well-coordinated eye movements, the two eyes converge on the object of attention even when one eye is covered. However, in many people, one or the other of the eyes wanders from the correct angle of vergence when it is covered. Such a drift of a covered eye is known as a **phoria**, and may be in the direction of increased convergence (esophoria), decreased convergence (exophoria), or may involve a torsional movement of the eye (cyclophoria). Strabismus or phoria are sometimes accompanied by a condition known as **latent nystagmus**. In this condition the patient's eyes execute nystagmic movements, with the fast phase directed towards the open eye, when either eye is occluded (Abadi, 1980). For further details about phoria, see Emsley (1946).

The types of eye movement listed in Table 5.1 will be discussed in more detail in the next chapter. The rest of this chapter is concerned with the structure and function of the extraocular muscles.

5.2 The geometry of eye movements

An eye does not have a fixed centre of rotation, either with reference to the eye itself or with reference to the orbit (Park and Park, 1933). In other words, the eye translates a little as it rotates. However, for most purposes it can be assumed that the eye rotates about a fixed centre 13.5 mm behind the cornea. The direction of gaze is specified with respect to the visual axis or principal visual line, that is, the line joining the point of fixation, the nodal point of the eye and the centre of the fovea. The straight-ahead, or **primary position** of the eye is not easy to define precisely, because the head and the eye lack clear landmarks. For the moment, I shall define the primary position of an eye as the direction of gaze when the head is held in an erect posture and the visual axis is horizontal and perpendicular to the interocular axis. The **interocular axis**, or base line, is the line joining the centres of rotation of the two eyes. When the visual axis of an eye moves from the primary position in a purely vertical direction, the eye may be raised (**elevation**) or lowered (**depression**); and when it moves laterally, or in a purely horizontal direction, the eye may turn nasally (**adduction**) or temporally (**abduction**). In either case, the eye moves into what is called a **secondary position**. The terms 'horizontal' and 'vertical' are not precise because they refer to planes oriented with respect to the head and are horizontal and vertical only when the head is upright. The term 'lateral' is preferable to 'horizontal', but there is no other single term for vertical in this context. When the visual axis moves from the primary position along any meridian other than a vertical or horizontal one the eye moves into a **tertiary position**.

An eye has no fixed axes of rotation. Therefore, in order to specify the position or movements of an eye, one may adopt any one of several axis systems. The choice is arbitrary although for a given purpose one system may have practical advantages. In the **Helmholtz system** the horizontal axis, about which vertical eye movements occur, is assumed to be fixed to the skull, and the vertical axis is assumed to rotate gimbal-fashion about the horizontal axis. The position of a short reference line on the surface of the eye is expressed in terms of elevation (λ) and azimuth (μ) (see Figure 5.1a). In the **Fick system** it is the vertical axis which is assumed to be fixed to the skull, and the position of a short reference line is expressed in terms of latitude (θ) and longitude (ϕ). The Fick system is simply the Helmholtz system turned to the side through 90° (see Figure 5.1b). In each system, rotation of the eye about the visual axis is known as **torsion**. Torsion is measured with respect to the 'vertical' axis in the Helmholtz system and with respect to the 'horizontal' axis in the Fick system.

The polar-coordinate, or **perimeter system**, expresses movements of an eye in terms of the angles of eccentricity (π) and meridional direction (κ) of a point at the centre of the pupil in relation to a spherical scale fixed with reference to the head

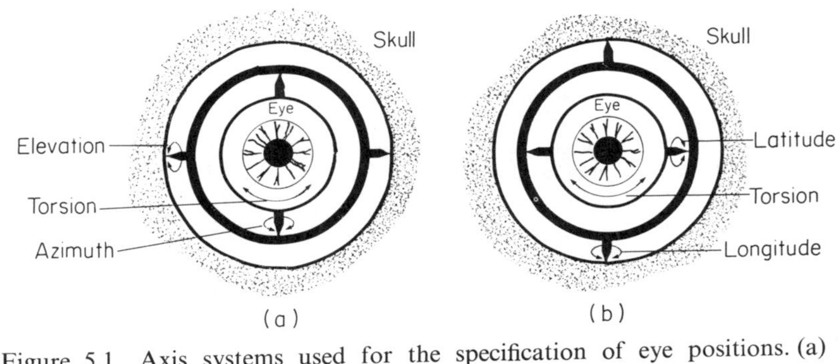

Figure 5.1 Axis systems used for the specification of eye positions. (a) The Helmholtz axis system. (b) The Fick axis system

and having its zero point on the visual axis of the eye when the latter is in its primary position.

Listing proposed that any movement of an eye may be regarded as occurring about an axis in a plane known as **Listing's plane**. Listing's plane is a plane fixed with respect to the head, and is defined as the mid-frontal or equatorial plane of the eye when the eye is in its primary position (plane HD′D in Figure 5.2). For elevations and depressions of the eye the axis in Listing's plane is the horizontal axis, for lateral movements it is the vertical axis, and for oblique movements it is between the horizontal and vertical axes. More precisely, the axis of rotation in Listing's plane is at right angles to the plane containing the initial and final positions of the visual axis (plane OPB in Figure 5.2). Thus, in any given movement of an eye, only one axis is assumed to exist; it always lies in Listing's plane, but its position in that plane depends on the meridian along which the

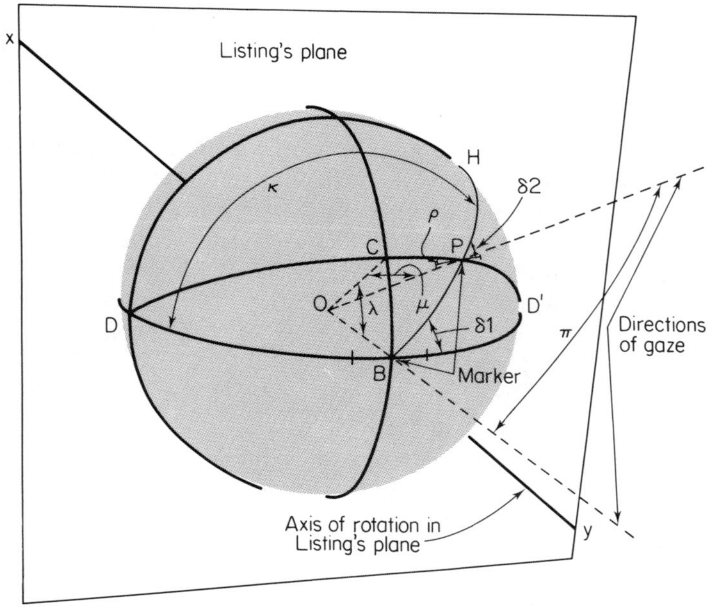

Figure 5.2 Diagrammatic representation of the geometry of eye movements. The direction of gaze is assumed to have moved from the primary position OB to an oblique position OP through an angle of eccentricity π. The movement has occurred along the meridian BH, which is at an angle K to the original horizontal meridian DBD′; which is equivalent to it having occurred about axis xy in the equatorial frontal plane (Listing's plane). The horizontal marker (between the small vertical bars) initially makes an angle $\delta 1$ with the meridian along which the visual axis moves and, according to Listing's law, this angle remains constant ($\delta 1 = \delta 2$). The eye can also be regarded as having moved on Helmholtz's axes, through an angle of elevation λ and an angle of azimuth μ. For angle $\delta 2$ to remain equal to $\delta 1$, the eye (and marker) must, in this system, undergo torsion through angle ρ relative to the final plane of regard DCD′

movement occurs. The extent of an eye movement is given by the angular separation of the initial and final positions of the direction of gaze (i.e. the angle of eccentricity π in the case of movements from the primary position); its direction is given by the angle which the meridian joining these two positions makes with an objectively horizontal meridian (δ, or its supplement κ, in Figure 5.2).

It follows that if a two-axis system such as Listing's system is to provide a comprehensive description of an eye movement, then torsional rotations of the eye about the visual axis, with reference to the meridian along which the visual axis travels, must be assumed not to occur. Helmholtz called this **Listing's law**. If an initially horizontal, short reference line is substituted for the point marker, then this law may be tested by seeing whether the angle δ remains constant. The law appears to be correct for conjugate eye movements (Quereau, 1954), but when the angle of convergence between the eyes changes, the law does not hold and torsion as defined by a change in the angle δ, does occur (Allen, 1954). The law does not hold in certain other circumstances, as we shall see later.

On the other hand, where Listing's law holds, and if movements of an eye are referred to either the Helmholtz or the Fick axis system, the angular positions which a line marker assumes during oblique version movements can be accounted for only by assuming that torsion, as defined in these systems, has occurred. But this merely means that torsion has been redefined. If torsion occurred in Listing's system, it would be indicated by a change in the angle δ of Figure 5.2, that is, by a change in the angle which a horizontal marker makes with the meridian (BH) along which the eye moves. This angle remains constant according to Listing's Law. Torsion in Helmholtz's system is the angle which the marker makes with the plane of regard (D′PD), that is, angle ρ. In Fick's system, torsion is the complement of angle ρ. These angles (ρ in Figure 5.2, or its complement) are a function of the angle δ and the angle of eccentricity π, and are not constant, even when the eyes move according to Listing's law. Clearly, also, the Helmholtz and Fick systems, being three-axis systems, can encompass torsion which occurs without any change in the position of the visual axis, whereas Listing's system was not designed for this purpose, because it is a two-axis system. One could adopt a modified Listing's system in which the eyes are said normally to rotate as if about an axis in Listing's plane, but in which any torsion induced by one or other of the stimulus conditions listed in Table 5.3 is indicated by changes in the angle δ.

The eyes do not obey Listing's law during sleep, which suggests that they are not mechanically constrained to move in this way. However, the eyes do obey Listing's law when moving in the dark, so that visual feedback is not required to cause them to move in this way. Therefore, when the eyes obey Listing's law it must be because of the nature of the neural programmes that control their movements (Nakayama, 1975). Listing's plane, within which axes of rotation are assumed to exist, is a plane fixed with reference to the head, so that it is as if the commands from the central nervous system that control eye movements are also referenced with respect to the head, rather than to a coordinate system fixed with

Table 5.3 Types of eye torsion and their inducing conditions. Those marked with an asterisk were reported in the literature for the first time only in the last few years. Numbers in brackets refer to sections in the book where that type of torsion is discussed

Name	Inducing conditions
Oblique gaze torsion ('false' torsion)	Function of direction and eccentricity of gaze and of axis system of measurement (5.2)
Countertorsional nystagmus	Active head tilt (canal stimulation) (9.2.8)
Static countertorsion	Active or static head tilt (utricular stimulation) (9.2.8)
Optostatic torsion*	Static tilt of visual field (Goodenough *et al.*, 1979) (10.3.4)
Optokinetic torsional nystagmus	Rotation of visual field about the visual axis (6.1.3)
Cyclovergence*	Cyclodisparity (Kertesz and Sullivan, 1978)
Cyclophoria	Absence of cyclofusional stimuli
Torsion during vergence	Change in vergence (see Allen and Carter, 1967)
Spontaneous fluctuation of torsion	Unknown
Pathological torsion	Brain pathology (see Sabin and Poche, 1969)
Voluntary torsion*	Training (Balliet and Nakayama, 1978a) (10.3.4)

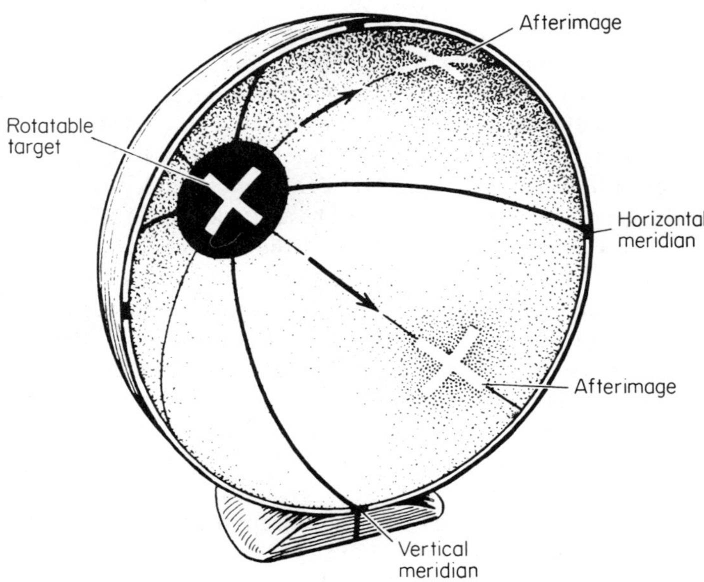

Figure 5.3 If the eye, in its primary position of gaze, is placed at the centre of curvature of a concave hemisphere and an afterimage of a cross is impressed upon it, the afterimage of the cross will remain in line with meridia passing through the original cross as the eye is rotated to different positions. This indicates the validity of Listing's law (From Noorden, G. K. von, *Burian–von Noorden's Binocular vision and Ocular Motility*, 2nd ed., 1980, St. Louis: The C. V. Mosby Co.)

respect to the eye. Listing's law has the important consequence that, as the gaze travels along any line in the visual field, the retinal image of the line remains self-congruent, and therefore continues to stimulate cortical orientation detectors with the same preferred orientation. This fact should be clear from an inspection of Figure 5.3. It follows from these relationships that an afterimage of a line imposed on the normally vertical meridian of the eye remains congruent with one of a set of circles parallel to the mid-vertical meridian of a sphere centred on the eye. These circles project as a series of hyperbolic arcs in a frontal plane (see Figure 5.4a). This latter fact enables one to construct a more precise definition of the **primary position of gaze** than that given at the beginning of this section. The primary position of gaze may be defined as the centre of the set of hyperbolic arcs which have just been described. Nakayama (1978) has proposed a rapid procedure for determining this point, which is illustrated in Figure 5.4 (b).

It should also be clear from these relationships that any torsion which occurs when the eye is in a primary or secondary position of gaze has the same angular value in whichever system of coordinates it is measured, and is considered to be 'real' torsion. When an eye moves into a tertiary position, it normally shows no torsion in Listing's coordinate system, but it does show torsion in the Fick or Helmholtz system. Such torsion is often, rather misleadingly, referred to as **false torsion**. Donders' law originally stated that the torsional position of the eye depends only on the direction of gaze. Donders (1848) wrote, '. . . to each position of the line of sight belongs a definite orientation of the horizontal and

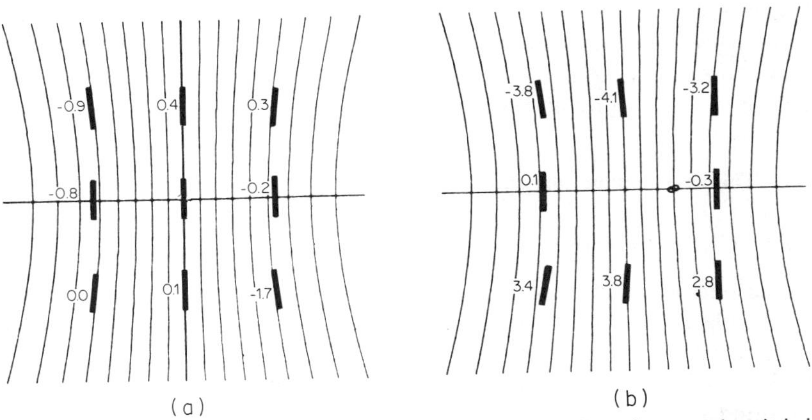

Figure 5.4 (a) The afterimage of a vertical line impressed on the eye when it is in the primary position of gaze remains approximately tangential to one of a family of hyperbolic arcs in the frontal plane as the eye is rotated into other positions of gaze. (b) If the eye is not in the primary position of gaze when an afterimage is impressed upon it, the afterimage does not remain tangential to the family of hyperbolic arcs centred on the original position of gaze. The set of angular deviations of the afterimages from the hyperbolic arcs provides a way of determining how far the eye was from the primary position of gaze when the afterimage was impressed upon it (From Nakayama, 1978. Reproduced by permision of the American Academy of Optometry)

vertical meridians relative to the coordinates of space'. Donders revised this statement when he became persuaded by his assistants that torsional movements of the eyes occur when the head is tilted (see Section 9.2.8). **Donders' law** is now taken to mean that the torsional position of the eyes for a given position of gaze does not depend on the path that they followed in reaching that position. Another way of stating Donders' law is to say that eye movements are commutative, or free from hysteresis. It is a more general form of Listing's law. Donders' law holds for eye movements which do not involve a change in vergence, but it does not hold if version and vergence movements are combined (Allen, 1954). Westheimer and McKee (1973) also found that Donders' law does not hold for transitional positions of the eyes in the course of a smooth version.

Torsion, as measured in any system, occurs under several conditions which are listed in Table 5.3. In all cases except the first, Listing's law does not hold, because such eye movements cannot be described as occurring about an axis in Listing's plane. It had been generally thought that torsion occurred as an entirely involuntary response to one or other of these stimulus conditions. However, Balliet and Nakayama (1978a) recently showed that torsion can be brought under voluntary control. The training procedure consisted of asking the subject to keep an afterimage of a line of light parallel to a stimulus line as the line was gradually rotated. After about 25 hours of training, one subject was able to voluntarily induce a conjugate torsion of his eyes with a range of up to 20°, in the absence of a visual stimulus.

In spite of the existence of these exceptions, Listing's law is still a useful general description of the way the eyes normally move. For further details about the geometry of eye movements, see Helmholtz (1962, Vol. 3, p. 37), Fry *et al.* (1945), and Fry *et al.* (1947). For details about how eye movements are measured, see Young and Sheena (1975).

5.3 The extraocular muscles

5.3.1 Gross structure and lines of action

The position and movement of each eye in its socket is controlled by six **extraocular**, or **extrinsic eye muscles**. These muscles form into three antagonistic pairs: a vertical pair—the superior and inferior recti; a horizontal pair—the medial and lateral recti; and the superior and inferior obliques. Their positions in relation to the orbit are shown in Figure 5.5. All the muscles except the inferior oblique have their **origins** in the circular tendon of Zinn at the apex of the orbit. The superior oblique passes over a pulley-like structure at the margin of the orbit known as the **trochlea**, which deflects the action of the muscle on the eye into an oblique direction. The inferior oblique has its origin in the medial wall of the orbit. Each muscle is joined to a tendon which blends into the sclera at a junction known as the **muscle insertion**. It can be seen in Figure 5.7 that each muscle is wrapped around the globe for some distance before its insertion. The **functional insertion** of a muscle is the point, known as the **point of tangency**, at which the

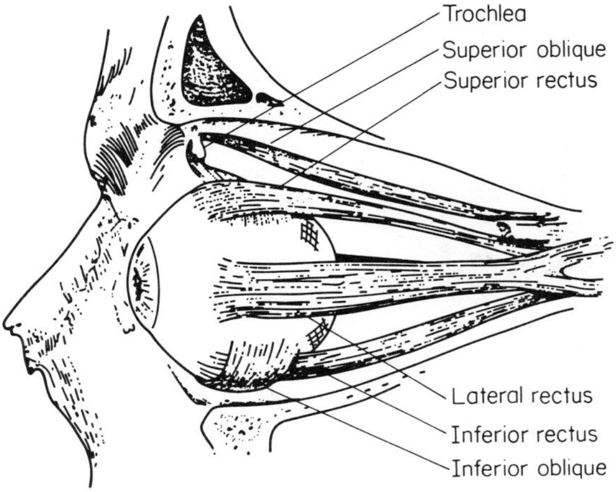

Figure 5.5 Origins and insertions of the extraocular muscles of the left eye (From Cogan, D. G., *Neurology of the Ocular Muscles*, 1956, Courtesy of Charles C. Thomas, Publisher, Springfield, Illinois)

muscle first touches the globe. This pulley-like arrangement ensures that the muscle acts tangentially on the globe within its normal range of action.

At first glance it would appear that the vertical recti elevate and depress the eye, the horizontal recti abduct and adduct it, and the obliques rotate it about the visual axis (intorsion and extorsion). But these statements are only approximately true. The horizontal recti lie in the normally horizontal plane of the eye, so that when these act alone, with the eye in the primary position, they produce an almost pure horizontal movement of the eye. The vertical recti are inserted at an angle of 23° to the median plane of the eye when the eye is in its primary position (see Figure 5.6). Thus, while the **primary action** of these muscles is to elevate or depress the eye, they also have what are known as **secondary actions**, which in this case are adduction and torsion. The primary action of the oblique muscles is an intorsion or extorsion, and their secondary actions are shown in Figure 5.7.

The above statements refer to movements of the eye from the primary position produced by a given antagonistic pair of muscles acting alone. When the eye departs from the primary position, the line of action of a given pair of muscles changes. For instance, if the eyes are elevated, the horizontal rectus muscles acquire a secondary elevating action (see Figure 5.7). Similar displacements of the **line of action** of other muscles occur for given departures from the primary positon. However, the muscles have a broad insertion and, as Helmholtz (1962, Vol. 3, p. 47) pointed out, this reduces the secondary effects that a given muscle produces when out of its normal plane of action (see Figure 5.8). For instance, when the eye is in an elevated position the line of action of the lateral rectus muscle is displaced, so that instead of merely abducting the eye, it also elevates it. However, when the eye is elevated, the lower fibres of this muscle are stretched

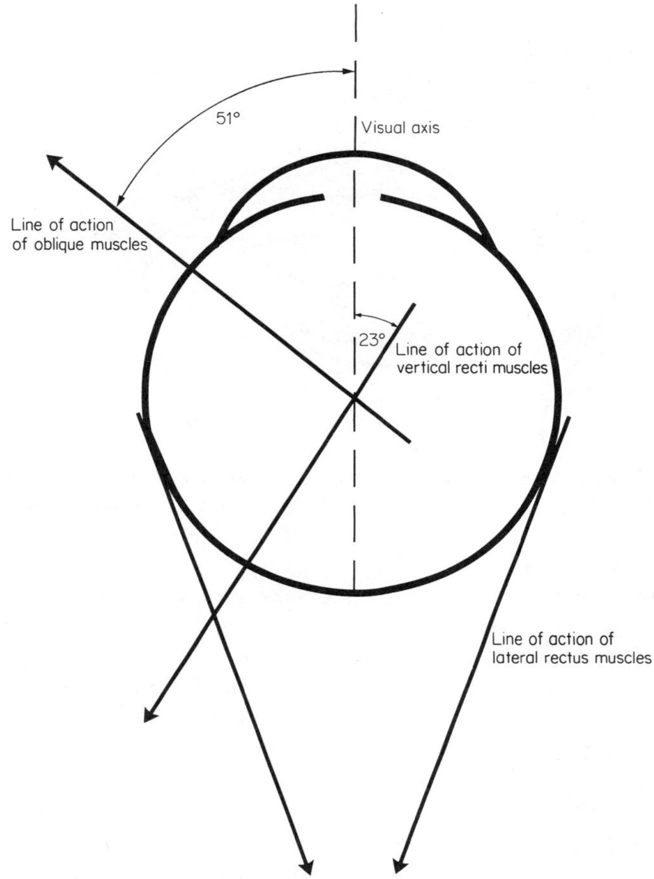

Figure 5.6 The angular relationships between the visual axis and the lines of action of the vertical recti and oblique muscles. The lines of action of the horizontal recti lie in the mid-transverse plane of the eye

more and therefore exert more tension than the upper fibres. This lowers the effective point of insertion and hence keeps the line of action nearer to the position it has when the eye is in the primary position. Support for this idea of an automatic compensatory mechanism comes from a study by Jampel (1966) in which he electrically stimulated the nerve supplying the superior oblique muscle and found that the axis of the eye's rotation changed very little in spite of changes in the horizontal deviation of the eye. This effect simplifies the relationship between the vestibular canals and the **planes of action** of the extraocular muscles (see Section 8.2.2). The broad insertion of muscles and the way they overlap each other prevents them from slipping over the surface of the eyeball when they are acting outside their primary plane of action.

One way to think of the effect that a given muscle has on the movements of the

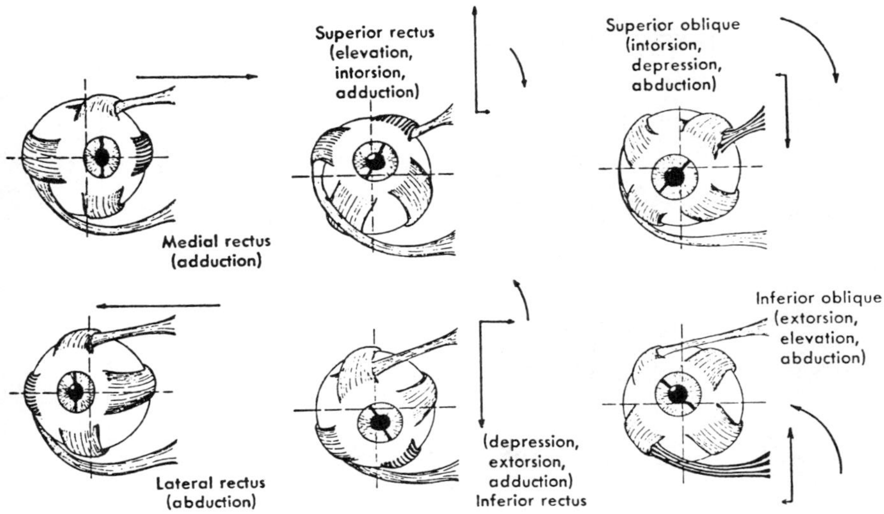

Figure 5.7 Diagrams illustrating the actions of the extraocular muscles of the right eye (From Moses, R. A. (Ed.), *Adler's Physiology of the Eye*, 5th ed., 1975, St. Louis: The C. V. Mosby Co.)

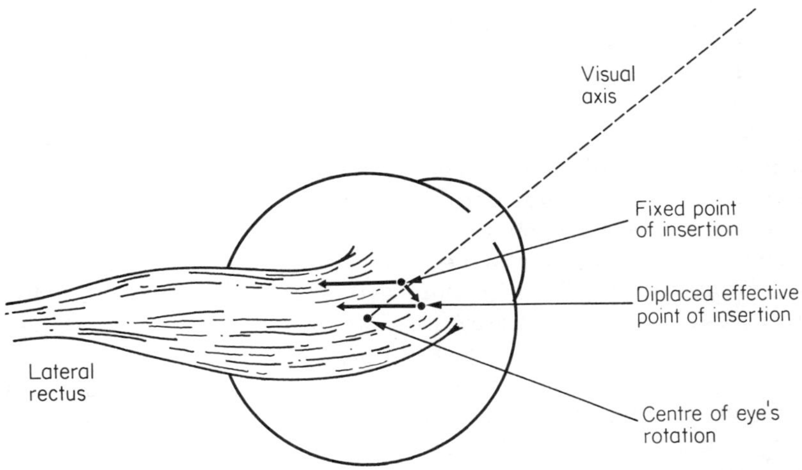

Figure 5.8 If the points of insertion of the horizontal rectus muscles were fixed, contraction of one of these muscles when the gaze is elevated would result in both a horizontal and vertical motion of the eye. Because of the muscles's broad insertion, the effective point of insertion is displaced towards the side of the muscle that is most stretched, and the elevating effect is reduced

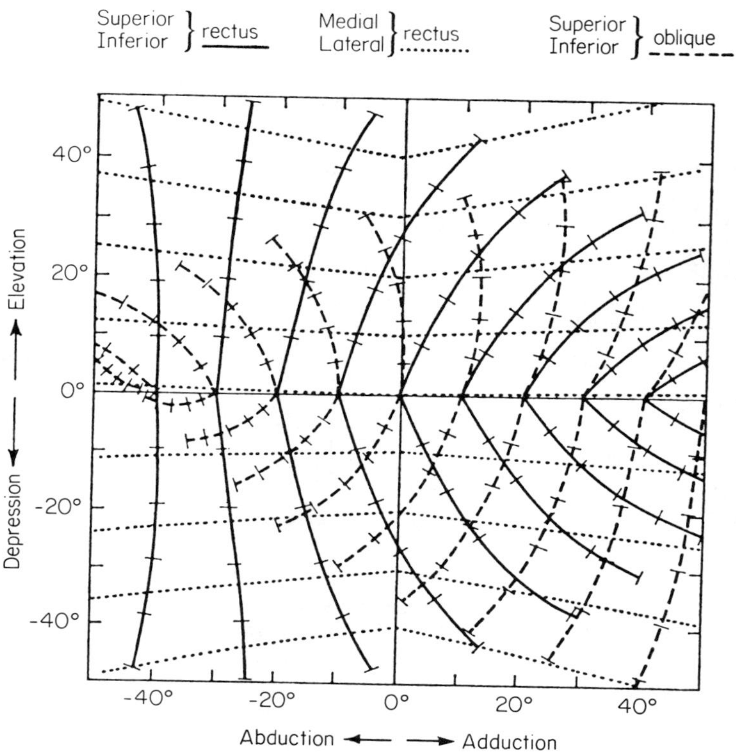

Figure 5.9 Traces of the point of fixation on a sphere concentric with the eye when the vertical muscles act individually from positions of adduction or abduction and the horizontal muscles from positions of elevation or depression (From Boeder, 1961. Reproduced by permission of the *American Journal of Opthalmology*)

eye is to consider the trace that the point of fixation would make on the inside of a sphere concentric with the eye when this muscle alone contracts. Boeder (1961) calculated the set of such traces for the vertical recti and oblique muscles for different positions of adduction and abduction, and also the set of traces for the horizontal muscles for different positions of elevation and depression. These sets of traces are shown in Figure 5.9. Boeder (1962) also calculated the length of each muscle for each position of the eye (see Davson, 1969, p. 37). By taking account of the established relationship between the length and tension of extraocular muscles and their degree of innervation, Robinson (1975) calculated the innervation that each muscle must receive for each position of gaze.

Pairs of muscles, one muscle in each eye, which act together in carrying out a particular conjugate movement of the eyes are known as **yoke muscles** or **synergists**. The lateral rectus of one eye is clearly the synergist of the medial rectus of the other eye. The superior rectus of one eye is considered to be the synergist of the inferior oblique of the other eye because, for instance, the left superior rectus and the right inferior oblique move the two eyes from the primary position to the upper right. The inferior rectus is the synergist of the opposite superior oblique. However, these statements are at best only approximations, and in any case, just as the lines of action of the muscles vary with the position of the eyes, so also do their synergistic relationships.

Each of the six extraocular muscles is under about 12 g of tension when the eye is in its primary position. As tension in an agonist increases the muscle contracts, and at the same time the stimulation of the antagonist is reduced or, in the case of rapid eye movements, cancelled entirely. This coupled action of the antagonistic pair of muscles is known as **reciprocal innervation**. The range of positions from the primary position out to the limit of deviation of an eye produced by contraction of a given muscle is known as the **field of action** of that muscle. Any movement of an eye which is brought about by the contraction of a given muscle is said to be in the **on-direction** of that muscle; movements in the opposite direction are in the **off-direction**. In a young person, the eyes can rotate from the primary position about 45–50° to left, to right, and downwards and about 40–45° upwards. Each degree of rotation involves about 0.2 mm of contraction in the agonistic muscle.

Several claims have been made that, when the tendon insertions of extrocular muscles are reversed on the eyeball of human patients or experimental animals, there is a gradual return of normal control of the eyes, even in the absence of visual feedback. This suggests that there is plasticity in the neural processes which control the extraocular muscles. However, Metz and Scott (1970) reviewed this evidence and, on the basis of their own studies, concluded that the pattern of innervation does not change after transplantation of the tendon insertions of extraocular muscles. They explained the recovery of function in terms of the development of mechanical linkages between the overlapping transplanted muscles. They did not deny that when visual feedback is provided new patterns of oculomotor control may be learned which compensate for disturbances of the muscle insertions. This question is discussed in Section 12.5.

5.3.2 The physiology of extraocular muscles

The four rectus muscles are between 40 and 50 mm long and each contains between twenty and thirty thousand muscle fibres. The oblique muscles contain just under twenty thousand fibres. The smallest muscle fibres are only 5 μm in diameter, the largest 50μm. Muscle fibres in the extraocular muscles, like those in other voluntary muscles, fall into two major groups, the **fast** and **slow** fibres. The fast fibres have a large diameter and are most prevalent in the inner, or **global** layers of the muscles. About 45% of the fibres are fast fibres. Each large-diameter fibre has a well-developed sarcoplasmic reticulum in which are embedded regularly spaced muscle fibrils (*Fibrillenstruktur*). Each muscle fibre receives a single specialized nerve ending, known as an **en plaque** ending, from which the post-synaptic impulse is propagated over the sarcoplasmic reticulum to each of the muscle fibrils. This ensures that all the fibrils contract rapidly and more or less simultaneously. Fast muscle fibres are thus designed to produce an all-or-none rapid and strong contraction for each nerve impulse. For this reason, they are referred to as **phasic** or **twitch** fibres. Each axon of the motor nerve innervates 10 or more fast muscle fibres. The fast fibres of the extraocular muscles are the fastest-acting muscles in the body, with contraction times of about 10 msec, although the effects of such a rapid contraction are not fully manifested at the tendon until after a delay of about 100 msec because of the viscous damping and inertial effects of the muscle tissues. Several subtypes of fast fibre have been described but their functional differences are obscure (see Carpenter, 1977, for details).

The slow muscle fibres have a small diameter and are concentrated largely in the outer, or **orbital** layers of the eye muscles. Each small fibre receives innervation from a branch of only one axon of the motor nerve, which terminates in multiple unspecialized neuromuscular synapses, known as **en grappe endings**. Each axon of the motor nerve innervates about three or four slow muscle fibres. The slow muscles fibres contain irregular masses of fibrils (*Felderstruktur*), and have a poorly developed sarcoplasmic reticulum. The post-synaptic impulse is not propagated in slow fibres, but each group of fibrils is stimulated by its own local synaptic ending. Thus, the slow fibres do not respond in an all-or-none fashion, but are designed to produce contractions of varying strength depending on the frequency of the stimulus. The slow fibres have a biochemical mechanism that enables them to sustain a continuous contraction without fatigue, and for this reason they are often referred to as **tonic** muscle fibres.

It was once thought that the slow muscle fibres are responsible for slow movements of the eye and for holding the eye in steady fixation, and that the fast fibres are responsible for rapid, or saccadic eye movements. This idea of a simple division of labour is now known to be wrong. Collins (1975) simultaneously monitored the activity of a large number of fast and slow muscle fibres by inserting a multiple electrode into the eye muscle of a human subject as the subject moved his eyes in prescribed ways. Collins found that the large, fast muscle fibres are essentially inactive out of the field of action of a muscle, that is, when the muscle is stretched beyond the length that it has in the primary position of the eye. The muscle is kept taut outside of its field of action by its own elasticity and by the contraction of the small, slow muscle fibres. As the eye rotates into the field of action of a given muscle, more and more fast and slow fibres are recruited into action until, at a deviation of 25°, all the fast fibres are fully contracted. Further deviation, up to the maximum deviation of about 50°, is brought about by the continued contraction of the slow fibres. The angular position of the eye at which a given muscle fibre is brought into action is known as its **threshold**. No fibres have thresholds greater than 25° in the on-direction, so that deviations of gaze more eccentric than this are determined by increasing firing rates rather than recruitment of muscle fibres. About one-third of the slow muscle fibres have thresholds lower than the angle of most extreme gaze in the off-direction (−45°). These fibres are therefore actively contracted at all angles of gaze from the most extreme off-position to the most extreme on-position. The other slow fibres and the fast fibres have thresholds evenly distributed over the range of action of the muscle. These relationships, which are summarized in Figure 5.10, describe the relative contributions of fast and slow fibres to holding the eye in various positions of fixation, and they also describe their relative contributions to slow movements of the eye.

When the eye moves rapidly, or saccadically, both the slow and fast fibres in the agonist receive a burst of innervation which is above the saturation frequency of the muscles. The muscles rapidly reach a state of maximum tension and the eye rotates ballistically at a high velocity. At the same time, all innervation to the antagonist is cancelled. In the second half of the movement, the innervation to the two muscle types in agonist and antagonist settles down exponentially to the pattern required to hold the eye in its new position.

The muscle fibres which have thresholds lower than zero degrees, that is, fibres which are under active contraction when the eye is in its primary position, tend to be slow tonic fibres. Thus, apart from brief pauses during saccades, these muscle fibres are innervated and contracted during the whole lifetime of the animal. This is the significance of their tonic characteristics. A phasic fibre is active only during on-direction saccades or when the eye moves slowly through or is held at angles of gaze above its threshold. Such states are transitory; people do not hold their eyes at eccentric angles for long, but rotate the head and restore the gaze to the primary position, that is, to a position below the threshold of most phasic fibres.

A certain proportion of cells in muscle tissue contain sensory endings and specialized contractile fibres, or **intrafusal** fibres. These cells are known as **muscle**

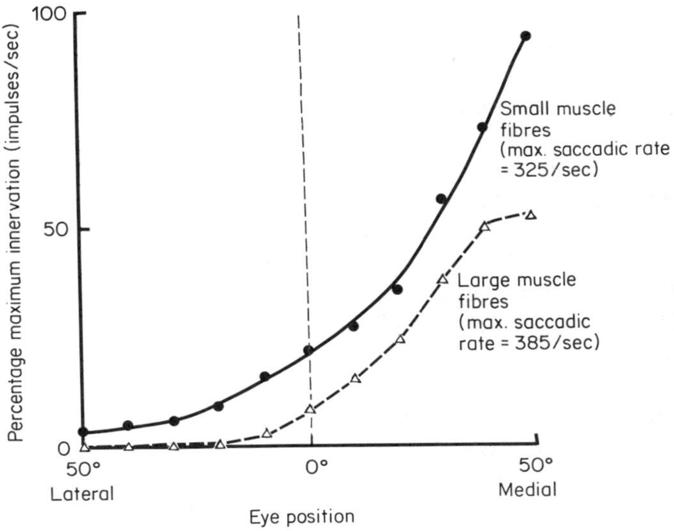

Figure 5.10 Innervation amplitude of large and small muscle fibres in the medial rectus as a function of lateral position of gaze. The recruitment of more muscle fibres as the eye moves into the medial position is reflected in the parabolic shape of the curves (From Collins, 1975. Reproduced by permission of Pergamon Press)

spindles. The sensory endings discharge when a muscle spindle is stretched and this is thought to help in the regulation of smooth muscle contraction. Contraction of the intrafusal muscle fibres is controlled by specialized motor nerves, known as **γ fibres**. Contraction of the intrafusal muscle fibres prevents the muscle spindles from becoming flaccid when the main muscles relax. Muscle spindles are responsible for initiating the **stretch reflex**, which is a reflex contraction of a muscle induced by stretching the muscle, particularly when it is actively contracting (see Matthews, 1972, for a detailed review of this subject).

Muscle spindles have been demonstrated in the extraocular muscles of several mammals, including man. The eye muscles of other mammals, like the dog and cat, do not possess muscle spindles, but they do have other types of stretch receptors (see Bach-y-Rita, 1971). Electrophysiological recordings from the single afferent neurones from muscle spindles in the extraocular muscles of the pig have revealed that muscle spindles respond to static muscle tension, and to changes in muscle tension, in much the same way as muscle spindles in skeletal muscles (Lennerstrand and Bach-y-Rita, 1974). The first-order afferent neurones from muscle spindles in the extraocular muscles of sheep pass through the ophthalmic branch of the trigeminal (5th cranial) nerve to the semilunar ganglion, from where second-order neurones arise and pass into the brainstem through the trigeminal root, and project to various parts of the trigeminal nucleus (Manni *et al.* 1975). However, there is some indication that the afferent pathways differ in different mammals (see Bach-y-Rita, 1975).

Responses to stretching of extraocular muscles have been recorded in single units of the various layers of the superior colliculus (Batini and Horcholle-Bossavit, 1977; Donaldson and Long, 1980), the vermis of the cerebellum (Tomlinson *et al.*, 1978), the visual cortex (Buisseret and Maffei, 1977), and the frontal eye fields (Dubrovsky and Barbas, 1977). The routes and functional significance of these projections are obscure.

Keller and Robinson (1971) were unable to record any activity in motoneurones of the abducens nucleus of the monkey while the eyes of the monkey were moved by the experimenter. They concluded that the monkey does not have a stretch reflex of the extraocular muscles. But there are two reasons why this conclusion is unjustified. In the first place, skeletal muscles show a much weaker stretch reflex to externally applied movement than to self-produced movement (Granit, 1971), so that the extraocular reflex may show when the eyes move actively. Bach-y-Rita (1972) demonstrated an inhibitory stretch reflex in extraocular muscles of the cat when the motor nerve was stimulated. In the second place, Keller and Robinson recorded only from larger phasic neurones in the abduceus and admitted that they may have overlooked stretch-induced changes in the firing rates of small tonic neurones. Marek and Markel (1971) detected changes in the electromyogram (potentials recorded from the surface of the extraocular muscles) during externally applied phasic stretch of the extraocular muscles of the cat. These responses were too weak to cause the eye to move, which suggests that they are involved in the tonic rather than the phasic system.

5.3.3 *Mechanical factors in eye movements*

The eye is a sphere with negligible moment of inertia, restrained in its rotation by elastic and viscous elements in the eye socket. Elasticity is defined as resistance to motion which is a function of the displacement of a body from some reference point. Viscosity is resistance to motion which is a function of the velocity of a body. Inertia is resistance to motion which is a function of the acceleration of a body. The total resistance R to motion through a given angle θ is given by:

$$R = k_1\theta + k_2\frac{d\theta}{dt} + k_3\frac{d^2\theta}{dt^2}$$

where k_1, k_2, and k_3 are the coefficients of elasticity, viscosity, and inertia, respectively. As the inertia of the eye to rotary motion may be neglected, the total resistance to motion of the eye is indicated by the sum of the first two terms of the above equation.

There are three types of elasticity in the eye: the passive elastivity of the non-muscular tissues of the orbit, the elasticity of the extraocular muscles to passive stretch, and the elasticity due to the active contraction of the extraocular muscles. Robinson (1964) was able to measure the contribution of each of these three types of elasticity in the human eye by pulling on a contact lens which was firmly placed onto the eye. By pulling on the normal eye, he measured the total elastic resistance; by pulling on an eye in which the eye muscles were paralysed, he measured the total passive elasticity; and by pulling on an eye in which the lateral muscles had been severed during eye surgery, he measured the elasticity of non-muscular tissues. For a given position of the eye he found that the orbital tissue contributed 20%, the passive elasticity of the muscle 30%, and the tonic activity of the muscle 50% to the total elastic tension. The increase in total resistance to movement per degree of rotation defines the slope of the elastic load function and is known as the **stiffness** of the eye. Collins *et al.* (1981) recently measured the stiffness of the eye to lateral movements and found that it varied from subject to subject and was slightly greater for movements in the nasal direction than for movements in the temporal direction. For normal subjects, it was found to be about 1 g/deg and to be remarkably constant over the entire range of eye movements. It was also found to be independent of the active force produced by activity of tonic muscles; such activity merely added to the elastic tension developed at each position.

When the eye moves rapidly, the restraining force increases because the viscous resistance, which is provided by the muscle tissue itself, increases with increasing velocity. The viscosity thus **damps** the motion of the eye. Objects moving under elastic tension tend to oscillate unless damped. The degree of damping required to just prevent oscillation is known as **critical damping**, and damping less or more than this is known as underdamping or overdamping. Robinson has claimed that the eye is overdamped, because when he pulled an eye to one side and then released it, it drifted exponentially back to its previous position over a period of 700 msec (see Robinson, 1968). However, Bengi and

Thomas (1968) found evidence of a mechanical resonant frequency of about 40 Hz in the motion of the eyeball. Furthermore, the eye tends to rebound after making a saccade in a way suggestive of underdamping (see Section 6.4.4). The situation is complicated by the fact that certain elastic elements are in parallel with the contractile muscle fibres and others are in series. It is also complicated by the fact that the elastic tension of the eye muscles is about 10 g greater at each position of stretch after the muscle has been stretched to a greater length, than when it approaches that length from a shorter length (Collins *et al.*, 1981) Collins (1975) has modelled the mechanical elements of the eye-movement **plant** and successfully simulated many features of actual eye movements (see Figure 5.11).

The best evidence we have on the tension developed in human extraocular muscles is from a study by Collins *et al.* (1975). They implanted a minute force transducer in series with the tendon of the horizontal recti of patients undergoing surgery for strabismus. The tensions required to hold the eye in various positions of lateral gaze are shown by the solid line in Figure 5.12. This is a parabolic function with a minimum at approximately 15° out of the field of action of the muscle. Tensions recorded at each position during smooth pursuit movements were slightly greater than those for static fixation. Tensions recorded at each position during saccadic movements of the eye also paralleled the locus of tensions for steady fixation, but were at a level equivalent to a position of the eye which was between 15 and 20° more eccentric than its actual position (see Figure 5.12). This extra tension is the tension required to drive the eye at a high velocity and overcome the inertial (acceleration-dependent) and particularly the viscous (velocity-dependent) resistance of the tissues which restrain the motion. At the end of the saccade the tension settles down to the value required to hold the eye in its new position.

Collins *et al.* also measured the tension developed in the tendons of the lateral recti when the eye was mechanically prevented from moving. In this experiment

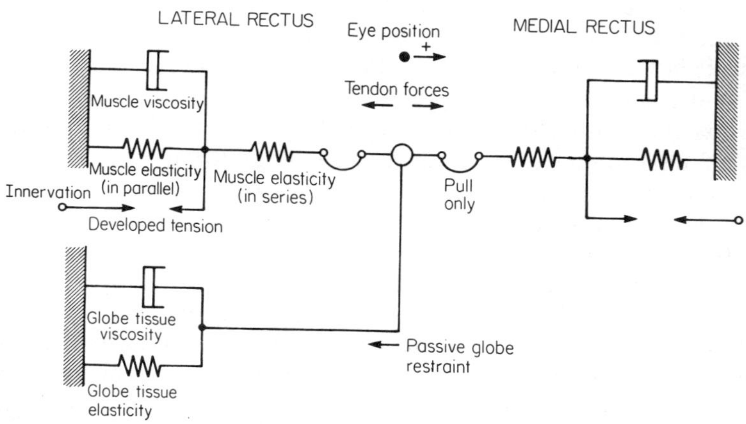

Figure 5.11 A simplified version of a model of the human oculomotor plant (From Collins, 1975. Reproduced by permission of Pergamon Press)

the subject was asked to direct his gaze to targets in various positions with the unrestrained eye, it being assumed that, because of Hering's law of equal innervation, the innervation to the restrained eye was the same as that to the unrestrained eye. This was done for various positions of the restrained eye. In this way the set of curves shown by dashed lines in Figure 5.12 was produced. Each of these curves shows the muscle tension in the restrained eye as a function of the position of the restrained eye while the degree of innervation, as determined by the position of the unrestrained eye, was held constant. Each curve therefore measures the length–tension characteristic of a given muscle as it is stretched and its innervation remains constant. It can be seen that, for tensions of more

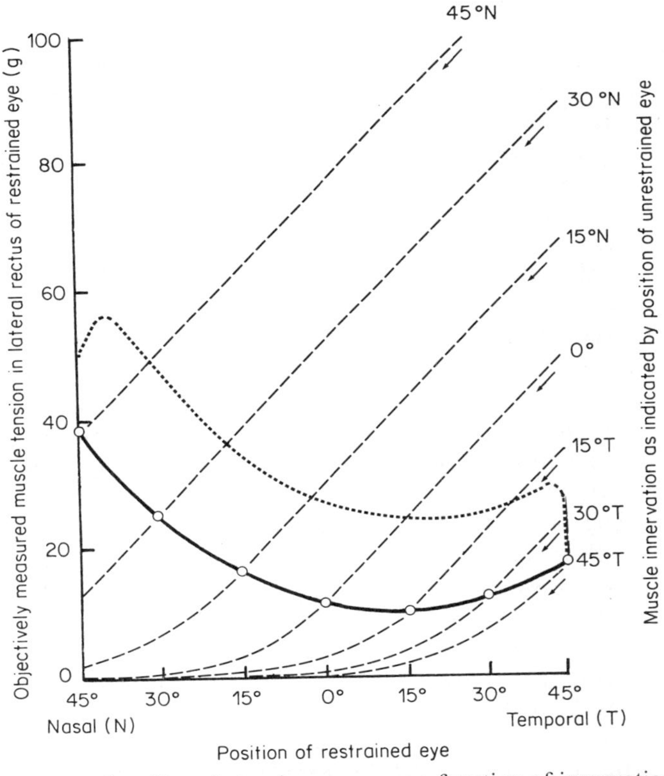

Figure 5.12 Family of length–tension curves as a function of innervation for the lateral recti (large dashed lines). Each open circle is the intersection of a given eye position with the length–tension curve of the corresponding innervation, signifying the force required to hold the eye at this position of gaze. The solid line is thus the locus of fixation forces for steady lateral gaze. The small dotted line running parallel to the fixation locus is the locus of force exerted when the eye is moving saccadically through each position of gaze. Thus, for a given position of the eye, the muscles exert about 10 g more tension when the eye makes a saccadic movement through that point than when the eye is at rest (From Collins, 1975. Reproduced by permission of Pergamon Press)

than 10 g, the increase in tension due to passive stretch of the muscle is a linear function of the eye position, that is, of the extent to which the muscle has been stretched. It is as if increasing the innervation to the muscle shifts the linear function along the muscle length axis. Below the 10 g level the muscle begins to go slack and eventually becomes quite limp where the curves are horizontal. However, in the unrestrained eye, the muscles are always under at least 10 g of tension, so that they always operate within the linear portion of the length–tension characteristic. That point on each curve where the innervation is appropriate to the position of the restrained eye represents the muscle tension required to hold the eye in that position under natural conditions. By joining up these points, one obtains the function that describes the total muscle tension as a function of the direction of gaze, which is the parabolic function already mentioned.

Because the length–tension functions of extraocular muscles with constant innervation are linear, the non-linearity of the active length–tension functions of Figure 5.12 must be due to the non-linearity of the active component of muscle contraction. This non-linearity is doubtless due to the recruitment of more and more muscle fibres into activity as the muscle shortens, but perhaps also to myogenically induced changes in innervation caused by the activity of muscle spindles and tendon stretch receptors.

The tensions plotted in Figure 5.12 are for one of the lateral rectus muscles. At each position of steady unrestricted fixation, the tension in the agonist must be balanced by an equal tension made up in part by the active and passive elasticity of the antagonist and in part by the elasticity of the orbital tissues.

For each position of the eye, the difference in innervation-induced muscle tension between agonist and antagonist equals the passive elastic forces that tend to restore the eye to its position of rest. The **position of rest** may be defined as that position where the innervation-induced forces are equal and the elastic forces are equal. According to recent measurements made by Collins *et al.* (1981), muscle forces are equally balanced at a position of about 10° of temporal gaze.

When an eye moves, tension in the agonists increases while that in the antagonists decreases. The two sets of muscles act in what is known as a 'class A push–pull' mode. It is a feature of push–pull devices, such as push–pull amplifiers, that they produce a linear difference output in spite of non-linearities in the opposing components. In the eye, differences in innervation-induced tension are required to be a linear function of eye position because the passive elastic forces that must be overcome increase linearly with increasing eccentricity of gaze.

The most significant feature of the mechanical properties of the eye is that the elastic load at any one position of gaze has a unique and constant magnitude and direction. No other muscular system concerned with the external world operates under such consistent load conditions. For instance, the force necessary to hold the arm in a given position depends on the weight that the arm is carrying and the direction in which gravity is acting on the limb. Thus, the innervation to the arm muscles bears no constant relationship to arm position, and this is why joint receptors are required to code limb position. The eye never carries loads that vary unpredictably and is immune to the effects of gravity because of its spherical shape. Thus, *to hold the eye in each position in the orbit, a unique pattern of innervation is required to the six extraocular muscles.* There is therefore no need for proprioceptors in the eye for signalling eye position; all that is required is some central registration of the pattern of innervation. This central registration of innervation has been called the **sense of innervation** by Helmholtz, the **corollary discharge** by Sperry, and the **efference copy** by von Holst.

5.3.4 The primary innervation of eye muscles

The two vertical rectus muscles, the medial rectus muscle, and the inferior oblique muscle are all innervated by the third, or **oculomotor** cranial nerve, the superior oblique muscle by the fourth, or **trochlear** cranial nerve, and the lateral rectus muscle by the sixth, or **abducens** cranial nerve. Each of the three cranial nerves originates in a brainstem nucleus of the same name. The three nuclei are embedded in the reticular formation and are collectively referred to as the **oculomotor nuclei**, which is confusing because that is also the specific name of one of them.

In recent years, several investigators, including Fuchs and Luschei (1970), Robinson (1970), Schiller (1970), Luschei and Fuchs (1972), and Henn and Cohen (1973), have recorded from single motoneurones in the abducens nucleus of the alert monkey while the animal made various kinds of lateral eye movements. All investigators have agreed that there are cells which begin to discharge at about 400 spikes/sec about 7 msec before the onset of a saccade in a given direction. Some units discharge at rates as high as 800 spikes/sec. This is well above the rate of about 250 spikes/sec required to contract the muscle to its maximum extent (the **saturation frequency**). This burst of activity is maintained at a high level until near the middle of the saccade, after which the unit's rate of discharge falls towards a steady rate (see Bahill, Clark and Stark, 1975a). The steady rate to which the discharge falls when the eye has reached its new position is a linear function of the position of the eye, although the slope of this function varies from unit to unit (see Figure 5.13). The function relating discharge rate to eye velocity is also linear, even though the highest discharge rates are above saturation frequency. Increases in innervation frequency above the saturation level produce higher *rates* of muscle tension development and hence a higher velocity of eye movement, even though the maximum tension when fully developed does not exceed its saturation value (Barmack *et al.*, 1971).

The behaviour of motoneurones in the abducens nucleus is very regular, as one would expect of neurones which control such a well-behaved organ as the eye, which always operates under predictable load conditions. The performance of each of these cells is approximated by a first-order equation of the form:

$$D = k_1 \theta + k_2 \frac{d\theta}{dt}$$

where D is the neural discharge rate, θ the eye position, k_1 the position-dependent

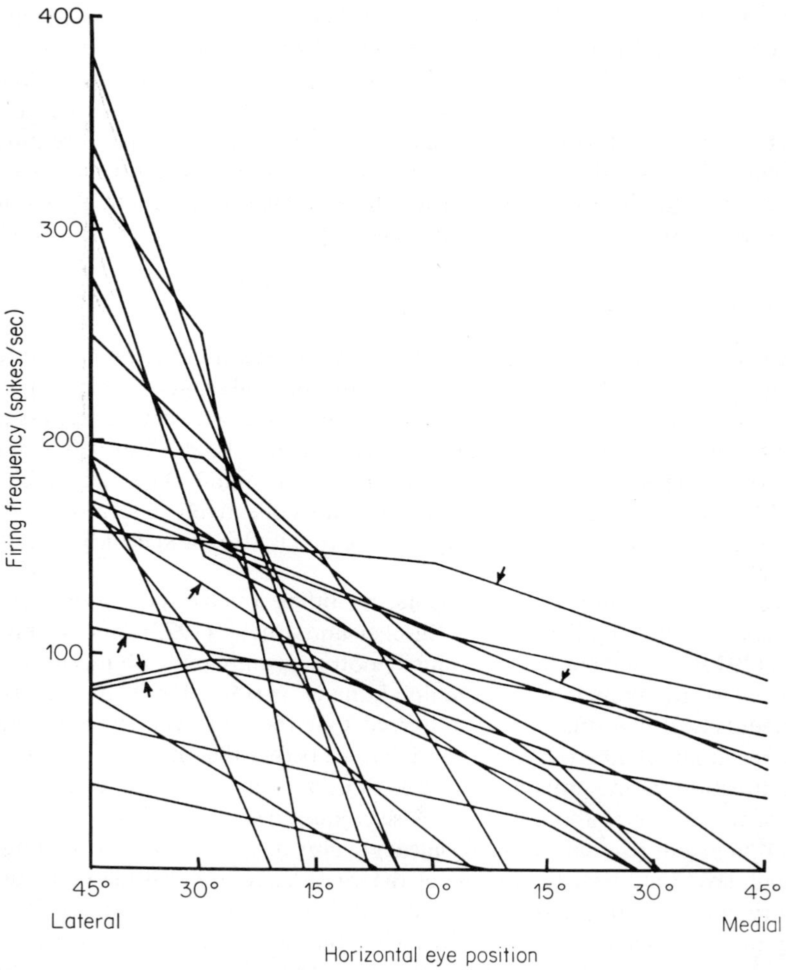

Figure 5.13 Relationship between firing frequency and horizontal eye position for 21 tonic units, having an ipsilateral on-direction, in the oculomotor nuclei of the monkey. Curves with arrows are from units whose tonic rate was related to vertical as well as to horizontal eye movements (From Luschei and Fuchs, 1972. Reproduced by permission of the *Journal of Neurophysiology*)

resistance (elasticity), and k_2 the velocity-dependent resistance (viscosity). The firing of such a cell is completely inhibited while the eye moves saccadically in the off-direction of the muscle which the unit innervates. The returning innervation to the antagonist muscle towards the end of a saccade probably provides some active breaking action, although the natural viscosity of the tissues also helps to prevent serious overshoot of the target position. The firing rate of these units is essentially unaffected by eye movements orthogonal to the on–off axis of the unit.

So we have a linear relationship between steady discharge rate and eye position

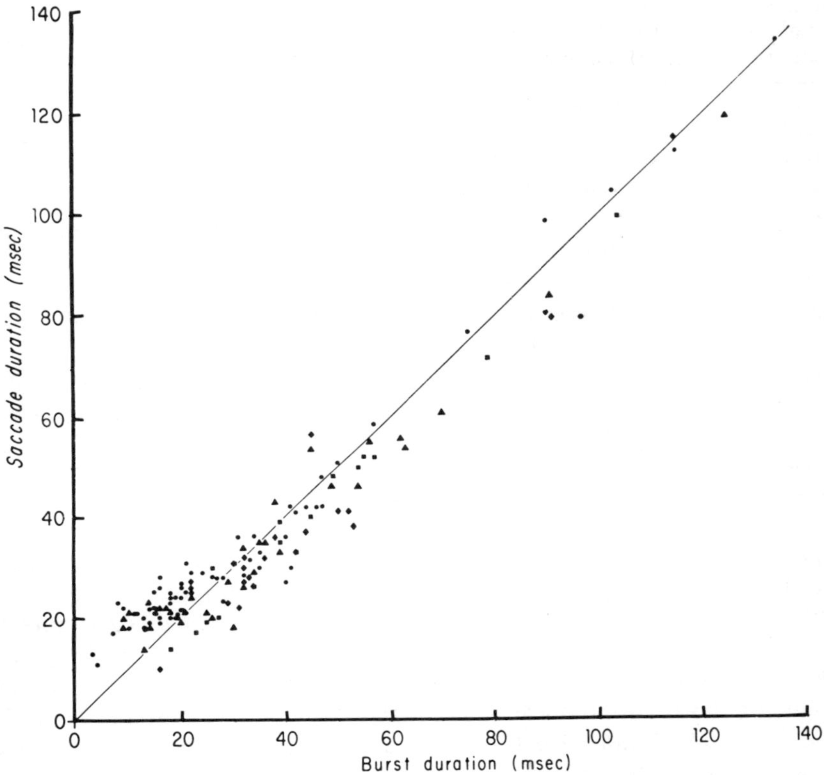

Figure 5.14 Relationship between saccadic duration and burst duration for four saccadic burst units in the brainstem of the monkey (From Luschei and Fuchs, 1972. Reproduced by permission of the *Journal of Neurophysiology*)

for single units in oculomotor nuclei, and it is known that tension in a muscle fibre is proportional to innervation rate for the range of tensions involved in holding the eye steady. These facts seem to contradict the report by Collins *et al.* of a parabolic relation between total muscle tension and eye position (Figure 5.12). The non-linearity in the length–tension function of the whole muscle is probably due to the fact that, for every 10° of rotation of the eye in the on-direction of a muscle, about 10% more muscle fibres are recruited into action (Robinson, 1970). Therefore, in looking at the length–tension function of the whole muscle, we see the cumulative effect of the superimposition of more and more linear units.

Several investigators have reported that all units in the oculomotor nuclei are essentially the same kind and participate in all types of eye movement in the on-direction. However, Reinhart and Zuber (1971), recording from the abducens in the cat, found evidence of units with distinct patterns of discharge associated with the slow and fast phases of nystagmus. One type of unit discharged at a higher frequency only during the fast saccadic phase of the eye movement;

Table 5.4 A review of the main types of motor units found by Luschei and Fuchs (1972) in the abducens nucleus (burst-tonic type only), the dorsomedial and pontine reticular formations, and the medial and superior vestibular nucleus of the monkey

Type of unit	Characteristics
Burst-tonic	High-frequency discharge 5–6 msec before and during saccade, and maintained slow discharge during pursuit and fixation
Burst	High-frequency discharge during saccade, sometimes beginning before and sometimes during the saccade. Burst duration related linearly to saccadic duration (see Figure 5.14)
Tonic	Lower-frequency discharge related linearly to static eye position (see Figure 5.13)
Pausing	Constant firing rate during fixation, stopping during saccades, some related to ipsilateral, some to contralateral, and some to both movements
Miscellaneous	

another type discharged at a higher rate only during the slow or pursuit phase; and a third type combined the properties of the first two types.

Although Luschei and Fuchs (1972) could not confirm the presence of different types of neurones in the abducens nucleus of the monkey, they did find types similar to those described by Reinhart and Zuber in the regions of the brainstem surrounding the abducens nucleus. Those they found are reviewed in Table 5.4 and in Figures 5.13 and 5.14. The significance of these types of neurones will become clearer in later sections.

5.3.5 Hering's law of equal innervation

Hering (1942) proposed in the nineteenth century that the movement of one eye is accompanied by a movement of the other eye of equal amplitude and velocity, either in the same or in the opposite direction. This has come to be known as **Hering's law of equal innervation**. Up to now, the movements of the eyes have been specified in terms of the magnitude, direction, and velocity of each eye separately. When considering coordinated eye movements, it is more convenient to specify eye movements in terms of version and vergence. **Version** is a binocular eye movement in which the rotations of the two eyes are equal in magnitude and direction. **Vergence** is a movement in which the rotations are equal in magnitude but in opposite directions. A **Vieth-Müller circle** is the path along which the gaze must move in order to keep vergence constant while version changes. A set of Vieth-Müller circles for different angles of vergence is shown in Figure 5.15. A **hyperbola of Hillebrand** is a path along which the gaze must move in order to keep version constant while vergence changes. Any movement of the two eyes may be described as the sum of a version and a vergence. Let θ be the version component with movements towards the right signed positive and let μ be the vergence

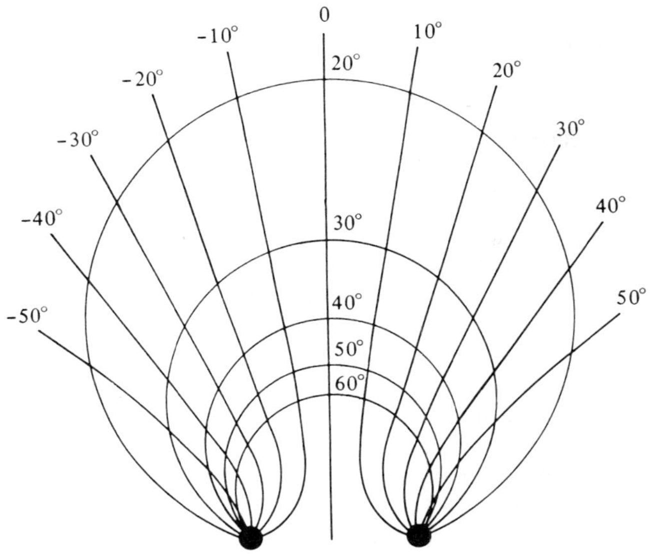

Figure 5.15 Vieth-Müller circles, or paths along which the gaze moves when vergence is constant, and hyperbolae of Hillebrand, or paths along which the gaze moves when version is constant (After Luneburg, 1948)

component with convergence signed positive. When the movement is a pure vergence, $\mu/2$ equals the movement of each eye. In general (after Ono, 1980).

$$\theta + (\mu/2) = \text{rotation of the left eye} \tag{1}$$

$$\theta - (\mu/2) = \text{rotation of the right eye.} \tag{2}$$

Hering's law would be a tautology if it merely stated that all binocular movements may be described as the sum of a version and a vergence component, and that each component may be regarded as the sum or difference of two equal movements in the two eyes. A stronger expression of the law is that the *velocity* of the two eyes is always equal. Many people have interpreted Hering's law in this way, and Hering himself wrote,. 'The two eyes are so related to one another that one cannot be moved independently of the other; rather, the musculature of both eyes reacts simultaneously to one and the same impulse of will' (Hering, 1942, p. 152). But this is not what Hering really meant, for in the same place he went on to write, '. . . it is possible for us to move both eyes simultaneously about different angles and with different speeds . . . and even to move one eye outward or inward while the other remains still'. It is clearly not the amplitude or velocity of the movements of the two eyes which are equal in Hering's law—but *the amplitude and velocity of the vergence component in each eye and of the version component in each eye, considered separately*. The actual movement that these two components produce in a given eye may be zero because, for a given eye, they may be directed in opposite directions and therefore cancel. In other words, Hering's law states that there are two centres in the nervous system, one for vergence and one for version, and that, from each of these centres, the innervation to one eye is always the same as the innervation to the other eye. When these innervations are pooled, the movement of each eye is the algebraic sum of the innervations from the two centres. Hering believed that there were distinct motoneurones for version and vergence movements, and that the pooling of the effects of the two types of innervation occurred in the muscles themselves. He thought he detected these opposed contractions in the muscles of the stationary eye during a change of fixation along the visual axis of that eye, and more recently, investigators have claimed to have detected them electromyographically (Tamler *et al.*, 1958). However, others have found no changes in the electrical activity of muscles of a stationary eye under these conditions (Breinin, 1955; Blodi and Van Allen, 1957). The electromyographically detected movements that were reported by Tamler *et al.* may have resulted from the slight torsional movements that both eyes undergo when there is a change in vergence (see Allen and Carter, 1967). A recent study by Enright (1980) has revealed that there is also a slight lateral translation of an eye when there is a vergence change along its visual axis. But these torsional and translatory movements of the eye are probably incidental to the version and vergence components, and it is more in accord with the physiological evidence reviewed in Section 5.4 to believe that there are no separate pathways for version and vergence in the oculomotor nerves, and that integration of version and vergence signals is achieved in the oculomotor nuclei. Hering's law should perhaps be called the **Law of equal component innervations**. Consider the special case where the gaze shifts from point A to point B, which both lie on the visual axis of the left eye, as in Figure 5.16. This movement involves a change of gaze of the right eye of ϕ. It can be seen in the figure that ϕ must equal the change in the angle of vergence (μ) which the movement produces, which in this case is negative (divergence). Hering's law requires that the two eyes receive the same vergence signal, which in this case must move each eye through angle $-(\phi/2)$. The other half of the total movement of the right eye must equal the magnitude of the vergence signal: in this case a rightward version of $+(\phi/2)$. When these values are put into equations (1) and (2) it is seen that they cancel for the left eye and add for the right eye. The right eye therefore does all the moving and the left eye remains stationary. It is in this way that Hering's law is compatible with the occurrence of unequal movements of the two eyes. However, the eyes do not normally move in this way, because this type of movement requires that the outputs of the vergence and version centres are simultaneous. As we shall now see, this is not usually true.

Versions in response to a stepped stimulus are normally saccadic and occur with a latency of about 200 msec and a velocity of up to 800°/sec. Vergences, on the other hand, have a latency of about 175 msec and a peak velocity of only about 10°/sec. One would expect therefore that, in a movement such as that depicted in Figure 5.16, the vergence movement would start before the version, then the version would quickly be achieved, leaving the rest of the slow vergence to be completed. If we assume for the sake of simplicity that the vergence mechanism is inhibited during the time that version is occurring, the resulting eye

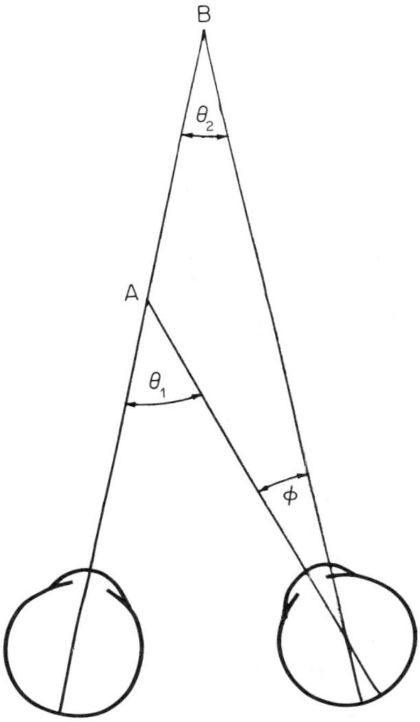

Figure 5.16 A change in gaze from point A to point B involves a vergence change of $\theta_1 - \theta_2$. But $\theta_1 - \theta_2 = \phi$, so that each eye (or the centre controlling it) receives a vergence signal of $-(\phi/2)$ and a version signal of $\phi/2$. In the left eye these two signals cancel and in the right eye they add

movements will be as depicted in Figure 5.17. A pure vergence will occur at the start, that is, the gaze will travel along a Hillebrand hyperbola. Then the pure version will take the gaze along a Vieth-Müller circle. Finally, the residue of the vergence will occur, again carrying the gaze along a Hillebrand hyperbola. Note that the law of equal component commands holds throughout this sequence. Furthermore, with a purely sequential programming of version and vergence, the amplitudes and velocities of the two eyes are always equal, not merely the 'amplitudes' and 'velocities' of the component commands. Actual records of eye movements during a rapid change in gaze, which involves both vergence and version components, show that the two components are achieved in the sequential manner just described (Miller *et al.*, 1980). However, it is not clear from the records whether vergence is inhibited during version or whether the singnals act concurrently. Miller *et al.* attempted to clarify this point by measuring the velocity of an eye movement when the version and vergence components to an eye were directed in the same direction and when they were directed in opposite directions. If the vergence signal is inhibited during version, these two velocities should be the same, but if the vergence and version signals act simultaneously so that they add algebraically, the two velocities should differ by

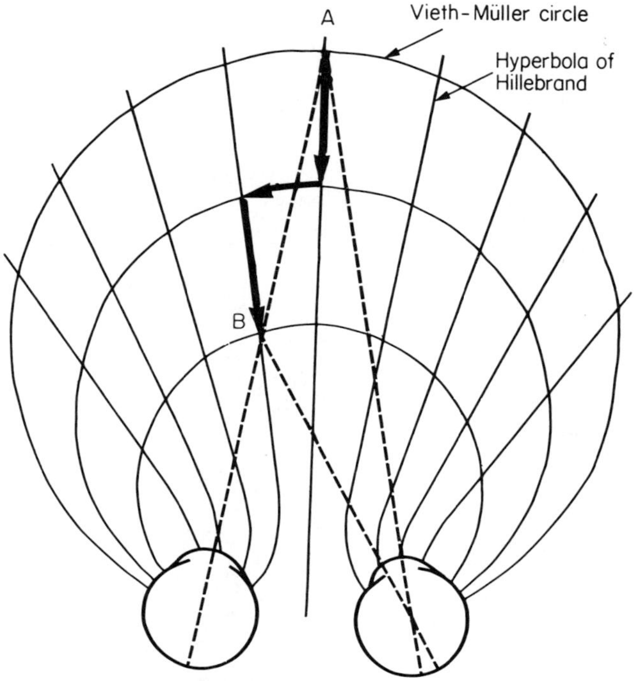

Figure 5.17 The hypothetical path followed by the point of fixation in a change of gaze from A to B. Part of the required convergence is first executed, followed by a rapid change of version along a Vieth-Müller circle, and then the rest of the required vergence is completed, along a Hillebrand hyperbola. It is assumed that the vergence response is inhibited while version is occurring

a predictable amount. In fact, if was found that the two velocities did differ, but not by the full amount predicted by the additivity of the two component innervations. This lack of complete additivity may be interpreted as a departure from the law of equal component commands, or one might still say that the component commands are equal but do not add in a linear fashion when feeding into the motoneurones perhaps because of saturation effects.

One might wonder what evidence would constitute a departure from the law of equal component commands. In general, the law would be violated if it could be shown that the two eyes receive unequal version or unequal vergence signals. It is not possible to detect such differences directly, because the neurophysiological channels have not been identified. However, eye movement recordings have revealed that microsaccades involved in the maintenance of fixation do not necessarily occur with the same amplitude in the two eyes (Krauskopf *et al.* 1960). It has also been reported that the two eyes do not move exactly in phase during large saccades, especially during states of fatigue (Bahill *et al.*, 1976). These are therefore instances where the law of equal innervations is violated.

If a visual target moves slowly along the visual axis of one eye, the other eye is able to pursue it while the first eye remains essentially stationary. However, as I have already shown, this is not necessarily a violation of the law of equal component commands because it can be argued that the immobile eye responds to equal and opposite commands, one from the slow vergence system and one from the equally slow versional pursuit system. One might argue in a similar way to account for the relative immobility of one eye during **accommodative vergence**, that is, the vergence response of an occluded eye when a stimulus is moved along the visual axis of the open eye (see Kenyon *et al.*, 1978, for a recent discussion of accommodative vergence).

5.4 Supranuclear control of eye movements

The oculomotor nuclei are the immediate source of the motor commands which travel to the extraocular muscles along the oculomotor nerves. Centres in the brain which directly or indirectly feed into the oculomotor nuclei are known as supranuclear centres. I shall briefly describe those supranuclear centres about which most is known. Fuller reviews are provided by Baker and Berthoz (1977), Raphan and Cohen (1978), and Bender (1980). The superior colliculus, which is one of these supranuclear centres, is described separately in Section 5.5.

5.4.1 The reticular formation

The supranuclear structures closest to the oculomotor nuclei are the **paramedian pontine reticular formation**, or **PPRF**, and the **mesencephalic reticular formation**, or **MRF**. It seems that the PPRF on each side is concerned mainly with conjugate ipsilateral movements in the horizontal plane, and that the MRF on both sides is concerned with conjugate eye movements in the vertical plane. It has been suggested that the vertical and horizontal components of oblique eye movements are coordinated in the PPRF (Büttner, Hepp and Henn, 1977).

Monosynaptic pathways exist between the PPRF and the oculomotor nuclei, particularly the abducens nucleus, which is concerned with conjugate lateral eye movements (Büttner-Ennever and Henn, 1976). Pathways from other supranuclear structures, including the vestibular nuclei, superior colliculi, and frontal eye fields, converge in the PPRF, which suggests that lateral eye movements of all types, slow, fast, voluntary and involuntary, are processed in this region. This suggestion is confirmed by the fact that lesions in the PPRF cause paralysis of all types of conjugate eye movements towards the side of the lesion (Bender and Shanzer, 1964).

Eye movements similar to naturally occurring saccadic lateral eye movements may be induced by electrical stimulation of cells in the PPRF, although pursuit movements have not been induced in this way. Furthermore, the eyes tend to remain in their deviated position for a fraction of a second after stimulation ceases, which suggests that the signal is integrated over time, a suggestion which has some bearing on the discussion in Section 9.2.2 (Cohen and Komatsuzaki, 1972). Activity in single cells of the PPRF as the eyes move in various ways has been recorded by a number of workers (see Keller, 1974, 1977). It is generally agreed that the cells fall into the following types: **medium-lead burst units**, which fire only with a brief high-frequency discharge about 20 msec before and continuing into the saccade; **long-lead burst units**, which begin a gradual increase in firing rate as much as 250 msec before a saccade before finally discharging like medium-lead burst units; **medium-lead burst-tonic units**, which behave like medium-lead burst units, except that they also fire steadily when the eyes are not moving, at a rate proportional to the eccentricity of gaze in a particular direction; and finally, **tonic units**, which fire at a steady position-dependent rate when the eyes are not moving, but show a complete pause in firing during saccades. A few cells fire with a burst just after the start of a saccade and some fire intermittently during saccades. The activity of all these types of cells has been found to be the same for all types of saccade. For instance, the activity of a burst unit is the same for a saccade forming the fast phase of vestibular or optokinetic nystagmus as it is for a voluntary saccade.

Henn and Cohen (1976) described how the activity of different burst units is related to different parameters of the saccadic eye movement. Some units showed a close correspondence between the number of spikes in a burst and the amplitude of the saccade, regardless of its direction. Other units showed a correspondence between the frequency of their discharge and the direction of the saccade, regardless of its amplitude. For units which coded the direction of the eye movement, the preferred directions were either in the lateral plane or in a plane tilted 10–20° off the vertical plane. These planes coincide with the planes of action of the eye muscles. Movements of the eyes along other planes involved the simultaneous activity in the two sets of neurones. The activity in each unit was found to be related to the cosine of the angle between the direction of the eye movement and the line of action of the relevant set of muscles. The coding of saccadic amplitude and direction is therefore closely tied to the actions of the muscles, although the final innervation to each particular muscle is determined only when the outputs of the various types of burst unit in the PPRF are combined in the oculomotor nuclei.

Saccadic amplitude is normally proportional to saccadic velocity (see Figure 6.5), so that Henn and Cohen did not consider saccadic velocity as a separate parameter. However, in the drowsy monkey, saccades of a given amplitude are slower than normal, and Keller (1974) was able to capitalize on this dissociation of the normal relationship between saccadic amplitude and velocity and show that the mean frequency of spikes in saccadic burst neurones of the PPRF is related to saccadic velocity rather than to saccadic amplitude.

Büttner, Büttner-Ennever and Henn (1977) recorded from cells in the mesencephalic reticular formation (MRF) of the monkey during eye movements and found units with properties similar to those of units in the PPRF. However, in this case, the burst units which coded the direction of movement responded most favourably when the eyes moved in a vertical direction, which reinforces the idea that the MRF is concerned with vertical eye movements. King and Fuchs (1979) looked at the coding relationships between parameters of the discharge of saccadic burst neurones in the MRF and parameters of saccadic eye movements. They found burst duration to be correlated with saccadic duration regardless of the direction of the saccade and burst frequency to be correlated with the peak velocity of the vertical component of the saccade.

Lateral eye movements of a given eye seem to be controlled by the PPRF and other centres on the same side of the brain, whereas vertical eye movements of a given eye seem to be controlled by the simultaneous action of centres on both sides of the brain, particularly the MRF and the region known as the pretectum (Pasik, Pasik and Bender, 1969). If the subcortical commissures are sectioned, the bilateral cooperation of centres controlling vertical eye movements is interfered with and vertical eye movements are disrupted (Pasik *et al.*, 1971). Upward and downward movements are organized separately, as shown by the fact that lesions in one part of the brainstem produce paralysis of upward gaze while lesions in other, more rostral parts cause paralysis of downward gaze (Jacobs *et al.*, 1973; Cogan, 1974).

Neurones associated with slow pursuit eye movements have not been identified in the PPRF or MRF. However, Eckmiller and Mackeben (1980) recently recorded from cells just below the abducens nucleus, which responded in synchrony with ipsilateral pursuit movements, irrespective of the absolute position of the eyes.

Neurones from the PPRF which are responsible for the control of conjugate lateral gaze project to the ipsilateral abducens nucleus. This nucleus contains motoneurones innervating the ipsilateral external rectus, and also **interneurones** which send axons across the midline, through the median longitudinal fasciculus, to motoneurones in the opposite oculomotor nucleus. From here, axons arise which innervate the medial rectus muscle on the same side (Baker and Highstein, 1975). The direct and crossed pair of outputs from the abducens nucleus is responsible for the reciprocal innervation of external and medial rectus muscles which is involved in the control of conjugate lateral eye movements.

5.4.2 *The nucleus prepositus hypoglossi*

The **nucleus prepositus hypoglossi**, so called because it is near the hypoglossal nucleus in the brainstem, has recently been shown to be intimately related to the oculomotor system. I shall refer to this nucleus as the prepositus. The prepositus receives inputs from the retina, the pontine reticular formation, the superior colliculus, the vestibular nuclei, the cerebellum, and directly or indirectly from receptors in the neck (see Baker, 1977, for the relevant references). There are excitatory outputs to the ipsilateral oculomotor nuclei (Baker *et al.*, 1977) and to the cerebellum (Baker and Berthoz, 1975). Baker, Gresty and Berthoz (1976) induced eye movements in the cat by vestibular and optokinetic stimulation and recorded from cells in the prepositus. They found burst-tonic units which responded just before and during saccades and had a steady firing rate related to eye position. The activity of tonic units was related to the direction of an eye movement; different cells responded to movement upwards, downwards, to the left, and to the right. They also found tonic units for which the response was related to the position of the eyes in the orbit. Gresty and Baker (1976) reported that cells in the prepositus which respond to visual stimuli, cells which respond to eye movements, and cells which respond to movements of the head could be found in close proximity to each other. This strongly suggests that the prepositus is concerned with the coordination of eye movements with head movements (see Section 9.2.6). This idea is supported by the fact that lesions in the region of the prepositus in monkeys have been found to cause disturbances of eye–head coordination (Uemura and Cohen, 1973).

5.4.3 The vestibular nuclei

The vestibular nuclei may be regarded as the brainstem nuclei for the vestibular system, although some afferent fibres from the vestibular system also go directly to the cerebellum and to the reticular formation (see Section 9.5.5). On each side of the brainstem there are four principal vestibular nuclei, the **inferior**, the **superior**, the **lateral** and the **medial** nuclei. Together, they form a structure which sits astride the cerebellar peduncle on the floor of the fourth ventricle (Brodal, 1974; Gacek, 1974). The canals send fibres mainly to the superior and medial nuclei, the utricle mainly to the medial and inferior nuclei, and the saccule mainly to the inferior nucleus. However, the details of these projections are not yet fully worked out. The vestibular nuclei also receive fibres from the cerebellum, the pontine and mesencephalic reticular formations, and the spinal cord. There are also direct and indirect commissural connections between the vestibular nuclei on the two sides of the brainstem (see Section 8.2.2).

Efferent pathways from the vestibular nuclei provide the largest single source of fibres to the oculomotor nuclei. Each pair of synergistic canals is approximately parallel to the plane of action of a pair of extraocular muscles, so it is not surprising that it has been found that inputs from each pair of canals project predominantly to those oculomotor nuclei which control the movements of only one pair of eye muscles (Szentágothai, 1950; Cohen *et al.*, 1964). Electrophysiological stimulation of the vestibular nerve on one side has revealed the following pattern of excitatory connections in the rabbit: from the anterior canal to the ipsilateral superior rectus and contralateral inferior oblique, from the horizontal canal to the ipsilateral medial rectus and contralateral lateral rectus, and from the posterior canal to the ipsilateral superior oblique and contralateral inferior rectus. There are also patterns of inhibitory control between specific canals and specific extraocular muscles (see Baker *et al.*, 1973; Baker and Berthoz, 1974; Highstein, 1972; Ito *et al.*, 1976a, b,). For instance, the anterior vestibular canal on one side of the head and the posterior canal on the other side are stimulated in a reciprocal fashion by to and fro rotation of the head in a given plane (see Section 8.2.1). One would therefore expect sensory inputs from such a pair of canals to have a corresponding reciprocal effect on the antagonistic pairs of extraocular muscles which lie in the same plane. In line with this expectation, it has been found that the excitatory and inhibitory inputs which the superior rectus motoneurones derive from the anterior and posterior vestibular canals are reciprocally related to the inputs which the inferior rectus motoneurones derive from the same canals (Uchino *et al.*, 1980).

The pathways along which these excitatory and inhibitory impulses travel form the three-neurone vestibulo-ocular reflex arcs which are discussed in Section 9.2. Experiments involving acute lesions have revealed that the projections from the horizontal vestibular canals to the lateral and medial recti pass through the ipsilateral **medial longitudinal fasciculus** (Carpenter, 1971) and that, at least in the rabbit, the projections from the anterior canals pass through the **brachium conjunctivum** and those from the posterior canals pass through the contralateral longitudinal fasciculus (Ito *et al.*, 1976c). Other, less direct, although no less important pathways run from the vestibular nuclei to the oculomotor nuclei through the pontine and mesencephalic reticular formations and the tract of Dieters (Reisine and Highstein, 1979). Other efferent fibres from the vestibular nuclei are described in Section 8.2.2.

Electrophysiological recording in the vestibular nuclei from single neurones which respond to canal stimulation has revealed two types of cell, which have been called **type I** and **type II** (Duensing and Schaeffer, 1958). The side of the head towards which a rotation of the head is occurring is the ipsilateral side; the other side is the contralateral side. When the head rotates, ipsilateral type I cells increase their rates of firing above their resting values and ipsilateral type II cells decrease their rates of firing below their resting values. At the same time, on the contralateral side, type I cells decrease their rates of firing and type II cells increase their rates of firing (Waespe *et al.*, 1977). There are about equal numbers of each type of cell in the monkey (Fuchs and Kimm, 1975). When the head rotates, all vestibular afferents on the ipsilateral side increase their rates of firing (Keller, 1976), so that the response of ipsilateral type II cells is not due to the effects of ipsilateral inputs. It is generally believed that their response is due to inputs arriving from the contralateral vestibular nuclei along commissural fibres (Kasahara and Uchino, 1974; Abend, 1977). This system of reciprocal connections provides the basis for the push–pull system that is described in Section 8.2.2.

Melvill Jones and Milsum (1970, 1971) recorded from cells in the cat's vestibular nucleus while the animal was rotated sinusoidally in a horizontal plane. The firing of some units was found to be modulated sinusoidally in phase with the firing rate of afferent fibres in the vestibular nerve, that is, they fired in phase with the velocity of head rotation. The firing of other fibres ceased during part of the head rotation cycle, thus behaving as if the primary input had been rectified. However, the distinction between cells showing a fully modulated response and cells showing a rectified response was not clearcut, because the firing of some cells was fully modulated for small amplitudes of head rotation and rectified for large rotations.

Recently, recordings from single cells in the vestibular nuclei have revealed the presence of units which fire just before or during eye movements of any kind, not only eye movements evoked by vestibular stimulation. The vestibular nuclei must therefore be part of the general eye movement control system and not merely a part of the vestibulo-ocular system (Waespe and Henn, 1979). This question is discussed further in Section 5.4.5.

Miles (1974) recorded from single units in the vestibular nuclei of alert monkeys and found several types of cell which fired in relation to eye movements. Some cells fired tonically in relation to the position of the eyes in the orbit, others fired with a burst of activity just before or during saccades, with a duration which corresponded to the amplitude of the saccade. Other cells showed both tonic and burst activity in relation to saccades. Some of the burst units responded only when the eyes moved in a certain direction, others were not directionally specific. Some of these neurones responded only in relation to eye movements, but others responded both when the eyes moved and during vestibular stimulation. Fuchs and Kimm (1975) found similar types of cell in the vestibular nuclei of the alert monkey, but their sample contained no cells which responded only with a burst of activity before or during eye movements. Keller and Kamath (1975) also confirmed the existence of eye-movement related cells in the vestibular nuclei of the alert monkey. The latency of response of the eye-movement related cells, which also responded to vestibular stimulation, revealed that these cells receive a disynaptic input from the ipsilateral vestibular nerve. Keller and Kamath also revealed that the response of these second-order eye-movement related cells was phase-shifted by about 70° with respect to the velocity of head rotation. The significance of this fact will be discussed in Section 9.2.2. These authors found only the burst-tonic type of cell. More work is called for in order to reveal clearly just how many types of eye-movement related cells there are in the vestibular nucleus and their role in the initiation and control of eye movements.

The eye-movement related cells of the vestibular nuclei have intimate connections with the cerebellum. The significance of these connections will be discussed in Section 9.5.5.

5.4.4 *The cerebellum*

Certain areas of the cerebellum are closely concerned with eye movements. These areas include the **vermis** and regions which are collectively known as the **vestibulocerebellum**, which includes the uvula, nodulus, and flocculus. The afferent and efferent connections of the vestibulocerebellum are described in Section 9.5.5. Electrophysiological stimulation of single cells of the cerebellum has revealed that different types of eye movement are evoked by stimulating different regions (Ron and Robinson, 1973). Stimulation of cells in the vermis evokes saccades, the direction of which varies with the placement of the electrode. Stimulation of other lobes of the cerebellar cortex evokes both saccades and smooth pursuit movements, and stimulation of the vestibulo-cerebellum evokes nystagmus. Furthermore, the recorded activity of cells in the flocculus of the monkey has been found to be associated with saccades and pursuit movements (Noda and Suzuki, 1979a, b, c) and activity of cells in the vermis with saccades or eye position (Kase *et al.*, 1980). The cerebellum is associated with eye movements involved in vestibular nystagmus and other reflex eye movements, and also those involved in voluntary saccades and pursuit.

The importance of the cerebellum for the control of pursuit eye movements was revealed in a study by Westheimer and Blair (1974). They found that unilateral ablation of the cerebellum in the monkey leads to an inability to hold the gaze in the half of visual space on the side of the lesion. The eyes drift back to the centre from this side and the animal continually makes corrective saccades in an attempt to hold the gaze on an object. Furthermore, the animal is unable to pursue smoothly a visual target moving away from the midline towards the side of the lesion. Total cerebellectomy in monkeys was found to result in permanent loss of slow pursuit movements. Impairments of slow pursuit eye movements are also found in humans with cerebellar atrophy (Baloh *et al.* 1975; Nemet and Ron, 1977). Electrophysiological evidence for the involvement of the vestibulo-cerebellum in pursuit eye movements was reported by Miles and Fuller (1975). They recorded from single cells in the flocculus of the monkey and found that the firing rate was related to the velocity of a visually tracked object. But the interesting point is that when the target was mounted on the chair and oscillated with the animal, so that the eyes remained fixed with respect to the head, the cells' rate of firing was now related to the velocity of the head. When the fixation target was stationary and the monkey rotated, the firing rate of the cells was no longer modulated. In other words, these cells respond to the motion of the object relative to the fixed background, rather than to its motion relative to the head. The significance of these results will be discussed in more detail in Section 9.5.5.

Westheimer and Blair (1974) did not find that ablations of the cerebellum produced any abnormality in the accuracy of saccades, but they did not make precise measurements of the amplitudes of saccades to well-defined targets. They did find, however, that the eyes drifted back to the midline after a saccade to the side of the lesion. Others have found that saccades towards the side of a cerebellar lesion in monkeys and humans tend to undershoot severely the target and, therefore, require a series of corrective secondary saccades, a condition known as **ocular dysmetria** (Aschoff and Cohen, 1971; Ron and Nemet, 1977).

Bilateral lesions of the human cerebellum cause ocular dysmetria and oscillation about the target position, and similar symptoms occur in the movements of the arm when a cerebeller patient attempts to reach for an object (Higgins and Daroff, 1966). Ritchie (1976) has shown that the magnitude and sign of saccadic error in monkeys with bilateral lesions of the cerebellum are related to the initial position of the eyes in the orbit and to the direction in which the saccade is made. For instance, a saccade away from the primary position of gaze to an eccentric visual target is smaller in amplitude than a saccade from that eccentric point to the primary position. The general rule is that saccades made against greater restraining forces of the extraocular muscles are smaller than saccades of equal intended amplitude made against smaller elastic resistance. This suggests that what is lacking is information about the elastic tension in the muscles.

The increase in static innervation required to move an eye away from its primary position is equal to the decrease in static innervation required to move it back over the same path. Each position of an eye is associated with a particular pattern of static innervation and if this set of associated positions and innervations is coded in the central nervous system, the computation of the appropriate step component is easily achieved for any saccade. The pulse must be stronger for an eccentrically directed saccade than for a centrally directed one if the normal amplitude–velocity ratio is to be maintained. Cerebellectomy has been found to destroy the capacity of monkeys to adjust slowly the saccadic pulse in compensation for surgically induced saccadic dysmetria (Optican and Robinson, 1980). This suggests that the cerebellum is concerned with the adaptive control of the saccadic pulse. Presumably the long-term repair of the pulse depends on visual feedback, but its moment-to-moment adjustment as well as its long-term repair could also depend on information about muscle tension provided by inputs from tendon receptors and muscle spindles in the extraocular muscles. There is no shortage of evidence that inputs from the extraocular muscles project to the cerebellum (Fuchs and Kornhuber, 1969; Baker *et al.*, 1972a; Schwarz and Tomlinson, 1977; Tomlinson *et al.*, 1978). Information about the initial state of tension could be used for both current and parametric adjustment. Information about the rate of development of tension could not be processed in time to affect a current saccade but could be used in parametric adjustment.

Robinson (1974) completely removed the cerebellum of the cat and observed that the eyes tended to slip back to the midline, which is the same symptom that had been reported after unilateral cerebellectomy in the monkey. Robinson concluded that one function of the cerebellum is to integrate the signal which determines the velocity of the eye movement, to produce a signal for holding the eyes in the position they reach when the motion has stopped. He concluded from the nature of the return drift caused by cerebellectomy that only part of this integration is performed by the cerebellum and the rest by the pontine reticular formation. Perhaps this integration function of the cerebellum is achieved by the use of feedback from receptors in the proprioceptors of the eye muscles, which are known to feed into the cerebellum. Loss of integration may be due to an interruption of this feedback system. I shall make further comments on Robinson's idea of brainstem integrators in Section 9.2.2.

Pellionisz and Llinás (1979) have proposed a precisely worked out mechanism whereby a smoothly changing signal entering the cerebellum may be transformed into one which predicts the future course of the input signal. Inputs arriving on mossy fibres, which are known to be associated with saccadic eye movements, are distributed along granule cells to a series of Purkinje cells. The series of cells forms a tapped delay line (Freeman and Nicholson, 1970). The key idea of the model is that members of the first group of Purkinje cells in the series discharge with a frequency which is the zero time-derivative of the input, those of the second group of Purkinje cells discharge with a frequency which is a first time-derivative of the input, and so on down the series. The end result is the generation of a power series (Taylor series) expansion of the input signal. It is a mathematical fact that the integral of a Taylor series expansion of a time function produces a function which phase-leads (predicts) the original function. Therefore, the outputs of the series of Purkinje cells, when summated, produce a phase-advanced signal. In fact, a useful phase-advanced signal is produced by summing the first few terms of only a crudely derived Taylor series. It is perhaps the breakdown of this predictive mechanism that leads to ocular dysmetria after cerebellar damage.

Pellionisz and Llinás (1980) have developed their model of the cerebellum to account for the way in which an internal representation of intended eye movements in terms of the movement of the line of gaze is transformed into the multiparameter output that determines the contraction of the three pairs of extraocular muscles. It is suggested that each intended movement is first decomposed into a number of abstract covariant vectors, defined with reference to the principal meridia of space. These vectors are then transformed into contravariant physical components of the final motor output by a cerebellar metric tensor. The authors draw attention to the resemblance between the microstructure of the cerebellum and a device which could embody the matrix of the required metric tensor.

5.4.5 The nucleus of the optic tract

It has been known for some time that there are direct retinal inputs to the lateral geniculate nucleus and to the superior colliculus. It is less well known that, at least in the cat, there are also direct inputs of ganglion-cell axons to the pretectum, (Laties and Sprague, 1966). (A fourth direct visual input to the accessory optic tract is discussed briefly in Section 9.5.5.)

Collewijn (1975a) found that in the rabbit optokinetic nystagmus was abolished by small lesions in the lateral pretectum. Ablation of the visual cortex or superior colliculus did not abolish this response. Furthermore, Collewijn found that although optokinetic nystagmus could be elecited by stimulation of the visual cortex or superior colliculus, the most efficient trigger zone was in the lateral pretectum, in an area which he identified with the **nucleus of the optic tract** (NOT). Stimulation within this region induced on optokinetic response with the slow phase towards the side of stimulation. In a second paper Collewijn (1975b) recorded from single cells in the necleus of the optic tract of the rabbit. These cells were excited by a visual pattern presented to the contralateral eye and moving in a temporonasal direction, and inhibited by a pattern moving in the opposite direction. He suggested that there is a convergence of direction-selective retinal axons on to cells of the nucleus of the optic tract and that this nucleus is an essential first link in the optokinetic reflex arc. Similar direction-selective cells have been found in the nucleus of the optic tract of the cat (Hoffmann and Schoppmann, 1975) and the rat (Cazin *et al.*, 1980b). Cells with this type of selectivity have not been found in the superior colliculus, which therefore seems not to be implicated in the optokinetic response.

In submammalian species and lower mammals, which have no fovea, or poorly developed foveae, covering one eye abolishes optokinetic nystagmus towards the seeing eye. In other words, in these animals, the slow phase towards the right is controlled by the left eye and vice versa (Tauber and Atkin, 1968). This so-called **directional preponderance** is explained by the fact that cells in the nucleus of the optic tract in such animals respond only to temporonasal movement of stimuli in the visual field. Directional preponderance is not evident in animals with well developed foveate eyes. It looks as though the primitive subcortical optokinetic mechanism, although still present in foveate animals, is supplemented by a complementary pursuit mechanism controlled by the visual cortex. Furthermore, the evidence which I shall review in Section 6.2.4 suggests that the primitive mechanism is controlled by retinal-slip velocity and that the cortical mechanism is controlled by fixation error and by perceived velocity.

Evidence that the primitive unidirectional mechanism is still present in higher mammals is provided by the fact that human patients with visual pathologies may show a directional preponderance in their optokinetic response. For instance, Baloh *et al.* (1980) reported that patients with congenital achromatopsia (maldevelopment of cone receptors and abnormally developed foveae) showed a marked temporonasal directional preponderance of their optokinetic response when tested monocularly. They also reported that these patients showed a slow build-up in the optokinetic response, which is also a feature of nystagmus in afoveate animals.

It is to be expected that foveate animals, such as the cat, which do not normally show directional preponderance of the optokinetic response, will show it when the visual cortex is removed. Wood *et al.* (1973) found that complete removal of the visual cortex in cats produced a prominent directional preponderance, similar to that seen in afoveate animals. Harris, Leporé, Guillemot, and Cynader (1980) showed that cats which had been reared in dark surroundings had complete directional preponderance of the optokinetic response. Presumably dark-rearing prevents the development of the cortical mechanism, leaving only the unidirectional subcortical mechanism intact.

Harris *et al.* also investigated the effects of sectioning the optic chiasma. This procedure severs the crossed visual inputs from each eye to each contralateral nucleus of the optic tract but leaves the uncrossed inputs to the geniculo-striate system intact. These cats showed a monocular optokinetic response only to visual movement in a nasotemporal direction. This is the inverse of the usual direction preponderance. These findings can be explained if it is assumed that the optokinetic system receives signals indicating temporonasal visual movement over crossed pathways which ascend through the nucleus of the optic tract, and signals indicating nasotemporal movement over uncrossed pathways through the geniculo-striate system. According to this view, dark-rearing coupled with sectioning of the optic chiasma should abolish the optokinetic response entirely, and this is what Harris *et al.* found.

The gain of the slow phase of optokinetic nystagmus is reduced in strabismic humans, whichever eye is tested (Schor and Levi, 1980). Cynader and Harris (1980) found the same defect in strabismic cats. They recorded from single cells of the nucleus of the optic tract of a strabismic cat and found that there were marked differences in binocular connectivity among these neurones compared with those in normal cats. It is tempting to assume that the same deficiency is responsible for the behavioural deficit in strabismic humans.

It has been pointed out in Sections 5.4.3 and 6.1.2 that the vestibular nuclei are an essential component of the optokinetic response of nystagmus and are induced by vestibular stimulation (see also Cazin *et al.*, 1980a). Pretectal lesions abolish OKN and the response of cells in the vestibular nucleus to visual motion (Precht and Strata, 1980). Thus, the nucleus of the optic tract probably conveys information about visual motion to the vestibular nuclei. This pathway also involves the nucleus reticularis tegmenti pontis. Cazin *et al.* (1980b) found cells in this nucleus with direction-selectivities similar to those of cells in the nucleus of the optic tract. The cells of the tegmentum responded to pure vestibular stimuli induced by angular acceleration of the animal's head in the dark, in the same direction as the movement of visual stimuli to which these cells also responded. It looks as though the integration of retinal-motion signals and head-motion signals is at least partly achieved at this site. Integration of these two motion signals is required if the animal is to coordinate optokinetic and vestibular nystagmus when the head rotates in the presence of illuminated visual surroundings (see Sections 6.1.2 and 9.2.5).

5.4.6 The frontal eye fields

Over a century ago, Ferrier (1874) stimulated each side of the frontal cortex of a monkey, in what is now known as Brodmann's area 8, and observed eye movements directed towards the contralateral side. The two areas 8, one on each side of the brain, are now known as the **frontal eye fields**. In other early investigations, it was reported that stimulation of the frontal eye fields produced various types of eye movements. However, in these studies, the animals were anaesthetized and the eye movements were not objectively recorded. It is now known that with alert monkeys only saccades are elicited by stimulation of the frontal eye fields.

There are inputs to the frontal eye fields from the visual cortex (Kuypers *et al.*, 1965). Outputs from the frontal eye fields project directly to the basal ganglia, thalamus, superior colliculus, and the pontine and mesencephalic reticular formations. However, there are no direct connections to the oculomotor nuclei (Astruc, 1971).

Unilateral lesions of the frontal fields in the monkey produce a consistent deviation of the eyes and head towards the side of the lesion; furthermore, the animal ignores things seen in the contralateral visual field, and fails to pursue an object which moves across the midline into the contralateral field (Latto and Cowey, 1971a). After about a week, these gross abnormalities of gaze disappear, but reappear in exaggerated form in the other direction if the other frontal eye field is subsequently removed (Latto and Cowey, 1971b).

Electrophysiological recordings from single cells of the frontal eye fields of the alert monkey have revealed a variety of types of cell. Mohler *et al.* (1973) found that about half the cells which they contacted responded only to visual stimuli. The receptive fields of these cells were very large and never less than 20° in diameter. About one-tenth of the cells in the frontal eye fields responded only to movements of the monkey's hands, or to moving sounds, and about one-quarter of the cells responded to movements of the eyes. Bizzi and Schiller (1970) found two types of cell in the frontal eye fields which responded during eye movements. The first type fired during saccades, whether voluntary or reflex, and the second type fired when the eyes were in a given position or when they moved slowly through that position during a voluntary or reflex pursuit movement. We thus see the same division between burst and tonic types that is found in other centres concerned with eye movements. However, these cells in the frontal eye fields fired *after* the eyes had started to move and are therefore not in the chain of command which is responsible for initiating eye movements. Mohler *et al.* (1973) confirmed that eye-movement related cells in frontal eye fields fire up to about 30 msec after the movement has started, and never before. They reported two types of saccade-related cells: one type fired whether the eye movement occurred in darkness or in illuminated surroundings, but the other type fired only if the saccade occurred in response to a visual stimulus. The latter type still fired when the stimulus was extinguished just after the movement had started, which shows that these cells were responding to the eye movement and not to the motion of the retinal image. Bizzi and Schiller (1970) found cells which responded only when the head moved, and they showed also that the response of the eye-movement related cells was not affected by movements of the head.

The fact that eye movements start before the firing of cells in the frontal eye fields seems to contradict the findings of Robinson (1968b) and Robinson and Fuchs (1969) that stimulation of the frontal eye fields can evoke saccadic eye movements with a latency of about 25 msec (see Section 6.4.3). The answer may be, as Robinson himself suggested (see Roucoux and Crommelick, 1977, p. 487), that electrical stimulation of the frontal eye fields may evoke a response in the pontine reticular formation through axons which normally convey signals the other way. The precise role of the frontal eye fields in functions related to eye movements remains a mystery.

5.4.7 Other cortical areas

Eye-movement related activity has also been electrophysiologically recorded in the visual cortex. Changes in potential may be recorded from surface electrodes over the occipital lobes in man. Periodic changes in the evoked potential, known as **lambda waves**, are recorded after the onset of eye movements over a patterned scene. Lambda waves do not occur when the scene is devoid of luminance contrasts, or when the subject is in the dark (Scott *et al.*, 1967), which suggests that they are related to movements of the retinal image produced by eye movements, rather than to eye movements as such. However, this cannot be the whole story because lambda waves occur up to 200 msec before voluntary eye movements (Barlow and Cigánek, 1969). In the cat, evoked potentials in the lateral geniculate nucleus, pontine reticular formation, and visual cortex have been recorded following saccadic eye movements. These changes in potential have been called **ponto-geniculo-occipital waves**, or **PGO waves**. Unlike lambda waves, PGO waves persist when the eyes are closed. Jeannerod and Sakai (1970) recorded the presence of PGO waves from the visual cortex of a cat when its oculomotor system was excited by vestibular stimulation but its eye muscles were paralysed. PGO waves are also found in cats which are in the REM (rapid eye movement) phase of sleep (Brooks, 1968). The functional significance of these waves remains obscure.

Wurtz (1969a, b) recorded from single cells in the visual cortex of the alert monkey while the animal moved its eyes over a stationary scene or held its eyes still while a stimulus was moved across the receptive field of the cell being investigated. Cells were found that were excited by these two procedures and others that were inhibited by them but both types of cell were affected equally by the two types of image motion and no cells were found that differentiated between them. Nor were any cells found that responded to eye movements made in the dark. Wurtz concluded that it is the movement of the image over a cell's receptive field that causes its activity to be affected during eye movements, and not efferent or proprioceptive signals associated with the eye movement itself (Wurtz, 1976). However, Wurtz recorded only from cells which had receptive fields in the central region of the retina. Cohen and Feldman (1971) recorded from the calcarine region of the visual cortex of the monkey, which receives inputs from the peripheral retina, and found cells which responded to eye movements made in evenly illuminated surroundings. Noda *et al.* (1972) found that about 10% of cells in the visual cortex of the cat did not respond to stimulus motion across their receptive fields when the eyes were stationary, but were either excited or inhibited when the animal made saccadic eye movements across a stationary patterned scene. Some of these cells also responded when eye movements were made in darkness. These cells are, therefore, capable of distinguishing between image movement produced by an object moving relative to a stationary eye and image movement produced by a movement of the eye past a stationary object. The significance of this point is discussed in Section 7.8.5. It has still to be determined whether such cells occur in the primate visual cortex.

There are also cells in the inferior region of the parietal lobe which discharge in relation to eye movements. This region of the parietal lobe is known as the **suprasylvian gyrus**, or area 7, and receives inputs from the visual cortex, which are probably relayed through the thalamus. Straschill and Schick (1974) recorded from single cells in the suprasylvian gyrus of the alert cat and found that most cells responded to visual stimuli moving within a given region of the visual field and about half of them were directionally selective. Most of these visually responsive cells responded just before or just after saccadic eye movements. Those responding before eye movements did so when the eye movements occurred in the dark. Those responding after the initiation of the eye movement probably did so in response to motion of the retinal image.

Lynch *et al.* (1977) found large numbers of cells in the suprasylvian gyrus of the monkey which discharged just before the animal turned its gaze towards an object in which it was interested. Such cells continued to discharge while the animal's attention was engaged by the object. For instance, if the animal expected a reward on presentation of a stimulus, the discharge of such a cell continued until the reward was delivered. After that, the cell no longer responded, even though the animal may still have been looking at the stimulus object. The cells remained inactive when the animal casually inspected its surroundings. These cells were called **fixation cells**. Most fixation cells responded only when the gaze was in a particular quadrant of the total region within which the animal could move its gaze when the head was fixed. Other cells were found which responded while the eyes slowly pursued a visual object, and yet others which responded before and during visually evoked saccades. Sakata *et al.* (1980) confirmed the existence of fixation cells in area 7 of the parietal lobe of the monkey. Each fixation cell had a preferred direction of gaze and its discharge rate decreased as a function of the angle of deviation from this direction. In addition to, or instead of, a direction selectivity, many fixation cells had a selectivity for a particular convergence of the two eyes. About half the cells showed the same selectivity to eye position in light as in dark surroundings; the others showed a decrease in discharge rate in the dark. Fixation cells thus provide signals which indicate the position of the point of fixation in three-dimensional space. We shall see in Section 11.7.3 that area 7 of the parietal lobe is also concerned with the control of limb movements towards visual objects.

5.5 The superior colliculus

5.5.1 General structure

Lower vertebrates do not possess a cerebral cortex. The primary visual centre in the brains of such animals is known as the **optic tectum**. In higher vertebrates, the primary visual centre is in the occipital pole of the cerebral cortex, and the homologue of the optic tectum is a pair of organs in the midbrain, one on each side, known as the superior colliculi, or sometimes by the singular form, **superior colliculus**. In mammals, the superior colliculus sits above the mesencephalic reticular formation, just behind the pretectum and just in front of the inferior colliculus. It seems to be involved in the process of coding the oculocentric location of visual objects which interest the animal and in eliciting saccadic eye movements towards such objects. The following brief review is based on data from primates, as far as possible. Fuller reviews are provided by Sparks and Pollack (1977) and by Goldberg and Robinson (1978).

There are three principal layers of neural tissue in the superior colliculus: the **superficial**, **intermediate** and **deep** layers. Each of these layers is further divided into sublayers, each with its own name. I shall not burden the reader with these. Ganglion cell axons arising from the periphery of the retina send collateral branches directly to the superficial layer of the superior colliculus. Most inputs to the superior colliculus from the central region of the retina, about 10° of visual angle on either side of the fovea, reach the superficial layer of the colliculus after having been relayed through the lateral geniculate nucleus and visual cortex. There must also be some direct inputs from the central retina, even from the fovea, because cells in the superficial layer of the colliculus in the monkey have been found to respond to foveal stimulation after removal of the visual cortex (Schiller *et al.*, 1974). More recently, autoradiographic and staining techniques have revealed these direct foveal inputs (Hubel *et al.,* 1975; Cowey and Perry, 1980). Stimulation of cells in the superficial layer of the superior colliculus inhibits those cells in the visual cortex from which the collicular cells receive their input. This effect is thought to be due to recurrent collateral inhibition (Hayashi, 1969). The retinotopic organization of cells in the visual cortex is preserved in the regions of the superior colliculus to which they project. In the monkey, the region of the retina within 5° of the fovea occupies 30% of the collicular surface (Wilson and Toyne, 1970; Cynader and Berman, 1972). Inputs to the colliculus on one side are derived from the nasal half of the contralateral retina and from the temporal half of the ipsilateral retina (Hubel *et al.,* 1975; Graybiel, 1976). Thus, the colliculus on each side receives inputs which originate in the contralateral hemifield of visual space.

The intermediate and deep layers of the superior colliculus have no direct inputs from the retina or from the visual cortex. However, these layers receive inputs from centres associated with vision in the temporal, parietal and frontal lobes, particularly from the frontal eye fields. They also receive inputs from subcortical structures, including the inferior colliculus (a structure concerned with auditory functions). As we shall see later, many of these cells are bimodal or multimodal, that is, they are excited by visual stimulation and by stimulation of one or more sense organs in other modalities, including the auditory, tactile and proprioceptive modalities and perhaps also the vestibular system.

The outputs from the superficial layer ascend to more rostral centres in the brain, including the lateral geniculate nucleus (see Section 2.5) and the thalamus (Benevento and Fallon, 1975; McIlwain, 1978). There is also an ascending output from intermediate and deeper layers to various centres in the thalamus, the medial geniculate nucleus, and the olivary nucleus. Descending outputs from the intermediate and deep layers project in two major tracts to the pontine and mesencephalic reticular formations, the inferior colliculus, and various motor nuclei in the upper brainstem. There are apparently no direct connections between the superior colliculus and the cerebellum or the oculomotor nuclei of the brainstem. The influence that the colliculus exerts on eye movements is therefore indirect (see Sparks and Pollack, 1977, for references on all these afferent connections of the superior colliculus).

5.5.2 *Visual inputs to the superior colliculus*

Most, if not all cells in the superficial layer of the superior colliculus respond to visual stimulation in an appropriate region of the retina. A map of the retinotopic projection on the surface of the right colliculus of the monkey is shown in Figure 5.18. Only Y and W cells project to the colliculus. X cells project only to the visual cortex, and their inputs are not relayed to the colliculus (Schiller, Malpeli and Schein, 1979). The diameter of the excitatory region of the receptive fields of collicular cells is about 1° for cells serving the foveal region and as large as 10–30° for cells serving the retinal periphery. According to Goldberg and Wurtz (1972a), there is little spatial summation within the excitatory region, so that the size of the stimulus is not an important factor in determining the response of a collicular cell. However, Updyke (1974) found evidence of spatial summation in collicular cells. There is general agreement that the shape and orientation of the stimulus are not important factors in determining the response of collicular cells (Schiller and Koerner, 1971; Goldberg and Wurtz, 1972a). Certain cells in the superficial layer of the colliculus respond best to stimuli moving between 0.5 and 30°/sec, but the response declines if the same stimulus is repeated several times. Other cells respond to velocities as high as 800°/sec. Only about 10% of cells in the monkey's superior colliculus have a preferred direction of movement, the rest are *non-directional*. Directionally selective cells are particularly sensitive to flashing stimuli which move in the preferred direction (Syka *et al.*, 1979). Most of the non-directional cells have inhibitory regions surrounding the excitatory region of their receptive fields. Those cells which do not have an inhibitory surround are particularly sensitive to changes in general

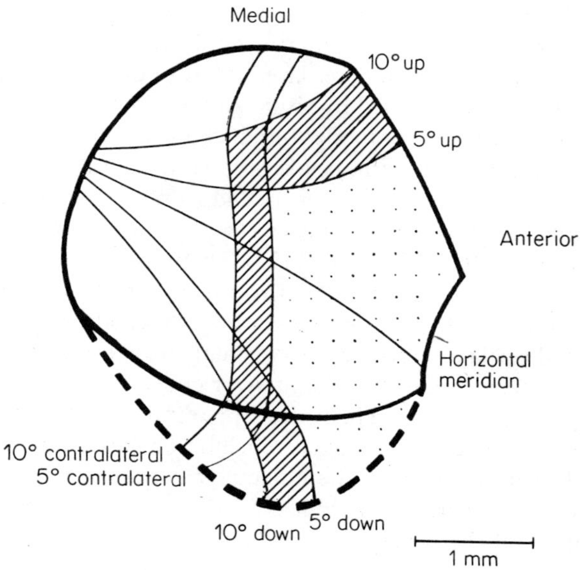

Figure 5.18 Map of the visual field on the surface of the right colliculus of the monkey. The stippled area represents parts of the contralateral field within 5° of the fovea. The striped and stippled areas together represent parts of the contralateral field within 10° of the fovea (From Cynader and Berman, 1972. Reproduced by permission of *Journal of Neurophysiology*)

illumination as well as to stimulus movement. Some cells are responsive to stimuli which approach the animal's head from a distance (Cynader and Berman, 1972; Goldberg and Wurtz, 1972a; Updyke, 1974). I shall return to this topic in Section 7.9.3. Many cells are binocularly driven, that is, they respond to stimulation of either eye.

Most visually driven cells in the superficial layer of the cat's colliculus are directionally selective and binocularly driven, but they do not develop these characteristics until the cat is about three weeks old (Norton, 1974). Furthermore, these cells lose their directional selectivity and binocularity if the visual cortex is removed (Wickelgren and Sterling, 1969). There are, nevertheless, collicular cells in the decorticate cat which respond to motion and which retain their retinotopic organization. It looks as though directional selectivity and binocularity are features of only those collicular cells which receive their input from the geniculostriate system; cells with a direct visual input are non-directional and monocularly driven. When the visual cortex is removed unilaterally in the neonatal kitten, the collicular cells which would have normally received inputs from the ablated hemisphere do not develop directional specificity or binocularity (Stein and Magalhães-Castro, 1975; Berman and Cynader, 1976; Flandrin and Jeannerod, 1977). Therefore, these collicular cells have no capacity to compensate for the early loss of cortical tissue, that is, they do not develop directional selectivity or binocularity in the absence of cortical tissue. Furthermore, the collicular cells of kittens which are reared in the dark do not acquire directional selectivity or binocularity (Flandrin and Jeannerod, 1977). Kittens reared in stroboscopic illumination, in which visual motion signals are absent, also lack directionally-specific cells in the colliculus (Flandrin *et al.*, 1976). Hoffmann and Sherman (1974, 1975) also found a lack of directional selectivity in collicular cells of cats reared in darkness and showed that the deficit was due to an absence of inputs from Y channels which reach the colliculus through the geniculostriate system. The direct Y- and W-cell inputs to the colliculus were intact in these animals. All the cells in the retina were found to be normal in dark-reared cats (Kratz *et al.*, 1979). Furthermore, stimulation of cells in the visual cortex of dark-reared cats produced a response in collicular cells. Thus, the absence of the indirect Y-channel input to the colliculus in dark-reared cats must be due to a deficit between the eye and the cortex.

Many of the collicular cells which respond to rapid stimulus motion do not respond when a saccadic eye movement sweeps the retina over a field of stationary objects. This suppression of discharge is evident when the eye movements are made in near total darkness. This fact, and evidence from the time course of the suppression effect, led Robinson and Wurtz (1976) to conclude that the signal responsible for this suppression is either the efference signal driving the eye movements or proprioceptive signals from receptors in the extraocular muscles. Cells which are suppressed in this way differentiate between image movements produced by the movement of objects in the visual field (exafference) and image movements produced by rotation of the eye (reafference) (see Section 7.7). That proprioceptive inputs from extraocular muscles may be involved in this suppression is indicated by the fact that passive stretch of extraocular muscles has been found to evoke responses in visually responsive neurones in the superior colliculus (Abrahams and Rose, 1975a, b). Furthermore, Rose and Abrahams (1975) found that the response evoked in a collicular cell by a passive movement of the eye occurred as the eye traversed a particular position in its orbit. Donaldson and Long (1977) found cells in the cat's superior colliculus which showed a suppression of their response to a visual stimulus soon after the eye was passively rotated. Proof that efference is also involved in the suppression of collicular cells would require an experiment *in which the response of collicular cells to visual movement is recorded in an animal which attempts to move its paralysed eyes.*

There are apparently no cells in the superior colliculus which respond to chromatic differences equated for luminance; in other words, the superior colliculus is colour-blind (Michael, 1972; Schiller and Malpeli, 1977).

Cells in the intermediate and deep layers of the superior colliculus have larger receptive fields than cells in the superficial layers. The receptive fields of cells which lie along a line at right angles to the collicular surface are roughly concentric, if these fields are plotted in the coordinate system of the collicular mapping of the visual fields (McIlwain, 1975). These relationships are illustrated in Figure 5.19. The cells in the intermediate and deep layers have a longer latency and a preference for more slowly moving stimuli than do the cells in the superficial layer (Marrocco and Li, 1977). Cells in these deeper layers habituate very rapidly if a stimulus is repeatedly presented to the same area of the retina. They respond best to a stimulus which is presented in a given position for the first time. These cells have therefore been called **novelty detectors** (Cynader and Berman, 1972). Furthermore, the responsiveness of a given cell is reduced as soon as second stimulus is introduced somewhere else in the visual field (Rizzolatti *et al.*, 1973; Camarda *et al.*, 1974).

Bisti *et al.* (1974) found that about 25% of the non-directional cells in the cat's superior colliculus varied their response to a visual stimulus as the animal was tilted about its longitudinal body axis, up to 5° from its normal position. Further tilt of the body had no effect. The effect of body tilt was abolished by sectioning of the eighth nerve, which suggests that vestibular inputs were responsible. As the collicular cells involved were not selectively responsive to either the direction of movement or the orientation of stimuli, it is difficult to see what role this relation between response and body position could play in a mechanism concerned with compensating for the visual effects of changes in posture. Perhaps these cells were associated with the countertorsion of the eyes which would have occurred had the animals not been anaesthetized.

Goldberg and Wurtz (1972b) trained monkeys to make a saccade to a spot of light sometimes but to ignore the light at other times. They then recorded from single cells in the superior colliculus as either one or the other type of stimulus was exposed in the receptive field of each of those cells. Several cells showed an enhanced response, but only when the stimulus was one towards which the animal was going to make a saccade. This enhancement was not due to general

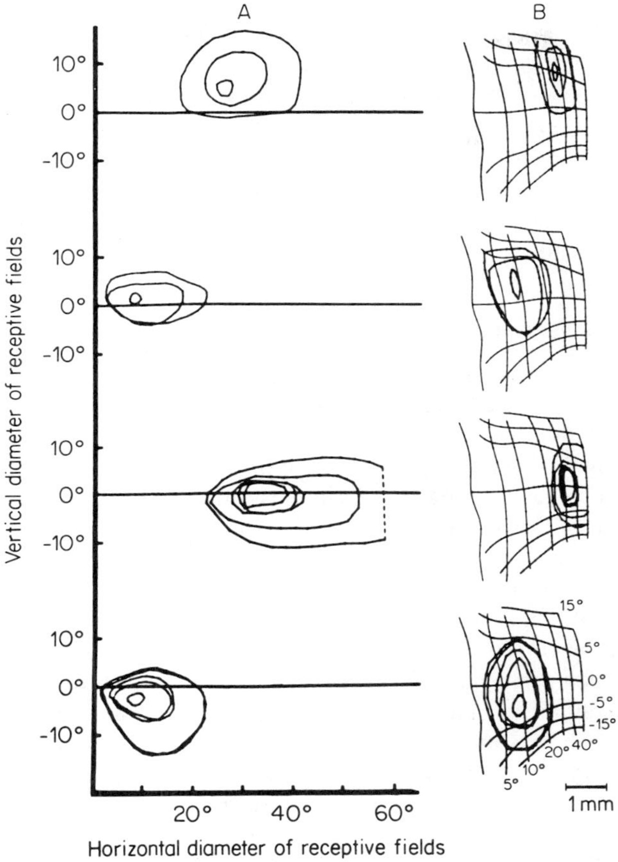

Figure 5.19 Nested maps of different retinal receptive fields in the monkey superior colliculus. Each nested map is of a receptive field at each of several penetrations into deeper layers of the colliculus. On the right, each map is replotted in the retinotopic coordinate system of the colliculus (From McIlwain, 1975. Reproduced by permission of the *Journal of Neurophysiology*)

arousal, because it was specific to cells receiving inputs from the target area. The enhancement was not evident when the animals made a saccade towards a given location in the dark, which shows that the cell's response was not part of the signal which initiated the eye movement. On the other hand, it did not occur when the animal moved its hand to a given target rather than its eyes, nor did it occur when the animal merely attended to a given area of its visual field without making an eye movement towards that area (Wurtz and Mohler, 1974, 1976). The enhancement was not mediated by the visual cortex, because the effect was increased rather than reduced when the visual cortex was ablated (Mohler and Wurtz, 1977). Mohler and Wurtz suggested that the effect is due to eye-movement related activity, ascending from deeper layers of the colliculus, facilitating visually induced activity in the superficial layer. Presumably, cortical areas, such as the parietal lobes, are involved, because of the attentional aspects of the response.

Thus, the superior colliculus seems to embody a mechanism which brings the animal's visual attention to bear on stimuli towards which it will make saccadic eye movements. However, these cells, whose firing is only loosely related to the occurrence of a saccade, are not the only cells in the colliculus related to eye movements. As we shall now see, there are other cells which always fire before and during a saccade.

5.5.3 Eye movements and the superior colliculus

Adamük (1870) first showed that electrical stimulation of the superior colliculus in the cat evoked conjugate eye movements towards the contra-lateral side. Although several investigators have confirmed this observation in the cat and in other animals, it was only recently that Robinson (1972) and Schiller and Stryker (1972) produced similar results in the monkey. There is agreement on the following points. Electrical stimulation of certain cells in the intermediate and deeper layers of the superior colliculus of the alert monkey results in a conjugate saccade (never any other kind of eye movement) towards a specific area of the contralateral hemifield. Movements in any direction into the contralateral hemifield may be produced by unilateral stimulation, except purely vertical movements, which presumably require the activation of cells on both sides. The region of the oculocentrically defined visual field towards which an eye moves when a given cell is stimulated is the **movement field** of that cell. Simultaneous stimulation of two regions of the colliculus evokes an eye movement in a direction between those of movements evoked by separate stimulation of those same regions. The retinotopic mapping of movement fields in the deeper layers corresponds to the retinotopic mapping of the receptive fields of cells lying above in the superficial layer of the colliculus. Each cell has a definite threshold of stimulus intensity below which stimulation produces no effect. This threshold is lower for cells in deeper layers. The latency period from stimulation to the beginning of the evoked saccade is between 20 and 30 msec. The amplitude and direction of the saccade is largely independent of the strength or duration of stimulation at above-threshold values, and depends only on the position of the electrode with respect to the retinotopic map over the surface of the colliculus. Cells at different depths within the same region produce saccades of similar amplitude and direction. The important point is that, in the monkey, stimulation of a given cell produces a saccade which is independent of the initial position of the eyes. In other words, each cell produces a saccade which carries the gaze to an oculocentrically defined point in the visual field, which I shall call an **oculocentric** saccade, and not a saccade to a point defined with respect to the head, which I shall call a **headcentric** saccade (see Figure 5.20). The two types of saccades can be distinguished in the following way. When a stimulus is repeatedly applied to the same collicular cell in quick succession, an oculocentrically coded system produces a staircase pattern of similar saccades in the same direction, but a headcentric system produces only one saccade because, after the initial response, the eyes are on target.

There has been some dispute about whether cells in the cat's colliculus are coded oculocentrically or headcentrically. Straschill and Rieger (1973) found that stimulation of the cat's superior colliculus produced headcentric saccades. This would mean that, in the cat, the retinotopic mapping of visual fields in the colliculus is adjusted in the colliculus itself by information regarding the position of the eyes at the start of the saccade. Stein, Goldberg and Clamann (1976) could not confirm this finding. However, more recent studies suggest that the cat has both an oculocentric and a headcentric coding system. Roucoux and Crommelinck (1976) and Crommelinck *et al.* (1977) claimed that, in the cat, collicular cells with movement fields within about 25° of the fovea are oculocentrically coded, whereas cells with more peripherally placed movement fields are headcentrically coded (see also Guitton *et al.*, 1980). Thus, repeated stimulation of cells with centrally placed movement fields produced a series of similar saccades, and repeated stimulation of cells with peripheral movement fields produced only one saccade. Eye movements evoked by cells with peripheral movement fields were found by Roucoux and Crommelinck to be often accompanied by movements of the head in the same direction. The reason why these investigators obtained only one saccade when cells with peripheral movement fields were stimulated may have been that the eyes were near the limit of their range of movement. Harris (1980) was not able to elicit headcentric saccades by the stimulation of any region

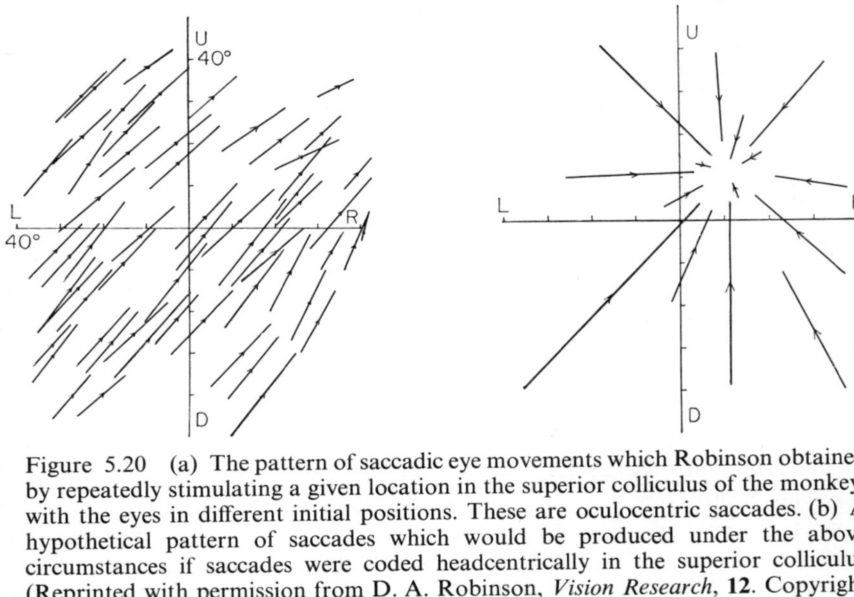

Figure 5.20 (a) The pattern of saccadic eye movements which Robinson obtained by repeatedly stimulating a given location in the superior colliculus of the monkey, with the eyes in different initial positions. These are oculocentric saccades. (b) A hypothetical pattern of saccades which would be produced under the above circumstances if saccades were coded headcentrically in the superior colliculus (Reprinted with permission from D. A. Robinson, *Vision Research*, **12**. Copyright 1972, Pergamon Press, Ltd.)

of the cat's superior colliculus, although he did notice the constraining effects due to the limited range of eye movements. He confirmed that saccades evoked by cells with peripheral movement fields were accompanied by head movements. On balance, the case for headcentrically coded cells in the cat's colliculus is not proven.

Primates, including man, certainly possess the ability to make saccades towards targets defined with respect to headcentric or exocentric positions, irrespective of the oculocentric position of the target. Evidence for this statement occurs at several places in the next two chapters, but I shall briefly summarize some of it here. Hallett and Lightstone (1976a) found that human subjects could make a saccade towards a target flashed on during a previous saccade. In this case the oculocentric position of the target with respect to the fovea did not define the required amplitude of the second saccade; its amplitude could be derived only by combining oculocentric and headcentric (eye position) information. This experiment does not prove what it has been thought to prove, namely, that the subject has accurate information about eye position during a saccade (see Section 7.8.2 for details on this point). The results of an experiment by Miller (1980), which is described in Section 7.8.2, showed that people can make fairly accurate saccades in the dark to remembered positions of visual targets, even though they may have moved their eyes after the target was presented. Finally, Mays and Sparks (1980a) found that monkeys could make an accurate saccade towards a briefly exposed visual target, even though a saccade was induced by collicular stimulation in the interval between the exposure of the target and the commencement of the voluntary saccade. The site of the mechanism which controls headcentrically or exocentrically defined saccades is not known, but other evidence produced by Mays and Sparks, which I shall review later in this section, suggests that cells exist in the colliculus which are at least related to the computation of the headcentric positions of visual targets.

A cell which when stimulated evokes a saccade of a given direction and amplitude also discharges when the animal makes a voluntary saccade of the same kind. In other words, the movement field of a cell determined by stimulation corresponds to its movement field determined by recording from the cell. Cells discharge at least 20 msec before the start of the eye movement and at least 10 msec before the first signs of activity in motor neurones of the extraocular muscles. They discharge to saccades made in the light or in the dark (Wurtz and Goldberg, 1971, 1972a; Schiller and Koerner, 1971; Sparks, 1978).

The literature on the effects of collicular lesions on eye movements is rather contradictory. Part of the problem has to do with ensuring that the lesion is restricted to the superior colliculus. There is general agreement that monkeys with bilateral collicular lesions tend to have a fixed gaze and those with unilateral lesions are more attentive to stimuli in the hemifield on the same side as the lesion. However, there seem to be no deficits in optokinetic or vestibular nystagmus after unilateral or bilateral lesions (Pasik, Pasik and Bender, 1969; Anderson and Symmes, 1969). Wurtz and Goldberg (1972b) produced general and localized lesions in the superior colliculus of the monkey and found that the only obvious effect was an increase in the latency of voluntary saccades towards visual targets in the movement field of the ablated area. Even this deficit declined over a period of time. A similar transient defect has been reported in humans with lesions in the superior colliculus (Heywood and Ratcliff, 1975). This suggests that mechanisms in other regions of the nervous system can substitute for the functions of the colliculus. Mohler and Wurtz (1977) found that monkeys could learn to make saccades into the part of the visual field corresponding to a local lesion of the visual cortex. However, if lesions were made in retinotopically corresponding regions of both the colliculus and the visual cortex, the animal could no longer detect visual objects or make saccades towards them. It looks as though the superior colliculus and the visual cortex may substitute for each other in the control of voluntary saccades.

Other evidence does not support this conclusion. Schiller (1977) found colliculectomy produced a persistent reduction in the number, size, and velocity of voluntary saccades. Furthermore, Kurtz and Butter (1980) reported a persistent inability of colliculectomized monkeys to direct their gaze voluntarily to peripheral targets. These latter investigators suggested that Wurtz and Goldberg, and Mohler and Wurtz failed to find persisting defects because their lesions were confined to the superficial layers of the colliculus. Schiller found that eye movements evoked by stimulation of the visual cortex were absent when the corresponding region of the colliculus had been destroyed, even after a six-month recovery period. On the other hand, Schiller, True and Conway (1979) found that combined destruction of the colliculus and frontal eye fields of the monkey permanently abolished the ability of the monkey to make organized saccades. This suggests that the visual cortex plus the frontal eye fields are required to compensate for a loss of the superior colliculus.

Robinson (1972) proposed that visual stimulation excites a retinotopically defined region of the superficial layer of the superior colliculus and that impulses then spread downwards into the intermediate and deeper layers through the columnar organization which was described by Michael (1970). From these deeper layers efferents descend to the pontine reticular formation, where messages concerning the oculocentric location of a saccadic target are converted into neural discharges with a frequency and duration appropriate to the required saccade (see Grantyn *et al.*, 1977, Grantyn *et al.*, 1979, and Raybourn and Keller, 1977, for details about colliculo-oculomotor pathways). Mohler and Wurtz (1976) pointed out that there are problems with this scheme. Superficial cells fire after, rather than before, deeper cells, and deeper cells often fire when superficial cells are quiet. They proposed a 'visual-movement convergence model' in which visually evoked activity moves down from superficial layers and combines in intermediate layers with movement-related activity moving upwards from deeper layers. This scheme would account for the properties of those cells in the intermediate layer, described in Section 5.2.2, which fire only when an eye movement is made towards a visual target.

Further evidence against any close coupling between cells in the superficial layer and eye-movement related cells in the intermediate layer is provided by the results of a study by Mays and Sparks (1980b). They trained monkeys to make two successive saccades towards two briefly presented targets. Thus, by the time the monkey came to make the second saccade, it had to rely on its memory of where the second target had been. This would involve registration of the initial oculocentric position of this target and an adjustment for the change in eye position due to the first saccade. By recording from cells in the intermediate layers of the colliculus they found cells which they called **quasi-visual cells**. Each of these cells responded when a target appeared in its receptive field whether or not a saccade was made towards that target. But, such a cell responded just before the eyes made the second of two saccades to a previously seen target. The cell which fired under these circumstances was not the one whose receptive field contained the originally seen second target, but the one whose receptive field contained the visually unstimulated oculocentric location defined with respect to the position of the eyes after they had completed the first saccade. In other words, the activity of quasi-visual cells reflected the **eye-position error**, defined as the difference between actual and desired eye positions, rather than **retinal error**, defined as the difference between the original oculocentric position of the target and the fovea. Furthermore, these cells seem to be capable of holding this information in store until a saccade occurs or is cancelled.

Cells in the deep layers of the cat's superior colliculus were found to respond when the eye was passively moved through a given position (Abrahams and Rose, 1975b). More recently, such cells have also been found in the superficial layers of the cat's superior colliculus (Donaldson and Long, 1980). Cells in the superficial layers were found to respond both to passive eye movement and to visual stimuli. The two types of stimuli, when presented simultaneously or in close succession, were found to be facilitative for some cells and inhibitory for others.

5.5.4 Head movements and the superior colliculus

Eye movements to visual targets off to one side are often accompanied by head movements (see Section 9.2.6), so that somewhere in the nervous system there must be a centre which is concerned with the joint programming of eye and head movements. Several lines of evidence suggest that the superior colliculus may be that centre. Other evidence suggests that this is not the case. I shall briefly review this conflicting evidence.

There is anatomical and electrophysiological evidence of projections from the superior colliculus, along the tectospinal tract, to motor centres in the cervical region of the spinal cord. Most of these projections involve at least one relay through synapses in brainstem nuclei (Anderson *et al.*, 1971). Furthermore, afferent fibres from muscle spindles in muscles of the neck project back to the same cells in the superior colliculus from which the tectospinal tract originates (Abrahams and Rose, 1975a).

Stimulation of the superior colliculus in rabbits and cats has been found to produce combined movements of the eyes and head (Hess *et al.* 1946; Syka and

Radil-Weiss, 1971; Schaefer, 1972). More recently, Stryker and Schiller (1975) stimulated single cells in the superior colliculus of the monkey. The animal was free to move its head in a horizontal plane and stimulation was confined to collicular regions associated with horizontal eye movements. The eye movement evoked by stimulation of a particular cell was the same each time stimulation occurred. Furthermore, these responses had a definite latency and threshold and were independent of the position of the eyes. Stimulation of a given cell sometimes also evoked a head movement, but not in a regular fashion. A head movement was more probable in a given direction when the eyes were already displaced in that same direction. Head movement responses had a very variable latency and threshold. Thus, in the monkey, the regular and simple relationship between stimulation of a particular cell and movement of the eyes is not found for movement of the head.

Robinson and Jarvis (1974) could find no neurones in the superior colliculus of the alert monkey that discharged in relation to head movements. Even when the amplitude of a saccade was reduced because it occurred with a head movement, the eye-movement related discharge from the colliculus was the same as it would have been had the saccade occurred at full amplitude. The adjustment of saccadic amplitude was probably accomplished at some other site. More recently, Straschill and Schick (1977) found cells in the superior colliculus of the cat which discharged either before or during contralateral head movements. Only a few cells behaved in this way, and most of them discharged only when the head movement occurred in the presence of a visual target. Passive movements of the head did not affect these cells, which rules out proprioceptive or vestibular inputs as the cause of their activity. Electrical stimulation of the cells induced contralateral head movements, with amplitudes which depended on the duration of stimulation.

There is evidence that the coupling between eye movements and head movements is more regularly programmed in the superior colliculi of animals, like the rabbit and cat, which try to maintain their eyes in a primary position to a greater extent than do primates (Harris, Blakemore and Donaghy, 1980). The coordination of eye and head movements is discussed further in Section 9.2.6.

5.5.5 *Non-visual inputs to the superior colliculus*

Cells in the deeper layers of the superior colliculus of the cat have been found to respond to auditory stimuli. Most of these cells respond to visual and to auditory stimuli and show the same selectivity for the direction of stimulus motion in the two modalities (Wickelgren, 1971). Similar cells have been found in the monkey (Cynader and Berman, 1972; Updyke, 1974). In the case of bimodally responsive cells in the cat, the positions of visual stimuli to which the cells were responsive corresponded to the positions of auditory stimuli which were effective in causing the cells to respond (Gordon, 1973). This spatial correspondence between visual and auditory inputs cannot be maintained when the animal holds its eyes in eccentric gaze, because the visual inputs are retinotopically organized and the auditory inputs are organized with respect to headcentric space. This is not a problem in cats, which prefer to turn the head towards the object of interest rather than hold the eyes in eccentric gaze for any length of time. It would be a problem in primates, and especially in humans, who often looks at things with an eccentric angle of gaze. *There seem to be no studies on intermodal mapping in the primate superior colliculus.*

Other cells in the deep layers of the superior colliculus of the cat respond to tactile stimulation of the whiskers, face, and paws. Many of these cells also respond to visual stimuli, and some respond to tactile, visual and auditory stimuli. Although the receptive fields of these multimodal cells are large, there is apparently some correspondence between the position of the cell's visual receptive field and the part of the visual field which normally contains the image of that part of the whiskers, face, and paw which evokes a response in that cell (Jassik-Gerschenfeld, 1965; Stein and Arigbede, 1972; Stein, Magalhães-Castro and Kruger, 1976).

It seems most likely that these multimodal cells are part of a mechanism for directing the gaze towards objects of interest which may be detected by vision, by audition, or by touch (see Section 11.7).

The basic structures and functions of the eye movement system have now been reviewed, and we are in a position to look in more detail at the types of eye movement that were briefly outlined in Section 5.1 and listed in Table 5.1.

Chapter 6

Eye Movements II Behavioural Studies

The basic types of eye movements and their functional significance were described in Section 5.1 and are listed in Table 5.1. I shall now review behavioural studies which have been done on each type of eye movement except vergence movements, which are not relevant to the subject of this book. The order outlined in Table 5.1 will be followed except that eye movements made in response to vestibular stimulation will be dealt with in Chapter 9.

6.1 Optokinetic nystagmus

6.1.1 General features

When the visual scene is moved in one direction across the field of view, there results an involuntary, conjugate, rhythmical movement of the eyes. Slow movements in pursuit of the moving scene alternate with quick return saccades. The alternating sequence of movements is known as **optokinetic nystagmus,** or OKN. This response may be observed easily in a person who is looking at the passing scene out of the window of a moving train. Helmholtz (1962, Vol. III, p. 247) noticed it under these circumstances, and Bárány (1908) called it *Eisenbahn* (railway) nystagmus. It is one of the most basic and phylogenetically primitive orienting responses and occurs in mammalian and many sub-mammalian species. Functionally, the slow phase keeps a stable image on the retina during movements of the whole field of view and the quick phase returns the eyes ready for another sweep. OKN in an appropriate direction may be elicited by a visual display moving in any direction across the visual field. Torsional movements of the eyes are elicited by a rotation of the visual scene in the frontal plane about the fixation point, although continued rotation of the scene produces a more or less steady deviation of the eyes, with the quick phase not much in evidence. OKN persists for about 20 sec after the stimulus has been replaced by a blank or dark field. This is known as **optokinetic afternystagmus,** or simply OKAN.

Nystagmic responses of the eyes are usually recorded by **electrooculography** (EOG). This method of recording eye movements depends on a naturally occurring difference in electrical potential, or dipole, between the cornea and retina. Movements of this dipole, which accompany eye movements, produce changes in potential which are picked up by electrodes placed round the orbit of the eye. This method introduces no extraneous visual stimuli, allows free movements of the head, and gives a record of the amplitude, speed, and frequency of nystagmus in the vertical and horizontal directions simultaneously. The method has a limited bandwidth; it suffers from low-frequency drift and introduces distortions when recording high-frequency responses. A procedure which overcomes this limitation is one in which the reflection of an infrared source from the iris–scleral boundary is detected by a photodiode attached to a spectacle frame (Young and Sheena, 1975).

The subject whose nystagmus is being recorded is usually placed at the centre of a hollow vertical cylinder which rotates around him and the inside of which is covered with vertical black and white stripes. It is important that the moving pattern fills the visual field, or, if it does not, that the stationary part of the field does not contain inhomogeneities upon which the subject can anchor his gaze. The area of the moving pattern need subtend only a few degrees, although larger fields are more effective, and peripheral stimuli need to be larger than centrally placed stimuli in order to elicit a response (Körner and Schiller, 1972). The **direction** of any nystagmus is designated according to the direction of the fast phase, and other important parameters are **frequency**, **amplitude**, the **velocity** of the slow phase, and the **symmetry** of the eye movement in relation to the primary position of the eyes. The **gain** of OKN is the ratio of the slow-phase velocity to the stimulus velocity.

Optokinetic nystagmus is difficult to inhibit consciously unless a stationary object is present upon which one may fixate. However, the response may be inhibited, at least for short periods, even without a fixation point, if one focuses beyond the moving stimuli and makes no effort to watch them (Honrubia *et al.*, 1968). Several investigators have confirmed that nystagmus with the fast phase downwards is particularly difficult to inhibit (Hulls and Stockwell, 1974). Guedry and Benson (1971) suggested that this is because, as a terrestrial animal moves about, stimuli evoking downbeat nystagmus occur less often than stimuli evoking nystagmus in other directions, and that the most frequently occurring responses are the ones which animals learn to inhibit. Optokinetic afternystagmus is more pronounced for downbeat nystagmus than for upbeat nystagmus (Matsuo *et al.*, 1979), and this may be related to the fact that upbeat nystagmus is more easily inhibited than downbeat nystagmus. Visual pathologies of long standing, such as amblyopia and cataract, reduce the gain of OKN, particularly that driven by temporalward and upward target motion. This asymmetry of gain is present in both the affected and unaffected eye, which suggested to Schor and Levi (1980) that it is due to incomplete development of binocular vision.

Some people can voluntarily produce nystagmic eye movements in the absence of the normal stimuli, and some are able to voluntarily produce nystagmus in one eye while the other remains still (Blair *et al.*, 1967). OKN cannot be induced during sleep (Gardner and Weitzman, 1967).

OKN occurs in neonates and even in premature neonates, although the gain of the slow phase is not as high as in adults (Kiff and Lepard, 1966; Kremenitzer *et al.*, 1979). It is present in cats and rabbits which have been reared in the dark for the first few months of life, although the velocity of the slow phase is not well matched to that of the stimulus (Vital-Durand and Jeannerod, 1974a,b; Collewijn, 1977; Kossut *et al.*, 1978). Vital-Durand and Jeannerod exposed dark-reared kittens to a visual environment moving in one direction for one hour a day. After 20 weeks, these kittens exhibited a more adequate OKN to stripes moving in the direction to which they had been exposed than in other directions. This selective effect of experience was probably due to the asymmetrical development of direction-sensitive motion detectors in the visual cortex (see Tretter *et al.*, 1975, for evidence on this point). However, optokinetic nystagmus is also present in cats which have had the visual cortex totally removed (Smith, 1937). Ter Braak and Van Vliet (1963) found it in decorticate monkeys, although Pasik and Pasik (1964) could not confirm this. Ter Braak *et al.* (1971) observed it in what they claimed was a completely destriate human patient.

The study of OKN has practical uses. For instance, it has been used in the objective determination of acuity in infants (Smith and Bojar, 1938); as a test for hysterical blindness (Snell, 1939); and as a test of eye dominance (Enoksson, 1963).

6.1.2 *The stimulus control of optokinetic nystagmus*

Although the slow phase of OKN is normally induced by visual stimuli moving continuously in one direction, this is not a necessary stimulus. If there are two superimposed stripe patterns moving in opposite directions, the subject may change the direction of his OKN by shifting his attention from one pattern to the other (ter Braak and Buis, 1970). OKN may be induced by a target that moves in discrete jumps, rather than continuously, as long as time intervals between target positions are short enough for the sequence to be seen as a continuous movement. As the time interval between successive target positions is increased, it is just at the point where the impression of continuous movement breaks down that the pursuit movement becomes interrupted by saccades in the same direction as the pursuit (Morgan and Turnbull, 1978). Ter Braak (1972) displayed two stationary out-of-phase stripe patterns in alternation so that the subject sometimes saw apparent movement one way, sometimes the other way, and sometimes as to-and-fro motion. To some extent the subject could voluntarily control which movement he saw and, while the apparent movement occurred consistently in one direction, the OKN usually occurred with its slow phase in the same direction. There are good reasons for believing that real and apparent movements activate the motion detectors of the visual system in the same way, so that there is nothing very surprising about that fact that OKN is induced by apparent movement.

The fact that a perceptibly smooth movement of the visual scene is required for OKN to occur suggests that motion information is monitored continuously by the oculomotor system, rather than discontinuously as was suggested in an earlier **sampled data model** of the pursuit system (Young and Stark, 1963). Further evidence in favour of a continuous monitoring of visual motion signals was reviewed by Brodkey and Stark (1968). For instance, if the quick phases in the record of an OKN movement are eliminated, the slow phases join into a smooth continuously varying trace rather than into a set of discontinuous segments that would result from a system which sampled the input intermittently.

It is possible to combine binocularly two random-element patterns in a stereoscope in such a way that a pattern is perceived which is not present in the retinal image of either eye (Julesz, 1971). A centrally synthesized image of this kind is called a **cyclopean image**. Steinbach and Anstis (1977) and Fox *et al.* (1978) reported that OKN is induced by a moving cyclopean image. This would seem to demonstrate that retinal motion is not required for OKN. However, one must be careful in drawing this conclusion, because the OKN which Fox *et al.* reported was of larger amplitude than normal OKN, and this suggests that it was a voluntary pursuit movement rather than involuntary OKN (see Section 6.2.4).

The velocity of the slow-phase OKN is close to the speed of the stimulus up to a velocity of about 30°/sec, that is, the gain is close to 1. At faster speeds, the eyes lag more and more behind the stimulus until, at a velocity of about 100°/sec, the regular pattern of the optokinetic movement breaks down completely (Honrubia *et al.*, 1968). Monkeys are able to pursue moving stimuli adequately at two to three times the velocity that human beings can manage (B. Cohen *et al.*, 1977). Higher velocities of pursuit require larger visual displays (Grüttner, 1939) and any pursuit movements are less adequate when the moving stimulus is small (Körner and Schiller, 1972). It is not the overall area of the display that counts, but rather its angular subtense in the direction of the eye movement (see Dichgans, 1977). Furthermore, as we shall see later, the gain and amplitude of the pursuit phase of OKN in animals with foveae also depend on whether the image of the moving stimulus falls on the fovea or on the periphery of the retina, and on whether or not the animal is attending to the stimulus. The maximum velocity of the slow phase of OKN is far less than, and is therefore not limited by, the velocity at which the eyes can move, which is over 800°/sec. Nor is it limited by the ability of the eyes to detect the separate elements of the moving stimulus. In other words, the flicker fusion frequency of the visual system is well above the frequency at which stripes of an optokinetic stimulus pass the eye when they are moving at the break velocity for OKN. The velocity of the slow phase of OKN is presumably limited by the rate at which the visual system can detect and respond to the visual consequences of a mismatch between stimulus velocity and eye velocity.

The simplest view of the signal controlling the pursuit phase of OKN is that it is the mismatch between eye velocity and target velocity. This **velocity error signal** is equal to the velocity of the movement of the retinal image, which is picked up by motion detectors in the visual system. The **gain** of a velocity-matching servosystem is defined as the pursuit velocity divided by the target velocity, which, for perfect stabilization of the image, equals unity. Such a system may be represented by the diagram in Figure 6.1. In normal operation, it is said to be a **closed-loop** system under negative feedback control. When visual feedback is disconnected from the input, it is an **open-loop** system. Opening a control loop of any servo-control system reveals the characteristics of the mechanism in its simplest form. There are two ways to open the control loop of the pursuit system of the eye. One is to present the moving stimulus to an eye which is mechanically prevented from moving and to record the movements of the other eye which is kept in the dark. In this way the stimulus is isolated from the response. The method works because, according to Hering's law, a stimulus to one eye is sufficient to drive both eyes. The other way to open the control loop of the eye's

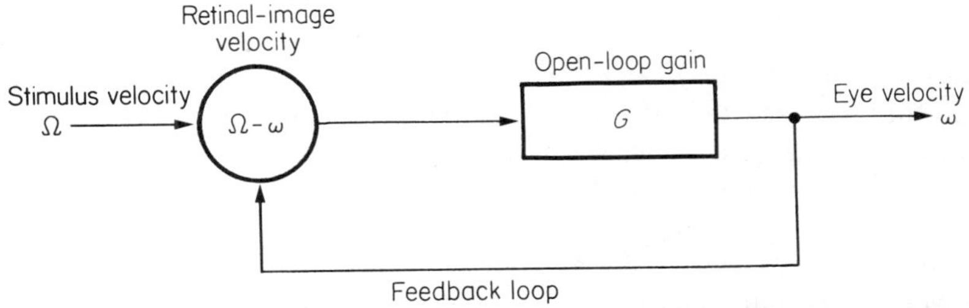

Figure 6.1 A stimulus moving with velocity Ω is tracked by the eye at velocity ω. The velocity of the retinal image ($\Omega - \omega$) generates a signal which modifies ω in the direction of making $\Omega - \omega$ zero. If G is the open-loop gain of the system, $G/(G+1)$ is the closed-loop gain

pursuit system is to stabilize the image on the retina, so that the movement of the eye has no effect on the movement of the image. Although both these procedures open the control loop, they have different consequences. In both procedures the image can be made to move over the retina at a constant velocity, no matter what the moving eye does. In the first case, the eye receiving the stimulus does not move, so that unchanging non-zero velocity and position errors are constantly present. In the second case, the eyes and the image move together, so that the velocity-error signal is zero unless the degree of image stabilization is manipulated.

The first procedure for opening the control loop of the pursuit system was applied to monkeys by Körner and Schiller (1972). When the stimulus velocity was between 2 and 5°/sec, the velocity of the pursuit phase of the unconstrained eye increased in steps from one sweep to the next until it was about 10 times that of the stimulus. The ratio of the eye velocity to the target velocity under these circumstances is the **open-loop gain**, or G. At lower and higher stimulus velocities, G was less than 10, and for velocities of over 30–40°/sec it remained constant at a value of just below 1. Therefore the open-loop gain is about 10, at best. The gain of a closed-loop servo system with unit negative feedback is given by $G/(G+1)$. It can be seen from this expression that the gain of the intact system approaches the desired value of 1 as G becomes larger, and we thus have an explanation for the high open-loop gain.

In afoveate animals the optokinetic response seems to be controlled by a subcortical velocity-matching servo system of the type just described. In this mechanism a leftward tracking optokinetic response is controlled by inputs to the right eye and a rightward tracking response by inputs to the left eye. In other words, each eye when tested alone responds only to stimuli moving in a temporonasal direction. Foveate animals supplement this simple subcortical mechanism with a cortical mechanism and do not show this directional preponderance unless the visual cortex has been destroyed. Details about these relationships were described in Section 5.4.5.

Several lines of evidence suggest that the visual system has **sensory integrators** each of which accumulates, or integrates, the retinal-velocity signal for one direction of motion. There are presumably at least four such integrators, one for each of the main directions of movement. The fact that optokinetic afternystagmus is weakest after upbeat nystagmus suggests that there is least velocity integration for upbeat nystagmus (Matsuo *et al.,* 1979). The possible sites of the integrators are discussed in Section 9.2.2. Velocity integrators would account for the way in which the eyes accelerate when exposed to a constant error signal in the first open-loop condition just described. Furthermore, integrators account for the fact that eyes continue their optokinetic response after the stimulus is removed, that is, they account for optokinetic afternystagmus. Afternystagmus decays after a short time, and the integrators presumably leak their charge at the same rate as the decay rate. Recent evidence suggests that the decay is linear, rather than exponential, as had been assumed previously (Collewijn *et al.,* 1980). Raphan *et al.* (1977) were able to estimate the time constant for charging integrators in the monkey by measuring the duration of exposure to a moving stimulus required for the eyes to reach a steady-state optokinetic response. As one would expect, this duration turns out to be the same as the duration of exposure required to produce a maximum period of afternystagmus. This reinforces the idea that the same integrators are responsible for maximizing pursuit velocities during OKN and for producing the afternystagmus. In the monkey, the afternystagmus discharges in darkness over a period of 20–60 sec, doing so more rapidly after higher stimulus velocities than after slow stimulus velocities. The afternystagmus is inhibited if the animal is allowed to see stationary objects and does not resume when the animal is returned to darkness after vision has been continued for a few seconds (B. Cohen *et al.*, 1977). This suggests that integrators are rapidly discharged by stationary visual inputs.

A coherent functional explanation of these features of the optokinetic system may be provided as follows. OKN is most commonly evoked by motion of the visual scene produced by movement of the head. Signals also arise in the vestibular system which indicate the velocity of head motion and trigger off vestibular nystagmus. These vestibular signals are fast-acting and initiate compensatory eye movements before any visually-evoked eye movements have had time to come into action. If the head continues to move at a steady velocity, the vestibular system ceases to respond, but by this time the optokinetic system will have taken over to generate adequate pursuit movements of the eyes. When the head comes to rest, the inertia of the fluids in the semicircular canals causes a deflection of the cupula in a direction opposite to that in which it was deflected during head acceleration. This, together with neuronal aftereffects of vestibular stimulation, induces postrotatory vestibular nystagmus, which, if unchecked, would destabilize the retinal image. However, the stored 'charge' in the integrator of the optokinetic system induces optokinetic afternystagmus. The postrotatory nystagmus and the optokinetic afternystagmus are opposite in sign and therefore cancel out. Recent evidence shows that the two aftereffects decay at the same rate, probably because they both depend on the same integrator mechanism (Raphan *et al.*, 1979; Collewijn *et al.*, 1980). They thus continue to cancel until the system has returned to its resting state.

The OKN and vestibular systems thus cooperate to stabilize the retinal image when the head moves, which suggests that they share certain neural components. Recent physiological investigations have revealed that one of these shared components is situated in the vestibular nuclei. Recordings from single cells in these nuclei in the monkey have shown that certain cells discharge at a rate which is correlated with the velocity of the slow phase of optokinetic afternystagmus and cease firing when the afternystagmus is inhibited by stationary stimuli (Waespe and Henn, 1977b). Furthermore, these same units respond transiently when the animal's head is rotated (Waespe and Henn, 1977a). Robinson (1977) constructed a computer simulation on the basis of Waespe and Henn's data (see Figure 6.2) and showed that it is reasonable to suppose that the optokinetic signals arising from retinal-image velocity and the vestibular signals add in a

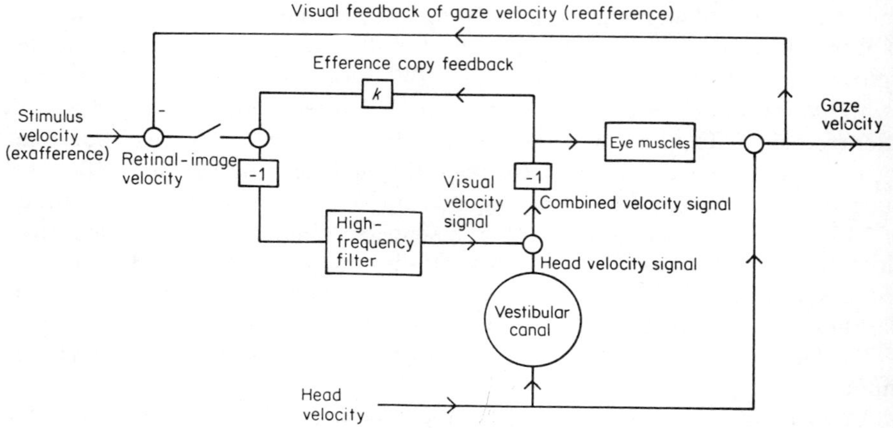

Figure 6.2 A simplified version of the model proposed by Robinson (1977) to account for the way visual and vestibular signals cooperate in producing eye movements which stabilize the retinal image

linear fashion to produce the signal which controls the velocity of pursuit movements of the eyes when the head is rotated with eyes open. See Section 5.4.5 for a discussion of the neurology of this process, and see Henn *et al.* (1980) and Robinson (1981) for a review of other models of the eye-movement system.

Further evidence for a common control system for the slow phases of OKN and vestibular nystagmus is provided by studies on the effects of labyrinthectomy. B. Cohen *et al.* (1973) showed that bilateral labyrinthectomy in the monkey abolishes optokinetic afternystagmus. Human beings with loss of the vestibular system since childhood also lack optokinetic afternystagmus and show a measurable defect in the ability to pursue stripes moving at high velocities (Zee, Yee and Robinson, 1976). In primates, effects of labyrinthectomy on the adequacy of OKN may be largely compensated for by voluntary pursuit. In order to reveal the capabilities of the involuntary system after labyrinthectomy, it is necessary to study an animal lacking the voluntary pursuit system. To this end, Collewijn (1976) performed a bilateral labyrinthectomy on the rabbit and found, in addition to the loss of optokinetic afternystagmus, a severe reduction in the velocity of the OKN slow phase. The maximum slow-phase velocity was a few degrees per second. In other words, the animal behaved as if it had only a very inadequate direct signal from the retinal motion detectors, unaided by an integrator. Collewijn (1972) developed an analogue model of the rabbit's optokinetic system. The retinal motion detectors in the rabbit have only four preferred directions of movement, and these approximately coincide with the lines of action of the four rectus muscles. Oyster *et al.* (1972) have shown electrophysiologically that the outputs of these retinal motion detectors have a strong relationship to the velocity of optokinetic nystagmus. It is most unlikely that man has a system as simple as this.

Optokinetic afternystagmus may be obtained after inspection of moving stimuli confined to the fovea (Brandt *et al.*, 1974; Muratore and Zee, 1979). This suggests that the velocity integrator operates for eye movements driven by foveal stimuli as well as by eye movements driven by peripheral stimuli. It has yet to be decided whether involuntary as well as voluntary pursuit eye movements feed through this same integrator.

There is evidence of secondary integrators which counteract the action of the primary integrators. The presence of the secondary integrators was inferred from the frequent occurrence of a second phase of afternystagmus following the first, but with opposite sign (Brandt *et al.*, 1974). The primary and secondary integrators are apparently distinct, because when monkeys are exposed to a period of repetitive whole-field optokinetic stimulation, the duration of the first phase of afternystagmus decreases, whereas the period of the second phase increases (Waespe and Henn, 1978b). Furthermore, whereas vision inhibits optokinetic afternystagmus, it has been found to enhance the second phase of afternystagmus (Waespe *et al.*, 1978).

The point at which the quick phase of optokinetic nystagmus is initiated is not related in any systematic fashion to the position of the eyes in the orbit, which suggests that the quick phase is not triggered by proprioceptors in the extraocular muscles. There is evidence of an inherent regular periodicity in the occurrence of the quick phase, but a person also has some voluntary control over the frequency of nystagmic beats (Cheng and Outerbridge, 1974b).

6.1.3 Optokinetic torsion

Brecher (1934) observed a subject's conjuctival blood vessels through a telescope and saw torsional motion of the eye when the gaze was fixed on the centre of a sectored disc rotating in a frontal plane. The eye rotated about the visual axis by a more or less constant amount as the stimulus motion continued, with an occasional saccadic jerk in the opposite direction. This motion of the eyes was confirmed by Howard and Templeton (1964b). The mean amplitude of the movement is normally between 1 and 3°, although Crone (1975) reported a maximum amplitude of 6°. Kertesz and Jones (1969) found the amplitude of the movement to be a function of the angular velocity of the sectored disc and the number of sectors drawn on the disc.

Optokinetically induced eye torsion is normally regarded as the torsional equivalent of vertical and horizontal optokinetic nystagmus. The fact that, in torsional nystagmus, the deviation of the eyes is more evident than an alternation of slow and quick phases may be partly due to the fact that investigators recorded the torsional position of the eye by intermittently photographing the iris or episcleral blood vessels, rather than by taking a continuous record of the motion. Petrov and Zenkin (1973) made a continuous record of the motion of a light attached to a contact lens while the subject tilted his head from side to side. The torsion which they recorded was induced by the tilt of the head, but they did show that when the subject could see the visual scene the deviation was accompanied by a nystagmic oscillation and when he was in the dark the deviation movement was very much smoother. It looks as though sight of the rotating retinal image induced an optokinetic nystagmic motion which was superimposed on the response induced by head tilt (see Figure 6.3). Petrov and Zenkin's records show that the velocity of the pursuit phases of the nystagmic motion was more or less matched to the velocity of the head motion. *What is needed is a continuous record of the movements of the eyes in response to a sectored rotating disc.*

I suggest that, in torsion induced by head tilt or by stimulus rotation, the nystagmic component is more important for vision than the steady deviation, because the nystagmic component allows the image to be momentarily stabilized. In man, the deviation component produced by tilting the head is grossly inadequate to keep the normally vertical meridian of the eye upright. This

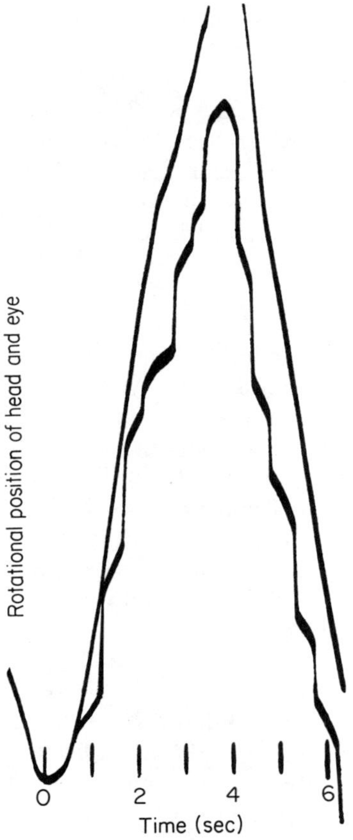

Figure 6.3 The smooth curve is a record of the head tilting towards the left shoulder through 50° and back. The discontinuous curve is a record of rotation of the eye around the visual axis during the head movement (Reprinted with permission from A. P. Petrov and G. M. Zenkin, *Vision Research*, **13**. Copyright 1973, Pergamon Press, Ltd.)

response may be an evolutionary relic in man. The deviation component in torsion induced by a rotating visual display may occur because the retinal image rarely rotates unless the head tilts, so that when the image does rotate, it is taken as a signal that the head is tilting and the appropriate response is evoked. We shall see in Section 9.4.4 that rotation of the visual scene does indeed give rise to the impression that the body is being tilted.

Eye torsion is also induced by inspection of a stationary tilted line or tilted grating. Howard and Templeton (1964b) could find no evidence of such **optostatic torsion** for a line which subtended only 10° to the eye. They used a line of this length because they were interested in the possible contribution of torsion to the tilt aftereffect in which lines of this length are used. Crone (1975) used a 25° disc-shaped grating at various angles of tilt and recorded a maximum optostatic torsion of 1° when the lines were at a fixed angle of tilt of about 15° from the vertical and no measurable effect for inclinations of the lines near the horizontal (see Figure 6.4). Goodenough, Sigman, Oltman, Rosso and Mertz (1979) used a tilted visual framework which subtended 28°, and also recorded torsion of up to 1° of amplitude. It is not clear whether the torsional response to these large displays was due to their size or to the fact that they stimulated the periphery of the retina. *Further experiments are needed to clarify this point.*

The eyes respond with oppositely directed torsion when presented with stereotargets which have a cyclodisparity. This **cyclofusional** response of the eyes serves to keep corresponding parts of disparate images on corresponding retinal meridia (Kertesz and Jones, 1970). It has been found to be reduced when the angular size of the stimulus is reduced (Kertesz and Sullivan, 1978), but here again, it is not clear whether it is size, rather than the eccentricity of the stimulus, which is important.

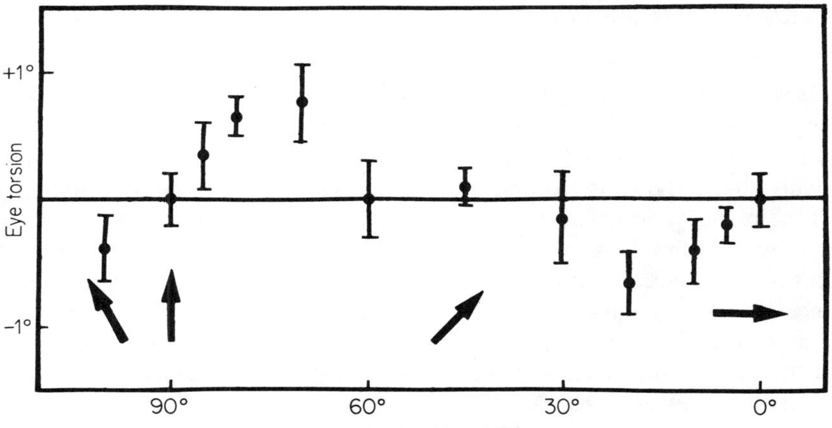

Figure 6.4 Optostatic eye torsion induced in one subject by inspection of a grating tilted at various angles in the frontal plane. The arrows represent the orientation of the stimulus. Bars are standard deviations (From Crone, 1975)

6.2 Voluntary pursuit eye movements

6.2.1 *Introduction*

The innate, largely involuntary, nystagmic eye movements which have just been described occur in most vertebrates, both in species with foveate eyes and in species with afoveate eyes. Human beings, and other animals with foveate eyes, possess the additional ability to maintain fixation on a moving object which interests them. Such animals can switch their gaze at will from an object moving in one direction to one moving in a different direction (Murphy *et al.,* 1975). Humans can also make a saccade to an object seen moving against a complex stationary background in the far periphery of the retina and, with almost no delay after the saccade, initiate an adequate pursuit movement (Atkin, 1969). Animals with foveate eyes therefore decide which of several motion signals is to drive the pursuit movement of the eyes. As we shall see, the motion signal need not even be visual; a person is able to track smoothly with his eyes a non-visual object, such as his own hand in the dark, or a sound source, or even an imagined moving object. Such abilities go far beyond the level of an involuntary reflex. Normally, a person is unable to move his eyes smoothly over a stationary scene. Such movements would be maladaptive, because they would blur the retinal image; the eyes naturally move in a series of saccades when looking over a stationary scene, thus affording a series of clearly seen stationary images.

It was at one time thought that voluntary visual pursuit is absent in the first two weeks of life (McGinnis, 1930), but it is now generally agreed that new-born infants are capable of fixating and pursuing a moving object, although not at such high speeds as adults (see Barten *et al.,* 1971; Kremenitzer *et al.,* 1979).

6.2.2 *Temporal characteristics of pursuit eye movements*

When the target motion occurs at an unpredictable time, a pursuit movement occurs with a latency of about 125 msec (D. A. Robinson, 1965). If the luminance contrast of the moving target is reduced, the pursuit latency increases (Haegerstrom-Portnoy and Brown, 1979). If the velocity of the target motion is suddenly changed during the first 90 msec of this latency period, the velocity of the subsequent eye movement may reflect this change (Barmack, 1970). In other words, up to a few tens of milliseconds before the eye movement occurs, the system is able to accept information relevant to how fast the eyes should move.

Once started, the pursuit movement accelerates to the velocity of the target over a period of 100 msec or more, depending on the velocity of the target. Smooth pursuit movements have been recorded in response to a target movement as slow as 5′/sec (Murphy *et al.*, 1975). As the velocity of the target increases, the pursuit velocity of the eyes increases by an equal amount, although, even at low velocities, the smooth movement is sometimes interrupted by small saccades, usually in the direction of the pursuit but sometimes in the other direction (Mackeben, *et al.*, 1980). At a target speed of about 30°/sec, the pursuit movement begins to lag consistently behind the target and an occasional saccade is made to recentre the target on the fovea. The maximum velocity of a normal voluntary pursuit movement is about 40°/sec. This is somewhat less than the maximum velocity of the slow phase of OKN induced by motion of the whole visual field. Pursuit velocity increases as the luminance contrast of the target is increased above threshold, but soon reaches its maximum value, after which further increases in contrast have no effect. The motion aftereffect has been found to be influenced in a similar way by changes in contrast (Keck, *et al.*, 1976). This similar dependence of pursuit eye movements and motion aftereffects on contrast presumably stems from the fact that they both depend on oculocentric motion detectors (Haegerstrom-Portnoy and Brown, 1979).

In any linear system, that is, a system in which the response to a complex input can be predicted by adding the effects of the components, there is a predictable relationship between the phase lag and the gain of the system to sinusoidal inputs. However, the phase lag of the visual pursuit system is rarely as large as one would predict from the assumption that it is a linear system, especially when the target motion is regular. There are two ways to think of this discrepancy. One is to continue to believe that the system is basically linear, but that its performance is supplemented by the ability of the system to predict the future position of the target from the current input. The other way is to discard the notion that the system is linear. St. Cyr and Fender (1969b, c) adopted the second view on the basis of evidence which they reviewed. They argued that, because the system is non-linear in the first place, the basis for inferring that its performance is supplemented by a predictor mechanism collapses. St. Cyr and Fender argued that the short phase lags that have been observed with regular sinusoidal target motion are not due to the operation of a predictor, but merely to the shorter time required by the system to process repetitive information. However, this theory cannot account for the existence of pursuit movements with zero lag, or for eye movements which anticipate the future position of a moving target; that is, the theory can account for short phase lags, but it cannot account for zero lags or for phase leads. If the motion of the target is predictable, because it is repetitive or because it is under the manual control of the subject, the latency of response to a change in target motion can, with practice, become vanishingly small. Michael and Melvill Jones (1966) found that the extent to which subjects could reduce the mean phase lag of their tracking eye movements varied with the degree of regularity, or predictability, in the motion of the visual target. We shall see later that several investigators have reported phase leads in pursuit eye movements (see also Kowler and Steinman, 1979a, b).

I suggest that St. Cyr and Fender are correct in questioning the existence of a predictor mechanism in which the future position of the target is computed on a moment-to-moment basis, but I think they overlook the possibility of another kind of predictor mechanism. In this second kind of predictor mechanism, eye movements are preprogrammed according to a learned pattern, and the system merely checks now and again to see that it is keeping in phase with the stimulus. In other words, when the system operates in this way it does not compute the future position of the target on a moment-to-moment basis, but neglects the current input and runs on its 'memory'. It improves its gain because it no longer processes the input, not because it processes it in a predictive mode, or because it has shortened the time required for processing.

A target that is being pursued may move horizontally and vertically at the same time, and one might ask whether each component of the eye movement is driven only by the appropriate component of the stimulus motion or whether horizontal eye movements are affected by vertically moving stimuli. Goodwin and Fender (1973a) used a target motion consisting of a sinusoid in the vertical direction and a low-frequency random motion in the horizontal direction. They found high cross-correlations between corresponding components of target and eye movements and concluded that the oculomotor system has independent vertical and horizontal channels with a relatively small degree of mutual cross-talk. Furthermore, subjects were aware of the fact that the vertical motion was sinusoidal and the horizontal motion was random. These same results were also found when the components of the target motion were in the two oblique directions (Goodwin and Fender, 1973b), which suggests that all orthogonal eye movements are independent, not merely vertical and horizontal movements.

One may also ask what happens when a person visually pursues a smoothly moving target that suddenly jumps and then continues moving smoothly. One possibility is that a saccade is superimposed on the pursuit movement to produce a saccade with higher than normal velocity for its amplitude, and another possibility is that the pursuit movement is suspended while the saccade occurs. Jürgens and Becker (1975) investigated this question by superimposing jumps of various sizes into the movement of a target that oscillated smoothly back and forth. The jump was sometimes with and sometimes against the smooth movement. The duration of each saccade made in response to the stimulus jump was used to indicate the intended amplitude of this saccade in accord with the relationship depicted in Figure 6.5. If saccadic and pursuit movements are superimposed so that their velocities add, then 'with' saccades and 'against' saccades that have the same duration (intended amplitude) should have different manifest amplitudes, because one will be quicker than the other by twice the velocity of the superimposed pursuit movement. If the two types of movement are not made simultaneously, the amplitudes of the 'with' and 'against' saccades should be equal, and this is what Jürgens and Becker found. The pursuit system thus closes down while a saccade is in progress, or at least pursuit signals do not gain access to the oculomotor nerves.

The results of this experiment also showed that the duration of the saccade was adjusted to compensate for the motion of the target during the time taken to make the saccade. The resulting saccadic duration was still related to the saccadic amplitude in the way characteristic of normal saccades. It looks as though the saccade was adjusted to the anticipated position of the target at the end of the saccade, which is to say that it took into account the movement that the pursuit system would have made had it not been shut off. This compensation suggests that the system extrapolates the future position of the moving spot on the basis of velocity information. Even if this is so, one need not assume that this extrapolation is done on a moment-to-moment basis, because the target in the Jürgens and Becker experiment moved back and forth at a steady velocity, which would allow the observer to achieve a reasonable degree of compensation by simply adding or subtracting an interval from the saccadic duration. But some degree of compensation of saccadic duration can be achieved even though not based on an estimate of velocity. We shall see in Section 6.4.3 that the saccadic system is able to use information relevant to the amplitude of a saccade up to about 80 msec before the onset of the saccade. In the case of a saccadic latency of 250 msec, this is 170 msec after the target step which invokes the saccade. Some degree of adjustment of saccadic duration can therefore be achieved if the saccade is directed to the position of the target as it is seen 80 msec before the saccade occurs. Heywood and Churcher (1981) recently demonstrated that responses to a step-ramp stimulus are best explained by a late position-sampling process of this kind rather than by the more complicated process in which velocity information is extrapolated.

I have already described how optokinetic nystagmus persists for a while after the stimulus evoking it has been removed. Muratore and Zee (1979) reported a similar persisting nystagmic eye movement following a period in which a subject pursued a point of light moving to and fro in a repetitive ramp-quick-return (sawtooth) pattern. They called this aftereffect the **pursuit afternystagmus**. But it may have been the quick return eye movements rather than the slow pursuit movements which generated this aftereffect. *It should be possible to determine the relative contributions of the slow and quick phases of a to-and-fro eye movement to the aftereffect by using a stimulus moving slowly to and fro in one condition and a stimulus moving rapidly to and fro in the other condition.* Moreover, optokinetic nystagmus may be elicited by a small stimulus, such as was used by Muratore and Zee (Dichgans, 1977), so that the pursuit movement that they studied may have involved the involuntary optokinetic system as well as the voluntary pursuit system, and the aftereffect may have been generated by the involuntary component. The fact that voluntary pursuit does not reach such high velocities as does involuntary pursuit suggests that the voluntary system does not operate through the velocity integrators. If the voluntary system does not operate through the velocity integrators, one would not expect it to be accompanied by aftereffects.

6.2.3 The stimulus for pursuit eye movements

It has been generally believed that the stimulus generating and controlling a voluntary pursuit eye movement is a visually detected motion of the retinal image, which signifies a mismatch between eye velocity and target velocity. It was not believed that a **position mismatch** between the target and the point of fixation, was an effective stimulus. This view was founded on the results of an experiment conducted by Rashbass (1961). He reported that when a visual target is stepped to one side and then moved slowly in the opposite direction, the eyes make the first movement in response to the slow return movement rather than to the initial step, even though the pursuit movement does not bring the image of the target onto the fovea. The simplest view is that pursuit movements aim to reduce the velocity of the image of the target, the **retinal slip velocity**, to zero. This may be the signal that drives involuntary pursuit, such as that involved in optokinetic nystagmus, but recent evidence demonstrates that this simple view must be rejected in the case of voluntary pursuit of particular targets. Pola and Wyatt (1980) have reported that pursuit movements were initiated in response to a series of to-and-fro sudden displacements, across the fovea, of an otherwise stabilized retinal image of a visual target. The displacements produced no actual or apparent smooth movement at the rates of alternation that were used. Furthermore, it was found that the pursuit velocity of the eyes was a linear function of target offset from the fovea, at least within the range 0.25 – 3°. It has also been shown that, once a pursuit movement has been initiated, small sudden displacements of the target are corrected by smooth pursuit movements, and saccadic movements are invoked only if the target gets too far from the fovea (Eckmiller and Mackeben, 1978). Other evidence, which will be reviewed below, indicates that pursuit eye movements may also be initiated and maintained by the perceived motion of a visual target which does not actually move over the retina or become displaced from the fovea. It may be concluded that voluntary pursuit eye movements may by driven by a retinal velocity mismatch signal, by a position mismatch signal, and by a perceived velocity signal.

Some idea of the adequacy of a pursuit eye movement driven by a velocity signal without a positional mismatch signal is obtained by a procedure devised by Williams and Fender (1979). The visual stimulus consisted of an endless moving string of closely packed luminous dots on a dark background. Saccades had no effect on the appearance of this display because the dots were all the same and only about 2′ of visual angle apart. It was found that there was typically a velocity mismatch of ± 10–20% between the target motion and eye motion for target velocities of 2°/sec. Furthermore, the velocity of one eye was not the same as that of the other eye. Although comparable figures were not presented for the pursuit of a single dot, for which both position and velocity cues would have been present, it was assumed that, at this velocity, the match would have been nearly perfect. It would seem that a velocity signal unaided by positional information does not allow the eyes to pursue accurately. However, these data must be interpreted with caution, because the endless display of dots removed not only the position cue but also the cue for binocular fusion. This lack of binocular fusion would have two effects. In the first place, the eyes would execute hunting vergence movements in search of a stimulus that would lock vergence, and secondly, the apparent distance of the display would become indeterminate. These two effects would certainly disturb the accuracy of pursuit and would also account for why the eyes were not in step.

Kowler and Steinman (1979a, b) found that when a subject could anticipate a sudden or gradual displacement of a visual target, the eyes made involuntary slow movements in the direction of the expected target motion. They also found that when a slowly moving pursued target suddenly stepped one way or the other at an anticipated time, the velocity of pursuit speeded up or slowed down in anticipation of the sudden displacement. These anticipatory eye movements were not found before saccades made in response to expected non-visual signals, which shows that they are caused by the anticipation of target displacement and not by the anticipation of a saccade. It is clear that they are not evoked by the positional retinal error, because they occurred before the target displacement and were not affected by the size of the displacement. They are presumably associated with the mechanism that allows the phase lag of pursuit to be improved with familiar target motions.

Although pursuit eye movements are normally made in response to a moving visual object, there are several circumstances under which people are able to make slow smooth eye movements in the absence of moving visual stimuli. These circumstances are listed below.

Pursuit of apparent visual movement Apparent visual movement may induce involuntary OKN, but it may also trigger voluntary pursuit movements. An impressive demonstration has been described by Lamontagne (1973). The subject tracked a small continuous light as it moved over a row of illuminated dots which flashed on and off in unison. As this was done, the whole row of dots jumped into apparent smooth movement in the same direction as the moving light. With a little practice, the subject was able to continue smoothly pursuing the apparent movement after the continuous light had been removed. The subject was, in effect, tracking his own eye movement, because the velocity of the apparent movement was determined by the velocity of the eye movement. A similar phenomenon was reported by Ward and Morgan (1978). They found that when subjects inspected a patch of dynamic visual noise they could smoothly pursue an illusory patch corresponding to the foveal projection.

Pursuit of induced movement A stationary visual object surrounded by a frame which is moving in a certain direction appears to move in the opposite direction. This is known as **induced motion**. Wyatt and Pola (1979) asked subjects to pursue a small visual target in dark surroundings. The motion of the target was controlled in such a way that it oscillated sinusoidally with an amplitude of a few degrees across the fovea of the subject. The velocity and amplitude of this movement with respect to the fovea were independent of the movement of the subject's eyes. In other words, the tracking was in open-loop mode. In a second condition, the target was surrounded by a frame which oscillated across the retina in couterphase to the motion of the target. The target appeared to be moving faster and further when the frame was present, and the amplitude and velocity of the pursuit eye movements were increased accordingly. In other words, the open-loop gain of the pursuit movement (maximum eye velocity divided by maximum retinal target velocity) was increased by an increase in apparent target velocity, with retinal velocity held constant.

Pursuit of pendular motion Westheimer and Conover (1954) reported that a person is able to move his eyes sinusoidally back and forth between two points, 30° apart, either in the dark or in the light. However, this was possible only when the eyes moved at about 3 Hz. This is much faster than ordinary pursuit movements, which tend to become irregular above about 0.5 Hz (Noorden and Mackenson, 1962). Deckert (1964) had subjects follow a pendulum bob moving through 36° at 0.5 Hz, and reported that, after they had closed their eyes, they could continue to move them to and fro in a reasonably smooth fashion. They could not pursue an imaginary pendulum with the eyes open and things in view.

Pursuit of afterimages Several investigators have reported that people can smoothly pursue an afterimage in the dark (Mack and Bachant, 1969; Heywood and Churcher, 1971). If the afterimage is on the fovea, the eyes tend to move sinusoidally from side to side with a frequency of 0.5–1.0 Hz. With an eccentric afterimage, the pursuit movement occurs in the direction of eccentricity with an occasional return saccade. Each pursuit phase remains essentially uninterrupted by small saccades as long as the afterimage is no more than 2° away from the fovea (Kommerell and Klein, 1971). When the afterimage is more eccentric than 2°, the pursuit phase is a mixture of smooth pursuits and small saccades, although the subject can improve the smoothness of the eye movements by making an effort (Puckett and Steinman, 1969; Heywood and Churcher, 1972). Steinbach and Pearce (1972) found that movements in pursuit of a centrally placed afterimage are initiated by slow spontaneous drifts of the eyes rather than by microsaccades. What is probably happening here is that, when a slow drift occurs, the afterimage moves as if it were an object moving in space with the eyes. The subject pursues this apparently moving object, but as he does so its apparent velocity increases. Eventually, the eyes settle down to a steady pursuit velocity. The gain is always 1, because the velocity of the afterimage in relation to a stationary background remains matched to the eye velocity at whatever steady state is selected.

This evidence suggests that retinal motion signals are not required to initiate pursuit movements. The essential stimulus would seem to be perceived headcentric motion of the stimulus. In the case of a foveal afterimage, the 'stimulus' for perceived egocentric movement is apparently the command signal to the eye muscles coupled with the absence of retinal motion.

Pursuit of proprioceptive and tactile stimuli Some people are able to track smoothly one of their own arms with the eyes as they move it back and forth in complete darkness (Gregory, 1958; Noorden and Mackenson, 1962). Steinbach (1969) found that only about one person in three could do this successfully, and Glenny and Heywood (1979) found performance was degraded if the subject knew there was an opaque screen between the eyes and the moving hand. Steinbach and Held (1968) reported that eye movements in pursuit of a small visible target attached to the hand were smooth when the arm motion was self-produced and interrupted by corrective saccades when the arm was moved passively by another person (see also Angel and Garland, 1972). Steinbach (1969) also showed that subjects could adequately track a flashing light attached to the moving hand, particularly the self-moved hand, even though they could not track the intermittent light adequately when it was moved without being attached to the hand. He concluded that visual pursuit is aided by proprioceptive inputs from a passively moved hand, and that it is helped even more by efference associated with an actively-moved hand. The extra advantage of self-produced movement is probably due to the fact that efferent signals provide information about the future motion of the hand and therefore enable the pursuit to be executed with reduced lag between target motion and eye motion. The advantage of self-produced movement over passive movement was evident in Steinbach's experiment because he used a rate of target motion (varying between 1.5–2.5 Hz with a variable amplitude of up to 20°) which is near the upper limit for adequate visual tracking. At rates of sinusiodal target motion below about 0.3 Hz, it has been shown that pursuit eye movements are equally good whether or not the stimulus motion is predictable (Michael and Melvill Jones, 1966; St. Cyr and Fender, 1969b), so that it is not surprising that at these low rates the advantage of active over passive movement has been found to be absent (Jordan, 1970). Not everyone is agreed that there is an advantage for self-produced movement over passive movement of the hand at high frequencies of oscillation (Mather and Lackner, 1980a).

Gauthier and Hofferer (1976) removed afferent signals from a moving arm by applying an ischemic block at the level of the triceps. The subject was no longer able to pursue adequately his moving arm in the dark, showing that efference alone is insufficient to instigate adequate pursuit.

Steinbach (1969) showed also that the beneficial effect of self-produced motion of the hand was available to subjects when the flashing light was attached to one end of a pivoted lever which was moved from the other end by the subject. This shows that it is not the motor efference to the arm which triggers the pursuit eye movements, but rather the information that efference provides, in the context of visual information, about the movement of the target being pursued. Angel *et al.* (1973) had subjects track a visual target held in the hand but allowed them to see only a videotaped image of the target after a delay of between about 0.2 and 1.5 sec. The pursuit movement was smoother with the self-produced movement than it was with similar externally driven target movements, as long as the delay was not greater than about 0.7 sec. This suggests that there is a short-term memory for arm movements which is available to the pursuit system. Lackner (1975) induced illusory movements of the arm by stimulating muscle spindles with a mechanical vibrator. The subject was able to track smoothly the apparent movement of the unseen arm, which he reported experiencing. People have also been shown to be capable of making reasonably adequate eye movements in the dark in pursuit of a tactile stimulus moved to and fro over the surface of the arm (Lackner and Evanoff, 1977).

Pursuit of hidden parts of visual objects Steinbach (1976) demonstrated that people are able to visually pursue hidden parts of a moving object. In one experiment, a square was moved to and fro behind a screen so that only the top half of the square could be seen. Subjects were able to pursue adequately a lower, hidden, corner of the square. In effect, the subject was pursuing the visible part of the square that was imaged outside the fovea. Winterson and Steinman (1978) confirmed that subjects are able to pursue a visible target that remains imaged outside the fovea. Steinbach also found that subjects could adequately pursue an object as it moved from side to side behind a narrow slit so that the only visible parts of the object were seen moving up and down in the slit. In a third experiment, subjects adequately pursued the unseen centre of a rolling wheel when all that could be seen was the cycloid path of a light attached to the rim of the wheel.

Thus, a person can visually pursue a 'perceived' object which does not itself provide a visual motion signal. Furthermore, Mack *et al.* (1979) reported that a pursuit eye movement may be invoked by a movement that is undetected. They showed that an isolated visual stimulus, moving so slowly that the motion is undetected, is pursued by the eyes just as adequately as when the stimulus is flanked by stationary stimuli which make its motion detectable. The voluntary pursuit system is clearly a very flexible one which may be driven by a variety of stimuli and which involves processes at several levels in the nervous system. While attempts to model, or simulate the more primitive and involuntary optokinetic response of the eyes have met with some success, attempts to model the voluntary pursuit system are probably premature (Young, 1977).

6.2.4 *Types of pursuit eye movement*

Pursuit eye movements, both those that occur in OKN and so-called voluntary pursuit movements, may be classified according to whether they are controlled by (a) a mechanism which seeks to match the eye velocity to the velocity of the stimulus, and which is under the control of an error signal from the velocity of the retinal image, (b) a velocity-matching mechanism under the control of perceived velocity, or (c) a mechanism that seeks to stabilize the image of a given feature of the stimulus on the fovea. Another way to classify pursuit eye movements is according to whether or not they are under voluntary control. There are no agreed criteria for deciding whether a function is or is not voluntary. The question could be decided on several grounds.

(a) Whether the movement can be evoked at will in the absence of an external stimulus.
(b) Whether the movement can be inhibited at will in the presence of the normal stimulus.
(c) Whether the active attention of the subject is required.
(d) Whether the movement requires the activity of centres in the cerebral cortex.
(e) Whether the subject is aware of the movement.

The characteristics of velocity-matching and foveation pursuit are listed below, together with evidence upon which the distinction is founded. At the same time comments are made bearing on the various aspects of the voluntary–involuntary distinction.

Pursuit in foveate and afoveate eye Velocity-matching eye movements under the control of an error signal from the velocity of the retinal image are very primitive, being present in animals which lack a well-developed fovea, such as many fish and reptiles. These pursuit movements stabilize the retinal image as the visual scene moves relative to the stationary animal, or as the animal moves relative to the scene, and are probably involuntary on all five criteria. Foveate animals, such as primates, also possess the involuntary velocity-matching system, but in such animals, this system is dominated by a more recently evolved and cortically controlled system which is invoked by the perceived velocity of the target. The evidence for this statement was presented in Section 6.2.3.

Foveate animals also possess a foveation mechanism which is invoked when the image of the object which the animal is attempting to fixate has wandered from the fovea. When the eyes fixate a stationary target, they often drift off-centre in a more or less random fashion and make small corrective saccades and sometimes corrective slow drifts (see Section 6.3.1). Subjects are normally not aware of these microsaccades and drifts, although subjects may control their frequency by voluntary effort, and they do not occur unless the person is attending to the visual target. If the target begins to move, the same foveation mechanism operates, except that the corrective drifts and microsaccades tend to be in the direction of target motion. The foveation pursuit system therefore involves a mechanism which has something in common with that which stabilizes the gaze on a stationary target. If the gaze wanders too far from a moving target, the more or less smooth pursuit movement is interrupted by a large corrective saccade.

Pursuit of central and peripheral stimuli Hood (1967) has argued that optokinetic nystagmus evoked by moving stimuli confined to the periphery of the retina is controlled by an involuntary velocity-matching mechanism, and that nystagmus evoked by stimuli confined to the central region of the retina is controlled by a voluntary mechanism which seeks to maintain the object of interest on the fovea.

Hood reported that if the direction of motion of a full-field visual stimulus is reversed, the slow phase of the human optokinetic response is interrupted by a saccade in the direction of the changed stimulus motion. This suggests that full-field OKN is dominated by the foveation mechanism, which aims to keep a given feature of the stimulus on the fovea. On the other hand, Hood reported that when the optokinetic stimulus is confined to the visual periphery, a reversal of stimulus motion induces a change in the direction of the ongoing pursuit phase. This suggests that peripherally driven OKN is dominated by a velocity-matching mechanism. Hood also reported that the velocity-matching mechanism is evident in patients with central scotomata. When vestibular nystagmus is induced by rotating the subject in the dark, and the direction of rotation is reversed, the initial response is a reversal of the direction of a slow phase. Vestibular nystagmus, which is known to be involuntary and under the control of a velocity signal from the semicircular canals, thus behaves like peripherally driven OKN. This suggests that velocity-matching pursuit is the more primitive and involuntary system. Hood also reported that patients with central scotomata achieve higher velocities of pursuit than do normal people. He concluded that the pursuit phase of involuntary, velocity-matching nystagmus may be driven to a velocity of about 90°/sec, whereas the maximum velocity of voluntary foveation pursuit is about 40°/sec.

Hood's distinction between centrally driven and peripherally driven optokinetic nystagmus has not been accepted by all other investigators. Dubois and Collewijn (1979) could not detect the difference between centrally and peripherally driven nystagmus which Hood had reported. Furthermore, several investigators have found that the velocity of the slow phase of optokinetic nystagmus is reduced for subjects with an artificial central scotoma, produced by blanking out the central region of the moving stimulus, rather than increased, as Hood had reported in patients with a natural scotoma (Körner and Schiller, 1972; Cheng and Outerbridge, 1975; Dubois and Collewijn, 1979).

It seems to me that all this evidence, on both sides of the argument, is problematical. The artificial scotoma which were used by Körner and Schiller and others must have had an oculocentrically fixed boundary to the blanked-out region, and it may have been this stationary boundary which caused the reduction in the velocity of the slow phase of nystagmus, rather than the fact that the central region was blanked out. This interpretation is supported by the fact that these investigators found that a moving stimulus confined to the visual periphery did not evoke nystagmus unless the subject attended to the peripheral stimulus. The subject's behaviour was clearly dominated by the oculocentrically fixed boundary of the central region. The subjects in Hood's study had natural central scotomata, and had probably learned not to regard the boundary of such a scotoma as an object to be fixated. In any case, the boundary of a natural scotoma is probably less sharp than the boundary of a central blanked-out region. On the other hand, the stimulus which Hood used was probably inappropriate also, because, as Dubois and Collewijn pointed out, he used stripes at intervals of 15° of visual angle, so that with stimulation confined to a small central region, the effectiveness of the stimulus would in any case be reduced. *The question of the relative contributions of velocity-matching pursuit and foveal pursuit to a person's ability to pursue foveal stimuli is clearly one which requires more study.*

Evidence of a different kind suggests that visual pursuit, whether driven by foveal or by peripheral stimuli, may be under the control of either an involuntary, velocity-matching mechanism, or a voluntary mechanism driven by perceived velocity, or a mechanism which attempts to keep a particular stimulus on a particular part of the eye. For instance, the evidence reviewed in the last section indicates that people are able to voluntarily pursue peripheral targets when they can infer the location of the fixated part of the moving pattern. Moreover, the ability to voluntarily pursue an off-centre afterimage also suggests that the peripheral velocity-matching system may be brought under voluntary control, or at least supplemented by a voluntary system of pursuit. Which mechanism predominates at any time may have more to do with the attitude of the subject than whether the stimulus is central or peripheral. It may be that people are more likely to invoke voluntary control when they view a central stimulus than when they view a peripheral stimulus.

Velocity integration There is evidence that Hood was correct in concluding that the involuntary velocity-matching pursuit system is more capable of following stimuli moving at a higher velocity than is the foveation pursuit system, unaided by a velocity-matching system. It was shown in Section 6.1.2 that the higher velocities reached by the velocity-matching system are due to the operation of a velocity integrator associated with the vestibular system. Labyrinthectomy in afoveate animals abolishes the integrator and leaves only a slower direct visual control of OKN. Foveate animals can to some extent compensate for the loss of the integrator by using the foveation mechanism and possibly by a velocity-matching mechanism driven by perceived velocity which may not feed through the integrators.

However, the voluntary pursuit systems cannot achieve such high velocities, because one of them (the foveation mechanism) requires positional information to be coded into a velocity signal and the other requires the higher-order processing of perceived velocity. These are inevitably more indirect, and therefore more time-consuming processes than that involved in a more direct mechanism that operates on the retinal slip velocity.

Passive and active optokinetic nystagmus Although OKN in humans is normally dominated by the foveation pursuit system, which is to a large extent under voluntary control, people can relax their voluntary control, leaving the OKN response under involuntary control (Rademaker and ter Braak, 1948). The voluntary type of OKN manifested when the subject consciously attends to the moving stimuli has been called **active** or **'look' nystagmus**, and OKN manifested when attention is relaxed has been called **passive** or **'stare' nystagmus**. Passive nystagmus has all the properties of involuntary velocity-matching nystagmus and active nystagmus has all the properties of nystagmus dominated by the foveation system which are listed above (Honrubia *et al.*, 1968; Hood and Leech, 1974).

The pursuit phase of passive, stare nystagmus has a mean duration of about 0.3 sec. When this is supplemented by voluntarily following, some of the quick phases tend to be suppressed, leading to an increase in the mean duration of the pursuit phase (Cheng and Outerbridge, 1974a, b). The extent to which quick phases are suppressed in voluntary pursuit is under voluntary control (Puckett and Steinman, 1969).

It may be concluded from all this evidence that, at least in the higher mammals, a primitive mechanism of visual pursuit, driven by a mechanism which automatically nulls retinal-image motion, is supplemented by a pursuit mechanism under the control of higher centres which is capable of being controlled by a variety of stimuli and which is highly adaptable.

6.3 The control of fixation

6.3.1 Eye movements during fixation

The most sensitive method of recording eye movements is to record the movements of a beam of light reflected off a mirror attached to a tightly fitting contact lens. When a person fixates his gaze on a small object, this procedure reveals three types of eye movement. There is a high-frequency noise-like tremor with a median amplitude of less than 1′ of arc and a frequency of up to 100 Hz. Superimposed on this tremor are **microsaccades**, with a mean amplitude of about 5′, which occur between one to three times each second, and slow **drifts** of the eye, with a mean velocity of about 4.0′/sec, which cause the gaze to wander with a standard deviation of between 2 and 5′ from the mean fixation position (Steinman *et al.*, 1973). During the pause between microsaccades, the image is held still with a standard deviation of only 0.25′ (Barlow, 1952).

Bengi and Thomas (1968) measured the high-frequency tremor of the eye when its inertia was increased by weights attached to a contact lens. They found two peaks in the velocity spectral density function, one at 40 Hz, which they identified with the inertia of the eyeball, and the other at 70 Hz, which they identified with fluctuations in muscle tension.

The stability of the eye may be indicated by the area of the region on the retina within which the image of a small point occurs a given proportion of the time. Sansbury *et al.* (1973) measured the stability of fixation with a foveal target and with targets which formed concentric rings round the fovea at various angles of eccentricity up to 15° of arc. Stability declined with increasing eccentricity of the target contours, which is what one would expect, because visual acuity declines with increasing eccentricity. However, eye stability declined with increasing eccentricity at a much slower rate than did acuity, which suggests that eye stability depends partly on a non-visual mechanism. This argument may not be as conclusive as it seems, because *the eye may be better at judging the centre of a symmetrical visual display than one would predict from performance on an acuity task*. This is a topic for an experiment.

Some attempt has been made to measure the separate contributions of the visual and the non-visual mechanisms of fixation stability. Timberlake *et al.* (1972) suggested that the first saccade which brings a newly exposed target to the fovea is triggered only by the visual-error signal. They found that the precision of this first saccade varied with the eccentricity of the target, in the same way as visual acuity varies with eccentricity. They argued that, once the fixation position has been found, the ability to hold it depends, not only on visual-error signals, but also on non-visual signals such as those from receptors in the extraocular muscles or signals associated with efference. The effectiveness of the non-visual signals in controlling eye stability is indicated by the ability of a person to hold a steady gaze when there is nothing in view. Skavenski (1971) found that the gaze could be kept to within about 2° of arc of a target position for more than two minutes after the light was turned off. It seems to me improbable that this relatively crude system could contribute to the much higher stability achieved with a visible target, even when the target is 15° into the periphery, but until we know how precisely a person is able to judge the centre of a symmetrical visual display, this issue will remain unsettled.

6.3.2 Theories of eye movements during fixation

Theories that have been proposed to account for eye movements during fixation will be listed and discussed under the following headings.

Prevention of image fading There are several procedures which enable an experimenter to arrest the motion of the retinal image during fixation (see Howard and Templeton, 1966, p. 47). When the image is **stopped** by one of these procedures all visual objects fade from view after a few seconds, because the receptor system adapts to the constant stimulation. Intermittent fading may be noticed when a pattern is fixated for some time, even without stabilizing the image in any other way. Clearly, some eye movements are required for the maintenance of vision, and it may be argued that fixation movements evolved as a mechanism for preventing image-fading. There is no evidence for or against the notion that the fine tremor has evolved specifically to serve this purpose. It might be worthwhile to look at the tremor in different species and see whether it is related to acuity. I think the most reasonable view is that tremor is simply the irreducible noise of the oculomotor system, which incidentally serves to prevent rapid adaptation of the image. If microsaccades or small drifting movements served to prevent fading, one might expect to find their frequency and amplitude increase when the 'stopped' image fades. However, Cornsweet (1956) found no such relationship, and concluded that image-fading is not the stimulus for these movements. In any case, animals rarely hold the gaze on one object for more than a few seconds, so that image-fading would not occur under normal circumstances, even if the eyes did not move while fixating.

Improvement of visual acuity Marshall and Talbot (1942) suggested that visual acuity is improved by the constant motion of the image over several receptors. The tremor of the eyes has a mean amplitude equal to the distance across three receptors, and it is possible that this movement improves acuity, but there is no evidence that it does. There is evidence that, if anything, acuity is degraded by microsaccades (see Section 2.3.4) and nobody has suggested how the slow drifts of the eyes might improve acuity.

I shall assume in what follows that tremor is the random irreducible noise of the oculomotor system and confine my attention to microsaccades and slow drifts. These are certainly not random or irreducible because, as we shall see, their directions and amplitudes are influenced by attention, learning, and other factors. Microsaccades and slow drifts seem to serve a purpose in addition to any role they may play in the prevention of image-fading. There are at least three theories about what that purpose might be.

Correction for drift of fixation Cornsweet (1956) argued that the slow drifts of the eyes are due to an inherent instability in the oculomotor system which causes the gaze to wander off the fixation target and that the resulting positional error triggers a microsaccade which brings the gaze onto target again. In a similar vein, Krauskopf *et al.* (1960) argued that corrective microsaccades of unequal amplitude in the two eyes correct for errors of vergence. This theory was brought into question when Rashbass (1961) claimed that a displacement of a visual target of less than 0.25° does not trigger corrective eye movements. This range of uncorrected fixation error is known as the **oculomotor dead zone**. The concept of a dead zone has been challenged by Wyman and Steinman (1973a), who found that subjects can consistently make corrective eye movements in response to a target displacement as small as 3.4. However, the theory that microsaccades correct for errors of fixation runs into other difficulties. Nachmias (1959) found that drifts of gaze along certain meridia were often corrected by return drifts rather than by saccades, and Boyce (1967) found that only about 30% of microsaccades were in a direction which corrected for fixation errors. St. Cyr and Fender (1969a) also found that drifts often correct for fixation errors, and for vergence errors. They suggested that the earlier conflicting evidence was due to the use of inappropriate measures of what constitutes a corrective movement. The measure they recommended was fixation error after an eye movement minus the error before the movement. St. Cyr and Fender also drew attention to intersubject variability in the pattern of fixation eye movements. This variability suggests that differential experience may play a part. These authors also pointed out that the shape and disposition of the target can have an effect, at least with some subjects. It thus seems that drifts as well as microsaccades both cause and correct errors of fixation.

Scanning strategies involved in pattern vision It has often been suggested that the small movements of the eyes are related to a strategy for extracting pattern information (see, for example, Gaarder, 1967). The gross movements of the eyes certainly allow the gaze to centre on important details of a visual scene, but nobody has produced any good evidence that the pattern of small eye movements is linked in any significant way to pattern recognition. A more reasonable conclusion was drawn by Barlow (1952, p. 306), when he stated:

> The eye might either work like a camera, taking a succession of snapshots, or the image might be moved deliberately over the retina in order to facilitate the extraction of certain types of information. The evidence that has been produced indicates that the first alternative is

correct, in which case the fixation pauses must be regarded as the units from which visual sensations are built up in the same way that a cinema film is built up of individual frames.

Small-scale pursuit and search strategies Perhaps the most satisfactory theory of the microsaccades and drifts which occur in fixation is that they are miniature versions of voluntary saccades and pursuit movements. Steinman *et al.* (1973) have made out a convincing case for this theory. The velocity–amplitude characteristics of microsaccades are similar to those of larger saccades (Zuber and Stark, 1965). Furthermore, the rate at which microsaccades occur is similar to the rate at which larger saccades occur (Cunitz and Steinman, 1969). The interval between them is about 300 msec, which is about as long as it takes to recognize a visual object. Haddad and Steinman (1973) found that subjects could make microsaccades voluntarily, and Gaarder (1966) and Steinman *et al.* (1967) found that subjects could voluntarily inhibit microsaccades during fixation, an ability which was not due merely to a relaxation of accommodation (Steinman *et al.,* 1969). But it has been reported that a person's ability to execute fine visuomotor skills, such as threading a needle, is not affected by whether or not he inhibits microsaccades (Winterson and Collewijn, 1976). Microsaccades probably reflect our normal habits of shifting our gaze over an object of interest once we have inspected it for about half a second. We do not normally look at one small location for longer than this, so that the microsaccades and drifts seen in studies of prolonged fixation may be mere laboratory curiosities rather than behaviourally significant events.

When a person attempts to hold a primary position of gaze in the dark, the mean value of the average deviation of the position of gaze from the initial position increases approximately monotonically with time in the dark, at least during the first few seconds (Cornsweet, 1956). This relationship suggests that eye position wanders in a random walk when there is no visual target to provide error feedback. Matin *et al.* (1970) confirmed that the drift from the primary position of gaze in the dark is substantially a random walk, although they did find a small but significant negative correlation between the direction of eye movements and the direction in which the gaze had wandered from its initial position. If there is a mechanism which corrects for the random drift of gaze, then the mean drift of the gaze from the primary position should eventually stabilize at some limiting value, rather than go on increasing in the way predicted by a purely random walk. In line with this suggestion, Skavenski and Steinman (1970) found that, after about 100 sec in the dark, the mean error of fixation stabilizes at about 2°. It is as if the system which controls eye position in the dark is inherently noisy but is capable of detecting when the eye has wandered too far off target and then initiating corrective movements.

Non-randomness in the drift of the gaze is much more in evidence when a person attempts to maintain an eccentric position of gaze. Skavenski and Steinman (1970) asked subjects to hold fixation in the position of a visual target, 10° from the primary position of gaze, after the target had been extinguished. The gaze did not show any overall tendency to return back to the primary position until the subjects were asked to voluntarily inhibit saccadic eye movements, whereupon the gaze drifted towards the primary position. This suggests that saccades correct for a mean drift towards the primary position. This suggestion has been confirmed by the results of an experiment conducted by Becker and Klein (1973). They asked subjects to maintain eccentric positions of gaze in the dark for angles of eccentricity up to 70°. The records of eye movements showed that the slow drifts of the eyes tended to carry the gaze back towards the primary position, but that saccades of up to 10° in amplitude corrected for this drift. The subjects were unaware of making these corrective saccades unless they were larger than about 5°. The velocity of the slow drift towards the primary position increased with the eccentricity of the gaze, up to a maximum velocity of about 10°/sec.

It is not clear where the signals originate that inform the system that the gaze has wandered from its original position. It may be concluded that they do not originate in the conjunctiva, because Skavenski and Steinman found that anaesthetizing the conjunctiva had little or no effect on the stability of gaze in the dark. They could arise from proprioceptors in the extraocular muscles or from the sense of muscular effort (corollary discharge) associated with holding the gaze in its original position.

6.4 Saccades

6.4.1 General features

Saccades are the most rapid movements of the eyes. In their most primitive form they terminate each slow pursuit phase of optokinetic and vestibular nystagmus, and in this form they are involuntary and occur in both foveate and afoveate eyes. In foveate animals, saccades also serve to bring the image of an object of interest onto the fovea. Human beings adopt characteristic scanning strategies when inspecting natural scenes and objects. Periods of steady fixation, which tend to be concentrated on regions of high information content, are interspersed with saccades (Yarbus, 1967; Rayner, 1978). When there is a simple stimulus on both sides of an initial fixation point, saccades tend to occur towards the stimulus nearest to the fovea, or, if the stimuli are equidistant, to the one which is flickering or moving (Findlay, 1980a).

Saccades may reach speeds of 700°/sec or more, and even a 40° saccade lasts only about 100 msec, much too short a time for there to be any guidance of the movement by error feedback from the retina or from proprioceptors. Saccades must be preprogrammed, or ballistic. The neural centres for generating saccades already existed as part of the nystagmic system before foveate eyes evolved. When the fovea did evolve, it required a saccadic system for rapidly and accurately centring the gaze on an object of interest. Changes of gaze must be rapid, so as to shorten the period of time during which vision is degraded by image motion. It is reasonable to suppose that the already existing saccadic pulse-generating mechanism in the brainstem would be harnessed to serve the requirements of voluntary shifts of gaze in species with foveate eyes. All the evidence suggests that there is indeed only one set of saccadic pulse generators which programme bursts of excitatory and inhibitory impulses in the oculomotor nerves for all saccades, both voluntary and involuntary. What varies is the signal which provokes the pulse-generating mechanism into action. The signals for voluntary saccades have precedence over those which trigger involuntary saccades. This fact was demonstrated by Judge (1973) in the following way. A subject was instructed to fixate a visual target which appeared to the left while his eyes were executing optokinetic nystagmus with the fast phase to the right. It was found that, during the 100 msec latency period before the voluntary saccade occurred, there were no rapid phases of nystagmus. It was concluded that, when the stimulus for a voluntary saccade is received, the optokinetic saccadic system is inhibited and remains inhibited for the remaining 100 msec that it takes for the voluntary saccadic movement to occur.

The most convincing evidence that all saccades are controlled by the same brainstem mechanism is that all saccades have the same characteristics (Guitton and Mandl, 1980a). A saccade is initiated by a high-frequency burst of activity in neurones innervating the phasic muscle fibres of the agonists, closely followed by a burst of activity in neurones innervating the tonic muscle fibres of the agonists, and accompanied by complete cancellation of activity in neurones innervating the antagonists. The magnitude of the sudden acceleration of the eye which this so-called **pulse** of neural activity produces is determined by the mechanical properties of the eye and is therefore the same for saccades of all amplitudes. The peak velocity of the eyes depends on the **pulse width** (how long the pulse is maintained) and the **pulse height** (number of muscle fibres recruited and the maximum firing frequency attained). Each of these factors increases with saccadic amplitude up to an amplitude of about 15°, and it can be seen from Figure 6.5 that up to this amplitude peak velocity is an approximately linear function of saccadic amplitude. For saccades larger than 15° most of the muscle fibres are recruited and nerve fibres are firing at their maximum frequency, so that beyond this point the peak velocity no longer increases at the same rate with increasing saccadic amplitude.

Bahill, Clark and Stark (1975a) have argued that for saccades under 15° the pulse of neural activity which drives the saccade maintains its full strength to about only halfway through the saccade, that is, up to the time that the peak velocity occurs. They argued that the earlier evidence of Robinson (1964) and Luschei and Fuchs (1972) that the pulse continued at full strength to near the end of the saccade was based on a misleading analysis of pulse width, in which the exponential decay of the pulse in the second half of the saccade was overlooked. As the pulse begins to decay, the innervation of the antagonists begins to return and, finally, the pattern of innervation to phasic and tonic muscle fibres in the agonist and antagonist settles down to that required to hold the eye in its new position.

The change in innervation brought about by a saccade is referred to as the **step**

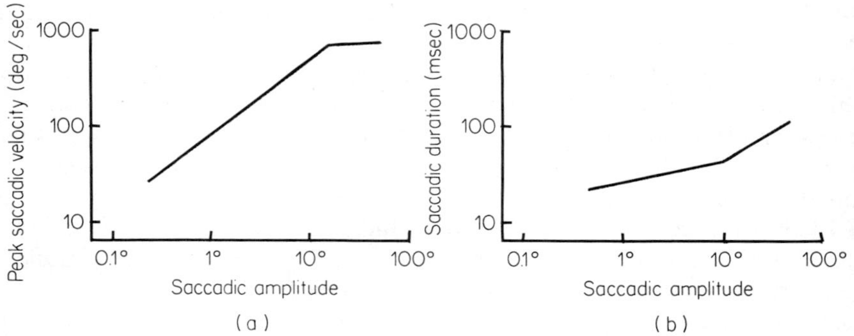

Figure 6.5 The saccadic main sequence. Normal saccades confrom to the peak velocity–amplitude characteristics shown in (a) and to the duration–amplitude characteristics shown in (b), and are said to be main sequence saccades (These plots are derived by approximation from data presented in Bahill, Clark and Stark, 1975a. Reproduced by permission of Elsevier North-Holland, Inc.)

component of the innervation. The innervation decays rapidly to its final level in the phasic system and more slowly in the tonic system (Eckmiller, 1975). The terminal portion of the tonic response sometimes reveals itself as a slow glide, or **glissade**, of the eye to its new position of rest after the main rapid phase of the saccade is over (see Bahill *et al.*, 1978).

If the oculomotor system were a linear second-order control system, the mean velocity of saccades would be proportional to saccadic amplitude, and saccadic duration would therefore be the same for saccades of all amplitudes. However, as Figure 6.5(b) shows, saccadic duration increases slightly with increasing amplitude up to an amplitude of about 15°, and more rapidly for saccades larger than 15° (Bahill, Clark and Stark, 1975a). Clearly, the oculomotor system is not strictly linear.

The relationships between saccadic duration, peak velocity, and amplitude revealed in Figures 6.5(a) and (b) are remarkably consistent for saccades of all types. Any saccade which fits these functions is said to be a **main sequence saccade**. All normal saccades made in illuminated surroundings are main sequence saccades, whether theyare the quick phases of optokinetic or vestibular nystagmus or voluntary changes of fixation, and it is this fact that leads one to suppose that all saccades funnel through the same pulse generators. Saccades made without the aid of a visual target, that is, in complete darkness or in an homogeneous illuminated visual field *(Ganzfeld)*, whether induced by vestibular stimulation or voluntary effort, are slower than main sequence saccades (Becker and Fuchs, 1969; Ron *et al.*, 1972).

A saccade made in normal visual surroundings that does not fit the main sequence may be a symptom of fatique (Bahill and Stark, 1975), alcohol intoxication (Wilkinson *et al.*, 1974), or spinocerebellar degeneration (Zee, Optican and Cook, 1976). There is evidence that when subjects attempt to read a word that lies along the path of a saccade, they can, with very little training, produce saccades voluntarily that are slower than normal during part of their trajectory (Viviani and Berthoz, 1977).

One would expect a consistent relationship between saccadic duration, peak velocity, and amplitude for saccades of different sizes made from the primary position of gaze, because there is a consistent increase in the elastic forces restraining the motion of the eye as it moves further into an eccentric position. What is surprising is that duration and peak velocity are the same for a saccade of a given amplitude, made from the primary position to an eccentric position, as they are for a saccade made in the reverse direction. In one direction the eyes work against an elastic load, and in the other direction they work with the load. The system must take load into account in order to maintain a constant duration and velocity of motion under these different load conditions. Ritchie (1976) has suggested that the information required for load compensation is derived from the afferent inputs from muscle spindles in the extraocular muscles, which feed into the cerebellum. He removed areas of the cerebellum related to saccadic eye movements in monkeys which had been trained to make accurate saccades to visual targets. The saccades were found to be inaccurate after the operation, and the sign and magnitude of the error depended on the initial position of the eyes, that is, on the initial balanced tension in the muscles. In spite of the fact that the saccadic amplitude varied, the duration of the saccades was not affected by the operatior These findings show that saccadic duration is related to the retinally detected position of the target and not to saccadic amplitude, and that saccadic amplitude is determined by a computation which involves taking the initial tension of the muscles into account. The close relationship between saccadic amplitude and duration which was described in Figure 6.5 is therefore not such a basic feature of the oculomotor system as many people have supposed.

About 85% of saccades have an amplitude of less than 15° (Bahill, Adler and Stark, 1975). If the gaze shifts more than this, the head usually moves also. Saccades of less than 15° usually fall short of the target by about 12% of the saccadic amplitude (J. M. Miller, 1977). However, whether or not an eye is on target after the saccade is complete, it often does not remain in the position which it initially reaches, but rebounds immediately through a small angle before coming to rest in its final position. This effect has been called **dynamic overshoot**, but I shall refer to it as **saccadic rebound**. Saccadic rebound is not a secondary corrective movement, because it is an integral part of the initial saccade and occurs without latency and without regard to the accuracy of the saccade. The first explanation that comes to mind is that saccadic rebound is due to the eye being slightly underdamped. Evidence cited in Section 5.3.3 suggests that the eye is overdamped. However, Adler and Fliegelman (1934) and Bengi and Thomas (1968) have detected a mechanical resonance in the eyeball at a frequency of about 40 Hz, and Barlow (1952) noticed that the rebound of the eye after a saccade was often followed by a secondary oscillation about the final position of rest, in a manner suggestive of an underdamped oscillation. Rebound could also be due to a mismatch between the ballistic, or pulse, component of the saccadic innervation and the steady-state, or step, component. For instance, if the ballistic component carried the eye beyond the position indicated by the steady-state component, the eye would rebound into its final position under the influence of the steady-state component. This aspect of saccadic rebound is discussed in detail by Bahill *et al.* (1978). It has also been suggested that saccadic rebound is caused by a burst of neural impulses in the antagonist muscles which occurs just after the muscles are released from inhibition, an effect known as **postinhibitory rebound** (Bahill, Clark and Stark, 1975b). Care must be taken to ensure that saccadic rebound is not an artefact of the instrument used to measure the eye movement.

Most of the work on the basic features of saccades has been done on saccades in a horizontal direction which involve activity in only one agonist at a time. Oblique eye movements involve the combined action of at least two agonists, each of which receives a burst of innervation from separate oculomotor nuclei. The horizontal and vertical recti presumably have distinct pulse generators and the question arises as to whether they work independently during an oblique saccade. The only evidence from man suggests that the horizontal component fires first, because the trajectory of an oblique saccade is curved in such a way that the initial part is deflected towards the horizontal (see Figure 6.6).

Bahill and Stark (1977), claimed that the velocity and duration of the horizontal component of an oblique saccade is the same as the velocity and duration of a purely horizontal saccade of the same amplitude as the component. More recent evidence suggests that the components of an oblique saccade are not as independent as was reported by Bahill and Stark. In all but oblique saccades in a 45° direction, the horizontal and vertical components have unequal amplitudes. Because of the relationship between saccadic amplitude and duration, these components will also have unequal durations. If the two components are

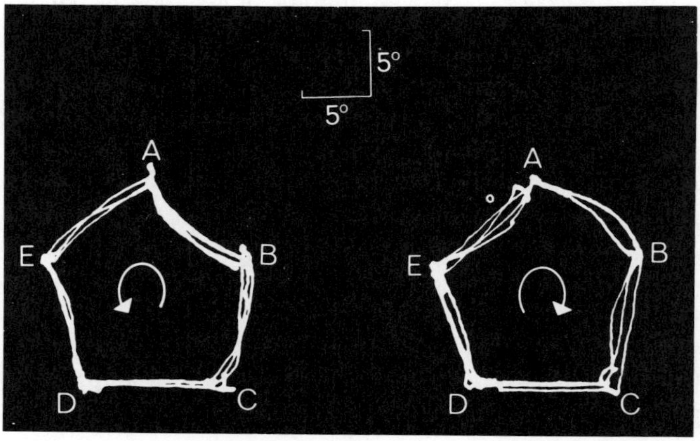

Figure 6.6 Saccadic scan paths between the points of a pentagon in (a) anticlockwise and (b) clockwise direction (Reprinted with permission from P. Viviani, A. Berthoz, and D. Tracey, *Vision Research*, **17**. Copyright. 1977, Pergamon Press, Ltd.)

independent, this difference in duration should result in a curved trajectory for such saccades, unless there is a compensating mechanism by which the durations of the components are made equal. Evinger and Fuchs (1978) and Guitton and Mandl (1980b) found that in the cat the duration of the shorter component of an oblique saccade is prolonged to make it more like the duration of the long component. But they found that this compensation is not complete and that in many saccades there is no compensation, which may account for the results of Bahill and Stark's experiment and perhaps for the results shown in Figure 6.6.

These basic features of saccades are largely immune to the effects of practice or voluntary effort. Saccades are very similar in different mammalian species and occur in neonate humans (Harris and MacFarlane, 1974; Aslin and Salapatek, 1975).

6.4.2 *Saccadic latency*

The latency, or reaction time, of a saccade is the time it takes the eyes to start moving after the fixation point has been suddenly stepped to a new location. The latency is normally about 250 msec, although it may be as low as 150 msec under certain circumstances. Reaction time increases slightly with increasing amplitude of the saccade (White *et al.*, 1962). It also increases as amplitude is reduced below 0.5° (Wyman and Steinman, 1973b). Latency for upward saccades has been reported to be shorter than for downward saccades, and latencies for leftward and rightward saccades have been reported to be equal (Heywood and Churcher, 1980). If a target moves regularly back and forth a person may learn its motion and produce saccades with vanishingly small reaction times (Fuchs, 1967).

The reaction time of an arm movement towards a visual target has been found to vary according to the number of target positions (N) and the probability that the given target will appear (P_i). The quantitative expression of this relationship is known as **Hick's law** (see Hick, 1952; Fitts and Peterson, 1964):

$$\text{R. T.} = a - b \sum_{i=1}^{N} P_i \log P_i$$

where a is a constant and b is the reciprocal of the channel capacity of the system, assumed constant. The right hand side of this equation provides a measure of the mean uncertainty (entropy) associated with each target; it reduces to $\mathrm{a + b \log N}$ when the N alternative signals are equally probable. One might expect a similar relationship to hold for saccadic eye movements. However, it has been found that for saccades in a given direction the degree of uncertainty about the position of a fixation target along the horizontal meridian does not affect saccadic reaction time (Saslow, 1967b; Miller, 1969). It may be that the act of making a decision about the amplitude of a saccade is exempt from Hick's law because it is an overlearned skill; overtrained arm movements have also been shown to be exempt from Hick's law (Fitts and Posner, 1967). Another reason might be that variations in decision time for the amplitude of a saccade are masked by the longer time it takes to decide about and programme the direction of the movement. Saccadic reaction time is longer when there is uncertainty about the direction in which a saccade is to be made than when there is prior knowledge about direction. Megaw and Armstrong (1973) varied the relative probability of a fixated target appearing to the left or right and found that reaction time was a linear increasing function of direction uncertainty. Norcia *et al.* (1980) conducted a two-dimensional version of this experiment. The fixation target jumped a fixed distance in one of 16 directions, but the probability of a jump in one of these directions was greater than that of a jump in any of the other directions. After the subject had become familiar with these different probabilities, the mean reaction time for saccades in the biased direction was reduced, and so was the reaction time for saccades in neighbouring directions, although not by as much. The reaction times of saccades in directions opposite the bias were increased. One possible explanation of this cost-benefit effect is that the directional programme for the biased direction is held in a buffer store. When the target appears in the biased direction, this ready-made programme is used, but when the target appears in the opposite direction, the ready-made programme is discarded and a new one substituted.

6.4.3 Saccadic stimulus refractoriness

A whole set of questions concerning the refractoriness of the visual system to information used in programming saccades may be investigated by recording the way the eyes respond to a visual target that steps twice in rapid succession (Westheimer, 1954a). The first step (**step 1**) is usually about 5° and is followed by **step 2** after an **interstep delay** of between 50 and 500 msec. Step 2 may be in the same direction as step 1 or it may be in the opposite direction, and it may be smaller than, equal to or larger than step 1.

There are two basic things that the saccadic system could do when presented with a double step: it could respond to each step and thus make two saccades in rapid succession, or it could ignore step 1 and aim straight for the position that the target occupies after step 2. In either case, as soon as step 1 occurs, the system will start to programme its response to it and the longer the target stays in this position, the more complete this programme will become. *The question is, must this programme run its course and produce a corresponding saccade once the programme is initiated, or is it possible for step 2 to modify or revoke the programme before the eyes start to move?*

Several investigators have shown that there is an increasing likelihood that step 2 will influence the initial saccade if the stimulus for step 2 occurs early within the period between step 1 and the start of the first saccade (Wheeless *et al.*, 1966; Becker and Fuchs, 1969; Becker and Jürgens, 1975). For instance, Becker and Jürgens (1975) found that if step 2 takes the target back towards its initial position more than 80 msec before the eyes have made their initial response, there is a high probability that the eyes will move directly to the target's final position. In other words, the amplitude of the saccade being planned is shortened in response to stimuli arriving well into the latency period, although the direction of the saccade is unaltered. The reaction time for this type of amended saccade, *counting from the second step*, was found to be shorter than the reaction time of a normal saccade (Komoda *et al.*, 1973). It was suggested that this is because the decision about direction involved in planning the saccade to the first step is retained in planning the response to the second step. The reaction time for this type of amended saccade, *counting from the first step*, was found to be slightly longer than a normal saccade, which presumably reflected the extra time required to incorporate new information from step 2 regarding the amplitude of the movement.

When step 2 returns the stimulus to a point further back than the position of the initial stimulus, an amended saccade must be reversed in direction as well as be changed in amplitude. It is not surprising, therefore, that saccades of this sort do not take place unless the second step occurs at least 170 msec before the eyes make their initial response, and that they have longer reaction times than amended saccades in which only amplitude is changed. The initial programme must now be scrapped and an entirely different one substituted (Komoda *et al.*, 1973; Becker and Jürgens, 1975; Hou and Fender, 1979). An essentially similar state of affairs holds when step 2 is at right angles to step 1 (Lisberger *et al.*, 1975; Hou and Fender, 1979).

There are two ways to think about these results. One way is to suppose, as did Hou and Fender, that programming of direction is complete early in the saccadic latency period and is then refractory to further information, but that the programming of amplitude takes longer and remains open to new information. The other possibility is that programming of direction takes a long time and is refractory to new information while it is in progress, but that programming of amplitude takes a short time and, although it too is refractory to new information while in progress, there is time for a second amplitude programme to be set up during the time required to plan direction. I suggest that the second possibility is the correct one, because it is more parsimonious and also because eye movements obey Hick's law with regard to direction but not with regard to amplitude (see Section 6.4.2).

These findings demonstrate that, during the time information about one stimulus event is being processed, the oculomotor system will accept other stimulus information until about 80 msec before the response occurs. Any stimulus presented later than this is too late to affect that saccade. However, information arriving too late to affect a saccade is registered by the visual system and may affect the next saccade. We know this because it has been shown that, when step 2 occurs just before or during the initial saccade and is extinguished before that saccade is complete, the eyes complete the initial saccade as planned, but then execute a second saccade with a normal latency towards the now extinguished second point (Hallett and Lightstone, 1976a, b). This indicates that the visual system is never refractory to new spatial information, not even just before or during a saccade. If information is too late to affect one saccade, it may be stored and used to control another saccade. However, in normal circumstances, visual information is degraded during saccades because of the smearing of the image across the retina, so that the visual system is effectively refractory to information during saccades. The sampled data theory proposed by Young and Stark (1963), according to which information for the control of eye movements is sampled only at fixed intervals of about 200 msec, is incompatible with the evidence just reviewed.

Normally, the visual system programmes only one saccade at a time, even though it may store information relevant to a subsequent saccade while a saccade is proceeding. This means that successive saccades are separated by at least one reaction time. Westheimer (1954b) demonstrated this by showing that if a person responds to both components of a target motion which steps out and then steps back, the return saccade does not commence until about 200 msec after the outgoing saccade, even though the target is in its displaced position for a shorter period than this. A similar intersaccadic delay is also evident when step 2 is at right angles to step 1 (Feinstein and Williams, 1972; Lisberger *et al.*, 1975). Even microsaccades, that occur when one is attempting to fixate a stationary object, are separated by about 200 msec. If a microsaccade must be separated from a larger saccade by this same interval, then this confounds the measurement of the latency of larger saccades, because it will be a matter of chance whether the stimulus is applied before or after a microsaccade has occurred. Saslow (1967a) showed that the latency of a saccade to an eccentric target turned on some time after the central fixation light had been turned off was only about 150 msec. Saslow claimed that microsaccades do not occur in the dark, so that in this experiment the eye was always in a maximum state of readiness. This suggests that 150 msec is a true measure of saccadic latency (see Carpenter, 1977, p. 63, for a discussion of this point).

The question may be asked whether the oculomotor system is basically incapable of organizing more than one saccade at a time, or whether it is merely a matter of strategy that it normally does not do so. After all, there would normally be no reason to plan more than one saccade at a time. After a saccade has occurred, it is an advantage to have the eyes stationary for a while so that visual information may be picked up. During this pause there is plenty of time to decide where the next saccade should be directed. Nevertheless, evidence which I shall now review shows that the eyes are capable of executing successive saccades with virtually no interval between them, showing that the saccades may be programmed simultaneously, or in parallel. It may be concluded that the normal serial nature of saccades is a question of strategy and does not result from a basic incapacity of the oculomotor system to plan saccades in parallel. Let us look at some of the evidence for this conclusion.

If it could be shown that the interval between successive saccades to double-step stimuli is shorter than a normal saccadic latency, this would provide evidence that the second saccade was in some sense preplanned before the first saccade was complete. Komoda *et al.* (1973) reported that the second saccade made in response to step 2 of a double step, with 50 msec step interval, had a latency of only 125 msec if step 2 was in the same direction as step 1. They argued that this unusually short latency was due to the direction component of the programme for the second saccade being retained from the first saccade. However, the 'saving' on the second saccade latency may not have been as great as Komoda *et al.* thought. In the first place, there would be no microsaccades in the intersaccadic interval, so that the proper comparison should be with the 150 msec latency of saccades made in the dark, rather than the 250 msec of a normal single saccade. In the second place, the latency for the second saccade should perhaps have been measured from the beginning of the first saccade rather than from its end, as Hallett and Lightstone (1976a) have argued.

When a person changes his gaze to a target that is more than about 5° to one side, he usually makes two saccades, one bringing the eye near the target and the second bringing the eye on target. These second, so-called **corrective saccades** follow the main saccade with a latency of about 125 msec, which is also less than the latency of a normal saccade. But here again it cannot be concluded that the second saccade is in any sense preprogrammed before the first saccade. Nevertheless, there is evidence that successive saccades may occur with considerably less than one latency period between them. Several investigators have reported double saccades separated by as little as 25 msec. These are called **overlapping saccades**, because the second saccade has started before the eye has come to rest. Bahill and Stark (1975) reviewed the literature on overlapping saccades and showed that they occur in only one eye at a time, and that fatigue increases the frequency of their occurrence. This suggests that they are really single saccades that have become fractionated. One possibility is that overlapping saccades result from a lack of synchrony in innervation of the horizontal and vertical agonists, which would explain why they occur frequently in oblique eye movements (Bahill and Stark, 1977). So this evidence does not prove that two saccades can be programmed together.

More convincing evidence that intersaccadic latency is not due to a basic refractoriness in the oculomotor system comes from studies on saccades evoked by stimulation of the frontal eye fields (Robinson, 1968). If a given location of the frontal eye field on one side of the brain is stimulated, a saccade of a given amplitude and direction is evoked. If the first stimulus is followed by a second stimulus in as little as 25 msec, two similar saccades are evoked. If the second stimulus is applied to the opposite frontal eye field, the second saccade occurs in the opposite direction, and no refractory period is evident; the second saccade occurs so soon after the first that it mechanically interferes with it. More recent evidence of the same kind has been reported by Marrocco (1978). He found that by stimulating the frontal eye field of a monkey, he could not evoke a saccade in a given direction within 50 msec after a naturally occurring saccade in the *same* direction. Furthermore, the electrically evoked saccade was not at full strength until 100 msec after a naturally occurring saccade. However, a saccade could be evoked electrically at any time during or after a naturally occurring saccade in the *opposite* direction, or at any time during or after a natural pursuit movement in either direction.

This evidence suggests that there are distinct pulse generators in the brainstem for each agonist, and that each pulse generator has a refractory period in the region of 25–50 msec. But this evidence still does not tell us whether two saccades are ever planned in parallel. The Hallett and Lightstone evidence mentioned above suggests that the visual system may register and store information relevant to a second saccade before or during a primary saccade, but that the initiation of a second saccadic programme is postponed at least until the primary saccade has started.

The most convincing evidence that saccades may be planned in parallel is the following. Täumer (1975) presented subjects with a visual target that made two 5° steps in the same direction with an interstep interval of 100 msec. He found that most subjects on some trials made two saccades with an intersaccadic latency of about 120 msec, which indicates that these saccades were planned in series, or one after the other. However, all subjects on some trials produced two saccades with less than 40 msec between them, which indicates that these saccades had been planned in parallel, or simultaneously. Lévy-Schoen and Blanc-Garin (1974) and Becker and Jürgens (1975) also reported double saccades with very brief, and even zero intersaccadic intervals, which they also interpreted as indicating the ability of the oculomotor system to preplan two saccades at the same time. Furthermore, the accuracy of the second saccade was found to be unrelated to the accuracy of the first, which suggests that in planning the second saccade, the system knows where the first saccade is going to bring the eye, even when the first saccade is destined to be inaccurate.

6.4.4 Saccadic accuracy

The accuracy of a saccade is normally indicated by the signed difference between the amplitude of the saccade and the distance between the initial fixation point and the position of the visual target to which the saccade is directed. This accuracy may be partitioned into a component in the direction of the target and a component at right angles to this. Care must be taken to ensure that the subject has no prior knowledge about the amplitude of the saccade he will be required to make. This precaution has not always been taken; for instance, Weber and Daroff (1972) measured saccadic accuracy while subjects glanced back and forth between two stable target positions, thus allowing the subjects to have prior knowledge about the positions of the targets. As Henson (1979) has pointed out, the only satisfactory procedure is to present each visual target at a new, randomly determined position along the meridian over which the saccades are being made.

If the initial saccade misses the visual target, a second, short-latency, corrective saccade is initiated which normally brings the eye very close to the target. Thus, the amplitude of these corrective saccades indicates the error of the initial saccade. Becker (1972) measured the amplitude of corrective saccades for different angular separations between the initial fixation point and the refixation target. He measured saccades which were symmetrical about the midline, saccades which started at the midline, and saccades which returned to the midline. His results are shown in Figure 6.7. It can be seen that saccades of over 10° generally undershoot the target, and that the probability of undershooting

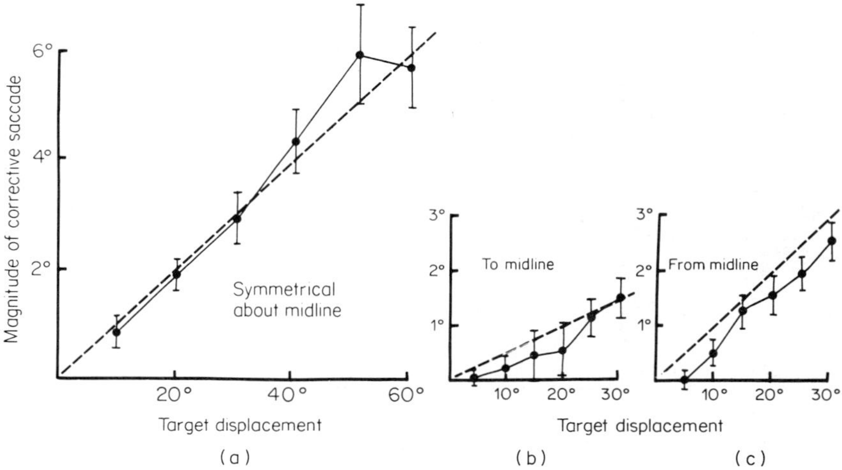

Figure 6.7 Magnitude of corrective saccades for different magnitudes of target eccentricity. In each case the initial saccade undershot the target by the amount indicated on the ordinate. This error invoked a corrective saccade of the same magnitude. Each data point is the mean of 24 trials for one subject. Bars indicate 95% confidence intervals. The dashed lines indicate the loci of 10% (a and c) or of 5% (b) saccadic undershoots (From Becker, 1972. Reproduced by permission of S. Karger AG, Basel)

increases with saccadic amplitude. The size of the error of the initial saccade is greater for outward directed saccades than for saccades which return the eye to the primary position (see Section 6.4.1 for a discussion of related phenomena). For saccades symmetrical about the midline, the error of the initial saccade is about 10% of the target distance, normally in the direction of undershoot. In other words, the initial saccade usually covers the first 90% of the distance to the target, leaving the second 10% to be taken up by a corrective saccade.

The accuracy of saccades under 10° was investigated by Henson (1979). His results are shown in Figure 6.8. It can be seen that he found a large mean magnitude of undershoots for 10° saccades and a small mean magnitude of undershoots for 4° saccades. When an undershoot occurred it was about 8% of the target distance for saccades of 10% or less.

The main characteristics of saccades which undershoot the target, that is, most saccades over about 5°, are as follows:

1. The initial saccade has a measured latency of about 250 msec and covers about 90% of the distance to the visual target.
2. The system anticipates having to make a corrective saccade of about 10%, so that if, at the end of the first saccade, the target is off-centre by about this amount, a corrective saccade is initiated with a latency of about 150 msec. This latency is independent of the magnitude of the correction for corrections of between 3 and 10°, and is shorter than the latency of a normal saccade. The latency may be

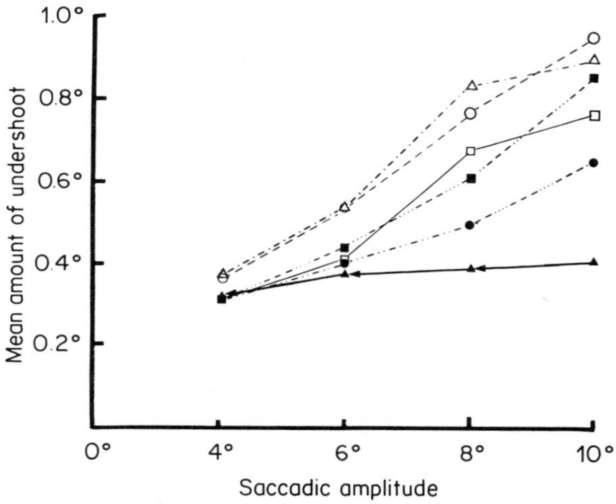

Figure 6.8 Mean extent to which a saccade undershoots its target as a function of saccadic amplitude for amplitudes under 10°. Each saccade was made from the primary position of gaze to one of four randomly determined target positions on either side. Each curve is the result for one subject. The curve with arrows is for saccades made back to the midline by one subject (Reprinted with permission from D. B. Henson, *Vision Research*, **19**. Copyright 1979, Pergamon Press, Ltd.)

shorter because of the absence of preceding microsaccades, but it may also be because the direction of the corrective saccade is preprogrammed. It has also been suggested that retinal signals regarding the location of the target may be taken into account before the first saccade is complete, and this may also contribute to the short latency of the corrective saccade (Prablanc *et al.*, 1978).

3. If the error of fixation after the first saccade is larger than anticipated, either the preprogrammed secondary saccade occurs, to be followed by a third saccade, or the preprogrammed saccade is cancelled and replaced by a second saccade of normal latency (Becker and Fuchs, 1969).

4. If the visual target is switched off before the end of the first saccade, and the error of the initial saccade is greater than 3°, a corrective saccade is initiated with a latency longer than that of a normal corrective saccade and which is less accurate in bringing the eyes onto the position of the target than is a corrective saccade made when the target remains in view (Barnes and Gresty, 1973). However, if the initial saccade in the dark brings the eyes to within about 3° of the target position, a corrective saccade is not initiated, as it would have been had the target been in view (Prablanc and Jeannerod, 1975; Becker, 1976). The system is thus capable of detecting the mismatch between the position of the eyes and the position of the absent target with a sensitivity of about 3°. It is not known how the system detects this mismatch, but it must rely either on proprioception from receptors in the muscles or tendons or on a comparison between the efferent command and the remembered headcentric location of the target (Shebilske, 1976).

When the error is less than 3°, the system is apparently incapable of detecting any mismatch between the position of the target seen briefly before the saccade and the direction of gaze after completion of the saccade towards that target. Thus, when a subject is asked to match the position of a light probe seen just after a saccade and the position of the target seen just before the saccade, he does so by placing the probe on the visual axis. In other words, he acts as though his gaze had reached the intended position, even though an objective measure of his position of gaze indicates that it has not (Lennie and Sidwell, 1978). *An experiment is needed to determine whether or not the oculocentric position of a peripheral target is underestimated when a saccade is not executed towards it.* If it is, at least part of saccadic undershoot would be a consequence of oculocentric underestimation.

In summary, it can be said that the eyes deliberately undershoot the target, especially for larger distances, and get ready to make a corrective saccade in the same direction as the main saccade. The visual information is quickly sampled just before and just after the first saccade, and the short-latency corrective saccade adjusted accordingly, as long as the error is in the excepted range. If the error is more than expected, the system either makes two corrective saccades or one long-latency corrective saccade. If the target is not in view after the first saccade, the system makes a rather inaccurate corrective saccade if it finds itself too far from its anticipated position.

Undershooting can be seen as a deliberate strategy adopted by the oculomotor system. Henson (1978) altered the position of the visual target during a saccade in such a way as to reduce the amount of undershooting. After a few trials, the system 'learned' to undershoot the displaced target by about the same amount that it previously undershot a normal target. This demonstrates how persistent the tendency to undershoot is. McLaughlin (1967) and Weisfeld (1972) also demonstrated the capacity of the saccadic system to adjust the amplitude of saccades when the visual target is repeatedly displaced during the saccade by a fixed amount in the direction of initial fixation. The persistence of undershooting in the Henson study suggests that the system will adjust less readily to targets which are displaced away from the initial fixation point than to targets displaced the other way. This suggestion is borne out by the results of an experiment conducted by J. M. Miller *et al.* (1979).

In all these experiments, training lasted only a short time. There is evidence that an adjustment for undershooting occurs after several days of experiece with an eye in which saccades are hypometric. For instance, Optican and Robinson (1980) found that monkeys learned over a period of days to compensate for hypometria induced by surgically weakening the horizontal recti. The capacity to adapt to hypometria was abolished after cerebellectomy (see Section 5.4.4).

An overshoot error readily invokes a parametric adjustment of the system but an undershoot error invokes an adjustment only after an extended period. It looks as though undershoot errors are not only tolerated, but desired. Several good reasons have been suggested to explain the preponderance of undershooting, which is particularly marked in saccades over 10° in amplitude (Frost and Pöppel, 1976).

1. By making undershooting the rule, the system economizes on the decision about the direction of the corrective saccade, which allows its latency to be shortened.
2. If the eyes undershoot the target, it remains in the same half of the visual field as it was to begin with, which means that the same side of the brain is involved in computing its position.
3. The total distance travelled by the eyes is less when they undershoot the target than when they overshoot it.
4. Large saccades are often accompanied by a head movement in the same direction, so that undershooting may be part of a strategy for anticipating the effects of this head movement (see Section 9.2.6).

Sometimes a saccadic eye movement ends in a slow drift of the eye onto the target. This slow drift is known as a **glissade** and normally occurs in only one eye at a time (Weber and Daroff, 1972). The glissade is a continuation of the saccadic movement and has no perceptible latency. It is therefore not a corrective movement made in response to an error signal. The most reasonable explanation is that a glissade occurs when the pulse of innervation fails to carry the eye to the new position specified by the maintained firing rates of motoneurones; the eye will then glide into its new stable position from the unstable position in which the pulse left it. The fact that the pulse can be mismatched to the new maintained firing rate for one eye while being matched for the other eye suggests that it is programmed separately for the two eyes (Easter, 1973).

So far in this discussion of saccadic accuracy it has been assumed that only one small visual target is in view at the same time. When a person is asked to make a saccade to a large visual object or to one of a cluster of visual targets, the first saccade typically brings the gaze to the centre of the display and one or more extra saccades are required before the desired fixation is achieved (Coren and Hoenig, 1972; Findlay, 1980b). One possible explanation of this behaviour is that only the low spatial-frequency components of the peripherally positioned visual display are detected before the first saccade is executed and that the higher spatial frequencies, which allow for the discrimination of parts of an object or of separate objects in a cluster, are detected only after the gaze has been centred on the display.

This completes this review of behavioural studies of eye movements. As we have seen, eye movements serve to stabilize the retinal image or to direct the gaze towards an object of interest. But the mobility of the eyes complicates the task of judging the headcentric or bodycentric direction of a visual object. This is the issue which will be discussed in the next chapter.

Chapter 7

Visual Direction with Respect to Head and Body

7.1 Basic concepts

This chapter is about how people judge the directions of objects with respect to the head or the body. There are no conventional terms for these two systems of reference but I shall refer to them as **headcentric** and **bodycentric** directions. Many investigators have failed to make clear which system of reference they were studying, and, as we shall see, this has resulted in confusions. Lateral headcentric directions may be objectively specified with respect to the median plane of the head and the extension of the spinal axis. Up and down headcentric directions may be specified with respect to the transverse plane through the centres of the eyes (which is the horizon plane if the head is upright) and the point where this plane cuts the spinal axis. Lateral bodycentric directions may be specified with respect to the median plane of the body and the spinal axis. Up and down bodycentric directions have little behavioural significance.

These definitions are rather vague because the human head and body do not have precise landmarks. But most experiments have to do only with the variability of directional judgments or with comparing constant errors under different conditions and, as long as other conditions remain the same, precise difinitions of the median planes are not required for these purposes.

A judgment about the headcentric direction of a visual object requires the subject to register the position of the images on the two retinae and the position of the eyes in the head. These two components of a headcentric judgment will be referred to as the **oculocentric component** and the **eye-position component** respectively (see Table 7.1). The object whose direction is being judged will be referred to as the **target**. If the subject fixates the target, the image is normally registered very precisely on the centre of the fovea in each eye. Under these conditions, uncertainty about the oculocentric position of the retinal image contributes very little to the variance of a headcentric judgment. Therefore, with foveal fixation, a headcentric judgment reduces essentially to the task of registering the position of the eyes in the head. Information about the position of the eyes in the head could theoretically be provided by proprioceptors in the extraocular muscles or tendons or by the efferent command to those muscles. I shall discuss this issue in Section 7.7. Information about the position of the eyes could also be provided visually by the position of the images of the tip of the nose and orbital ridges. However, it has been found that the variability of judgments of straight ahead is not affected by whether or not the nose is in view. It seems that the nose is too eccentrically placed in each monocular visual field to be effective, since an external reference object fixed with respect to the head has been found

Table 7.1 Types of directional judgments and their components

Type of judgment		Components
Proprioceptive		
Egocentric	Oculocentric	Sense of retinal position of target's image
	Headcentric	Sense of eye position in the head
	Bodycentric	Sense of head position on the body
Exocentric (allocentric)		

to reduce the variability of judgments if it is reasonably close to the target (Shebilske and Nice, 1976; Wetherick, 1977).

A judgment about the bodycentric direction of a visual object involves the oculocentric and eye-position components, but it also requires the subject to register the position of the head on the body—the **head-position component**. The position of the head on the body is indicated by proprioceptors in the neck, but it may also be indicated visually if the body is in view.

People habitually judge the lateral visual direction of an object in terms of its position to left or right of the apparent median plane of the head or body, so that these planes are the norms for these judgments. One cannot determine the position of an apparent median plane by asking the subject to place a single visual stimulus straight ahead of the head or body. This procedure indicates only one point in a median plane for that particular fixation distance; a point can never specify a plane. In spite of these limitations, single-point determinations are useful for comparing the effects of various factors on visual judgments of direction, and the results are probably valid as a basis for generalization at least for the fixation distance used. It is not legitimate to describe these judgments as 'median-plane judgments'; a better description is 'straight-ahead judgments'.

7.2 Laws of headcentric visual direction, and the egocentre

The analysis of headcentric visual direction must start with the basic concept of oculocentric direction, namely, the visual line. In Section 2.1, a visual line was defined as any straight line passing through the nodal point of the eye. *Any one visual line is the locus of all points fixed relative to the eye which stimulate a given point on the retina.* The visual line through the centre of the fovea is the **principal visual line**, or visual axis. Any other visual line may be specified in terms of its angle of elevation with respect to the eye's median plane and of its azimuth with respect to the eye's mid-transverse plane or in terms of its angle of eccentricity and meridional angle, as described in Figure 2.1.

All objects which lie on the same visual line, whatever their distance from the observer, are judged to be in the same oculocentric direction. Furthermore, no other objects are judged to be in this same direction. This is the **law of oculocentric direction**. When looking with one stationary eye, the observer will say that all objects on the same visual line are in the same headcentric direction. Thus, the *first law of headcentric visual direction is that, in monocular viewing, objects lying on the same visual line, for a given position of the eye in the head, are judged to be in the same headcentric direction, which is unique to that visual line.*

When a person sights with one eye along a rod, the rod lies along the principal visual line and is experienced as pointing directly at the self. One may also align a rod with other visual lines by maintaining fixation on a point in the median plane while sighting along an eccentrically placed rod, and in this case, too, the rod appears to point directly at the self. Note that a rod which lies on a visual line for one position of the eye will not lie on a visual line after the eye has rotated to another position. This is because the centre of rotation of the eye is behind the nodal point, so that a rotation of the eye displaces the nodal point in the same direction as the eye movement (Figure 7.1). The general rule is that, in monocular viewing, stimuli which lie on any one visual line appear to be in alignment with the self. The **visual egocentre** is the location in the head towards which visually aligned objects appear to point. A more precise definition will be given in what follows. Note that the egocentre is not necessarily the point towards which monocularly aligned objects *actually* point, which is of course the nodal point of the eye. In fact, as we shall see, the egocentre is normally in the median plane of the head. The second law of headcentric direction is that, *in monocular viewing, all visual lines appear to be directed towards a common visual egocentre.*

The next stage in the analysis is to assume that for every retinal point in the binocular portion of the field of one eye there is a corresponding point in the retina of the other eye, such that when images fall simultaneously on both points there is judged to be an aligned pair of objects, or, if the images are similar

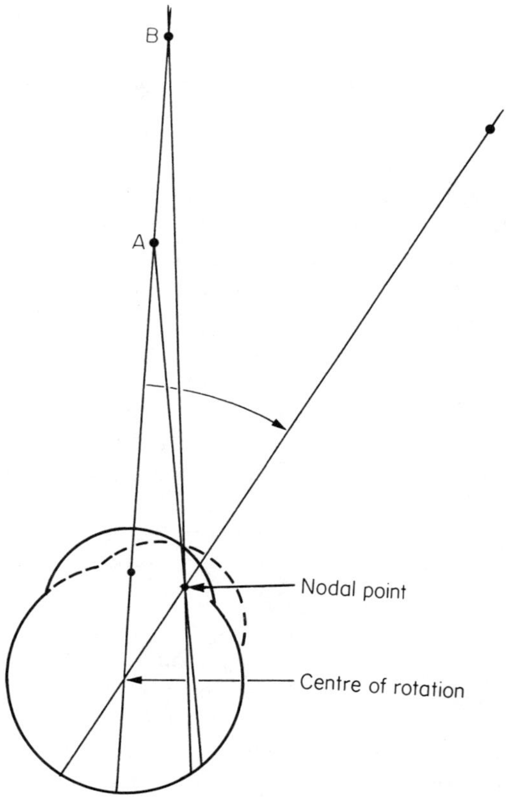

Figure 7.1 Showing how two objects, A and B, are no longer aligned with the nodal point of the eye after the eye has changed its angle of gaze

enough, a single 'fused' object. **Corresponding retinal points**, therefore, have identical headcentric space values, which means that objects on the visual lines of a pair of corresponding points will be reported to be in spatial alignment with each other and also with the egocentre. The simplest way to define corresponding points is to say that they are the pairs of points in the two retinae which would be brought into coincidence if one retina were swung in an arc about the fixation point until it came to lie over the other retina. If corresponding points conformed to this simple **geometrical** definition of correspondence, the members of each pair of corresponding points would have the same oculocentric space values as well as having the same headcentric space values. In fact, binocular correspondence cannot be accurately defined by this simple method. There are two reasons for this. In the first place, if the normally horizontal retinal meridia are brought into congruence by transposing one retina onto the other, the two corresponding meridia in the vertical direction are extorted about 2° from the true vertical (see Figure 7.3b). In the second place, points lying along horizontal meridia are not in precise geometrical correspondence. Only if oculocentric directions were defined in terms of these departures from geometrical correspondence could one say that corresponding retinal meridia have the same oculocentric space values.

For any point of binocular convergence, there is a region in space within which stimuli must be placed if they are to stimulate corresponding retinal points. This region is known as the **horopter**. The images of any stimulus object outside the horopter stimulate non-corresponding points in two eyes, and are said to be **disparate images**. If the binocular disparity is not too large, the images experientially **fuse** into a single image, and if the disparity is large, the viewer experiences **diplopia**, that is, he sees two images. An object between the horopter and the observer is said to project **crossed images**, because, of the pair of disparate images which such an object produces, the one which the observer sees on the left is in the right eye and the one seen on the right is in the left eye. Objects beyond the horopter are said to produce **uncrossed images**, for a similar reason. If a person has a divergent squint, or strabismus, of recent onset, corresponding visual lines cannot be converged onto a visual object, and therefore the images of all objects fall on non-corresponding retinal points and all objects are reported as double. In squints of long standing double vision is avoided because the input to one eye is **suppressed** when both eyes are open. The suppressed eye often has reduced acuity and other visual defects, a condition known as **strabismic amblyopia**.

The horopter is best understood by considering it in two parts, the **longitudinal horopter**, or the part lying in the plane containing the two visual axes, and the **vertical horopter**, or part lying at right angles to this plane. Ideally, the longitudinal horopter is a circle, known as the **Vieth-Müller circle**, which passes through the fixation point and the nodal points of the eyes. If a person's gaze is converged on point F, as in Figure 7.2, the images of F will fall on corresponding points (the foveae in this case) on the two retinae. While convergence is maintained on point F, a necessary condition for any other point, P, to stimulate corresponding points is that the two angles marked ϕ in Figure 7.2 are equal.

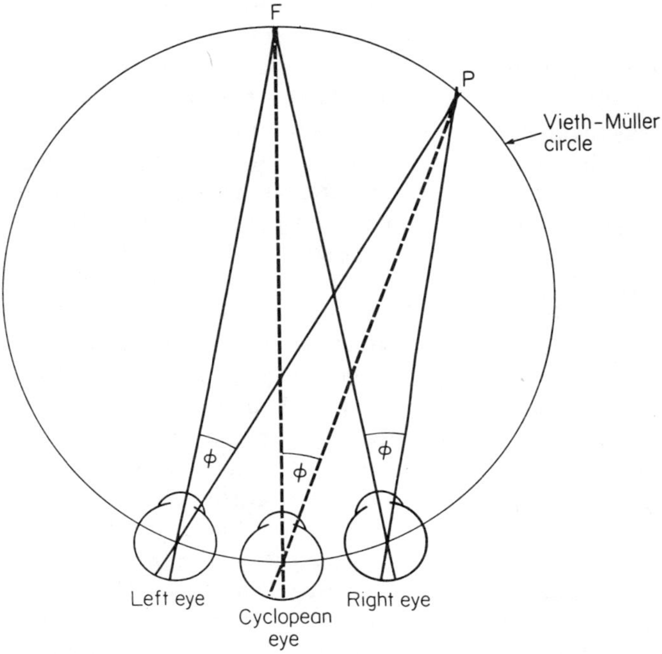

Figure 7.2 The theoretical longitudinal horopter, or Vieth-Müller circle. Any two points on the Vieth-Müller circle subtend the same bipolar azimuth, ϕ, to the two eyes. The same angle is subtended by these points at a cyclopean eye positioned on the Vieth-Müller circle, midway between the eyes. If this cyclopean eye is defined as the egocentre, the biploar azimuth defines the headcentric azimuth of one point with respect to the other

From elementary geometry we know that these angles will be equal if the line joining F and P is the chord of a circle which passes through the nodal points of the two eyes, that is, if F and P fall on the Vieth-Müller circle. For any two points in visual space, the angle ϕ is known as the **bipolar azimuth**. If the horizontal locus of points in space which produce single images for a given point of fixation of the eyes is empirically determined, it is found to deviate from the ideal Vieth-Müller circle (see Figure 7.4). This so-called **Hering–Hillebrand deviation** occurs because the distributions of corresponding points along horizontal meridia in the two retinae are not geometrically identical (see Ogle, 1964, for more details on this topic).

With the eyes in symmetrical convergence, the ideal vertical horopter is a single vertical line passing through the point of binocular convergence. In practice, the vertical horopter was shown by Helmholtz (1962), and more recently by Nakayama *et al.* (1980), to slant away from the observer (see Figure 7.3). This is a consequence of the fact, already mentioned, that corresponding retinal meridia in the vertical direction are extorted about 2°. Thus, in order for a line in space to stimulate these corresponding extorted meridia it must slant away from the

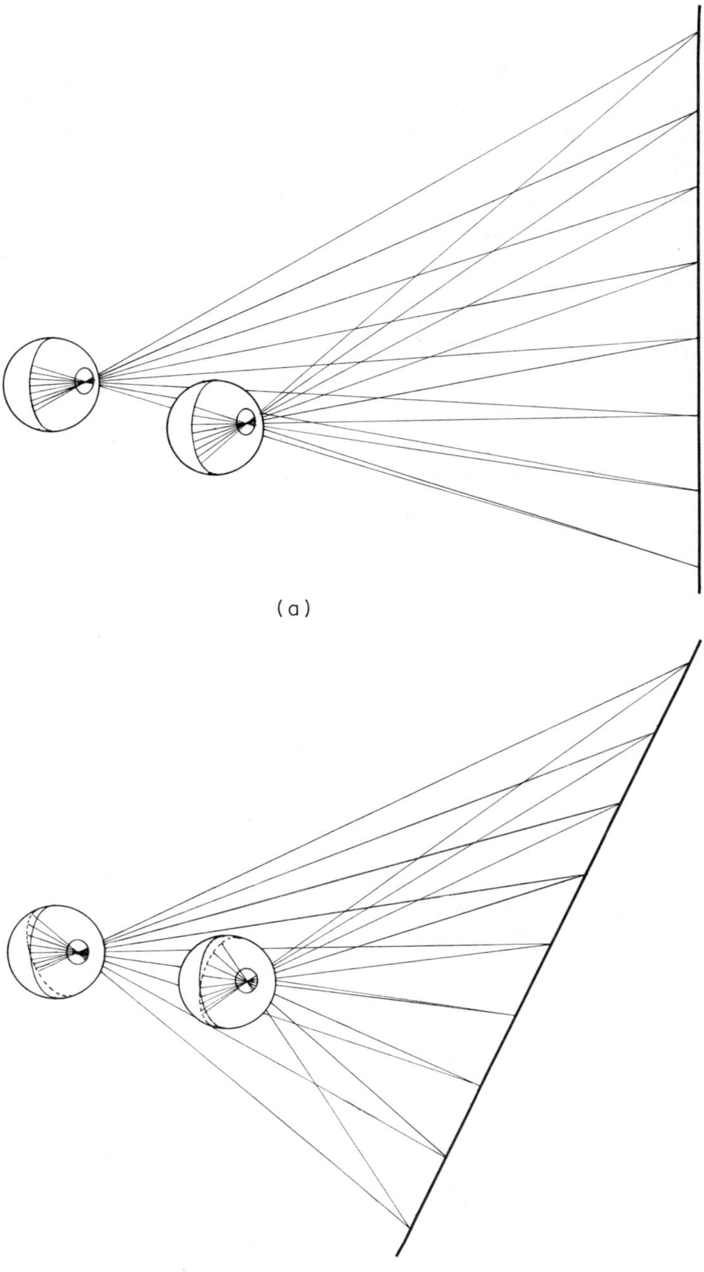

Figure 7.3 (a) The theoretical vertical horopter is a vertical line only if the corresponding retinal meridia are vertical. (b) The actual vertical horopter is a line slanting away from the observer, which implies that the corresponding retinal meridia (dashed lines) are extorted with respect to the vertical meridia (solid lines) (From Tyler and Scott, 1979. Reproduced by permission of Harper & Row, Publishers, Inc.)

observer. It is a fact of geometry that the degree of slant required increases with viewing distance, so that the vertical horopter becomes more slanted with increasing distance, until at infinity it is a line parallel to the ground. The relation between the inclination of the vertical horopter, i, and the angle between the corresponding meridia, δ, is given by $\tan \delta = a/b \tan i$, where a is the interpupillary distance and b is the observation distance (Ogle, 1964). The slant of the vertical horopter is probably related to the fact that we more often look at surfaces which slant away from us than at vertical surfaces, although it is not known whether the pattern of binocular correspondence responsible for the slant of the vertical horopter is innate or experientially determined.

The longitudinal and vertical horopters are ideally lines, but in practice they have volume, because a certain amount of binocular disparity is tolerated before diplopia is apparent (see Figure 7.4). For a given point of convergence, there are no other positions of a point-like object which produce images which

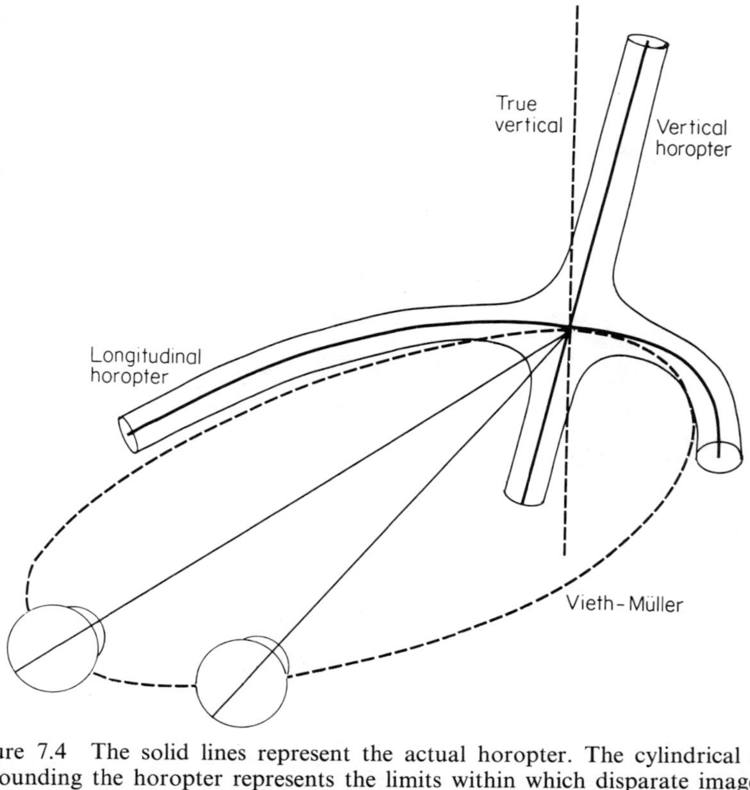

Figure 7.4 The solid lines represent the actual horopter. The cylindrical zone surrounding the horopter represents the limits within which disparate images of point-like objects are perceived as fused (Panum's fusional areas). The dashed lines represent the theoretical horopter based on the assumption of exact geometrical correspondence between the two retinae (From Tyler and Scott, 1979. Reproduced by permission of Harper & Row, Publishers, Inc.)

correspond in both horizontal and vertical directions simultaneously. I have considered only the horopter for symmetrical convergence. With asymmetrical convergence, the situation is more complex. Readers are referred to Tyler and Scott (1979) for a discussion of this topic, which is beyond the scope of the present volume.

Hering stated the facts of binocular correspondence as the **law of identical oculocentric visual directions** in the following words: 'For any given two corresponding lines of direction, or visual lines, there is in visual space a single visual direction line upon which *appears* everything which *actually lies* in the pair of visual lines' (Hering, 1942, p. 41). The truth of this statement was demonstrated by Hering in the following way: 'Let the observer stand about 1/2 metre from a window which affords a view of outdoors, hold his head very steady, close the right eye, and direct the left to an object located somewhat to the right. Let us suppose it is a tree which is well set off from its surroundings. While fixing the tree with the left eye a black mark is made on the window pane at a spot in line with the tree. Now the left eye is closed and the right opened and directed at the spot on the window, and beyond that to some object in line with it, for example, a chimney. Then with both eyes open and directed at the spot, this latter will appear to cover parts of the tree and chimney. Both will be seen simultaneously, now the tree more distinctly, now chimney, and sometimes both equally well, according to which eye's image is victor in the conflict. One sees therefore, the spot on the pane, the tree and the chimney in the same direction' (Hering, 1942, p. 38). Hering used Figure 7.5 to illustrate this situation; the figure shows only one of each of the disparate images in each eye. Another way to illustrate the same point is to hold a horizontal card up to the eyes with a line line extending from the centre of each pupil to an apex, as in Figure 7.6. A thin vertical separator down the centre of the card ensures that each eye sees only its own line. If the two lines are made visually distinct, for instance by drawing one line in one colour and the other in another colour, and if fixation is maintained on the point where they meet, the two lines will appear to extend straight out from this point. Hiroshi Ono recently drew my attention to the fact that W. C. Wells, a London physician, first described these principles of visual direction in 1792, 42 years before Hering was born.

There is certainly a problem when we judge the direction of an object with both eyes open, because the two eyes are not in the same place. If we are to have a unified sense of visual direction, some way must be found to reconcile impressions derived from these two vantage points. The observations just described demonstrate that, at least for symmetrical convergence, we judge visual directions with reference to a point midway between the eyes. This common point of reference for the two eyes is known as the **cyclopean eye**, or **visual egocentre**. It is as if the directions of objects on any pair of corresponding visual lines were judged as though the objects were seen by an eye midway between the eyes (see Figure 7.5).

We can think of the visual fields of the two eyes fusing to form a unified **cyclopean field**. The oculocentric directions of points within this cyclopean visual

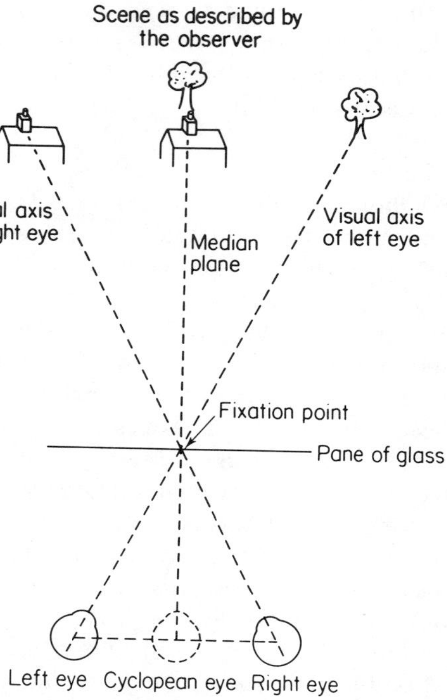

Figure 7.5 Diagrammatic representation of Hering's law of identical binocular directions

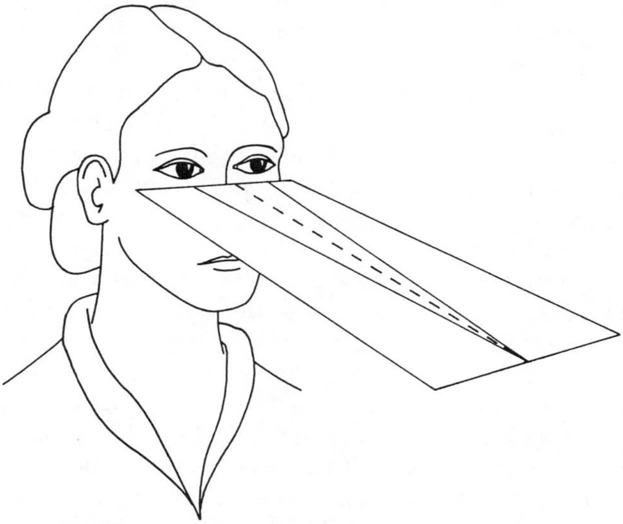

Figure 7.6 A simple way to demonstrate that visual directions in the two eyes are referred to a common egocentre. Each line must point accurately to the pupil of an eye and fixation maintained on the point where the lines meet. The two lines will then appear to be superimposed along the median plane of the head

field are formed by the superimposition of the sets of corresponding points from each retina. However, not only corresponding images but also slightly disparate images are experientially fused in the cyclopean field. One may then ask what determines the oculocentric direction of two fused but disparate images within the cyclopean field. The cyclopean oculocentric direction of a fused image could be that of the image from one or other eye or it could be the average of the oculocentric directions of the two monocular images. To determine the answer to this question, Ono *et al.* (1977) presented subjects with the stereogram shown in Figure 7.7. When the eyes were converged on the two circles so that the circles stimulated corresponding points in the two retinae, it was found that the fused image of the enclosed dots appeared centred in the fused image of the circle. This indicates that the cyclopean oculocentric direction of fused disparate images is the average of the oculocentric directions of the two monocular images. With large disparities, the experience of fusion gave way to an impression of two images in rivalry, with one dot being suppressed by the other. When this happened the apparent direction of the dot was determined by whichever of the two disparate images was dominant at the time. If disparity between the monocular images was very large, the images neither fused nor rivalled but appeared as two images. Sheedy and Fry (1979) also found that the cyclopean oculocentric direction of fused disparate images is essentially an equally weighted compromise between the two monocular directions, and Kertesz (1980) has reported that the perceived orientation of the fused image of stereograms consisting of gratings which have a cyclodisparity is the mean orientation of the disparate images.

A corollary of the law of identical visual directions is that two objects at different distances, which appear in alignment when viewed with one eye, will not appear to be in alignment when viewed with the other eye. This fact is demonstrated when one sights a distant object through a ring with one eye and then transfers the gaze to the other eye. When both eyes are open, there is conflicting information about the alignment of the ring and the object. In this situation there is a very large binocular disparity and most people accept the information in one eye, their sighting eye, and ignore what they see with the other eye. The *sighting eye is therefore that eye which is used preferentially for making judgments about the alignment of disparate external objects.* Great care is needed here to avoid confusion. The tasks must be carefully defined. Care must be taken to distinguish between a judgment about when things are in alignment, a judgment about when things are pointing at the self, and a judgment about which

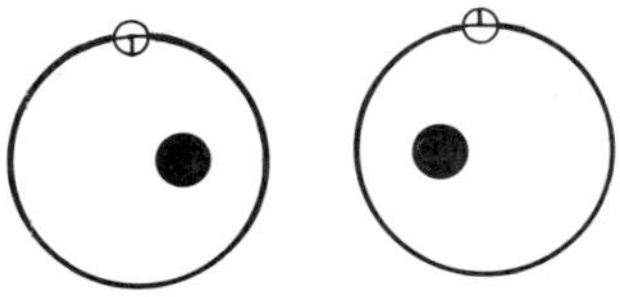

Figure 7.7 Stereogram used by Ono *et al.* (1977)

part of the self they are judged to be in alignment with. These points can be illustrated by sighting down a pencil with only the left eye open. If you point the pencil directly at the self, I think you will find that it points to the nodal point of the eye, although it seems to point to somewhere between the two eyes. This illusion is more evident if the pencil is held in the nasal hemifield. If the observation is repeated with the right eye, the pencil will still appear to point to the self when in fact it is pointing to the nodal point of the right eye. If the pencil is pointed at the self with both eyes open, most people with normal binocular vision will point it at the bridge of the nose, especially if they look back and forth along the pencil.

A horizontal line in the median plane of the head stimulates non-corresponding points in the two retinae, except in the region of binocular fixation. Outside the region of fixation the disparities are too large for binocular fusion or rivalry, so that two lines are seen intersecting at the fixation point. This cross is easily observed by taking a card with a line drawn on it and holding it horizontally just below eye level so that one end of the line touches the bridge of the nose. It is as if the space before one eye had rotated scissor-fashion about the fixation point over the space before the other eye, apparently transferring the objects on each visual axis to the median plane and, for each eye, objects in the objective median plane of the head to the visual axis of the opposite eye.

With symmetrical convergence, all visible objects which are imaged on the foveae are judged to have the same headcentric direction, which lies approximately in the median plane of the head. Furthermore, with symmetrical convergence, objects on the visual axis of an eye are usually judged to be in the median plane of the head even when only one eye is open or when, because of an obstruction, the object can be seen by only one eye. This was demonstrated by Hering in the following manner. Several centimetres in front of the right eye, a card was held with a pin-hole at its centre (see Figure 7.8). A black dot, *P*, on a pane of glass was fixated directly by the left eye and by the right eye through the pin-hole. A small object *A* was placed beyond the glass on the visual axis of the left eye. Although the object was seen by only the left eye and was situated to the right of the subject's median plane, it was reported to be in the median plane behind the point *P*. If the right eye was closed, the subject's report remained the same; only if the eyes changed their position did the apparent position of *A* change.

Suppose that fixation is maintained on point F in the median plane of the head, as in Figure 7.2. The bipolar azimuth of any other point P on the horopter with respect to F will be the same for both eyes. But according to Hering's law of cyclopean projection, the direction of P with respect to F will be judged as if both eyes were in the median plane of the head. In the ideal case, the point in the median plane of the head to which F and P subtend this same angle is the point where the Vieth-Müller circle cuts the median plane. This point provides a theoretical definition of the egocentre. Furthermore, the headcentric direction of point P with respect to such an egocentre is the direction of the line joining the point to the egocentre with respect to the median plane. In other

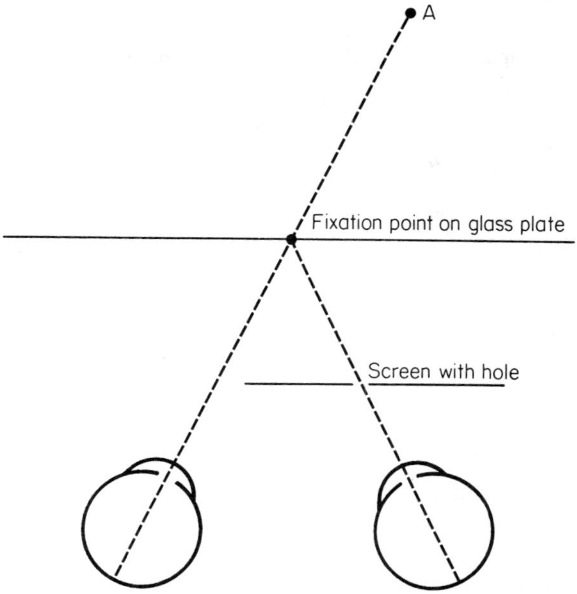

Figure 7.8 Arrangement used by Hering to demonstrate the law of cyclopean projection for an object on the visual axis of only one eye

words, the headcentric direction of any point on the horopter may be defined as the bipolar azimuth of that point with respect to the egocentre and median plane of the head.

Instead of having the gaze held on point F, consider the case in which the gaze is directed to any other point on the same longitudinal horopter, that is, where convergence is asymmetrical but still in the horizontal plane. If the centre of rotation of the eyes were in the same location as the nodal point, the geometry of the situation would be the same as that just given for asymmetrical visual stimuli. But the nodal point is in front of the eye's centre of rotation, so that the geometry will in fact be more complex than this. If the eyes move into a tertiary position, the geometry of the headcentric directions of visual targets becomes even more complicated by the fact that the eyes obey Listing's law (Section 5.2).

The fact that most people consistently use the same eye for sighting or aligning things led Walls (1951a) to suggest that the dominant eye is the egocentre. But most people judge directions as if both their eyes are located in the median plane of the head, no matter which eye they use in a sighting task. When only one eye is open, sighting along a rod and pointing the rod at the self generally give the same result. But when both eyes are open these two tasks give different answers, and it was probably a confusion over this point that led Walls to propose that the dominant eye is the egocentre. An amblyope will probably continue to point a rod at his good eye when both eyes are open—but even he will believe that the rod points to a location somewhere between his eyes.

Here then are five laws or principles of visual direction.

1. *The law of oculocentric direction* states that objects on a given visual line appear to be aligned, or superimposed, that is, they appear to have the same oculocentric direction. A corollary of this law is that non-coincident retinal images give rise to a judgment of spatial separateness. There are occasions when this law does not hold; a single object seen monocularly sometimes evokes a judgment that there are two objects in view. This condition is known as **monocular diplopia** and is discussed in Howard and Templeton (1966).
2. *The first law of monocular headcentric direction* states that, in monocular viewing, objects on the same visual line are judged to be in the same headcentric direction, which is unique to that visual line.
3. *The second law of monocular headcentric direction* states that all visual lines appear to point to the same visual egocentre.
4. *The law of binocular headcentric visual directions* states that every visual line in the binocular field of one eye has a corresponding visual line in the other eye, with an identical apparent headcentric direction. These sets of identical headcentric directions are superimposed to form the cyclopean visual field. This principle is supplemented by the following principles. Within the cyclopean field the headcentric direction of slightly disparate fused images is the average of the headcentric directions of the component monocular images. The cyclopean direction of rivalling images in the two eyes is that of the dominant image. Widely disparate images are experienced as diplopic images in the cyclopean field.
5. *The law of cyclopean projection* states that objects on the visual axes of the two eyes when in symmetrical convergence are judged to be in the median plane of the head. A more general form of this law for asymmetrical stimuli and asymmetrical convergence is that objects on any corresponding pair of visual lines are judged to be in a plane passing through the egocentre and the point in the horopter contained in both visual lines. Complications due to phoria and strabismus are discussed in Section 7.6.3. Ono (1979) has proposed a similar set of principles.

One final qualification must be mentioned. According to this analysis, objects on corresponding visual lines are judged to be in line with the egocentre and to be lying in the plane which passes through this centre and the point of fixation (see Figure 7.5). However, the plane can be said to pass through the objective position of the point of fixation only if it is known that the position of this point is accurately judged by the observer.

This analysis of visual direction accounts for the way in which directional judgments are made about objects at different distances. The fact that the directions of objects at different distances along visual lines are judged in relation to a common egocentre means that the directions of all objects other than those in the horopter are misjudged (see Ono and Angus, 1974). This would seem to be intolerable from a behavioural point of view; however, one usually fixates on an object to which one is attending, so that the illusory directions of other objects are of little consequence.

7.3 Determining the location of the visual egocentre

There are three basic procedures for determining the location of the egocentre.

1. *Multiple alignments with the visual egocentre.* This method was suggested by Howard and Templeton (1966). The subject is required to binocularly fixate the near end of a long rod in a horizontal plane at eye level and rotate the rod until he judges it to be pointing directly at himself. This is repeated for different positions of the rod. The egocentre is then defined as the intersection of the projections of the axis of the rod in its various positions.

Mitson *et al.* (1976) suggested a modification of this method which removes the distracting double images of the non-fixated parts of the rod. In this version the rod is replaced by two stimuli at different distances from the subject. These stimuli may be thought of as replacing the front and back ends of the rod (see Figure 7.9a). Binocular disparity is avoided by arranging that only one stimulus is visible at any given time. The subject is asked to switch on the near and far stimuli in alternation and to move the near stimulus until the imaginary axis joining the two stimuli is judged to be pointing directly at himself. The procedure is repeated for different directions, and the intersection point of the projected axes is taken to be the egocentre.

2. *Pointing to targets in the horopter.* This method was proposed by Funaishi (1926). The subject fixates a point in the median plane at a distance of two metres and is shown other non-fixated targets in the same frontal-parallel plane. He is asked to align his unseen finger with each of the non-fixated targets, with his finger first in a frontal plane at one distance and then in a frontal plane at another distance. The lines joining each pair of finger positions are extended back towards the subject (see Figure 7.9b). The egocentre is the point where the lines intersect. Mitson *et al.* suggested that the subject should be allowed to fixate each visual target in turn rather than keep the gaze fixed on the central target.

3. *Judging points of alignment in the face plane.* In this method, which was devised by Roelofs (1959), the subject sights down a tube with one eye and indicates the place on his face where the tube appears to be aimed. This is repeated for different positions of the tube, and the egocentre is the intersection of the set of lines joining the front of the tube to each location on the face to which it appears to point.

Ono *et al.* (1972) showed that a monocularly viewed point may be mislocalized if there is a phoria in the occluded eye. This could contaminate the results obtained by Roelofs' method, and for this reason Mitson, Barbeito and Ono modified the method. The subject binocularly fixates a near visual stimulus. A far stimulus is seen, by only one eye, in line with the near stimulus (see Figure 7.9c). The subject points with unseen finger towards the apparent position of the far stimulus. This is repeated for several directions of the two stimuli. The egocentre is the point of intersection of the set of lines each of which joins the fixation point to the point which the subject indicated.

Mitson *et al.* (1976) found that each of the three methods produced results with high retest reliability but found no correlation between them. This lack of

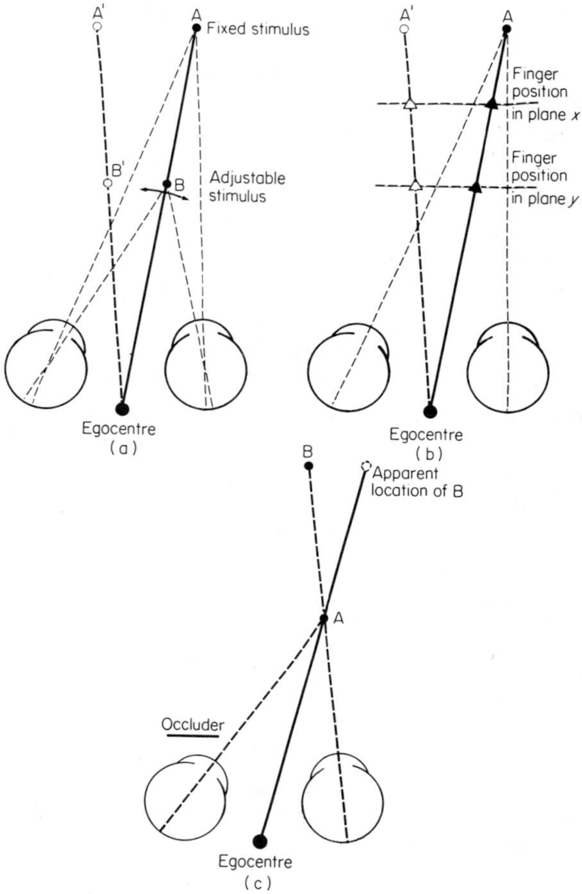

Figure 7.9 Illustrations of the three tasks used by Mitson *et al* (1976) to measure the location of the visual egocentre. (a) Modified Howard and Templeton's method. Subject sees A and B in alternation and adjusts B to be in the same direction as A. This is repeated for other positions of A and B, such as A′ and B′. (b) Modified Funaishi's task, in which the subject fixates A and points with unseen finger in planes *x* and *y* towards A. This procedure is repeated for other positions of A, such A′. (c) Modified Roelofs' method. The subject fixates A with both eyes and points to the apparent location of B. This procedure is repeated for other positions of A and B (Copyright, 1976, Canadian Psychological Association. Reprinted by permission)

correlation suggests that the procedures were not measuring the same thing, or at least that they are subject to errors of measurement which vary across methods and across subjects. Barbeito and Ono (1979) found that the three methods gave different results for a given subject, although each of them gave results which were consistent both within and between testing sessions. They tested the ability of each method to predict a subject's performance on other spatial tasks which depend on the location of the egocentre. These criterion tasks were (1) judging straight ahead, (2) setting a point at one distance between two other points at another distance, and (3) judging the extent of apparent movement of visual targets during accommodative vergence. Only the first method, involving the multiple alignment of targets with the egocentre, successfully predicted performance on these criterion tasks. This is the only purely visual method; the others involve pointing to the apparent position of visual stimuli, and this may be why they relate less well to other purely visual tasks. The first method was also the most precise method, which again is probably due to the fact that it does not involve highly variable pointing responses.

In Section 7.2 I defined the egocentre as the point on the median plane of the head to which bipolar azimuths are projected unchanged. In the ideal case, this definition places the egocentre just behind the corneal plane of the eyes. On the other hand, one might have predicted on other theoretical grounds that the egocentre is on the axis of rotation of the head. Directional judgments made in relation to such a centre would not be affected by rotations of the head on a vertical axis, as they would be if the egocentre were situated outside this axis. However, the bipolar azimuths of points perceived by each eye could not be transferred unchanged to such an egocentre.

Funaishi (1927) and Roelofs (1959) reported that the egocentre lies on the centre of rotation of the head. On the other hand, Fry (1950) placed it several centimetres in front of the corneal plane. Barbeito and Ono (1979) found that all methods of determining the position of the egocentre are critically dependent on accurate control of fixation. In the earlier methods, fixation was difficult to control, and this was probably why such divergent results were reported. Barbeito and Ono controlled fixation in a way that has already been indicated. They found that, in the three methods that they used, the egocentre was on average very close to a point in the corneal plane midway between the eyes.

7.4 Procedures for studying headcentric and bodycentric visual localization

If an experiment is being planned on headcentric or bodycentric visual localization, several basic decisions must be made about the best procedure to adopt.

1. It must be decided whether the experiment is to be about headcentric or about bodycentric direction. If it is about headcentric direction, the instrument must be calibrated with respect to the median plane of the head and the subject's head must be clamped, preferably with the aid of a wax bite plate. The experimenter may wish to take measurements with the head in more than one position, which is all right as long as measurements are made with reference to the median plane of the head. The important thing is that the head should not be allowed to move inadvertently, because any such movement disturbs the calibration of the measuring system. If the experiment is about bodycentric direction, the body must be clamped and the instrument calibrated with respect to the median plane of the body.

2. Some way must be found to ensure that the subject is making the correct judgment. It is no use expecting subjects to follow verbal instructions, because they may not understand them, and even if they do, their conscious understanding may not guide their behaviour. Some behavioural test of the criteria used by subjects must be embedded in the experiment. As an illustration of how this may be done, consider an experiment to determine the precision of headcentric directional judgments of a visual target in various locations. If the investigator is interested in only the eye-position component of the task, the subject should be required to fixate each target, and this should be sufficient guarantee that he will make headcentric and not oculocentric judgments. But suppose the experimenter is interested in headcentric judgments for different retinal locations of the target as well as different positions of the eye. In other words, suppose he is interested in both the eye-position and oculocentric components of headcentric localization. The experimenter will present the target in various retinal locations for each of several positions of the eyes. This will normally require a fixation stimulus in addition to the target, and the presence of two stimuli may predispose the subject to make oculocentric judgments in spite of instructions which he may be given. However, if the subject is required to make judgments about the direction of a fixed target with the eyes in several positions, his performance will reveal whether or not he has interpreted the instructions correctly. If he is making oculocentric judgments of the target, he will give very different answers for different positions of his eyes, and if he is making headcentric judgments, he will tend to give the same answer whatever the position of his eyes (insofar as direction constancy holds). So this is the kind of thing I mean when I talk about the need for a test of the subject's criterion to be embedded in the experiment.

If the experiment is about bodycentric directions, decisions must be made about whether the oculocentric, the eye-position, or the head-position components are to be varied, which means that decisions should be made about whether each target is to be fixated and whether the head is to be allowed to centre on each target. If a pointing response is used, this is sufficient guarantee that bodycentric judgments are being made. But if verbal estimates of direction are used, all combinations of target position with respect to the three components must be presented, so that a check may be kept on the frame of reference that the subject is using.

3. Great care must be taken in the design of the visual display, special attention being paid to the boundaries of the visual field. The stimuli may be presented on a dark field, in which case the visual field will have no boundaries, or they may be presented in an homogeneously illuminated field (*Ganzfeld*), in which case the boundaries of the field may be the orbital ridges and nose or an artificial boundary. If a boundary is in view, it must be realized that this may affect performance.

4. Subjects should not be given knowledge of the accuracy of their responses, because this may lead them to make use of stimulus cues which they do not normally employ.

5. The response or judgment that the subject is required to make must be

carefully selected. Any judgment of direction necessarily involves a standard of reference, which may be any one of the following:

(a) a second stimulus presented at the same time
(b) a second stimulus presented previously
(c) a natural norm such as straight ahead
(d) the position of the unseen finger.

If possible, only one target should be in view, because subjects will be prone to judge the relative oculocentric direction of one object relative to another if they can see more than one thing at once. Even if subjects are required to judge the position of a target in relation to another stimulus object presented previously, they may still be prone to make oculocentric rather than egocentric judgments. Furthermore, this procedure introduces memory for the location of the first target. Memories fade at a certain rate, so that interstimulus intervals must be carefully controlled.

If the subject is required to judge the position of targets in relation to a natural norm, such as straight ahead, there are various scaling procedures that may be used. He may be asked to report verbally the angular position of the visual target in terms of degrees or some other arbitrary scale. People tend to round off scale values in the neighbourhood of the ends of the scale (Saltman and Garner, 1950), and so the pattern of constant errors revealed by this method may be an artefact. This problem may be avoided by asking subjects to place a target in specified locations, or by asking them to successively partition a space.

Pointing with unseen hand to visual targets is a procedure which should be used with care in experiments on headcentric direction. Any response which involves the neck or anything below the neck requires the subject to register the position of the head on the body, and the task is then no longer a headcentric one. The fact that the head is clamped may not be sufficient precaution because the clamp fixes only the objective position of the head; it does not fix the apparent position of the head, and that is what matters if the response involves pointing with any organ below the neck. When bodycentric judgments are being investigated, the task should preferably involve pointing with the hand because this will ensure that the subject is making bodycentric judgments. However, it must be realized that this introduces another source of variance, namely, the sense of position of the arm on the body, and it may also introduce constant errors, because the arm is asymmetrically placed on the body. It may help to have the subject point with both arms at once.

Experiments which will be reviewed in Section 7.8.2 required subjects to localize transient stimuli as they moved their eyes to and fro along a fixed path. Once a subject in such an experiment has become familiar with the duration of the motion of the fixation point, he may judge the positions of transient stimuli on the basis of their time of arrival rather than on the basis of their spatial position, and this may lead the experimenter to reach false conclusions about the subject's spatial abilities.

These then are some of the precautions required in conducting experiments on headcentric or bodycentric direction; precautions which have often been neglected. The rest of this chapter is devoted to the subject of headcentric direction. The earlier literature on bodycentric localization is reviewed in Howard and Templeton (1966). Other studies related to bodycentric localization are reviewed in Section 11.3.

7.5 The precision of headcentric visual localization

7.5.1 Psychophysical determinations

Merton (1961) conducted a series of experiments on the precision of directing the eyes in the dark to targets previously seen. The head was clamped as it should have been for a headcentric judgment. The oculocentric component was minimized because the subject was allowed to fixate the target. The subject fixated a target presented in dark surrounds in one of several positions between 20° to the left and 20° to the right. The dark surrounding eliminated any contamination due to the boundaries of the visual field. The target was switched off, the subject lowered and shut his eyes, and then opened them and attempted to fixate the position of the extinguished target. Standard deviations of lateral fixation ranged between about 2 and 4° for different subjects, with a mean of about 3°. (Similar results were obtained by Heywood, 1973.) This performance reflects the variability of the initial registration of the horizontal position of the eye, the loss due to 'forgetting' while the eyes were closed, and the variability of refixations. However, the time intervals, and hence the amount of 'forgetting', were not controlled and so the significance of these results is obscure. The variability of refixations was less in the vertical direction, but this may have been due to the fact that the vertical position of the target did not vary. As we shall see, others have not confirmed this difference. The variability of refixations to the straight-ahead position was less than to lateral positions, although data for only one subject were analysed in this way. In a second experiment, Merton allowed the subject to fixate a target, which was then extinguished, moved to left or right through several degrees, and put on again. The subject had to replace the target to its original position. The standard deviation of one subject was 0.6°. The time taken to make the judgment was not controlled, as it should have been. Furthermore, it is not clear what the subject did with his eyes in this experiment. He may simply have kept his eyes in the position of the initial target and restored the displaced target to the centre of vision, in which case the experiment measured the ability of the subject to maintain a given position of fixation without a fixation point, rather than his ability to relocate his eyes.

Matin and Kibler (1966) conducted a study of what they referred to as '. . . the ability of an observer to specify an egocentric direction' (p. 407). The subject, in a dark room, fixated a light which was either straight ahead or 3.4° to left or right. This light was then extinguished for a period of 3 sec, during which time the subject attempted to maintain the same fixation position. At the end of this period a 100 msec flashed target was exposed at one of several positions on one side or the other of the extinguished fixation light and the subject had to report '. . . whether the flash appeared to the left, to the right, or in the same location as the fixation target' (p. 410). A similar experiment was done in which the subject had to judge the vertical position of the flashed target with respect to fixation lights above and below straight ahead. The standard deviations of the psychophysical functions were about 20′ of arc for the straight-ahead position of gaze and significantly more for lateral positions of gaze. The lateral and vertical gaze positions gave similar results. This precision is two orders of magnitude less than acuity measured when the two visual stimuli are present simultaneously. Two things could have been happening in the 3 sec interval between stimuli: the subject may have 'forgotten' the retinal location of the first stimulus, or his eyes may have wandered from the fixation point, thus altering the relative retinal location of the two stimuli. We shall see in Section 7.5.2 that the eyes do in fact drift in a random walk when a person tries to maintain fixation in the dark. Nobody has suggested a way of investigating whether people forget the retinal location of a stimulus. The effects of eye drift were investigated by Matin *et al.* (1966), who took the same measurements as Matin and Kibler but in addition recorded the position of the eyes during the 3 sec interval between the presentation of the fixation target and the flash target. Involuntary movements of up to one degree in either direction occurred in this interval, and most of the variability in the subject's judgments could be accounted for in terms of the consequent change in the oculocentric position of the targets. The obvious conclusion is that the subjects had been making oculocentric judgments, and not headcentric judgments as Matin and Kibler had thought. This does not seem to be the conclusion drawn by Matin, Pearce, Matin and Kibler, because they wrote that their results '. . . showed no trace of compensation for the ocular movements from proprioceptive sources' (p. 416). The implication here is that the subjects were making headcentric judgments but failed to take proper account of the drift of their eyes. But I suggest that when a subject is presented with two visual stimuli on different parts of the retina, he will, whatever the instructions, judge their relative oculocentric directions. To investigate headcentric judgments, Matin *et al.* should have allowed the subjects to fixate the flash target, thus forcing them to make headcentric judgments. They might then have found that eye position is taken into account; not necessarily proprioceptively, but perhaps by registered efference.

Kinchla and Smyzer (1967) used a signal-detection procedure to measure their subject's sensitivity to a sudden change in the position of a small light in dark surroundings. In this study, and in a subsequent study by Keller and Kinchla (1968), it was shown that the ability of a subject to detect a displacement of the stimulus declined as the interstimulus interval increased. Sensitivity to a displacement provides a measure of the extent to which the subject has become uncertain about the position of the initial stimulus. They found that the amplitude of displacement which was reliably detected increased with the interstimulus interval at a greater rate than the extent to which the eyes are known to drift from a previous position of gaze, and concluded from this that drift of the eyes is responsible for only part of the increasing uncertainty about image position. However, the crucial factor in eye drift may not be the extent to which the eyes wander from the initial fixation of gaze, but rather an imbalance in the registration of slow drifts and saccades (see Section 7.5.2). Kinchla and Allen (1969) developed a stochastic model of the way in which egocentric positional information is lost during the interval between the presentation of two comparison stimuli (see also Kinchla, 1976).

Findlay (1974) measured the precision of directional judgments with time intervals of 100–500 msec between the initial and final position of a displaced visual target. He also found that the precision fell off more rapidly than the rate predicted from the extent of eye drift. In a second experiment he found that the reported position of the second stimulus relative to the first tended to be in the same direction as a microsaccade which occurred just before the judgment was made. Findlay argued that this is opposite to the effect one would expect on the basis of the relative oculocentric positions of the targets. He suggested that microsaccades in one direction suppress the signals indicating image motion. It is not clear to me how this would work, because the second stimulus was not in view when the saccade occurred.

The only other recent experiment on the precision of headcentric visual localization of which I am aware was conducted by J. W. Miller and Hall (1963). They set out to measure '. . . the capability of the eye to position itself in the absence of retinal stimulation. . .' (p. 14). The subject was positioned at the centre of an illuminated spherical *Ganzfeld* and asked to return a visual target to the straight-ahead position from various off-centre positions. The mean absolute error (a poor measure of precision) was 2–3° with reference to the median plane of the subject's body. However, the subject's head was free to move, so that drifts in head position would contaminate the measure that was taken. These results therefore tell us nothing about the '. . . capability of the eye to position itself'. Furthermore, the subjects were free to follow the displaced target with their eyes or remain fixated at the centre, so that we do not know whether this task was performed as a headcentric or as an oculocentric one. This study was replicated by Katz *et al.* (1965), but the head and eyes were not restrained in their study either.

More carefully conducted experiments are called for on the precision of headcentric judgments of visual direction.

The precision of headcentric visual localization is also indicated by the extent to which a stationary visual object appears to change its position as fixation is maintained upon it. This is the topic which I shall now review.

7.5.2 Autokinetic movement

In 1799 the astronomer von Humbolt noticed that stars seen through a telescope appeared to drift about. He was convinced that the effect was due to physical factors outside the observer, and it was not until 1858 that Sweizer demonstrated the subjective nature of the phenomenon by noting that different people gave differing reports of movements of the same star at the same time. Aubert rediscovered the phenomenon in 1887 and called it **autokinetic movement**, by which name it is now generally known. For detailed reviews of this subject, see Royce *et al.* (1966) and Levy (1972).

There are many circumstances which induce apparent movement in a visually fixated stationary point of light seen in dark surroundings. If a display moving in a certain direction has been inspected for some time, a stationary point of light seen subsequently appears to move in the opposite direction. This is the well-known **movement aftereffect**. If a point of light is surrounded by a frame which moves in a certain direction, the point appears to move in the opposite direction. This is known as **induced movement** (see Section 7.6.2). If the vestibular canals or otolith organs are stimulated, apparent visual movements are induced which are known as the **oculogyral** and **oculogravic illusions**, respectively (see Sections 9.3 and 10.4.8). Finally, if the eyes are held in an eccentric posture for some time and then returned to fixate a point of light in the straight-ahead position, the light usually appears to be displaced in a direction opposite to that of the previous eccentric posture (see Section 7.6.4). Autokinetic movement occurs when a point of light in dark surroundings is visually fixated in the straight-ahead position when none of the above conditions applies. The light appears to drift slowly hither and thither in a frontal plane. The movement does not usually start until the light has been fixated for several seconds and its amplitude usually increases as fixation is continued.

There are two sensations involved in autokinetic movement. One is the sensation that the point of light is moving, and the other is the sensation that it has changed its egocentric location. These are not necessarily related in the way that logic might lead one to expect. For instance, Graybiel and Clark (1945) noted that subjects often reported that the apparent distance covered by a point of light was much less than the product of the duration and apparent speed. A mismatch of this kind between apparent displacement and apparent speed is known as **paradoxical movement**. The sensation is of an object moving but not changing its location, or changing its location by less than the movement indicates.

It is difficult to measure autokinetic movement. It cannot be measured by providing the subject with a visual frame of reference, because the introduction of other visible objects disturbs the effect. The subject may be asked to give a verbal estimate of the apparent movement, but because the distance of the target from the subject is indeterminate, so will be the estimates of the linear distance moved. It is not easy to make judgments in terms of visual angle. Guilford and Dallenbach (1928) asked the subject to trace the course of the apparent movements on paper, but such a procedure assumes that people can draw accurately in the dark. Bridges and Bitterman (1954) provided the subject with a lever which permitted him to move the autokinetic stimulus in a frontal plane. The subject was told that the light would be moved and he was to return it to its starting point. A permanent record of the direction and extent of the subject's tracking of his own autokinetic movement was thus obtained. To what extent this procedure contaminates the results is not known: it certainly depends on the ability of subjects to remember the starting point. The method indicates the apparent displacement of the light rather than its apparent speed. Pearce and Matin (1966) used a nulling method which was designed to indicate the apparent speed of autokinetic movement. In this procedure a point on an oscilloscope was moved by the experimenter so as to null the reports of movement which the subjects made.

Levy (1972) described a signal-detection procedure for measuring autokinetic movement, but did not put it into practice. The idea is that autokinetic movement may be regarded as a false positive response in a visual-movement detection task. The subject is presented with stimuli which actually move in various directions and at various speeds and is asked to report the presence and direction of the movement. On catch trials, the stimulus does not move. With such a procedure it should be possible to decide to what extent autokinetic movement is influenced by factors which cause the subject to change his criterion (β) as opposed to factors which change his ability to detect movement (d'). The evidence that will be reviewed suggests that autokinetic movement is due to 'noise' in the visual mechanisms responsible for assessing egocentric direction, but that whether or not a subject interprets this noise as a real movement of the visual object depends on the criterion level which he sets.

Several theories of autokinetic movement have been proposed, but it may be misleading to regard them as alternatives: they are best considered possible coexisting contributory mechanisms. There are two types of theory. In the first type, autokinetic movement it said to be caused by movements of the retinal image which occur as a result of the small drifts and saccadic movements of the eyes during fixation. In the second type, autokinetic movement is thought to be caused by instability in the sense of direction of gaze, due either to eye movements which are unregistered by the perceptual system or to fluctuations in the pattern of innervation of the extraocular muscles. I shall deal with each type of theory in turn.

Autokinesis and motions of the retinal image Any involuntary eye movement, such as nystagmus or passive movement of the eye by the finger, certainly produces an illusion of movement of the visual scene. The crucial question is whether during autokinetic movement eye movements occur which are large enough to move the retinal image by the required amount. Guilford and Dallenbach (1928) photographed the eye during autokinetic movement and found no evidence of eye movements in excess of 1°, which was the smallest movement that could be detected with their apparatus.

A new approach to the problem of the role of eye movements in autokinetic movement has been reported by Matin and MacKinnon (1964). They compared the amount of autokinetic movement of a point of light in each of the eight principal directions with normal viewing, and again when horizontal retinal-image movements consequent upon eye movements were compensated for by the 'stopped-image' technique. The horizontal autokinetic movement was markedly reduced in the stopped-image condition compared with the normal viewing condition. This suggests that the small drifts and saccades which are known to occur even during steady fixation contribute to autokinesis.

There are at least two ways of accounting for the fact that eye movements as small as those involved in visual fixation can cause apparent movements of a visual target of several degrees. One possibility is that, in a given period of time, there is a predominance of slow drifts of the eyes in one direction. The eyes do not necessarily wander very far, because microsaccades may occur in the other direction. The key idea is that the motion of the retinal image produced by the slow drifts is detectable by the visual system but the very rapid image motions produced by the microsaccades go undetected as movements. As far as the motion-detection mechanism of the visual system is concerned, it is as if the image had moved through the total extent of the slow drifts occurring in that interval of time. As far as the position-detecting mechanism is concerned, the image has not moved. Crone and Verduyn Lunel (1969) proposed another theory to account for why small eye movements would cause autokinetic movement. They argued that when a moving visual object is pursued by the eyes, the eyes tend to lag slightly behind the object, thus causing the image to be displaced to one side of the centre of the fovea. A persisting small eccentricity of a fixated object normally indicates to the visual system that the object is moving, and provides a stimulus to tracking movements of the eyes. If during fixation of a stationary point of light the image becomes eccentric for a while, this will be interpreted as due to a movement of the object which is being pursued by the eyes. Crone and Verduyn Lunel produced evidence in support of this theory by showing that autokinetic movement at any time tended to be in the direction of the eccentricity of the retinal image at that time.

Autokinesis and the sense of the position of gaze Both the theories just described are based on the idea that autokinetic movement is due to the way in which movements or displacements of the retinal image are interpreted. Autokinetic movement may also be explained in terms of the way in which movements of the eyes are interpreted. For instance, if, during fixation, the slow drifts of the eyes are registered by the visual system but the microsaccades are unregistered, the net effect will be that several slow drifts in one direction will be interpreted as a movement of the eyes through the total extent of those slow drifts. Over a longer period of time, as the eyes drift randomly in different directions, there will be a random walk of the apparent position of gaze and hence of the apparent position of the point of light. A similar argument would apply if the slow drifts were unregistered and the microsaccades were registered. The important point is that the sense of direction of gaze will be unstable if there is any difference in the way drifts and microsaccades are registered. There is evidence, which I shall now review, that autokinetic movement is associated with changes in the sense of position of the eyes, although this evidence does not isolate the precise cause of these changes.

It has been found that autokinetic movement occurs in the same direction as a previous fixation of the eyes (Carr, 1910; Gregory and Zangwill, 1963) and in the same direction as a maintained eccentricity of gaze (Adams, 1912; Luchins, 1954a; Levy, 1973). In Section 7.6.4 it will be shown that eccentric fixation alters the sense of position of the eyes, and it is presumably this which induces autokinesis under these circumstances.

The role of changes in the felt position of the eyes in autokinesis is also indicated by the results of an experiment by Jordan (1968) in which subjects were asked to place their eyes in a straight-ahead position while observing autokinetic movement. In response to such instructions subjects moved their gaze in the opposite direction to that in which the autokinetic light had just been seen to be moving. DeSisto and McLaughlin (1968) obtained the same result.

There should be one position of the eyes which, if maintained for some time, would produce an equal balance of innervation in the extraocular muscles and a symmetrical distribution of drifts and microsaccades in the various directions. When the eyes are in such a position, autokinetic movement should be equally likely to occur in each direction. Matin and Kibler (1966) asked subjects to make judgments about the position of a point of light in relation to a previously seen point of light and found that there was a position of gaze for which errors in different directions were equally probable. Vaegan (1976) also found that there was a position of gaze for which autokinetic movements in different directions were equally probable, and that this position was correlated with the position of gaze determined by Matin and Kibler's method. He suggested that this position be characterized as the **physiological position of rest**.

The involvement of changes in the sense of direction of gaze in autokinetic movement is also suggested by the fact that afterimages are subject to autokinetic movement (Piggins, 1965; Brosgole, 1968). If autokinetic movement depended only on movements of the retinal image, it should be absent when the target is an afterimage.

Autokinetic movement is probably the result of several factors: retinal movements induced by small eye movements, slight changes in the retinal eccentricity of the fixated light, unmonitored movements of the eyes, and imbalance in the innervation of the extraocular muscles. These factors may all be subsumed under the general title of noise in the headcentric visual-direction system. But one question still remains. Why are the effects of noise in the headcentric visual system evident only when a small object is viewed in dark or homogeneous surroundings? Why do not ordinary visual scenes appear to drift about? I think that the general answer to this question is that we assume ordinary visual scenes do not move, and set a high criterion level (β level) when assessing visual-motion signals with a full field of view. In other words, visual-motion signals from the retina or from eye movements must be stronger when there is a full field of objects than when there is only one isolated object before we are willing to interpret these signals as an actual movement. Let us take a brief look at the evidence on the role of the visual frame and of cognitive factors in autokinesis.

Royce *et al.* (1962) studied the role of the visual frame quantitatively. As the number of lights in a multiple-light configuration was increased, autokinesis was less often reported. A concentric circular band of light also reduced the apparent movement of a central point, irrespective of the radius of the circle. The brighter the circle, the more the reduction of the movement. Luchins (1954b) controlled the amount of visual field by gradually illuminating the inside of a box in the centre of which the autokinetic light was situated. As the illumination was increased, the apparent movement became less. Larger single patches of light are reported as moving less than smaller patches (Honeyman *et al.*, 1946; Edwards, 1954). The effect of size may be due to the increase in the amount of light flux rather than the area of the stimulus: increasing the brightness of a light of constant area has been found to decrease the autokinetic movement (Karwoski *et al.*, 1948; Spigel, 1963). Black points in an homogeneous light field also show autokinetic movement but of smaller magnitude than light points on a dark field (Luchins, 1954b). This reduced effect could be due to the fact that when the whole field is bright the edges of the field come into view and provide an anchor or standard of reference.

Although the amount of autokinetic movement decreases as the visual frame is strengthened, apparent movements may be observed even in a full and varied field of view if steady fixation is maintained for some time (Marks, 1949; Honisett and Oldfield, 1961). The whole field may appear to move, or parts may move relative to other parts. These effects are particularly marked if one steadily fixates a large pattern of regular black lines. The lines bend and contort, and at the same time appear to drift as a whole across the visual field. Stable visual directional values would seem to require constant eye movements as well as a full field of view.

There is a lot of evidence that suggestion and expectation can affect the amount of autokinesis, even when only a single point of light is in view (Corteen, 1970). Furthermore, in the case of visual targets which have an implied direction of motion, such as a picture of an automobile, autokinetic movement has been found to occur more often in the direction of implied movement than in other directions (Honeyman *et al.*, 1946; Borresen, 1973). The magnitude of autokinetic movement has been found to decrease when the observer was allowed to touch the light source or to hold a tube through which the light was seen (Worchel and Burnham, 1967; Lackner and Zabkar, 1977).

It seems reasonable to conclude that autokinesis results from noise in the headcentric localization system but that whether this noise is interpreted as an actual movement of the stimulus depends on cognitive criteria.

7.6 Constant errors in headcentric visual localization

7.6.1 The effect of oculocentric stimulus asymmetry

The position of the apparent visual straight ahead is affected by asymmetries in the visual field. Dietzel (1924), and later Roelofs (1935), found that if, for instance, the right-hand edge of a luminous figure seen in the dark is placed so that it is in the objective median plane, this edge of the figure is judged to be displaced to the right. In other words, the apparent straight ahead is displaced towards the centre of an asymmetrical visual display. This has come to be known as the **Dietzel–Roelofs effect**. The pre-1966 literature on this effect is reviewed in Howard and Templeton (1966) and, as far as I am aware, no significant studies have appeared since that time. The only well-formulated theory of the Dietzel–Roelofs effect was proposed by Bruell and Albee (1955). They postulated that a fixated object is seen to the right in headcentric space if fixation is maintained by voluntary innervation of the dextro-rotators of the eye and to the left if fixation is maintained by the innervation of laevo-rotators. They suggested that the pattern of voluntary oculomotor innervation at any time could be determined if certain assumptions were made. The first of these assumptions is that an involuntary fixation reflex causes eye turning towards the centre of the object to which one is attending; this has recently received some experimental support (Kaufman and Richards, 1969; Coren and Hoenig, 1972). The second assumption is that when the eyes are turned out of the primary position, involuntary reflexes tend to restore them to that position. These assumptions were then applied to the analysis of the Dietzel–Roelofs effect. With a stimulus extending into the left periphery of vision, a reflex to turn the eyes to the left is invoked which the subject has to overcome voluntarily in order to maintain his central fixation. According to the first hypothesis, such voluntary innervation of the dextro-rotators results in the judgment that the fixated edge of the stimulus is to the right. The fixated part of the figure has to be moved to the left in order to be judged straight ahead, but this causes the eyes to move out of the primary position. Restoring reflexes become increasingly active until they cancel out the fixation reflex. At this point of 'reflex equilibrium', fixation of the right edge of the stimulus can be maintained without voluntary control and therefore the fixated edge will be judged straight ahead.

According to this theory, anything which disturbs the normal balance of voluntary innervation required to hold the eyes in a given location should cause an apparent shift in headcentric directions. We shall see in what follows that there is a lot of experimental support for this idea.

7.6.2 The effect of oculocentric motion

If a stationary object is seen against a moving framework, the object appears to move in the opposite direction. This is the well-known phenomenon of induced movement first described in detail by Duncker (1929). In addition to this apparent motion, the object appears to be egocentrically *displaced* in a direction opposite to the motion of the background. The apparent motion and displacement could be due to the increasing asymmetry of the moving framework, in which case they would be instances of the Dietzel–Roelofs effect. However, this cannot account for the whole of these effects, because induced motion and displacement are present when the background consists of an array of moving spots seen within the confines of a stationary and centrally placed framework (Sugarman and Cohen, 1968). The effects are also present when a complete panoramic array of stripes is rotated round the subject (Fischer and Kornmüller, 1930). Brecher *et al.* (1972) found that the apparent straight ahead was displaced in the direction of rotation of a striped drum and that the magnitude of the effect increased with increasing velocity of the drum up to a maximum displacement of about 10° of visual angle at a drum speed of about 80°/sec. With higher drum speeds the displacement effect became smaller (see Figure 7.10).

This effect is clearly not due to an asymmetrical distribution of visual stimuli, but rather to the asymmetry of visual motion. One could still think of the effect as being due to asymmetrical fixation tendencies, because the moving stripes invoke optokinetic nystagmus which is inhibited by fixation on the target which the subject is asked to set to the straight ahead. Thus, an explanation similar to that which Bruell and Albee offered for the apparent displacement of the straight ahead with asymmetrical static displays may account for the effect obtained with moving but otherwise symmetrical displays. But there are two other possible contributory causes of the displacement effect with moving displays. The first is that it may be due to contrast effects within the oculocentric visual-motion detection system. Recent electrophysiological evidence has revealed the presence of motion detectors in the visual system which are sensitive to the relative occulocentric movements of objects, in addition to the better-known detectors that respond to local retinal motion (Burns *et al.*, 1972), and it may be contrast between these detectors that causes apparent motion and displacement effects. The second possible contributory cause of the motion and displacement effects

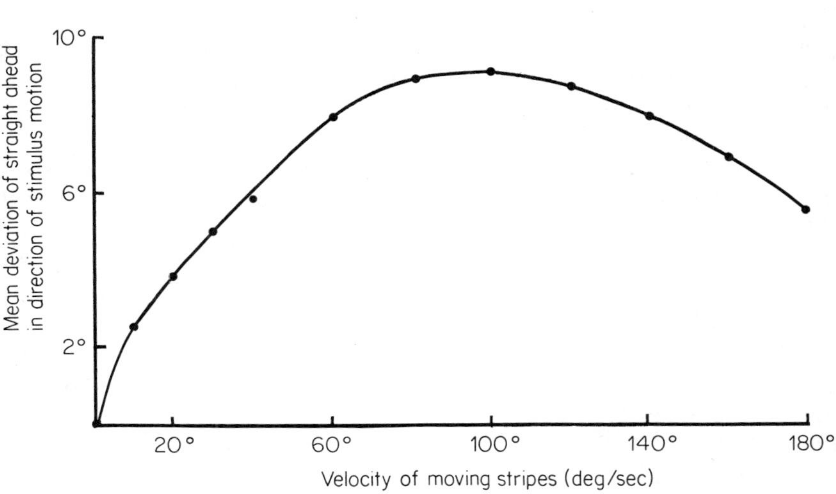

Figure 7.10 Mean signed deviations of the apparent visual straight ahead for a subject viewing horizontally moving stripes at various velocities (Based on data presented in Brecher *et al.*, 1972. Reproduced by permission of Taylor & Francis, Ltd.)

produced by moving stripes is the illusory self motion of the subject induced by the visual movement (see Section 9.4 for a discussion of this possibility).

Gogel (1977) reported that two frames moving apart from each other induced apparent inward motion and displacements of stationary objects contained within each frame. It is difficult to see how eye-movement tendencies could cause this effect, because there was equal and opposite motion of the frames. The effect could have been due to local motion contrast effects or to an illusion of forward body motion induced by the impression of looming created by the parting frames. *Further experiments are required in which measurements are made of apparent self motion and visual displacement induced by looming or contracting visual displays* (see Section 7.9).

7.6.3 The effects of phoria and strabismus

The law of cyclopean projection given in Section 7.2 states that objects on a corresponding pair of visual lines are judged to be in a plane passing through the egocentre and the point in the horopter contained in both visual lines. It is assumed in this analysis that the eyes are correctly converged upon a defined visual object. There are two conditions which complicate the situation. One of these is **phoria**, which is a deviation of an eye inwards (esophoria) or outwards (exophoria) from the intended point of convergence when that eye is occuluded. The other condition is squint or **strabismus**, which is a deviation of one or other eye from the intended point of convergence when both eyes can see the visual display. Some strabismics show a constant convergent (esotropic) or divergent (exotropic) deviation of one eye, while others, so-called **alternating strabismics**, show a deviation sometimes of one eye and sometimes of the other.

Consider first the effects of phoria on egocentric localization. According to the law of cyclopean projection, an object upon which the eyes are converged is perceived as lying on a line drawn through the cyclopean eye and the point of convergence. As long as convergence is maintained on the target, the above principle will define its apparent location (assuming that its distance is accurately judged). However, if one eye is esophoric, that eye will deviate inwards when it is occluded, and the perceived location of the target should shift away from the occluded eye, as shown in Figure 7.11(a). In the case of a person with an occluded exophoric eye the perceived location of the target should shift towards the occluded eye, as shown in Figure 7.11(b). In either case the subject should experience an apparent movement of the target in the direction of his phoria as the phoric eye is occluded, and a return movement as that eye is uncovered. These predictions were confirmed in studies by Ono *et al.* (1972) and Ono and Gonda (1978). These findings confirm that egocentric directions are judged in terms of pooled information about the positions of the two eyes and not merely the position of the dominant eye or of the eye that happens to be seeing the target.

A person with a constant strabismus of long standing learns to suppress the visual input from the deviating eye when both eyes are open. When only the deviating eye is open, that eye is able to fixate on an object and see it, although the

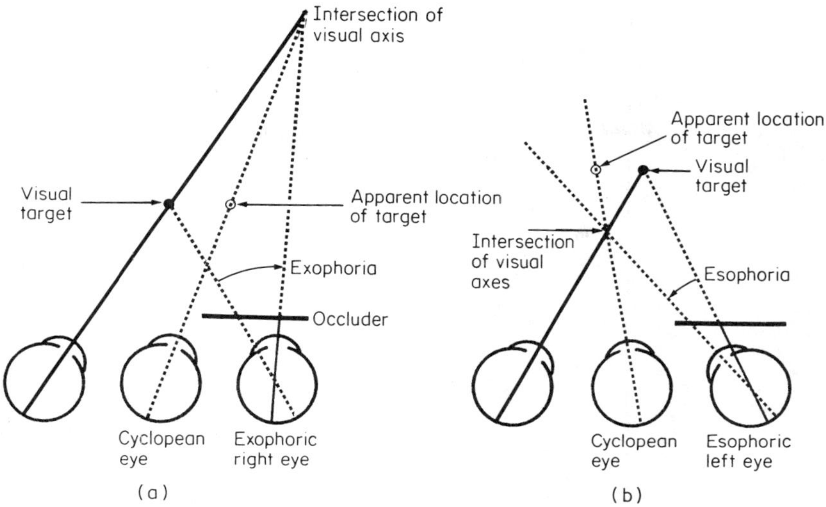

Figure 7.11 The predicted effect of occluding (a) an exophoric eye and (b) an esophoric eye on the apparent location of a visual target. It is assumed that the distance of target is perceived correctly (Adapted from Ono and Gonda, 1978. Reproduced by permission of Pion Ltd.)

covered normal eye will not be converged on the object being inspected. The question arises as to whether, under these circumstances, the direction of the object fixated by the strabismic eye is judged in terms of the position of that eye, in terms of the position of the covered normal eye, or in terms of their combined positions. Mann *et al.* (1979a) showed that the second possibility is true: constant strabismics learn to use the direction of their normal eye whether or not that eye is viewing the visual target. Alternating strabismics, on the other hand, judge the direction of a visual target in terms of whichever eye is open: these subjects learn to dissociate the two eyes in making directional judgments.

7.6.4 The effects of previous eye and head positions

Several related distortions of headcentric space are produced when the eyes are held in an off-centre position for a short time. While the eyes are deviated, there is a shift of several degrees in the apparent straight-ahead position of a visual target in the direction of deviation of the eyes. In addition to this perdeviation effect, there are two postdeviation effects, or **postural aftereffects**. The first of these is that when the observer attempts to return his gaze to the straight-ahead position his eyes settle in a position which is displaced in the direction of previous deviation. The second aftereffect is that when a person attempts to point to visual targets with unseen hand he misses in the direction of previous eye deviation.

Figure 7.12 illustrates the perdeviation displacement of the apparent straight-ahead position of a visual target while the eyes are fixated on a stimulus at various angles of eccentricity. It can be seen that the shift in the apparent straight ahead is

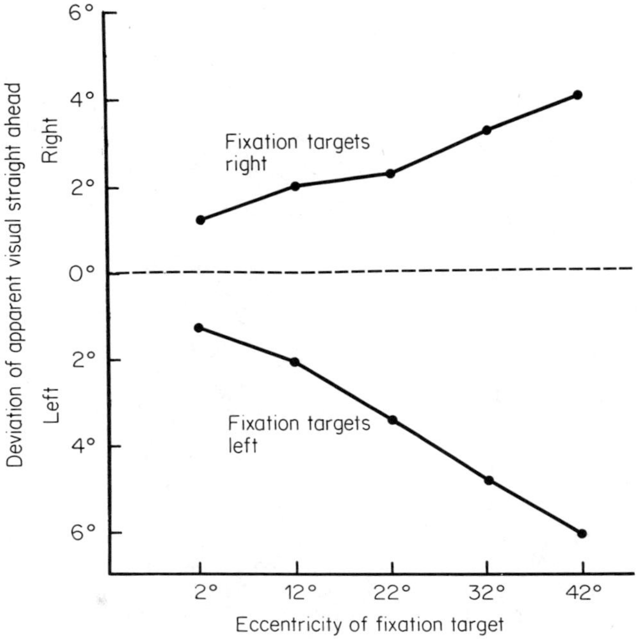

Figure 7.12 Deviation of the apparent visual straight ahead as a function of eccentricity of gaze. Data based on means of six subjects (From Morgan, 1978. Reproduced by permission of Psychonomic Society, Inc)

a more or less linear function of the eccentricity of gaze, reaching a maximum of 6° at a deviation of 42°. Hill (1972) and Morgan (1978) conducted experiments which showed that the above effect is due to the asymmetrical posture of the eyes rather than to the asymmetrical position of the visual target.

The aftereffect of asymmetrical eye posture on egocentric localization seems to have been first reported by MacDougall (1903), and has been confirmed recently by Park (1969) and Craske *et al.* (1975). It has also been known for some time that an object seen in dark surroundings appears to drift in the opposite direction to that of a previous deviation of the eyes, but this topic was discussed in section 7.5.2.

Paap and Ebenholtz (1976) produced the data shown in Figures 7.13 and 7.14. It can be seen from Figure 7.13 that the change in apparent straight ahead increases with both the eccentricity and the duration of the previous eye position. The maximum aftereffect is similar in magnitude to the maximum perdeviation effects. Figure 7.14 shows that the aftereffect of holding the eye at an eccentricity of 32° decays more rapidly for the shorter durations of eye deviation. It can also be seen that with the shorter induction periods, the initial aftereffect is followed by an aftereffect of opposite sign. The aftereffect for 8 min of induction lasted for nearly 4 min and may also have reversed in sign if testing had been extended over a longer time.

The initial aftereffects of asymmetrical eye posture can be understood as a

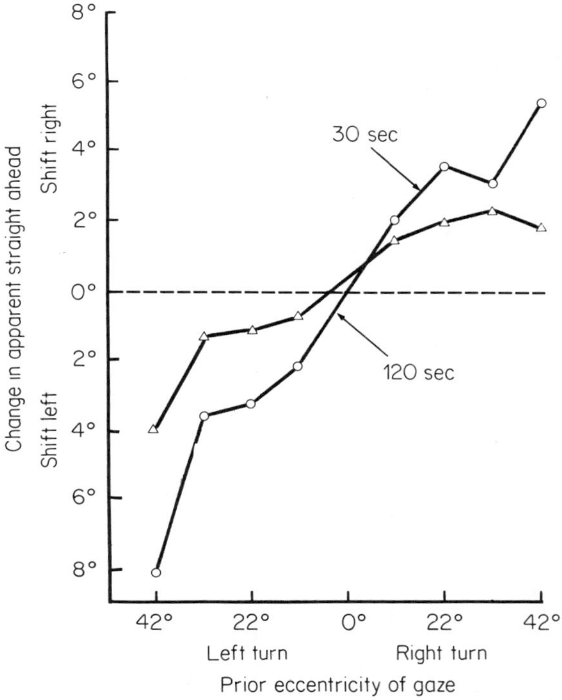

Figure 7.13 Changes in the apparent visual straight ahead 2 sec after holding the eyes at various angles of eccentricity for 30 and for 120 sec (From Paap and Ebenholtz, 1976. Copyright, 1976, by the American Psychological Association. Reprinted by permission)

persisting change in the felt position of the eyes away from a position of eccentric gaze. One would expect such large effects to be revealed in physiological studies. Eckmiller (1974) recorded from single units in the region of the oculomotor nuclei of the monkey. The firing rate of each unit was related to the position of the eyes. However, the firing rate of a given cell was higher after the eyes had approached a given position from a less eccentric direction than it was when the eyes had approached the position from a more eccentric direction. In other words, the cells showed a hysteresis effect. The difference between the ingoing and outgoing impulse rates was between 5 and 22 impulses/sec, which corresponds to a difference in eye position of several degrees. Tension hysteresis has been measured directly in extraocular muscles (see beginning of Section 5.3.3).

These physiological data are compatible with the psychophysical findings, and both types of effect could be due to any of the following interrelated sources of hysteresis.

1. Muscles become more responsive to a given level of innervation after a period of active contraction. This effect is known as **posttetanic potentiation** (Hughes,

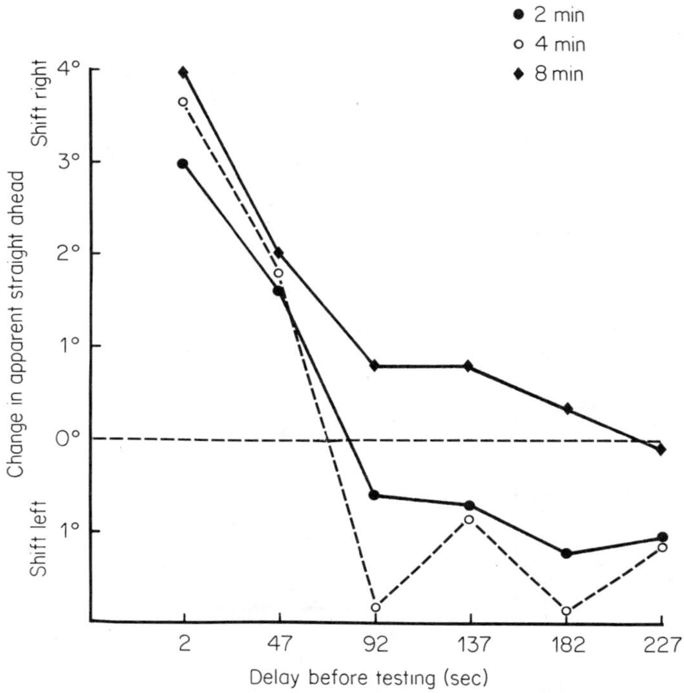

Figure 7.14 Progressive changes in the apparent straight ahead after holding the eyes at an eccentricity of 32° right for various periods of time (From Paap and Ebenholtz, 1976. Copyright, 1976, by the American Psychological Association. Reprinted by permission)

1958; Olson and Swett, 1971). However, it is counterbalanced by the fact that the elastic tension in a muscle is increased by prestretching (Bosco and Komi, 1979).
2. There is a persistent sensory discharge from muscle-spindle receptors after the contraction of a muscle, which is probably due to an alteration in the contractile state of extrafusal and intrafusal muscle fibres (Hutton *et al.*, 1973).
3. The discharge from Golgi tendon organs and ligament receptors adapts when these structures are subjected to steady tension (Houck, 1967 Ferrell, 1980).

An asymmetrical posture of the head leads to per- and postdeviation effects which are analogous to those produced by a deviation of the eyes (Howard and Anstis, 1974: Ebenholtz, 1976; Shebilske and Fogelgren, 1977; Shebilske and Karmiohl, 1979). Howard and Anstis found that the apparent straight-ahead position of the head was displaced about 6° in the direction of a previous 24° deviated position of the head about a vertical axis. However, holding the head straight, against a load tending to rotate it to one side, produced no measurable aftereffect on the apparent position of the head. It was concluded that, at least for small loads, the aftereffect of head posture is not due to the aftereffect of asymmetrical muscle tension, but to adaptive processes in the joint receptors of the neck. The neck differs from the eye in two crucial ways—it has joint receptors, whereas the eye has none, and the head may be held in an asymmetrical posture with relaxed muscles, whereas the eye cannot. It makes sense therefore to suppose that aftereffects of head posture are due to adaptation in joint receptors and that aftereffects of eye posture are due to neuromuscular factors.

7.7 The roles of efference and afference in headcentric visual localization

The studies which have just been reviewed demonstrate that an imbalance in the extraocular muscles disturbs judgments of headcentric direction. However, it is not clear whether the effects are due to imbalances in motor innervation or imbalances in proprioceptive inputs from muscle spindles and tendon organs. The debate between those supporting a motor innervation, or efference, theory and those supporting a proprioceptive, or afference, theory of headcentric localization has been going on for over a hundred years. The early literature is reviewed in Howard and Templeton (1966), and I shall concentrate on the recent literature here.

The investigation of the relative contributions of efference and afference to egocentric localization requires techniques for altering each factor independently. One technique used for changing afference without changing efference has been to displace the eye passively. Brindley and Merton (1960) anaesthetized the conjunctiva in one eye and deflected the eye passively up to 40° either way by pulling on the rectus muscles with forceps. With both eyes occluded, the subject remained unaware that one eye had been moved. Furthermore, the other eye did not move, which demonstrated an absence of a stretch reflex to passive stretch of extraocular muscles (Section 5.3.2). When the eye not being deflected was open the subject saw no apparent movement of the visual scene, which demonstrated that afference from the stretched muscles was not affecting judgments of direction. These results seem to provide conclusive evidence against any role of afference in spatial localization. However, passively induced afference is not the same as afference induced by active muscle contractions, because only in active contraction are the γ-nerve fibres activated. These nerve fibres innervate the contractile tissues within the sensory muscle spindles, and it is only when these sensory organs are so innervated that the spindles operate effectively.

Skavenski (1972) objected to Brindley and Merton's crude psychophysical procedure. In his own experiment he pulled on a tightly fitting contact lens and found that two subjects, in the dark, could usually indicate which way the eye was being pulled, even when the conjunctiva was anaesthetized. He also found that subjects could hold an eye in a given position in the dark while a load was being applied to the eye, and argued that these skills are controlled by afference from muscle spindles or tendon organs. These experiments should be replicated, but in any case they do not establish a direct involvement of afference in normal spatial judgments. This question was investigated in another experiment (Skavenski *et al.*, 1972). The subject's task was to judge the position of a target seen with a passively moved eye in relation to a fixation target seen with the other eye when the two targets were presented in alternation. It was argued that, because of Hering's law of equal innervation, the efference to the two eyes was the same, although the afference from one eye varied with the applied load. The results failed to reveal any effects of the anomalous afference on spatial judgments, and it was concluded that headcentric visual localization depends only on efference. This conclusion is open to the same objection that was raised in the discussion of Brindley and Merton's experiment, namely, that afference from passive stretch of a muscle is not the same as afference from active contraction.

A procedure for changing efference without changing afference is to surgically deafferent the extraocular muscles. Hein *et al.* (1981) dark reared six kittens until they were 10 weeks old. During this time one eye was removed and the other was deafferented by sectioning of the ophthalmic branch of the fifth cranial nerve. Even after six weeks in an illuminated environment these kittens collided with obstacles and could not place their paws on moveable prongs. Spatial skills developed in kittens with a deafferented seeing eye and a normal sutured eye, and were retained in cats which were deafferented after they had developed them. This is strong evidence that afference from eye muscles is involved in the development of visual-motor skills. It could be involved directly in the early calibration of headcentric space, or its absence could cause deficits in the way that the eyes move.

A second technique for varying efference without changing afference is to ask a subject to report visual impressions while trying to move his paralysed eyes. It is well known that when patients with paralysed eye muscles try to move their eyes they experience a displacement of the visual scene and reach too far when trying to fixate and touch an off-centre visual target. This is known as **past-pointing**, and is what one would expect if eye positon is judged in terms of efference. Perenin *et al.* (1977) studied patients with paralysis of one rectus muscle in one eye. When a visual target was presented to only the paralysed eye, on the side of the non-paralysed muscle, pointing movements with unseen hand were either accurate or showed a slight undershoot, but pointing movements to objects on the side opposite the paralysed muscle showed a large overshoot (see Figure 7.15). However, if the patient made no effort to fixate the visual targets, his reaching movements were accurate on both sides.

Kornmuller (1931) artificially paralysed the extraocular muscles of one eye with novocaine and reported that his visual straight ahead was apparently displaced in the direction of an attempted eye movement. Several other investigators have confirmed this result which, together with the clinical data, suggests that sense of eye position depends on efference. However, in these earlier studies the eyes were probably not completely paralysed.

Siebeck (1954) induced complete paralysis of the extraocular muscles of both eyes with curare. When the subjects attempted to move their eyes they did not experience an illusion of spatial displacement, but were aware only that their eyes

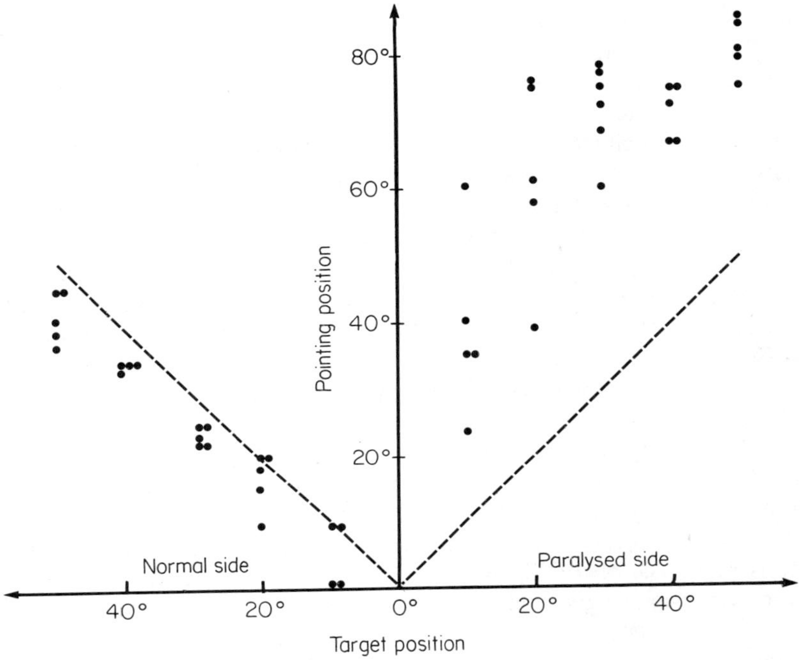

Figure 7.15 Pointing errors of a subject required to look at and point to a target at various positions on the side of paralysed gaze and on the side of normal gaze. The dashed lines indicate the loci of correct pointing (From Perenin *et al.*, 1977. Reproduced by permission of S. Karger AG, Basel)

were paralysed. Brindley *et al.* (1976) induced complete paralysis of the extraocular muscles with lignocaine and confirmed Siebeck's findings. When the eyes were partially paralysed, they reported that their subjects past-pointed with unseen hand. But when paralysis was complete, they pointed accurately to objects which they attempted to fixate. We are not told whether the head was clamped, nor about what was in the subjects' visual field, so that the significance of these findings is difficult to assess. These authors reported a striking effect of complete paralysis. An afterimage in the non-paralysed eye appeared to move with the eyes, as one would expect on either an efference or afference theory, but the afterimage in the paralysed eye did not appear to move during attempted eye movements. This is evidence that afference is involved in the sense of eye position. However, one could argue for an efference-theory account of this result in the following way. Perhaps, when the eye is completely paralysed, the total absence of proprioceptive inputs convinces the subject that the eye has not moved, and this impression may be confirmed by the immobility of the image of whatever stationary objects are in view. Under these circumstances the subject may simply ignore efference and continue to believe that the eye is in its primary position (see Matin, 1976a for a discussion of other possibilities).

Stevens *et al.* (1976) also compared the effects of partial and complete paralysis. With partial paralysis, the visual scene disappeared and reappeared in another apparent location when the subject attempted to make a saccadic movement. Slow apparent drifts of the scene were reported which were coupled with slow movements of the eye back to its primary position. When pointing with unseen hand, subjects overshot visual objects in the direction of intended fixations. When paralysis was complete, the apparent displacements and movements of the scene were not reported. Subjects could not point to targets because the curare affected the whole body. Again the subject's head was not clamped and we are not told what was in his visual field. Like the completely paralysed subjects in other studies, the paralysed subjects in this study were aware that their eyes were not moving. This may have been because of the unmoving images of the nose, orbital ridges, and other objects in view, or it may have been because of the complete absence of proprioception. Whatever the cause, the feeling that their eyes had not obeyed their command may have persuaded them to disregard efference under these circumstances, even though efference is what normally informs them about eye position.

The crucial importance of the visual background in this type of experiment was demonstrated recently by Matin *et al.* (1981). They found that errors in pointing to eccentric visual targets increased with increasing degrees of eye paralysis, but only when the targets were seen in dark surroundings. With illuminated surroundings, there were no pointing errors. In this experiment care was taken to control for the contaminating effects of head movements. Complete paralysis was not induced so that there is still some uncertainty about what happens to headcentric spatial judgments when all afference and all visual surroundings are absent (Section 11.6 contains further comments on this topic).

7.8 Headcentric visual localization during and about the time of eye movements

7.8.1 Introduction

It was shown in the previous section that people can make reasonably accurate and precise judgments of the headcentric location of an object seen in dark surroundings with the eyes in different directions of gaze. Such judgments are possible only if the position of the eyes is taken into account together with the retinal location of the image. This ability to combine information about eye position and retinal-image position in judging headcentric direction is referred to as **direction constancy**. One question to be dealt with in this section is whether there is direction constancy while the eyes are moving. If there is, there must be accurate information about the retinal location of images and about the position of the eyes during eye movements. But if constancy breaks down, then retinal and/or oculomotor information must be degraded during eye movements. The basic procedure for investigating this question is to present a visual target very briefly at various times during an eye movement and ask the subject to judge its headcentric direction. The task most frequently used is that of comparing the position of a flashed target with the position of a second target presented just before or just after the eye movement. This is not a good procedure because it predisposes the subject to make relative oculocentric judgments rather than headcentric judgments. This problem of measurement, and other problems which will be mentioned later, has made it difficult to interpret the results of experiments on this question. Nevertheless, there is a general consensus that direction constancy breaks down just before and during saccades. This is not surprising, because there must be a change in the directional significance of retinal loci when there is a shift in the direction of gaze, and this change must take some time to accomplish, and may even start before the eyes begin to move. Furthermore, under normal circumstances, the images of stationary objects are smeared across the retina during a saccade, which severely degrades the capacity of the visual system to extract information of any kind from the visual input. We are just not used to taking any notice of visual events that occur during a saccade, so that one would not expect subjects to be able to localize briefly flashed targets presented at such a time in an artificial experimental situation.

The question of visual localization during pursuit eye movements is more complicated, for there are now two kinds of visual object, the object that is being pursued and the image of which remains relatively fixed on the retina, and objects in the stationary background, the images of which sweep across the retina. It seems most likely that the object being pursued is the one that the person is interested in and which his visual system is geared to deal with. Things in the stationary background are of little or no interest—if they were, the person would stop moving his eyes and look at them. Pursuit eye movements have evolved to stabilize the images of moving objects, so as to simplify the processing of information. I suggest that people have reasonably accurate awareness of the headcentric location of an object which they are pursuing but a poor ability to localize objects in the stationary background. Yet almost all experiments have had to do with the trivial question of the background and virtually none with the more fundamental question of the pursued object. But what does it mean to say that a person can accurately localize an object egocentrically? I suggest that the only significant meaning is that he can accurately reach with an unseen part of his body towards the seen object. Accurate headcentric localization would be indicated by the ability to set the median plane of the head (the nose) in line with the object. Accurate bodycentric localization would be indicated by the ability to reach towards the object with unseen hand. *Sensorimotor accuracy and precision define the accuracy and precision of egocentric localization.* But in very few studies purporting to investigate headcentric or bodycentric localization during pursuit eye movements has a sensorimotor task been used. The task normally used has been that of relating the apparent position of one visual target with respect to another, and this predisposes subjects to use an *oculocentric* rather than a *headcentric* or *bodycentric* criterion. *What we want to know is how accurately people can point some part of the body to the position of a pursued target at a time indicated by some transient signal*. Only one such experiment has been done, as we shall see in Section 7.8.4.

A related issue, which has often been confused with the question of egocentric direction constancy, has to do with the perceived *stability* of the visual world during eye movements. An object may appear to move without appearing to change its position. For instance, if moving stripes are viewed for a while, a stationary scene seen subsequently appears to move in the opposite direction, and yet the scene does not seem to change its position. In other words, the visual motion-detection system is not closely tied to the visual direction system, and that is why the question of direction constancy should not be confused with that of visual stability.

With head and eyes stationary, a headcentric movement of an object produces an oculocentric movement of its image, which I shall refer to as an **exafferent oculocentric movement**. A movement of an image produced by a self-produced movement of the head or eyes will be referred to as a **reafferent oculocentric movement**. If the eyes pursue a moving object, and nothing else is in view, the movement of the object is detected in terms of whatever non-retinal information about the movement of the eyes is available. This source of information regarding eye movements is often referred to as the **extraretinal signal**, without implying that it is either exclusively a motor signal or exclusively a proprioceptive signal.

The traditional question is whether the world appears to be stationary during eye movements. Almost every discussion of this issue in the context of saccadic eye movements starts with the assertion that the world does not appear to move when we make voluntary saccades, in spite of the fact that there are retinal-motion signals. The debate then centres on the question of whether these reafferent retinal-motion signals are 'cancelled' by the concomitant oculomotor signals (Holst and Mittelstaedt, 1950; Holst, 1954; Gregory, 1958), or whether such signals are 'interpreted' in terms of the intention to move the eyes, rather than cancelled (MacKay, 1973). This debate, insofar as it is about the stability of the visual world *during* saccadic eye movements, is based on a misleading premise—that there are more or less precise retinal-motion signals during a saccade. But this is not true. A typical saccade has a velocity of about 800°/sec, which produces an image motion that is much too fast to be seen as anything but a smear. Although there are cells in the superior colliculus which respond to image motions at this velocity (Section 5.5.2), there is no evidence that they allow a person to discriminate between one velocity and another at this high range. Furthermore, object perception is completely degraded during a saccade, mainly because of the smearing of the image but also because visual sensitivity is reduced when the image moves. We do not perceive a stable or a moving scene during a saccade, we perceive a smeared image, and we perceive this only if we attend to it.

During a pursuit eye movement we certainly have reafferent retinal-motion signals from stationary objects, so that it makes sense to enquire whether such signals are interpreted correctly to produce the impression of an egocentrically stable visual world. The answer should be easy to find. Consider whether the stationary background appears to move when you pursue your own finger as it moves from side to side. Does the background appear to move or doesn't it? I get a strong impression of movement, but am I responding to the oculocentric movement of the image or do I see the background moving relative to my head? The apparently simple question, 'Do I see movement?' is ambiguous; the answer depends on what criterion is being used. The problem is also confounded by the fact that we know that ordinary backgrounds do not move. That is why experimenters have typically used abstract visual stimuli rather than real backgrounds for investigating this issue. But whenever there is movement of one thing relative to another there remains an essential ambiguity about which thing is interpreted as moving and which thing is interpreted as stationary, no matter which object is moving relative to the retina. Duncker (1929) demonstrated this ambiguity many years ago and showed that the relative sizes of the moving objects, enclosing relationships, and visual eccentricity have more to do with determining which object appears to move than has objective or retinal motion. More recently, Johansson (1977) has extended the study of the interpretation of relative motion, and has shown how the grouping and meaningfulness of relative motions affects the way in which motion signals are resolved into relative and absolute components. In spite of these difficulties associated with enquiring into the stability of the visual world during pursuit movements, the experiments that have been done suggest that stability is not achieved, but that people are subject to consistent illusions of motion for objects in the stationary background during pursuit eye movements. But the more fundamental question has to do with how well people judge the motion of the object being pursued. This is the object in which the observer is interested and the one towards which his behaviour is directed. For this reason, I suggest that the accuracy of reaching or tracking with unseen hand is the most relevant measure of headcentric and bodycentric localization during pursuit eye movements, in spite of the difficulties mentioned in Section 7.4. But the only experiments of this kind that I have come across are those concerned with ball-catching (see Sharp and Whiting, 1974). There is a large literature on visual-motor tracking but these experiments are not concerned with the ability to track visual targets with unseen hand, but rather with the ability to control the movements of one visual target in relation to another.

The other question to be discussed in this section has to do with the perception of the change in the location of an object sensed, not *during* the eye movement, but just before and just after the eye movement. In other words, does a stationary object appear to change its headcentric location when it can be seen before and after an eye movement, but not during the eye movement? This question is relevant only for saccadic eye movements, and I know of only one study that tackles it. This is odd, because it is surely the more important question. It is not what we see during saccades that matters, but rather how well and how soon we reassess directional information after the eye movement is complete.

7.8.2 *Visual localization around the time of a saccade*

An image at a given location on the retina must come to signify a different headcentric direction when the eyes execute a saccade to a new position of gaze. Several studies have been done to determine whether this change occurs during the latency period of the saccade, during the saccade, or just after the eyes have come to rest in their new position, and whether the change occurs suddenly or gradually. However, as we shall see, the answers to these questions vary with the procedure used to measure the subject's localization of the target.

In a series of studies, Matin and his associates used the following procedure (Matin and Pearce, 1965; Matin, Matin and Pearce, 1969; Matin, Matin and Pola, 1970). The subject fixated a central visual target for 4 sec, this target was then extinguished, and 300 msec later a second target was exposed for 70 msec at a variable distance to the right of the first target. The subject was asked to move his gaze to the second target as soon as it appeared. At some time during the interval between the appearance of the second target and 1.5 sec after the beginning of the saccade, a 1 msec test flash was presented at one of several the path positions along of the saccade. The subject was asked to report whether the test flash occurred to the left or to the right of the first target in one set of experiments, and to the left or right of the second target in another set of experiments. By combining the subject's reports with information from an eye-position monitor, Matin and associates determined the retinal location which the test flash occupied, before, during and after the saccade, when the subject reported that it was in the same direction as the comparison target. Changes in this point of subjective equality were interpreted as changes in the apparent headcentric direction of the test flash. The results from several experiments in which the first target was the reference stimulus indicated that the apparent headcentric direction of the test flash began to move in the direction of the saccade 100 msec before the saccade and continued to move until several hundred milliseconds after the saccade. In other words, the subject behaved as if his eyes had started to move during the latency period of the saccade, but he did not take full account of the new position of the eyes until some time after they had reached their final position. Thus, whatever information was being used to register the change in the position of the eyes, it was not closely tied to the time course of the actual eye movement. Quite different results were obtained when the position of the test flash was compared with that of the second fixation target. Under these conditions, judgments of direction were based solely on retinal location. In other words, there was no evidence of a change in the apparent headcentric direction of the test flash. However, 500 msec or more elapsed between the first target and the test flash and only about 200 msec between the test flash and the second target, and Matin suggested that the influence of changing eye position may not affect the relative visual direction of stimuli separated by only 200 msec. This suggestion was supported by the results of another experiment in which subjects judged the relative directions of two test flashes occurring during the same saccade (L. Matin, 1976b). It was found that if the flashes occurred within a period of 200 msec or less their relative directions were judged entirely on the basis of their relative retinal locations. This duration is in the same neighbourhood as the visual persistence time and the suggestion is that stimuli arriving within this period are bracketed together in the same headcentric direction.

The precise interpretation of these results is made difficult because of several factors. In the first place, the use of an intravisual comparison task in which the direction of one visual object is judged with respect to another may predispose the subject to make oculocentric rather than headcentric judgments. Oculocentric judgments were certainly made when the objects were near together in time, as I have just explained, but even when they were well separated, a tendency to make oculocentric judgments may have contaminated the results. Secondly, the task required the subject to remember the headcentric direction of the comparison target, and the method relies on the unsupported assumption that the memory of its location remained stable during the eye movement. Thirdly, the eyes always moved the same way in a given experiment, and always from the primary position to a position of eccentric gaze. One wonders what would have happened if the subject had been unable to anticipate the direction of the eye movement, or if the movements had been towards the primary position. Fourthly, few subjects were used, and there were large unexplained individual differences.

Another approach to the question of visual localization during saccades, that overcomes the problem of memory for the comparison stimulus, was used by Bischof and Kramer (1968). Subjects were asked to judge the location of a test flash presented in a fixed direction at different times during or after a 16° saccade. The flash was seen above the centre of a continuously visible numbered scale that extended across the subject's field of view in line with the direction of the saccade. The investigators reported that during the early part of the saccade the apparent direction of the flash moved in the direction of the saccade with respect to its actual position above the scale, and then moved slightly in the reverse direction before finally being judged in its correct location 50–100 msec after the end of the saccade. Mateeff (1978) repeated the experiment but presented test flashes before as well as during and after the saccade. It was reported that the apparent location of the flash began moving in the direction of the saccade about 100 msec before the saccade began, and again was not judged in its correct location with respect to the scale until 50–100 msec after the saccade (see Figure 7.16). These results substantially confirm Matin's results, except that the system reached its postsaccadic stable state sooner with respect to the visible scale than with respect to the remembered comparison target in Matin's procedure. This is not surprising, because a stable visible scale provides a compelling comparison for judging relative oculocentric directions. The subject would soon correctly locate the flash on the numbered scale when both were clearly in view, and his success tells us nothing about the apparent *headcentric* location of either the flash or the scale. During the eye movement the numbers on the scale would be blurred, so that it is not surprising that the flash was mislocalized at that time. It remains to be explained why the flash was mislocalized with respect to the

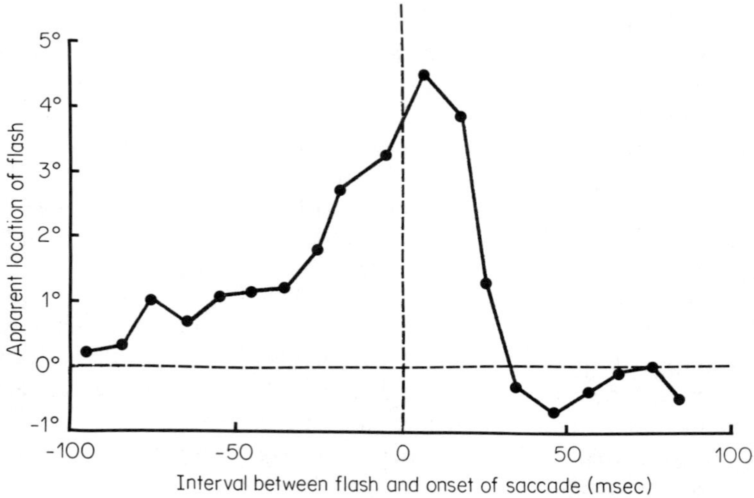

Figure 7.16 The apparent location of a test flash on a scale as a function of time interval between flash and onset of saccade. The flash was always presented at zero on the scale. Results for one subject (From Mateeff, 1978. Reproduced by permission of Psychonomic Society, Inc)

continuously visible scale before and just after the saccade. Mateeff suggested that this mislocalization represents the beginning of a change in the apparent direction of the flash. But why was the direction of the scale not similarly affected, leaving the comparison between the two unaltered?

The answer to this problem has been supplied by Sperling and Speelman (1966). They measured the apparent location of a test flash relative to a stationary background during a saccade and compared this with the apparent location of a test flash seen with stationary eyes against a background moving at saccadic velocities. The retinal velocity of the background was similar in the two cases. The apparent shift in the relative location of the flash was the same in the two conditions! MacKay (1970b) repeated this experiment and obtained the same result (see Figure 7.17). It seems therefore that the apparent shifts in relative location of test flash and scale reported by Bishof and Kramer and by Mateeff had nothing to do with eye movements and were not even headcentric shifts, but were apparent oculocentric mislocations of flash relative to scale induced by the retinal motion of the image of the scale. MacKay suggested the following explanation of why this oculocentric shift is apparent before the scale begins to move. It takes a certain time for a person to relate the position of the flash to the scale. If the scale moves during this period, the new information about its position arrives in time to bias the judgment about the position of the scale in relation to the flash. The shorter the interval between the flash and the scale movement, the greater the biasing effect of information from the displaced background. When the flash occurs just after the background movement, only the residue of the earlier information remains to bias the decision about the

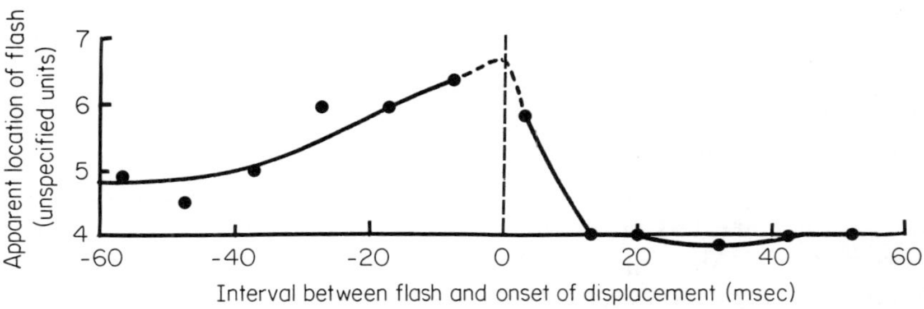

Figure 7.17 Perceived location of a test flash on an 11° luminous scale as a function of time interval between flash and onset of a 4° displacement of the scale. The actual location of the flash was always at point 4 on the scale. Results for one subject (After MacKay, 1970b. Reproduced by permission of Macmillan Journals Ltd.)

relative location of flash and scale, and this accounts for the small 'overshoot'.

To investigate the change in the directional significance of a retinal locus, which is associated with a saccade, a procedure is required that forces the subject to make headcentric rather than oculocentric judgments. One way to do this is to ask the subject to point with unseen hand to the location of a test flash which occurs before, during, or after a saccade. Templeton and Anstis (1975) did an experiment of this type. Subjects made 10° saccades to left or right between green fixation targets which were presented briefly in sequence, 5° on either side of the primary position of gaze. They were asked to point to the apparent location of a red test flash that occurred in various positions and at various times with respect to the saccade. The headcentric shift began about 100 msec before the saccade and reached its new value about 100 msec after the saccade, which is in tolerable agreement with Matin's results. Templeton and Anstis found similar shifts in localization for test flashes presented in the saccadic latency period in a control condition in which subjects were asked not to move their eyes. It is as if the appearance of the displaced fixation point triggered the programming of a saccade which the subject then cancelled.

The only evidence that spatial information is accurately registered *during* a saccade comes from an experiment by Hallett and Lightstone (1976b) in which they showed that a person can make an accurate secondary saccade to the position of a flash which was visible only momentarily during a saccade. This suggests that the headcentric location of the flash is correctly registered while the eyes are in motion, which implies that the visual system has accurate up-to-date information about the position of the eyes during a saccade. There is no need to come to this conclusion if Hallett and Lightstone's procedure is considered in detail. The flash always occurred at the same time during the saccade, and after each secondary saccade the position of the flashed target was revealed to the subject. The subject made many trials and the results presented were the overall means. With a stable temporal sequence and knowledge of results, the subjects could easily learn to make the appropriate secondary saccade. The more significant question is whether subjects *can perform accurately when there is uncertainty about the time and location of the flash and when they are not given knowledge of results.*

The studies reviewed in the earlier part of this section reveal a relatively clear picture of the time course of the eye-position component in headcentric localization when a saccade occurs. A somewhat different question is how adequately eye-position information adjusts the headcentric significance of a retinal locus when this information has settled to its postsaccadic value. An experiment by J. M. Miller (1980) provides some data on this question. The subject made single saccades with an amplitude of his own choosing in a range of about 2–15° from a fixation point into the dark. A few hundred milliseconds after a saccade, he attempted to localize the point fixated before the saccade, which was now not illuminated. The subject did this either by moving a visible spot or by making a saccade back to that point. Thus, oculocentric judgments were used to measure the precision and accuracy of the eye-position component for a situation in which this component could be determined only by extraretinal information. Extraretinal information about eye position was found to be highly accurate and precise, underestimating the actual saccade amplitude by less than 4% on average, with a standard deviation of about 16%. Accuracy and precision were even better if the initial saccade was made to a visible target which provided oculocentric information about the 'intended' amplitude of the saccade.

In summary, it may be concluded that the directional significance of a retinal locus begins to change when a saccade is being programmed and continues to change until about 100 msec after the saccade is complete, at which time it has reached a value appropriate to the size of the preceding saccade. The oculocentric direction of a stationary object which is continually in view during an eye movement is assessed in terms of a weighted mean of its retinal location over a period of at least 50 msec. But it is clear that the directional shifts which occur in the neighbourhood of a saccade are not tied to the movement itself, and can therefore not be due to the direct effect of either efferent or afferent-proprioceptive signals. There is no reason why the change in headcentric space should be closely tied to the actual movements of the eye. Any distortion in visual direction at about the time of a saccade has no behavioural significance. In the first place, the object towards which the eyes are turning will have been correctly localized before the saccade was initiated and it is this object which interests the animal at the time—that is why the eyes are being directed to it. Secondly, the retinal image is blurred during a saccade, so that pattern vision is severely degraded (Gresty *et al.*, 1976). The best strategy for an animal or a person is to ignore the visual input about the time of a saccade.

7.8.3 Judgments of movements and displacements occurring during and around the time of a saccade

The issues to be dealt with under this heading relate to the traditional question of whether the visual world is perceived as stable during a saccade. There are two crucial distinctions which are rarely made clear. In the first place, it is rarely made clear whether the issue under study refers to the impression of movement sampled *during* the saccade (persaccadic judgments) or the comparison of the locations of things sampled *before and just after* the saccade (pre- and postsaccadic judgments). Secondly, it is rarely made clear whether the movement being studied is the relative motion of objects in the visual field or the motion of an object with respect to the observer. On the basis of these two distinction, the following four tasks may be defined.

1. *Persaccadic judgments of visual motion.* The issue here is how precisely and accurately reafferent oculocentric movement of the image of an object is distinguished from exafferent movement of the image when the movement is sampled during a saccade. It is important that the object whose movement is being judged is not visible both before the saccade and after it. If it is, the subject may compare its location before and after the saccade, and the task is then no longer one of distinguishing movements during the saccade. Furthermore, there must be no stationary background in view which would provide information about the movement of one object relative to another.

2. *Persaccadic judgments of relative motion.* The question here is how precisely and accurately the movement of one object relative to another is detected when the movement is sampled during a saccade. There must now be one object moving relative to another, but again, they must not be in view both before the saccade and after it.

3. *Pre-postsaccadic judgments of visual displacement.* Here we want to know how precisely and accurately an actual displacement of whatever is in view is detected when the movement is sampled before and after the saccade. If a person makes a saccade towards a visible target, the initial saccade may not accurately bring his gaze to the desired location. Call this a **normal saccadic error**. On the other hand, the target may move during the saccade and this will also cause a saccadic error. When everything in view is displaced, including the target to which the saccade is directed, the question we are considering becomes, *how far does the target have to be displaced before the subject ascribes the saccadic error to objective displacement, as opposed to normal saccadic inaccuracy?* In order to investigate this question, the target should not be in view during the saccade, because we are not interested in how the subject discriminates retinal motion signals but only in how he assesses the saccadic error when the saccade is complete.

4. *Pre- postsaccadic judgments of relative displacement.* The issue here is how precisely and accurately a displacement of one object relative to another is detected when sampled before and again after a saccade.

We can thus investigate how people discriminate between headcentric and reafferent movement sampled during a saccade and how they discriminate between headcentric and reafferent displacement sampled before and after a saccade. The experimental literature on these questions is difficult to review, because it is not often made clear which issue is being investigated, and even when it is, the procedures are often inappropriate.

The evidence bearing on the first three of these issues will be reviewed next; there seems to be no evidence bearing on the fourth issue.

1. *Persaccadic judgments of visual motion, or the discrimination between*

reafferent oculocentric movements and headcentric movements sampled during a saccade. Orban *et al.* (1973) asked the subject to shift his gaze back and forth between two lamps placed symmetrically 10° on either side of the straight-ahead position. After some practice the lamps were switched off and the subject was asked to continue making 20° saccades. In the middle of the saccade a spot on a cathode-ray tube moved at a velocity of 40° /sec through 1° of arc one way or the other across the centre of the screen. The spot was visible all the time, but nothing else was. The subject had to report which way the spot moved. Subjects reported that the spot moved in the same direction as the saccade whichever way it actually moved. In other words, the subject could not detect the direction of movement and assumed that the stimulus moved in the same direction as the eyes. This is not surprising, because the retinal velocity of the spot was about 800° /sec $\pm$ 40° /sec. A velocity of 800°/sec is too high for the discrimination of a 5% difference in velocity. The subject would have two spurious cues for making the required discrimination. The first is the length of the path traversed by the spot across the retina. This would be 20° $\pm$ 1°. This, too, is a 5% difference, probably below the difference threshold for length discrimination. The second cue is the change in the headcentric direction of the spot as sampled before and after the saccade. But this, too, would probably be below the difference threshold. In any case, the study was not about pre- postsaccadic judgments, so that the spot should not have been visible after the saccade.

Holly (1975) improved on Duysens' and Callen's procedure by presenting the spot whose movement was being judged only during the saccade and by varying the velocity of the spot from 10 to over 4000° /sec. It is unfortunate that the fixation targets and other things were always in view. The subject had to draw on a piece of paper the apparent path of motion of the spot seen during the saccade. Notice that this task is no longer one of judging velocity. When the velocity of the spot was greater or less than the velocity of the eyes, the length of the perceived path of motion was predictable from that produced on the retina. When the spot moved at the same velocity as the eyes, it was drawn as being stationary and in the position in which it first appeared. Holly argued that subjects did not experience headcentric motion under this condition, because they interpreted the spot as an afterimage produced by a stationary flash. It could also be argued that the subjects bracketed all stimuli into the same headcentric direction when they impinged on the same retinal location within 200 msec (see Section 7.8.2).

Bridgeman *et al.* (1975) presented the subject with 13 fixation points spaced horizontally at 1° intervals and asked them to make saccades at will from one to another for a period of time. At unpredictable times the whole display was rapidly displaced 1, 2 or 4° to left or right, and the subject had to report whenever he saw the stimulus jump. If the jump occurred just before, during or just after a saccade, the probability of its being detected was very low, although it was easily detected at other times. It was concluded that there is a suppression of motion detection during saccades (see also Bridgeman and Stark, 1979). This conclusion is unjustified as regards the during-saccade comparison. During saccades the imposed velocity of 900° sec was superimposed on a retinal-image velocity of about 700°/sec, but between saccades the imposed velocity was seen in comparison with zero image velocity. The difficulty of detecting the imposed velocity during saccades as compared with between saccades was probably due to this difference in comparative retinal velocity, and may have had nothing to do with eye movements. If one wanted to know whether the threshold for motion detection is affected by eye movements, a stabilized retinal image should be used as the baseline. That is, *the image of a spot should be stabilized on the retina, and its velocity varied from that baseline during an eye movement; the spot should not be still in view when the eye comes to rest. The threshold for motion detection obtained this way would be compared with the threshold for motion detection of a spot with eyes stationary.* In this way it could be decided whether motion detection is suppressed during eye movements. As this book was going to press a report appeared of an experiment along these lines by Brooks, Yates, and Coleman (1980). They found that small, unpredictable displacements imposed upon a target moving at saccadic velocity were detected equally well when the basic movement was caused by an eye movement and when it was caused by an externally imposed movement. There is thus no specific suppression of velocity sensitivity during a saccade.

A phenomenon that is often cited in support of the contention that motion detection is suppressed during saccades is that one does not notice the movement of one's own eyes when looking in the mirror. Campbell and Corfield (1977) investigated this phenomenon by watching the movements of their own eyes on a television monitor. The movements were not seen whatever the magnification or orientation of the monitor. When a video recording of the same movements was viewed, they were easily detected. Wallach and Lewis (1966) conducted a similar experiment with similar results. But again, these experiments prove nothing about the suppression of motion detection during saccades because the movements would also be difficult to detect if the videotaped display were displaced at 700°/sec while the image of the eyes moved. It is simply that superimposed movements are hard to detect when the whole display is moving at a high velocity (Stark *et al.*, 1976; Brooks *et al.*, 1980).

Another possible reason why motion is difficult to detect during saccades is that the threshold for light detection is elevated during saccades—an effect referred to as **saccadic suppression** (see E. Matin, 1974a, for a general review of this topic). This term suggests that the visual input is suppressed by signals associated with a saccadic eye movement. These signals could be (a) efference signals, or corollary discharge, associated with the activity of the extraocular muscles, (b) visual signals arising from the rapid motion of the retinal image, or (c) proprioceptive signals from the extraocular muscles. Recent evidence suggests that visual signals are responsible for saccadic suppression within the geniculostriate system. On the other hand, there is electrophysiological evidence that saccadic suppression within the superior colliculus is caused by the corollary discharge as well as by visual signals, although the behavioural significance of this suppression is unclear. I shall now briefly review this evidence.

Electrophysiological evidence for saccade-related changes in the response of cells in the lateral geniculate nucleus was reviewed in Section 2.5. Recent studies have shown that these changes are the same for movements of the retinal image induced by saccades as they are for externally produced image motions, which suggests that it is not the eye movement as such which leads to such changes but motion of the retinal image produced by a saccade (Jeannerod and Chouvet, 1973; Noda and Adey, 1974). Judge *et al.* (1980) recorded from cells in the visual cortex of the alert monkey trained to fixate and make saccadic eye movements. The response of all cells to visual stimulation was attenuated during a saccade, but only cells which were visually excited just before the saccade. It was concluded that forward visual masking is the main cause of saccadic suppression in cortical cells.

Psychophysical evidence for the importance of retinal-image motion, rather than the movement of the eyes, in saccadic suppression has been provided by Wurtz (1969b), MacKay (1970a, b, c) and Mitrani *et al.* (1971, 1973). In these studies it was shown that image motion at saccadic velocities in the stationary eye produces the same effect as image motion during saccades. It was also shown that there is no threshold elevation during saccades when the background is dark; the degree of suppression increases as background illumination is increased. An evenly illuminated background is sufficient to produce suppression, a patterned background is not required (Brooks, Impelman, and Lum, 1980). Kulikowski and Leighton (1977) found no change in contrast threshold when a subject attempted to make a saccade with an eye in which the extraocular muscles had been paralysed. There would have been an effect in this case if corollary discharge had been involved.

There are several reasons why a rapidly moving image should be more difficult to detect than a stationary one. (a) A rapidly moving image excites a given retinal location for too brief a time to initiate a normal response in the receptors. (b) The rapid succession of different images at any one retinal location produces visual masking (Breitmeyer and Ganz, 1976; Volkmann *et al.*, 1978). (c) The response of sustained cells, which carry pattern information, is inhibited in the lateral geniculate body by the simultaneous activity in transient cells, which respond to rapid motions of the retinal image (see Section 2.5). (d) It has been reported that when a stationary grating is replaced by a grey field for as long as it takes a saccade to occur, the interruption of the grating is not noticed, even though the eyes are not moving (Corfield *et al.*, 1978). It looks as though the visual system is designed to ignore a temporary 'grey-out' of the stimulus because, when the grating was interrupted by other patterns, the event was noticed.

Richmond and Wurtz (1980) detected a reduction in the firing rate of cells in the superficial layers of the monkey's superior colliculus which were sensitive to visual movement when the animal attempted to move its paralysed eyes. This must be due to the effects of efference or corollary discharge. It looks as though the colliculus contains a mechanism for distinguishing between image motion due to eye movements (reafference) and image motion due to object motion (exafference). However, this mechanism may have nothing to do with the perception of a stable visual world during saccades, but rather with preventing the rapid-motion detectors of the colliculus from adapting when the eyes move. Collin and Cowey (1980) found no clear evidence that colliculectomised monkeys were unable to distinguish between exafferent and reafferent visual motion. Wurtz *et al.* (1980) found that the response of each visually responsive cell in the colliculus was attenuated when the cell was visually excited just before the presentation of the test stimulus. In other words, these cells showed evidence of forward visual masking, which would presumably operate during saccades. The colliculus seems to be endowed with two mechanisms which produce visual suppression during saccades, one driven by the corollary discharge and one driven by motion of the retinal image.

Even though there is an elevation of threshold during saccades the effect is not large and is unlikely to contribute to the experience of visual stability during saccades made under well illuminated conditions. Some people are able to execute a rapid series of saccades alternating in direction, in the course of which they experience a very unstable visual world. By comparison, there is little, if any, experience of loss of stability during a series of normally spaced saccades. Nagle *et al.* (1980) found that saccadic suppression was the same for both types of saccade. If suppression were a cause of stability it would produce stability for rapidly alternating as well as for normal saccades. I suggest that stability is lost with rapidly alternating saccades, not because experience during each saccade of such a series is any different from experience during a normal saccade, but because, in the rapid series the eye does not come to rest between saccades for long enough to register a stable scene. We tolerate and ignore the transitory experience of visual motion during a series of normal saccades because our attention is anchored on the stable scene seen during intersaccadic intervals. If these intervals are reduced below a critical duration, we are left with only a series of moving images. *Just what this critical interval might be must be determined by further experiments.*

I suggest that the whole question of the perception of movement *during* saccades is devoid of theoretical interest, because the velocities are much too high and the duration is much too short for variations in velocity under natural conditions to have any significance.

2. *Persaccadic judgments of relative motion.* There seem to be no studies on this question. In the study by Orban *et al.* mentioned under heading 1, there was a condition in which subjects had to report the direction in which a spot moved relative to a stationary background during a saccade. The subjects could correctly identify the direction of movement and it was concluded that relative movements may be detected during a saccade. But the conclusion is unjustified, because the spot was visible before and after the saccade, so that the subject could detect the motion by noting the changed position of the spot on the background after the saccade.

3. *Pre- postsaccadic judgments of visual displacement.* The study which bears most directly on this question is an unpublished report by Braddick (1969). The subject was asked to make saccades to and fro between opposite sides of a circle which subtended 100′ of arc at the subject's eye. During each saccade the circle

was blanked out and reappeared 300 msec later, after the saccade was complete. While the circle was blanked out it was displaced by a varying amount, and the subject had to report the presence and direction of this displacement. It was found that displacements of 10′ of arc were detected, which is 10% of the saccade amplitude.

A very similar experiment was reported by Whipple and Wallach (1978). Again subjects were asked to make saccades to and fro across a circle, which in this case subtended 7° of arc, and after each saccade report on the direction in which the circle was displaced during the saccade. When the circle was displaced along the axis of the saccade it had to be moved between 9 and 10% of the saccadic amplitude for the movement to be detected. This is in good agreement with Braddick's results, considering that Whipple and Wallach used a larger saccadic amplitude and did not blank out the stimulus during the saccade.

These experiments indicate that when the centre of gaze misses its intended target by more than about 10% of the saccadic amplitude the error is ascribed to motion of the scene rather than to an inaccuracy of the saccade. The normal accuracy of a saccade is about 10% of the amplitude of the intended movement (see Section 6.4.4) so that it is reasonable to suppose that inaccuracies of more than this would be ascribed to motion of the scene, at least in conditions where the subject is set to expect such a motion.

Mack (1970) did an experiment that relates to this issue. During a saccade to a briefly flashed target, the initial fixation target, which remained visible, was displaced by some variable fraction of the saccadic amplitude, either to the left, to the right, or up, or down. Movements of the initial fixation target greater than 10% of the saccadic amplitude were detected by subjects on a significant number of trials. Because of the arguments already given, it is reasonable to suppose that performance was based on a before–after comparison rather than on a sampling of velocity information during the saccade. Interpreted this way, the results suggest that a person can judge with 10% precision that the distance from the present position of gaze to a previous position of gaze is equal to the amplitude of the just-completed saccade.

In another experiment, Mack *et al.* (1978) demonstrated that the threshold for the detection of an upward target displacement and the threshold for a downward displacement occurring during a horizontal saccade were affected in different ways according to whether the subject had been exposed to an upward or to a downward target displacement. In other words, the baseline against which shifts are assessed is subject to modification through experience.

In Braddick's study the stimulus was blanked out during the saccade, and in the other studies just mentioned the stimulus remained in view. Although the results of these studies were similar, a direct comparison of the effects of blanking, as opposed to not blanking, on the *threshold* for detection of displacement has not been made. However, Festinger and Holtzman (1978) did compare the effect of blanking and not blanking on the *apparent size* of a displacement of the scene during a saccade. The subject made a 2.5° saccade from a spot in the centre of a square to the edge of the square and gave an estimate of the extent to which the dot and square were displaced during the saccade. In fact, the square was displaced either through the same distance as the saccade or half that distance in one of four orthogonal directions. In one condition the displacement coincided with the end of the saccade, so that during the early part of the saccade the retinal image smear was like that in a normal saccade; in a second condition the image was stabilized on the retina, so that there was no image smear; and in a third condition there was also no image smear because the stimulus was blanked out during the saccade. There was also a control condition in which the displacement occurred while the eyes were stationary. The main results were that a target displacement with or against the saccade was judged to be smaller when it occurred with normal smear during the saccade than when it occurred with eyes stationary. The displacement was judged to be smaller still when there was no retinal image smear, that is, when the target was stabilized or blanked out. The judged extent of displacements orthogonal to the saccade was not affected by these factors. Note that even the smaller displacement in this experiment was well above the 10% threshold for displacement detection, so these data tell us nothing about whether retinal smear affects the *threshold* for displacement detection. Festinger and Holtzman interpreted their results as demonstrating a bias towards perceiving a stable visual world during saccades. They concluded that even if the world is perceived to have been displaced during a saccade there is a bias towards perceiving less displacement than actually occurs, especially if retinal smear is not present.

Bridgeman *et al.* (1979) caused a visual stimulus to step unpredictably to left or right through 2° of visual angle while the subject made a 20° saccade. It was found, in conformity with the results of the other experiments which have been reviewed, that normally this stimulus displacement was not detected. However, it was discovered that subjects could point accurately towards the visual target soon after the saccade was completed. Bridgeman *et al.* concluded that the mechanism responsible for pointing has access to information regarding the position of the displaced target which is not available to the 'cognitive' system. This conclusion is a little misleading, because the motor and cognitive tasks were not strictly comparable. The cognitive task involved a before–after comparison of the position of the visual target, whereas the motor task involved only the appreciation of the postsaccadic position of the target. It may thus have been the presence or absence of this memory factor, rather than a difference between cognition and a motor response, that was responsible for the difference that Bridgeman *et al.* found. The other possibility is that subjects were predisposed to make oculocentric judgments in the cognitive task and headcentric judgments in the pointing task.

7.8.4 *Visual localization during pursuit eye movements*

The only published report of an experiment on the visual estimation of direction during visual pursuit is by Ward (1976). He found that a target flashed on during a pursuit movement was mislocalized in the direction of the eye movement relative to a continuously visible scale. This is the same kind of effect that was discussed in Section 7.8.2, and which was shown to have nothing to do with eye movements.

Stoper (1967) conducted an experiment which throws more light on the question, although the results have not been published. The subject was asked to report the location of two targets with respect to each other when they were flashed on at different times during a pursuit movement. It was found that for time intervals up to about 300 msec the judgment of relative location was made almost entirely on the basis of the retinal locations of the flashes. A similar experiment by Matin, described at the beginning of Section 7.8.2, which was done with saccades, gave similar results. Thus, during saccades and pursuit movements, the relative directions of brief stimuli arriving up to about 300 msec after each other are judged entirely in terms of their relative retinal locations. This is not surprising. If the system were to evaluate the relative headcentric locations of sequentially flashed targets during an eye movement it would have to match precisely the moment-to-moment changes in the signals arising from the eye-position detectors with the signals from the image-position detectors. The time it takes to register and process the two types of signal would have to be the same, which would require a very intricate process for phase-locking the changing signals. Such a mechanism would serve no purpose because we do not attend to things that are unstable on the retina. The background which sweeps across the retina when we move our eyes in pursuit of something or because we are changing fixation is of no immediate interest; if it were we would not be moving our eyes. The whole point about eye movements is that they allow us to stabilize the retinal image of whatever we wish to look at, so that the visual system may more easily process the information. The visual system is geared to decoding stationary images, and that is why we need to pursue moving objects. Having evolved this system it would be pointless to also have a system for dealing with reafferent retinal motion. The best way to find out whether we can judge accurately the headcentric direction of objects during a pursuit movement is to ask subjects to point to the position of a target flashed on during the movement.

Hansen (1979) has done the only experiment of this kind. Subjects were required to pursue a light in dark surroundings as it moved horizontally to and fro over a fixed 10° path at various speeds between 15′ and 30°/sec. A second light was briefly exposed at some randomly varying time and position within the pursuit path. The subject was required to hit the briefly exposed target with an unseen hammer. No knowledge of results was given. The mean signed error was within 3′ of the position of the target for the two subjects, with a standard deviation of about 0.5°. Performance was as good as when the subjects aimed at targets with the eyes not moving. Furthermore, the constant error was the same when the target occurred in a leftward pursuit movement as when it occurred in a rightward pursuit movement, which demonstrates that there was no directional lag in the estimation of eye or target position. In another experiment the subject pursued a light which moved horizontally with a sinusoidally varying velocity over a fixed 4° path and, on command, aimed the hammer at a second continuously visible light as it moved in phase with the first light but over a path in one of several orientations with respect to the horizontal pursuit path. The oculocentric path of the untracked light was thus the resultant of the motion component supplied by its headcentric motion and the component supplied by the reafferent motion signal. Nevertheless, the subject was able to aim the hammer accurately; blows fell along the actual path of motion rather than along the oculocentric path, showing that the subjects had allowed for the motion of the eyes in determining their direction of aim. In a third experiment subjects adjusted the path of the untracked light so that it appeared vertical. Subjects were able to do this task with a mean error of no more than 8° of tilt, in spite of the fact that when the untracked light was actually moving vertically its oculocentric direction was at 45°. Indeed, the subjects visually experienced the 45° path of the untracked light, but were nevertheless able to extract the headcentric component of the movement and make a correct vertical setting. If the experiment had predisposed the subjects to make oculocentric judgments, they would have indicated that the path of the vertically moving light appeared inclined at 45°, and one may have concluded that people cannot make allowance for eye position. As it was, Hansen's subjects were disposed to make headcentric judgments by the nature of the task and instructions, and Hansen concluded from this, and from his two other experiments, that people have accurate information about the position of the eyes during visual pursuit.

I have only one complaint about this conclusion. Hansen allowed the subjects to see the tracked target several times over the same fixed path. It may well be that, once the subjects had got used to the duration of each to-and-fro motion, they interpolated the time of occurrence of the briefly exposed target (or the command to aim at the untracked target) and used this as a basis for judging the position of the target with respect to the fixed scan path of the eye movement. They would then have to make an adjustment for the retinal eccentricity of the target before finally judging its position in space. *The experiment should be repeated with pursuit movements which follow a target moving at a different speed and over a different scan path on each trial.* Only then will we know for sure whether eye-position information is accurately monitored during a pursuit eye movement.

J. M. Miller (1980) asked subjects to return a visual target back to its starting position after completing an eye movement in pursuit of the same target seen in dark surroundings. He found that the settings fell short by an average of 11%, which suggests that the amplitude of the eye movement had been underestimated by this amount. If the subject pursued the target back and forth over the same track several times before the return setting was made, the underestimation jumped to 33%. This was a significantly greater error than in the single-sweep case even after allowance had been made for differences in tracking accuracy. Miller argued that tracking of repetitive, predictable target motions involves a subsystem whose operation is automatic and not consciously available, hence the large error. It is possible, however, that some sort of adaptation to repetitive motion might explain the larger error in the repetitive case.

There are several other reports in the literature with titles that suggest that they are about visual localization during pursuit eye movements. But in fact they deal only with the perception of movement during pursuit. It was implied in these reports that if an object appears to move during a pursuit movement, this reflects an instability in its egocentric direction. But this does not follow because, as I have already argued, the question of whether an object appears to move during an eye movement is ambiguous—it can refer to exocentric, headcentric or oculocentric movement. In any case, motion detection, at least oculocentric motion detection, and judgments of direction depend on different mechanisms.

7.8.5 Judgments of movements occurring during pursuit eye movements

Only two of the four tasks listed at the beginning of Section 7.8.3 have been investigated in the case of pursuit movements. These will be dealt with in turn.

1. *Discrimination between reafferent oculocentric movements and egocentric movements sampled during pursuit.* Fleischl (1882) observed that an object moving at a certain objective velocity appears to be going faster when it is observed with the eyes stationary than when it is visually pursued. Aubert (1886) confirmed this observation, and it is referred to as the **Aubert–Fleischl phenomenon**.

The usual explanation of this effect is that when we pursue a moving object with the eyes we underestimate the velocity of the eyes and hence the velocity of the object. But another way of looking at it is to suppose that we overestimate the velocity of objects which sweep across the retina and correctly estimate the velocity of eye movements. This question can be resolved only by eliciting a response which indicates the subject's estimate of the actual velocity of the moving object. *One possibility would be to have the subject track the motion with his unseen hand both while he fixates on a stationary point and while he follows the motion with his eyes.* This would at least indicate whether retinal movement or movement pursued by the eyes is more accurately integrated with reaching movements, and that is what visual localization is all about. Until such an experiment is done we cannot be sure whether the velocity of an object is more accurately perceived when we pursue it or when we don't. I predict that pursued motions will be tracked with the hand more accurately than non-pursued motions, because we normally pursue an object that we are interested in catching. People certainly catch balls more accurately when they follow the path of the ball with their eyes than when they hold their gaze fixed (Whiting, 1969).

A series of experiments from Freiburg demonstrates that the Aubert–Fleischl phenomenon may have nothing to do with retinal motion as such. Dichgans *et al.* (1969) asked subjects to scale perceived velocities of a horizontally moving pattern of vertical stripes. A given exafferent oculocentric motion was rated to be about 1.6 times faster than the same motion when it was visually pursued. In a subsequent study by Dichgans *et al.* (1975), it was found that this ratio of perceived velocities decreased as the spatial frequency of the moving pattern was reduced, and when only a single moving contour was seen the perceived velocities were the same (see Figure 7.18). Thus the Aubert–Fleischl phenomenon depends on the presence of a temporal sequence of stimulation – not merely on movement across the retina. They accounted for their finding in terms of the results of another study in which it was shown that the apparent velocity of a moving striped pattern viewed with stationary eyes is a joint function of its velocity and the temporal intermittency generated by its spatial frequency (Diener *et al.*, 1976). When there is only a single moving edge, the temporal intermittency is zero and apparent velocity depends only on the actual velocity. But there is a zero intermittency when a periodic pattern is pursued accurately by the eyes, so that in this case also, apparent velocity depends only on the actual velocity.

A further question having to do with the perception of motion during pursuit eye movements is whether the experience of apparent motion, generated by stimuli flashed in sequence, depends on the relative retinal, or the relative headcentric positions of flashed stimuli. Stoper (1973) presented flashed stimuli sequentially during a pursuit movement, either so that they stimulated the same retinal location but occurred in different headcentric locations, or so that they stimulated different retinal locations but occupied the same headcentric location. Apparent movement was reported only when the retinal locations differed. This experiment demonstrates that apparent motion is a property of the oculocentric

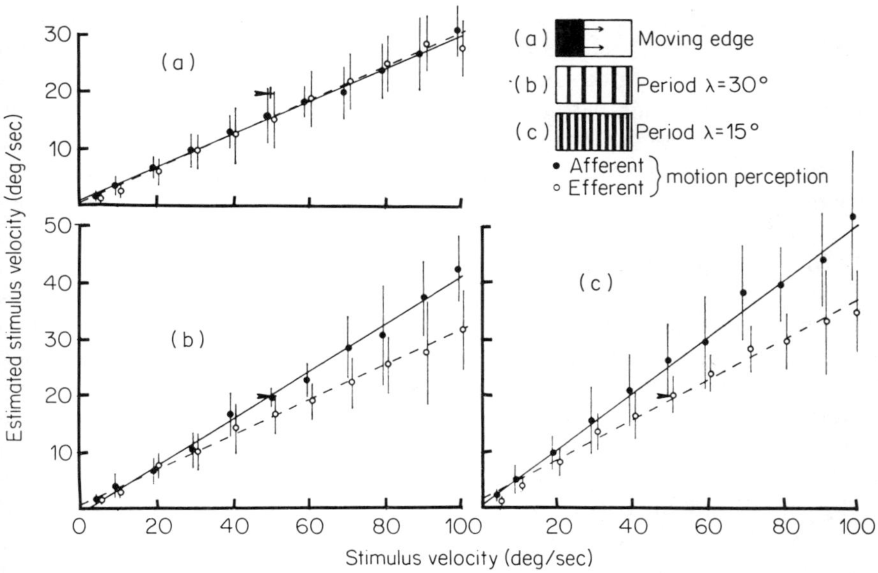

Figure 7.18 The apparent velocity of a moving display as a function of actual velocity when viewed with stationary eyes (solid lines) and when visually pursued (dashed lines). When the display is a single moving edge, as in (a), the two functions are the same. When the display is a moving grating, as in (b) and (c), the two functions diverge as the spatial frequency of the display increases. The bars represent standard deviations (From Dichgans *et al.*, 1975)

motion-detecting system. It tells us nothing about visual stability during eye movements, because the subjects were not asked whether they perceived the apparent movement as oculocentric or headcentric. Stoper suggested that apparent movement was not reported when the sequential stimuli fell on the same retinal location because the duration of the motion was too brief to evoke a sensation of motion within the headcentric system even though it was not too brief to evoke a sensation of motion within the oculocentric motion-detection system.

If a moving object seen in dark surroundings is visually pursued for some time it appears to move. In this case there is no oculocentric motion to confound the judgment. In a similar way, if an afterimage is 'pursued' for some time it appears to move objectively (Mack and Bachant, 1969). This apparent motion of the afterimage seems to depend on oculomotor command signals, or efference, because an afterimage appears stationary when the eye is displaced passively by pressure applied to one side of it. However, we cannot conclude from this that receptors in the extraocular muscles do not contribute to egocentric localization, because the proprioceptive input generated by a passive eye movement is not the same as that generated by active movement.

2. *Discrimination of relative movement sampled during a pursuit eye movement.* Filehne (1922) observed that when the eyes follow a moving object seen against a stationary background, the background appears to move in the opposite direction. This has become known as the **Filehne illusion**. The 'pendular whiplash' illusion is a related phenomenon (see Ford, 1910). There is no doubt that the reafferent oculocentric motion of the image of the background excites the oculocentric motion detectors, and, as I argued in the last section, this will surely give rise to sensations of seen movement. The question is whether the sensations are interpreted as an actual motion of the background. If the background is an ordinary scene, such as a room, subjects know that there is no actual motion, whatever sensations they may experience (Gibson, 1968). That is why Mack and Herman (1973) investigated this question using a background consisting of a horizontal band of abstract shapes. The Filehne illusion was experienced by about half the subjects, but there is still the problem of deciding whether the subjects were reporting oculocentric or headcentric motion. Mack and Herman explained the illusion in terms of an underestimation of the velocity of pursuit eye movements which causes the velocity of the pursued object to be underestimated. According to this view, the subject then accounts for the motion of the pursued object relative to the background by assuming that the background is moving the other way. In other words, the background gains what the pursued object loses, keeping the relative velocity constant. Mack and Herman also conducted an experiment which they claimed demonstrated that the velocity of a non-pursued target in dark surroundings is more accurately perceived than the velocity of a pursued target and they suggested that this evidence supports their theory of underestimation of eye velocity. However, their data were based on an arbitrary comparison of relative velocities and not on

estimates of headcentric velocities, such as might be provided by the accuracy of tracking objects with unseen hand. Their data do not provide unequivocal support for their theory, and indeed the evidence reviewed in this Section suggests that the velocity of a pursued object is more accurately perceived than that of a non-pursued object. This would certainly make more sense, because it is more important for an animal to judge the velocity of an object which it is pursuing than of one which it is not pursuing. It is the object being pursued that the animal is interested in and perhaps trying to catch. I suggest that the Filehne illusion is not due to an underestimation of pursuit velocity but to an overestimation of the velocity of reafferent oculocentric motion generated by the stationary background.

The importance of retinal motion signals in the Filehne illusion was, in fact, demonstrated in a further study by Mack and Herman (1978). The subject pursued a short vertical line moving horizontally at 5° /sec while the experimenter adjusted the velocity of a background, consisting of a single luminous point, until the subject judged the point to be objectively stationary. The mean of background velocities which were judged to be stationary was 1°/sec in the direction of pursuit when the background was visible for 1.2 sec, and 3.35°/sec when the background was visible for only 0.2 sec. Thus, subjects based their judgments mainly on relative oculocentric velocity with brief exposures. This result is compatible with the result of Stoper's experiment reported above. In another condition, in which Mack and Herman blanked out the pursuit target while the background point was in view, the subjects selected a much slower background movement as stationary. The experimenters suggested that the Filehne illusion is due in part to a tendency to base the estimate of motion on the relative oculocentric motion of background and target, and this comes closer to the view that I have put forward. There is electrophysiological evidence that the visual system contains cells which are specifically sensitive to the motion of one stimulus relative to another rather than to motion of an isolated stimulus (Burns *et al.*, 1972; Bridgeman, 1972). Thus, when there is relative retinal motion, there is a stronger excitation of motion detectors than when there is motion of an isolated stimulus, and this enhances the Filehne illusion.

I have argued that when we pursue an object, we are interested in detecting its headcentric or bodycentric location and motion so that we may accurately catch or track it. When we pursue something with our eyes we ignore reafferent motion signals arising from stationary objects. If, as is often the case, the object we are following has parts which move relative to each other, then it is the appreciation of this relative retinal motion that is important—not the headcentric or bodycentric motion of the parts. I suggest that for this reason we normally attend to the oculocentric rather than to the headcentric motion of objects which generate retinal motion signals during pursuit eye movements. For instance, the experiments of Johansson (1973, 1974) demonstrate how, when observing lights on a rolling wheel or attached to the limbs of a walking figure in dark surroundings, we perceive the pattern of relative motion of the parts of the object superimposed on a general egocentric motion of the object. We do not have to move our eyes to achieve this partitioning of motions into relative and absolute components, but if we do track the motion of the figure the task is simplified, because the headcentric motion is now coded as an eye movement and the relative motion of parts as retinal motion. Normally, when we can see only a light on the rim of a rolling wheel, the motion is perceived as a series of cycloids rather than rolling motion. If a light is added to the centre of the wheel, the relative motion of the lights is now seen as a rotation, and the shared translatory motion of the lights as a progression of the wheel. But Stern and Emelity (1978) showed that a centre light is not necessary for the perception of rolling motion if the eyes track at the velocity of the translatory motion of the wheel. When the eyes do this, the eye motion provides an egocentric framework in terms of which the residual retinal motion of the rim light is seen as circular motion. This is a nice illustration of the point that I am stressing, namely, that when we pursue an object we more or less adequately keep track of its headcentric direction, ignore the reafferent motion of the background, and interpret other retinal motion signals as occurring relative to the changing egocentric location of the pursued object.

7.9 Judging whether approaching objects are likely to hit us

7.9.1 Introduction

Up to now in this chapter we have considered the directions of objects with respect to axes and planes fixed in relation to the head or torso. These are true egocentric directions. There is a class of **semi-egocentric directions** which are defined, not with respect to the median plane of the head or body, but with respect to the plane containing the stimulus and the egocentre of the observer (see Section 1.3.2). Consider the task of deciding whether a rod is pointing towards the self (the egocentre) irrespective of its direction with respect to the median plane of the head. According to the requirements of this task, a rod anywhere round the head has the same direction if it points to the egocentre. If the rod does not point to the egocentre, its semi-egocentric direction is given by the angle between the rod and the line joining the midpoint of the rod to the egocentre. The position of the rod relative to the median plane of the head is not involved.

Semi-egocentric directional judgments are involved when we judge whether someone's gaze is centred on us (see Anstis *et al.*, 1969) or when we judge whether someone is pointing a gun at us. Such judgments are also involved when an animal judges whether an object which is approaching it is likely to hit it, or when a moving animal judges whether it is likely to collide with a stationary object. These types of judgment are obviously very important for survival and the avoidance of injury.

There are two types of stimulus cue that indicate the direction in which an object is approaching: monocular cues and binocular cues. In order to simplify the discussion of these cues, I shall confine my attention to movements of an object within the plane containing the two visual axes.

7.9.2 Monocular cues to the direction of an object's approach

Consider first a stimulus object with no discernible size, such as a point of light, which is viewed monocularly. As such an object moves, the observer can detect only that component of its motion which is projected into the frontal plane. The direction of this projected path of motion may be detected either by the changing position of the retinal image or, if the object is pursued by the eyes, by the changing position of the eyes. In the absence of any other information, this cue is ambiguous because the projected path of motion produced by an object moving at an appropriate speed and in an appropriate direction within a frontal plane is the same as that produced by an object travelling at an angle towards the observer. Under certain circumstances, however, the projected path of motion of a point source does provide unambiguous information. Suppose, for instance, that an observer is steering a vehicle towards what he knows to be a stationary light in otherwise dark surroundings. For instance, he may be a pilot of a plane flying towards a lighthouse on a dark night. Under these circumstances, the observer knows that any perceived movement of the light is due to the fact that his plane is not flying directly towards the light. In other words, the sideways motion of a small visible object which a moving observer knows to be fixed provides unambiguous information that the observer is not heading towards that object. If the observer knows the distance of the object and his own velocity, the direction and velocity of relative sideways motion of the object provides unambiguous information not only about whether or not the observer is heading towards it, but also about his relative direction of approach. Llewellyn (1971) found that the precision and accuracy of judgments about heading based solely on the apparent sideways drift of a single small target were just as good as the precision and accuracy of judgments based on seeing the expanding pattern produced by an array of objects. Llewellyn did not investigate the relative potency of these two types of display under conditions in which the subject had no basis for assuming that the visual scene was fixed and all the stimulus motion was due to his own movement.

Consider now an extended stimulus object or an array of objects, viewed monocularly. The image of such a stimulus expands as the stimulus approaches the observer or the observer approaches the stimulus. This expansion is known as **looming**, and if the observer knows that the object is not actually varying in size it provides unambiguous information about the rate and direction of approach (Gibson, 1950, 1958). With monocular viewing, there are three features of looming to be considered; its pattern, rate, and focus.

Consider first the **pattern of looming**. One important aspect of this pattern is the **symmetry of looming**. An object approaching along the visual axis or any other visual line of an eye produces an image which expands symmetrically, and this signifies that the object will ultimately hit the observer if no avoidance response is made. If the object approaches along any path other than a visual line, its image expands asymmetically and if this asymmetry is large enough it indicates that the object is destined to miss the head. Note that this cue depends only on the direction of the path of the approaching object relative to the station point of the observer; it is not substantially affected by rotations of the observer's eye.

Schiff *et al.* (1962) found that monkeys showed avoidance responses to a looming image seen on a screen. Schiff (1965) observed avoidance behaviour in response to accelerated looming of shadows cast on a screen in a variety of species, including crabs, chicks, and humans. He produced some evidence that animals could detect the 'direction' of the simulated approach from the degree of asymmetry in the looming pattern.

These avoidance responses to looming could be innate or develop only after the experiential coupling of looming and impact. Ball and Tronick (1971) claimed to have observed avoidance responses to symmetrically expanding shadows in human infants only two weeks old. The avoidance responses were not observed when the shadows expanded asymmetrically or contracted. Yonas *et al.* (1977) found no evidence of avoidance responses to symmetrically looming shadows, or even to real approaching objects, in infants under about four months of age. They suggested that the responses observed by Ball and Tronick were due to the attempt of the infant to keep the gaze fixed on the top of the approaching object, as it rose higher in the visual field, and not to the movement of the object in depth. Dunkeld and Bower (1980) challenged this assertion and produced evidence that infants between three and four weeks of age show avoidance responses which are triggered specifically by the approach of an object under conditions where responses to rising edges are controlled for.

Another important aspect of the pattern of looming is the **divergence and straightness of looming**. If an observer moves rapidly past or towards a large surface its image may be reduced to a blurred flow pattern. Harrington *et al.* (1980) have argued that the pattern of 'blur lines' in motion-produced images provides information about the nature of the observer's motion with respect to a surface. Figure 7.19 shows some of the blur patterns that are visible to an observer looking down on a surface over which he is moving. Harrington *et al.* showed that people are reasonably precise and accurate in judging the degree of divergence in simulated blur patterns.

The **rate of looming** per unit distance indicates the velocity of the object's approach and is proportional to the time to impact. A rapidly looming image produced by an object at any distance is alarming because it signifies that impact is imminent.

If an observer approaches a large surface, the **focus of expansion** of its looming image indicates where on the surface impact will occur. J. J. Gibson has stressed the importance of this information for a pilot landing an aircraft (Gibson *et al.*, 1955). According to a study by Johnston *et al.* (1973) observers are able to estimate the focus of expansion of a looming pattern with a mean error of 10°, which is larger than one would expect if this were a critical factor in locomotor control.

Whether responses to looming are innate or learned, adult animals which show such responses must possess a mechanism for the detection of looming and

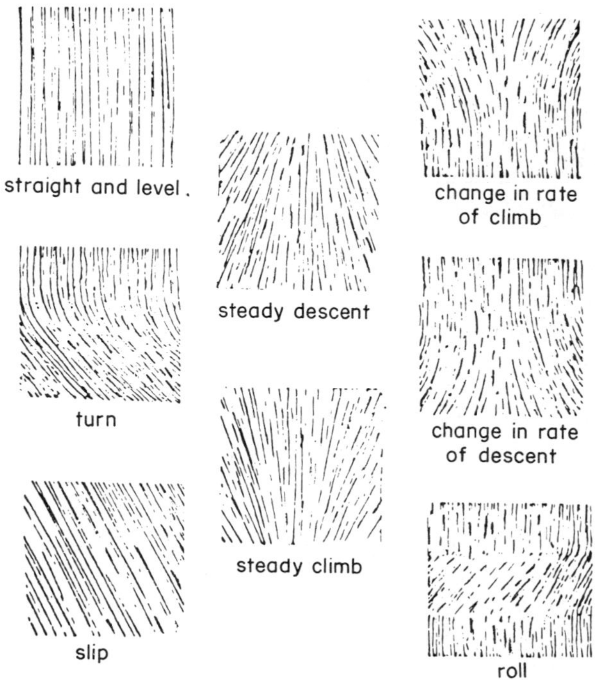

Figure 7.19 Blur patterns produced in the eye of an observer looking down on a surface above which he is moving (From Harrington *et al.*, 1980. Reproduced by permission of the Psychonomic Society)

degree of symmetry in the looming pattern. Looming might be expected to manifest differential adaptation and aftereffects like those shown for other visual attributes, such as lateral motion and orientation. Regan and Beverley (1978) reported evidence of this kind. They asked the subject to inspect for 10 min the outline of a 0.5° square, which either pulsated in width or oscillated sinusoidally from side to side with a frequency of 1 Hz. It was found that inspection of pulsation specifically elevated the threshold for the detection of pulsation in a test square. If the effect had been due to adaptation of detectors for lateral motion, the oscillating and pulsating inspection squares would have had equal effects on the pulsation threshold. When the pulsations and oscillations were slow one way with a quick return (ramp waveform) the threshold elevation was evident only when ramp motion of the test square was in the same direction as that of the inspection square. The effect was restricted to the region of the retina stimulated by the inspection square, but showed when a test square was presented to the eye opposite to that exposed to the inspection square.

Regan and Beverley concluded that there are channels in the visual system specifically tuned to looming. Presumably the detectors responsible receive inputs from motion detectors with opposite preferred directions. It was concluded that each looming detector collects motion signals from an area of no more than about 1.5° of visual angle, because of the fact that oscillating squares larger than this did not produce aftereffects of the type just described (see also Beverley and Regan, 1979). In a subsequent paper Regan and Beverley (1979) exposed subjects for some minutes to an inspection pattern which loomed symmetrically from a fixed point and measured the elevation that this produced in the sensitivity to pulsating motion of a subsequently exposed test square. The degree of threshold elevation declined with increasing distance between the centre of the test square and the focus of the looming inspection pattern. This was true even when the focus of the looming pattern was not on the point of fixation. It was concluded that channels for the detection of looming are specifically tuned to the degree of symmetry in the looming stimulus, which is what one would expect if these detectors are designed to detect the direction of approach of an object on the basis of looming cues.

Regan and Cynader (1979) recorded from single cells in the cat's visual cortex in the hope of finding cells which responded selectively to a looming stimulus. Most cells responded to changing size, but that alone is not sufficient to indicate that they are specifically tuned to looming, because changing the size of a stimulus also involves local changes in brightness which all cortical cells detect. Furthermore, most of the cells which responded to looming responded best to expansion in one part of their receptive field and to contraction in another part, and this reversal of preference is not what one would expect of a true looming detector and can in any case be explained in terms of the organization of receptive fields of known classes of cortical cell. Nevertheless, Regan and Cynader did find a few cells which responded consistently to looming, irrespective of the brightness polarity of the stimulus or the position of the stimulus in the cell's receptive field. These cells seem to have been true looming detectors.

7.9.3 Binocular cues to the direction of an object's approach

The nature of the binocular cues to the direction of a small approaching object depends on whether the observer maintains a fixed position of gaze or whether he tracks the approaching object. If he maintains a fixed position of gaze, the rate and direction of image motion across the retina of one eye differs with respect to the rate and direction of image motion across the other eye, according to the direction of the object's approach. Consider an object moving within the plane containing the two visual axes. If it moves along the median plane of the head, the two binocular images move symmetrically; outwards if the object approaches and inwards if it recedes. We can say, after Regan *et al.* (1979), that the ratio of movements is 1 : 1 in antiphase. If the object moves along a visual line of one or other eye, the image in only one eye moves, and the ratio of movements is 1 :0. If the object moves along a path which misses the head, the images move in the same direction, that is, in phase. The limiting case for in-phase motion is motion of the object along the horizontal horopter, when the ratio of movements is 1 : 1 in phase. Thus, these rates and phases of relative image motion provide unambiguous information about the direction of an object's approach relative to the observer.

Psychophysical evidence for the existence of visual mechanisms specifically tuned to the relative rates of movement of the images in the two eyes was provided by an experiment which I conducted (Howard, 1961). If a three-dimensional rotating wire framework is inspected for some time, it undergoes an apparent reversal of perspective and direction of rotation. I found that the time taken for such an apparent change in direction to occur, with binocular viewing, was on average about 2 min. If, at the end of this latency period, the actual direction of rotation was reversed, the delay before the next apparent reversal was about 4 min. It is as if the effects produced by an initial direction of rotation in depth have to 'unwind' before the system adapts to the new direction. The visual system is specifically adapted to a particular pattern of rotation in depth, because the purely lateral movements produced by the front and back of the framework were the same whichever way the framework was rotating. The effect depends on binocular cues to depth, because it was not evident with monocular viewing. It is restricted to the region of the retina stimulated by the inspection stimulus. Further evidence of this kind was provided by Beverley and Regan (1973). Subjects viewed an irregular pattern of dots with the left eye and a matching pattern of dots with the right eye. The central 2° areas of the two patterns were sinusoidally oscillated from side to side. When the two displays were stereoscopically fused, the central area appeared to oscillate back and forth in depth. The direction of the path of motion in depth was varied by changing the phase and relative amplitudes (symmetry) of the movements of the central discs. After subjects were exposed for 10 min to such a display, it was found that their sensitivity to simulated movement in depth was reduced. That is, there was an increase in the amplitude of relative image motions required to evoke an impression of movement in depth. This change in sensitivity was greatest when the test movement was in the same direction in depth as the inspection movement and fell to zero if the directions of the two movements differed by more than a certain amount. The data suggested that there are four visual channels, each tuned to a different range of directions of approach: one for directions of approach between the nose and the left eye, one for approach between the nose and the right eye, and one each for directions of approach on either side of the head.

There is also physiological evidence for the presence of binocular visual mechanisms specifically tuned to the direction of approaching objects. Updyke (1974) recorded from cells in the superior colliculus of the monkey and found cells which responded best to objects approaching the animal's head, although it was not clear whether the crucial stimulus was the symmetry of looming or the relative movements of binocular images. Zeki (1974a) found cells in the visual cortex of the monkey which responded to oppositely directed stimulus motion in the left and right eyes. However, he did not stimulate both eyes simultaneously. Pettigrew (1973) did stimulate both eyes simultaneously, and found a few cells which responded best to oppositely directed movements in the two eyes.

Regan and Beverley (1973) recorded evoked potentials from the scalp over the visual cortex in human subjects as the subjects watched a stereoscopic display of simulated motion in depth, and obtained responses that were specifically related to the appearance of motion in depth. More recently, Cynader and Regan (1978) recorded from cells in area 18 of the cat's visual cortex as the cat viewed a stereoscopically simulated display of a bar oscillating in depth along various directions of motion. They found cells which were tuned to the direction of movement in depth with a directional range of 2–3°, which included trajectories that would hit or narrowly miss the cat's head. Other cells were tuned to trajectories which missed the head, and yet others were tuned to movements within a frontal plane.

It may be concluded that the mammalian visual system contains mechanisms specifically tuned to looming and others specifically tuned to binocular differences in image motion. These two mechanisms presumably pool their outputs into a system which directs the animal to make appropriate responses.

Up to now it has been assumed that the body and head of the subject were stationary. When a person attempts to judge the directions of objects while the head or the whole body are moved about they can succeed only to the extent to which these movements are taken into account. The next two chapters will be devoted to a discussion of this issue.

Chapter 8

The Vestibular System

8.1 General structure and function

The vestibular system is a set of five sensory epithelia embedded within the bony labyrinths on each side of the skull in close proximity to the auditory cochlea. These sense organs, together with the cochlea, evolved from ancestral organs in primitive aquatic vertebrates. The ancestral organs consisted of cells covered with sensitive hairs, or cilia, lying in fluid-filled pits in the skin. The cilia were affected by movements of the fluid caused by external disturbances or by movements of the animal. The sensory pits became specialized and linked by canals under the skin to form the lateral-line and vestibular organs of fish and, by further specialization, the vestibular–cochlear organs in the bony cavities of the inner ears of mammals. Three subdivisions evolved, which in mammals are represented by (1) the **spiral cochlea**, sensitive to acoustic vibrations (2) the sack-shaped **utricle** and **saccule**, sensitive to gravity and to linear acceleration of the head, and (3) the three **canals**, sensitive to rotations of the head (see Figure 8.1). A complete set of organs occurs on each side of the head. Ducts join the utricle-canal complex to the saccule, and the saccule to the cochlea. Another duct runs from the utricle to a lymphatic reservoir, the **endolymphatic sac**. The whole system is filled with a fluid, the **endolymph**. The cavities containing the sensory epithelia and endolymph are lined with a membrane which is separated from the bony walls of the labyrinth by a fine network of connective tissue bathed in **perilymph** fluid.

The sensory cells of the vestibular system, like those of other sense organs, may

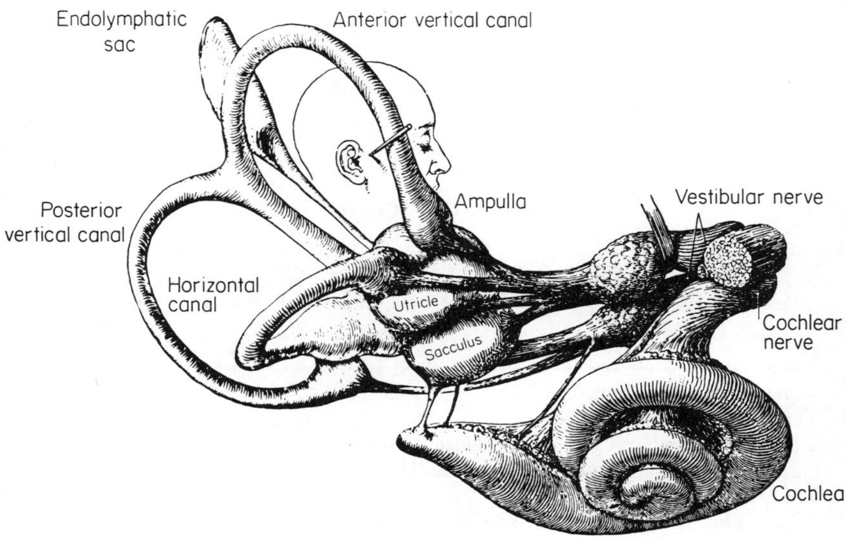

Figure 8.1 Diagram of the inner ear, showing the semicircular canals, utricle, saccule, and cochlea, together with the nerves innervating them (From Hardy, 1934)

be artificially stimulated by a variety of stimuli including thermal, chemical, mechanical and electrical stimuli. But they are maintained in a dark environment which is thermally and chemically constant, and are normally protected from forms of energy other than mechanical. Furthermore, the nature, frequency, direction, and amplitude of the mechanical energy to which any sensory cell of the vestibular system normally responds is determined largely by the mechanical properties of the organ in which the cell lies and by the mechanical arrangement of the sensitive hair cells, or cilia, on the surface of the cell.

8.2 The vestibular canals

8.2.1 The mechanics of the vestibular canals

The canals are usually called the semicircular canals, but the prefix 'semi' is misleading; although the canals share a common cavity in the utricle, each one is functionally a complete and independent fluid circuit (see Figure 8.2). When a person holds the head in a natural erect posture, the horizontal canals are almost parallel to the earth-horizontal plane (see Wilson and Melvill Jones, 1979, p 19). A more precise landmark for specifying the position of the canals is provided by the 'horizontal' reference plane of the Horsley Clark stereotaxic instrument. This plane passes through the lower margins of the orbits of the eyes and of the auditory canals, and is inclined forwards 25° with respect to the horizontal canals in man. When the head is in a natural erect posture the posterior canal lies in an almost vertical plane which forms an angle of 45° to the frontal plane of the head,

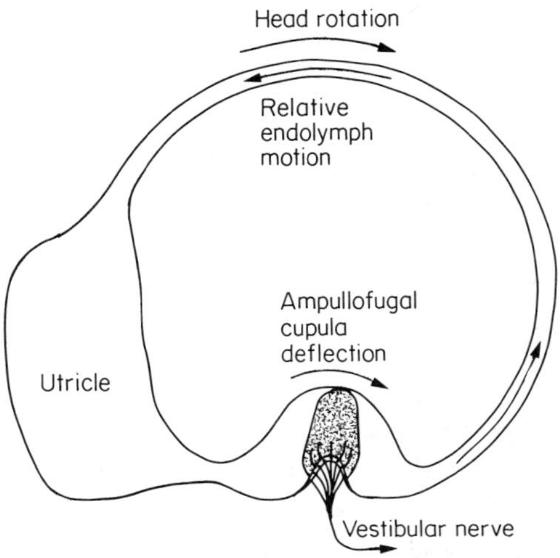

Figure 8.2 Diagram of a vestibular canal showing the complete fluid circuit. The arrows depict the consequences of a clockwise rotation of the head in the plane of the canal. At ordinary levels of stimulation the cupula does not deflect as shown in the diagram, but acts like a diaphragm, with the apex not moving (Adapted from Milsum and Melvill Jones, 1969. Reproduced by permission of the New York Academy of Science)

and the anterior canal lies in a plane inclined at an angle of about 15° to the vertical which also forms an angle of 45° to the frontal plane. Each canal has a contralateral partner to which it is most nearly parallel. Each pair forms a functional unit—a **synergistic pair**. Thus, the two horizontal canals, the left anterior and right posterior canals, and the left posterior and right anterior canals form synergistic pairs. The three canals are not exactly at right angles and synergistic pairs are not exactly parallel, so that there is no plane of head rotation which stimulates only one pair of synergists; any rotation gives rise to a pattern of stimulation in several canals (see Blanks *et al.*, 1975, for details of the dispositions of the canals).

At one point of each canal, near its junction with the utricle, is a swelling or **ampulla**, which contains the sensory epithelium, or **crista ampullaris**. The structure of a crista is shown in Figures 8.3 and 8.4. It is a ridge of epithelium protruding into the cavity, or **lumen**, of the ampulla, and carrying many multiciliated sensory cells interspersed with structural cells. The cilia of all the cells project into a common gelatinous mass, or **cupula**. The cupula reaches to the far side of the ampulla, which is arched at this point. It is a difficult structure to see, because it has the same refractive index as the endolymph. For this reason, the early theorists did not realize that the cupula seals the lumen of the ampulla; they thought that the endolymph flowed over the cristae. It was not until 1931 that Steinhausen observed the functioning of the cupula *in vivo* by injecting Indian ink into the endolymph of a fish. He showed that the cupula forms an

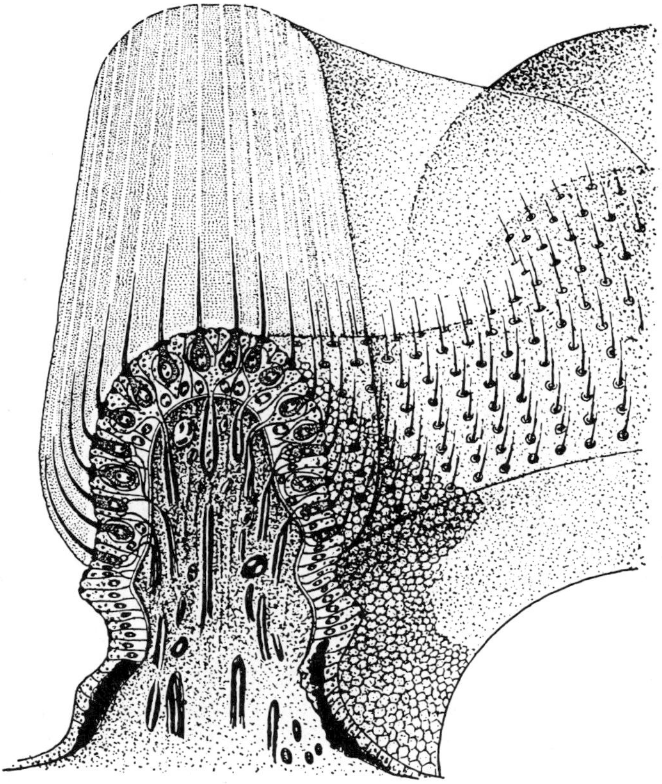

Figure 8.3 Schematic drawing of one half of a crista ampullaris, showing innervation of its epithelium. Thick nerve fibres forming nerve calyces round type I hair cells at the summit of the crista; medium calibre fibres innervating type I hair cells on the slope of the crista; medium calibre and fine nerve fibres forming a nerve plexus innervating hair cells of type II. The sensory hairs pass from the hair cells into fine canals in the cupula, which is separated from the epithelium by a narrow subcupular space (From Wersäll, 1956. Reproduced by permission of *Acta Oto-Laryngologica*)

effective seal which prevents leakage of endolymph past that point. Thus, the endolymph and cupula are always displaced as one. When the head is turned in the plane of a canal the endolymph and cupula lag behind because of their inertia. The fluid circulates relative to the wall of the canal in a direction opposite to the direction of rotation of the head, and displaces the cupula by the same amount. It was thought until recently that the cupula swings from side to side like a swingdoor, but it is now known that normal displacements of the endolymph are not sufficient to displace the apex of the cupula, so that the cupula normally behaves like a diaphragm (McLaren and Hillman, 1979).

The force acting on the cupula is equal to the product of the angular acceleration of the head, α, in the plane of the canal and the moment of inertia of

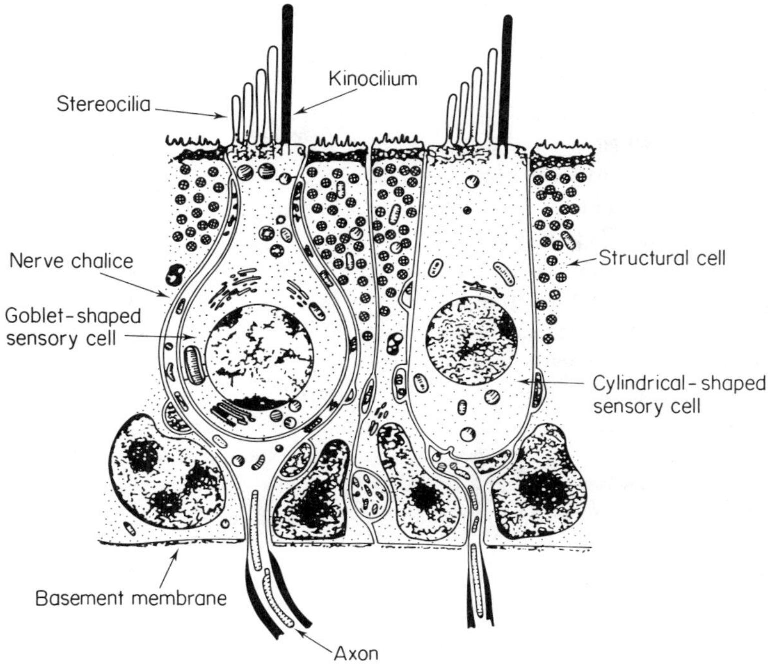

Figure 8.4 Diagrammatic representation of the detailed structure of the crista ampullaris (From Engström, 1961. Reproduced by permission of *Acta Oto-Laryngologica*)

the cupula and endolymph, H. This force is related to the angular displacement of the endolymph and cupula, θ, by the following differential equation:

$$\alpha H = k\theta + r\frac{d\theta}{dt} + H\frac{d^2\theta}{dt^2}$$

where k is the coefficient of elastic, or position-dependent, resistance, r is the coefficient of viscous, or velocity-dependent resistance, and H the moment of inertia, or coefficient of mass-dependent resistance of the cupula and endolymph (Mayne, 1950). But the diameter of the cross-section of the canal is only about 1/3 mm in man, which means that the mass of the endolymph is small and that the viscous resistance is high for even moderate velocities. Furthermore, the elasticity of the cupula is small compared with the viscous resistance. So that, approximately,

$$\alpha H = r\frac{d\theta}{dt}, \quad \text{or } \alpha \text{ is proportional to } \frac{d\theta}{dt}.$$

If both sides of this relation are integrated with respect to time, we can say that the angular velocity of the head, ω, is proportional to the angular displacement of the cupula, θ. In other words, the semicircular canal is an integrating angular accelerometer, that is, an angular speedometer. When the head rotates very slowly, the viscous forces become smaller than the inertial forces, and the response of the system becomes proportional to the acceleration of the head (Melvill Jones, 1976). However, as far as the velocities and frequencies of normal head movements are concerned, the canals operate as velocity detectors. The dimensions, and hence the viscous resistance, of the canals of other species are apparently adjusted to suit their natural range of velocities and frequencies of head rotation (Melvill Jones and Spells, 1963). The high viscous resistance of the canal has several advantages.

(a) It converts the system into a velocity detector for the normally encountered, or **physiological range,** of frequencies of head oscillation, which in a human being lies between about 0–1 Hz and 5 Hz. Over this range of frequencies of head oscillation the **gain** or sensitivity of the system (cupula displacement per unit of head velocity) is constant, and the cupula deflection keeps in phase with head velocity, neither lagging nor leading. At lower frequencies of head oscillation, sensitivity decreases and cupula displacement *leads* head velocity. When this phase advance reaches 90°, the cupula displacement is in phase with head acceleration, and it is then that the system acts as an accelerometer. At higher frequencies of head oscillation, sensitivity decreases again but cupula displacement *lags* behind head velocity (see Wilson and Melvill Jones, 1979, p. 53, for details on these points).

(b) It keeps the flow of endolymph smooth, or non-turbulent. The Reynolds number is the ratio of the rate of flow of fluid in a pipe to the product of the diameter of the pipe and the viscosity of the fluid. When the Reynolds number is small, as it is in the vestibular system, the flow of the fluid is laminar, or smooth. When the number is large, the flow is turbulent.

(c) It provides viscous damping which prevents the system from oscillating.

(d) It reduces the flow of the endolymph and hence keeps the deflection of the cupula within a small range, where its characteristics are most linear. Calculations show that the cupula deflects only about 10 millimicrons for an acceleration of the head of $0.1°/sec^2$, which is about the smallest acceleration of normal head movements (Oman and Young, 1972). These are extremely small movements, and explain why it has been difficult to observe the cupula deflection directly.

(e) It reduces the latency of the cupula displacement. We have seen that the response of the cupula is determined by the elastic, viscous and inertial constants, k, r, and H. The ratio H/r is known as the **inertial time constant**, and the ratio r/k is known as the **elastic time constant**. The inertial time constant is about 3–5 msec and is known as the **short time constant**. It is the time taken for the cupula displacement to reach $1/e$ of its maximum value after a sudden change in angular velocity of the head. The elastic time constant, as indicated by the discharge characteristics of primary vestibular afferents in the monkey, varies from about 3.8 to 5.5 sec, depending on whether the cell has phasic or tonic properties (Goldberg and Fernández, 1971a). This is known as the **long time constant**. The long time constant is related to how long it takes the cupula to return to its midposition after the accelerative stimulus is removed. Estimates of the long time constant in man are based on the persistence of oculomotor responses to vestibular stimulation, and are about three times as long as those I have just mentioned. Wilson and Melvill Jones (1979) suggested that the behaviourally determined time constant is longer than the time constant at the peripheral organ because of the presence of a positive feeback loop in the vestibulo-ocular control system. A recent review of vestibular dynamics has been provided by Barnes (1980).

It is clear from what has been said that the cupula is deflected when the canal walls move relative to the endolymph. Because of the elasticity of the cupula and the friction between the endolymph and the canal walls, the endolymph catches up to the movement of the canal walls when a person is rotated for a while at constant angular velocity. When this has happened, the cupula returns to its resting position. If the person is now decelerated and brought to rest, the cupula displaces in the opposite direction, and it takes up to about 30 sec before the behavioural effects of this deflection are dissipated, although the cupula itself is restored to its central position in a shorter time than this. During the 30 sec recovery period the subject behaves as if he is decelerating. He falls to one side, his eyes move nystagmically, and he experiences turning sensations and nausea. If an acceleration is followed immediately by a deceleration, as is usually the case when a person voluntarily turns his body, the two opposed deflections of the cupulae cancel out, leaving no residual deflection and no aftereffects.

All this assumes that the subject does not move his head relative to his body while he is being rotated. If he does, he will modify the effects which the rotation has on his vestibular canals. If he rotates his head about the same axis as the applied rotation, there is a momentary increase or decrease in its total angular acceleration. If he turns his head about any other axis, his own angular velocity interacts with the applied velocity in a complex way. Anyone who has tried to turn the axis of a spinning wheel or top at right angles to itself will have noticed that the axis tends to rotate at right angles both to itself and to the manually applied torque. This resultant motion is known as **precession**, and the forces which produce it are called **Coriolis forces.** It is also sometimes convenient to talk about the rotations being **cross-coupled**. Consider the events on one side of the head when a person is rotated to the left at a steady angular velocity, ω, in the plane of the horizontal canal (see Figure 8.5). After some time, the cupula in the horizontal canal will restore to the central position and the whole system will stop responding. If the head is then quickly inclined forward through an angle ϕ, the two vertical canals project themselves as narrow ellipses in the plane of the original rotation and each becomes subject to a momentum component of that rotation of $\omega \sin \phi$. The horizontal canal moves slightly out of the plane of rotation and loses a fraction of its original momentum, equal to $\omega \cos \phi$. These three changes in momentum add like vectors to produce a resultant momentum proportional to:

$$\omega[(1 - \cos \phi)^2 + 2 \sin^2 \phi]^{1/2}$$

This resultant momentum acts in the plane of head tilt in a direction towards

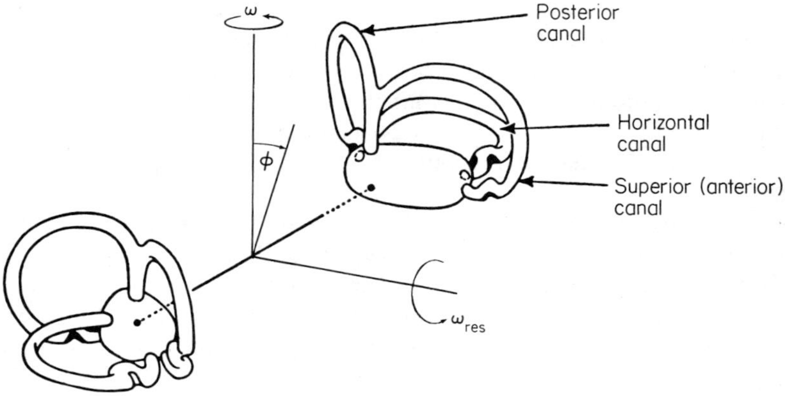

Figure 8.5 Demonstrating cross-coupling between head tilt and body rotation. The head is rotating round a vertical axis, angular velocity ω. The lateral canals are in a horizontal plane. The head is suddenly inclined forwards over the angle ϕ. A resultant momentum drives the endolymph clockwise; hence a sensation that the head is being tilted to the left, equivalent to a stimulus intensity:

$$\omega[(1-\cos\phi)^2+2\sin^2\phi]^{1/2}$$

(From Groen, 1961. Reproduced by permission of S. Karger AG, Basel)

the right shoulder and hence produces a sensation that the head is being tilted to the left. After a while the cupulae will again be restored to the central position, but any other similar movement of the head will induce another illusory sensation of tilting. These illusory sensations are very disturbing and result in illusory movements of the visual surroundings, nausea, and extreme discomfort (Collins, 1968b; Melvill Jones, 1970; Reason and Graybiel, 1971). Coriolis inputs from the canals should not in themselves induce nausea, because they resemble the inputs from a true rotation of the head in an equivalent direction. The nausea must result from the conflict between the canal inputs which produce a non-veridical signal of head tilt and the otolith, visual and kinaesthetic inputs which signal the veridical direction of head inclination. If movements of the head are confined to the plane of the rotation of the body, these effects are avoided.

8.2.2 The afferent signal from vestibular canals

The microstructure of the sensory epithelium of the cristae is shown diagrammatically in Figure 8.4. There are two types of sensory **hair cell**. **Type I cells** are flask-shaped and almost totally enclosed, singly or several together, in a goblet-shaped nerve chalice innervated predominantly by the larger diameter fibres in the cristal nerve. Any one nerve fibre may branch to supply several type I cells. **Type II hair cells** are cylindrical, and innervated mainly by smaller diameter nerve fibres, each with several knob-shaped nerve endings (Wersäll and Bagger-Sjöbäck, 1974). A single type II cell may be innervated by several nerve fibres and a given nerve may innervate several sensory cells, sometimes a mixture of type I and type II cells (see Wilson and Melvill Jones, 1979, p. 35). Both types of cells are also innervated by efferent neurones which are capable of inhibiting their response (Klinke and Galley, 1974). The functional significance of the efferent neurones remains obscure.

Each sensory cell projects about 70 fine cilia into the cupula mass; these **stereocilia** are distinguishable from a single, larger **kinocilium** in each cell. The stereocilia are graded in length towards the kinocilium, as shown in Figure 8.4. The membrane over the cilia is continuous with the cell membrane and a resting potential of about 160 mV exists across it; the cell interior is negative with respect to the contents of the cupula. This potential is due partly to a polarization of the cell membrane and partly to a concentration of positive ions in the endolymph. Microelectrode studies have revealed that a deflection of the cupula in one direction is accompanied by a **depolarization**, or decrease, of the resting potential across the cilia membrane, and that a deflection in the other direction is accompanied by a **hyperpolarization**, or an increase in the resting potential (Trincker, 1962). It is thought to be the shearing movement of the cilia produced by the deflection of the cupula which alters the resting potential; if the cilia are deflected at right angles to their normal plane of action the resting potential is not affected (Flock, 1965). There must be an asymmetry in the structure of the cilia to account for this asymmetry of action within a specific plane. Lowenstein and Wersäll (1959) found that the kinocilium is always situated on that side of the stereocilia that points in the direction of depolarizing (excitatory) cupula deflection. In any one canal each kinocilium lies on the same side of its group of stereocilia and the stereocilia are graded in increasing length in this direction. Here then is the required asymmetry. A mechanical model of how this structural asymmetry may produce the bidirectional but asymmetrical response of the hair cell has been proposed by Malcolm (1974).

By recording from single afferent fibres arising from the hair cells of the vestibular system of the squirrel monkey, Goldberg, Fernández and their collaborators have revealed the way in which cupula deflection is coded into a neural response (Goldberg and Fernández, 1971a, b; Fernández and Goldberg, 1971). Vestibular afferents have a potential dynamic range of 0–400 spikes/sec. Most, if not all of them discharge even when the cupula is not deflected. In the monkey, this **resting discharge** has been found to average about 110 spikes/sec, and to vary from cell to cell over a range of 18–160 spikes/sec (Keller, 1976). The resting discharge is very regular from moment to moment in some cells which are therefore called **regular cells**, and variable in others which are called **irregular cells**. The large-diameter fibres from type I (chalice) hair cells tend to fire irregularly and the thin fibres from type II (cylindrical) hair cells tend to fire regularly. It has been suggested that the more stable firing rate of thin fibres is due to the fact that these fibres innervate more hair cells than do thick fibres (Goldberg and Fernández, 1977). Furthermore, the firing rate of regular cells is closely related to the amount of deflection of the cupula, whereas the firing rate of irregular cells is related to the velocity of cupula displacement as well as the magnitude of its deflection. This difference reflects the fact that the firing rate of the irregular units adapts to steady displacement of the cupula. Adaptation is equivalent to a differentiation of the cupula displacement. These differences between the two types of cell may be summed up by saying that the regular units are static, or **tonic** receptors and the irregular units are dynamic, or **phasic** receptors. Regular tonic afferents have longer time constants and lower sensitivity than irregular phasic afferents (Malcolm and Melvill Jones, 1970).

The function relating frequency of firing of primary afferents to cupula deflection is S-shaped (Groen *et al.*, 1952). Thus, this function is approximately linear only over the middle of the range of stimulus values to which the cell is sensitive. However, the resting discharge rate and the linear portion of the curve occur at different positions along the cupula deflection axis, so that there is a continuum of types which differ in the value of a 'bias', analogous to the bias on a thermionic valve. The net result is that the frequency of discharge of the whole set of primary afferents is a more or less linear function of cupula deflection (head velocity) over a wide range of deflection amplitudes.

There is very little phase lag in the response of primary afferents to sinusoidal rotations of the head over the frequency range of natural head movements, that is, over the range 0.1–1 Hz. At frequencies below and above this range the rate of response of primary afferents shows an increasing phase advance relative to the velocity of the head (Fernández and Goldberg, 1971).

The fact that most afferent units in the vestibular system have a resting discharge confers several advantages on the system.

1. It provides a continuous influx of excitation into the motor centres concerned with the maintenance of posture. This helps to keep the skeletal muscles under steady tonic contraction. Loss of vestibular input results in prolonged muscular flaccidity (Lowenstein, 1975).
2. The resting discharge enables the system to work bidirectionally, that is, it helps to produce signals which are different for opposed directions of rotation.
3. The resting discharge renders the vestibular receptors virtually thresholdless (Goldberg and Fernández, 1971a).
4. The bidirectionality of the output from most units means that, when the head rotates in the plane of a canal in a given direction, there is an increase in rate of firing from that canal on one side of the head and a related decrease in rate of firing from the synergistic canal on the other side of the head. We shall see later that primary vestibular afferents on each side feed into the ipsilateral vestibular nucleus in the brainstem, where they synapse with secondary neurones. Commissures directly and indirectly connect one vestibular nucleus with the vestibular nucleus on the other side.

The net result is that each input to the vestibular nucleus on one side inhibits the corresponding contralateral input. In other words, the inputs from the two sides combine reciprocally in the vestibular nuclei to form a differential, or push–pull system. It is a feature of push–pull detectors that any non-linearities in one half of the system are cancelled by non-linearities in the other half, so that the 'difference signal' produced between them is linear. Furthermore, noise, or an aberrant input, from only one canal will produce a difference signal only half as strong as a true stimulus of the same intensity occurring in both canals. In other words, the push–pull arrangement should improve the signal-to-noise ratio of the vestibular system, although whether it does so in practice has yet to be determined experimentally. Any disturbance of the neutral balance point of the vestibular push–pull system creates an impression of head rotation when the head is not moving. The fact that animals slowly adapt to constant levels of vestibular stimulation suggests that they possess a mechanism which automatically adjusts the balance of the system (Malcolm and Melvill Jones, 1970).

Consider now the sequence of events in the right horizontal canal when the head is rotated to the right in the plane of the canal. The inertia of the contents of the canal causes the endolymph to lag behind the canal walls and flow in the opposite direction relative to the walls of the canal, and the cupula to be deflected towards the utricle (ampullopetally). This results in a depolarization of the hair-cell membrane and a consequent increase in the frequency of firing in the sensory neurones. There occurs simultaneously an ampullofugal deflection of the left-hand cupula (see Figure 8.2) which induces an increase in the resting potential (hyperpolarization) and a decrease in the frequency of the afferent discharge (see Table 8.1). The effect on each side is reversed for a head rotation to the left.

In the vertical canals, depolarization, and hence increased afferent discharge, is produced by ampullofugal deflection. One might conclude that a deflection of the cupula one way drives nystagmic eye movements in one direction and a deflection the other way drives nystagmus in the other direction. But this is not how it works. The bidirectional output from a given canal is combined in push–pull fashion with the output from the synergistic canal on the other side of the head. It is this **difference signal**, produced in the vestibular nuclei of the brainstem, that drives the behaviour associated with vestibular stimulation, such as postural balance and nystagmus. Consider what happens when the horizontal canal on one side of the head is destroyed. With the head stationary, the resting discharge from the intact canal is uncancelled by the now absent resting discharge from the other side. In other words, the lack of input from the damaged side has the same effect as an extreme ampullofugal cupula deflection on that side. The resulting central imbalance of the push–pull system induces a spontaneous and persistent nystagmus with its quick phase directed to the intact side. If two intact synergistic

Table 8.1 Summarizing the consequences in the horizontal canals of rotating a person to the right in the plane of the canals

Right ear (ipsilateral)	Left ear (contralateral)
Ampullopetal deflection of cupula	Ampullofugal deflection of cupula
Decreased resting potential (depolarization)	Increased resting potential (hyperpolarization)
Increased frequency of afferent firing	Decreased frequency of afferent firing

canals are electrically or thermally stimulated simultaneously in the same way, either both will be hyperpolarized or both will be depolarized. But this will not disturb the central balance and nothing will happen (Fluur and Mellström, 1971a).

The neural discharge from afferent neurones diminishes or adapts under conditions of constant cupula deflection produced by a prolonged period of steady angular acceleration (Goldberg and Fernández, 1971a). The rate of adaptation is particularly rapid for the normally silent units and for units with irregular rates of resting discharge. This adaptation of the input is observed after the vestibular nerve is cut, showing that it is not dependent on inhibitory impulses from the central nervous system transmitted along efferent fibres (Lowenstein, 1975). Thus, the second-order differential equation given at the beginning of Section 8.2.1 must be modified if it is to encompass the process of adaptation. Young and Oman (1970) have suggested an appropriate modification and used it with success to predict behavioural consequences of vestibular adaptation. Another theoretically predictable consequence of adaptation in the irregular afferents is that they exhibit a diminished phase lag and an enhancement of sensitivity (gain) to changes in acceleration as the frequency of oscillation is increased (Louie and Kimm, 1976).

The response of afferent nerve fibres from the canals is not modified by visual stimulation, not even by a movement of the whole visual scene, even though visual movement induces illusory sensations of head rotation (see Section 9.4). Nor is the response of canal afferents modified by movements of the eyes (Keller, 1976).

There have been several reports that responses in afferent neurones from the canals can be detected when the head is tilted or subjected to other linear accelerative stimuli. Very small differences in density between cupula and endolymph would be sufficient to cause such responses, and Goldberg and Fernández (1975) have shown that such density differences may arise from thermal gradients and other disturbances introduced by surgical exposure. By plugging the canals they were able to abolish these thermal effects and, at the same time, abolish the response of the canals to angular acceleration. However, the blocked canals still responded to linear acceleration, which was thus seen as a normal response of the system, and not merely due to thermal gradients. I shall return to this question in Section 9.2.7.

Alcohol intoxication induces nystagmus when the head is tilted, and it is thought that this **positional nystagmus** is due to a disturbance of the specific gravity of the cupula caused by the differential absorption of alcohol into different parts of the canal system. If alcohol is mixed with heavy water (deuterium oxide) the density of the mixture can be made equal to that of water. A person intoxicated on 'heavy' alcohol does not suffer from positional nystagmus and its attendant dizziness (Money *et al.*, 1965; Money and Myles, 1974). The only snag is that a shot of 'heavy' whisky costs about thirty dollars!

The response of vestibular afferents in the monkey has been found not to be affected by anaesthesia of the central nervous system (Keller, 1976). Thus, the efferent fibres, which are known to carry impulses from the central nervous system to the vestibular system, seem to have nothing to do with the normal maintenance of the system's response. Nevertheless, Goldberg and Fernández (1980) found that electrical stimulation of efferent fibres slightly increased the gain of irregularly discharging primary afferents, and speculated that efference may serve to extend the dynamic range of the phasic units.

On each side of the head the primary afferents from the three canals terminate in specific regions of the vestibular nuclei which sit astride the cerebellar peduncle in the upper brainstem (see Section 5.4.3). Direct innervation is confined to the ipsilateral nuclei, but commissural fibres provide indirect connections between the two sides. Electrophysiological recordings from single cells in the vestibular nucleus of the monkey have revealed that about half of the cells are excited by acceleration of the head to the ipsilateral side and about half are excited by a contralateral acceleration. Units excited by ipsilateral acceleration are known as **type I cells** and those excited by contralateral acceleration are known as **type II cells**. It seems that type II cells receive their innervation indirectly through commissural fibres from the contralateral vestibular nucleus, and exert inhibitory influences on type I cells. Some primary afferents from the canals go directly to the cerebellum and others directly to the reticular formation (Blanks *et al.*, 1977).

Secondary neurones responsive to canal stimulation leave the vestibular nuclei to enter the pair of medial longitudinal fasciculi, each of which terminates bilaterally in the oculomotor nuclei (Tarlov, 1975). Other fibres from the vestibular nuclei feed into polysynaptic pathways which project through the pontine reticular system and also finally reach the oculomotor nuclei. Stimulation of each ampullary nerve causes contraction of a particular extraocular muscle in one eye and a relaxation of the antagonistic muscle in the other eye (see Section 5.4.3). Other secondary neurones responsive to canal stimulation leave the vestibular nuclei and terminate in the cerebellum, particularly in the nodulus, uvula, flocculus, vermis and fastigial nuclei, mainly ipsilaterally (Pompeiano, 1974). There is also a projection back from the cerebellum to the vestibular nucleus (Walberg, 1975). Other nerve fibres leave the vestibular nuclei on each side and descend to all levels of the spinal cord as the **lateral vestibulospinal pathways** and to the upper spinal cord as the **medial vestibulospinal pathways**. The **reticulospinal pathways** also receive vestibular inputs and they too descend to all levels of the spinal cord. The lateral vestibulospinal and the reticulospinal pathways mediate spinal reflexes and exercise a reciprocal control over flexor and extensor motoneurones throughout the body (Gernandt, 1974; Pompeiano, 1975). The shorter medial pathways mediate reflexes concerned with the postural control of the head (Nyberg-Hansen, 1975). Although doubted for a long time, it is now established that the vestibular apparatus projects to the cerebral cortex, principally to the contralateral motor and somatosensory cortex (Fredrickson and Schwarz, 1977; Büttner and Lang, 1979).

The complex ramifications of the outputs from the vestibular nuclei suggest that the behavioural consequences of canal stimulation are equally complex. However, the paucity of cortical projections from the vestibular system suggests that these behavioural consequences are largely at the unconscious reflex level. The contents of Chapter 9 will demonstrate the truth of these suggestions.

8.3 The otolith organs

8.3.1 The mechanics of the otolith organs

The utricle and saccule are sack-shaped cavities at the junction of the three vestibular canals. Each cavity is filled with endolymph and contains a sensory epithelium, the **macula**, which responds to the magnitude and direction of linear acceleration, including the magnitude and direction of gravity. Organs sensitive to position with respect to gravity occurred early in the evolutionary scale. They are represented in every animal phylum, and their general mode of functioning is remarkably constant. In invertebrates they are known as **statocysts**. One of the neatest ways of studying their function is provided by the crayfish's habit of shedding the linings of its statocysts every so often. The crayfish uses its claws to replace grains of sand in the new cavity. These grains serve to stimulate the sensory cells as the crayfish moves about. If iron filings are the only small particles available the animal places these in the statocyst cavity and when a magnet is moved about, the crayfish alters the orientation of its body (see Lowenstein, 1975).

Precise data on the orientation of human maculae do not seem to be available. The utricular macula is approximately in the same plane as the horizontal canal, although it turns upwards at its anterior end. The saccular macula lies on the medial wall of the saccule and is tilted inwards from the vertical by at least 20°. The surface of each macula is slightly concave (see Figure 8.6) and is covered with an epithelium of ciliated cells, very similar to those found in the cristae of the canals (see Figure 8.7). The cilia protrude into a gelatinous substance containing calcite crystals—the **otoliths**, or otoconia. The specific gravity of the otoliths is about three times that of the surrounding endolymph. Such an organ is clearly suited to respond to displacements of the heavy otoliths induced by changes in the extent and direction of the linear acceleration of the head, brought about by changes in the linear velocity of the body or in the direction of motion of the body or by a tilt of the head with respect to gravity.

Trincker (1962) recorded the changes in potential inside the sensory cells of the macula of the guinea pig. Only shearing forces produced by displacements tangential to the macula surface were effective in producing changes in potential. Pressure or tension applied at right angles to the macular epithelium had no effect. Békésy (1953) had found a similar state of affairs in the organ of Corti in the cochlea. On the basis of shearing forces, the utricular response should be a sine function of the angle of tilt whereas, on the basis of pressure forces, it should be a cosine function of tilt. Trincker was able to demonstrate the sine function in actual recordings from the tilting macula of the guinea pig.

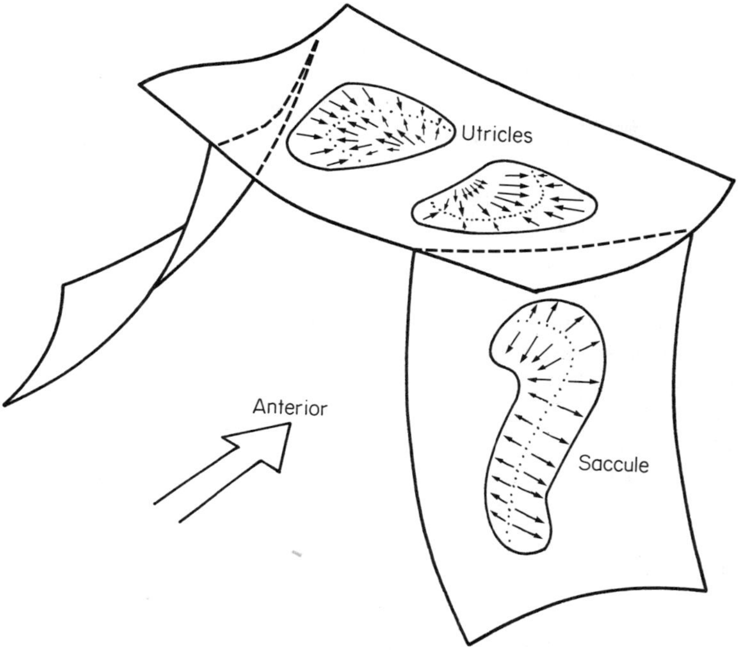

Figure 8.6 Dispositions of the otolith organs showing the polarization maps of their macular surfaces. The arrows point in the direction of depolarization of the hair cells, that is, in the direction of the kinocilium (Adpated from Malcolm and Melvill Jones, 1974. Reproduced by permission of *Acta Oto-Laryngologica*)

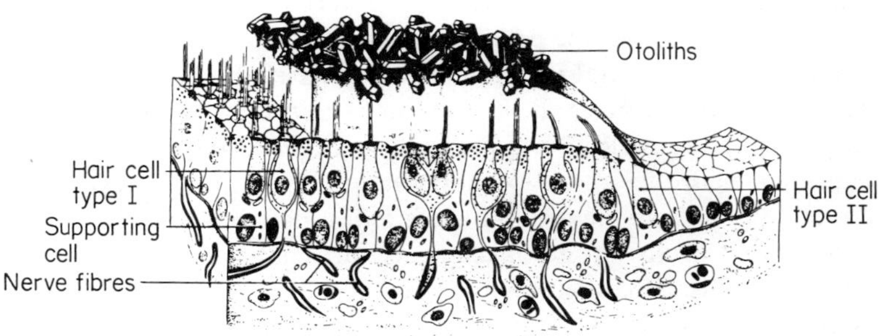

Figure 8.7 Diagrammatic representation of the detailed structure of the macular epithelium (Reprinted with permission from Iurato, 1967. Copyright, 1967, Pergamon Press Ltd.)

The shearing force acting on an otolith organ is the product of the excess mass of the otoliths, β, and the linear acceleration, α, acting in the plane of the macula. The excess mass of the otoliths is the mass of the otoliths minus the mass of the same volume of endolymph. If the density of the endolymph is assumed to be 1, then

$$\beta = \frac{m(\rho - 1)}{\rho}$$

where ρ is the density of the otoliths and m their mass. The shearing force is opposed by elastic, or position-dependent resistance k; viscous, or velocity-dependent resistance r; and mass, or acceleration-dependent resistance m. Thus,

$$\beta\alpha = kx + r\frac{dx}{dt} + m_e \frac{d^2x}{dt^2}$$

where m_e is the effective mass of the otoliths, that is, the total mass of tissue and fluid that is accelerated, and x is the linear displacement of the otoliths. This is the same well-known equation of motion used to describe the mechanics of the cupula. de Vries (1950) obtained direct measurements of these physical constants in a fish. In mammals, the constants which have been inferred from behavioural measurements reflect properties of neural elements as well as those of the peripheral organ (Young and Meiry, 1968).

The human threshold for the perception of linear acceleration has been determined as approximately 0.005 g (Parker *et al.*, 1978). This represents an estimated displacement of the otoliths of 0.15 millionths of a metre and puts the sensitivity of the organ at 33 μm/g.

8.3.2 The afferent signal from otolith organs

The cellular structure of the maculae is very similar to that of the cristae. There are about 33,000 hair cells in the human utricle and about 19,000 in the saccule. Some of these are **type I** sensory cells enclosed in a chalice and innervated by larger-diameter nerve fibres and some are **type II** cylindrical cells innervated by small-diameter fibres. Each cell has about 70 stereocilia and one kinocilium, with a resting potential across the cell membrane.

The membrane potential is maximally depolarized when a force shears the stereocilia in the direction of the kinocilium, and is hyperpolarized when a force acts in the opposite direction. Depolarization increases the resting discharge rate of the afferent neurone, and hyperpolarization decreases the resting discharge. The direction of most effective stimulation of a hair cell is its **polarization axis**. Histological and electrophysiological procedures have been used to map out the distribution of cells with different polarization axes on the surface of the utricular and saccular maculae (Flock, 1964; Spoendlin, 1965; Rosenhall, 1972). Figure 8.6 is a representation of such polarization maps. In each case, oppositely polarized regions are separated by a central 'parting' known as the **striola**. Synergistic and inhibitory relationships between different regions of the macula have been worked out (Fluur, 1970). In particular, the oppositely directed regions inhibit each other and this may explain why otolith organs have been found to be unaffected by even intense angular acceleration (Goldberg and Fernández, 1975). Consider the case where a person rotates about his body axis so that the utricular maculae lie in the plane of rotation. The **rotary component** of the motion should stimulate equal numbers of cells with opposite polarization axes in each utricle, producing a zero output. The **linear**, or centrifugal component of the rotation should produce equal and opposite stimulation of the utricles on the two sides of the head. Presumably these opposed inputs from the two sides are cancelled in the vestibular nuclei. Similar arguments, which are illustrated in Figure 8.8, show that both the rotary and centrifugal components of rotations of the head about any other axis passing through the centre of the

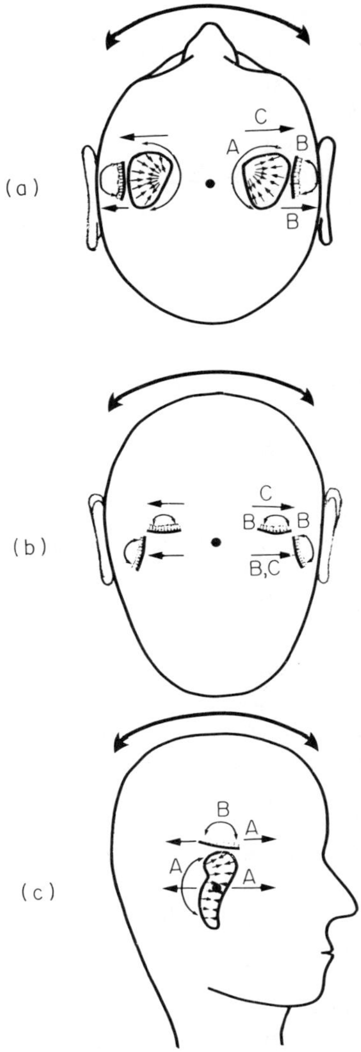

Figure 8.8 Showing how the rotary and centrifugal components of rotary accelerations of the head about each of three axes passing through the centre of the head produce effects on the utricles and saccules which either cancel within each end organ (cases marked A), act only radially on the macula (cases marked B), or are opposite on the two sides of the head (cases marked C). The straight arrows represent centrifugal forces and the curved arrows rotational forces produced by the head acceleration

head should also produce effects in the utricles which cancel within the utricles or within the vestibular nucleus.

Fernández and Goldberg (1976a, b, c) recorded from single afferent fibres in the vestibular nerve of the squirrel monkey as the monkey was placed in various tilted positions or subjected to centrifugal forces in various planes. They found that each nerve fibre was excited most when the head was tilted at a particular angle in a particular plane and least when the head was tilted at 180° in the opposite direction in the same plane. This axis of maximum and minimum excitability defines the polarization axis of the nerve fibre and corresponds to the polarization axis of the hair cells from which that afferent neurone receives its innervation. Shearing of the hair cells in a direction orthogonal to the polarization axis within the plane of the macula had very little effect on the discharge rate of the afferent fibre, and changing the magnitude of the force acting perpendicular to the macular surface had no significant effect. Units in the superior region of the nerve were found to have polarization axes in the plane of the utricular macula, and those in the inferior region were found to have axes in the plane of the saccular macula. When the hair cells of the macula serviced by a given afferent unit were in line with the direction of gravity, either standing up or hanging down, the discharge frequency was, when averaged over a sample of cells, about 60 spikes/sec, with a standard deviation of about 27 spikes/sec. This is the **resting discharge rate**. By exposing the animal to forces of up to 5*g* in one direction or the other along the polarization axis of a unit, S-shaped force-response curves were derived. Acceleration in the inhibitory direction reduced the discharge rate to a minimum residual level of between 5 and 20 spikes/sec, which was reached at a force of about 2*g*. Stronger inhibitory forces had no further effect. Acceleration in the excitatory direction increased the discharge rate in a more or less linear fashion up to a force of about 3*g*, after which the response levelled off and saturated at about 5*g* (see Figure 8.9). It can be seen from the figure that the physiological range of $\pm 1g$ is situated in the lower portion of the force–response function. Fernández and Goldberg identified three response parameters to account for the differences in the force–response characteristics of different units: the **resting discharge rate**, **sensitivity**, or change in discharge rate produced by a shear force of 1*g*, and an **asymmetry coefficient**, or the position of the resting rate with respect to the curve's inflection point.

Most afferent neurones were found to have a regular resting discharge in the resting positions; these are designated **regular units**, to distinguish them from **irregular units**, which have an irregular resting discharge. It can be seen in Figure 8.6 that, when the head is erect, some cells in the sacculus are deflected in an excitatory direction and others are deflected in an inhibitory direction. Nevertheless, the discharge rates of these two types of cells were found to be similar when the head was erect. This means that these cells have an opposite bias in the function which relates cilia deflection to discharge rate, so that each of them can respond most effectively to the range of tilt about the normal vertical position of the head.

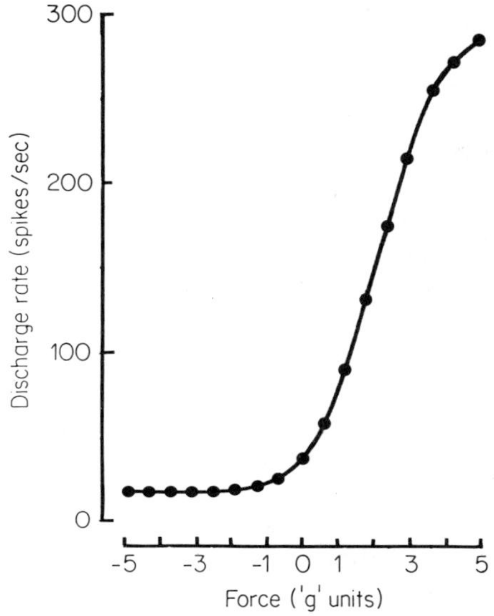

Figure 8.9 A force–response curve for an otolith neurone of the squirrel monkey. Other units produced curves of the same form but which varied in resting discharge rate, sensitivity (steepness), and the positions of the inflection point and saturation levels (From Fernández and Goldberg, 1976b. Reproduced by permission of the *Journal of Neurophysiology*)

The **sensitivity**, or gain, of a unit is defined as the change in the rate of discharge produced by a shear force of 1*g*. Fernández and Goldberg found that the sensitivity of most units was between 30 and 40 spikes/sec/*g*. The sensitivity of regular units was proportional to their resting discharge. During a period of constant stimulation, the discharge rate of units tended to return towards the resting level. This **adaptation** was more prominent among the irregular than the regular units. The irregular units are therefore referred to as **phasic** receptors and the regular units as **tonic** receptors, as was the case for canal afferents. Because of their high rate of adaptation, the phasic units differentiate the input, which gives them a higher sensitivity (gain) to high-frequency changes in head position. Adaptation also causes these units to show a phase lead of 20–40° as frequency is increased.

The primary afferents from the otolith organs project to the ipsilateral vestibular nuclei in the brainstem, particularly to the lateral and medial nuclei (Gacek, 1975). There is no direct input to the contralateral vestibular nucleus, but commissural fibres connect the two sides. Excitation of primary afferents from otolith organs on one side facilitates the response produced in the contralateral vestibular nucleus (Wilson *et al.*, 1978). The otolith inputs are thus not connected in a push–pull fashion, as are the inputs from the canals. But the geometry of the canals is such that each canal is a simple **bidirectional** detector with a single output which increases or decreases according to the direction and magnitude of the rotary force acting in its plane. A rotation of the head in one direction has opposite effects on synergistic canals, so that as far as the canals are concerned, it is the difference, not the sum of the synergistic inputs that varies with the stimulus. The geometry of the otolith organs is more complex. Each otolith organ is a **multidirectional** detector with several outputs, each of which increases or decreases according to the linear vector acting along its polarization axis. Furthermore, a tilt of the head in a given direction has the same effect on synergists, so that, in the case of the otolith organs, it is the summed output from regions with the same polarization axis on the two sides of the head which is the signal for linear acceleration. There could be push–pull inhibitory relationships in the vestibular nuclei between inputs from opposed units on the two sides of the head, but such interactions have not been described. Perhaps otolith inputs do not reciprocate in this way because a maintained input from the system is required for all postures of the body, including the normal erect posture, for the maintenance of postural tones. There is a natural zero value of rotary acceleration because we are not always turning our heads, but there is no natural zero value of linear acceleration, because the utricles are normally inclined several degrees to a constant force field of 1*g*, and a maintained input is needed to register the posture of the body at all times.

There is a fundamental ambiguity in the signals from the otoliths. For instance, the effect of a horizontal linear acceleration, *a*, acting in the plane of the utricles is indistinguishable from the effect of tilting the head through the angle whose sine is *a*. One well-known consequence of this ambiguity is the experience pilots have when there is a change in the speed of the aircraft. If an aircraft gains speed, the otolith organs are stimulated in the same way that they would be by an appropriate upward pitch of the aircraft. If the aircraft is in cloud, the pilot may correct for this apparent upward pitch by bringing the nose of the aircraft down, and he may then fly the aircraft into the ground (Wolfe and Cramer, 1970; Kirkham *et al.* 1978). The fact that the ambiguous input from the otolith organs is interpreted as a tilt of the body rather than a linear acceleration suggests that the primary function of these organs is the detection of head tilt. However, the canals may help to disambiguate the otolith inputs, as we shall see in Section 9.4.4.

Secondary neurones arising from the vestibular nuclei convey information of macular origin to the cerebellum (Marini *et al.*, 1976), the reticular formation (Spyer *et al.*, 1974; Kubin *et al.*, 1980), and the motoneurones of the neck (Wilson *et al.*, 1977). These connections presumably mediate postural reflexes. In addition, secondary neurones pass to the oculomotor nuclei where they mediate eye-movement reflexes associated with head tilt and linear acceleration (Baker *et al.*, 1973; Schwindt *et al.*, 1973).

Chapter 9

Visual–Vestibular Interactions

9.1 Vestibular psychophysics

9.1.1 Methods of stimulating and recording

The most obvious way to stimulate the vestibular canals is to use the 'adequate' (i.e. normal) mode of stimulation, namely, rotary acceleration. To avoid the difficulty of observing a subject while he is rotating, use is made of the fact that vestibular responses outlive a period of deceleration. In the original procedure

Figure 9.1 A rotational device at the Defence and Civil Institute of Environmental Medicine, Toronto, which is capable of rotating a human subject about any axis with reference to the body or gravity.

used by Bárány, the blindfolded subject is accelerated in a rotating chair, then rotated for 20 sec at a constant speed so that the cupula restores to its resting position, and finally decelerated quickly to rest and his responses noted. In the postrotatory period a variety of behavioural indices of the vestibular response is used, such as nystagmic eye movements, sensations of rotation, vertigo, and illusions of visual motion (the oculogyral effect). Since the introduction of electrooculography, the movements of the eyes may be recorded while the subject is rotating, and perrotatory procedures are now in common use. When applied in this way electrooculography is referred to as **electronystagmography** (Aschan, 1956). On-line computer programs have been devised for the rapid extraction of information from the eye-movement record (Allum *et al.*, 1975).

Modern rotating chairs and centrifuges allow the subject to be rotated while his body is held in any one of various attitudes, and provide for the separate or mixed stimulation of the canals and otolith organs (see Figure 9.1). For specific research purposes, elaborate devices have been designed for exposing subjects to well-defined linear or angular velocity stimulus profiles, such as those shown in Figure 9.2. The Bárány procedure involves a sudden, or **step** deceleration which, in its original form, was excessively massive. In the more recent development of this method, known as **cupulometry**, the velocity reached before deceleration is systematically varied and the duration of nystagmus, or of other vestibular responses following rapid deceleration, is plotted as a function of the velocity reached before deceleration. The resulting graph is known as a **cupulogram** (Cawthorne *et al.*, 1956; Groen, 1957). The **trapezoidal** velocity profile allows for a greater refinement of the acceleratory and deceleratory phases and keeps their effects separate by the interposition of a period of constant velocity. The **triangular** velocity profile has been recommended for studying normal vestibular functions because it resembles the stimulus profile of ordinary head movements (Guedry, Stockwell, Norman and Owens, 1971). The **sinusoidal** velocity profile also resembles natural stimulation and is a standard stimulus for investigating

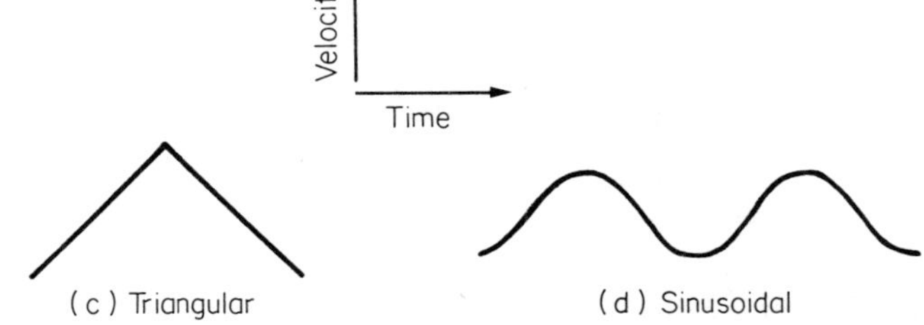

Figure 9.2 Rotary velocity profiles used in the study of vestibular functions

the dynamics of any control system (see Section 2.2.2). One convenient method for varying the period and amplitude of the sinusoid is to adjust weights on a **torsion swing** as illustrated in Figure 9.3.

The cupulogram obtained by the method of cupulometry is ideally a linear function. Insofar as the duration of postrotatory nystagmus represents the time which the vestibular system takes to return to equilibrium, the slope of the cupulogram equals the long time constant of the canal which was stimulated (see Section 8.2.1). However, the slope of the cupulogram has been found to change with practice (see Section 9.5.2). Furthermore, the slope of the cupulogram obtained by plotting the duration of postrotatory vertigo or the oculogyral illusion may not be the same as the slope obtained from the duration of postrotatory nystagmus (Groen, 1960). It looks as though behavioural measures reflect the time constants of neural processes rather than the time constant of cupula restoration. That this is so is indicated by the fact that the long time constant of discharge of primary vestibular afferents in the monkey, which probably indicates the time constant of the cupula, has been found to be only 6 sec, which is shorter than the time constant of decay of vestibular nystagmus in the monkey (Goldberg and Fernández, 1971a).

Blair and Gavin (1979a) tested monkeys with a modified Bárány procedure and also by using sinusoidal rotation. The long time constant of the canal system from the Bárány procedure was indicated by the time required for the smooth phase velocity of vestibular nystagmus to drop to 37% of its initial value after an impulsive stimulus, and the time constant from the sinusoidal procedure was indicated by $\frac{1}{2}\pi f_c$, where f_c is the frequency at which the phase lead of nystagmus with respect to the movement of the head is 45°. The two procedures were found to give similar estimates of the long time constant of the neural response driven

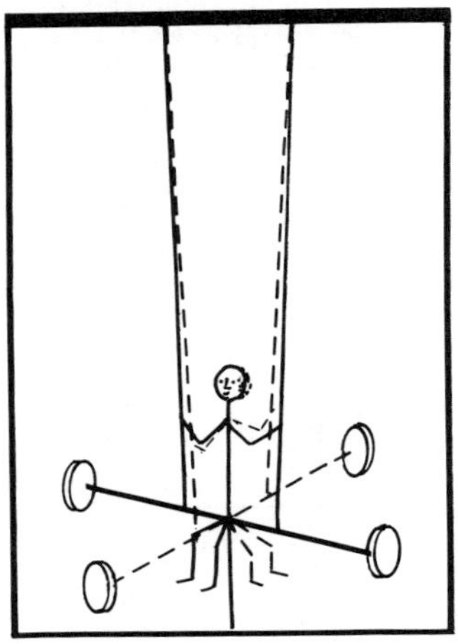

Figure 9.3 The torsion swing used to vary the period and amplitude of sinusoidal rotation of the body (From Jongkees, 1975. Reproduced by permission of Academic Press, Inc.)

by the canal system, although the Bárány procedure was recommended because it is much quicker and, since it involves only one acceleration and deceleration, is less subject to the effects of repetitive stimulation. When the Bárány test was repeatedly applied, the long time constant of the nystagmic response declined to an asymptotic value of about 7 sec, which is close to the physical long time constant of the cupula. It looks as though repetitive stimulation reduces the neural component of the long time constant to zero. This decline in the long time constant of vestibular nystagmus with repeated testing was not found in the rabbit or in man (see beginning of Section 9.5.2), and Blair and Gavin suggested that this was because the time constant in these experiments was already close to the limiting value set by the physical time constant of the cupula. But another factor may be that Blair and Gavin did not keep their animals alert during testing (see Section 9.5.2 for a discussion of this factor).

The vestibular canals may also be stimulated by inadequate (i.e. unusual) stimuli. If the outer ear is irrigated with water at above blood temperature, heat is conducted into the vestibular apparatus, reaching the most lateral part of the horizontal canal first and then the lateral parts of the other canals. Temperature gradients, and hence specific gravity gradients, arise across the ampullae. If a particular canal is not in a horizontal plane, any specific gravity gradient induces convection currents in the endolymph, which causes a cupula displacement. If cold water is used, a temperature and specific gravity gradient is induced in the opposite direction and, for a given posture of the head, this deflects a given cupula the other way. Nystagmus induced in this way is known as **caloric nystagmus**, and this theory of the action of thermal gradients has been confirmed by the absence of caloric nystagmus in the weightless conditions of parabolic flight (Kellogg and Graybiel, 1967; Oosterveld and van der Laarse, 1969). With the caloric procedure it is possible to stimulate the vestibular system on either side of the head, and any one canal in either direction. The procedure of caloric stimulation has been standardized, for both clinical and research use, by Fitzgerald and Hallpike (1942). The fast-phase saccades in caloric nystagmus are slower than normal saccades of the same amplitude (see Figure 6.5a). Animals also execute slow saccades when they are not alert, and Ron *et al.* (1972) suggested that the slow saccades in caloric nystagmus are due to a loss of alertness. Another factor might be that caloric nystagmus is induced by the excitation of a canal on only one side of the head, so that there is an absence of activity that stimulation of the synergist would normally produce.

An electric current, either a.c. or d.c., may be passed through the skull or to inserted electrodes. Faradic (a.c.) stimulation produces a single-phase nystagmic response. Galvanic (d.c.) stimulation produces nystagmus with the slow phase directed towards the anode. Reversing the direction of the current reverses the direction of the nystagmus (Spiegel and Scala, 1943; Weiss and Tole, 1973). In animals, the cristae may be stimulated directly by inserting a small piston or pneumatic hammer into the lumen of the canals. All these unusual modes of stimulation have the advantage that they allow the various cristae and maculae to be stimulated separately.

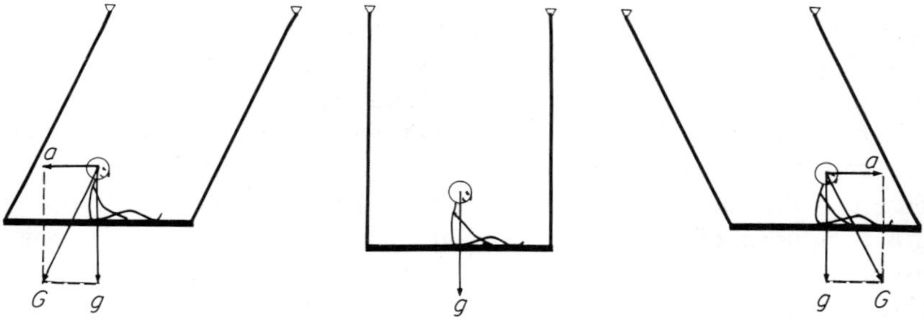

Figure 9.4 The parallel swing used to stimulate the otolith organs sinusoidally. G is the resultant of the applied acceleration, a, and the acceleration due to gravity, g (From Jongkees, 1975. Reproduced by permission of Academic Press, Inc.)

There is a vast clinical literature on the use of these various procedures for the diagnosis of diseases of the peripheral and central mechanisms of the vestibular system (Hood, 1978; Kornhuber, 1974a, b; Naunton, 1975).

Stimulation of the otolith organs is best produced by a well-defined tilt or linear acceleration of the whole body. The whole body should be moved or tilted, and not merely the head, because movements at the neck may induce some of the same responses as stimulation of the otolith organs. Tilting has the disadvantage that it stimulates the canals while the body is being moved into position. One way round this problem is to use the **parallel swing** as shown in Figure 9.4. This excites the otolith organs rhythmically while leaving the canals essentially unaffected.

If accelerations in excess of 1g are required, the subject is rotated in a human centrifuge (see Section 10.4.8). In a centrifuge the otolith organs are exposed to a force which is the resultant of gravity and the centrifugal force generated by the rotation (see Figure 10.8). The direction of this resultant force with respect to the body of the subject is controlled by varying the attitude of the body. A centrifuge imposes an unwanted rotary acceleration on the subject until after the machine has reached a constant angular velocity. If the subject is counterrotated about his own body axis at the same angular velocity as the centrifuge the effects of the rotary acceleration are cancelled, but the linear component of the motion now rotates with respect to the subject's body axis. The subject is said to be subject to a **rotating linear vector**. I refer to this type of stimulation as **ferris wheel rotation**, and there are certain ways in which it is put to use, as we shall see in Section 9.2.7.

9.1.2 Thresholds for rotary acceleration

There are three manifestations of vestibular canal activity which have been used to determine threshold values. These are: (1) reports of feelings of rotation, (2) nystagmus, and (3) the oculogyral effect (the apparent movement of a point of light in the dark, induced by vestibular stimulation). It is misleading to regard these thresholds as measures of a single underlying end-organ threshold; they each reflect threshold functions at various levels of partially distinct neural systems. Some of the many determinations of these three thresholds are shown in

Table 9.1. It can be seen that they vary between 0.02° /sec^2 and 2° /sec^2. On the whole, the oculogyral illusion seems to be the most sensitive indicator (Clark, 1967; Clark and Stewart, 1968a; Miller and Graybiel, 1975).

It is not surprising that there is so much variation between the different determinations even of the same threshold. It is not easy to accelerate a human subject smoothly, avoiding all extraneous sources of stimulation, and experimenters have differed in the extent to which they were successful. The threshold here, as in other modalities, is a function of the signal-to-noise ratio: reduce the 'noise', and one reduces the threshold. In any case, there are many other factors which modify the results. One of these is the psychophysical method used. For instance, a perrotatory double staircase method apparently yields lower thresholds than cupulometry, which suggests that the staircase method is more sensitive (Clark and Stewart, 1968b; Dockstader, 1971). Furthermore, the duration of the stimulus affects vestibular thresholds for at least two reasons. The first is that vestibular reactions adapt like other sensory functions. The second is that for shorter times of application greater accelerations are required to deflect the cupula to the same degree. The original statement of this relationship was that the product of acceleration and stimulus duration for threshold stimuli is constant. This constant is referred to as **Mulder's constant**, and the relationship is analogous to Bloch's law in vision. More recent studies have confirmed the inverse relationship between acceleration and duration but the product has been found to be not constant but a linearly increasing function of stimulus duration (Doty, 1969).

There seem to be no published reports of a direct comparison between acceleration thresholds in different planes of rotation. Clark (1967) reviews an unpublished experiment by Meiry in which it was shown that the threshold for rotation about the Z axis (head erect, axis of rotation vertical) was between 0.1 and 0.2°/sec^2 and the threshold for rotation about the X axis (head supine, axis of rotation vertical) was 0.5°/sec^2. This difference is consonant with the different time constants of vestibular responses in yaw and pitch (see Section 8.2.1).

Table 9.1 Human thresholds for the detection of rotary acceleration, as indicated by various responses

Response	Investigator	Threshold acceleration (deg/sec^2)
First reports of rotation	Mach (1875)	2.0
	Tumarkin (1937)	0.2
	Clark and Stewart (1962)	0.12
Nystagmus	Dohlman (1935)	1.0
	Buys and Rijlant (1939)	0.8
Oculogyral illusion	Christian (1940)	0.13–0.33
	Groen and Jongkees (1948)	0.28–2.0
	Graybiel *et al.* (1948)	0.12
	Miller and Graybiel (1975)	0.02–0.95

Gundry (1978) compared the threshold for rotation of a vertically seated person about the *X* axis (roll) with that for rotation of a prone person about the same body axis. The threshold and time for detection were lower for the erect person and this was probably due to the fact that with this posture the canals and the utricles would be stimulated, whereas in the prone posture only the canals would be stimulated (see also Guedry, 1965).

Several investigators have used magnitude estimation procedures to derive functions relating the perceived velocity of rotation to various parameters of the stimulus (Brown, 1966, 1968; Clark and Stewart, 1968c, 1972; Guedry, Stockwell, and Gilson, 1971; Elsner, 1971).

9.1.3 Thresholds for linear acceleration

Thresholds for sinusoidal linear motion, as indicated by the first reported sensation of motion, have been obtained for motion along horizontal and along vertical tracks with the subject's *x*, *y* or *z* body axis parallel to the direction of motion. Some of these determinations are shown in Table 9.2. It can be seen that the thresholds hover about the value of 10 cm/sec^2, which is equivalent to a shear force of about 0.01 *g*. It is not easy to compare these different determinations because the direction and period of the oscillatory motion varied from study to study. Thresholds for sustained horizontal linear acceleration are easier to interpret and compare than are thresholds for sinusoidal linear motion. However, sustained linear acceleration requires the construction of long smooth tracks, and these are expensive and difficult to make. The time it takes the subject to report correctly the direction of movement on a given percentage of trials after

Table 9.2 Human thresholds for the detection of a linear oscillation of the body

Investigator	Posture of subjects	Direction of oscillation	Period of oscillation (sec)	Max. accel. at threshold (cm/sec^2)
Mach (1875)	Upright	Vertical	7	10–12
Travis and Dodge (1928)	Seated facing movement	Horizontal	2–8	20–25
	Standing facing movement	Horizontal	2–8	8
	Standing sideways to movement	Horizontal	2–8	5
Walsh (1962)	Prone or supine	Horizontal	3	7–8
	Prone or supine	Horizontal	9	6–12
Walsh (1964)	Prone or supine	Vertical	3	6
	Prone or supine	Vertical	9	18
Parker *et al.* (1978)	Prone	Horizontal	2.6	3–9

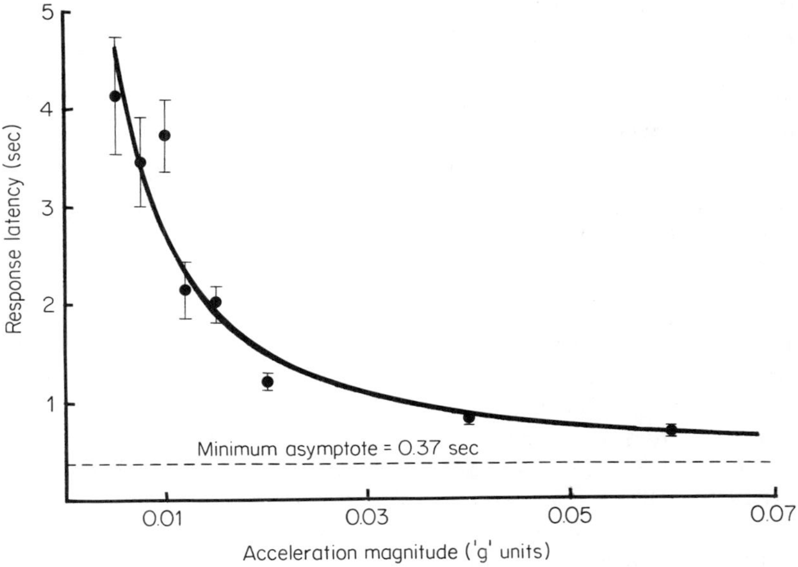

Figure 9.5 Latency of detection of vertical acceleration as a function of rate of acceleration. Based on means of eight subjects (bars give standard errors) (From Melvill Jones and Young, 1978. Reproduced by permission of *Acta Oto-Laryngologica*)

the acceleration has been applied is a useful measure of the threshold. This measure indicates how long a given acceleration must be applied in order to produce a threshold deflection of otoliths. The absolute threshold is that acceleration for which the response latency becomes very long. Figure 9.5 shows the response latency as a function of linear vertical acceleration for subjects in an erect seated posture (Melvill Jones and Young, 1978). It can be seen that the latency becomes long at an acceleration of less than 0.01 *g*. It was inferred from the equation of the function relating acceleration to latency that a consistent linear *velocity* had to be achieved before a sensation of movement was reported. The mean value of this velocity was 21.6 cm/sec. Melvill Jones and Young compared their results for acceleration of a seated subject along a vertical path with previous determinations for subjects moved along a horizontal path (Young and Meiry, 1968) and found that they were similar. This means that the utricles have about the same sensitivity as the saccules. However, for some unknown reason, subjects are more likely to be confused about the direction of a horizontal motion than about the direction of a vertical one. (Malcolm and Melvill Jones, 1974), even though the direction of the reflex oculomotor response is unambiguously related to the direction of vertical linear acceleration (Melvill Jones *et al.*, 1980).

The threshold for linear acceleration temporarily increases following a period of sustained oscillating linear acceleration (Parker *et al.*, 1978). It is not clear whether this is due to adaptation of phasic receptors or to fatigue. It was apparently not due to adaptation of tonic receptors, because there was no change in the threshold after 10 min of maintained unidirectional linear acceleration in a centrifuge.

The whole question of vestibular psychophysics has been thoroughly reviewed by Guedry (1974).

9.2 Vestibular nystagmus

9.2.1 General features

When a person is rotated about his body axis the eyes execute rhythmic, conjugate, reflex movements. They move slowly in the opposite direction to the rotation (the slow phase) and quickly back (the quick phase). Vestibular nystagmus may be elicited in a horizontal direction, a vertical direction, or about the visual axis (torsional nystagmus), depending upon which canals are in the plane of rotation. Many people show some nystagmic responses when the body is held in a tilted position. Such responses are known as **positional nystagmus**, and if they are large in amplitude or persistent they are a sign of pathology or alcohol intoxication (Barber and Wright, 1973). As we shall see later, **steady excitation** of the otolith organs produces a constant deflection of the eyes and there is some uncertainty about whether it also produces nystagmus. Positional nystagmus may be due to some effect of linear acceleration on the canals. However, nystagmus is produced by **rhythmic stimulation** of the otolith organs. Turning of the torso with head held stationary may also induce deviation of the eyes interrupted by nystagmic beats. But this so-called **cervico-ocular response** is prominent only in infants or in pathological states in adults (see Section 9.2.4).

Some people when subjected to sinusoidal rotation of the body in the dark show nystagmic movements of the head. This is known as the **vestibulocollic reflex**, and is very pronounced in animals, like the owl, which do not move their eyes far (Outerbridge and Melvill Jones, 1971; Steinbach and Money, 1973; V. J. Wilson *et al.*, 1979).

Vestibular nystagmus is basically under the control of innate mechanisms in the brainstem, although, as we shall see later, the response may to some extent be controlled by higher centres. The response occurs in neonates (Tibbling, 1969) and in 12-week-old kittens reared in total darkness (Berthoz, Jeannerod, Vital-Durand and Oliveras, 1975). The amplitude of each beat of the response is larger in the infant than it is in the adult, probably because it is less under visual control in the child (Ornitz *et al.*, 1979). Vestibular nystagmus is weakened in states of relaxed attention (Sokolovski, 1966). During light sleep it is replaced by maintained ocular deviation in the compensatory direction, with an absence of the quick component, and during deep sleep the response is entirely absent (Melvill Jones and Sugie, 1972).

If a person is decelerated after a period of rotation at constant angular velocity, the acquired momentum of the endolymph–cupula of the canal in the plane of the rotation causes a deflection of the cupula in the same direction as the previous rotation. This induces **postrotatory nystagmus** which is opposite in direction to the perrotatory nystagmus. This response lasts for up to 30 sec in the dark (see Figure 9.6). Its duration depends on the strength of the stimulus, attention and practice (see Section 9.5.2) and also on the axis of rotation (Melvill Jones *et al.*, 1964). It tends to be inhibited when stationary objects are in view (see next section).

It has been mentioned already that during a period of constant rotary acceleration (constant cupula deflection) the firing rate of phasic vestibular afferents returns towards the resting level. In other words, the vestibular system adapts to constant acceleration. This adaptation reveals itself as a decrement in the velocity of the slow phase of vestibular nystagmus and a decrease in the perceived velocity of body rotation (J. H. Brown and Wolfe, 1969; Luster, 1972). If the body is brought to rest after the system has adapted in this way, it takes time for the system to adapt back to its normal state, and during this time the perrotatory response continues. In the initial postrotatory period this in-phase nystagmus is overshadowed by the anti-phase, postrotatory nystagmus induced by the deceleration of the body. However, the time constant for recovery from adaptation is longer than the time constant for cupula restoration. One estimate puts their respective values at 82 and 21 sec (Malcolm and Melvill Jones, 1970). This means that the in-phase nystagmus will reveal itself after the anti-phase, postrotatory nystagmus has worn off. This post-postrotatory nystagmus is known as **secondary nystagmus** (see Figure 9.7). Models of these processes have been developed (Young and Oman, 1969; Stockwell *et al.*, 1973).

If the experimenter rotates the head of the subject to one side with high angular velocity, the eyes make a brief slow-phase movement followed by a large saccade to the *same side* and a slow return to the primary position as the head turns. Under ordinary circumstances, rapid head movements are executed in order to bring the gaze towards an object situated off to one side. The large saccade can thus be understood as part of the strategy of fixating such an object (see

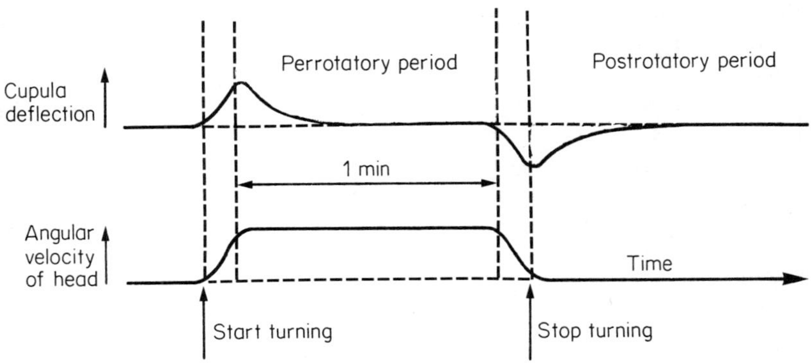

Figure 9.6 The pattern of cupula deflection during (top curve) and after (bottom curve) a rotation of the body (After Melvill Jones, 1976. Reprinted by permission of Lawrence Erlbaum Associates, Inc.)

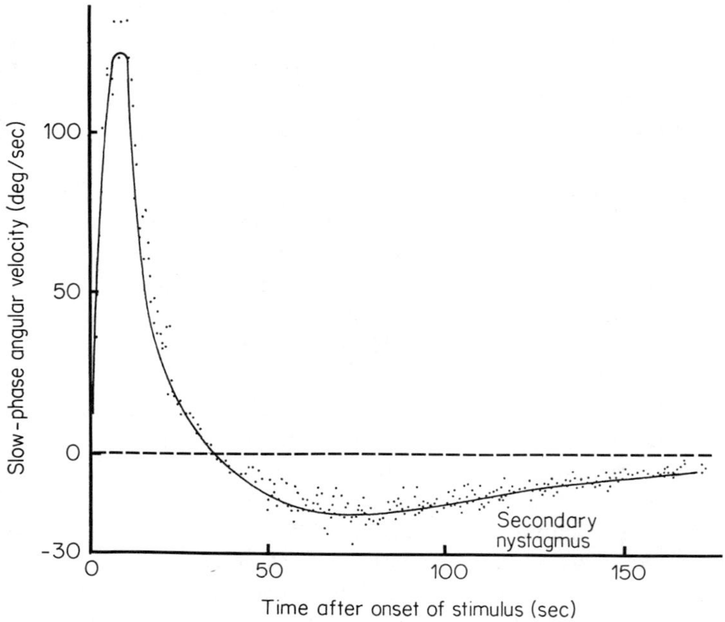

Figure 9.7 The angular velocity of the slow phase of vestibular nystagmus plotted beat by beat after a sudden change in the angular velocity of the head (From Malcolm and Melvill Jones, 1970. Reproduced by permission of *Acta Oto-Laryngologica*)

Section 9.2.6). However, it is apparently controlled by vestibular inputs, because it occurs in the dark as well as when objects are in view (Melvill Jones, 1964).

If an upright person is slowly oscillated sinusoidally about a vertical axis, there can be seen, superimposed on the nystagmic response, a slow change in the mean position of the eye relative to the skull. This drift in mean eye position has the same frequency as the oscillatory motion of the body but its phase is advanced by 90° with respect to the change in head position. In any sinusoidal motion, displacement and velocity are 90° out of phase, so that the drift in eye position is approximately in phase with the angular velocity of the head. This, and the fact that the drift occurs in the dark, suggests that it is driven by cupula deflection (Mishkin and Melvill Jones, 1966).

The pathways which convey signals from the vestibular nuclei to the oculomotor nuclei are described in Section 5.7.3. These pathways have been reviewed in detail by Cohen (1974) and by Wilson and Melvill Jones (1979).

9.2.2 Temporal characteristics of vestibular nystagmus

The reaction time, or latency of vestibular nystagmus made in response to an externally applied rotation of the body varies with the intensity of vestibular stimulation. With a rotary acceleration of the body of $1.0°/sec^2$ the latency to the completion of the first saccadic beat is several seconds, and with an acceleration of $8°/sec^2$ it can be as short as half a second (Fluur and Mendel, 1966). Once the response is initiated and the subject continues to be accelerated at a steady rate, it takes several seconds for the response to reach its maximum frequency. This so-called **crescendo time** is shorter with higher rates of acceleration (Fluur and Mendel, 1970). In the case of self-produced head movements the latency may be zero, which means that the head and eye movements are initiated together. On this account, these eye movements have been called **coordinate compensatory eye movements**.

The **gain** of vestibular nystagmus induced by unidirectional rotation of the head is the ratio of the angular velocity of the eyes during the slow phase to the angular velocity of the head. This determination must be done with the subject in the dark so as to avoid the intrusion of optokinetic nystagmus. If the gain is 1, the slow phase compensates perfectly for the head rotation, so that during this phase the image of a distant stationary scene would be stabilized on the retina. (We shall see in Section 9.5.4 that the stabilization of near scenes requires a gain of more than 1.) If the gain is less than 1, the eyes fail to keep up with the moving image of a stationary scene. If the body is oscillated sinusoidally about a given axis, the gain of vestibular nystagmus is indicated by the peak velocity of the slow phase of nystagmus divided by the peak velocity of the head. The difference in phase between these two points of maximum velocity when the subject is in the dark is a measure of the **phase lag** of the vestibulo-ocular system.

When the subject is in the dark, the gain of the slow phase varies with the frequency of sinusoidal oscillation of the body. The gain is only about 0.6 when subjects are oscillated at 0.1 Hz, and rises to 1 at about 2 Hz and to above 1 at higher frequencies in humans (Benson, 1970; Meiry, 1971) and in monkeys (Keller, 1978). It is significant that, where the gain is 1 at a frequency of 2 Hz, the acceleration of the head is about the same as it is in natural head movements. When subjects can see a stationary background, the gain is 1 over a wide range of frequencies of head oscillation; up to 6 Hz in the monkey (Keller, 1978). But the gain is 1 up to about 5 Hz when the head is oscillated voluntarily in the dark (Tomlinson *et al.*, 1980), presumably because of the contribution of efference. The gain of nystagmus with slow phase downwards is less than that of nystagmus with slow phase upwards (Darlot *et al.*, 1981).

I have not seen data on the phase lag of human vestibular nystagmus, but for the monkey, in the dark, it has been found that the position of the eyes is approximately in phase with head position over the frequency range 0.04–3.5 Hz, after which there is an increasing phase lag (Skavenski and Robinson, 1973; Keller, 1978). Note that the input from the canals over this range of frequencies is proportional to head velocity, not to head position. In any sinusoidal motion, velocity and displacement are 90° out of phase. Thus, if the vestibular input were to drive the slow phase of nystagmus directly, one would expect a 90° phase lag between changes in eye position and changes in head position when the head is oscillated sinusoidally in the dark. The absence of such a phase lag led Skavenski and Robinson to postulate the existence of a neural integrator between the vestibular nuclei and the oculomotor centres in the brainstem. The operation of integrating a velocity signal is equivalent to converting it into a position signal, which in turn is equivalent to the required 90° phase shift.

Carpenter (1972) found that deactivation of the cerebellum of the cat by local cooling produced changes in vestibular nystagmus which were compatible with the idea that integration of vestibular signals is achieved in the cerebellum. Robinson (1974) repeated this experiment and, although his findings confirmed that removal of the cerebellum increases the phase lage of nystagmus, the increase which he found accounted for only part of the 90° phase shift which an adequate integrator must achieve. Gardner and Fuchs (1975) recorded from single cells in the cerebellum which fire in relation to the slow phase of vestibular nystagmus. The responses of some of these cells showed a phase lag intermediate between that of canal afferents and cells in the oculomotor nuclei, which supports Robinson's conclusion that only part of the required integration is achieved in the cerebellum.

Recent electrophysiological recordings from secondary and tertiary cells in the vestibular nucleus of the alert monkey have revealed that the required 90° phase shift is already evident in the later stages of processing within the vestibular nucleus (Keller and Kamath, 1975; Buettner *et al.*, 1978; Pola and Robinson, 1978). It is possible that this integration is achieved by recurrent interactions between the vestibular nuclei, the cerebellum, and the reticular pontine formation (see Section 4.7.1). The idea that the reticular pontine formation is involved is supported by the work of Cohen and Komatsuzaki (1972) on the effects of electrical stimulation of the reticular formation and by electrophysiological recordings from single cells in that region (Luschei and Fuchs, 1972).

Integration of vestibular signals may also be partly achieved by the mediation of proprioceptive inputs from the extraocular muscles, as indicated by the finding of Allum and Graf (1977) that anaesthesia of these muscles in the goldfish reduced the long time constant of the response of cells in the vestibular nuclei to vestibular stimuli. Keller and Precht (1979b) found no such effect in the cat, and in any case, such an effect would shift the phase of only low-frequency oscillations, whereas a phase shift is required over the whole waveband of the system.

Robinson has suggested that the brainstem integrator is responsible not only for bringing head and eye movements into correct phase, but also for the ability of a person to hold his gaze in a desired direction after an eye movement. It is known that people can maintain an eccentric position of gaze in the dark for many seconds (Skavenski and Steinman, 1970; Becker and Klein, 1973). I find it difficult to believe that the same brainstem integrator is responsible for this behaviour, and suggest that a system higher in the neuraxis is involved—a system in which patterns of static innervation to eye muscles are coded in terms of eye position. Perhaps this higher-level coding system is also involved in the more menial task of bringing head and eye movements into phase. It is known that the phase shift evident in cells of the vestibular nuclei is present only in the alert animal (Keller and Kamath, 1975).

9.2.3 *The quick phase of vestibular nystagmus*

It is generally believed that the quick phase of vestibular nystagmus is under the control of a mechanism which is distinct from that driving the slow phase. One reason for this belief is that the two phases are affected in different ways by drugs and by brain lesions (Carpenter, 1972; Haciska, 1973). It is known that the quick phase is not initiated by proprioceptive signals from the extraocular muscles, because the rhythmic discharge of motoneurones can still be detected during vestibular stimulation after the eye and its muscles have been removed (McIntyre, 1939). Nor is the quick phase initiated by the motion of the retinal image, because nystagmus occurs in the dark. Cheng and Outerbridge (1974a) have shown that the quick phase of vestibular nystagmus tends to occur at regular intervals of time, although this regularity may not be apparent because of interaction between this inherent periodicity and the periodicity of the rotational stimulus. Electrophysiological evidence suggests that the signals for the quick phase originate in the pontine reticular formation (Berthoz *et al.*, 1974; Hikosaka and Kawakami, 1977). It is apparently the same centre which controls the quick phase of optokinetic nystagmus.

9.2.4 *The cervico-ocular reflex*

A rotation of the torso about the *z* axis relative to the stationary head produces small-amplitude horizontal nystagmic movements of the eyes and a deviation of gaze in a direction opposite to the movement of the torso. This response is more prominent in infants, in pathological states in adults, and in certain animals than it is in the normal human adult (Jongkees, 1969). A forward inclination of the torso with respect to the head produces vertical eye movements, although a sideways tilt of the torso in the absence of utricular stimulation does not induce eye torsion (Scott, 1967). The horizontal and vertical responses are induced by stimulation of the joint receptors in the neck and are known as **cervico-ocular reflexes** and also as **spino-ocular reflexes** (Meiry, 1971; Takemori and Suzuki, 1971). Dichgans (1974) concluded, in a review of the evidence, that the cervico-ocular reflex and the vestibulo-ocular reflex cooperate in achieving visual stability during head movements. However, Barnes and Forbat (1979) found that the mean gain of the human cervico-ocular response was only between 0.04 and 0.06 at frequencies of passive body oscillation from 0.2 to 1.3 Hz. Furthermore, it has been found that the gain and even the direction of the reflex varies from animal to animal and from trial to trial in the same animal (Fuller, 1980). These facts suggest that the reflex has little functional significance and does not contribute significantly to stabilization of the retinal image during natural movements of the head. However, before this conclusion can be accepted there are other factors to be considered.

In many animals the head stays in a stationary attitude with respect to space if the body is oscillated about any of its three orthogonal axes. This reflex is particularly pronounced in the owl; the head remains stubbornly in a fixed position even when the body is oscillated rapidly from side to side through 360°.

Understandably, the cervico-ocular reflex is absent under these circumstances its presence would destabilize rather than stabilize the retinal image. Presumably the cervico-ocular reflex is inhibited by the head-stabilizing reflex. Perhaps when an animal's torso is moved with respect to its mechanically restrained head, the head-stabilizing reflex is sometimes evoked and sometimes relaxed, and this would explain the variations in gain and direction of the cervico-ocular reflex in the conditions under which it is normally tested. One would expect the reflex to be evoked consistently only when an animal actively moves its head with respect to a stationary body, because this is when eye movements are needed to stabilize the retinal image. But under these circumstances it is difficult to distinguish from the vestibulo-ocular response.

There are two way to dissociate these two reflexes. One is to use animals lacking a vestibular system. Primates with chronic vestibular lesions have indeed been shown to exhibit effective compensatory eye movements when the head is actively moved (Dichgans, Bizzi, Morasso, and Tagliasco, 1974) and the same has been shown for labyrinthine-defective humans (Kasai and Zee, 1978). Presumably, these eye movements are invoked by the cervico-ocular reflex. The other way to dissociate the cervico-ocular and vestibulo-ocular responses is to *ask a subject sitting on a freely moving chair to rotate his head while it is anchored.* This experiment has not been done. But Tomlinson *et al.* (1980) have shown that the gain of vestibular nystagmus in the dark is unity over a wider range of head velocities when the head is moved actively compared with when it is moved passively. The improvement could be due either to the contribution of efference or to the recruitment of the cervico-ocular reflex.

Neurones originating in proprioceptors in the neck project directly or indirectly to the superior colliculus, the nucleus prepositus hyperglossi, and the frontal eye fields, where they converge with visual and extraocular muscle afferents (Abrahams and Rose, 1975a; Gresty and Baker, 1976; Dubrovsky and Barbas, 1977). There is also a projection from neck proprioceptors to the contralateral vestibular nucleus (Hikosaka and Maeda, 1973).

Excitation of stretch receptors in the extraocular muscles of the cat gives rise to discharges in those brainstem motor nuclei which control movements of the head and forelimbs (Easton, 1972). These connections presumably serve to initiate reflex postural adjustments in response to movements of the eyes. Such reflexes would be significant in the hunting behaviour of cats and may not be present in man.

9.2.5 The interaction of vestibular nystagmus and visual pursuit

It has been mentioned already that the gain of the slow phase of vestibular nystagmus varies with the frequency of head oscillation when the subject is in the dark. When a stationary background is in view, the gain of the slow phase is close to 1 over a wide range of frequencies in the monkey and in man (Keller, 1978; Koenig *et al.*, 1978). However, when subjects are asked to make voluntary head movements in lighted surroundings, the slow-phase gain has been found to vary from person to person and from eye to eye. Steinman and Collewijn (1980) found that it was very close to 1 in both eyes of one of their subjects but that in another subject it was 0.87 in one eye and only 0.66 in the other. In spite of the imperfect image stabilization in some subjects, they all reported that the visual scene appeared stable and binocularly fused when they moved their heads. Thus, even though the visual and vestibular reflexes may not achieve perfect stabilization of the retinal image, they are adequate for the maintenance of normal visual performance. One indication of the adequacy of vision during head movements is the fact that a person can read a stationary display of digits whilst being oscillated about a vertical axis up to frequencies of about 10 Hz. When the head is stationary, the gain of a visual pursuit eye movement with respect to the velocity of stimulus motion shows a rapid decline at frequencies about 1 Hz, just at the frequency at which the gain of vestibular pursuit is improving. Thus, when a display of digits is oscillated in front of a stationary observer, his ability to read them is impaired even at a frequency of 0.5 Hz, and at 2 Hz, reading speed is only about 10% of normal (Benson and Barnes, 1978). This is in marked contrast to the adequacy of visual pursuit of stationary objects during head oscillation mentioned above.

When a subject fixates his gaze on a visual object which rotates with the head the vestibulo-ocular response is inhibited. Benson and Barnes (1978) found that nystagmus was effectively suppressed up to a frequency of sinusoidal head rotation of about 0.5 Hz, after which the eyes began to move in spite of the efforts of the observer to maintain fixation. At a frequency of 2 Hz, the amplitude of the nystagmus was found to be almost as great as when the eyes were closed. This loss of fixation stability in tracking a display rotating with the head revealed itself as a progressive decrement in the ability to read a display of digits which rotated with the head (Barnes *et al.*, 1978). The most effective image stabilization is achieved for an objectively stationary display, rather than for one which rotates with or faster than the subject (Guedry *et al.*, 1979). In other words, the most effective visual stabilization is achieved when vestibular nystagmus and optokinetic nystagmus are concordant. The adequacy of suppression of nystagmus for forward rotation of the body is less than for backward rotation or rotation about a vertical axis (Benson and Guedry, 1971; Darlot *et al.*, 1981). The ability to visually suppress vestibular nystagmus may be impaired in people with lesions in the cerebellum, brainstem or cerebral hemispheres (Dichgans *et al.*, 1978).

All this evidence suggests that a stationary visual stimulus improves the gain of the vestibular slow phase and that, within limits, a visual stimulus which rotates with the subject allows him to inhibit vestibular nystagmus (see Section 9.5.5). However, Barr *et al.* (1976) found that when subjects were asked merely to imagine a stationary visual scene when they were oscillated at 0.3 Hz, the gain of the slow phase approached unity, and that when they were asked to imagine a scene rotating with the head, the gain dropped to 0.35. This demonstrates that vestibular nystagmus can be controlled by high-level processes (see also Yasui and Young, 1975).

If the head and body rotate at a constant velocity for some time, the cupula is restored to its central position and the stimulus for vestibular nystagmus ceases. However, if the eyes are open, optokinetic nystagmus takes over and maintains the nystagmus with a slow-phase velocity more or less adequately matched to the rotation of the head. So that between them, vestibular and optokinetic nystagmus stabilize the image of the stationary visual scene as the head turns. After a sudden deceleration of the head, the cupula is deflected in the direction of previous rotation, which induces a postrotatory nystagmus opposite in phase to the perrotatory response. This lasts for several seconds if the subject is in the dark, but is soon inhibited if stationary objects are in view after rotation has ceased. Postrotatory nystagmus is inhibited in this case, because the ability of the subject to fixate visual objects overcomes the effects of the vestibular stimuli. But postrotatory nystagmus is also inhibited if a scene was in view while the body was being rotated, even though the subject is in the dark after coming to rest. As I pointed out in Section 6.1.1, optokinetic stimulation during body rotation generates optokinetic afternystagmus which is opposite in phase to postrotatory vestibular nystagmus. The two postrotatory effects therefore cancel, leaving the image of the stationary scene stabilized on the retina. This is yet another example of the way in which the characteristics of the vestibular system nicely mesh with those of the visual pursuit system to maintain effective visual performance as the observer moves about. These relationships are discussed in more detail in Section 9.4.

When sinusoidal linear acceleration of the body along the interaural axis is combined with optokinetic stimulation, the slow phase of the optokinetic response is periodically modulated by the accelerative stimulus applied to the body (Tokunaga, 1977; Buizza *et al.*, 1980). A review of visual-vestibular interactions has recently been provided by Henn *et al.* (1980).

9.2.6 Coordinated head and eye movements during changes of visual fixation

When a person saccadically redirects his gaze to an object which is some distance off to one side, the natural reaction is not merely to move the eyes but also to move the head. The end result of this coordinated movement of the eyes and the head is that the image of the object of interest becomes centred on the foveae and that the eyes return to their primary position in the head. Although the motor commands to the head are initiated before those to the eyes (Bizzi *et al.*, 1971), the low inertia of the oculomotor system causes the saccade to start about 20 msec before the head movement. The saccade is completed before the head has reached its peak velocity. The eyes then make a slow return movement with respect to the head while the head continues to rotate towards the visual target. In this way the image of the scene is stabilized on the retina during the last part of the head movement. Detailed measurements of sequences of eye and head movements are provided by Uemura *et al.* (1980).

Morasso *et al.* (1973) found that the amplitude of a saccade made in conjunction with a head movement compared with a saccade made without a head movement is reduced by just the amount that the head has moved by the time the saccade is complete. This could be interpreted as evidence that the saccade is modulated during its execution by the motion of the head, but a more likely interpretation is that the saccade is preprogrammed in such a way that the anticipated motion of the head is taken into account. Nevertheless, there is a general tendency for saccades to undershoot the target even when the head is not moving (see Section 6.4.4), and the extent of this undershooting is large for movements towards targets which are offset more than 60° (Barnes, 1979). Normally, any inaccuracy of a saccade is corrected by a secondary saccade, but in a combined movement of eyes and head there is very little delay between the end of the initial saccade and the start of the slow compensatory return motion of the eyes. For this reason, the corrective saccade does not occur until after the slow motion of the eyes has begun. This means that the image of the object towards which the gaze is being directed is not accurately foveated until after the slow compensatory eye motion has begun, which in turn explains why digits which are exposed briefly immediately after the initial saccade are not recognized (Bartz, 1966).

The question arises as to what extent this sequence of eye and head movements is under vestibular as opposed to visual control. Several pieces of evidence suggest that the basic sequence is under vestibular control, but that it is modified by visual inputs and perhaps also by inputs from joint receptors in the neck.

Melvill Jones (1964) described how the eyes execute a saccade in the same direction as a rapid, passively induced movement of the head in the dark just after the head has started to move. This effect has also been reported for voluntary movements of the head in the dark, and also in illuminated surroundings when the subject makes no attempt to fixate a visual object (Henriksson *et al.*, 1974; Barnes, 1979). The fact that the saccade occurs *after* the onset of the head movement under these circumstances suggests that it is a response to vestibular stimulation and/or to stimulation of receptors in the neck. On the other hand, when the subject is attempting to direct his gaze towards an eccentrically placed object, the saccade occurs *before* the head movement and must therefore be preprogrammed along with the head movement.

The compensatory return motion of the eyes is apparently not preprogrammed, but is triggered by vestibular, visual or by proprioceptive inputs from the neck. Bizzi *et al.* (1971) demonstrated this by unpredictably applying a brake which stopped the motion of the head just after it had begun to move towards a new point of fixation. When this was done, the initial saccade occurred but the compensatory return movement did not. If the return movement were preprogrammed, a sudden application of a brake would not interfere with it. The brake did of course forestall stimulation of the vestibular, visual and neck receptors. When the head is rotated voluntarily or passively in the dark, the slow return motion of the eyes occurs, showing that vestibular and/or proprioceptive inputs are sufficient to induce it (Barnes, 1979). Nevertheless, in the dark, the slow return movement of the eyes lags behind the head motion. If objects are in view, the gain of the slow return movement is improved by the recruitment of the optokinetic pursuit system and corrective saccades may interrupt the motion (Gresty, 1974; Lanman *et al.*, 1978).

The coordinated pattern of eye and head movements which has just been described is particularly evident when a person directs his gaze towards an object which is suddenly presented in an unexpected position. If a person knows when and where the target is going to appear, he may adopt a predictive strategy of eye and head movements. In this pattern of movements the head moves before the target is presented and a corrective saccade occurs when the target appears, followed by a secondary head movement and a compensatory eye movement (Bizzi *et al.*, 1972).

When subjects are required to pursue visual targets moving to and fro along a horizontal track, they adopt different strategies of eye and head motion, depending on whether the motion is fast or slow and whether it is random or regular. Gresty and Leech (1977) showed that when the motion is random at any speed or when it is regular and fast, subjects pursue mainly with their eyes, presumably because of the lower inertia of the eyes but also to avoid the intrusion of the vestibulo-ocular reflex which would be triggered by a head movement. With regular sinusoidal motion below 1 Hz, subjects often use a more or less adequate combination of pursuit movements of the head and eyes. At higher frequencies of regular target motion, the pursuit eye movements are interrupted by saccades, partly of vestibular origin and partly in response to inaccuracies of visual fixation.

People with complete loss of labyrinthine function adopt a variety of strategies of eye and head movement when directing the gaze to off-centre visual targets. They also show a compensatory enhancement of eye movements triggered by neck proprioceptors (Kasai and Zee, 1978).

9.2.7 Interactions between vestibular nystagmus and stimulation of the otolith organs

Electrical stimulation of afferents from a utricle or saccule induces a steady deviation of the eyes (Suzuki *et al.*, 1969; Janeke *et al.*, 1970). Fluur and Mellström (1970a, b, 1971b) electrically stimulated different regions of the utricle and saccule in the cat and mapped out the corresponding deviations of the eyes. It can be seen from Figure 9.8 that, in general, the direction of the eye deviations corresponds to the polarization axis at each position on the maculae. In addition to these coordinated eye movements, uncoordinated deviations and torsional movements of the eyes were observed. Nystagmic responses were sometimes observed, although these were probably due to incidental stimulation of the canals, because cats with sectioned ampullar nerves but intact afferents from utricle and saccule were found not to exhibit nystagmus (Fluur and Siegborn, 1973a). Observations of the deviations of the eyes in humans subjected to linear acceleration in various directions have produced maps of the utricular and saccular maculae similar to those observed in the cat (Fluur, 1970). Also in

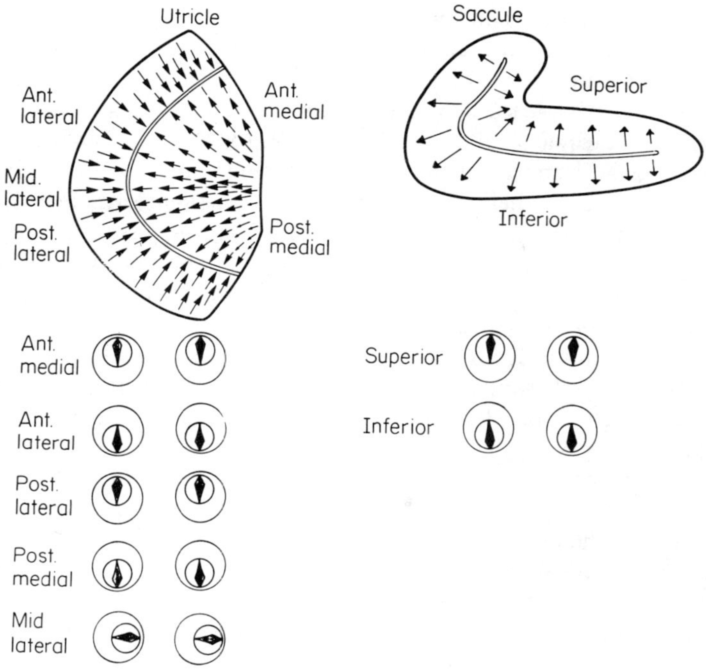

Figure 9.8 Coordinated eye movements produced by electrical stimulation of the utricle and saccule of the cat (From Fluur and Mellström, 1970a, b. Reproduced by permission of *The Laryngoscope*)

humans, natural stimulation of the otolith organs in the dark by a constant linear acceleration of the body along the interaural axis has been found to induce nystagmus with an amplitude of a few degrees. However, the magnitude of the accelerative stimulus required to produce this response is well above that required for the perception of acceleration (Niven *et al.*, 1966; Buizza *et al.*, 1980). It is not clear whether this response is due to stimulation of the otolith organs or to some indirect effect of linear acceleration on the canals.

Although there is some doubt about whether steady stimulation of the otoliths induces nystagmus, there seems to be no doubt that ablation or stimulation of these organs may facilitate or inhibit nystagmus induced by stimulation of the canals. These relationships between the otolith organs and the canals have been investigated in some detail in the cat by Fluur and Siegborn (1973a, b, c, d, 1974a, b). More recently, Igarashi *et al.* (1977) ablated the otolith organs in the squirrel monkey, first on one side and then on both sides. For a short time after the unilateral operation, some animals showed a spontaneous lateral nystagmus when they were held stationary in the dark. But after that effect had disappeared, there remained an apparently permanent reduction in the velocity of the slow phase of rotational nystagmus induced by rotation towards the operated side. After bilateral ablation of the otoliths, this slow-phase reduction was present in nystagmus to both sides. This modulation of vestibular nystgmus by the otolith ograns may be seen in the effect of body tilt on the spontaneous nystagmus in patients with unilateral vestibular damage. When such a patient tilts his head towards the affected side the nystagmus frequency increases, and when he tilts it the other way it decreases (Fluur, 1973a, 1974a, b). This is probably the reason why such patients sleep with their good side down (Fluur, 1973b). A nystagmus induced or increased by head tilt is referred to as **positional nystagmus**, and it is generally concluded that it is due to the defective interplay between inputs from the canals and otolith organs, except in the case of alcohol positional nystagmus, which is probably due to a disturbance of the specific gravity of the cupula.

A recent electrophysiological study has revealed a possible neural basis for the interaction of inputs from the otolith organs and the canals. Favilla, *et al.* (1980) recorded from cells in the fastigial nucleus of the cerebellum and found some which responded to both linear and rotatory acceleration of the animal, in such a way as to suggest that they receive inputs from otolith organs and canals. This result was probably due to a convergence of otolith and canal inputs onto these cells, rather than to interactions at the level of the end organs, because most of the cells responded to only one or other type of stimulation.

If a person with head erect is rotated on a centrifuge in the dark, in such a way that his body always faces in the same direction relative to the stationary surroundings, he is subjected to a rotating linear acceleration in a horizontal plane, in the absence of any angular movement. Humans rotated in this way, which I suggest may be called **ferris-wheel rotation**, manifest a horizontal nystagmus (Benson, 1968). This response probably represents the sequential eye deviations induced by the changing direction of the linear vector. The **ferris-wheel nystagmus** was found to persist when the six canals were presumably put out of action by blocking the endolymphatic ducts (Correia and Money, 1970). Benson and Barnes (1973) presented a theoretical model of the motion of the otolith membranes when subjected to ferris-wheel rotation and found that it described the observed nystagmic responses.

Evidence against the view that responses to ferris-wheel rotation are due only to otolith stimulation comes from a study by Benson *et al.* (1970). They recorded from canal-driven neural units in the vestibular nuclei of the cat while it was subjected to ferris-wheel rotation. Those units that were normally excited by a rotary acceleration in the same direction showed increased firing rates, even though the canals were not, in any obvious way, being subjected to a rotary stimulus. Goldberg and Fernández (1975) recorded responses from canal afferents in a monkey subjected to linear acceleration with the canals plugged so that they could not respond to rotary accelerations. This response of the canal afferents is unlikely to be due to a direct effect of linear acceleration on the cupula because objective measurements have failed to reveal any difference between the specific gravity of the cupula and the endolymph (Money *et al.*, 1971). Benson and Bodin (1966a) suggested that, because of the density difference between the perilymph and endolymph, a rotating linear vector causes a cyclic motion of the membranous duct within the bony cavity. This motion would cause a continuous cyclic redistribution of the endolymph in the manner of a roller pump. This mechanism may well operate even in plugged canals. However, before it is concluded that the canals respond to such a stimulus, it would have to be confirmed that canal-driven units of the vestibular nucleus receive no inputs from otolith organs, that there is no direct neural coupling between the cristae and maculae, and that canal afferents are not affected by efferent neurones responding to stimulation of the otolith organs.

A type of linear acceleration which is most likely to stimulate only the otolith organs is sinusoidal linear motion in a purely vertical direction. Melvill Jones *et al.* (1973) exposed human subjects to such a stimulus in the dark, and found weak eye movements which varied systematically with the frequency of the stimulus. With this type of stimulus, oppositely polarized units in the maculae are alternately stimulated, and this presumably evoked the alternate eye deviations which Melvill Jones *et al.* observed.

When a person is rotated about the mid-body axis with the body axis horizontal **(barbecue rotation)**, he is subjected to a rotary acceleration in the plane of the horizontal canals and a rotating linear acceleration vector in the plane of the utricles. Under such circumstances, nystagmus persists, even during an extended period of constant velocity of rotation in the dark (Guedry, 1965; Benson and Bodin, 1966a). Presumably the canal-instigated response ceases during constant velocity, just as it does when the head is rotated at constant velocity about a vertical axis, so that the persisting nystagmus must be due to the otolith contribution. The duration of postrotatory nystagmus is less after barbecue rotation than it is after rotation about a vertical axis (Benson and Bodin, 1966b). Benson (1974) has suggested that this reduction is due to a modification of the adaptation time constant of the canals by stimulation of the otolith organs. That the otolith organs are responsible for all these modifications of nystagmus during barbecue rotation is confirmed by the fact that, after ablation of these organs, nystagmus in barbecue rotation has been found to be similar to that induced by rotation of the erect body about a vertical axis (Janeke *et al.*, 1970).

In summary, it may be stated that stimulation of the otolith organs by constant linear acceleration induces deviations of the eyes in a direction which depends on the direction of the stimulus; whether it also induces nystagmus is not finally settled. If the balance of the vestibular system is disturbed by unilateral damage, the resulting spontaneous nystagmus is modified by otolith stimulation or a nystagmus is induced by such stimulation. If a person is subjected to a ferris-wheel rotation, the changing otolith-induced eye deviation produces a nystagmic response. The canals also respond to this ferris-wheel rotation, possibly because of a pumping action produced by the cyclic displacement of the endolymphatic canal. In barbecue rotation the effects of rotary acceleration interact with the effects of the rotating linear acceleration vector in a complex way.

9.2.8 Countertorsion

When the head is tilted to one side, the eyes rotate in their orbits about the visual axes in the opposite direction. This response is known as **countertorsion**. Because of countertorsion the normally vertical meridian of the retina does not tilt as much as the head. In certain species countertorsion enables the animal to keep the same retinal meridian aligned with gravity, but in man countertorsion is only about 20% of the head tilt for angles of tilt up to about 60° (see Figure 9.9). In the dark the response is fairly smooth, with an occasional nystagmic beat which probably arises from stimulation of the canals. In illuminated surroundings the nystagmic beats are much more evident (see Figure 6.3), and are probably due to optokinetic stimulation from the moving retinal image (Petrov and Zenkin, 1973).

The amplitude of countertorsion has been reported to be a function of the sine of the angle of head tilt. This suggests that it is the shearing force acting on the utricles which drives countertorsion (Schöne, 1962; Miller and Graybiel, 1971).

The amplitude of countertorsion has also been found to increase as the velocity of head tilt is increased, although gain and phase plots show too much scatter for an accurate determination of the transfer function (Hannen *et al.*, 1966). A more accurate determination of the transfer function has been obtained by electromyographically recording the response of the muscles that are involved in eye torsion. Anderson and Precht (1979) took this approach and were able to determine the phase lag of the muscular response to sinusoidal tilting movements of the head over the frequency range of 0.025–0.25 Hz by averaging the response over several cycles. Their results indicate that for low-frequency tilting movements of the head, torsion is a displacement-related response. At the higher frequencies of tilt which were tested, they found a significant phase lag in the torsional response, even though the primary output of the utricle showed a phase lead at these same frequencies. Anderson and Precht concluded that there is a 'leaky' integrator in the utriculo-ocular pathway, perhaps the same integrator that is known to operate in the canal–eye movement system.

The amplitude of countertorsion produced by a tilt of the head alone is the same as that produced by a tilt of the whole body (Krejĉová *et al.*, 1971). This suggests that receptors in the joints of the neck do not contribute significantly to

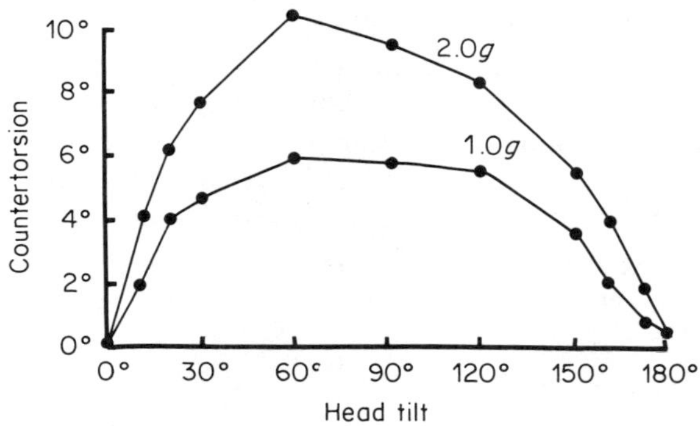

Figure 9.9 Countertorsion as a function of head tilt under 1.0*g* (lower curve) and under 2.0*g* (upper curve) (From Schöne, 1962)

countertorsion. This is supported by the fact that countertorsion does not occur when the head is tilted towards the shoulder with the body in a supine posture (Scott, 1967).

After loss of labyrinthine functions on one side countertorsion towards the defective side is reduced, and after bilabyrinthectomy countertorsion is reduced in both directions (Krejĉová *et al.*, 1971; Nelson and Cope, 1971). The small residual countertorsion that persists after bilabyrinthectomy is present when the head and torso are tilted together and is therefore not due to inputs from the neck. This residual torsion is also found when the subject is in the dark and is therefore not induced visually. It must be due either to incomplete loss of vestibular functions or to somato-sensory afferents responding to changes in body posture (Smiles *et al.*, 1975).

The amplitude of countertorsion is not reduced if the head is held in a tilted position for several hours (Miller and Graybiel, 1974), which provides evidence that countertorsion is driven by non-adapting utricular receptors.

9.3 Visual effects of rotary acceleration: the oculogyral illusion

Everyone will have noticed that, just after spinning oneself about the body's longitudinal axis, stationary visual objects appear to move in the opposite direction. Graybiel and Hupp (1946) gave the name **oculogyral illusion** to any such illusion of visual movement seen during or just after rotary acceleration or deceleration of the body.

The best stimulus for studying the oculogyral illusion is a small luminous object seen in dark surroundings and fixed with respect to the subject so that, as he is rotated, it maintains a constant headcentric direction. The subject is able to fixate such a stimulus and thus eliminate complications due to nystagmus. One would expect such an object to appear to move exocentrically in the same direction and at the same angular velocity as the body, because that is what it is really doing. Such a sensation of visual movement would not be an illusion, but is rather the baseline from which the oculogyral illusion should be measured. This point has not usually been made clear in discussions of the oculogyral illusion. If there is an illusory apparent change in angular velocity of the body, the velocity of the visual object relative to the imagined unseen stationary surroundings should appear to change accordingly. This should give rise to a predictable misperception of the objective velocity of the visual object. This is an illusion but not an interesting one, because it simply means that the subject has correctly judged the headcentric stability of the target and estimated its objective motion relative to his apparent body rotation. A much more interesting effect is the experience of headcentric motion of a visual target which rotates with the subject. This is of course an illusion because the target has a fixed headcentric location. It is not always clear whether subjects in experiments on the oculogyral illusion were asked to judge the headcentric or the absolute (exocentric) motion of the visual target, but most accounts leave little doubt that subjects in fact made headcentric judgments. Misperception of the velocity of body rotation should not be reflected in this measure of the oculogyral illusion. Thus, when the subject is asked to report only when the headcentrically fixed visual target appears to be moving faster or slower than himself, any effect must be due to a misperception of the movement of the target relative to the head, not to a misperception of the movement of the head. In other words, it must be due to some disturbance of the systems which register headcentric motion, that is, it must be due to misregistration of oculocentric motion or of eye movements (see Table 7.1). For this reason I shall define the oculogyral illusion as an illusion of headcentric visual motion of a headcentrically stationary object induced by stimulation of the vestibular canals. An analogous illusory movement of a sound source has been called the **audiogyral illusion** (Mayne, 1952; Lester and Morant, 1970). All these illusory movements are, like the motion aftereffect, paradoxical, because they are not accompanied by an impression of change of location–the object seems to be moving without getting anywhere. This paradox is understandable when one remembers that the neural signals for movement are different from the signals for displacement.

The most systematic determination of the oculogyral illusion has been carried out by Parsons (1970). He asked subjects to estimate the velocity of a visual target using a rating scale. These estimates were made every five seconds during a rotary acceleration of the erect body about its mid-body axis and during the subsequent period of rotation at a steady velocity. Accelerations of 2, 3, 6 and $9°/\sec^2$ were used, each extending over periods of 1, 3, 6 and 9 sec. The mean results of 10 subjects for an acceleration of $9°/\sec^2$ and for two durations of acceleration are shown in Figure 9.10. A control group, in which all measurements were made during a period of rotation at a steady velocity, showed no evidence of an oculogyral illusion. It can be seen that in those conditions in which the acceleration was continued for 9 sec, the illusory visual movement in the direction of body rotation increased during the period of acceleration, although a true measure of the **crescendo time** of the effect is not revealed by these data because the acceleration was not continued for long enough. It can also be seen that after a steady velocity was reached, the illusory movement declined to zero in about 40 sec, which is about the same that postrotatory nystagmus normally persists. After the longer period of body acceleration the illusion passed into a reversed phase which lasted for approximately another two minutes. This phase corresponds to the period when the vestibular apparatus is recovering from the effects of neural adaptation to the preceding acceleration. This reversed phase was not shown when the acceleration of $9°/\sec^2$ was continued for only 1 sec, as can be seen in Figure 9.10, and it was not shown for durations of even 9 sec when acceleration was only $2°/\sec^2$. Comparison of Figures 9.7 and 9.10 is informative.

The oculogyral illusion cannot be due to movements of the retinal image resulting from vestibular nystagmus, because the gaze is fixed on the visual target, and in any case the illusion occurs when an afterimage is used as the target (Göthlin, 1946). The illusion could be due to the effects of the visually inhibited oculomotor signals. This is unlikely to be the only cause of the illusion, because the illusion occurs during rotary accelerations that are too weak to induce

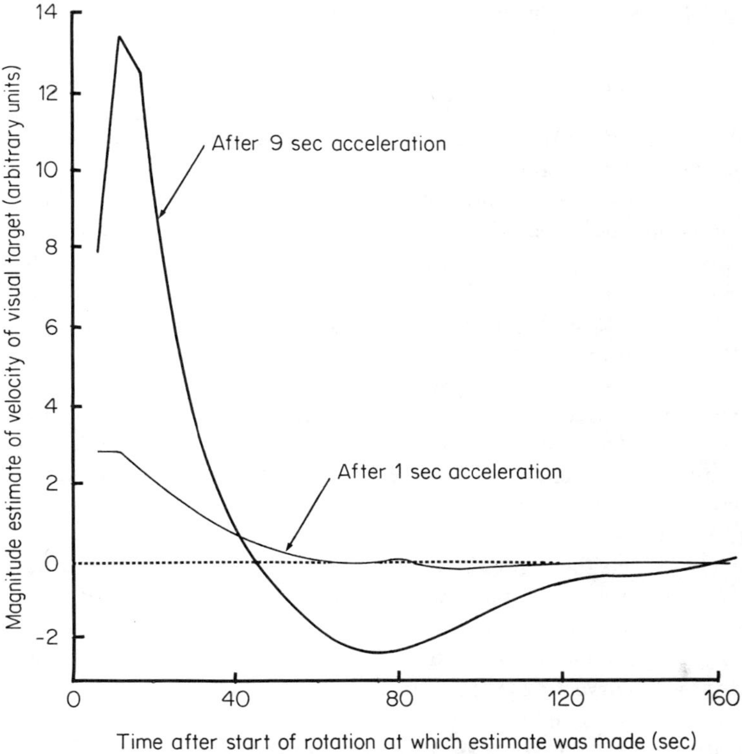

Figure 9.10 Mean magnitude estimates of the oculogyral illusion for 10 subjects rotating at constant velocity after being subjected to an acceleration of $9°/sec^2$ for periods of 9 and 1 sec. Positive apparent movement is in the direction of body rotation compare with Figure 9.7. (Adapted from Parsons, 1970. Copyright, 1970, by the American Psychological Association. Reprinted by permission).

nystagmus and lasts longer than nystagmus (Guedry *et al.*, 1961). Thus, on logical grounds, the oculogyral illusion cannot be due to any misjudgment of the velocity of body rotation, and it seems not to depend on anything to do with eye movements. It looks as though the illusion is due to a direct effect of vestibular stimulation on the visual centres concerned with the assessment of headcentric visual motion (Clark and Stewart, 1969). Interaction between vestibular and visual inputs seems to be a two-way process, as we shall see in Section 9.4.

The only general conclusion to be drawn at the moment is that if there is an input from the canals which is normally associated with a rotation of the body in a given direction at a given speed, a visual object in a fixed headcentric direction appears to move in the same direction but at a higher velocity than the body seems to be moving. Visual illusions of movement are also produced by subjecting a person to linear acceleration. These are known as **oculogravic illusions**, and are discussed in Section 10.4.8.

9.4 Visually-induced illusions of self motion

9.4.1 Introduction

During the nineteenth century there was a fairground device known as the haunted swing. People entered a boat-shaped chamber and artificial scenery was slowly swung backwards and forwards outside the windows. This gave a compelling illusion that the chamber was rocking, and the people felt all the body sensations of real motion, including loss of postural stability and vertigo (Wood, 1895). Similar visually-induced sensations of self motion are experienced when one is sitting in a stationary train while watching the neighbouring train pull out of the station, and sometimes in the wide-screen cinema.

Illusions of self motion and nystagmus may also be induced by displacing sound sources relative to a stationary observer in the dark (Hennebert, 1960; Lackner, 1977). They may also be induced when a subject in the dark keeps his hand on an object which moves round him (Brandt *et al.*, 1977) and even when tactile stimuli are moved over a large part of the body surface (Dichgans and Brandt, 1978).

Mach (1875) was the first to investigate visually-induced self motion, which he did by surrounding the subject with a rotating cylinder painted with vertical black and white stripes. Fischer and Kornmüller (1930) used the same method and gave the name **circularvection** to visually-induced illusions of self rotation and the name **linearvection** to illusions of visually-induced self translation. These early investigators concluded that illusions of self movement are due to the central effects of the efference associated with optokinetic nystagmus. Brandt *et al.* (1973) revealed the inadequacy of this theory by showing that the illusion of self rotation induced by a large moving display was little affected by a change in the direction of optokinetic nystagmus triggered by a change in the motion of a small display confined to the centre of the larger moving display.

All observers are agreed that visually-induced sensations of body motion are indistinguishable from sensations arising from actual motions of the body, which suggests that motion of the visual scene has the same effect on the nervous system as stimulation of the vestibular sense organs. This equivalence between the effects of induced and real vestibular stimulation is dramatically illustrated by the fact that if a person inclines his head forward while experiencing visually-induced rotation about the mid-body axis, he has the same sensations of Coriolis vertigo that are induced when the head is tilted during a real rotation. These sensations are known as the **pseudo-Coriolis effect** (Dichgans and Brandt, 1973). The equivalence between real vestibular inputs and visually 'simulated' inputs makes sense because, during a prolonged period of real passive rotary or linear motion of the body at a constant velocity, the vestibular input ceases and the only indication of motion is the sight of the countermoving surroundings. Natural visual surroundings do not move, so that if they are perceived to be moving relative to the self, the best bet is that the motion is due to self motion. It is, therefore, not surprising that a large moving visual display induces an illusion of self motion—it is the natural cue for such motion. When a person moves along a straight path through a natural visual environment, the optical array undergoes transformations: a parallactic gradient of motion over horizontal surfaces on either side, translation of vertical surfaces on either side, and expansion outwards from the front (Gibson, 1968); and when a person rotates, the optical array rotates relative to the head in the opposite direction about the same axis. This complex optical flow pattern contains information regarding the observer's velocity of motion, his heading (Warren, 1976), and the relative distances of surfaces and objects. Nakayama and Loomis (1974) have suggested that these optical flow patterns are processed by specialized cells in the visual system.

If, during a real motion of the body, the subject's gaze is fixed on an object which moves with him, against a stationary background, the image of the stationary background moves over the retina. On the other hand, if the subject's attention is directed to the stationary surroundings rather than to the object which moves with him, the eyes execute nystagmic movements which stabilize the image of the surroundings during the slow phases. In either case, the surroundings move relative to the head of the moving subject, and it is this headcentric motion that provides the only cue to the subject that he is moving after the movement has continued for a while. Headcentric motion of the stationary scene is assessed by summing the motion of the retinal image of the scene and the motion of pursuit eye movements. For a given type and velocity of body motion, the more there is of one, the less there is of the other. It follows that if illusory self motion is induced by the perceived headcentric motion of the scene, the illusion should not depend on whether or not the eyes move, but on the accuracy with which combined image motion and eye motion are perceived. We shall see that these predictions are substantiated by the evidence which will now be reviewed.

9.4.2 Temporal features of visually-induced sensations of self motion

Consider first the case where an upright subject is placed at the centre of a rotating vertical cylinder painted on the inside with black and white stripes. When the subject is instructed to open his eyes, he is suddenly confronted by an array of vertical stripes moving across his entire field of view. For an initial period of 3–4 sec the subject experiences an objective movement of the scene with himself as stationary. He then begins to experience illusory sensations of rotation of his own body about its mid-vertical axis. Gradually the apparent velocity of self rotation increases while the apparent velocity of the scene decreases, until, after a variable period of up to 30 sec or more, only self rotation is experienced and the scene appears stationary (Brandt *et al.*, 1973). These relationships are revealed in Figure 9.11, which shows the results of a study by Wong and Frost (1978) in which they obtained magnitude estimations of the velocities of stimulus motion and of self motion at different times after the initial exposure to a rotating scene for different velocities of rotation of the scene.

These findings may be interpreted in the following way. Consider first what happens when a person is actually rotated. During the initial phase of. a real

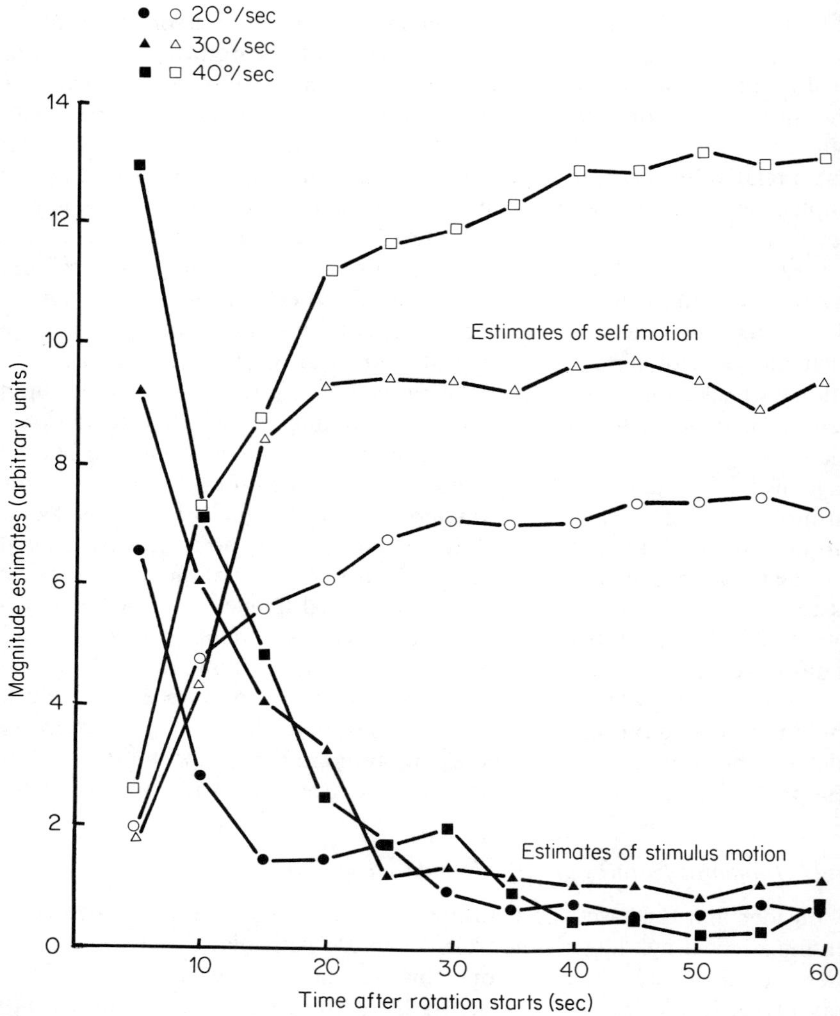

Figure 9.11 Mean magnitude estimates of velocity of stimulus motion (bottom curves) and self motion (top curves) of four subjects at different times after a striped drum is set in motion round them at each of three velocities (From Wong and Frost, 1978. Reproduced by permission of Psychonomic Society, Inc)

rotation of the body, the vestibular system is excited. From the time that the rotary acceleration becomes a constant rotary velocity, it takes about 10 sec for the cupula to become restored to its central position, and up to 30 sec for associated neural activity to cease. During the period of cupula deflection, pursuit eye movements and sensations of body rotation are largely under the control of vestibular inputs. During the period of rotation at constant angular velocity, the vestibular input weakens as the cupula returns to its central position and visual sensations become more dominant as the visual pursuit system is recruited, until finally, sensations of body rotation are governed wholly by visual inputs. Consider now what should happen when a stationary person observes a scene rotating at constant velocity. During the initial period he will know that it is the scene that is moving and not himself, because the vestibular inputs that a natural acceleration of the body would produce are absent. But gradually, the restraint imposed by the absence of vestibular inputs will lessen, finally leaving the visual motion of the scene to be interpreted wholly as self rotation. This exactly describes the empirical results shown in Figure 9.10.

If the moving scene is observed for a long time, the sensation of body rotation sometimes reverses direction and then the scene seems to be moving with the observer but at a higher velocity than him. Brandt *et al.* (1974) related these reversals to habituation of vestibular sensations, which is a topic discussed in Section 9.5.2.

The latency of the sensation of self rotation is shortened by an actual imposed impulsive rotation of the body in the same direction (Brandt *et al.*, 1974). The latency of illusory self motion is also shortened when induced by visual motion towards the side of spontaneous nystagmus in patients with a unilateral peripheral vestibular lesion, at least before they have learned to compensate for the deficit (Hökendorf *et al.*, 1977). The shortening of latency in both these cases is what one would expect; any actual vestibular stimulation counter to the scene motion will encourage the system to interpret the scene motion as due to body rotation and hence shorten the latency of that response. An actual rotation of the body in the opposite direction to an already present sensation of body rotation abruptly destroys the illusion (Young *et al.*, 1973; Teixeira and Lackner, 1979). This too is just what one would expect, if it is assumed that the sensation of body rotation is dominated by vestibular inputs when these signal an acceleration or deceleration and by visual stimulation only when the vestibular inputs are neutralized after a period of rest or steady velocity.

The estimated angular velocity of self motion has been found to be a linear function of stimulus velocity up to about 90°/sec. At higher stimulus velocities apparent self motion is accompanied by apparent motion of the scene and there are interludes when the subject no longer feels himself rotating. For stimulus velocities above about 180°/sec, there is no further increase in the velocity of apparent self motion. Such high velocities of continued real self motion are rarely met with, so that it is not surprising that the system shows this indecision in how it interprets such signals.

The function relating the velocity of the scene to the apparent velocity of self rotation is complicated by the fact that the apparent velocity of a moving visual display depends on the spatial frequency of the visual pattern (see Section 6.1.2). It is also complicated by the fact that the apparent linear velocity of a moving display varies with its apparent distance from the observer. An object travelling past an observer at a certain linear velocity subtends an image which moves across the retina at a much higher velocity when it is nearby than when it is far away, so that if an object which produces a given retinal velocity appears to get further away, its linear velocity will seem to increase. Its apparent *angular* velocity should of course be unaffected by its apparent distance. Wist *et al.* (1975) found that the apparent angular velocity of visually-induced self rotation is determined by the apparent linear velocity (apparent distance) of the moving scene, rather than by its angular velocity. This suggests that we apply size–distance scaling to rotary as well as to linear motion. In this sense we treat the rotary motion as if it were a linear motion even though the self motion that it induces is a rotary rather than translatory one.

If the lights are put out after a moving scene has been inspected for some time there is typically a continuation of the illusory self motion in the same direction, designated the **positive aftereffect**, followed by a period of illusory self motion in the opposite direction, designated the **negative aftereffect**. One reason for the positive aftereffect must surely be that there are no vestibular signals to inform the system that the body has stopped rotating. But there is probably another reason that has to do with the visual aftereffects of observing the rotating scene. We have seen that the initial increase in the apparent velocity of induced self-rotation may be related to the time it would have taken the cupula and neural elements to restore to their resting level, had the subject actually been rotating, and to the recruitment time of optokinetic nystagmus. Brandt *et al.* (1974) found that the time course of the positive and negative aftereffects of apparent self rotation as a function of stimulus duration was similar to the time course of positive and negative optokinetic afternystagmus (see Section 6.1.1). The two positive aftereffects were found to increase in duration to about 36 sec as stimulus exposure increased up to 60 sec, and to decrease in duration with stimulus exposures longer than this. The negative aftereffects continued to increase with increasing stimulus duration, at least up to 15 min, the longest duration tested. Thus, when the moving scene is first exposed the illusory self motion follows the time course of recruitment of optokinetic nystagmus and after the stimulus is removed it follows the time course of the positive and negative phases of optokinetic afternystagmus.

Optokinetic nystagmus also resembles illusory self motion in another way. The aftereffects of both of them occur whether the subject pursues the moving scene or fixates on a headcentrically stable object during the exposure of the inducing stimuli. Illusory self motion is not caused by optokinetic nystagmus, but both effects seem to depend on the recruitment of the same underlying process. Optokinetic nystagmus is known to be associated with neural impulses in the vestibular nuclei (see Section 5.4.3), so that perhaps these same circuits are related to sensations of body rotation.

If the moving scene has been inspected for some time and the subject continues to view it after it has stopped moving, the sensation of self motion is immediately reversed in its direction. This negative aftereffect is due to the visual movement aftereffect induced by the oculoulocentric motion of the scene. This visually-driven negative aftereffect therefore outweighs the positive aftereffect arising from the vestibular and optokinetic systems.

9.4.3 *Effects of size and location of the visual stimulus on illusory self motion*

Illusory self motion depends on the presence of a moving display which stimulates the periphery of the retina rather than the central region. Thus, Brandt *et al.* (1973) found that a stimulus confined to an area of up to 30° in diameter in the central region of the visual field does not elicit illusions of self motion, whereas a moving display of the same area which stimulates the retinal periphery is very effective. When the moving display covers the entire visual field, masking the central region up to 60° in diameter does not reduce the illusion of self motion, and even a central mask of 120° in diameter reduces the illusion only slightly. Furthermore, if a peripheral display moves one way and a central display of equal area moves in the opposite direction, the illusion of self motion is determined by the motion of the peripheral display—even though the eyes may pursue the central display. So once again we come across a situation in which headcentric orientation is governed by the peripheral retina.

In a visual field consisting of objects at various distances from the observer, illusory self motion is induced by movement occurring in the visual background rather than by movement in the foreground. Thus, a moving background induces illusory self motion even if the foreground is stationary, but a moving foreground does not induce illusory self motion if the background is stationary (Brandt *et al.*, 1975).

The orienting responses triggered by peripheral vision are important at all times for general coordination and balance. It is therefore not surprising that illusory self motion has been found not to be influenced by a reduction in luminance to scotopic levels or by defocusing the image (Leibowitz *et al.*, 1979).

9.4.4 *Illusory self motion in different planes*

This discussion of visually induced sensations of self motion has so far been confined to the effects induced by vertical stripes rotating round an erect observer. This is the simplest case because the illusory self motion is one of simple rotation about the body axis uncomplicated by sensations of body tilt. I shall now consider the effects of scenes rotating about other axes or moving in a linear fashion.

A scene which rotates in a frontal plane around the observer's visual axis induces a sensation of body rotation in the opposite direction. This will be referred to as an **illusory sensation of self tilt**. The otolith organs and the somaesthetic senses supply information which conflicts with these roll sensations. These conflicting signals produce paradoxical sensations—the body seems to be rotating continuously and at the same time seems to be tilted only to a limited degree. Held *et al.* (1975) investigated the illusion of self tilt using a visual display consisting of a random texture rather than radial lines, so as to avoid complications due to visual cues to the direction of gravity. They found that the illusion of body motion started soon after the stimulus motion and reached a steady state after about 18 sec. The effect was found to depend on stimulus area, and peripheral stimuli were found to be more effective than central stimuli, just as in the case of rotation about the body axis. The amount of apparent body tilt increased with the angular velocity of the stimulus and reached an average value of 15° for seven subjects, with one subject showing an effect of 40°. Accompanying the sensation of body tilt was an apparent tilt of a vertical line in the same direction. This effect will be discussed in more detail in Section 10.4.

The magnitude of the sensation of self tilt induced by a scene rotating about the visual axis is increased if the head is inclined to 90° or inverted, and the sensation is decreased if the head is inclined so that the utricular maculae are horizontal (Young *et al.*, 1975). These relationships make sense in terms of the fact that the sensitivity of utricular maculae is greatest when they are horizontal and becomes progressively less the more they are inclined. In the horizontal position, the utricles therefore provide more effective information about the true position of the head.

Illusory sensations of self inclination (pitch) are produced by a scene which moves upwards or downwards across the field of view of the erect observer. The sensation of continuous body motion is now accompanied by a paradoxical sensation of the body being at a constant angle of pitch against the direction of stimulus motion. The illusion of pitch forwards is larger than the illusion of pitch backwards (Young *et al.*, 1975). Similar asymmetries which occur in optokinetic nystagmus were mentioned in Section 6.1.1.

Illusory sensations of self linear motion are produced by translatory motion of the visual scene. This illusion is sometimes referred to as **linearvection**. Although such effects have been known for a long time, a particularly effective apparatus for demonstrating them was recently devised by Lishman and Lee (1973). The subject is enclosed within a large illuminated box which swings back and forth parallel to the direction in which the subject is facing. At the same time he may stand on a trolley and be moved passively back and forth, either in phase or out of phase with the box, or he may walk actively forward while the box moves with him or against him. The sensations of self motion were found to be determined by the direction of motion of the visual scene relative to the observer. Even if the observer were being actually moved one way, he would experience a motion of the body the other way if that is what the relative motion of the scene indicated. When a person walked in one direction and the scene induced a sensation of self motion in the other direction because it moved at a faster rate than the subject, the subject had the sensation of walking on a moving treadmill. Subjects tended to sway in synchrony with the scene as they attempted to counter the sensation of body sway that the moving scene induced. Young children who had just learned to walk fell over as soon as the scene began to move (see Section 9.4.6).

Apparent linear motion of the body may be induced by a moving scene with a velocity at about the threshold for the visual detection of motion. It has also been found to have a latency and saturation velocity similar to those for illusory rotary motion about the body axis (Berthoz, Pavard and Young, 1975). Like illusory rotary motion, it is more readily induced by motion in the visual periphery than by motion in the centre of the visual field (Rock, 1968; Johansson, 1977).

These sensations of linear motion presumably occur because the visual input triggers some mechanism that is normally triggered by the otoliths. This makes sense because if, for instance, a person walks at a steady pace, the otoliths are not stimulated, and the only sensory cues associated with the motion are the relative motion of the scene and the motions of the legs. If a person is moved passively at a steady velocity, he has only the scene motion to tell him that he is moving. We have already seen that illusory rotary motion is followed by positive and negative aftereffects once the subject closes his eyes, and that these aftereffects may be related to the way the cupula responds after a real rotation and to optokinetic afternystagmus. When a real linear motion of the body is stopped the otoliths, unlike the cupulae, return to their resting condition very quickly (de Vries, 1950). Furthermore, linear motion forwards in a full-field stationary surround induces neither nystagmus nor afternystagmus if the gaze is directed in the direction of motion. Therefore, *if cupula restoration time and optokinetic afternystagmus are the cause of aftereffects of induced self rotation, such aftereffects should be absent for induced linear motion produced by a symmetrically moving scene*. There seem to be no data on this point.

9.4.5 *The neurological basis of illusory self motion*

All the evidence reviewed so far suggests that the motion of large visual displays relative to a stationary observer has an effect which is equivalent in many ways to the effect of stimulating the vestibular end organs. This suggests that visual and vestibular inputs converge at some site in the central nervous system, and the most obvious place to look for such a convergence is the vestibular nuclei. Direct electrophysiological recordings from single cells in the vestibular nuclei of the goldfish, rabbit, and cat have revealed neurones which respond both to stimulation of the vestibular end organs and to motion of the visual surroundings. References to studies on these subprimate species are given in Dichgans and Brandt (1978).

Henn *et al.* (1974) and Waespe and Henn (1977a) found that nearly all cells in the vestibular nuclei of the rhesus monkey which respond to canal stimulation also respond to moving visual fields. A unit which is excited by head rotation to the left is also excited by motion of the visual scene to the right, that is, by scene motion that would normally accompany that head motion. Activity in these units outlasts the removal of the moving scene, which accounts nicely for the positive aftereffect of illusory self rotation. When the animal was actually rotated in the direction of induced self rotation, the vestibular units were found to increase their rate of firing above the level recorded when only the scene rotated. When the animal was rotated in the other direction, the firing rates were found to fall below the level for scene rotation. These findings also fit in nicely with the behavioural data reported in Section 9.4.2. Furthermore, cells related to the otolith organs have been found in the cat's vestibular nucleus which respond to linear motion of the animal but also to translatory motion of the visual scene in relation to the stationary animal (Daunton and Thomsen, 1979).

The pathways serving visual–vestibular interactions are not fully known. There are no direct visual inputs from the optic tract or superior colliculus to the vestibular nuclei, but the flocculus of the cerebellum receives an input from the visual system and from primary and secondary vestibular neurones (see Section 9.5.5). Furthermore, lesions of the flocculus impair the suppression of vestibular nystagmus by vision (Section 9.5.5). For these reasons, it has been generally assumed that visual signals reach the vestibular nuclei through the cerebellum. However, Keller and Precht (1979b) found that the response of cells in the vestibular nucleus to moving visual stimuli was unaffected by complete removal of the cerebellum in cats. Visual signals must also reach the vestibular nuclei by a route which does not pass through the cerebellum. The crucial visual input could perhaps reach the vestibular nuclei through the superior colliculus (see Section 5.5). However, against this view is the fact that cells with the required movement sensitivities have not been found in the superior colliculus. Such cells have been found in the nucleus of the optic tract in the rabbit and cat (see Section 5.4.5). This is a band of tissue receiving visual inputs which lies along the dorsolateral edge of the pretectum in the midbrain (Hoffmann and Schoppmann, 1975). So perhaps this is where the visual inputs underlying illusory self motion originate.

9.4.6 Illusory self motion and postural stability

The potent effects that peripheral visual motion has on the sense of body motion suggest that vision plays an important role in the maintenance of postural stability. The experiments conducted by Lishman and Lee, that were described in Section 9.4.4, illustrate how postural stability is disturbed by movement of the visual surroundings. Postural stability is also controlled by vestibular inputs, by inputs from joint receptors in the hips and ankles, by stretch reflexes in the muscles of the legs and back, and by tactile inputs from the balls of the feet. Reviews of these non-visual postural mechanisms are to be found in Howard and Templeton (1966), Stein *et al.* (1973), and Roberts (1978).

The importance of vision for postural stability changes with age. Brandt *et al.* (1976) found that children between six and 12 months of age who have learned to sit upright are not disturbed by a movement of their visual surroundings. However, Butterworth and Hicks (1977) reported the opposite finding. There is no doubt, though, that once children have learned to stand, their postural stability is readily disrupted by visual motion. A movement of the whole scene in the frontal plane causes the child to sway to the side in the direction of scene motion, and movement of the whole scene forwards or backwards in a parasagittal plane causes the child to sway, and probably fall, forwards or backwards (Lee and Aronson, 1974). After the age of five, the effects of visual motion on postural stability slowly decrease. The normal adult shows only moderate body sway in the direction of visual motion (Brandt *et al.*, 1976).

If a person wears optical devices which reverse the direction of motion of the retinal image induced by sideways motion of the body, postural stability is severely disturbed. Gonshor and Melvill Jones (1980) had subjects wear reversing Dove prisms continuously for up to 27 days and tested their progress in regaining postural stability on a series of graded tests which involved walking along rails of various widths. From an initial period of catastrophically disorganized postural control, subjects continued to improve over a period of days. After the prisms were removed subjects soon regained stability when walking on the broader rails, but required days to regain full stability on the narrowest rail.

The sensations of illusory self motion induced by visual motion have a latency of several seconds. This is much too long a latency for an effective postural response. If a corrective response to an actual self motion were delayed this long, the person would fall before making the appropriate adjustment. Nashner and Berthoz (1978) looked at this question and, by measuring the electromyographic response in the leg muscles of a standing adult, were able to show that under favourable conditions the muscles respond to a movement of the visual scene with a latency of about 100 msec. Thus, the long-latency responses to visual motion, involving sensations of body motion and visual tilt, are distinct from the short-latency responses having to do with the maintenance of postural stability. The illusory sensations are delayed because the absence of changes in vestibular and proprioceptive signals informs the subject that he is not tilting. The subject interprets the visual signals as due to self motion only after the time when the other signals would have adapted. The postural responses to visual motion are presumably not subject to these restraints.

A latency of 100 msec is still rather long compared with the latency of postural responses to non-visual inputs. This suggests that visual inputs help to correct only for low-frequency spontaneous oscillations of the body when a person is standing in stationary surroundings, and that high-frequency oscillations are corrected by the shorter-latency, non-visual inputs. Dichgans, Mauritz, Allum and Brandt (1976) found that a sinusoidal to-and-fro motion of the visual surroundings destabilized posture in a standing adult most effectively when the scene oscillated at about 0.1 Hz. This frequency is at the lower end of the average power spectrum of spontaneous body sway for a standing adult. Lestienne *et al.* (1977) obtained similar results. Dichgans et al. found that in patients with postural ataxia, in whom the non-visual mechanisms of postural stability were defective, the destabilizing effects of sinusoidal visual motion extended to higher frequencies. Such patients clearly depend on vision for compensating for postural sway to a greater extent than do normal subjects. (See Dichgans and Brandt, 1978 for a more detailed discussion of these issues.)

The sensations of illusory motion induced by visual tilt require a threshold displacement of over 2°, whereas amplitudes of normal body sway are below 2° to either side of vertical (Dichgans, Mauritz, Allum and Brandt, 1976). The higher threshold for illusory self motion could, like its longer latency, be due to the restraining influence of the absence of vestibular and proprioceptive inputs which normally accompany body sway. Motion of the retinal image resulting from movements of the eyes does not induce postural instability, but similar image motion caused by movements of the visual surroundings does (White *et al.*, 1980).

9.5 The adaptability of the vestibulo–ocular system

9.5.1 Introduction

This section is about long-lasting changes in one or other of the perrotatory or postrotatory responses to vestibular stimulation. These responses include nystagmus, sensations of rotation, postural instability, the oculogyral illusion, and nausea. Long-term changes, which may last hours, days, or longer, must be distinguished from short-term changes, which are rapidly induced and last only minutes. Short-term changes may be due to adaptation of the vestibular end organ or to loss of alertness. For instance, Collins *et al.* (1961) demonstrated that when subjects are told to relax, vestibular nystagmus is weak, and when they are asked to solve arithmetic problems or observe external events, the nystagmus is vigorous. Several other investigators have confirmed the importance of arousal or alertness in vestibular responses and it has been suggested that lack of arousal is a causal factor in the weak nystagmic responses of schizophrenic and autistic patients. The whole subject of the effects of arousal on vestibular responses is thoroughly reviewed in Collins (1974a). The short-term effects of a loss of arousal are non-specific and are normally easily reversed if the state of arousal is changed. In extreme fatigue the state of arousal may not be so quickly reversed but the effects on vestibular responsiveness are still non-specific.

The long-term effects that are discussed in this section tend to be related specifically to the type of vestibular stimulation that induces them and are not due merely to a loss of alertness. A long-term change may involve a reduction in one or other vestibular response and the term **vestibular habituation** is usually applied to such changes. However, I hope to show that the term 'habituation' is inappropriate to describe all such response reductions. Some of them, and perhaps all, are special cases of a broader class of adaptive modifications of vestibular responses to atypical sensory stimulation. I shall distinguish four possible causes of long-term changes in vestibular responses in the next four subsections, beginning with the contentious mechanism of habituation.

9.5.2 Vestibular habituation

The term **habituation** was originally used to signify a type of simple learning in which a response to a potentially noxious stimulus gradually weakens if the stimulus is repeated in benign circumstances. For instance, a snail rapidly withdraws into its shell when lightly touched, but if it is touched repeatedly the withdrawal response gradually declines. The signal gradually becomes less arousing, but regains its previous potency if accompanied by some noxious event.

Atypical visual–vestibular stimulation is potentially noxious, because it signifies that the animal is in an unusual situation with which it may not be able to deal. But this would not explain the vomiting and depression which often accompany motion sickness. Treisman (1977) suggested that these symptoms arise because one of the first indications an animal has that it has been poisoned is a disturbance of the vestibular system. The vertigo-induced vomiting therefore serves to expel the 'poison' from the system and the depression serves to keep the animal still while it recovers. One could argue that when a person at sea gradually recovers from seasickness he habituates these responses as the potentially noxious stimulation is repeated without undesirable consequences (see Reason and Brand, 1975, for a review of the topic of motion sickness).

There has been a long, and often heated debate about whether postrotatory nystagmus decreases in duration, velocity or amplitude with repeated exposure to body rotation. This literature is reviewed in Collins (1974b). One would not expect nystagmus or sensations of rotation to habituate when a person is subjected continually to alternating brief rotations of the body. After all, these responses are well adapted to serve a useful purpose and any decrement would be maladaptive. Furthermore, ordinary living involves the continual stimulation of the vestibular system, so that whatever habituation may occur will already have occurred. Results of two studies seem to indicate that responses to sinusoidal rotation do not habituate. Rabbits subjected to a continuous sinusoidal oscillation for two days at a frequency of 0.17 Hz with eyes covered, showed no systematic decrement in the velocity or amplitude of nystagmus (Kleinschmidt and Collewijn, 1975). Men subjected to a similar sinusoidal rotation for one hour each day for three days have also been found not to show any sign of response habituation (Gonshor and Melvill Jones, 1973). Note that this type of stimulation does not induce aftereffects, so that it was the perrotatory nystagmus that did not habituate. Optokinetic nystagmus has also been found not to habituate when subjects were exposed repeatedly to stimuli moving in the same direction; there is only a transitory increase in gain (Clément, *et al.*, 1981).

In most studies claiming to demonstrate vestibular habituation, the subjects were repeatedly exposed to a succession of 10–20 periods of rotation or to caloric stimulation spread over an hour or two or, in some cases, over several days. During each period of exposure the subject was rotated in a given direction for a few minutes. In some studies all the periods of rotation were in the same direction, and in others periods of left rotation alternated with periods of right rotation. Earlier studies were complicated by inconsistencies in, or lack of control over (a) the subject's level of arousal, (b) the measures of nystagmus (duration, frequency, velocity, or amplitude), (c) whether or not the eyes were open during and after rotation, (d) the degree to which stimulation was typical of that to which the vestibular system is normally exposed, and (e) the frequency of head oscillation.

Two recent studies go some way towards resolving the contradictions in this literature. Clément *et al.* (1981) exposed cats to periods of unidirectional velocity steps for five consecutive days. This produced long-term reduction in the time constant and gain of nystagmus only in the trained direction. In spite of the resulting asymmetry in gain, the animals showed no spontaneous nystagmus when stationary. This suggested to Clément *et al.* that the habituation was confined to tonic receptors which have no resting discharge and respond to low-frequency oscillations of the head. This would explain why previous investigators found no habituation at the relatively high frequency of 0.17 Hz. Jäger and Henn (1981) confirmed that there is no habituation at frequencies of head oscillation above 0.1 Hz, but found it after monkeys were exposed to sinusoidal oscillations of between 0.002 Hz and 0.1 Hz. There was a persisting decline of the time constant of nystagmus to pulse angular accelerations, predictably accompanied by a reduction in gain and increase in phase lead. The habituation effects induced by one frequency of rotation transferred to nystagmus generated by sinusoidal oscillations over the whole frequency range, except that increased phase leads were not evident for very low-frequency test oscillations, which already showed a 90° lead. Perhaps continued rotation of the head in the dark is an atypical form of stimulation. *It would be interesting to know whether adaptation occurs after rotation in illuminated surroundings.* Other forms of atypical stimulation produce habituation.

In Coriolis stimulation, a deflection of the head is imposed on an existing rotation in another plane, which creates a resultant angular acceleration in a third direction. This contradicts the inputs from the otoliths, which respond only to the tilt of the head, and it also contradicts visual information if the eyes are open. A series of studies has been conducted at the Aerospace Medical Institute in Pensacola on the way people adjust to repeated Coriolis stimulation. Subjects lived for days in a rotating room, so that when they tilted their heads they were subjected to Coriolis stimulation which induced nystagmus, dizziness, and nausea. If head tilting was practised in a given direction these symptoms subsided, but only for the practised direction (Guedry, 1964). After the rotation of the room was stopped, they experienced an oppositely directed nystagmus when the head was tilted in the practised direction (Dowd and Cramer, 1967). This suggests that there had been a specific central recalibration of the vestibulo-ocular system, and it is perhaps this recalibration that was responsible for the reduction in dizziness and vertigo, rather than a more general and passive process of habituation. The active nature of the recalibration process is suggested by the fact that subjects deprived of sleep have been found not to adapt to Coriolis stimulation as readily as well-rested subjects (Dowd, 1974).

Unilateral caloric stimulation is another atypical vestibular input to which animals have been said to habituate. But Collins (1965) pointed out that in many instances the 'habituation' reported in earlier studies was due to a temporary lapse of alertness. Collins did find a persisting reduction in the amplitude of nystagmus induced by repeated unilateral caloric stimulation when the subject was allowed to see his surroundings. But in this case there are contradictory vestibular and visual inputs, so that the reduction in nystagmic amplitude can be understood as a recalibration of the system rather than habituation (see Section 9.5.4). We have seen in Section 9.4 that exposure to a rotating visual field leads to an illusion of self rotation. Repeated exposure to this type of stimulus has been found to lead to a 40% reduction in the velocity of the slow phase of vestibular nystagmus in the direction of the induced self rotation even though the subjects were kept alert (Young and Henn, 1974). This can also be understood as a recalibration of the vestibular response, due to anomalous pairing of vestibular and visual inputs.

Perhaps all cases of 'habituation' are best understood as recalibrations of the vestibulo–ocular system to atypical stimulation.

9.5.3 Acquired techniques for controlling vestibular responses

A second type of vestibular response decrement is typified by the ability of figure skaters and ballet dancers to spin for long periods without suffering disturbing aftereffects. At least part of this ability depends on the prevention of sustained cupula deflection by the habit of **spotting**. As the body rotates dancers keep the head pointing towards the audience for as long as possible, and then flick the head round faster than the body until it again faces the audience. They thus overcome the need to execute eye nystagmus by substituting head nystagmus, and at the same time subject their cupulas to alternating clockwise and counterclockwise stimulation as the head successively accelerates and decelerates. The conditions necessary for postrotatory nystagmus are therefore not allowed to develop (Osterhammel *et al.*, 1968a, b).

But skaters and ballet dancers can avoid disturbing aftereffects even without spotting; indeed figure skaters do not normally use the spotting technique. Postrotatory effects may also be inhibited by fixating on a visual object after spinning (Collins, 1966). If figure skaters keep their eyes closed after spinning, they show postrotatory nystagmus and lose their balance just like anybody else (Dix and Hood, 1969). Even a brief period of room illumination following a rapid rotation of the body is sufficient to forestall vertigo (Collins, 1968a, b; Schroeder, 1971). Perrotatory spotting and postrotatory visual fixation are therefore techniques that people may acquire to help them inhibit disturbing aftereffects of continued body rotation. Most people avoid such disturbances by not spinning round like skaters and ballet dancers.

9.5.4 Recalibration of the vestibulo-ocular system

The third type of long-term change of vestibular responses is illustrated by the capacity of a person to adapt the gain, and even the sign of the slow phase of vestibular nystagmus to atypical combinations of head and scene motion. These changes are normally referred to as adaptive changes; this is all right as long as they are not confused with sensory adaptation. A better term might be **recalibration** of the vestibulo-ocular system.

When spectacles which reverse the visual scene from left to right are worn, a sideways motion of the head causes the retinal image to move across the retina in a direction opposite to that in which it normally moves. The anomalous image motion is annulled if the subject pursues the image with his eyes, but this involves a reversal of the sign of the nystagmus that normally accompanies a rotation of the head. Whether or not the subject pursues the motion of the image, the visual scene is experienced as moving in the same direction as the motion of the head and at twice the angular velocity of head motion. Stratton (1897a, b) observed this phenomenon in his studies on the effects of continuously wearing an optical device which reversed and inverted the visual scene. By the third day of wearing this device, he noted that head movements were still accompanied by slight swinging of the scene. By the sixth day, he wrote:

> Movements of the head or of the body, which shifted the field of view, seemed now to be in entire keeping with the visual changes thus produced; the motion seemed to be towards that side on which objects entered the visual field, and not towards the opposite side, as the pre-experimental representation of the movement would have required. (Stratton, 1897b, p. 358)

This experience suggests that he had learned to reinterpret the relationship between head movements and image movements. However, a study by Hay (1968) suggests that image motion was not involved. In this study, the subject was asked to nod his head up and down while following the movement of a point of light on an oscilloscope. The movement of the light was coupled to the head movement in such a way that, as the head moved up and down, the spot moved from left to right. Thus, the subject's eyes had to move diagonally in order to keep the image of the light centred on the fovea. After exposure to this discordant situation, the subject was asked to judge whether a stationary point of light appeared to move as he continued to nod his head. After as little as one minute of exposure, the subject reported that the stationary object moved left and right in synchrony with his head, but with directions reversed with respect to the exposure condition. Controls showed that a movement of the head alone or of the light alone did not produce the aftereffect. In a later study, Hay and Goldsmith (1973) measured the movements of the eyes during exposure to anomalous head–image motion and found that the subjects soon learned to pursue the visual target. What was learned here was a new coupling between head movements and eye movements. Retinal-image motion was not involved, because the point of light was always fixated. It is a pity that Hay and Goldsmith did not measure the movements of the eyes during the testing procedure which followed exposure to the anomalous motion, but in their discussion they suggested that the eyes moved out of phase with the head and caused the retinal image to drift from the fovea. Wallach and Frey (1969) and Wallach *et al.* (1969) have done similar experiments.

A point which is easily overlooked is that the velocity of eye movements necessary to stabilize the retinal image of an object when the head rotates varies with the distance of the object. This is because the eyes undergo a translatory as well as a rotary motion when the head rotates. For a head rotation of θ, the actual compensatory eye movement required to stabilize the image of a stationary object at distance D is

$$\theta + \tan^{-1}\frac{d\sin\theta}{D}$$

where d is the distance from the axis of head rotation to the nodal point of the eye. For a distant object the effects of this translation are negligible, but for near objects the velocity of eye movements required to stabilize the retinal image becomes about double the velocity required for distant objects. In other words, the gain of the slow phase of vestibular nystagmus required for image stabilization increases from 1 for distant objects to 2 for near objects. Adjustments for this factor could be achieved in two ways. One way would be for information about the distance of the visual object, perhaps indicated by the convergence of the eyes, to be fed into the vestibulo-ocular control system and to control the gain of the nystagmus. The other way would be for the gain of the vestibular–ocular system to be set for infinity viewing and for the visual motion system to trim the velocity of the response when near objects are viewed. *One way to distinguish between these possibilities would be to see whether the gain of vestibular nystagmus in the dark is modified by a change in the state of convergence of the eyes.* This experiment has not been done, but Hay and Sawyer (1969) have shown that a change in the convergence of the eyes alters the velocity at which a visual scene must be moved in order to appear stationary when the head rotates.

Recently Gonshor and Melvill Jones (1976a) continuously exposed four adults to optical reversal of the visual scene for periods of between two and 27 days. The gain of vestibular nystagmus was measured periodically for each subject by rotating him sinusoidally from side to side about the mid-body axis in the dark. The experimenters had established that repetition of this testing procedure did not itself lead to any adaptation or habituation of nystagmus. All subjects showed a substantial reduction in the gain of the slow phase of nystagmus during the first two days of exposure to reversed vision, and at the end of the week the response was almost absent. During the second week, a nystagmus of reversed phase began to appear. The gain of this reversed movement gradually improved. After normal vision was restored, the nystagmus returned to normal in about two hours. During all this time, nystagmic responses to rotations of the head in the sagittal plane were unaffected. In another experiment Gonshor and Melvill Jones (1976b) exposed subjects to reversed vision for only 16 minutes on each of three consecutive days in order to show that only a small amount of experience is required to produce a change in the gain of vestibular nystagmus. During each 16-minute exposure subjects visually tracked a mirror-reversed target as they were rotated sinusoidally from side to side about the mid-body axis. Subjects in a control group were simply rotated sinusoidally in the dark. Subjects in the experimental group showed a 25% reduction in the gain of the slow phase of nystagmus measured in the dark. There should have been a control in which subjects tracked a non-reversed visual target; the gain of nystagmus might have been reduced by the experience of tracking a target when rotated sinusoidally even if the scene motion had not been reversed.

Melvill Jones and Mandl (1979) obtained evidence of adaptation of the gain of vestibular nystagmus in human subjects exposed to reversed vision when the stimulus field to which they were exposed was illuminated intermittently by a stroboscope. This result could be interpreted as showing that smooth slippage of the retinal image is not required for adaptation of the vestibular system. However, as Melvill Jones and Mandl realized, a stroboscopic sequence of images is capable of stimulating the motion detectors of the visual system, so that this experiment does not demonstrate that adaptation may occur when motion detectors are not stimulated. Furthermore, Mandl and Melvill Jones (1979) found that the subjects were able to track the stroboscopically illuminated scene visually as they were rotated, so that the adaptation of the vestibular ocular reflex was probably due to the anomalous direction of this visual tracking in relation to the direction of eye movements that would otherwise have been invoked by the vestibular system.

Gauthier and Robinson (1975) pointed out that adaptation of the gain of vestibular nystagmus has practical implications for underwater divers exposed to the magnifying effect of a face-mask and for patients with bilateral aphakia (loss of the lens), who wear spectacle lenses which inadvertently magnify the visual scene. Gauthier and Robinson fitted a normal subject with lenses which magnified by a factor of two. The experience of visual stability during head movements returned after four days. After the lenses were removed, visual stability was regained in about five hours. The gain of the slow phase of nystagmus in the dark increased by 70% during the five days of exposure to the lenses. Similar adaptive changes in the gain of vestibular nystagmus have been reported in the cat (Melvill Jones and Davies, 1976) and in the monkey (Miles and Fuller, 1974; Miles and Eighmy, 1980).

Repeated exposure to a display of vertical stripes rotating one way across the visual field of a subject whose head and body are stationary has been found to lead to a reduction in the gain of the slow phase of nystagmus induced by rotation of the body in the dark (Young and Henn, 1974; Pfaltz and Ohtsuka, 1975). The combination of movement of the visual scene with no movement of the body may be regarded as an anomalous pairing of visual and vestibular inputs, so that the reduction in the gain of vestibular nystagmus after this type of exposure may be seen as another instance of the recalibration of the vestibulo-ocular system. The experience of visually-induced self motion has also been found to be reversed after a period of several hours in which subjects walked about wearing left–right reversing prisms. Indeed, induced self motion was found to be reversed in sign before there was any evidence of a change in the vestibulo-ocular reflex (Oman *et al.,* 1980).

9.5.5. *The physiological basis of the recalibration of the vestibulo–ocular system*

The recalibration of the vestibulo-ocular system that has just been described involves more than the decrement of unwanted responses; it also involves the substitution of responses better suited to the anomalous relationship between visual inputs, eye movements, and vestibular inputs. It has been shown that there is little or no change in the response of vestibular afferents during recalibration of the vestibulo-ocular system, so that changes in the vestibular sense organs are not involved (Miles and Braitman, 1980; Lisberger and Miles, 1980). There are good reasons for believing that these adaptive changes are associated with changes in the response of the vestibulocerebellum, although whether this is the ultimate site of the adaptive changes or whether a changed pattern of response to vestibular stimulation is merely channelled through this structure is not yet established. A brief review of what is known about the neurology of the cerebellar control of the gain of vestibular nystagmus will now be given.

The most ancient part of the cerebellum, which includes the uvula, nodulus, and flocculus, is known as the **vestibulocerebellum**, because it has been known for some time that the evolutionary development of this region is connected with that of the vestibular organs. The inputs to the vestibulocerebellum are along **mossy** and **climbing fibres**. **Purkinje cell** axons are the sole efferents. Each climbing fibre excites its own group of Purkinje cells directly. Each mossy fibre makes initial contact with a granule cell which spreads the input to many Purkinje cells, each of which also receives inputs from other granule cells (see Figure 9.12). Both types of input also excite inhibitory interneurones, including **Golgi cells** and **basket cells**.

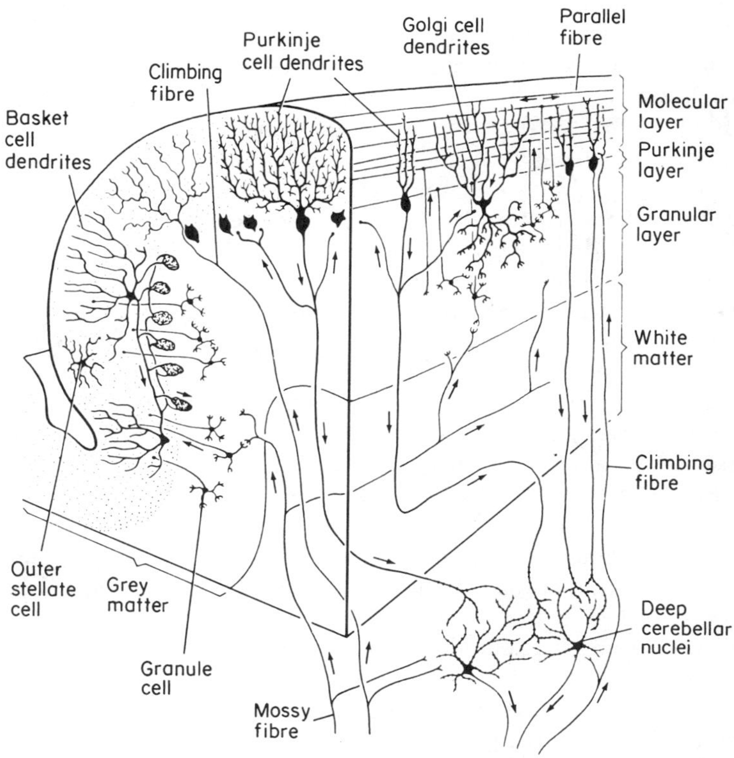

Figure 9.12 Diagrammatic representation of the connections of the principal types of neurones in the cerebellum, as described in the text (From Eccles *et al*., 1967. Reproduced by permission of Springer-Verlag New York, Inc.)

One major group of mossy fibres of the vestibulocerebellum of the monkey receives afferent signals directly from the vestibular organs and also indirectly through the vestibular nuclei. The firing rate of these **vestibular mossy fibres** is modulated in phase with rotations of the head. The firing rate of some of these cells is also related to the position of the eyes in their sockets, and the firing of some of them is inhibited during saccades. A second major group of mossy fibres, **eye-movement mossy fibres**, have been shown, in the monkey, to fire with a burst before and/or during saccades (burst types), at a steady rate determined by the position of gaze (tonic type), or with both a burst and tonically (burst-tonic type). These eye-movement mossy fibres are not modulated by changes in head position. During sinusoidal eye movements the firing rate of tonic and burst-tonic types is modulated sinusoidally with a phase lag of about 40° (Lisberger and Fuchs, 1978b). Recent electrophysiological evidence from the cat suggests that these cells receive signals directly from cells with similar response properties in the median and paramedian pontine reticular formation (Nakao *et al.,* 1980).

Climbing fibres, at least in the rabbit and cat, have been found to respond to signals arising from specific regions of the retina. These signals are conveyed through the accessory optic tract by way of the inferior olivary nucleus (Maekawa and Simpson, 1973; Simpson *et al.*, 1974; A. M. Graybiel, 1974; Baker, Gibson, Glickstein and Stein, 1976). The cat's cerebellar **vermis,** which is associated with the vestibulocerebellum, also receives visual inputs, which in this case are relayed through the superior colliculus and pons (Mower *et al.,* 1980). The vestibulocerebellum therefore receives information about the motion of the head, movements of the eyes, and movements of the retinal image. In other words, it receives all the information needed to assess the adequacy of vestibulo-ocular reflexes. These fuctional relationships are depicted in Figure 9.13.

On the output side of the vestibulocerebellum, electrophysiological recordings from Purkinje cells of the flocculus have revealed two distinct types of discharge from each cell. The first type, known as **simple spikes**, represents activity arising from mossy fibres, and the second type, known as **complex spikes**, represents activity arising from climbing fibres (see Pellionisz and Llinás, 1977, for a model of how these spikes may be generated). There is a steady rate of discharge of simple spikes in the absence of visual or vestibular stimuli. This steady discharge is modulated when the head rotates or when the eyes rotate. The firing rate of most Purkinje cells increases in phase with increases in the velocity of ipsilateral eye or head rotation and decreases in phase with increases in the velocity of contralateral eye or head rotation. An increase in the simple spike firing rate inhibits the primary ipsilateral vestibular inputs as they impinge on secondary neurones in the vestibular nucleus, and thus prevents vestibular signals from reaching the oculomotor system. A decrease in simple spike firing rate facilitates vestibular inputs. The pattern of in-phase and out-of-phase activity of complex spikes seems in general to be the converse of that of simple spikes (Ito, 1976), but the significance of these relationships is obscure.

During the slow phase of vestibular nystagmus the head rotates one way and the eyes move the other way at more or less the same velocity. There is thus an in-

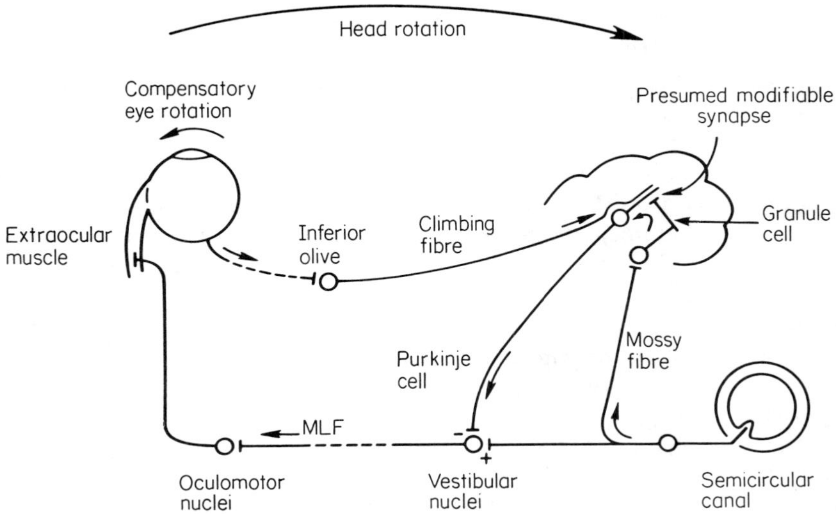

Figure 9.13 A simplified diagram emphasizing the flocculus as an inhibitory side-loop of the main vestibulo-ocular pathway, receiving mossy fibre – granule cell vestibular input and sending inhibitory fibres to the vestibular nuclei through its Purkinje cell output. In Ito's model, visual inputs through the inferior olive and climbing fibres indicate an error in the slow phase of nystagmus, and cause a gradual change in the vestibular-induced inhibitory control of the vestibular nucleus, which in turn adjusts the gain of the slow phase of vestibular nystagmus until it is again adequate (From Miles, 1977. Reproduced by permission of Plenum Publishing Corporation)

phase modulation of the simple spike discharge by the head rotation and an out-of-phase modulation by the eye rotation. The out-of-phase and in-phase modulations of the Purkinje cells therefore cancel (Lisberger and Fuchs, 1974, 1978a; Miles, Fuller, Braitman and Dow, 1980). There has been some debate as to whether, under these circumstances, there is a slight preponderance of in-phase (inhibitory) or out-of-phase (excitatory) signals passing from the flocculus to the vestibular nucleus (see Robinson, 1977; Dufossé *et al.*, 1978). But in any case, under these conditions the vestibular signals are allowed to pass through to the oculomotor system and initiate nystagmus.

When vestibular nystagmus is inhibited because the gaze remains fixated on an object which rotates with the head, the modulated output of the vestibulocerebellum, which is in phase with the head rotation, acts with full force on the secondary neurones of the vestibular nucleus and inhibits the primary ipsilateral vestibular output (Wolfe, 1969; Baker *et al.*, 1972b). Keller and Daniels (1975) and Waespe and Henn (1978a) found this inhibition of the vestibular output to be incomplete, even when the nystagmus was completely absent. Either this incompletely inhibited activity of the vestibular nucleus is not an appropriate signal for nystagmus, or there is a further stage at which the residual output of the vestibular nucleus is inhibited. But there is no doubt that fixation on a large visual display which rotates with the subject may completely inhibit vestibular nystagmus (Miles, 1977; Lisberger and Fuchs, 1978a). Monkeys and men with cerebellar lesions are not able to inhibit vestibular nystagmus visually (Takemori and Cohen, 1974; Dichgans *et al.*, 1978).

The final output of the flocculus component of the vestibulocerebellum is related to the algebraic sum of the head and eye velocities (Lisberger and Fuchs, 1978a). The output of this part of the system does not seem to be modulated directly by retinal-image velocity but is apparently part of a mechanism for controlling the velocity of eye movements in pursuit of objects which move at various velocities while the head is moving. The output of Purkinje cells in the vermal region of the vestibulocerebellum is related to the algebraic sum of the velocity of pursuit eye movements and the velocity of retinal-image motion (Kase *et al.,* 1979). This part of the system seems to be a mechanism for assessing the velocity of target motion relative to the head. It is odd, in this context, that monkeys in which the vermis was ablated have been reported to be still capable of modifying their vestibular nystagmus in the presence of discrepant visual inputs (Blair and Gavin, 1979b). Another possible function of the vestibulocerebellum is that it could be involved in correcting for any mismatch between the step and pulse components of saccades (see Section 6.4.4). It is also possible that the vestibulocerebellum is involved in adjusting the gain of vestibular nystagmus to the distance of the visual object being fixated. I showed in Section 9.5.4 that the gain required to stabilize the image of a near object is greater than the gain required to stabilize the image of a far object. Miles, Fuller, Braitman and Dow (1980) found cells in the flocculus of the monkey which responded to changes in the convergence of the eyes, so that the information required for adjusting the gain of nystagmus to the distance of objects is available within the vestibulocerebellum.

Ito (1972, 1976) suggested, on the basis of an idea put forward by Marr (1969), that the adequacy of the vestibulo-ocular reflex is assessed by the vestibulocerebellum in terms of the information about image motion which it receives along the climbing fibres. He also suggested that, if the velocity of the slow phase of vestibular nystagmus is inadequate for stabilization of the retinal image, the balance of in-phase and out-of-phase simple spike activity is shifted as a result of modification of synaptic connections within the vestibulocerebellum. It was further suggested that this shifted balance of simple spike activity plays upon the vestibular inputs as they reach the vestibular nucleus in such a way as to modify appropriately the gain of the head movement to eye movement system.

Evidence in support of the involvement of the vestibulocerebellum in the adaptive control of vestibular nystagmus is provided by the fact that this adaptive control is lost in the cat when this part of the cerebellum is ablated (Robinson, 1976). The importance of the accessory visual input through the inferior olive is indicated by the loss of adaptive plasticity after removal of the olive in the cat (Haddad *et al.,* 1980).

Recently, it has been shown electrophysiologically that the response characteristics of single cells in the vestibulocerebellum are changed as a result of the recalibration of the vestibulo-ocular system. Dufossé *et al.* (1978) recorded changes in simple spike activity from Purkinje cells in a rabbit and found that these changes were related to the behavioural changes induced by exposing the animal to anomalous visual–vestibular inputs. This does not prove that the site of the adaptive changes is in the vestibulocerebellum, because these recorded changes in the response of cells may merely reflect adaptive changes occurring at some other site, which funnel through the vestibulocerebellum and into the vestibular nuclei. What is required to prove the involvement of the cerebellum is the simultaneous recording of inputs and outputs to the cerebellum before and after adaptation.

Keller and Precht (1979a) recorded changes in cells of the vestibular nucleus of the cat after it had been subjected to a two-thirds reduction in the gain of its vestibulo-ocular reflex. The resting firing rates of cells were not altered, but the gain and phase relationships between head velocity and the modulation of the firing rates of cells were found to have been modified. However, these modifications of response characteristics showed only in certain cells, particularly in cells which normally have a high gain, and the modification was not large enough to account for the behavioural changes. Keller and Precht suggested that the floccular mechanism proposed to account for vestibulo-ocular recalibration may have to be modified or supplemented by adaptive changes outside the cerebellum. Miles, Braitman and Dow (1980) recorded changes in the flocculus of the monkey which were associated with increases and with decreases in the gain of the vestibulo-ocular reflex. However, they found that the change in gain in an animal adapted to anomalous visual inputs was evident in eye movements induced at short latencies after a passive rotation of the head before any changes were evident in the discharges of Purkinje cells leaving the flocculus. They also concluded that the floccular mechanism is not the primary site of the changes underlying adaptation of the vestibulo-ocular reflex. To finally settle this matter we need simultaneous recordings from cells entering the flocculus and from cells which leave it, before and after adaptation. One possible clue to the site of adaptive changes is the fact that dark-reared cats have been found to be incapable to modifying the gain of the vestibulo-ocular response (Harris and Cynader, 1981).

9.5.6 *Compensation for vestibular loss*

A final type of long-term change in vestibular responses is shown by patients recovering from unilateral vestibular damage. The initial reaction to unilateral damage is a spontaneous nystagmus with the quick phase towards the normal side, postural instability, a tendency to hold the head tilted towards the intact side, and nausea. A vertical line may appear tilted one way or the other depending on the type and extent of the unilateral damage (Friedmann, 1971). These symptoms result from the imbalance of the push–pull mechanism in the vestibular nuclei that was described in Section 8.2.2. In a patient with unilateral vestibular loss, the resting discharge from each canal of the intact labyrinth is not cancelled in the vestibular nuclei by the corresponding input from the opposite side of the head. The resulting difference signal when the head is stationary is thus equivalent to that normally produced by a rotation of the head in a direction which increases the inputs from the intact canals relative to those from the other side. In the case of the horizontal canal, this is equivalent to rotation of the head about a vertical axis towards the intact labyrinth. It is this imbalanced input that induces the spontaneous nystagmus. In the case of the vertical canals, considered as a pair, the imbalanced input is equivalent to that produced by tilting the head towards the intact side. This should induce torsional nystagmus, although I have not seen any mention of that symptom of unilateral vestibular damage. The loss of the otolith organs on one side is presumably responsible for the apparent tilt of vertical lines. But the otolith imbalance also induces a deviation of the eyes to the affected side. If the eyes are allowed to respond to this induced deviation, the spontaneous canal-driven nystagmus is reduced in intensity, but if the eyes are forced in the other direction, the otolith inputs enhance the canal-driven nystagmus (see Section 9.2.7). That is one reason why patients tilt the head towards the intact side and sleep on their sound side (Fluur, 1973a, b), but another reason is that a sideways tilt of the head is induced by the otolith imbalance (Hallpike, 1975).

The simplest view of recovery from unilateral vestibular loss is that it is due to a reappearance of spontaneous activity in neurones of the vestibular nucleus which are normally driven by the damaged canals. Such a reappearance of activity has been recorded in cats which have recovered from hemilabyrinthectomy (Precht *et al.,* 1966). But the more fundamental question concerns the origin of the activity in the deafferented vestibular neurones. One possibility is that these cells come under the control of excitatory impulses arriving from the intact labyrinth through commissural fibres from the vestibular nuclei on the intact side. However, spontaneous activity recovers on both sides in bilabyrinthectomised animals, so that this possibility seems to be ruled out (Ryu and McCabe, 1976). Another possibility is that adaptive changes occurring in the vestibulocerebellum are responsible for the restoration of vestibular balance. We saw in the last section that the cerebellum is involved in the restoration of other forms of imbalance in the visual–vestibular system. However, Haddad *et al.* (1977) found that spontaneous nystagmus declined in cerebellectomized cats at much the same rate that it declined in normal cats, so that this possibility seems to be ruled out also.

Recovery from unilateral hemilabyrinthectomy depends upon the integrity of other sensorimotor systems. For instance, recovery is delayed in immobilized baboons (Lacour, *et al.*, 1976). If a guinea pig is spinally transected after compensation has been acquired, the symptom of spontaneous nystagmus returns (Azzena, 1969). It is not clear whether these effects are due to the absence of motor efference or to the reduction in sensory inputs that motor inactivity entails. But in any case, vision is necessary for compensation, because cats kept in the dark after hemilabyrinthectomy do not recover postural or nystagmic stability, whereas cats kept in the light show significant recovery in a few days (Putkonen *et al.*, 1977; Courjon *et al.*, 1977). Furthermore, blindfolding an animal causes a reemergence of symptoms even after it has recovered (Schäfer and Meyer, 1974).

The source of recovery of the resting discharge of deafferented cells in the vestibular nuclei would therefore seem to be a potentiation of the normal effects that other modalities have on cells in the vestibular nuclei.

After both labyrinths are removed, the resting discharge from both sides is absent, which leaves the central push–pull system undisturbed. There is therefore no spontaneous nystagmus after bilabyrinthectomy, at least initially. However, there is a loss of muscular tonus, an increase in postural instability when the eyes are closed, and a gross inadequacy of compensatory eye movements during head movements. People with bilateral loss of vestibular function are easily disoriented when submerged in water (Padden, 1959) but they do not suffer from motion sickness. The disturbing symptoms of bilabyrinthectomy subside over a period of months. Dichgans, Bizzi, Morasso and Tagliasco, (1973, 1974) teased out two contributory factors related to this recovery process by recording nystagmus in monkey recovering from bilabyrinthectomy. The first factor was a recovery in the velocity of the slow phase of nystagmus induced by head turning. The slow-phase gain had returned to about 30% of its normal value in about two months. Such a recovery could be due to a potentiation of the normally weak reflex control of eye movements by joint receptors in the neck or to a potentiation of the effects of inputs from neck-joint receptors on the vestibular nuclei. The second factor was revealed when the monkeys moved their heads actively. The gain of the compensatory nystagmus was now higher than with passive movement, and Dichgans *et al.* found that the eye movement appeared when the intended head movement was unexpectedly prevented from occurring. This suggests that the eye movement was not reflexly triggered by the head movement but was preprogrammed along with the head movement. The monkeys had apparently learned to preprogramme these eye movements after being labyrinthectomized, since normal monkeys show no evidence of such behaviour.

Chapter 10

Visual Orientation to Gravity

10.1 Introduction

Gravity is the most pervasive and significant of all the features of man's environment to which he orients himself. Under normal conditions it is virtually constant, both in strength and direction. Humans have a mid-body axis which, in the standing posture, is kept in line with the direction of gravity; any disturbance is corrected by a complex of postural reflexes. Normally, the eyes are also oriented so that for each eye a particular retinal meridian is kept in line with the direction of gravity. This is referred to as the **normally vertical retinal**

meridian. When the head is in a normal erect position and the eyes are in their primary positions of gaze, the normally vertical retinal meridians are in their primary orientation with respect to the head. When the head tilts, the eyes counterrotate and hence deviate from their primary orientation in the head. They

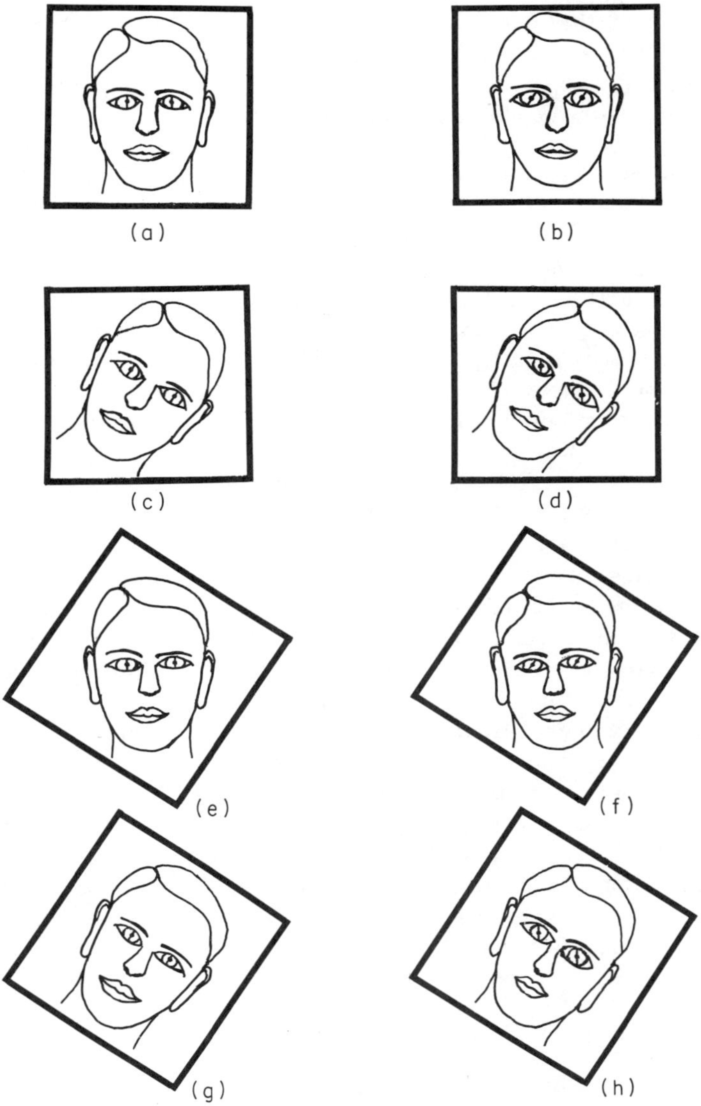

Figure 10.1 A classification of the relative positions of head, eyes, and visual frame. (a) Normal. (b) Eyes tilted, spontaneous torsion or cyclophoria. (c) Head and eyes tilted. (d) Head tilted, eyes erect, countertorsion. (e) Frame tilted. (f) Frame and eyes tilted, optostatic or optokinetic torsion. (g) Head, eyes, and frame tilted. (h) Head and frame tilted, eyes erect, countertorsion

also deviate from their primary orientation in response to one or other of the conditions listed in Table 5.3. Man lives in an environment of visible objects and surfaces which usually maintain a constant orientation to the direction of gravity. Such objects and surfaces constitute the **visual frame of reference for the vertical**. The first part of this chapter is about the precision and accuracy with which people judge the tilt of visual objects with respect to gravity when the body, each eye, and the visual framework are held in the vertical position. The second part is concerned with the effects produced by tilting the visual frame, and the third part deals with the effects produced by tilting the body.

The stimulus domain of interest in this chapter is the set of all possible combinations of body tilt, eye tilt (torsion), and visual-frame tilt and is schematically presented in Figure 10.1. The squares depict the visual frame. The orientation of the eyes is indicated by the lines through the pupils. All directions are with reference to gravity, which is considered to be parallel to the sides of the page. There are thus four stimulus factors: (1) body tilt, (2) eye torsion, (3) visible-frame tilt, and (4) the direction of gravity, but only three degrees of freedom since gravity is always considered fixed. Ignoring differences of extent and direction of tilt, there are eight possible states of the system. This is the basic plan of stimulus conditions in terms of which judgements of, and motor responses to gravity will be discussed.

The response domain covers any task which involves adjusting or estimating the orientation of an object with respect to the direction of gravity. The 'object' may be an external visible object, or the body of the subject, or part of his body.

Consider the task of adjusting or estimating the orientation of a visible line in a frontal plane with respect to the vertical when the orientation of the line, the eyes, and the body of the observer may vary independently, and where there is no visual frame of reference. The ideal perceiver for such a task is simply one which extracts the algebraic sum of the following three signed angles: (1) the deviation of the image of the line relative to the normally vertical retinal meridian, (2) the deviation of the eyes relative to their primary orientation in the head (torsion), (3) the deviation of the head and body relative to gravity.

If a visual framework, such as a room, which the perceiver knows is vertical in view, the task reduces to one of registering the angle between the line being judged and a vertical feature of the framework. If the visual framework is tilted when it is assumed to be vertical, illusions of orientation are to be expected and, as we shall see, such illusions do occur.

10.2 Judging visual tilt

10.2.1 The precision of visual-tilt judgments

Several investigators have reported on the precision with which a person is able to set a luminous line to the vertical. Reported average unsigned deviations from the true vertical are approximately 1° (Witkin and Asch, 1948; Mann *et al.*, 1949; Kaufmann *et al.*, 1949). Westheimer *et al.* (1976) found that on 75% of trials subjects could detect when a 0.5° line was as little as about 0.5° from vertical. Considering that the length of the line and other factors were not the same in the various studies, the results are in good agreement. Figures for the vertical and horizontal were found to be similar by Mann *et al.* There is a good deal of evidence that the vertical and horizontal positions of a line are more precisely judged and produced than other positions (Gibson and Radner, 1937; Volkmann and Pufall, 1972). The precision of estimates of the visual vertical is highest when the body is in its normal upright position and decreases with increasing tilt of the body (Miller, *et al.,* 1965). This decrease in precision is related to the decrease in the sensitivity of the utricles with increasing tilt of the body; it is particularly evident when the somaesthetic inputs are eliminated by immersing the subject in water (Lechner-Steinleitner and Schöne, 1980). Precision in setting a line to the vertical has been reported to improve up to the age of 10 years (Volkmann and Pufall, 1972).

10.2.2 The accuracy of visual-tilt judgments

The most extensive studies of the judgment of tilt have been carried out by Rogers, Volkmann, Reese, and Kaufman (see Reese, 1953, pp. 9–16). Each of 50 subjects was asked to estimate the tilt of stimuli distributed between 10° anticlockwise and 100° clockwise. The constant errors and variances were least at the vertical and horizontal positions and maximum at the 30, 75 and 100° positions, Intermediate constant errors and variances were obtained at the 45° position. Within-subject variances ranged from 2 to nearly 8°. The substitution of a 14 cm or a 4 cm line for the original 56 cm line did not affect the results.

When the method of production was used rather than the method of estimation, the constant and variable errors were smaller but the pattern of errors for the various inclinations was similar for the two methods. Muller *et al.* (1955), using an entire circular range of stimulus lines, found that identification of tilt was similar in each quadrant. They also found that accuracy was highest at the horizontal and vertical positions, but unlike Rogers *et al.* they did not find a point of intermediate accuracy at the mid-quadrant position. Keene (1963) has shown that the response categories allowed in judgments of radial position affect the distribution of errors within a quadrant. Three response scales were used: degrees, a letter–number code which had a category break at 90°, and a letter–number code with breaks at 60 and 120°. It was hypothesized that accuracy would centre round category breaks due to the imposed salience of these particular responses in the response series. The result in each case showed minimum errors about the natural vertical and horizontal points, demonstrating that these points have a natural salience. Although the 60–120° anchor scale did not give secondary-error minima at these within-quadrant points, it did interfere with the secondary minimum which otherwise occurred at the 45° position. In other words, the normal vertical–horizontal categories exert their influence in spite of imposed category breaks, even though imposed categories may affect the within-quadrant distribution of error magnitudes.

It has been reported that the apparent vertical is displaced to the right when the right eye is used and to the left when the left eye is used (Fischer, 1927; Schneider, 1966; Wade, 1969b). However, for a stimulus line in a given location, the direction of gaze is not the same for each eye and it may have been this difference in the direction gaze that caused the difference in constant errors, rather than the eye used. Direction of gaze has an effect, as we shall now see.

10.2.3 The effects of angle of gaze on judgments of visual tilt

When an eye moves into an oblique position of gaze the image of a vertical line no longer falls on the normally vertical meridian of the retina. This deviation is due to the fact that the eyes rotate according to Listing's law (Section 5.2). If the rotation of an eye about a horizontal axis is θ and about a vertical axis ϕ, then the orientation Ψ of the normally vertical meridian of the eye with respect to the vertical is given by:

$$\Psi = \sin^{-1} \frac{\sin\theta \sin\phi}{(1 + \cos\theta \cos\phi)}. \text{(Robinson, 1963)}$$

In the Fick axis system, angle Ψ is equivalent to a torsion of the eye about the visual axis. In Figure 10.2 the eye is fixated in an oblique angle of gaze on a point

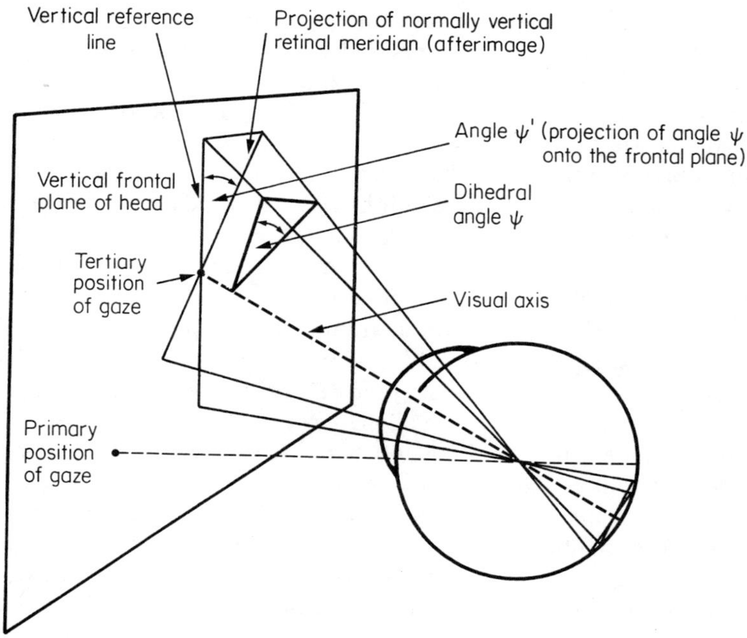

Figure 10.2 Showing how an afterimage imposed on the normally vertical meridian of the eye is projected at an angle Ψ' to the objective vertical when the eye is in a tertiary position of gaze. The angle Ψ' is the projection of the dihedral angle Ψ, contained in the frontal plane of the eye, onto the frontal plane of the head

on a screen erected in a frontal plane relative to the head of the observer. A vertical reference line is drawn through the fixation point, and an afterimage of a line, impressed on the normally vertical meridian of the retina, is projected on the screen through the same point. The angle Ψ is the dihedral angle between two planes; the first plane is the vertical plane through the centre of the eye and the vertical reference line, and the second plane is that through the centre of the eye and the afterimage. This dihedral angle is contained in a frontal plane of the eye, which is inclined with respect to the frontal plane of the head when the eye is not in its primary position of gaze. The angle Ψ' between the afterimage projected onto the vertical screen and the vertical reference line on the screen may be calculated. It is the projection of dihedral angle Ψ onto the vertical screen (Nakayama and Balliet, 1977).

By asking subjects to align a reference line to the projected afterimage impressed on the normally vertical meridian of the eye, several investigators have confirmed that the eyes obey Listing's law (Quereau, 1954). If a line must fall on the normally vertical retinal meridian in order to be perceived as vertical within a frontal plane, then angle Ψ' is the predicted deviation of the apparent vertical for an oblique angle of gaze. The question is whether vertical lines appear to tilt in this way when viewed obliquely or whether the vertical is judged veridically. The ability to judge the vertical veridically when the normally vertical meridian is tilted is referred to as **orientation constancy**.

Nakayama and Balliet (1977) presented subjects with a luminous line seen monocularly in dark surroundings at various positions in a frontal plane. The subjects fixated the centre of the line and set it to the apparent vertical. Figure 10.3 shows the mean setting at each position, together with the orientation of the projection of the normally vertical meridian of the eye at the same positions. It can be seen that for oblique angles of gaze the apparent vertical is a compromise between a veridical setting and the projection of the eye's vertical meridian. Thus, orientation constancy breaks down when only a single line in dark surroundings is in view. However, judgments of the vertical for oblique gaze were found to be veridical when the room lights were left on, so that under normal viewing conditions orientation constancy is presumably maintained by the simple process of referring objects to the visual frame of reference.

With binocular viewing the situation is more complex, because the two eyes view a given line with different angles of gaze. This means that the orientation of any line (except a horizontal line or one lying along the vertical horopter) with respect to a particular retinal meridian of one eye will differ from its orientation with respect to the corresponding meridian of the other eye. Any such disparity in the orientations of binocular images constitutes a **cyclodisparity**. The situation is complicated by the fact that the eyes may rotate about the visual axes so as to nullify a cyclodisparity (see Table 5.3). If we ignore this possibility, it can be stated that the degree of cyclodisparity produced by a line lying straight ahead in the median plane of the head with the eyes fixated on the centre of the line is a function of the angle of inclination between the line and the vertical horopter at that viewing distance. It was shown in Section 7.2 that the vertical horopter is

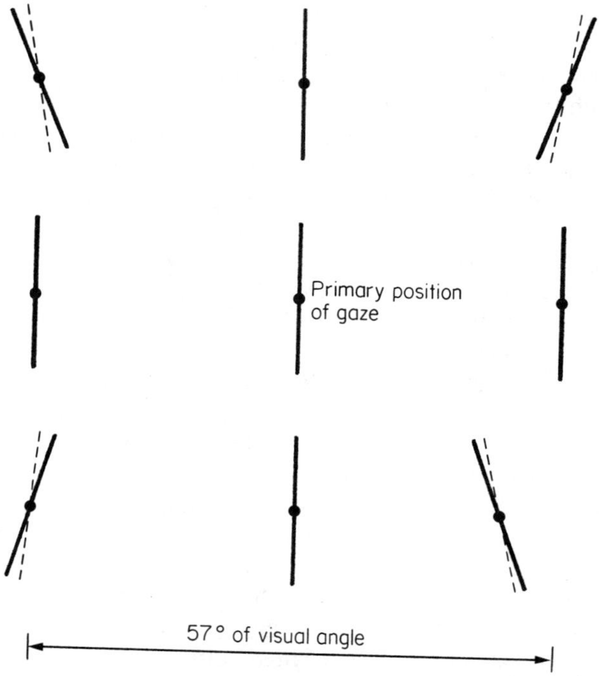

Figure 10.3 An afterimage of a vertical line is impressed on the eye while the eye is in its primary position of gaze. Each solid line shows the orientation of this afterimage on a frontal plane of the head for that position of gaze. The apparent vertical corresponds to the direction of the normally vertical meridian, except for tertiary (oblique) positions of gaze. At these positions the apparent vertical is represented by dashed lines and is a compromise between the normally vertical meridian and the true vertical. (Reprinted with permission from K. Nakayama and R. Balliet, *Vision Research*, **17**. Copyright 1977, Pergamon Press, Ltd.)

inclined within the median plane of the head at an angle that increases with increasing viewing distance. In spite of this inclination of the vertical horopter, it has been found that, on average, people correctly estimate the vertical position of a line within the median plane (Ebenholtz and Paap, 1973; Cogan, 1979).

Another directional judgment related to gravity is the estimation of the elevation of the visual horizon. The literature on this topic is reviewed in Howard and Templeton (1966, p. 183), and some relevant studies are mentioned in Section 10.4.8.

10.3 The effects of tilting the visual frame of reference on judgments of the visual vertical

10.3.1 Introduction

The tilt contrast and normalization effects discussed in Sections 4.5 and 4.6 are rarely greater than 2 or 3°, and do not depend on the meaning or significance of the patterns which induce them. It has been assumed that these effects are due to a direct effect of visual stimuli on orientation detectors at a basic or early stage in the processing of visual information. However, under certain conditions, the apparent visual vertical can be displaced through angles much larger than 3°. For instance, a person in an aeroplane perceives the inside of the plane to be upright even when it is tilted through a considerable angle. A pilot flying in a cloud may not even be aware that his plane is upside-down, and it is a common experience of sailors that a water surface or plumb line seen in the cabin of a listing boat seems to be slanted at a large angle (Dixon and Dixon, 1966). Wertheimer (1912) noticed that a room seen in a tilted mirror soon came to appear upright, and it has been reported that people living in buildings tilted 8° by an earthquake perceive the rooms as vertical (Kitahara and Uno, 1967).

All these phenomena have in common the presence of a tilted visual scene which is normally upright. A familiar, normally vertical scene is said to constitute a **frame of reference** for the visual perception of the vertical. A frame of reference may be defined in general as *an attribute of certain objects which does not normally vary, and in terms of which variations of the same attribute in other objects perceived at the same time are judged.* A frame of reference is not the same as a norm. A norm is a central point, or point of symmetry, in an oppositional sensory scale. The sides of a vertical object exhibit the norm in the stimulus dimension of tilt, but they constitute a frame of reference only if the object is one which the subject assumes is normally vertical.

The walls and floor of a room are normally vertical and horizontal, and the orientations of other objects in the room are judged with respect to that of the walls and floor. Whereas the tilt contrast effect discussed in Section 4.6 is small and limited to the location of the inducing figure, the effect of a frame of reference is larger and not necessarily limited to one location in the visual field. The effect produced by a tilted frame is like changing the zero point on a measurement scale, which leaves all relative values unaffected, while the tilt adaptation effect is like compressing one part of a scale relative to the rest. The tilt adaptation effect outlasts the presence of the inducing figure for several seconds. There are conflicting reports about whether frame of reference effects outlast the presence of the frame. Morant and Aronoff (1966) claimed that they do not, but Austin *et al.* (1969) and Wallace, *et al.* (1975) claimed that they do.

10.3.2 The rod-and-frame test

A quantitative study of the effect of a tilted frame of reference on the apparent vertical was conducted by Asch and Witkin (1948). The subject stood outside a model room which was tilted by 22° and, after looking into it for four minutes, set a rod seen inside the tilted room to appear vertical with respect to the normal room in which he was standing. In one condition, this was done with the subject viewing the tilted room through a tube which excluded the normal upright room from view. In this case, the rod was displaced on average 15° in the direction of the tilted room. In the second condition, settings were done without a tube so that the tilted room was seen in the context of the normal room. The rod was now displaced only 8.5° on average. Thus large displacements of the visual vertical are found when the visual field is filled with a tilted scene containing familiar objects, such as the walls of a room, which are normally vertical. However, a familiar scene is not necessary for the production of these effects. Witkin and Asch (1948) used a luminous square frame with 1 m sides, which was tilted about its centre 28° to the left or right in the subject's frontal plane. A 97 cm rod was pivoted about the same centre and the subject had to judge when it was vertical. Nothing else was in view. The mean constant error was 6° in the direction of the tilted frame, but there were wide variations between the performances of the various subjects. Some subjects placed the rod nearly parallel to one or other side of the frame even though they were instructed to set it to the gravitational vertical. Other subjects were little affected by the frame. Witkin developed the **rod-and-frame test**, which is a standardized form of the rod-and-frame apparatus with a standardized scoring procedure. A person was said to be more **frame dependent** the more he set the rod in line with the tilted frame. Frame dependency was found to correlate with scores on a variety of perceptual and personality tests (see Witkin *et al.*, 1973, for a bibliography on this topic).

The rod-and-frame test has been used widely for a variety of research purposes, so that it is unfortunate that crucial factors such as the stability of the head (Lester, 1968), the starting position of the rod (Nyborg, 1974), the rotation speed of the rod (Corah, 1965), and the type of score used have usually not been controlled. All these factors have been shown to affect the result. It is particularly unfortunate that the test is often scored in terms of the unsigned deviation from the vertical. Such a measure confounds the signed mean error, or constant error, which is the item of prime interest, with variability, which is best measured by the unsigned mean of deviation scores or by the variance (Wenderoth, 1974; Fine and Danforth, 1975). Attempts to apply signal-detection procedures (Schuck *et al.*, 1970) have only added to the confusion, because this procedure is totally inappropriate for measuring constant errors.

Children of both sexes are affected more than adults by a tilted frame. By the age of 13–16 years sex differences begin to show: men are less frame-dependent than women (Nyborg, 1980). These sex differences may have something to do with differential experience, although the results of earlier experiments on the effects of experience and training on frame dependency were contradictory (see Howard and Templeton, 1966, p. 208).

One would expect the size of the frame to affect the magnitude of apparent displacement of the visual vertical. Ebenholtz (1977) showed that larger frames have a greater effect, and that it is the size of the retinal image of the frame that is important—not the actual size of the frame. When the side of the frame subtended less than 10° at the eye, the effect on the apparent tilt of the rod was found to be less than 2°, but the effect rose to 8° when the frame subtended 40°. One interpretation of these results is that a frame projected onto the periphery of the retina is a more effective frame of reference than one projected nearer the fovea. This would be compatible with the fact that rotating peripheral stimuli are particularly effective in inducing illusions of visual tilt and self motion

(Section 9.4.4). However, *experiments must be done with small frames at different eccentricities within the visual field before the effects of changing size can be dissociated from the effects of changing eccentricity.*

Gogel and Newton (1975) reported that a frame at the same apparent distance as a rod had more effect on the apparent tilt of the rod than a frame at some other distance.

10.3.3 Varying the tilt of the frame

A square frame has no polar axis and therefore, when tilted at 45°, may be regarded as an erect diamond. In this position it should not affect the apparent vertical position of an enclosed rod, and of course the frame should have no effect when tilted at 90°. Beh *et al.* (1971) and Beh and Wenderoth (1972) measured the effect of triangular, square and hexagonal frames tilted at various angles and found that the apparent vertical position of a rod enclosed within the frame tended to be displaced towards the nearest axis of symmetry of the frame. In a subsequent study, Wenderoth and Beh (1977) studied the effects of frames with various sides omitted, and again stressed the importance of axes of symmetry. The maximum displacement of the apparent vertical was only about 1.5° in these experiments, which is not surprising, because the frames subtended only about 7° of visual angle. These small effects may have more in common with tilt adaptation than with the effects of a frame of reference.

If the frame is a room, it has polar features (a ceiling and a floor) and therefore appears vertical only in one position. If the polar axis is the frame of reference for judgments of the vertical, one would expect the apparent vertical position of the rod to be displaced in the direction of tilt of the room for all angles of tilt up to 180°, that is, until the room is upside-down. Singer *et al.* (1970) confirmed that the addition of polar features (furniture) to an otherwise bare box tilted at 45° caused the rod to be displaced 17° in the direction of tilt of the box. In this experiment the frame subtended about 53° of visual angle which explains the large effect. When the bare frame was tilted at 67° the rod was displaced towards the nearest side of the frame, as one would expect. However, at this angle, even when the furniture was added, the rod was still displaced towards the nearest side of the frame rather than in the direction of the polar axis of the room. Purcell *et al.* (1978) found that, in the case of tilted images of outdoor scenes, the apparent vertical was displaced in the direction of the polar axis up to a tilt of about 75°, at which point the displacement was zero. Perhaps in all these experiments the frame was not very compelling; after all, it was only either a box or a projected scene. Morant and Beller (1965) plotted the displacement of the apparent vertical as a function of the tilt of a real three-dimensional scene. They found that the displacement increased up to a tilt of 75°, the largest angle tested. This is not surprising because Morant and Beller had subjects walk about in the real world while looking through rotating prisms, whereas the subjects in Purcell's studies sat passively in front of a tilted box or projected pictures. I suspect that when people are totally enclosed in a tilting room, they accept the room as their frame of reference at extreme angles of tilt. In some instances, an inverted scene may be accepted as upright. For instance, pilots in an inverted aircraft may not realize that they are inverted, although of course, in this case, the pilot is also upside-down.

10.3.4 Causal factors in the rod-and-frame effect

In several studies a mean displacement of the apparent vertical of about 8° in the direction of a 22° simple tilted frame has been reported; some subjects showed greater effects, some less. Up to about 2° of this may be accounted for in terms of a tilt contrast effect of the sides of the frame on the rod (see Goodenough, Oltman, Sigman, Rosso and Mertz, 1979, for evidence on this point). Other possible factors are (a) eye torsion induced by the tilting frame, (b) an apparent tilt of the body induced by the tilting frame, and (c) a cognitive set to use the tilted frame as the vertical reference.

Howard and Templeton (1964b) found no evidence of eye torsion when subjects inspected a tilted line which subtended 10° at the eye. However, more recently, Goodenough, Sigman Oltman Rosso and Mertz (1979) found torsional movements of the eyes of about 1° when subjects inspected a tilted visual framework. Kertesz and Sullivan (1978) found that cyclotorsion is induced to a measurable extent with cyclodisparities, but only with large visual displays. Balliet and Nakayama (1978b) were able to train human subjects to make voluntary torsional movements of the eyes of up to 30° in total amplitude. Balliet and Nakayama (1978a) reported that the orientation of a truly vertical luminous line was apparently displaced in the opposite direction to voluntarily induced eye torsion in much the same way that it was apparently displaced by tilting the head (Section 10.4.2). They suggested that signals associated with eye torsion and signals from the vestibular canals associated with head tilt share common circuits in the vestibular nuclei (Section 9.4.5). However, normal optostatic torsion would not account for more than about 1° of deviation of the apparent vertical; we are still left with a mean effect of about 7° of visually-induced deviation of the apparent vertical to account for.

It has been suggested on several occasions that inspection of a tilted frame induces an apparent tilt of the body in the opposite direction. This apparent tilt of the body would then induce an apparent displacement of the rod. Sigman *et al.* (1979) found that the apparent vertical position of a rod was displaced on average 7° in the direction of a 28° tilted frame, and that the apparent vertical position of the body was displaced 2.5° in the same direction. There was a high correlation between the two settings. This suggests that part but not all of the effect of a tilted frame on the apparent orientation of a rod is due to a change in the apparent orientation of the body. In a similar study by Ebenholtz and Benzschawel (1977), the apparent tilt of the head in the presence of a tilting frame was measured. Here too, the effect, although in the right direction to account for the visual effect, was not large enough to account for it fully. If the instructions required subjects to set the rod in line with the body axis rather than in line with the direction of gravity, any effect due to an apparent tilt of the body should be eliminated. Sigman *et al.* conducted a second experiment in which they found that bodycentric instructions reduced the mean size of the effect of a 28° tilted frame on a rod by 2.4°. This is, as one would expect, about the same angle as the apparent tilt of the body which had been found in their first experiment.

It looks as though tilt adaptation, eye torsion, and apparent body tilt all contribute something to, but do not fully account for, the effect of a tilted visual frame on the apparent tilt of a rod. The other factor is presumably the subject's inability to dissociate non-visual impressions of the gravitational vertical from the cues provided by the visual framework. Even at the level of conscious knowledge, most adults do not have a clear concept of the horizontal and vertical, as we shall see in the next section.

10.3.5 The percept and concept of horizontal

The best-known indicator of whether a person has a clear understanding of the concept of horizontality is Piaget's water-level test. In this test, subjects are asked to draw the water surface in each of a set of pictures of a bottle tilted at various angles, and to point out the correct ones from among pictures of a bottle with water in it when the water and bottle are at various angles. Piaget and Inhelder (1956) reported that only children over eight years of age revealed by their performance on these tests that they knew that water remains horizontal in a tilted vessel. It was argued that perceptual experience alone does not lead to the development of the correct principle. Smedslund (1963) confirmed these findings using similar methods.

It would seem that children do not know what the orientation of the surface of a liquid in a tilted vessel should be and are not able to recognize the correct orientation. But these conclusions are based on faulty experimental procedures. In the first place, pictures of bottles may not evoke the correct percept or concept of the level of the liquid. The children may have regarded them as tilted pictures of upright bottles rather than pictures of tilted bottles, which would, of course, make a difference to where the water surface would be placed, even for a person who had the correct principle. Secondly, half of the test involved drawing in the water level. It is known that when children are asked to copy an abstract pattern, they make intersecting lines more perpendicular than those in the models (Ibbotson and Bryant, 1976). In any case, a 'production' test is never a valid procedure for diagnosing percepts or concepts. For instance, children cannot draw in perspective, but this does not prove that they do not perceive the world in terms of perspective. Young children often draw a 'man' with no body, but they would be very surprised to see such a creature in real life.

Several investigators have reported that a high proportion of adults fail Piaget's water-level test, and concluded that many adults also lack the appropriate percept and concept (Rebelsky, 1964; Thomas *et al.*, 1973). However, production methods were used in these investigations also so that the percept may not have been evoked. Ordinary experience of liquid in tilting vessels may lead people to recognize when the surface of a liquid is correctly oriented in a life-like situation, even though they may not be able to express the principle in words, or make correct judgments in less life-like situations.

I conducted an experiment to check on these possibilities (Howard, 1978b). The subjects saw a full-size, three-dimensional coloured stereoscopic picture, like that shown in Figure 10.4. The liquid was presented at various angles in the jug, which was tilted at 45°. Whenever one of these pictures was presented, the subject was asked to say whether there was anything wrong with the liquid. It was found that about 50% of the sample of college students of both sexes accepted pictures in which the liquid was on average anywhere within 15° either way of its correct orientation. On further questioning it was found that none of these subjects had any knowledge of the principle that liquid remains horizontal in a tilted vessel, whereas all the other students knew the principle and none of them made an error of more than 5°. The experiment was repeated using cinematographic sequences in which the jug was seen while being tilted and with the liquid moving either correctly or at various incorrect rates of incline. The results were the same. About half of students tested did not know the water-level principle and had only the

Figure 10.4 One of the scenes used by Howard (1978b) to investigate the ability of people to recognize the correct orientation of the surface of a liquid in a tilted vessel. The actual scenes were coloured, full size and stereoscopic. This orientation of the surface was one of those which was accepted as correct by subjects who did not know that water remains horizontal (Reproduced by permission of Pion Ltd.)

crudest percept about how liquid should look in a tilted vessel. Only those subjects who did know the principle had an adequate percept. One subject remembered the correct principle halfway through the testing procedure. Up to that point he behaved like those adults who did not know the principle, but after he remembered he behaved like those subjects who knew it. When forced to select the best scene or when asked to draw in the water level on a picture of the tilted jug, even subjects who did not know the principle performed correctly on average. In other words, these subjects passed Piaget's test. The imprecision of their percept was revealed only when they were asked to give an opinion about each scene, a procedure which measured the range of tolerance of their percept.

Their poor performance was not due to the effect of the sides of the tilted jug on the apparent horizontal, because the rest of the scene provided adequate reference to the true horizontal, and in any case, if this had been the cause of the difficulty, it would have shown in subjects who knew the principle. Furthermore, subjects who did not know the principle performed adequately after verbal instruction. Visual illusions of tilt are largely immune to the effects of knowledge.

Human beings certainly develop some accurate and precise percepts before they acquire conscious knowledge of the principles involved. For example, people make due allowance for distance when judging the sizes of objects, although the principle of perspective was not correctly stated before the fifteenth century. The perceptual schema for the water-level principle is represented in only a very crude form in people who do not know the principle.

10.3.6 The effects of rotating backgrounds on the visual vertical

In Section 9.4.4 It was shown that a large visual display in a frontal plane rotating about the visual axis induces a sensation of body tilt and body rotation in the opposite direction. It was also mentioned that such a display causes an apparent angular displacement of a stationary vertical line, also in the opposite direction. It is this latter effect which is of interest in the context of this section.

The effects of background rotation on the apparent visual vertical were first reported by Brecher (1934), and more recently by Bischof and Scheerer (1970). Hughes *et al.* (1972) found that the displacement of the apparent vertical position of a line in the direction of background rotation increased to a maximum value of about 3° as the angular velocity of the background was increased up to about 36°/sec. However, the background in this experiment subtended a visual angle of only 19°, which is rather small for an effect which is probably induced more by peripheral visual stimulation than by central stimulation. Dichgans *et al.* (1972) used as background a rotating display of random dots which subtended 130° of visual angle, and found that the apparent vertical position of a centrally placed edge was displaced a maximum mean value of 15° as the angular velocity of the background was increased to about 30°/sec (see Figure 10.5). At higher angular velocities the effect remained more or less constant. At each velocity the apparent tilt of the edge reached its steady level after a period of about 18 sec and returned to its normal position in about the same period after the rotation was stopped.

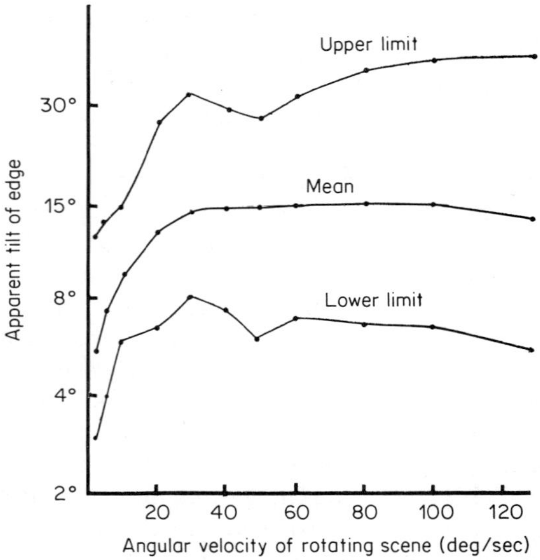

Figure 10.5 Apparent tilt of a stationary vertical edge as a function of angular velocity of a surrounding pattern of random dots. The curve represents the mean results for seven subjects (From Dichgans *et al.*, 1972. Copyright, 1972, by the American Association for the Advancement of Science)

The drift in the apparent angular position of the vertical edge was accompanied by a shift in the apparent angular position of the observer's body and by paradoxical sensations of continuous rotation of the body and of the visual line. In a later study, Held *et al.* (1975) found that an annulus-shaped display of a given area rotating in the frontal plane about the fixation point was more effective when it stimulated more peripheral regions of the retina. This confirms once more the importance of the peripheral retina for mechanisms of egocentric orientation.

The background display used by Dichgans and his associates consisted of a random pattern which contained no visual cues to the vertical. These induced shifts in the visual vertical are therefore not produced by tilted scenes or frames. Perhaps the best way to think of the motion-induced effects is in terms of mechanisms of postural stability. It is well known that postural sway is reduced when a stable visual scene is in view (Section 9.4.6). This stabilizing effect of vision is partly due to the visual cues to the vertical that most natural scenes contain, but it is also due to the motions of the retinal image of the scene induced by body sway. In other words, rotation of the images of visual scenes is a natural signal that signifies body sway. The importance of visual motion is illustrated by the fact that infants fall over as soon as the visual scene is moved (Section 9.4.6), and by the fact that the stabilizing effect of a visual scene is reduced if it is flashed on and off at rates below 6 Hz, that is, at rates where visual motion signals become ineffective (Amblard and Cremieux, 1976).

Normally, the body sways at amplitudes of less than 2°, so that one would suppose that visually-induced tilt would occur for background rotations of this amplitude. Mauritz *et al.* (1977), using sinusoidally rotating backgrounds, found that the induced visual tilt for this type of motion is a function of the amplitude of the oscillation rather than of its angular velocity or its frequency of oscillation. However, an oscillation amplitude of 5–8° was required to produce any effect at all. They explained this high threshold in relation to the amplitude of natural body sway by the fact that natural body sway is accompanied by vestibular stimulation, which is lacking in visually simulated sway. It is presumably this absence of vestibular stimulation in visually simulated sway that also accounts for the long latency of visually-induced apparent tilt. Insofar as inputs from the utricles act as a constraint on the magnitude of apparent displacements of the visual vertical, one would expect that the effectiveness of this constraint to be less when the head is tilted to one side into a position where the utricles are known to be less sensitive. In conformity with this prediction, the apparent displacement of the visual vertical has been found to be larger when the head is tilted in this way (Dichgans, Diener, and Brandt, 1974).

A visual display rotating about the visual axis induces a shift in the torsional position of the eyes of about 3° in the same direction (Section 6.1.3). If this torsional movement were unregistered it could account for an apparent displacement of the visual vertical of this amount. Torsion may well account for the small displacement of the visual vertical observed with small rotating displays. There is no evidence on this question. However, eye torsion could account for only part of the displacement of up to 15° observed with large rotating displays. The only evidence on this point is from a study by Finke and Held (1978), who found that the magnitude of apparent visual tilt was not correlated with the amplitude of torsion induced by a large rotating display.

10.4 The effects of body tilt on the visual vertical

10.4.1 The Aubert (A) and Müller (E) effects

Aubert (1861) happened one day to be looking at a vertical streak of light in an otherwise dark room and noticed that when he tilted his head to one side the streak of light appeared to tilt in the opposite direction. The room had to be completely dark; if the furniture became visible the effect disappeared. When darkness was restored, the effect returned after a few seconds' delay. The effect was equally strong with a horizontal line of light and did not depend on constant fixation.

Müller (1916) pointed out that with small tilts of the head a vertical line often appears to be tilted in the same direction as the head, rather than in the opposite direction, as in the Aubert phenomenon. Müller called his effect the **E-effect** and the Aubert effect the **A-effect**. I shall sometimes describe the A-effect as an apparent tilt of a truly vertical line in the opposite direction to the head tilt, and sometimes as a tilt of the apparent vertical in the same direction as the head tilt, depending on the technique of measurement used in the particular experiment being described. Similarly, the E-effect will be described either as an apparent tilt of the true vertical in the same direction as the head or as a tilt of the apparent vertical in the opposite direction to the head. Müller found that most people experience the E-effect with small angles of head tilt and the A-effect with larger angles, while some people experience only the A-effect. The E-effect tends to reach a maximum value of between 3 and 12° at a tilt of the whole body of between 30 and 40°. At greater angles of body tilt the E-effect declines and reaches zero at an inclination of about 60°. With further tilt of the whole body up to 180°, the A-effect increases, reaching a maximum value of up to 15° (Bauermeister, 1964; Miller *et al.*, 1965; Wade, 1969a). A sample of data is shown in Figure 10.6.

The A-effect is what one would expect if a person underestimated the extent of the head tilt, and the E-effect is what one would expect from an overestimation of the head tilt. The A- and E-effects normally reported would result if large tilts of the head were underestimated and small tilts overestimated. It can be seen from Figure 10.6 that the function relating the magnitude of the two effects to the degree of body tilt resembles a sine function. This prompted Schöne (1962) to suggest that the illusions are related to the shearing force acting on the utricular surface, which is also a sine function of body tilt. This cannot always be the case, however, because, as we shall see in Section 10.4.7, these illusions are present in people with loss of vestibular functions.

There are wide individual differences in the incidence and magnitude of A- and E-effects. One factor that may be important, and which has usually not been controlled, is the speed at which the subject tilts his head. Stockwell and Guedry (1970) reported that for small rapid tilts of the head the A-effect predominates and for small slow tilts the E-effect predominates.

The effects of body tilt on the visual vertical do not seem to be due to the subject being confused about what is meant by the gravitational vertical when the

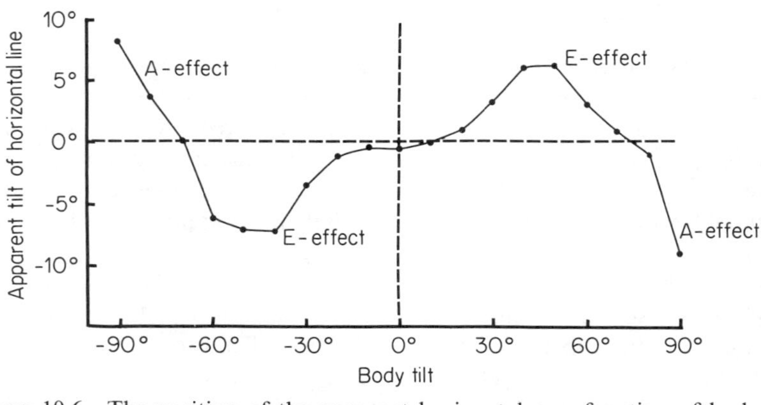

Figure 10.6 The position of the apparent horizontal as a function of body tilt. Means of three subjects (From a study by Miller *et al.*, 1965)

body is tilted, because Wade (1970a) obtained effects of about the same size with subjects who were carefully instructed about the meaning of the term 'vertical' as with subjects who were merely asked to adjust the line until it looked vertical. In another study, Wade (1970b) found that an E-effect was produced with a 30° head tilt whichever psychophysical method he used, although the magnitude of the E-effect was greater with some methods than with others.

One very important factor that has been almost completely overlooked in studies of the E- and A-effects is the length of the luminous line. Evidence is presented at many places in this book that stimuli which fall on the peripheral retina are assessed in terms of inputs from the vestibular and kinaesthetic system in such a way as to stabilize visual space during eye and head movements, and stimuli falling on the central retina have to do with pattern vision and judgments of the movements of one visual stimulus with respect to other visual stimuli. On this basis one may predict that the apparent orientation of a line of a given length confined to peripheral vision would be less affected by head tilt than a line of the same length in the centre of the visual field. Wade (1969c) showed that a line which subtended about 40° of visual angle produced a smaller E-effect than a line which subtended only 7°. Other evidence comes from a study by Wallach and Bacon (1976). The rotation of a radial pattern was linked to the sideways tilting of the subject's head by a system which allowed the ratio of the velocities of the two rotations to be varied. The radial pattern was centred in the visual field and subtended 40° of arc. It was found that the pattern had to rotate in the same direction as the head in excess of 5% of the head rotation in order for its change of orientation to be noticed. All rotations of the pattern in the opposite direction were noticed. Similar results were obtained when the central region of the pattern was blanked out, so that stimulation was confined to the periphery. When the radial pattern was confined to the central few degrees of the visual field it required a rotation of up to about 18% of the head rotation before subjects reported the motion. Both these settings of the no-motion ratio are equivalent in direction to an E-effect, and the results suggest that the E-effect is much smaller when the stimulus is confined to the peripheral retina than when confined to the central retina. However, this experiment was concerned with perception of motion during head tilt, not with the perception of the static tilt of visual objects. *An experiment is needed in which judgments of static tilt for foveal and peripheral visual displays are compared.*

If the body has been rotating about the visual axis for some time and is then suddenly stopped, there is an apparent displacement of a vertical line in the direction opposite to that of the previous rotation of the body. This effect decays over a period of a minute or two and is replaced by the typical E- or A-effect. The transitory effect is presumably due to the postrotatory deflection of the cupula. The magnitude and duration of this illusory effect are least if the head is brought to rest in a vertical position, where the otolith organs most effectively restrain the canal influences from exerting their effects, and greatest when the head is brought to rest at an angle of 150°, where the otoliths are least effective (Udo de Haes and Schöne, 1970).

Inclining the head backwards produces similar apparent deviations of a vertical line in the median plane (Ebenholtz, 1970a) although there are complicating factors in this type of judgment (see Ebenholtz, 1972). It seems that the effect of a sideways tilt of the head on the apparent tilt of a line in a vertical plane is not influenced by a simultaneous inclination of the head (Wade, 1972).

The *variability* of estimates of the vertical also increases with increasing tilt of the whole body (Miller *et al.*, 1965). This increase in variability is presumably related to the decrease in the sensitivity of the utricle with increasing tilt of the head, and it probably accounts for the fluctuations which have been observed in the apparent orientation of a luminous line when the head is held in a tilted position (Aubert, 1861; Miller and Graybiel, 1963).

The E- and A-effects must be due to one or more of the following sensorimotor changes which occur when the head is tilted or to changes in the way these events are interpreted: (1) countertorsion of the eyes, (2) changes in vestibular inputs, particularly those from the otolith organs, (3) changes in somaesthetic inputs arising from contact between the body and surfaces which support it, (4) changes in kinaesthetic inputs from the joints and musculature of the legs, back, and neck. In the sections which follow, experiments which have been done to disentangle the contributions of these various systems to the visual perception of the vertical will be reviewed.

10.4.2 Eye torsion and the A- and E-effects

When the head is in a tilted position the eyes are counterrotated with respect to the head, so that the normally vertical meridians of the eyes do not tilt as far as the head. This counterrotation is about 10% of the angle of head tilt (see Section 9.2.8). If the system failed to take this countertorsion into account, the resulting error would be in the same direction as the E-effect. However, the E-effect is not a simple function of eye torsion. Fischer (1930) measured countertorsion and the apparent tilt of a vertical line but found no parallel between the curves representing the two effects as a function of body tilt; there was not even a proportional relationship, although they were always in the same direction. Furthermore, Miller *et al.* (1968) found E-effects in people whose eyes did not tort, because of vestibular malfunction. Udo de Haes (1970) could find no relationship between the variability of judgments of the vertical and the intensity and duration of postrotatory torsional nystagmus. Merker and Held (1981) used a line which subtended 32° of visual angle and found no consistent relationship between static eye torsion and the amount or the variability of the apparent angular displacement of the line as the head tilted.

Eye torsion in animals, for example in the rabbit and in fish, has an amplitude large enough to stabilize effectively the retinal image as the head tilts through large angles. In man, eye torsion is only about 10% of the head tilt, which suggests that this function of countertorsion is vestigial in man. It may be that, in ordinary surroundings, humans can correctly judge the vertical when they tilt the head, not because of the stabilizing effects of eye torsion, but because they rely on the perceived stability of the visual framework. But torsion induced by head movements does serve an important function in man. As the head is tilted with objects in view, the motion of the retinal image is adequately stabilized during each slow phase of torsional nystagmus, and this prevents the degradation of vision that would occur if the image moved continuously over the retina (see Figure (6.3). This explains torsional nystagmus but not the residual torsion when the head is inclined, and it is this latter type of eye movement which could disturb the perception of the vertical. The fact that countertorsion has no consistent effect on the apparent vertical suggests that it is normally allowed for Balliet and Nakayama (1978a) found that voluntarily induced eye torsion was accompanied by a corresponding shift in the apparent vertical, so that not all types of torsion are allowed for.

10.4.3 The visual vertical with the body immersed in water

If a person is freely suspended in water in a state of neutral buoyancy, and if extraneous cues to the vertical such as thermal gradients, air bubbles, and visual cues are eliminated, the only cues to the vertical are provided by the otolith organs. A comparison of a person's ability to set a line to the vertical under these circumstances with his ability when he is seated in a chair in air should indicate the contribution that tactile cues make to the perception of the visual vertical. Early experiments of this type were unsatisfactory because subjects were tested just after they had been rotated several times, so that rotational aftereffects would disturb their judgments, and direct comparisons with performance in air were not made (see J. L. Brown, 1961). Schöne (1964) made careful measurements of the apparent visual vertical for the immersed subject at angles of tilt of the body between 0 and 180°. Up to 90° of body tilt the precision and accuracy of setting the line to the vertical were much the same under water as in air, but beyond 90° the variability of the settings under water became relatively larger, especially in the upside-down position, when the range of settings was about 30°. In the upside-down position subjects sometimes reported that their bodies were erect. Very similar results were reported by Wade (1973b). Ross *et al.* (1969) also confirmed the instability of orientation judgments when the immersed subject is upside-down. Graybiel *et al.* (1968) confirmed the similarity between immersed and dry performance for body tilts up to 30°, but did not test with large tilts.

The general conclusion from experiments on immersion is that the somaesthetic senses contribute little to the accuracy or precision of judgments of the visual vertical for small angles of body tilt, but do supplement the otolith organs at larger angles of tilt.

10.4.4 The visual vertical under conditions of weightlessness

In true zero-gravity conditions, such as are met with in orbiting spacecraft, the notion of gravitational vertical is meaningless, and of course the otolith organs no longer indicate the position of the head. If the inside of the craft is in view the occupants normally accept the surface nearest their feet as 'down', and this impression is reinforced if instruments and other familiar objects are oriented in the same way as the observer. If the astronaut is strapped into a seat, contact cues will provide a frame of reference for 'vertical', even in the dark. Graybiel *et al.* (1967) asked subjects during the time they were exposed to weightlessness for eight days in the Gemini spaceflights, to set a line of light in dark surroundings to be parallel to a remembered feature of the control panel which was in front of them. It was found that this task was perfomed just as precisely and accurately under weightless conditions as it was under normal gravity conditions, and it was concluded that '. . . lifting the gravitational load from the otolith organs did not result in any disturbance of . . . the perceived direction of space'. But the subjects in this experiment remained in the same position while being tested, so that they could succeed by simply setting the visible line with respect to the body axis, which is a task that does not depend on the otolith organs or on tactile cues. *What is needed is an experiment in which weightless subjects are asked to set a line to a remembered external reference when the direction of application of contact cues is varied.* One way to do this would be to secure the subject with cushioned straps against a hard board so that the tactile cues are more pronounced and directionally localized along one surface of the body. In one condition the subject could be positioned at various angles with respect to the board so that the direction of the tactile forces changes. At each angle he could be asked to position a visual line with respect to the board. This would test his ability to align a visual object to the direction of application of tactile stimuli. In another condition the subject and board could be rotated together and the subject asked to set the visual line to its original position in space. This would test the subject's ability to judge the extent and direction of his movement on the basis of inputs from the vestibular canals with somaesthetic inputs held constant.

It has been reported that Russian cosmonauts and American astronauts had a sensation of being upside-down when they closed their eyes while exposed to the weightless conditions of orbital flight (Graybiel and Kellog, 1967). This phenomenon has been called the **inversion illusion**, although in weightless conditions it is no more an illusion to believe that you are upside-down than it is to believe that you are right-way-up. Graybiel and Kellog found that labyrinthine-defective subjects do not experience this illusion when exposed to the same conditions that give rise to the illusion in normal subjects. They concluded that the illusion is due to the way normal subjects interpret the absence of inputs from the otolith organs in the weightless state. The Russian cosmonauts reported that the illusion was dispelled if they pushed themselves against the surface of their chairs.

10.4.5 The visual vertical and variations in somaesthetic stimulation

One way to investigate the role of tactile cues in the perception of the vertical is to anaesthetize the areas of skin in contact with surfaces. Garten (1920) achieved partial paralysis of the skin on the buttocks of a seated subject by cooling. This was found not to affect the ability of the subject to set the chair and his body to the vertical. Arndts (1924) anaesthetized standing and sitting subjects in those areas of skin in contact with the support surface, and again found that the ability to set the body to a vertical position was not affected. But these experiments do not tell us whether skin anaesthesia affects the ability to judge the inclination of the body at various angles of tilt, and it is this ability that is required for the accurate perception of the visual vertical when the body is tilted. A more direct approach was taken by Nyborg (1971). Each subject was supported in a framework in a standing posture either with rubber cushions, so that the tactile stimulation was diffused, or with four hard supports, so that there were areas of local contact. The subjects were upright or tilted 28° to left or right and had to set a rod to the visual vertical in the presence of a tilted frame. When the body was upright the settings were only slightly less variable with the local skin contact than with the diffuse contact, but when the body was tilted the mean unsigned error was 2° less with skin contact than with the diffuse contact. So again we find that somaesthesis becomes a factor only when the body is tilted.

10.4.6 The effects of tilting the head with erect and supine body

One way to eliminate the contribution of the otolith organs to the task of judging the visual orientation of a line is to place the subject in a supine position (on his back) and ask him to set a line in the frontal plane to be parallel with his body axis. This is less expensive than sending a man into the weightless conditions of space.

When the head is in line with the trunk, the headcentric orientation of a line has been found to be little affected by whether the subject is in an erect or supine posture (Rock, 1954; Brosgole and Cristal, 1967). This is not very surprising because when the head is in line with the trunk the sensorimotor system is symmetrical and, in any case, the otoliths and tactile–kinaesthetic cues are irrelevant to this task, in whichever posture the subject is held.

A more decisive issue is whether tilting the head relative to the torso has a greater effect on the ability to set a line to the body midline when the subject is supine than when he is erect. Consider first the erect observer. The eyes countertort as the head is tilted, and the torsion will have to be registered by the subject's perceptual system if it is not to disturb his judgments of the headcentric or bodycentric orientation of a line. The subject must also register the position of the head with respect to the body. The joint receptors in the neck provide this information, but it may also be provided by a combination of inputs from the utricles, from pressure receptors in the buttocks or feet, and from kinaesthetic receptors in the legs and back. Finally, the angle of the head to gravity is indicated by the otolith organs. By combining the sensations from his trunk and feet with the otolith inputs, he should be able to judge the position of his head with respect to his body. I shall refer to this set of inputs as the **somaesthetic–otolith complex**. Now consider the supine subject. His eyes will not have any residual torsion as he tilts his head, because the essential stimulus for residual torsion from the otolith organs is absent. He has information from joint receptors in the neck for registering the position of the head on the body, but that is all; the somaesthetic–otolith complex is absent. There are thus two differences between the erect and supine postures—the presence or absence of the somaesthetic–otolith complex and the presence or absence of eye torsion.

The contribution of the torsion factor may be independently assessed by asking the subject to set a visual line to the median plane of the *head*, with the head tilted at different angles to the torso, with the torso in the erect and supine positions. The somaesthetic–otolith complex is of course irrelevant to this task. Wade (1970c) did this experiment, although not with this purpose in mind. He found that the apparent median plane of the head was displaced about 4° in the direction of the torso when the head was tilted 40° with the torso in an erect posture, and that tilting the head had little or no effect when the torso was supine. The apparent tilt of the head in the erect posture is in the same direction as the E-effect, and is what one would expect from an unregistered eye torsion. The absence of the illusion in the supine position, where there is no eye torsion, strengthens the impression that torsion is the cause of the 4° apparent displacement found in the erect posture.

The combined contribution of eye torsion and the somaesthetic–otolith complex may be assessed by asking subjects to set a visual line to the median plane of the *torso* with the head tilted at different angles with respect to the erect and supine body. Templeton (1973) did this experiment. The constant errors were not significant for a 20° head tilt in the erect posture. In the supine posture the same head tilt produced a constant error of several degrees in the direction of the head. I suggest that in the erect posture the somaesthetic–otolith complex provided reliable information about the position of the head on the body, and that any error due to misregistered eye torsion would be only about 2° at 20° of head tilt. In the supine posture only the inputs from joint receptors in the neck were available for the task, and these receptors are prone to adaptation effects of several degrees, as we shall see in the next section. The effects of adaptation of joint receptors are in the right direction to have caused the constant errors that Templeton reported.

Comparative studies of visual orientation in erect and supine postures have thus provided strong evidence that eye torsion is incorrectly registered by the orientation system (at least for lines imaged on the central retina) and that the somaesthetic–otolith complex provides a better basis for more accurate judgments of head–body posture than do the joint receptors of the neck acting alone. We will look more carefully at the neck-joint receptors in the next section.

Lechner-Steinleitner (1978) approached the problem of the interaction between head tilt and body posture in a different way. The subject's trunk was strapped to a vertical board which was then tilted from side to side to various roll positions between vertical and 150° to the subject's right. For each of these roll positions, the subject's trunk was either in line with the head or deflected backwards into a prone position. When the head was deflected backwards, the

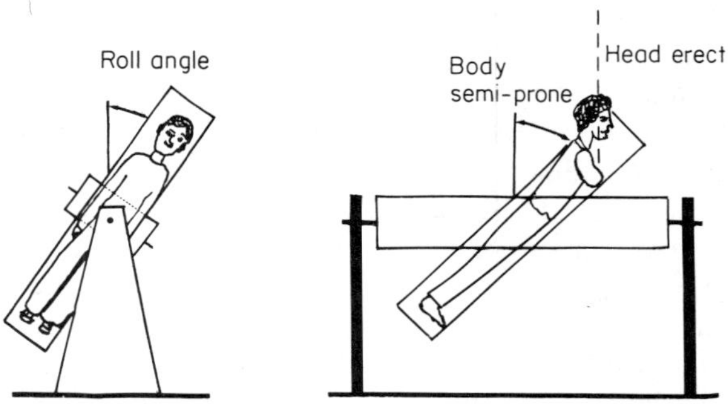

Figure 10.7 Apparatus used by Lechner-Steinleitner (1978) as described in the text

board was inclined forward through the same angle so that the coronal plane of the head was vertical (see Figure 10.7). In each position the subject was shown a luminous line in a frontal plane of the head and was asked to set it to the apparent vertical. Otolith stimulation (and eye torsion) would vary with the sideways tilt of the head and body. But the way it varied should not have been dependent on the position of the trunk with respect to the head—although the neck became flexed backwards, this movement was in a plane orthogonal to the plane within which movement of the luminous line occurred and would therefore be irrelevant. Thus, the only sensory factors that varied with the posture of the trunk, and which could have affected the task, were the changing somaesthetic sensations due to the contact of the body with the board. It was found that changing the posture of the trunk into the prone position increased the constant and standard errors of the visual judgment, but only for roll angles of the body greater than 90°. It was concluded that performance was limited by inputs from the otolith organs for smaller angles of body tilt, but that as the reliability of these inputs declined at higher tilt angles, the performance became limited by somaesthetic inputs. Schöne and Udo de Haes (1968) had conducted a similar experiment in the same laboratory and reached similar conclusions.

10.4.7 The visual vertical in subjects with loss of vestibular function

The only way to eliminate the influence of the labyrinths (otolith organs and canals) on the perception of the visual vertical is to use subjects with bilateral loss of vestibular function. In using such subjects one must be alert to the possible presence of residual function. Subjects with a recent vestibular loss should not be used because they will be subject to transitory disturbances during a period of recovery. On the other hand, it must be realized that subjects with a long-standing loss will probably have learned to use their other senses in a way that is not typical of normal subjects.

People with loss of vestibular functions have no difficulty balancing or orientating themselves when they can see their surroundings. But even in dark surroundings their constant and variable errors in setting a luminous line to the vertical are similar to those of normal people when the body is erect. The errors of the two types of subject are also similar when they are asked to set a luminous line to the vertical while standing with the trunk and head in various postures of sideways tilt (Clark and Graybiel, 1967). With such actively maintained postures neither group shows significant constant errors with the head, or head and trunk, tilted 20° to one side. Apparently the kinaesthetic and motor activity associated with actively maintained posture and the differential pressure on the feet allow both groups of subjects to appreciate the visual vertical accurately in the absence of a visual frame of reference. Normal people show about 2° of residual eye torsion when the head is tilted 20°, whereas there is no countertorsion in people without functioning labyrinths. That this difference was not reflected in the mean constant errors of either group of subjects suggests that subjects in both groups were able to assess the torsional positions of their eyes adequately, whether or not their eyes made countertorsional movements.

When normal and vestibular-defective subjects were tilted passively in a chair, and asked to set a luminous line to the vertical, both groups of subjects showed a mean E-effect up to about 60° of body tilt and an A-effect at larger angles of tilt. However, the defective subjects showed larger illusory effects and the variability of their settings increased with increasing body tilt at a faster rate than that of normal subjects (Miller *et al.*, 1968).

Submerged, blindfolded subjects with vestibular defects become more easily disoriented than normal subjects under the same circumstances (Padden, 1959).

10.4.8 The visual vertical in the human centrifuge: the oculogravic effect

A human centrifuge is a device which rotates at various speeds about a vertical axis and which carries a small cabin within which a person can be strapped in various postures. The cabin may be positioned at various distances from the centre of rotation, and it may be held in a fixed orientation or allowed to swing out so as to remain in line with the resultant force as the centrifuge speeds up.

The radially directed centrifugal force per unit mass acting on a subject at distance r from the centre of rotation of a centrifuge, rotating at an angular velocity ω, is given by $\omega^2 r$ (see Figure 10.8). This force and the force of gravity (g) form a resultant gravitoinertial force per unit mass of $\{(\omega^2 r)^2 + g^2\}^{1/2}$ which acts at an angle θ with respect to the vertical, given by $\tan\theta = (\omega^2 r/g)$.

Spinning a human being in a centrifuge has six main effects on his sense organs. (1) it displaces the direction of the force acting on his body, unless the cabin is hinged so that it is free to align itself and the subject with the direction of the resultant force. (2) it increases the force acting on his body. (3) it rotates him about his own body axis on every cycle of the centrifuge. (4) It produces a gradient in the direction and magnitude of the force acting on his body. Thus, if he faces in the direction of motion, his utricles will be stimulated unequally, and if he moves towards or away from the axis of rotation, he will experience complex changes in the direction and magnitude of the resultant force. (5) If he moves his head he will be subjected to Coriolis effects. (6) As the centrifuge changes its

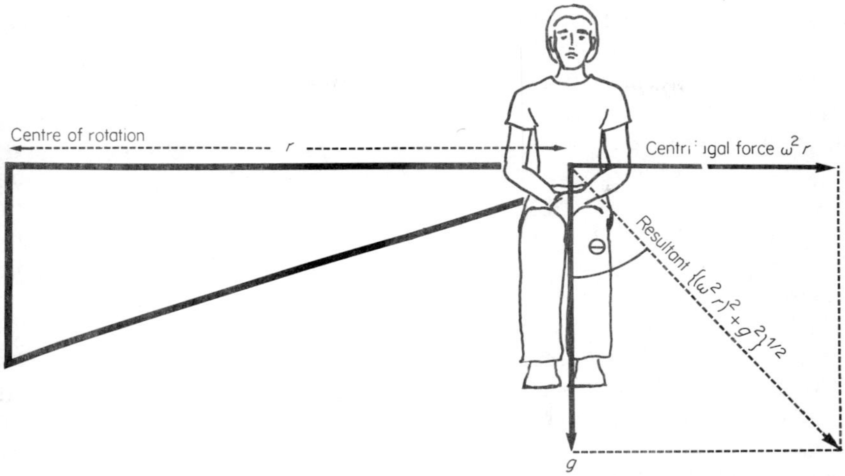

Figure 10.8 Showing how rotating a subject at angular velocity ω, in a horizontal plane at a distance r from the centre of rotation, produces a centrifugal force of $\omega^2 r$ and a resultant force of $\{(\omega^2 r)^2 + g^2\}^{1/2}$ acting at an angle θ to his vertical mid-body axis. $\mathrm{Tan}\,\theta = (\omega^2 r/g)$

angular velocity, the otolith organs and somaesthetic senses are stimulated as if the body were being tilted, but the vestibular canals do not register a rotation of the head in the same plane as the apparent tilt, as they would if the body were really being tilted.

The centrifuge has sometimes been used solely as a means of studying the effects on the judgment of the vertical of a displacement of the direction of the force acting on the body. I have argued previously that the cheapest, most convenient and least contaminated procedure for studying the effects of changing the direction of the force acting on the body is simply to tilt a person (Howard and Templeton, 1963). The centrifuge is useful for studying the effects of forces in excess of 1g and for subjecting subjects to rotating linear accelerations as described in Section 9.1.1. It also provides a way of changing the direction of the force acting on the otolith organs and somaesthetic system without stimulating the vestibular canals in the way they are normally stimulated when the head is being tilted.

Mach (1875) formulated the hypothesis that when an individual is rotated in a centrifuge he senses the direction of the resultant gravitoinertial force and regards this as the vertical if there is no visual frame of reference. If we disregard the fact that the resultant force is larger than 1g, and consider only the static situation, Mach's hypothesis simply states that a gravitoinertial force which is displaced θ degrees to the upright body has the same effect on the sense organs as a normal gravitational force acting on a body tilted through the same angle. We know from Einstein that there is no way of distinguishing between a gravitational field and a centrifugal field, so that it cannot matter whether we produce a change in the direction of the force on the body by tilting the subject or by rotating him. Hence Mach's hypothesis does not require experimental validation. Nevertheless, there have been several experimental validations of it in the last 100 years. For instance, Clark and Graybiel (1951), working at Pensacola, had their subjects set a luminous line to the horizontal in an otherwise dark cubicle rotated on a centrifuge. Their results fitted Mach's hypothesis fairly well. A subject in an enclosed chamber, subjected to a displaced direction of gravity, has no option (unless he reasons the situation out in a way that is not expected of him) but to adopt the new direction of 'gravity' as his standard of reference. If the subject faces towards the centre of the centrifuge, he experiences a backward inclination of the body equivalent to the change in the shear force acting on his utricles and an equivalent apparent elevation of a point of light at eye level (Schöne, 1964; M. M. Cohen, 1973a). If the subject faces in the direction of rotation of the centrifuge, he experiences a similar tilt of the body and an equivalent apparent tilt of a vertical luminous line positioned in his frontal plane. These effects are called **oculogravic illusions** (Graybiel, 1952). But strictly speaking they are not illusions when considered from the point of view of the subject. The point *is* elevated and the line *is* tilted *relative to the subject's gravitoinertial frame*, and that is the judgment that the subject is asked to make. It is no more an illusion than is the report that a person makes about the movement, in relation to gravity, of a line fixed with reference to his body as he is tilted. It would be an illusion only if the apparent vertical position of the line were *not* reported to be in line with the resultant direction of the gravitoinertial force. Such an illusion would be an Aubert or Müller effect.

However, there are two differences between displacing the resultant force acting on a subject in a centrifuge and tilting him in a chair, both of which should produce deviations from Mach's hypothesis. In the first place, the magnitude of the resultant force (F) in a centrifuge exceeds $1g$. Consider first the effect that this will produce on the otolith organs. Assume that the centrifugal force is $1g$, so that the resultant force of $(2g)^{1/2}$ acts at an angle of 45° to the utricular maculae of a subject facing in the direction of rotation. Bearing in mind that the shearing force on the utricles is proportional to $\sin\theta$, where θ is the angle between the force and the normal to the macula, then a force of $(2g)^{1/2}$ acting at an angle of 45° is equivalent to a shearing force of $1g$ acting at an angle whose sine is $\sqrt{2}\sin 45°$. This angle is 90°. Similar calculations show that, when the resultant force is displaced 20 or 30°, the shearing forces are equivalent to those produced by tilting the body 21.3 and 35° respectively. Thus, it is not until the resultant force is displaced more than about 30° that the deviation from Mach's hypothesis on account of the magnitude factor becomes severe. An extension of this argument shows that if the subject's body is always aligned with the resultant force, and if the head is positioned so that the utricular maculae are normal to the resultant force, then increasing the force should have no effect on the utricle, and hence no effect on the apparent tilt of the body. The increased force will deflect hair cells only on the saccular macula, but this should induce a sensation of being in a rising elevator, not a sensation of tilt. When the utricular maculae are not normal to the resultant force, increasing the force should increase the shearing force and induce the impression of increasing body tilt. The results of studies by Miller and Graybiel (1966) and Wade and Schöne (1971) are in approximate agreement with this analysis. The resultant force in these studies, although well over 1g, was not large enough to produce a shearing of the maculae beyond the normal limit of 1g acting at 90° to the macula surface. In other studies, in which supernormal shearing of the maculae was induced, it was found that increases beyond the normal limit did not produce impressions of increased body tilt or visual tilt (Schöne *et al.*, 1967; Schöne and Wade, 1971). So it looks as though 1g acting at 90° to the macula is the saturation point for impressions of body tilt.

Correia *et al* (1968) suggested another explanation for the fact that higher g values do not produce the apparent tilts that one would expect from the view that the effective stimulus for the otolith organs is proportional to sin θ. They proposed instead that the effective stimulus is proportional to the tangent of the angle of head tilt. By plotting the apparent tilt of a visual line as a function of $F \tan \theta$ they got a better fit for normal and supernormal stimulus strengths than by plotting apparent tilt against $F \sin \theta$. This suggests that compression forces come into play with higher g forces, although a precise mechanism was not presented.

Consider now the effect that a resultant force of over 1g has on the somaesthetic senses. With the body at a given angle to the resultant force, an increase in the force does not alter its direction of application on the surface of the body, and should therefore not induce a change in the apparent angle of tilt of the body. However, the tactile and kinaesthetic inputs should be enhanced and this should perhaps lead to a more veridical appreciation of the angle of tilt. On the other hand, the muscular effort required to maintain the body in its posture will increase with increasing magnitude of the resultant force and this should give rise to the impression of increased body tilt. There seem to be no experimental data bearing on these points.

The other difference between 'tilting' a person in a centrifuge and tilting him in a chair is that in the centrifuge the vestibular canals in the plane of the apparent tilt are not affected when the centrifuge is accelerating, that is, while the force is being displaced, whereas, when a person is tilted, the canals register the rotation of the head. Thus, in the centrifuge, 'tilt' information from the utricles is contradicted by the absence of the canal inputs that normally accompany a real head tilt. The impression of tilt should therefore be less in the centrifuge than when a person is being tilted in a chair. This reduced apparent tilt of a line deserves to be called a true oculogravic illusion. The cupulae do not restore to their central position until several seconds after the head has been tilted, so that the absence of cupula displacement in the centrifuge should delay the impression of body tilt for some time after the machine has reached a steady velocity. The experimental results accord with this analysis (Graybiel *et al.*, 1979).

A prominent feature of the oculogravic illusion is that it takes about 80 sec for the sensations of body tilt and visual displacement to reach their full magnitude after the centrifuge has reached its final velocity. The lower curve in Figure 10.9 is from Clark and Graybiel (1966a) and shows the gradual shift in

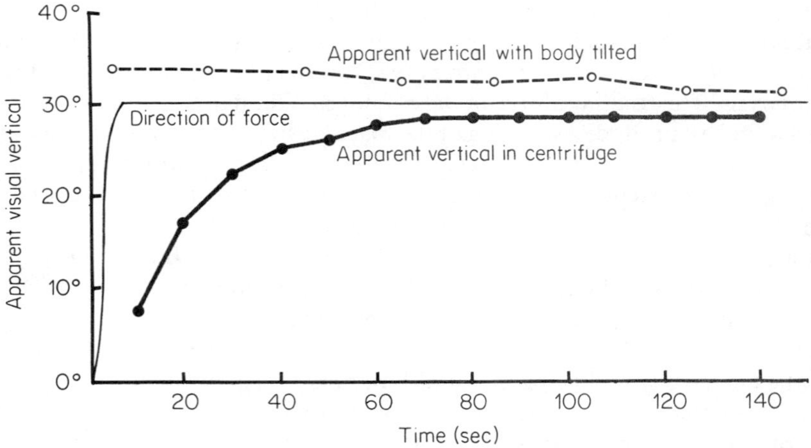

Figure 10.9 The mean position of the apparent visual vertical with respect to the body axis for various angles of sideways body tilt (top curve) and for various directions of gravitoinertial force imposed in a centrifuge (lower curve). The middle curve shows the actual direction of the force acting on the body in both cases (The upper curve is from Stockwell and Guedry, 1970, and the lower curve is from data from Clark and Graybiel, 1966a. Reproduced by permission of *Acta Oto-Laryngologica*)

the apparent visual vertical which occurs when a centrifuge is accelerated so as to displace the gravitoinertial force through 30°. It can be seen that the apparent vertical eventually comes to lie close to the direction of the displaced force as predicted by Mach's hypothesis, but only after about 60–80 sec. The other function in Figure 10.9 is from Stockwell and Guedry (1970) and shows the mean apparent vertical after subjects were actually tilted through 30° in a padded chair. In this case the apparent vertical is close to the gravitational vertical straight away. The delay in the centrifuge case must therefore be due to the restraining influence of the canals, which inform the subject that he is not tilting. This restraining influence decays after the new steady state is reached, with a time course that matches the decay of postrotatory vestibular responses (Stockwell and Guedry, 1970).

When a person is oscillated back and forth along a linear path, the illusory changes in body tilt and in the visual vertical are much less than when the same otolith shear is produced by actually tilting the person (Niven *et al.*, 1966; Schöne and Mortag, 1968). In this case the otolith stimulation is not accompanied by changes in canal stimulation that accompany body tilt, so that the subject should experience linear motion rather than body tilt. If a steady linear acceleration is continued for some time, the subject should gradually experience body tilt rather than linear motion. After a period of steady acceleration, the absence of rotational signals from the canals would no longer contradict the sensation of tilt because, with prolonged tilt, the cupulae would have been restored to their central position. In line with this suggestion it has been reported that subjects in a centrifuge subjected to a constant displacement of the force acting on their bodies experience an apparent tilt of the body and of their visual surroundings (Schöne, 1964). Furthermore, experiences of people riding in accelerating aircraft confirm that illusions of tilt do occur with prolonged linear acceleration. It has often been reported that when an aircraft is accelerated in cloud the pilot may interpret signals from the otolith organs as being due to an upward pitch of the aircraft, and may turn the nose of the plane down in an effort to correct the plane's attitude, only to drive the plane into the ground (Section 8.3.2).

10.4.9 The visual vertical and postural after effects

In section 7.6.4 it was shown how the sense of position of a limb, the head or the eyes may be temporarily affected when an asymmetrical posture is maintained for some time. It is therefore not surprising to find that such **postural after effects** disturb judgments of the apparent orientation of a luminous line seen in dark surroundings. There are two cases to consider. The first is the effect of tilting the head relative to the torso and the second is the effect of tilting the whole body.

Day and Wade (1966) asked subjects to set a luminous line to the vertical with head erect, before and after holding the head for two minutes in one of various angles of tilt up to 40°. After the head tilt, subjects set the line up to 6° in the direction of previous head tilt. The direction of the after effect was what one would expect if the joint receptors in the neck had adapted during the period of sustained head tilt and remained in the adapted state after the head was returned to the vertical. The other possibility is that the otolith organs had adapted. However, in a subsequent experiment Day and Wade (1968) found that when subjects were asked to align the line of light to the body axis, the after effect of head tilt was much the same when the head was tilted with the body erect as it was when it was tilted with the body in a supine posture. The joint receptors and not the otoliths thus seem to be the cause of the postural after effect. In any case, there is physiological evidence that the otolith organs do not adapt (Section 8.3.2). Wade and Day (1968a) found that the size of the visual after effect increased with the duration of the head tilt and that it decayed exponentially after the head was returned to the vertical.

One would expect that a prolonged tilt of the head would induce an apparent tilt of the head by the same amount and in the same direction as the apparent displacement of a truly vertical line of light when the head is restored to the upright position. Wade and Day (1968b) found that the two after effects were the same after subjects had maintained a 20° tilt of their heads for some time. However, with larger head tilts the visual after effect increased whereas the postural after effect decreased. The reasons for these different trends are obscure—perhaps eye torsion is involved. These experiments should be repeated.

After effects on the postural and visual vertical are also produced by maintained tilted postures of the whole body, and in this case the joint receptors of the neck cannot be implicated. Clark and Graybiel (1963) found that when subjects were asked to return themselves into an erect posture after being held in a tilted posture for some time, they positioned themselves at an angle in the direction of the previous, tilted posture. Recently, Schöne and Lechner-Steinleitner (1978) have reported a similar hysteresis effect in the way subjects set a luminous line to the vertical for various lateral tilts of the whole body, when each of these tilts was preceded by more clockwise, or more anticlockwise tilts. The direction of this effect is consistent with the after effects reported by Wade and by Clark and Graybiel.

The most reasonable interpretation of the postural and visual after effects of whole-body tilt is that they are due to adaptation of somaesthetic receptors. The idea that these receptors, rather than the otolith organs, are implicated is supported by the following evidence: (a) Clark and Graybiel (1966b) found that labyrinthine-defective subjects show a larger after effect than normal subjects; (b) the effect is least when the whole body is vertical, that is, in the posture where otolith organs are most sensitive, and greatest at angles of tilt of 90° or more, where the otolith organs are least sensitive (see Figure 8.1c); (c) with the posture of the head held constant, the effect is larger when the trunk is erect and the somaesthetic receptors are more heavily stimulated than it is when the trunk is prone and the somaesthetic receptors are less heavily stimulated (Schöne and Lechner-Steinleitner, 1978).

The visual vertical is therefore affected by adaptation of joint receptors in the neck and by adaptation of somaesthetic receptors. Wade (1968) found that, for a given degree of tilt, the after effects of head tilt on the apparent vertical are greater than the after effects of whole-body tilt. Wade also measured the visual after effect of holding the head vertical and the trunk in a tilted posture. In this case, the after effect from adaptation of the joint receptors of the neck is opposed to the after effect from adaptation of the somaesthetic receptors of the trunk. The after effect Wade obtained was in the same direction as that predicted from the adaptation of the joint receptors, which shows that this is a stronger after effect than that produced by adaptation of somaesthetic receptors.

The effects of sensory adaptation during a maintained posture should be revealed not only as an after effect, but also as a shift in the apparent visual vertical while the posture is held. Wade (1970d) showed that the mean E-effect of 30 subjects, evident when the head was tilted 30°, gradually declined from an initial value of about 6° to a mean value of about 4° while the tilted posture of the head was maintained for several minutes. He also found a similar but smaller reduction in the E-effect when the same tilt of the whole body was maintained for a few minutes.

In summary, it may be concluded that a stable visual frame is the most effective cue to the visual vertical when the body is tilted. An isolated line of light is subject to A and E effects. Although these illusions typically follow a sine function, resembling the shear-force function of the utricles, their cause is not known. At large angles of body tilt, with no visual frame, the utricles become less effective and changes in somaesthetic and kinaesthetic stimuli affect the precision and accuracy of judgments of the visual vertical.

Chapter 11

Intersensory and Multisensory Coordination

11.1 Introduction

Many sensory attributes are peculiar to one sense. Thus, colour is peculiar to vision, tone is peculiar to hearing, and pressure is peculiar to touch. On the other hand, all senses respond to intensitive and temporal attributes of stimuli, and the spatial attributes of stimuli are common to the senses of vision, audition, touch, and kinaesthesis. Intensity and time are one-dimensional stimulus attributes and time has the further restriction that observers are not free to move at will within it. These facts simplify decisions about the relative intensities and times of occurrence of stimuli in the same or in different senses. Space is three-dimensional and stimulus objects and perceivers may move independently in space. The task of relating experiences spatially in different senses is therefore complex. The accurate spatial coordination of the various spatial senses with each other and with motor control is a prerequisite of most overt behaviour and this chapter is concerned with the processes by which this coordination is achieved.

Consider the act of picking up a small handbell, looking at it, and ringing it. The direction of the bell is sensed visually, acoustically, and kinaesthetically. The visual, acoustic and kinaesthetic experiences all seem to originate from the same source—the three senses are spatially coordinated. Furthermore, as you move the bell, the change in location is sensed by each modality and is also associated with the motor commands which cause the movement. In other words, the sensory experiences and the intentional movements covary in characteristic ways. Underlying such common experiences is the fundamental question of how well, and in what ways, the different senses and the motor system are spatially coordinated. It is cumbersome to have to refer to intersensory and sensorimotor coordination; let it be assumed that the term 'intersensory', when used in a general sense, refers also to sensorimotor coordination.

It was shown in Chapter 7 how visual localization depends upon the integration of spatial information from the retina, from the eye-movement system, and from the head-movement system. These are integrative processes involving several sense organs, but they are not intersensory or multisensory in the sense used in the present chapter, because in the integrative processes the presence of the stimulus object is detected by only one sense organ. True intersensory or multisensory judgments involve the combination of spatial information from more than one sense organ when the stimulus object or objects are detected by more than one sense organ. This distinction will become clearer in what follows.

The first issue to be discussed in this chapter is that of the accuracy and precision of **intersensory spatial judgments**. An intersensory spatial task may be defined as *one in which the subject is required to bring an object which is detected by one sense organ into a specified spatial relationship with a second, perceptually distinct, object which is detected by another sense organ*. The 'objects' may be external objects or parts of the subject's body. The 'spatial relationship' may be that of identity of spatial position, parallelism, alignment, equality of length, or any other spatial quality in two or three dimensions. However, I shall be concerned only with intersensory judgments of direction and orientation in a frontal plane. The two sense organs involved in an intersensory judgment may be in the same modality. For instance, the task may be that of aligning a finger on one hand with a finger on the other hand in the dark, in which case the two sense organs are the joint receptors in the two arms. On the other hand, the two sense organs in an intersensory judgment may be in different modalities, as they are, for instance, when the task is that of aligning the unseen finger with a visual target.

An intersensory task is an 'open-loop' task, in the sense that the error of the intersensory judgment is not displayed. As we shall see, the sense organs involved in an intersensory task are 'in series', which means that, ideally, the variance of the total performance equals the sum of the variances of the spatial acuities of the senses considered separately.

A related but distinct issue is that of the accuracy and precision of **multisensory spatial judgments**. A multisensory spatial task may be defined as *one in which the subject is presented with one stimulus object which he can detect with two sense organs at the same time and is required to estimate its direction or its orientation in relation to his body or to some other landmark*. In this case, each sense organ receives independent information about the spatial disposition of the same object. The sense organs are therefore operating 'in parallel', which means that, ideally, the variance of the composite judgment should be less than the variance of judgments based on either modality alone. Intersensory and multisensory spatial judgments are discussed in Section 11.3.

The third issue concerns the effects of artificially distorting the input to one or other of the sense organs involved in a multisensory task. For instance, a person may view an object through prisms which apparently displace the object at the same time that he is holding the object or listening to the sound it makes. The questions are (a) whether the subject experiences a single object, and (b) if so, in what direction it seems to lie. This type of situation will be referred to as **intersensory discordance**, although only question (a) involves intersensory judgments; question (b) involves multisensory judgments. Intersensory discordance is discussed in Section 11.4, and also in Chapter 12. The present chapter is concerned with how people react to a discordance when initially exposed to it. Chapter 12 is concerned with how people learn to adapt to an intersensory discordance.

The fourth issue, which is discussed in Sections 11.5 and 11.6, concerns the development of intersensory behaviour in young children. Finally, in Section 11.7, a brief review of the cortical control of intersensory and sensorimotor coordination will be presented. But first some general theoretical points must be clarified.

11.2 The stimulus domain of intersensory and multisensory localization

Any investigator of intersensory or multisensory localization requires an objective metric with which to relate the direction of an object as detected by one sense organ to its direction as detected by another sense organ. For this purpose he must specify a reference point and reference planes which are common to all the sense organs being considered, which, in the present context, are the eyes, ears, and kinaesthetic sense organs of the neck and arm. A convenient reference point is any point on the mid-spinal axis of the human torso, and convenient reference planes are the median plane and the transverse plane through that point. The direction of any object in space may be given in terms of its azimuth, or horizontal direction and its elevation, or vertical direction with respect to this point and these reference planes. If the position of an object must be specified, and not merely its direction, then the distance of the object from the reference point must be known. One can also use a Cartesian system in which the median plane of the body is the x plane, the mid-frontal plane the z-plane, and the transverse plane through the point of reference in the body axis the y-plane (see Figure 1.4). In this system, the position of a stimulus is given by its x, y and z coordinates. A distal stimulus is any small object which can be seen, heard and touched.

Consider now the nature of the proximal stimulus for each of the three modalities we are considering. In all experiments on intersensory localization which have been reported, the subject was allowed to fixate the target visually so that it can be assumed that the position of the image on the retina was always accurately and precisely localized by the subject. The simplest assumption is that the direction of the visually sensed object is coded as the algebraic sum of the angle of lateral gaze and the angular position of the head. I shall ignore complications due to the fact that the centre of rotation of an eye is not on the same axis as the rotation of the head. Visual localization of a fixated object is thus, essentially, a kinaesthetic task, and involves very little purely visual processing. The object is, of course, *detected* visually, and that is why we talk of visual localization, but the direction of the object is determined kinaesthetically. The direction of a heard object is coded in terms of the various cues to auditory direction, such as interaural intensity and time differences. These cues are used to code the direction of an object in relation to the median plane of the head. In order to appreciate the direction of a heard object with reference to the torso, the subject must also know the position of his head on his body. Finally, the position of a felt object is coded in terms of the sense of position of the pointing finger in relation to the body. This is given by the algebraic sum of the angular positions of the various joints in the arm, with suitable allowance made for the length of each segment of the limb (Section 12.4.2). However, for the moment, I shall refer simply to the angular position of the arm on the body, as if the arm consisted of a single straight segment.

An intersensory localization task is essentially a discrimination task; it requires the subject to detect when there is a lack of correspondence between the direction of an object sensed by one sense organ and its direction sensed by another sense organ. A visual–kinaesthetic localization task requires the subject to make the algebraic sum of his angle of gaze and his head position equal to the angular position of the arm. The position of the head is irrelevant in a visual-auditory localization task, because the eyes and ears are both in the head; the subject is simply required to bring the angle of gaze into correspondence with the auditory direction of the object. If the subject is allowed to move his head rather than his eyes, the visual–auditory task is even simpler. For instance, if the subject is asked to place a visible object in the direction of an unseen sound source (by turning a knob), he may simply rotate his head until the sound is in the median plane of the head and then set the visible object in the direction in which he is gazing.

The ideal observer for an intersensory localization task may now be defined. Let the task be that of pointing with the unseen hand to a visual target, and let it be assumed that the ideal observer refers all directions to a common centre and a common median plane. Suppose that the variance of directional judgments for each separate modality is known. Let SD_v^2 be the variance of judgments of the bodycentric visual direction of a stimulus, and let SD_k^2 be the variance of the felt direction of a stimulus. Then, for an ideal observer, the variance of performance on the visual–kinaesthetic task would be the sum of the variances for the individual modalities (Merton, 1961; G. H. Fisher, 1960, 1962a). That is:

$$SD_{vk}^2 = SD_v^2 + SD_k^2. \tag{1}$$

This will be referred to as the **additive variance hypothesis** for intersensory localization.

Any constant errors in the visual and kinaesthetic modalities would, in the ideal case, add algebraically to give the constant error of the intersensory task, or:

$$CE_v + CE_k = CE_{vk}. \tag{2}$$

This will be referred to as the **additive constant-error hypothesis**.

Consider now the multisensory task in which the observer is asked to judge the bodycentric direction of the object he can both see and feel. In the ideal case there is a pooling of information, so that the variance of the multisensory judgment should be less than the variance of either component. Using R. A. Fisher's (1966, p. 239) measure of information, we can state that:

$$\frac{1}{SD_{(vk)2}} = \frac{1}{SD_v^2} + \frac{1}{SD_k^2}. \tag{3}$$

This will be referred to as the **additive information hypothesis** for multisensory judgments. The bracket indicates that it is a multisensory rather than an intersensory task. The observer *combines* information in a multisensory task and *compares* information in an intersensory task. It is assumed that the target is clearly visible. See Mulligan and Shaw (1980) for a discussion of multisensory signal detection.

An observer may judge an object to be in one direction by touch when he cannot see it, and judge it to be visually in another direction when he sees it without touching it. This could happen if there were constant errors in either visual or kinaesthetic judgments of direction or if the input to one or other sense were distorted. In such a case, if the observer were allowed to both touch and see the object, and if he were persuaded that there was only one object, he would receive conflicting information. The ideal strategy for dealing with a conflict of this kind would be to take the mean of the two sensed positions as the true position. This will be referred to as a **compromise judgment**. If the ideal observer were informed which judgment was in error, or if there were a reason to trust one sense more than the other, the observer would either ignore the information in one modality or recalibrate the weaker system to bring it into correspondence with the other. In both these cases, one modality is said to **dominate** the other. In general, for a multisensory task,

$$\mathrm{CE}_{(\mathrm{vk})} = \tfrac{1}{2}(W_1 \mathrm{CE}_\mathrm{v} + W_2 \mathrm{CE}_\mathrm{k})$$

where W_1 and W_2 are weighting factors assigned by the observer to each type of error, with values summing to one.

Let it be stressed again that, in a visual–kinaesthetic task, visual localization refers to visual localization with respect to the torso and involves the registration of the direction of gaze and the direction in which the head is pointing. Similarly, in an auditory–kinaesthetic localization task, auditory direction is specified relative to the torso. The direction of the head may be ignored only in a visual–auditory localization task.

This theoretical analysis has provided a set of hypotheses in terms of which the behaviour of actual observers can be assessed. These may be listed as follows.

1. The variance of performance in an intersensory localization task should equal the sum of the variances of localizations in each of the component modalities. The evidence relevant to this hypothesis is reviewed at the end of Section 11.3.2.
2. The constant error in an intersensory localization task should equal the algebraic sum of the constant errors of localizations in each component modality. Evidence bearing on this hypothesis is reviewed in Howard and Templeton (1966, p. 354).
3. The reciprocal of the variance of performance on a multisensory localization task should equal the sum of the reciprocals of the variances of localizations in each component modality. I know of no evidence bearing on this hypothesis. However, it is known that different senses do pool information under certain circumstances. For instance, the probability of detecting a stimulus that is both seen and heard is greater than the probability of detecting the stimulus by either modality alone (Brown and Hopkins, 1967).
4. The constant error in a multisensory localization task should equal the mean of the constant errors of the component modalities, unless one modality is known to be more reliable than the other, in which case directional judgments should be made entirely in terms of the dominant modality. The evidence bearing on this issue is reviewed in Section 11.4.

11.3 The precision and accuracy of intersensory localization

11.3.1 Procedures

The problems associated with measuring the precision and accuracy of intersensory localizations are much the same whichever pair of modalities is being considered. Therefore, for simplicity of exposition, the discussion of these problems will be confined to visual–kinaesthetic localization. Visual–Kinaesthetic localization has been studied in each of the three planes of egocentric space, the frontal, horizontal and median planes (see Figure 1.4).

Sandström (1951) worked in a frontal plane. He varied the position of a small luminous visual target within a frontal plane in otherwise dark surroundings and asked the subject to point to it with unseen hand (see also Holding, 1968).

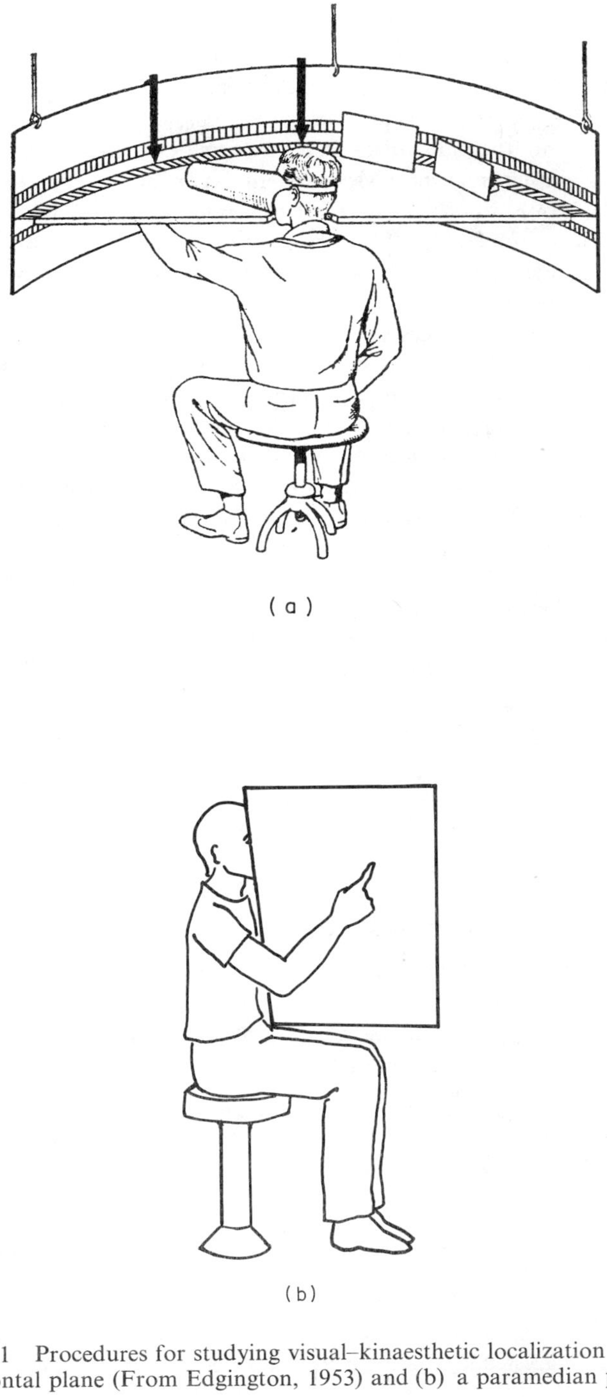

Figure 11.1 Procedures for studying visual–kinaesthetic localization in (a) a horizontal plane (From Edgington, 1953) and (b) a paramedian plane

Typically, an apparatus like that shown in Figure 11.1(a) is used to study intersensory localization in a horizontal plane. The visual target appears at various positions along an arc at a fixed distance from the subject. Therefore, in such experiments, localizations are measured not over the whole plane but only along one arc within the plane. Pre-1966 experiments of this type are reviewed in Howard and Templeton (1966). More recent experiments have been conducted by Churchill (1969), Auerbach and Sperling (1974), and Lackner (1974a). It seems that Foley and Held (1972) have conducted the only study of visual–kinaesthetic localization over the whole of a horizontal plane within reach of the subject, including various distances as well as various directions.

The only reference to intersensory localization in a median or paramedian plane that I have found is a study by Legge (1970). He used an arrangement like that shown in Figure 11.1(b) to measure variable and constant errors of pointing with unseen hand to visual targets in a paramedian plane.

It seems that nobody has studied visual–kinaesthetic localizations in three-dimensional space. One way of doing this would be to use the arrangement shown in Figure 11.2. The visual targets are positioned above the subject, who sees them reflected in the space in front of him. In this way the hands do not get in the way of the targets and at the same time remain hidden. Some way of registering the three-dimensional position of the fingertip would have to be devised. One way of doing this would be to slide a vertical calibrated plane forward on a track towards the tip of the subject's finger at the end of each pointing response and read the x and y coordinates of the finger from scales on the plane and the z coordinate from a scale on the track. A useful device for

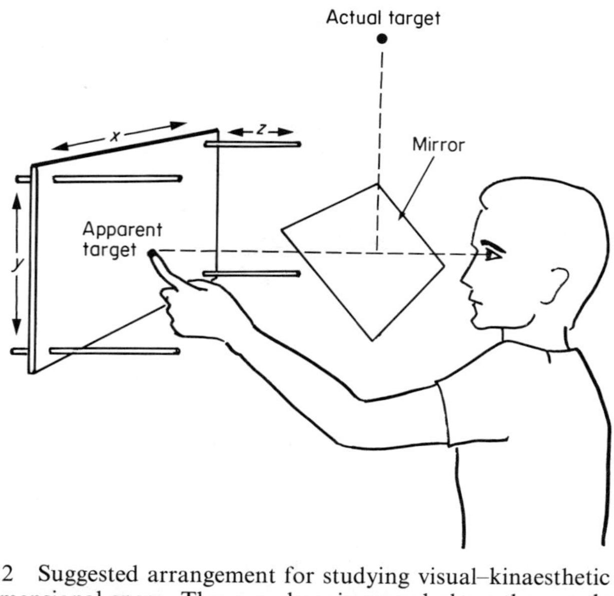

Figure 11.2 Suggested arrangement for studying visual–kinaesthetic localization in three-dimensional space. The x–y plane is moved along the z rack towards the finger after each localizing response

recording the position of the finger in a plane has been described by Bauer *et al.* (1969). A more convenient, though more expensive procedure is to place a click generator on the subject's finger and record its x, y, and z coordinates by measuring the acoustic delay to each of three linear microphones. This type of system is available commercially.

11.3.2 Factors affecting intersensory precision and accuracy

It would be tedious and theoretically pointless to review the variable and the constant errors that have been reported in the studies just mentioned, especially as most of the investigators themselves made no attempt to explain them. In any case, the results of different experiments cannot be compared because some were given in centimetres, others in degrees. Furthermore, the variability of responses was sometimes given as the variance and sometimes as the mean unsigned error relative to the point of judged equality, and sometimes as the unsigned absolute error. The last is a poor measure because it confounds variability with the constant error. Let me remind readers that I use the term 'precision' to refer to any measure of variability of responses about the point of judged equality, and the term 'accuracy', or 'constant error', to refer to the mean signed absolute error. Even if the data from different experiments were recomputed into a compatible form, the results would still not be comparable, because the experimental conditions varied from study to study.

The following are some of the factors that might be expected to affect the precision and accuracy of visual–kinaesthetic, visual–auditory and auditory–kinaesthetic localizations. These factors should be carefully controlled in experiments on intersensory or multisensory localization.

Asymmetry of stimulation Evidence that asymmetry of stimulation in any particular modality can produce large constant errors of localization within that modality was reviewed in Section 7.6. For instance, if an arm is held out to one side for a minute or so, the apparent straight-ahead position of the arm is displaced several degrees in that same direction. Consider a visual–kinaesthetic localization task in which the subject must position a visible object to coincide with the unseen index finger of an arm which is held out to one side of the body midline. If he succeeds, both the visual and the kinaesthetic stimulation will be asymmetrical. However, the apparent position of the arm would be affected most, because the arm is held in position for a longer period of time. The subject would therefore most likely position the visual target nearer to the median plane than the actual position of the finger. On the other hand, if the subject points his unseen finger to an off-centre visual target, the apparent position of the visual target would normalize towards the median plane. One would now expect the finger to be placed nearer to the median plane than the actual position of the visual target. Note that in the two conditions just mentioned, the constant errors of the intersensory tasks would be in opposite directions.

Suppose the experimenter ensures that the visual target and the arm are held in an asymmetrical position for the same length of time. If normalization is now the same in both modalities, it should result in a zero constant error of visual–kinaesthetic localization, whichever way round measurements are taken. But consider what would happen to the intersensory constant errors if the kinaesthetic sense of arm position normalized more than the apparent position of the visual target, even though the eyes and arm were held in an asymmetric position for the same length of time. In that case, if the subject pointed to an off-centre visual target, he would position his arm further away from the median plane than the visual target in order to bring the finger to the same apparent position as the target. On the other hand, if the subject were to place the visual target to correspond with the position of the finger, he would place it nearer to the median plane than the finger. The two intersensory constant errors would now be in the same direction (see Table 11.1). The situation would be even more complicated if the effects of asymmetrical stimulation in one modality transferred to the other modality. However, this apparently does not happen (Day and Singer, 1964).

Churchill (1969) reported that visual estimates of arm positions and kinaesthetic estimates of visual locations produced opposite constant errors, although he did not comment on the significance of this result. The experiment was not well designed, because the visual target was one of a whole array of simultaneously visible objects, whereas the kinaesthetic stimulus was a single position of the arm. However, the pattern of constant errors which he reported suggests that vision normalized more in one condition and kinaesthesis normalized more in the other condition.

Legge (1970) asked subjects to place an unseen stylus opposite visual targets arranged in a paramedian plane, as in Figure 11.1(b). He found that the stylus was generally placed further from the body than the visual target when the two were judged to be opposite each other. It is not possible to interpret this pattern of errors, because the visual and kinaesthetic stimuli were not exposed for the same length of time and the corresponding visual to kinaesthetic judgments were not recorded.

Table 11.1 Predicted constant errors due to either visual or kinaesthetic normalization when a light is set to the position of the finger of the off-centre arm and when the finger is pointed at an off-centre visual target. It is assumed that the arm and eyes are held in an asymmetrical posture for an equal period of time.

Localizing response	Light set to finger		Finger set to light	
Site of normalization	Visual	Kinaesthetic	Visual	Kinaesthetic
Resulting constant error	Light set further from median plane than the finger	Light set nearer to median plane than the finger	Finger set nearer to median plane than the light	Finger set further from median plane than the light

The modality of the localizing response We have just seen that care must be taken to make the duration of exposure to the stimuli equal for the two modalities. The best way to achieve this is to use the method of constant stimuli rather than the method of adjustment, and that is why the most carefully conducted experiments on intersensory localization have done this.

But even if the durations of the stimuli are made the same, the modality within which the localizing response is made may still affect the outcome. One reason for this is that usually, in a given trial, the position of one stimulus or limb is altered and the position of the other stimulus or limb is held constant. The effects of changes in the position of the stimulus on the constant errors may easily be controlled by the standard procedures of counterbalancing the extent and direction of stimulus motion. However, the effects of changes in stimulus position on the variable error are not necessarily controlled by such procedures. It is well known that the precision of the sense of arm position is much greater when the arm has just been moving than when it has been stationary for some time (Paillard and Brouchon, 1968), and the same is apparently true for the precision of the sense of eye position (Section 7.5.2). If the precision of position sense is more affected by lack of movement in one modality than in the other, one would expect the variance to be different according to which modality is used for the localizing response.

Temporal factors The accuracy and precision of intersensory and visual–motor localization should be adversely affected by any delay between the termination of the target stimulus and the initiation of the localizing response. Any such delay requires the subject to remember the position of the target. This question goes beyond the scope of this volume. Readers are referred to Posner (1967) and Holding (1968).

The reaction time of pointing to a visual target is about 350 msec. This is about 100 msec longer than the reaction time of any saccade that may be required to bring the image of the target onto the fovea (Prablanc *et al.*, 1979). These same investigators found that, when a peripherally positioned visual target was switched off at the onset of a saccade towards the target, there was no correlation between the saccadic errors and the pointing errors. They suggested that the two responses are planned in parallel.

The minimum duration of an accurate movement of a stylus over a surface from a given position to a visual target has been shown to be a function of the width of the target, W, and of the amplitude of the movement. Fitts (1954) expressed these relationships in quantitative form in a formula derived from information theory:

$$\text{Response time} = a + b \log\left(\frac{A}{W}\right)$$

where a and b are empirically determined constants.

Training If the subject receives knowledge of results in an intersensory localization experiment, it is to be expected that his constant errors would tend to zero. This topic is discussed in Chapter 12. One might also expect that the subject's variance would be reduced by training. There is a vast literature showing how the precision of visual–motor skills is improved by training (see Bilodeau, 1966; Holding, 1965; Welford, 1968). But this general topic is beyond the scope of this volume. Usually, in such studies, the subject can see his limb while aiming at or tracking a visual target (Beggs and Howarth, 1972). Lackner (1974a) prevented the subject from seeing his hand while pointing to visual targets arranged along a horizontal arc as in Figure 11.1(a), but allowed him to feel the position of the target at the end of each aiming motion. This type of feedback led to a significant reduction in the variability of pointing in a test situation in which no knowledge of results was provided. This suggests that people do not normally perform at their limit of precision when pointing with unseen hand at visual targets. The reason may by that the hand is usually in view and high precision may be achieved by visually guiding the finger onto the target.

Interactions between direction, distance, and the position of the egocentre It was shown in Section 7.2 that visual directions are judged with reference to an egocentre which is typically in the median plane of the head, somewhere behind the bridge of the nose. It was also shown that the accuracy of pointing with unseen hand to a visual target in a horizontal plane is affected by the position of the visual egocentre, the apparent distance of the visual target, and the angle of convergence of the eyes. Foley and Held (1972) reported that constant errors of pointing with unseen hand to visual targets in a horizontal plane tended to be to the side opposite the sighting (dominant) eye, when the only cue to visual distance was the convergence of the eyes, although this conclusion was qualified in a subsequent study (Foley, 1975). In the original study it was assumed that subjects judged visual directions with reference to a common visual-kinaesthetic egocentre displaced towards the sighting eye and the pattern of constant errors was explained by assuming that the subject's eyes were converged beyond the actual position of the target, as in Figure 11.3a. But this interpretation is valid only if it is assumed that both the visual and the kinaesthetic egocentres are displaced towards the dominant eye. If this assumption is discarded the same pattern of constant errors may be explained in at least two other ways without making any reference to the apparent distance of the visual target. If the visual egocentre were displaced towards the dominant eye and the kinaesthetic egocentre were in the centre of the head, as in Figure 11.3b, the subject would point towards the non-dominant eye in order to equate angles θ and ϕ. If the centre for visual space were in the median plane in front of the centre for tactile–kinaesthetic space, as in Figure 11.3(c), then the mean constant errors for intersensory localizations would be different depending upon whether the stimuli were presented to the left or right of the median plane. If the centre for visual space were in front of the centre for tactile–kinaesthetic space, and if visual and tactile–kinaesthetic stimuli were presented together to the left of the subject's head, then the tactile–kinaesthetic stimulus would appear to be to the

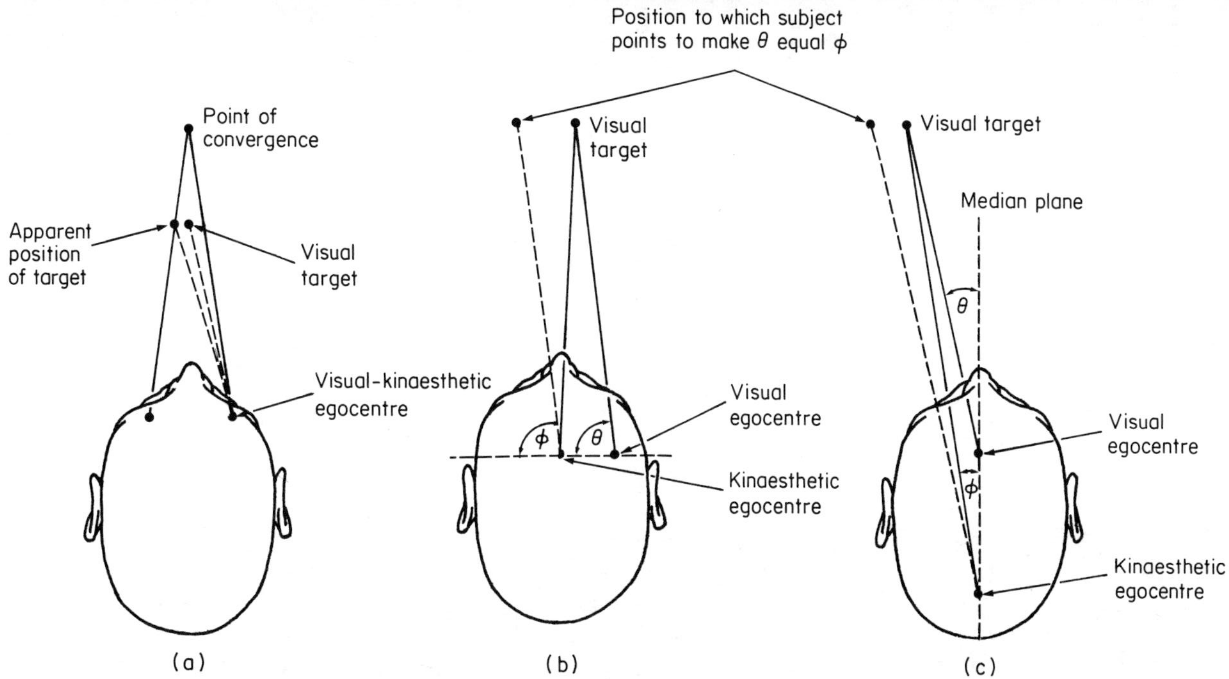

Figure 11.3 (a) The apparent direction of a visual target with respect to the right eye when convergence is beyond the target. This will predict the direction of the pointing error reported by Foley and Held if the visual–kinaesthetic egocentre is in the position of the right eye. (b) Showing how constant errors in pointing to a visual target result if the kinaesthetic and visual egocentres do not coincide laterally. (c) Showing how errors in pointing to a visual target result if the visual egocentre is in front of the kinaesthetic egocentre

right of the visual stimulus. That is, angle ϕ would be smaller than angle θ in Figure 11.3(c). However, if the stimuli were presented to the right, the tactile–kinaesthetic stimulus would appear to be to the left of the visual stimulus.

Having thus reviewed some of the factors that may affect the precision and accuracy of intersensory localizations, we are in a better position to look at the evidence bearing on the additive variance hypothesis for intersensory localizations that was mentioned in Section 11.2.

If SD_{VK}^2, SD_{KA}^2, and SD_{AV}^2 are the observed variances for visual–kinaesthetic, kinaesthetic–auditory and auditory–visual intersensory localizations respectively, and SD_V^2, SD_K^2, and SD_A^2 are the variances of localizations in each separate modality, then, according to the additive variance hypothesis:

$$SD_{VK}^2 = SD_V^2 + SD_K^2$$

$$SD_{KA}^2 = SD_K^2 + SD_A^2$$

$$SD_{AV}^2 = SD_A^2 + SD_V^2$$

Since the three intersensory variances may be measured, the values of the three single-modality variances which satisfy the additive hypothesis may be calculated by solving the three simultaneous equations. These hypothetical values may then be compared with experimental values. The apparatus that Fisher (1960) used for measuring intersensory and intrasensory variances is shown in Figure 11.4. A light, a small loudspeaker, and a small tactile–kinaesthetic 'button' were each mounted at arm's length on the end of a separate boom. Each boom could be rotated independently about the spinal axis of the subject. The subject was

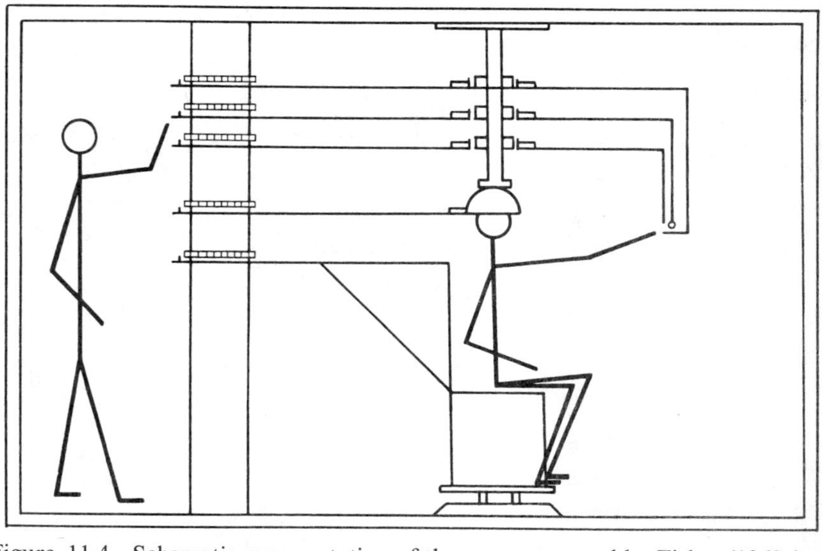

Figure 11.4 Schematic representation of the apparatus used by Fisher (1962a) to study intersensory localization

presented with pairs of stimuli in various positions and, for each pair, had to report which of the two stimuli within the pair was to the left of the other. The results of this experiment were inconclusive, but a similar experiment was done more recently by Auerbach and Sperling (1974). In this experiment, attention was confined to auditory–visual localization. The variances of auditory–visual localization and of localizations in vision and audition were determined. Care was taken to keep these three tasks as similar as possible, using a signal-detection procedure. The results supported the additive variance hypothesis, at least for auditory–visual localization.

11.4 Reactions to intersensory discordance: ventriloquism

11.4.1 Basic findings

Consider any object which can be seen and felt and which emits a characteristic noise. Normally, the cues which indicate the direction of the object to each of the three senses covary in a characteristic way when the object is moved. This sensory covariance may be disturbed by displacing the visual input with prisms, by displacing the auditory input with pseudophones, or by disguising the real source of the sound. For instance, in the cinema the person speaking can be seen on the screen, but the actual source of the sound is a loudspeaker to one side of the screen. It is common knowledge that the sound seems to come from the mouth of the actor. This is one example of the general principle that, when visual and auditory sources of spatial information are in conflict, the sound source is apparently displaced to coincide with the visible object with which it is associated. This phenomenon is known as **visual dominance** or **visual capture**. In general, in any intersensory conflict, if judgments are based more on one modality than on the other, the first modality is said to dominate or capture the other modality. Sensory dominance has been studied within three pairs of modalities, vision–audition, vision–kinaesthesis, and audition–kinaesthesis. Here are some examples of experiments on each pair.

Perhaps the best-known example of visual dominance over audition is ventriloquism, in which the voice of the ventriloquist seems to come from a doll. Jackson (1953) did an experiment to determine how far a visual object has to be separated from the position of an associated sound before the discordance becomes noticeable. Visual and auditory stimuli were made to conflict by the use of silent steaming kettles and hidden compressed-air whistles. Subjects reported that the whistling sound appeared to originate from the kettles until the steaming kettles and the whistles were separated in a horizontal plane by more than 30°. Separations of this order indicate clearly that we are concerned not with a simple problem of poor spatial acuity, but rather with a mechanism, or mechanisms, in which spatial information in one modality is partly or wholly interpreted in terms of information from the other modality. Thurlow and Rosenthal (1976) recently confirmed that vision dominates audition up to separations of about 30°. They found that hidden sound sources in the region behind a subject's head were also apparently displaced to coincide with associated visible objects. This is because sounds behind the head are, in any case, easily confused with sounds in front of the head.

A nice example of a conflict between vision and kinaesthesis is provided by a study by Tastevin (1937). He showed his subject a life-like replica of one of the subject's fingers protruding from under a screen, while his hidden real finger was some distance to one side. The subject was convinced that the model finger was his own, and felt that it was where he saw it to be. This is a case of vision dominating kinaesthesis.

Hay *et al.* (1965) confirmed that vision dominates kinaesthesis. They asked subjects to point with the hidden right hand to the apparent position of their left hand seen through 10° displacing prisms. Subjects pointed very close to the seen position of the displaced left hand. Over (1966) also confirmed that vision dominates kinaesthesis. In his experiment, subjects were asked to rotate a bar in a frontal plane until it was judged to be horizontal. Subjects could feel the bar with unseen hand while viewing it through prisms which apparently tilted it. With discordances up to 15° judgments were dominated by vision, and with larger discordances compromise settings were made. Vision has been found to dominate even when pitted against simultaneously presented discordant auditory and kinaesthetic information (Warren, 1979). Rock and Victor (1964) also found that vision is dominant over touch–kinaesthesis. In this experiment subjects handled a square object while looking through a lens which caused the object to appear rectangular. They were then asked to select a matching object from a set of objects they could feel but not see and from a set of objects they could see but not feel. The use of two types of judgment to measure the effect of the discordance, controlled for the possible contaminating effects of a response bias. Subjects were also asked to draw the standard object that they had seen and felt. In all conditions, most subjects selected an object which matched the shape as seen rather than the shape as felt, and few subjects were aware of any conflict.

One must be cautious in comparing the results of studies in which there is a discordance between the seen and felt *direction* of an object and studies in which there is a discordance between the seen and felt *shape* of an object. In the first situation, any bias in visual judgments would almost certainly be due to a bias in the apparent position of the eyes rather than a change in retinal local sign (oculocentric change). Indeed, a slight biasing of visual directional judgments by discordant kinaesthetic inputs has been reported (Pick, Warren, and Hay, 1969; Warren and Pick, 1970). In the second situation, any bias in visual judgments must necessarily be due to an oculocentric change. This is an improbable occurrence and it is not surprising that this type of visual bias has not been reported.

Finally, the conflict between audition and touch–kinaesthesis was studied by Pick, Warren, and Hay (1969). Each blindfolded subject was asked to point with his right hand to the felt position of his left hand while the latter hand was touching a loudspeaker which was emitting clicks. The two hands were not allowed to touch. At the same time, subjects wore a pseudophone which apparently displaced the source of the clicks by 11° to one side. Subjects pointed to the true position of the left hand and ignored the discordant auditory information. Willott (1973) confirmed this finding, using a testing procedure that did not involve pointing, showing that the results of Pick *et al.* were not due to the biasing effect of their kinaesthetic testing procedure.

Thus, there seems to be general agreement that vision dominates audition and kinaesthesis and that kinaesthesis dominates audition. Nobody seems to have mentioned the vestibular system in this context. In Section 9.4 it was shown that when vestibular and visual information regarding the tilt or rotation of the head are in conflict, strong visual illusions are often produced, showing that vision is not necessarily dominant over the vestibular system.

When two pieces of spatial information are discordant it is reasonable to suppose that the information which is normally least precisely perceived will be discarded. Fisher (1962b) carried out a series of experiments to test this idea. Subjects were told that two stimuli in two different modalities originated in the same object on a horizontal semicircular arc in front of them and at arm's length. In fact, the relative locations of the two sources of stimulation were varied and the subjects had to indicate by pointing where they judged the apparently single source to be. Visual–auditory, and tactile–kinaesthetic stimuli were used in various combinations. Visual stimuli were completely dominant over both tactile–kinaesthetic and auditory stimuli, and tactile–kinaesthetic stimuli tended to be dominant over auditory stimuli. This is the order of dominance which Fisher predicted from the order of precision of localization of individual stimuli within each of the modalities which he had established in another set of experiments. Pick, Warren and Hay (1969) confirmed this same order of dominance between the various pairs of modalities. However, they did not find that the degree of dominance of vision over audition was quantitatively predictable from the dominance of kinaesthesis over audition. In other words, they found that, although the dominance relationships between the three pairs of modalities were qualitatively transitive, they were not quantitatively transitive.

11.4.2 Procedures and precautions

In spite of the good agreement between the results of different investigations of the patterns of sensory dominance, the findings must nevertheless be treated with caution, because in many experiments the procedures were biased or poorly controlled. Before these problems can be discussed, the various procedures that have been used to measure sensory dominance must be described.

The methods may be divided into two groups. In one group, the subject matches the direction or orientation of the biomodally discordant 'object' with the direction or orientation of an internal standard, such as the apparent median plane, or of a neutral comparison stimulus during the time that the discordant 'object' is presented. In the other group of methods, the subject makes the match after the discordant 'object' has been removed. These will be referred to as **perexposure** and **postexposure** methods, respectively.

There are two perexposure methods. In the first method the subject is asked to adjust the direction or orientation of the discordant 'object' to match an internal standard, or norm. This method was used in the study by Over mentioned above. The subject felt a rod with unseen hand and at the same time saw it through prisms which optically rotated it. By asking the subject to set the 'rod' to the apparent horizontal it was possible to determine which modality was dominant. Another example of the use of an internal standard is provided in a study by Kinney and Luria (1970). The subject wore a face-mask under water, which magnified visual images. He could see and feel a series of metal discs and was asked to select those discs that matched well-known American coins in size. The fact that subjects selected discs which visually matched the normal images of coins, rather than those that were the same size by touch, demonstrated that vision was dominant over touch–kinaesthesis.

In the second perexposure method the subject is asked to match the discordant 'object' with one of a series of comparison stimuli presented to a sense organ which is assumed not to be affected by the discordance. For example, in the Hay, Pick and Ikeda study mentioned at the beginning of this section the subject was asked to point with his hidden right hand to the apparent position of the left hand seen through displacing prisms. This procedure is valid only if it is assumed that any apparent change in the location of the left hand does not affect the apparent location of the right hand. Another example is provided by a study by Radeau and Bertelson (1976) in which the subject was exposed to discordant visual and auditory stimuli and asked to indicate the position of the 'stimulus' by pointing.

Postexposure methods are more problematical than perexposure methods, because the interpretation of the results depends on whether or not it is assumed that the effects of intersensory discordance outlast the presentation of the discordant stimuli. As an illustration of this problem consider the procedure used by Rock and Victor (1964) and by McDonnell and Duffett (1972). The subject viewed a block through a cylindrical lens, which compressed the image to about half, while at the same time feeling the block with unseen hand. Following exposure to the discordant stimuli, the subject was required to select by feel, from a set of blocks, one that matched the size of the exposure block. The size of matching block selected would clearly depend on whether or not there was an aftereffect produced by the exposure to discordant stimulation, and the procedure would measure what it was supposed to measure only if there was no aftereffect. But aftereffects of discordance are known to occur, as we shall see in Chapter 12, so that this type of procedure would seem to be invalid, especially if the discordant stimuli are exposed for some time, as they were in the McDonnell and Duffett experiment. The other postexposure procedures, which are designed to measure the aftereffects of discordance, will be described in Chapter 12 and therefore need not be discussed here.

This section will end with a list of precautions that must be taken whichever method for measuring the effects of intersensory discordance is adopted. They are also factors that must be considered when assessing the significance of any experimental finding in this area.

1. The test used to reveal the way that the subject has resolved the conflicting

spatial information must not bias the outcome. It has already been mentioned that Rock and Victor (1964) controlled for this source of bias by giving tests in both the modalities involved, but not all investigators have taken this precaution. See Warren (1980) for a discussion of this issue.

2. The extent to which the subject explores the stimuli actively, rather than passively, must be controlled. Welch *et al.* (1979) found that if the subject's unseen hand was moved by the experimenter towards an optically displaced visual target, visual capture was enhanced compared with when the hand was moved actively by the subject. A person has a more precise idea of where a limb is when he is actively moving it than when it is moved passively (Paillard and Brouchon, 1968), and that probably accounts for the findings of Welch *et al.* The improvement in the sense of arm position produced by active movement must dissipate rapidly after the arm has come to rest, because Steinbach (1970) found that a hidden, passively moved arm was just as precisely localized by the gaze as was an actively moved arm, 5 sec after the arm had been brought to rest.

3. The synchrony of the two discordant stimuli must be controlled. As we have already seen, some investigators have added controls in which successive discordant stimuli were presented. Subjects tended to localize the source of stimulation accurately in each modality when this was done. However, the simultaneous presentation of stimuli is not sufficient to guarantee that discordant stimuli will be experienced as a single object, as we shall soon see.

4. The possible effects of the asymmetrical distribution of stimuli with respect to the body midline must be controlled for. If the stimuli in one modality are arranged symmetrically, then the stimuli in the other modality cannot also be symmetrical if there is to be an intersensory discordance. The effects of any such asymmetry must be assessed by appropriate control conditions. This has generally not been done.

5. Several studies have revealed that sensory dominance effects do not occur unless the subject is convinced that the stimuli which he feels, hears and sees originate in the same object. For instance, Jack and Thurlow (1973) found that a delay of as little as 0.2 sec between the mouth movements of a dummy speaker and speech sounds led to a large decrease in the probability that subjects experienced a ventriloquism effect (see also Thurlow and Jack, 1973). E. A. Miller (1972) obtained a greater degree of visual dominance when subjects were told that the visual and discordant tactile stimuli originated from the same object than when they were led to believe that there were two objects.

6. The results must not be generalized beyond the type of display used in the experiment. For instance, in many of the studies in which it was found that vision dominated audition, the visual stimulus was set in a full field of view containing other objects, while the auditory stimulus was the sound from a single source. Under these circumstances, it is to be expected that more reliance will be placed on visual information than on auditory information. Radeau and Bertelson (1976) found a greater degree of visual dominance over discordant auditory stimulation when a full field of objects was visible than when the visual stimulus was a luminous point in a dark surround. Furthermore, the stimuli in all those studies in which vision was found to dominate were stationary. There is some

evidence that things are different when moving stimuli are used. For instance, illusions of visual movement are produced by motion signals arising in the vestibular system (Section 9.3). Lackner and Levine (1978) reported that subjects fixating a light attached to the stationary hand saw it move when illusory motion of the arm was induced by vibration of the arm muscles.

All the studies of intersensory discordance which I have reviewed involved spatial discordance. Vision is the most precise spatial sense, and it is therefore not surprising that it is the dominant sense for spatial judgments. In a recent review of intersensory discordance, Welch and Warren (1980) drew attention to the possibility that vision may not dominate in other types of discordance. For instance, audition is particularly good for judgments of temporal intervals and may be expected to dominate over vision when associated sequences of visual and auditory signals are out of synchrony. They cited some unpublished studies on visual–auditory asynchrony in which it was found that a visual flash rate appeared to be driven by an intermittent tone.

7. The duration of exposure to the discordant stimuli should be controlled. If the experiment is designed to explore how people resolve conflicting spatial information in two modalities, the stimuli should be presented for only a short period and ideally only once to each subject. Rock and Victor took these precautions in their study of visual–kinaesthetic discordance. Other investigators have not always been so careful. The simplest explanation of the way in which one modality tends to be dominant under conditions of brief exposure is that the subject localizes largely or wholly on the basis of stimuli from one sense organ. A less likely explanation is that coding of signals from one sense organ is temporarily modified by the inputs from the other sense organ. But in either case there is no need to suppose that there will be any aftereffects of brief exposure to intersensory discordance. On the other hand, if the stimuli are presented repeatedly or for long periods of time, the discordance may begin to induce a more permanent change in the coding processes of one or other sense organ. Persisting changes of this kind are best measured as after effects, using localization tests applied after the discordant stimuli have been removed, because an aftereffect provides convincing evidence that there has been a persisting change in the way spatial information is coded. Experiments of this kind, which are reviewed in Chapter 12, have typically revealed that discordance gradually modifies the way in which spatial information is coded by each modality, in such a way as to nullify the effects of the discordance. This suggests that we possess intersensory covariance schemata which are subject to modification through experience.

11.5 The development of intersensory coordination

11.5.1 Introduction

The literature on the early development of intersensory and sensorimotor coordination in the infant has centred around the controversy between nativists and empiricists. Those predisposed to a nativist position stress the presence of spatially coordinated units of behaviour in the human neonate. For instance, Bruner and Koslowski (1972) observed arm and hand movements in the very young infant which were appropriate to the size of a visually perceived object. Evidence such as this has inclined Bruner (1973) to the view that visually-guided behaviour involves the temporal and spatial organization of innate constituents of the total performance, such as visually-triggered swipes of the arm and anticipatory grasping movements of the hand. The role of learning is to order and shape these constituent subroutines to meet the demands of particular tasks. Bower (1974) takes a more extreme nativist approach on the basis of his own evidence that infants only a few days old can reach towards visual objects and adjust their grasp to the nature of the object. On the other hand, those investigators disposed to an empiricist, or learning approach believe that the different modalities and the motor system are relatively independent at birth and that coordinated behaviour develops only after error-guided experience. Piaget (1954), for instance, argued that in the first phase of what he called the sensorimotor stage of development the senses function separately, so that if related events occur in two modalities, this fact is not reflected in the behaviour of the infant. It is only during the second phase of sensorimotor development that coordination schemata are acquired by learning experiences such as the child seeing its own hand in action.

Even the most ardent empiricist would surely not deny that there are certain innate spatially coordinated sensorimotor units of behaviour, such as the optokinetic response (Brazelton *et al.*, 1966; Barten *et al.*, 1971), postural reflexes, and behaviour patterns associated with feeding, such as the turning of the head towards the side which is touched (Prechtl, 1958). On the other hand, the most devoted nativist must agree that many spatially coordinated skills are learned. The real problem is to unravel the many factors that contribute to the development of particular patterns of behaviour.

11.5.2 Visually coordinated reaching in infants

One of the earliest identifiable movements of the arm occurs in the necktonic reflex, or ‘fencing’ reflex (see Howard and Templeton, 1966). Another reflex movement of the arm which occurs in the neonate is the **traction response**. This is a flexion of the arm and hand in response to a pull on the arm. When the infant is about two weeks old, flexion of the fingers occurs in response to contact stimulation of the palm. This is the well-known **grasp reflex**, which at first involves all the fingers but, after about the fourth month, becomes adapted to the shape and orientation of the object (Twitchell, 1970). These reflexes are not under visual control at first, but if their development is hindered by neuropathologies, the onset of visually-guided reaching is also delayed. They should therefore be regarded as part of the substratum out of which develops consciously directed behaviour towards external objects which, according to most investigators, occurs in the fourth month (see McGraw, 1943).

White *et al.* (1964) outlined a normative developmental sequence after observing infants during the first six months. During the first month, the child shows innate pursuit movements of the eyes and head, and by the second month these movements become more refined and show signs of predicting the future position of pursued objects. Arm movements are unrelated to vision at this stage. The grasp reflex is present but is wholly under tactile control. Infants under one month of age do not attend to objects within arm's reach, probably because of inadequate accommodation and convergence. In the second and third months, the infant visually attends to near objects and begins to take a visual interest in his own arms. The first visually-directed swiping movements of the arm develop, but the child grasps an object only if the hand touches it. In the third month, the swipe gives way to a more directed arm movement, and the child looks back and forth between object and hand. In the third and fourth months, the child watches the two hands as they contact and manipulate each other, thus producing a double feedback experience. In the fifth month, this double arm action comes under visual control and gradually gives rise to the ability to reach for and grasp an object rapidly. Like Piaget, White *et al.* stressed the importance of the infant's observation of its own moving hand. The results of experiments with animals done in the same department also stress the importance of visually observed self-produced movements of the limbs (Section 11.6).

White (1970) enquired whether the normative sequence could be speeded up by providing the infant with an 'enriched' environment. He reported that, where infants were nurtured in an environment enriched by a variety of objects hanging within reach, the onset of sustained hand regard occurred at a mean age of 50 days, rather than at 60 days as in infants reared in a 'normal' environment.

Bruner and Koslowski (1972) also observed what they called preadapted constituents of fully-fledged directed reaching in two-month-old infants. These consisted of visually-directed swipes of the arm, bringing the arms to the body midline, and grasping movements at the midline in the presence of a visual stimulus. The infants were more likely to make these responses when shown a small graspable ball than when they were shown a ball too large to grasp. This suggests that the preadapted constituents are in some sense directed at the visual object, even though they are not organized into an adequate reaching and grasping motion.

Bower *et al.* (1970) reported some observations that conflicted with all previous reports. They claimed to have observed infants only a few days old reaching towards visual objects, touching them, and occasionally grasping them. Furthermore, they showed infants between six and 11 days old the virtual image of an object and reported that the infants cried with frustration when their arm movements failed to contact the object, whereas infants successfully reaching for the real object did not cry. Neonates were also observed to shape the hand in anticipation before they touched the object, although not until the third month did the presence of an object in the hand elicit looking. This suggests that early movements of the arm are under visual rather than tactile control, which is the reverse of the usual view that early arm movements are under tactile–kinaesthetic control.

However, several investigators have failed to find any convincing evidence of visually-directed reaching in neonates (Dodwell *et al.*, 1976; Ruff and Halton, 1978). Di Franco *et al.* (1978) found many of the components of reaching in neonates, and some infants showed evidence of a much more rapid rate of development than others, but visually-guided reaching was not seen until between the second and third months. Field (1977) could find no evidence of distress even in three-month-old infants when they reached for virtual images.

Another debate in this literature has centred round the question of whether the earliest visually-directed reaching movements are under visual control while being executed. Reaching movements which are directed towards a visual object, but without visual control once initiated, are called **visually-triggered movements**. Responses which are modified during execution by visually perceived error feedback are called **visually-guided movements**. Piaget (1954) described early reaching behaviour in which the child constantly looks back and forth from hand to object. White *et al.* (1964) also observed such behaviour once the hand had been brought to within a few centimetres of the object. Bruner (1973), Bower (1974), and Hofsten and Lindhagen (1979) did not observe this type of early reaching, and concluded that sight of the hand does not seem to be required once the movement has been initiated. Bower and Wishart (1972) found that 12-week-old infants could grasp an object which they had seen when the lights were put out before the reaching movement was started.

One must be careful to distinguish between the necessity of viewing the hand when the child is learning to reach or learning to accommodate the accuracy of his reach to the changing dimensions of his body, and the necessity of viewing the hand on every reaching response once the reach is reasonably accurate. That is, one must distinguish between **parametric** visual control and **current** visual control. The crucial question is what the infant does when he fails to contact the object reached for: can he anticipate a misreach while the arm is still moving (current control), or does he only appreciate the error once the reaching movement is complete (parametric control)? Bower stated that infants under five months of age cannot correct a misreach, but must withdraw the hand and start again. McDonnell (1975) placed 30-diopter prisms in front of the eyes of four-month-old infants and reported that the infants switched from a mispath to a hitpath in midcourse. Hofsten and Lindhagen (1979) confirmed that infants of about this age correct for misreaches during a movement. These investigators also showed that, by the time infants could reach accurately for stationary visual objects, when about four months old, they could also reach for moving objects. Success in this task could be achieved by the infant reaching towards the current position of the moving target and then using visual guidance to home in on the target. This strategy would produce a curved trajectory of the hand. Alternatively, the infant could extrapolate the future position of the target and reach directly to that location. Hofsten and Lindhagen claimed that the shape of the arm trajectory showed that the infants used the second strategy. It would be more satisfactory if *this experiment on reaching towards objects were repeated under conditions in which subjects were not allowed to see the arm during the movement.*

These experiments demonstrate that, in one way or another, infants can correct for reaching errors induced by viewing objects through displacing prisms. McDonnell and Abraham (1979) went on to show that in six-month-old infants these corrected movements persist as an aftereffect when the prisms are removed, in the same way that prism-adapted responses persist in adults. Infants of this age are thus capable of both current and parametric control of visually-directed reaching.

Few studies have been done on the development of visual–motor coordination beyond the infant level. None of the earlier studies revealed any age trends in the accuracy or precision of kinaesthetic–visual or visual–kinaesthetic tasks, although the youngest age with which adult performance was compared was six years (Sandström and Lundberg, 1956; Pick *et al.*, 1967; Smothergill, 1973). More recently, L. Hay (1978) measured the mean unsigned error of reaching with unseen hand towards visual targets, for children between the ages of four and 11 years. The level of performance of children below seven years of age was similar to that of adults, but the mean error was greatest for seven- and eight-year olds. The decrease in the level of performance at these ages was put down to the disruptive effects of a transition from a predominantly ballistic style of reaching which younger children use to a visually guided style of reaching typical of older children and adults. Films of pointing movements in children of different ages revealed these age-related differences in reaching style (Hay, 1979).

11.5.3 Visual–auditory coordination in infants

Wertheimer (1961) reported that a human neonate reliably turned the eyes in the direction of an auditory click. Butterworth and Castillo (1976) and McGurk *et al.* (1977) could not replicate Wertheimer's observation with randomly presented clicks. However, Butterworth and Castillo were able to elicit eye movements which were related to the side upon which a steady tone was sounded, although these were towards the side opposite to the tone. Some resolution of these conflicting reports is provided by the results of a study by Turkewitz *et al.* (1966) in which it was found that whether neonates turned their eyes away from, or towards a tone depended on the loudness of the tone, the level of arousal of the infant, and the ear into which the tone was sounded. Muir and Field (1979) found that when a continuous sound was presented to the side of the head and when the head was free to turn, the majority of newborn infants that were tested turned their heads towards the sound. (see also Crassini and Broerse, 1980) It has been reported that even blind infants turn their eyes towards a sound during the first few weeks of life (Freedman, 1974).

Aronson and Rosenbloom (1971) reported that infants 30 days old showed distress when they saw their mothers speak to them while the voice was emitted from a displaced loudspeaker. McGurk and Lewis (1974) criticized the design of this experiment on several counts and failed to replicate the finding in what they considered to be a better designed experiment (see also Condry *et al.* 1977). The success of experiments of this type depends upon the assumption that the child has either an innate or a learned association between particular sounds and particular visual objects. Spelke (1976) put this assumption to experimental test by showing four-month-old infants two motion-picture sequences of familiar events side by side, while playing a soundtrack appropriate to only one of the sequences through a centrally placed speaker. The infants looked mainly at the sequence specified by the soundtrack, domonstrating that they could associate sights with appropriate sounds in the absence of spatial cues. In a subsequent experiment, Spelke (1979) presented four-month-old infants with a motion film of two toys, side by side, being bounced at different regular rates. A repetitive sound was presented mid-way between the two pictures with the same frequency and phase as one or other of the bouncing toys. It was found that the infants tended to look first at the toy with which the sound was synchronized. In other similar experiments, it was found that infants tended to look towards the toy which bounced at the same frequency as the sound, even though the two series of events were out of phase, and towards the toy which bounced in phase with the sound when one toy was out of phase and one was in phase with the sound, both bouncing at the same frequency.

Wishart *et al.* (1978) reported that four-month-old infants could reach towards the source of a sound in the dark. Younger infants were not tested, but it was found that the frequency of such reaching movements peaked at the age of six months and then declined during the next few months. Wishart *et al.* suggested that this decline in the frequency of reaching to sounds was related to the fact that the infants often failed to contact the object making the noise and were more likely to be discouraged by this failure as they got older. The frequency of reaching recovered as the infants approached the age of 12 months, presumably because they became more accurate and therefore more confident.

11.6 The effects of restricted rearing on visual–motor coordination

It should be possible to discover what kinds of sensorimotor experiences are required for the development of different visual–motor skills by rearing animals in environments which restrict the types of experience which they receive. This idea has inspired a series of experiments by Richard Held and his associates. In the most famous of these experiments (Held and Hein, 1963), pairs of kittens were reared in darkness, except for a certain period each day when they were placed in an illuminated carousel apparatus, as shown in Figure 11.5. One kitten of each pair was always placed in the box so that its feet did not touch the ground and the other member of the pair was always placed on the other end of the rotating lever so that it could walk and thereby cause itself and its passive partner to be moved round inside the striped drum. Both kittens had the same visual experience of moving stripes, but for the active kitten the visual movement was

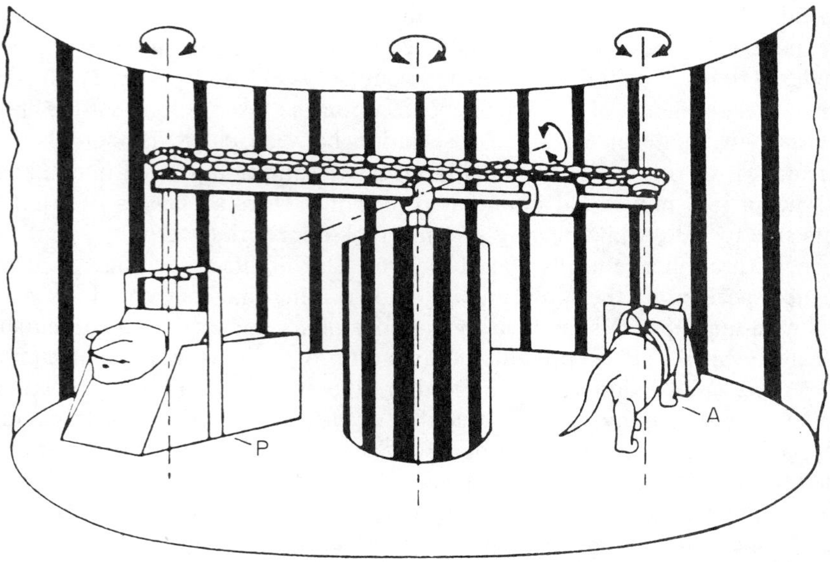

Figure 11.5 Apparatus used by Held and Hein (1963) for equating body and visual motion for an actively moving animal (A) and a passively moving animal (P) (Copyright, 1963, by the American Psychological Association. Reprinted by permission)

correlated with the efference signals associated with its walking and for the passive kitten the visual movement was not in any obvious way related to other experiences. Sensory stimulation which results from efference associated with self-produced movement is known as **reafference**, and sensory stimulation which occurs independently of efference is called **exafference**. Held and Hein found that only the active kitten developed the ability to avoid a cliff, blink at an approaching object, or extend its paw to a surface. They concluded that reafference is necessary for the development of visual–motor skills. I have pointed out (Howard and Templeton, 1966, p. 394) that the passive kitten was never given usable exafferent stimulation. All it saw was a moving display of stripes and there were no other features of its environment to which it could relate this visual input. Held and Hein never tried to teach it anything. But there is a vast range of possibilities for correlated exafferent inputs which could be tried. For instance a visible object could be placed in the path of the animal; on some occasions the object could collide with the animal, and on other occasions the object could miss. To provide reinforcement (another thing Held and Hein failed to supply to the passive animal), the object could be food and sometimes hit the kitten in the mouth, or it could give an electric shock, or make a noise, pleasant or unpleasant. Another possibility would be to cause the passive animal to bump down a step and thus receive strong vestibular and tactile stimulation. In all this one would merely have to ensure that the passive kitten could not move actively in such a way as to alter the visual signals. Eating movements, startle responses, and the like would be permitted, for these would not affect the critical spatial feedback.

It is known that the response of extending a paw towards a visual surface blinking to an approaching object, and cliff avoidance may be elicited in animals totally lacking in visual–motor experience, so that the absence of these responses in Held and Hein's passive kitten was a mystery. Hein, Held and Gower (1970) went some way towards explaining this mystery. They showed that the deficits evident in the passive kitten were not due to motor defects in the limbs by demonstrating that the response deficits could be specific to only one eye. They repeated the original carousel experiment and showed that after the experiment the passive kittens took 41 days to learn the paw extension response when allowed to move themselves in a visual environment, which was much longer than it took a kitten which had been totally deprived of vision for the same length of time. Furthermore, Hein, Gower and Diamond (1970) found that the paw-placing response developed in immobilized kittens which had experienced only diffuse light. The passive experience in the carousel must have interfered with the normal maturational process by which the response develops, even without reafferent experience.

Therefore, it cannot be claimed that reafferent stimulation is necessary for the

Figure 11.6 Apparatus devised by Hein and Held (1967) to test visually-guided reaching in the cat (Copyright, 1967, by the American Association for the Advancement of Science)

development of the three visual–motor skills that Held and Hein originally used. These responses develop without such experience. It can only be claimed that the peculiar conditions of passive experience in the carousel inhibit that development.

Hein and Held (1967) developed tests of other abilities which they claimed did not develop without reafferent visual experience. One of these is the reaching response of the forelimb towards a visual target in a specific direction. In the cat, this took the form of a task in which the cat was required to visually guide its paw onto one of a set of prongs, as shown in Figure 11.6. Hein and Held reared kittens in darkness until they were four weeks old after which they were allowed each day to walk about freely in an illuminated enclosure, but with opaque collars round their necks which prevented them from seeing their limbs (see Figure 11.7). After 12 days of this treatment they extended their paws towards the surface defined by the prongs but did not hit the prongs more frequently than one would expect by chance, and they could not accurately strike a ball that was dangled in front of them. Held and Bauer (1967) conducted a very similar experiment with dark-reared monkeys. They too wore a collar which prevented them from seeing their arms (see Figure 11.8). During the exposure phase of the experiment they were

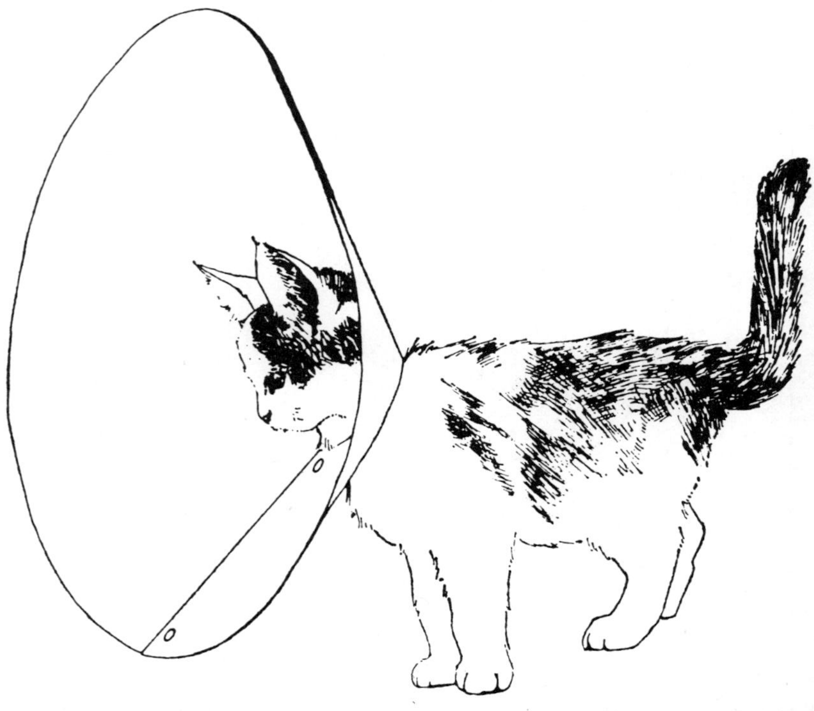

Figure 11.7 Procedure used by Hein and Held (1967) to prevent a developing kitten from seeing its own limbs (Copyright, 1967, by the American Association for the Advancement of Science)

Figure 11.8 Apparatus used by Held and Bauer (1967) to prevent a developing monkey from seeing its own limbs (Copyright, 1967, by the American Association for the Advancement of Science)

placed in the apparatus and shown a milk bottle towards which they directed their gaze. They were reinforced for extending the arms in the general direction of the bottle, *but were not allowed to touch it.* After 35 days one of the arms was allowed to come into view, and as soon as this happened the monkeys gazed at the hand and followed it as it moved from side to side. The monkeys did not move their arms accurately towards the bottle at first but, after several days, accurate visual guidance of the hand towards the bottle was achieved. It was concluded that '. . . an infant primate initially fails to reach accurately for attractive visible objects with a limb that it has never previously viewed.' The same conclusion was meant to apply to the cat also.

There is a serious flaw in this conclusion. The monkeys were not allowed to touch the bottle before the collar was removed, so that they had no experiential basis for learning to relate the seen object with anything that the arm did. The monkeys should have been allowed to touch the bottle with unseen hand while looking at it. Only if visually-guided reaching failed to develop under these circumstances would it be legitimate to conclude that sight of the arm is necessary. Walk and Bond (1971) replicated Held and Bauer's experiment with this control added. They allowed the monkeys to touch one end of a rod, the other end of which they could see projecting above the edge of the collar which occluded the hand. After this exposure the visual preoccupation with the previously unseen hand was absent and visually-guided reaching was tolerably accurate. Therefore, sight of the hand is not required for the development of visual–motor coordination.

Sight of the moving limb is no doubt an effective way to train visual–motor coordination, but a more general principle of visual–motor learning which probably accords with the facts is that a visual–motor skill will develop in the presence of information of any type which informs the animal about the accuracy of performance. The performance of humans with sensory and motor defects from birth certainly attests to the truth of this more general principle for humans. But visual–motor learning can be very specific to the conditions under which it is learned, as we shall now see.

Visual–motor skills learned under the guidance of one eye do not necessarily transfer to the other eye. This type of specificity was demonstrated by Hein, Held and Gower (1970) with kittens in the carousel, each of which was exposed alternately to (a) passive body movement when only one eye was open, and (b) active movement when only the other eye was open. In subsequent tests, applied with each eye open in turn, behavioural deficits showed only when the eye which had been exposed to passive movement was open. Van Hof-van Duin (1976) reared cats with one eye sutured for 10 months and found persistent defects of visually-guided reaching, obstacle avoidance, and visual tracking when vision was restored to the deprived eye and the experienced eye was closed. Pattern vision was not affected in either eye. Robinson and Fish (1974) reared cats in a similar way, but for 18 months. They found a similar pattern of eye-specific deficiencies in visually-guided behaviour, especially for objects imaged in the central part of the retina and in photopic levels of illumination. They also found a disruption of conjugate eye movements and defective pattern vision. The specificity of visual–motor learning was also revealed in an experiment by Hein and Diamond (1971). They reared cats binocularly in low levels of illumination and found that visually-guided behaviour was evident only in dimly-lit surroundings.

There thus seems to be no doubt that visually-guided reaching and obstacle avoidance do not develop in the light-deprived eye or even in an eye exposed to the conditions of passive movement in the carousel. More recently, Hein *et al.* (1979) surgically immobilized one eye in kittens from the time when they were first allowed to see. The kittens were given visual experience with this eye, alternating with visual experience with their normal eye. Later, the kittens did not show visual reaching or avoidance behaviour when using their immobile eye, even though they could execute these skills when using their normal eye. Controls showed that the deficit was not due merely to a loss of accommodation in the operated eye nor to a loss of vestibular reflexes. Immobilization of the eye in a visually experienced kitten did not produce these deficits. It was concluded that eye movements are necessary for the development of visual–motor skills, but the crucial factor in causing these defects may have been the inability to stabilize or foveate the retinal image. Hein *et al.* (1981) found that deafferentation of an eye without immobilizing it caused these same deficits in kittens (see Section 7.7 for details). This suggests that afference from eye muscles is necessary for visual–motor development, but it is possible that the defects were a secondary effect of abnormal eye movements in deafferented animals.

Experiments involving rearing animals under restricted sensorimotor conditions do not lead to any firm conclusions. Certain types of early deprivation have a general debilitating effect, but visual–motor learning is usually very specific to the conditions under which it occurs.

11.7 The neurology of intersensory coordination

11.7.1 Introduction

Very little is known about the neurology of intersensory coordination. For any pair of spatially coordinated modalities, there must presumably be cells at one or more sites in the nervous system that respond to spatially corresponding inputs from both modalities. Indeed, *intersensory conjuction of inputs* must occur at several levels, because intersensory behaviour also occurs at several levels. At the most basic level, there is a conjunction of sensory information in the reticular arousal system of the mesencephalon (see Bell *et al.,* 1964). This system is responsible for arousing subcortical and cortical centres to the presence of stimulation and has more to do with the intensity and general nature of stimuli than with their location in space. At the next level, the simple visual orienting response that an animal shows towards a source of sound, or the visual righting reflexes probably involve the conjunction of sensory inputs at subcortical levels (see, for example, Gattass *et al*., 1978). At a still higher level, the conscious control of directed reaching to visual or auditory targets involves relationships between visual and auditory inputs in subcortical structures, in the visual and auditory cortex, and in the parietal and frontal lobes, as we shall see in what follows. Finally, the construction of a mental representation of a familiar object or geographical region, in terms of which inputs from several modalities may be assessed, is a very high-level form of intersensory, or suprasensory spatial coordination, and must involve many parts of the brain. This last topic is beyond the scope of this book. When a neurophysiologist reports that he has found cells that respond to inputs from two sensory modalities, the level of behavioural integration that these cells serve may not be known, and care must be exercised in interpreting such findings.

11.7.2 Conjunction of visual and auditory inputs

Chalupa *et al.* (1975) reported that cells in the lateral geniculate nucleus of the cat respond only to visual and not to auditory stimulation. However, continued stimulation of lateral geniculate cells of the sustained type by flashes of light paired with auditory clicks led to a decrement in response frequency. This decrement occurred only when the flashes and clicks were separated by less than 0.1 sec. This effect may be mediated by inputs from the mesencephalic reticular system (Section 5.4.1) or by inputs from cells in the visual cortex which respond to both visual and auditory inputs. Chalupa *et al.* mentioned similar findings by others, but it is not clear whether or not any of these findings are relevant to the problem of intersensory localization.

Perhaps of more relevance are reports by several investigators that some neurons in areas 17, 18, and 19 of the cat's visual cortex respond both to visual and to auditory stimuli (Jung, *et al.*, 1963; Spinelli *et al.*, 1968; Morrell, 1972). Morrell, for instance, reported that the position in a horizontal plane of an acoustic stimulus that produced a response in a given cortical cell was the same as that of a visual stimulus that produced a response in the same cell. Here then is a true conjunction of spatially corresponding inputs from two modalities. Fishman and Michael (1973) found that approximately one-third of cortical cells in areas 17 and 18 of the cat responded to both auditory and visual stimuli. The responses were all excitatory; no inhibitory effects were seen. The **bimodal** cells had binocular inputs and some were simple cells and some complex cells. Like Morrell, these investigators found that many of the bimodal cells responded to visual and acoustic stimuli in corresponding positions. Furthermore, they reported that bimodal cells tend to occur in clusters within distinct cortical columns. Thus, some columns were found to contain mostly bimodal cells and other columns were found to contain none.

The various investigators have tended to assume that these bimodal cells have something to do with directing movements of the eyes and head towards auditory and visual stimuli. There is an obvious problem with this interpretation, as Pöppel (1973) has pointed out. The cells of the visual cortex respond to stimuli in a particular occulocentric direction, whereas auditory localization is related to headcentric directions. Thus, if the eyes were to move, the two inputs to a bimodal cells would be put out of register. But this is not a problem for the cat. When a cat is hunting it does not move its eyes far without rapidly directing its head towards any slight visual or auditory cue—ready to pounce. When it hunts quiet animals in dim light, a conjunction of inputs in the earlier stages of sensory processing will aid detection, because the signal-to-noise ratio is theoretically improved by a factor of $\sqrt{2}$ when two signals with uncorrelated noise are combined. One would not necessarily expect such an early and direct intersensory coupling of information in humans.

11.7.3 The cortical control of visual–motor coordination

There are two types of evidence bearing on the question of the cortical control of visual–motor coordination: that deriving from surgical and clinical studies on animals and men, and that deriving from the study of responses of single cells in the central nervous system. Only a very brief review of each of these areas of study can be given here, and attention will be confined to studies of eye–hand coordination in reaching.

A disorder of visually-directed reaching is produced by surgically removing either the visual cortex or the precentral motor areas of the arm, but in each case the disorder of coordination is a by-product of widespread visual or motor deficits and tell us little about the mechanisms underlying visual–motor coordination. Of more specific interest are studies of the effects of severing interhemispheric connections passing through the corpus callosum, which I shall now review.

Anatomical studies have revealed that the motor cortex on each side of the brain has control over coarse movements of limbs on both sides of the body, through the ventromedial pathways of the spinal cord. The finer movements of the arm, and particularly of the hand, are controlled mainly by the motor cortex on the opposite side of the brain, over neurones which descend in dorsolateral pathways of the spinal cord (Brinkman and Kuypers, 1973). If movements of an arm are to be controlled by vision, then pathways must connect the visual cortex in the occipital lobe with the motor cortex in the frontal lobe. Ipsilateral connections would be involved when an arm reaches for a seen object on its own side of the visual field, and contralateral connections would be involved when an arm reaches for something on the opposite side. Remember that the left half of visual space is projected onto the right hemisphere (Figure 2.22).

Downer (1959), and later Lund *et al.* (1979), severed the corpus callosum and optic chiasma of a monkey and at the same time restricted vision to one eye. In this way visual inputs were restricted to the ipsilateral hemisphere and the most probable route for contralateral connections from the visual cortex to the motor cortex was interrupted. It was found that, in a variety of tasks, the monkey could adequately control reaching movements of the hand contralateral to the seeing eye (controlled by the ipsilateral motor cortex), but groped awkwardly with the ipsilateral hand. This seems to confirm that the corpus callosum is indeed the route taken by pathways connecting the visual cortex on one side with the contralateral motor centres controlling fine movements of the limbs. It also strongly suggests that ipsilateral connections run through intracortical pathways. Downer found that after the operation the monkeys gradually improved their performance with the affected limb, which suggests that they were learning to use the coarser system of motor control in a more refined way although, as we shall see, there are other ways of accounting for this improvement.

Certain other investigators have failed to replicate Downer's findings (Myers *et al.*, 1962; Hamilton, 1967). But these investigators may have overlooked the motor deficit because of a failure to differentiate between coarse and fine limb movements. It is also possible that the deficit was obscured by the presence of tactile control of movements of the fingers, involving somaesthetic pathways to the contralateral hemisphere (Brinkman and Kuypers, 1973), or by kinaesthetic cues associated with movements of the eyes or the head (see later). In any case, Downer's results are consistent with results of studies on split-brain humans which I shall now review.

The corpus collosum of human beings is sometimes severed as a clinical procedure. In these so-called **split-brain** humans the optic chiasma is not severed but, by controlling fixation, it is possible to restrict a given visual stimulus to one hemisphere. When this is done, the arm contralateral to the visual stimulus can locate the stimulus only by groping, whereas the ipsilateral arm performs normally (Gazzaniga *et al.*, 1967) (don't forget, each hemifield projects to the contralateral hemisphere and each hemisphere controls the opposite arm). If the subject fixates his gaze on the visual target, by moving either the eyes or the head, he can perform adequately with both arms because a foveal image falls across both hemispheres. But Gazzaniga (1969) found that split-brain subjects can also reach accurately to the position of a visual target with either arm, even if the visual target is switched off before the centring movement of the eye or head is complete. In this case the kinaesthetic–motor cues associated with the already initiated eye or head movements are available to both hemispheres, and this enables the subject to adopt what Gazzaniga has called a 'cross-cueing strategy' in controlling the contralateral arm.

These studies tell us something about the interhemispheric control of visual–motor coordination. But perhaps more basic questions have to do with intrahemispheric processes. Haaxma and Kuypers (1975) severed the corpus callosum of the monkey so that a given arm was controlled by only one hemisphere. They then removed the *posterior pareital lobe* on one side. This produced a defect in visually guided reaching with the contralateral arm; animals missed when attempting to reach for food, and movements of the fingers in the act of grasping were impaired. When food was held behind a transparent barrier, the monkeys reached round the barrier with the ipsilateral arm, but when this arm was restrained, they tried to reach directly through the barrier with the contralateral arm, as if relying on a more primitive mechanism which was unable to take account of the presence of the barrier. Similar deficits caused by unilateral or bilateral ablation of the posterior parietal lobe in monkeys have been noted by several investigators. Additional symptoms that have been reported include a loss of ability to discriminate between objects by touch and an absence of spontaneous movements of the contralateral limbs. It is generally agreed that there are no gross sensory losses, such as loss of visual acuity or tactile sensitivity, and no motor paralysis (see Darian-Smith *et al.*, 1979 and Faugier-Grimaud *et al.*, 1978, for references). Although errors of misreaching towards visible targets are confined to the limb contralateral to cortical damage, the ipsilateral limb shows evidence of impairment when the animal attempts to reach for an object after the lights are put out (Ratcliff *et al.*, 1977; LaMotte and Acuna, 1978). Several investigators have reported that defects in reaching are manifest only with respect to objects in extrapersonal space; animals show no deficit when reaching towards parts of their own bodies.

Rondot *et al.* (1977) recently reviewed the literature on a clinical condition in man known as **visuomotor ataxia**, which is due to damage of the posterior parietal lobe. The symptoms of this condition closely resemble the defects in animals which have just been described. They do not necessarily include motor paralysis of the arm or eye muscles, or visual defects. These symptoms must also be distinguished from defects due to damage to the cerebellum or basal ganglia.

The patient has difficulty reaching for and grasping a visual object and resorts to groping randomly, both from side to side and in depth. In striking contrast, movements which do not require visual guidance, such as buttoning garments, or pointing to some part of the body, are unaffected (Damasio and Benton, 1979). The defect may be **unilateral**, that is, it may affect only one half of visual space, or it may be **bilateral**. If it is unilateral, it is called **direct** when it affects only the hand on the same side as the affected visual hemifield and **crossed** when it is confined to the other hand. It is both direct and crossed when both hands are affected. If the defect is bilateral, both hands may be affected whichever side they reach to or each hand may have difficulties only when reaching into the contralateral hemifield. When the defect is unilateral and affects only one arm, it normally does not show when the patient is allowed to fixate the visual target. The defect is distinguishable from spatial agnosia because spatial agnosia is not dependent on which arm is used. It is also distinguishable from motor apraxia because this would not be confined to one half of visual space. In all the cases reported by Randot *et al.* the damage affected the junction of the occipital and parietal lobes. It is assumed that the pathways from the visual areas to the motor areas of the frontal lobe pass through the parietal lobe, and that these can be both crossed and uncrossed.

Other literature on the behavioural effects of parietal lobe damage is reviewed by Critchley (1969), and a more recent review of the functions of the parietal lobes is provided by Mountcastle *et al.* (1975) and Lynch (1980).

The posterior parietal lobes are well placed to serve as centres for the higher control of coordination between vision, somaesthesis, eye movements, and limb movements. Although there are no direct sensory inputs to or motor outputs from the posterior parietal lobes, there are indirect corticocortical connections with primary sensory and motor area (Darian-Smith *et al.*, 1979). There are also connections with the limbic system and with subcortical structures, including the thalamus, basal ganglia, and pontine nuclei (see Lynch, 1980). The analysis of cerebral blood flow in man has revealed that the parietal lobes are particularly active during the execution of voluntary movements in relation to seen objects (Roland *et al.*, 1980).

Mountcastle and his associates have explored the trigger features of single cells in the posterior parietal lobe (Brodmann's areas 5 and 7) in the monkey (Mountcastle *et al.*, 1975; Lynch *et al.*, 1977; Yin and Mountcastle, 1977). one class of cells was found which responded to passive movement of the limbs or skin. In a second class, some cells responded when the animal reached for an object of interest. These were called **hand projection neurones**. Others responded when the animal grasped an object, and were called **hand manipulation neurones**. A third class of cells responded to visual stimuli, and were called **visually responsive neurones**. These cells had large receptive fields which never included the fovea and which were usually in the contralateral hemifield. Some of them increased and some decreased their rate of firing just before a saccadic eye movement, and their functional properties suggested to Mountcastle that they derive their afferent signals from the superior colliculus. A fourth class of cells responded in association with eye movements; **fixation neurones** fired when the animal fixated an object of interest within a given region, **saccadic neurones** discharged just before saccadic eye movements to objects of interest, and **visual tracking neurones** discharged during pursuit of an object (see Section 5.4.7).

Robinson *et al.* (1978) found essentially the same types of cell in the posterior parietal lobe of the monkey. However, these investigators found that, while all parietal neurones responded to appropriate sensory stimulation in the absence of active limb or eye movement, none responded in the presence of limb or eye movement when sensory stimulation was absent. They concluded that the posterior parietal lobe is concerned with directing the animal's attention towards stimuli of interest, rather than with organizing responses to such stimuli. Mountcastle, on the other hand, implicated the posterior parietal lobe in the initiation of movements of eyes and limbs.

Similar types of cells have also been reported in a series of studies from the laboratory of Leinonen and Hÿvärinen (see Leinonen and Hÿvärinen, 1980 for a review of this work). They found cells which responded only during active movements of the arms or eyes towards objects of interest, and this supports the notion of a command function for this region of the brain. Even Robinson *et al.* admit that many cells which respond to visual stimuli do so more vigorously if the stimulus triggers a saccade. Perhaps the debate about whether or not the parietal lobes have a motor command function is spurious. Even if parietal cells do respond when movements occur, this may be because of a corollary discharge from motor centres, rather than because the cells are directly involved in organizing motor responses. In any case, the idea of command cells, or cells that initiate movements, is misleading. We do not know where or how movements are initiated, or even what it means to say that a movement is initiated. The dichotomy between sensory and motor functions has meaning for the peripheral sensory and motor systems, but is probably inapplicable in structures like the parietal lobes, which are concerned with the high-level integration of diverse types of information.

One must be careful not to emphasize the role of the cerebral cortex in the control of visually guided movement to the exclusion of other centres, especially subcortical centres such as the cerebellum and basal ganglia. Beaubaton *et al.* (1980) report that paralysis of the globus pallidus (one of the basal ganglia) or of the dentate nucleus (in the cerebellum) in the monkey leads to a loss of accuracy in pointing to visual targets. Defects of this kind could be due to disruption of the motor command system rather than of the system which integrates visual signals with motor commands. Evidence that the basal ganglia are specifically implicated in visual–motor control is provided by a study by Aldridge *et al.* (1980). They took electrophysiological recordings from single cells in the basal ganglia of the monkey and detected cells which received highly processed information, probably from cortical association areas, and which were active during the time that the animal tracked a moving visual object with its hand. Damage to the basal ganglia of humans, such as is associated with Parkinson's disease, results in

ıkinesia, or an inability to initiate movements towards a visual object. This lisorder is manifest only in certain contexts, which suggests that the basal ganglia ıre concerned with organizing responses to specific visual information.

This brief review indicates that our knowledge of the neurology of visual–motor coordination is very fragmentary. Perhaps the development of ıutomata will help to provide theoretical insights which will guide future physiological work.

Chapter 12

Adaptation to Discordant Stimulation

12.1 Introduction

Subcortical sensory systems responsible for the initial coding of stimulus attributes are 'hard wired' in the human neonate. For instance, the retinotopic projection of ganglion cells to the visual cortex is fixed before birth and is normally immune to the effects of subsequent environmental influences. Other sensory mechanisms, more centrally placed in the nervous system, such as the orientation detectors of the visual cortex, pass through a critical period in early infancy. During this period they are readily modified by environmental influences but are then largely immune to such influences. In a similar way the basic machinery of the motor system is laid down in the embryo or in early infancy. Nevertheless, the spatially coordinated behaviour of adult human beings is remarkably flexible and readily adjusted to suit changes in stimulus conditions. This flexibility arises, not in the basic sensory or motor mechanisms, but rather in the processes which relate one sensory system with another, or which coordinate sensory events and motor responses. For instance, we saw in Section 9.2 that the mechanism of vestibular nystagmus, although basically innate, has built into it a process for adapting the gain of the response to changing relationships between image motion, eye motion, and head motion.

Flexibility is required even in involuntary systems of this sort, because the precise conditions under which such mechanisms will be required to work cannot be fully predicted. There are several reasons for this.

In the first place, the shape and size of body parts change as a person grows, and such changes would disturb the calibration of spatial mechanisms if these mechanisms did not possess some power of adjustment.

Secondly, human beings inhabit a variety of environments, each of which possesses its own peculiar stimulus features. For instance, the stimulus world of a Japanese pearl diver is very different from that of a Greenland Eskimo. Lower animals lack the intelligence to exploit a variety of environments, so that they have less need of flexibility in their basic spatial skills. They tend to possess highly specialized sense organs and specially shaped motor organs, such as claws, beaks, and wings, which are used in relatively stereotyped behaviour. Man, on the other hand, can use his intelligence to exploit almost any environment, and has dispensed with narrowly specialized sense organs and limbs. In doing so, he has increased the need for adaptability of the neural structures which serve these relatively unspecialized structures. This adaptability reaches its highest expression in learned skills.

Flexibility of spatially coordinated bahaviour is also required to compensate for the effects of injury. Many lower animals can grow a replacement limb if one is lost, but severely injured humans must learn to coordinate their remaining limbs and sense organs in a novel way.

Finally, flexibility is required in spatial behaviour, to correct for the tendency of any complex control system to drift from its peak level of performance.

There are several ways of studying the flexibility of spatially coordinated behaviour. One way is to anatomically rotate or transplant sense organs or tendon insertions. This method can be applied to humans only when radical surgery is required for medical reasons. A second procedure is to study animals that have been deliberately reared in anomalous sensory environments (see Section 11.6). Human beings cannot be deliberately manipulated in this way, but some people have severe sensory deficits from an early age, and their performance on selected tasks is often informative. Finally, the flexibility of spatially coordinated behaviour may be studied by temporarily distorting the normal relationship between one stimulus dimension and another, or between a stimulus dimension and associated responses, and measuring the adaptive changes that occur in selected perceptual and perceptual–motor tasks. This method is ideally suited for use with human subjects, and this chapter is concerned largely with experiments in which it has been used.

The classic study of sensory distortion in man was conducted by Stratton (1897a, b). For six days, he continuously wore spectacles which inverted and reversed the visual scene. Stratton gave himself no systematic tests, but recorded his experience in the context of his normal routine of life. He found that his actions gradually became more smoothly coordinated, although each one had to be mastered separately. Similar studies were done by Erisman and Kohler in Innsbruck (see Kottenhoff, 1957a, b, for an English summary of this work). Like

Stratton, they also found that there was little transfer of learning from one action to another. Apart from this rather general finding, these experiments revealed very little about the nature of visual–motor adaptation to distorted vision. This was because the exposure and testing procedures were unsystematic and because wearing reversing and/or inverting spectacles disturbed many aspects of behaviour simultaneously, so that it was difficult to isolate particular effects. Consider, for example, the effects produced by an optical system which rotates the visual scene through 180° in a frontal plane. There is first the *intravisual* disturbance produced by the upside-down appearance of familiar objects, an effect which I refer to as **reversed visual polarity**, and which is discussed in Section 14.4. Then there are the *intersensory* disturbances, such as the disturbed relationship between where an object, such as a ringing bell, is seen and where it is heard. Finally, there are the disturbances of *visual–motor coordination*. Each of these disturbances should be studied separately, with controlled conditions of exposure and systematic testing procedures. A **sensory discordance** is a disturbance (relative to some agreed standard) in the way in which stimuli within one sensory dimension covary with respect to those in another sensory dimension. If the two sensory dimensions are in the same sensory modality, we have an **intrasensory discordance**, and if they are in separate modalities, we have an **intersensory discordance**. Note that an intersensory discordance may not be noticed by a subject when both stimuli are present together, because impressions in one modality may dominate over impressions received in the other modality (Section 11.4). A **sensorimotor discordance** is a discordance between changes in some sensory attribute and the covarying changes in the self-produced movement of some part of the body. Examples of intrasensory discordance are provided in Table 1.1 and examples of intersensory discordance in Chapter 11. The present chapter will be devoted to the whole question of how human subjects adjust to intersensory and sensorimotor discordance. Other recent reviews of this area have been provided by Rock (1966). Ross (1974), Kornheiser (1976), Welch (1978), and Howard (1978a).

A discordance between two sensory dimensions may be produced in two basic ways. One way is to *distort the chain of events between the distal stimuli in one of the sensory dimensions and the neural events into which the stimulus energy is ultimately transduced.* For example, a discordance between the seen and felt position of an object may be introduced by optically displacing the visual image of the object. The same discordance may also be introduced by perturbing the sense organs which signal arm position, for instance by mechanically vibrating them. The second way in which a discordance may be produced is to *mislead the subject about the identity of the object from which stimuli arise.* For instance, a visual–auditory discordance is produced if a subject believes that the source of a ringing sound is a bell that he can see when in fact the sound originates from a hidden source in some other location. Another example is provided by the situation in which a person believes that he is seeing his own finger when in fact it is somebody else's finger in another position. A spatial discordance due to a mistake in the indentification of an object may be replicated by a distortion of one of the sensory inputs. For this reason I shall use the term 'sensory distortion' to refer to anything which is equivalent to a disturbance of the normal chain of events occurring between a distal stimulus and the neural events underlying judgments which a subject makes about the stimulus.

A perceptual system devoted to a given sensory dimension cannot, by itself, distinguish between a sensory distortion and a transformation of the distal stimulus. For instance, a wedge prism causes the retinal image of a visual display to be displaced to one side. But this same displacement of the image may also be produced by a truly displaced distal display, and the visual system cannot distinguish between these two causes of image displacement. In other words, the effects on a given sensory system of a sensory distortion may always be matched by a suitably selected simple transformation of the distal stimulus.

A sensory distortion can be distinguished from a simple distal transformation only by comparing the effects which the distal stimulus has on the distorted sensory system with the effects that it has on some other, undistorted sensory system. For instance, if a person views a scene through an optical device which displaces the image to one side, he will not be aware that vision is distorted (assuming of course that he is not told about the optical device and that he was not allowed to see the display beforehand). However, as soon as he reaches towards an object in his field of view, he will become aware that there is a sensory distortion, because he will see that his hand has not reached in the right direction. He may not know which sense has been distorted but, because of discordance between the visually-sensed direction of the object and the kinaesthetically-sensed direction of his hand, he will know that there is a sensory distortion and not merely a transformation of the distal stimulus. If the coordination of vision and the kinaesthetic–motor system is to be restored, some readjustment of the internal calibration of vision with respect to the kinaesthetic–motor system must occur. In this chapter I will outline what is known about the conditions leading to this kind of readjustment and what is known about the site of the processes responsible for it. But I must first consider types of visual distortion and the procedures used in studying adaptation to discordant stimulation.

12.2 Types of discordance

I have already explained that a spatial discordance may be an intrasensory, an intersensory or a sensorimotor discordance. Apart from inverted vision, which is discussed in Section 14.4, most of the intravisual discordances that have been studied have been between different cues to depth. For instance, by viewing the world through a pseudoscope which reverses the sign of binocular disparities, an intravisual discordance is produced between binocular and monocular cues to depth. Depth perception is outside the scope of this book, so that intravisual discordance will not be discussed. Intersensory discordance was discussed in Chapter 11, but that chapter was concerned only with the initial response to a discordance and not with the question of how people learn to adapt to discordant stimulation.

The concordance between vision and another spatial sense may be disturbed by distorting the visual input or by distorting the input to the other sense organ. However, it is much easier to distort the visual input by an optical device than it is to distort the inputs to other senses, so that most experiments have involved optical distortions. I shall now consider the types of optical distortion that have been used.

Most experimenters have used a **similarity transformation**, that is, a transformation in which angles are preserved so that the shape of the retinal image is not distorted. Similarity transformations include the **isometric transformations** in which angles *and* distances are preserved, namely, translations, reflections, and rotations (tilts or a 180° rotation), and those in which only angles are preserved, namely, magnification and minification. **Non-similarity transformations**, such as curvature, shear, and non-uniform magnification, have been used in a few studies. These similarity and non-similarity transformations are all *continuous* transformations; the only *discontinuous* transformation that has been used is a vertical displacement of the two halves of the monocular field in opposite directions. The types of transformation are depicted in Figure 12.1

Translation of the visual input, usually a lateral translation, is the most commonly used visual distortion, and the most commonly used device for producing such a distortion is the wedge prism. The degree to which a light beam is refracted by a prism depends on the refractive index of the material out of which the prism is made, the apical angle of the prism, and the angle of incidence of the light beam. Let the angular position of a prism be defined by the orientation of the plane which bisects its apex. The refractive power of a prism is measured in **prism diopters**. A 1 diopter prism produces a 1 cm optical displacement of an object at a distance of 1 m, when the prism is in its position of minimum deviation (See Figure 12.2). One centimetre subtends 1° of visual angle at a distance of about 56 cm, so that a 1 diopter prism displaces the middle part of the visual scene by approximately 0.5° of visual angle. However, the seen displacement produced by a prism depends on the position of the object with respect to the prism and the eye. Consider a prism mounted vertically in front of an eye with its base to the left. Such a prism displaces the visual scene to the right. But objects in visual space on the base side of the prism are displaced less than objects in visual space on the apex side of the prism. In other words, there is an uneven lateral magnification of the visual scene. Furthermore, vertical lines are apparently curved, and horizontal lines above the horizon slope one way and those below the horizon slope the other way (see Figure 12.3). The only undistorted line is the horizon line (assuming that the subject is erect, the prism is in a frontal plane, and the displacement axis is horizontal). A prism may also produce colour fringes round objects because of chromatic aberration. If the aim of an experiment is to investigate the effects of displacing the visual scene, these other distortions which a prism produces should be minimized. This is done by presenting the subject with only one visual target at a time in otherwise dark surroundings and by varying the position of the target only along the axis of zero curvature and slope and not too far from a position opposite the eye of the

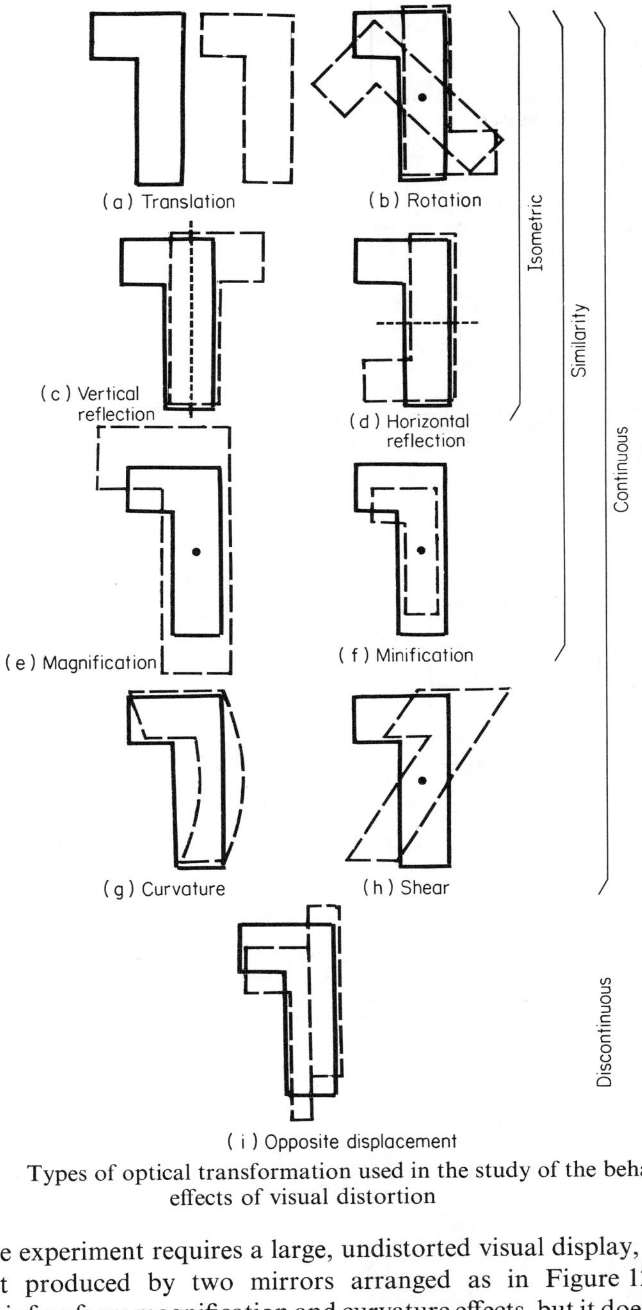

Figure 12.1 Types of optical transformation used in the study of the behavioural effects of visual distortion

subject. If the experiment requires a large, undistorted visual display, displacement is best produced by two mirrors arranged as in Figure 12.4. This arrangement is free from magnification and curvature effects, but it does increase the apparent distance of the scene.

The simplest way to invert or reverse the visual scene is to observe it in a mirror

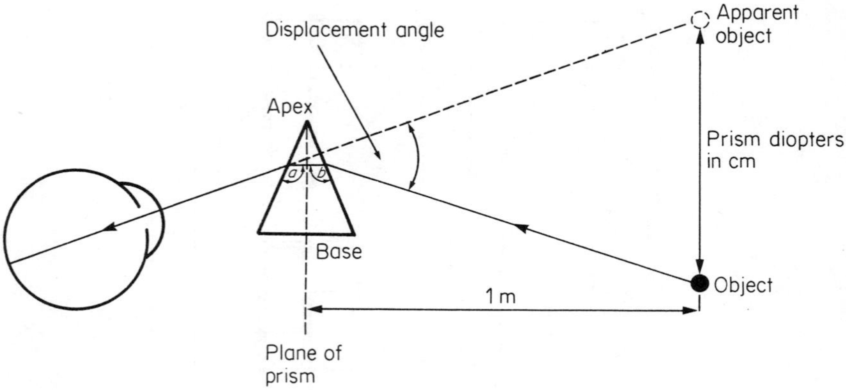

Figure 12.2 The displacement in centimetres produced by a prism at a distance of 1 m when angle *a* equals angle *b*, defines the dioptric power of the prism

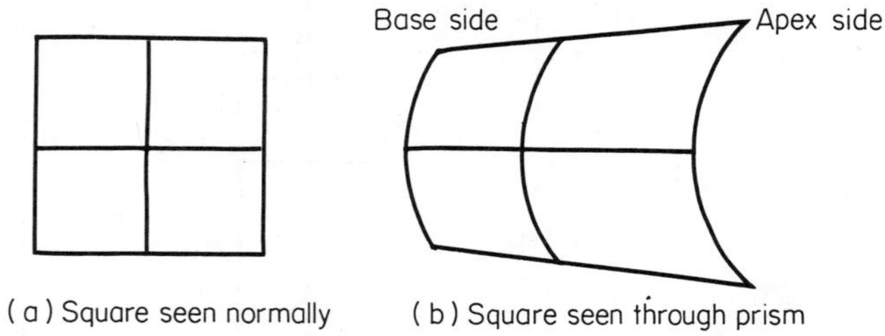

Figure 12.3 The distortion produced by a wedge prism

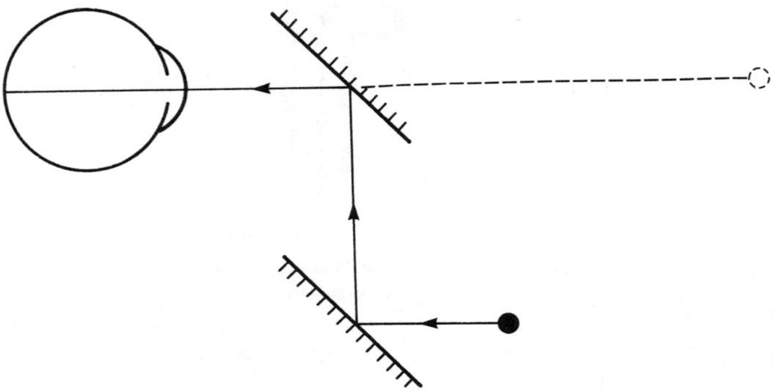

Figure 12.4 Showing how a displacement of the visual scene may be produced by two mirrors. The mirrors incidentally increase the path length of the light and hence the apparent distance of objects

held with the reflecting surface near and parallel to the visual axis. For instance, if the mirror is held horizontally just above the eyes, as in Figure 12.5, the visual scene is inverted, that is, reflected about a horizontal axis. If the mirror is held vertically to one side of the visual axis, the scene is reversed, or reflected about a vertical axis. These arrangements were used by Erisman and Kohler (see Kottenhoff, 1957a) in their studies on the effects of inverted and reversed vision. The subject's field of view is severely restricted by the screen which must be used to prevent him seeing the non-inverted scene, and the scene is vertically *displaced* as well as inverted, because the mirror is to one side of the visual axis. A Dove prism does the same thing as a mirror but gives a wider field of view and reverses or inverts the scene without displacing it. A Dove prism is essentially a mirror, which reflects the image, plus two refracting surfaces which prevent the image from being displaced (see Figure 12.6). The simplest way to understand what a Dove prism does to the appearance of the visual scene is to take a circular piece of paper with an L shape drawn through the centre of each side, so that the two shapes are congruent (see Figure 12.7). It will be noticed that, as the paper is turned over about any axis in the plane of the paper, corresponding to one of the reflection axes of the prism, the shape is reflected about that axis. But inspection will reveal that this is equivalent to a reflection of the shape about its own axis and a rotation of the shape in its own plane through twice the angle between the shape axis and the reflection axis of the prism. This means that, when

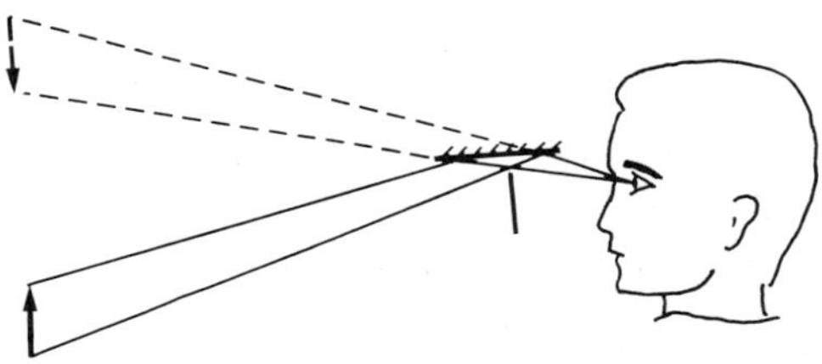

Figure 12.5 Showing how a mirror held above the eyes inverts the visual scene. The screen prevents the subject from seeing things not reflected in the mirror

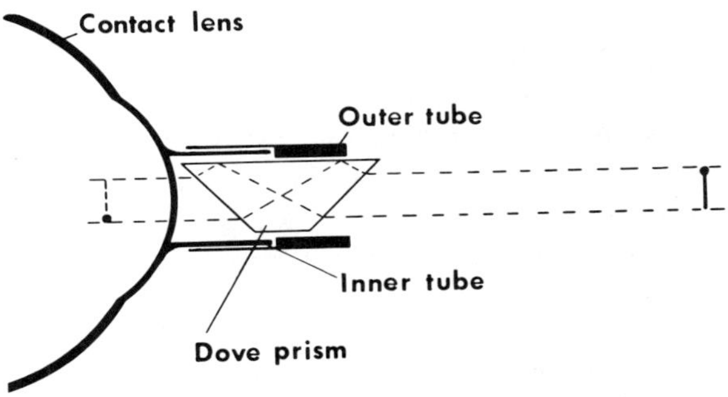

Figure 12.6 Showing a Dove prism mounted on a contact lens, and the optical effect produced by it (Reproduced by permission of Psychonomic Society, Inc)

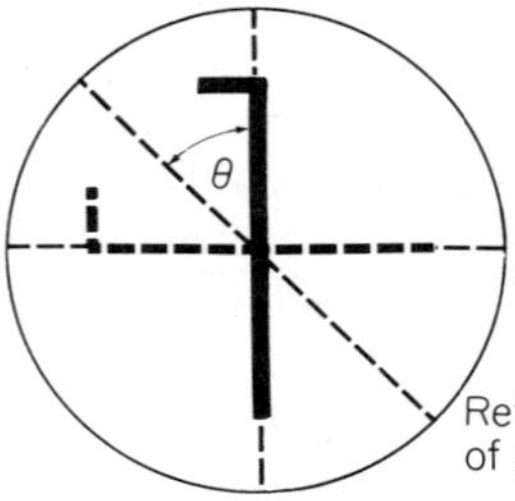

Figure 12.7 The dashed 'L' is the image of the solid 'L' as seen through a Dove prism with its reflection axis lying in the direction indicated. The transformation is equivalent to a reflection of the object about its own axis and a rotation of the object through 2θ, where θ is the angle between the reflection axis of the prism and the axis of the object

a Dove prism is rotated about the visual axis, the visual scene appears to rotate in a frontal plane at twice the angular velocity of the prism. It is possible to remove the reflection of the scene, leaving only its rotation, by placing a second Dove prism in line with the first, and it is possible to remove the appearance of rotation, leaving only the reflection, by placing the reflection axis of the prism at right angles, or parallel to the main lines of the visual display. The Dove prism is, therefore, a useful device for both reflecting and rotating the visual scene. In an animal, the visual scene may be oculocentrically rotated by surgically rotating the eyes.

The effects produced by an inverting or reversing device vary according to whether it is mounted on the subject's head or on a contact lens on his eye. For instance, if a reversing device is mounted on the head, any rotation of the head about a vertical or transverse horizontal axis causes a motion of the scene relative to the eye which is in the opposite direction to that in which it normally moves. This relative motion of the scene will appear to the subject as a real motion of the scene at twice the velocity of the head in the same direction as the head. I shall refer to this as **head parallax distortion**. If the subject pursues the anomalously moving scene with his eyes, he will restore the stability of his retinal image, but he will have reversed the sign of his vestibulo-ocular reflex. This question was discussed in Section 9.5.4. If the optical device is attached to the head, and the head is held still, any movement of the eyes is accompanied by a normal motion of the image on the retina, so that there is no distortion of **eye-movement parallax** under these circumstances. If the optical device is worn on a contact lens on the eye, there is a distortion of head-movement parallax when the head moves and a distortion of eye-movement parallax when the eyes move. *There have been no systematic studies of how people adapt to this situation.*

Magnification and minification may be produced by viewing the scene through lenses and non-linear distortions may be produced by prisms or curved mirrors. These then are the types of discordance used to study the adaptability of intersensory and sensorimotor coordination. I shall now discuss the experimental procedures which have been used in this type of study.

12.3 Procedures for studying adaptation to discordant stimulation

The general paradigm for studying adaptation to discordant stimulation is to pretest the subject on one or more selected perceptual or motor tasks under conditions in which he is not given knowledge of results, then to expose him to the discordant stimulation for a given period, and finally to repeat the same tests that were given before exposure, again without knowledge of results. The post tests may be interspersed with exposure trials if one wishes to plot the progress of adaptation or if one wishes to prevent the effects from wearing off. These three phases of an experiment will be referred to as **pretesting**, **exposure** (or training), and **post testing**. The amount of adaptation produced in a given exposure is the difference between the pretest and post test scores on that task. It is usually necessary to have one or more control condition in which the exposure phase is varied in some crucial fashion. I shall first consider the exposure phase by discussing the types of discordance that have been used and the methods of presenting them to the subject.

If the aim of an experiment is merely to see how people react immediately to discordant stimulation, one or other of the procedures described in Section 11.4.2 could be employed. But if the experiment is designed to find out whether exposure to discordant stimuli leads to some persisting modification in the way in which stimuli in one or other modality are responded to, the procedures described in Chapter 11 may not be appropriate, because people typically ignore spatial discrepancies between vision and other senses. It has been argued that adaptive changes which persist after exposure to intersensory discordance are simply a continuation of the process of visual dominance that occurs while the discordance is experienced. According to this view, adaptation would be greater if dominance were greater. But the converse is more likely to be true because, if there were more visual dominance, the discordance would be less likely to be detected. This second view is supported by the results of an experiment by Radeau and Bertelson (1977, 1978). They showed that the conditions which enhance the impression of ventriloquism, when associated faces and voices are spatially separated, are not the same as those which lead to a persisting change in the way subjects point towards visual and auditory targets. Warren and Schmitt (1978) and Welch *et al.* (1979) have also shown that the conditions which potentiate the immediate ventriloquism effect attenuate the effects of long-term adaptation to discordance.

A subject is more likely to recalibrate his sensory system if the discordance is vividly displayed, that is, under conditions in which dominance (ventriloquism) does not occur. One way to make the discordance more vivid is to ask the subject to make successive rather than simultaneous judgments about the location of stimuli in each modality and then reveal to him the magnitude of any discrepancy in his judgments. If the discordant stimuli are exposed at the same time, the discrepancy will probably be more apparent if the stimuli are moved rather than presented in a static display. Other factors governing the vividness of discordant inputs are listed in Section 12.4.4.

In 1968, I proposed a classification of procedures which have been used for displaying discordant stimulation in the exposure phase of an experiment. This classification has been generally adopted. I present this scheme in slightly amended form in Table 12.1. It can be seen from the table that procedures are classified basically in terms of orthogonal criteria. On the one hand, they are classified according to whether the subject is required to execute an explicit task or whether he is merely required to observe the discordance. On the other hand, they are classified according to whether the discordant stimuli are displayed concurrently, terminally or successively. In a **concurrent display**, the discordance is continuously presented. In a **terminal display**, the discordance is not made evident initially, either because one stimulus is presented before the other or because the feedback loop that allows the subject to detect the discordance is not closed until some specified time. For instance, if a person were to reach towards an optically displaced visual target and were allowed to see his hand only when he completed the aiming movement, this would be an example of terminal display, because, although the two senses receive stimulation during the whole movement, the feedback loop between kinaesthesis and vision closes only when the subject is allowed to see his hand. In a successive display the two discordant stimuli are presented one after the other, without overlap. In this case, the subject must rely on short-term memory in order to detect the discordance. Successive displays are rarely used.

I have further subdivided these categories according to whether the discordance is displayed exafferently or reafferently. The distinction between exafference and reafference is described in Section 12.4.4. Further, or alternative subdivisions of task and display categories are, of course, possible.

There is one further distinction which is not shown in the table. This concerns whether the discordance is introduced fully in one step or whether it is gradually introduced over several responses during the exposure phase. In most experiments the discordance was introduced in one step, which usually meant that, initially, the subject was aware of the discordance. In 1968 I described a procedure, which I called **prismatic shaping**, which prevents the subject being aware of the discordance. In this procedure the sensory distortion which causes the discordance is introduced very gradually during the exposure phase, so that the subject adapts to each small step as it occurs and never becomes aware that there is any discrepancy in his judgments or responses.

Perhaps the most prevalent and most commonly overlooked source of contamination in an experiment on sensory discordance is stimulus asymmetry. For instance, if one is investigating adaptation of reaching behaviour to displaced visual inputs, either the visual targets are displaced by the prism to an asymmetrical position in the visual field or, if they are displayed in an apparently symmetrical array, then the arm must point asymmetrically in order to reach them. Asymmetrical visual stimuli, asymmetrical positions of gaze, and asymmetrical arm postures each produce aftereffects which to the unwary may be accepted as evidence that the subject has adapted to the discordance (Section 7.6.4). There is no shortage of evidence that the effects of asymmetrical stimulation may contaminate the results of an adaptation experiment (Sekuler

Table 12.1 A classification of types of exposure to discordant stimulation, with examples of each type. Based on an experiment in which vision is displaced by a prism

	Involving no explicit task		Involving explicit task	
	Exafferent display	Reafferent display	Exafferent display	Reafferent display
Concurrent display of discordance	Watching passively moved arm	Watching actively moved arm	Judging when a light is coincident with a sound source	Viewing hand while aiming at a target
Terminal display of discordance	Looking at a bell, then both looking and listening to it	Moving hidden arm to a location, then viewing it	Aligning hidden, passively moved arm with visual target, then viewing arm and target	Actively pointing to visual target with unseen hand, then viewing arm and target
Successive display of discordance	As in terminal display but with one stimulus removed before the other is seen			

and Bauer, 1966; McLaughlin *et al.*, 1966; Collins and Singer, 1968; Ebenholtz, 1974; Willey *et al.*, 1978; Fogelgren and Shebilske, 1979). It is essential, in any experiment, that control conditions be run in which the effects of asymmetrical stimulation are assessed so that they may be allowed for in estimating the effects of discordance.

Another, almost completely neglected source of potential contamination in experiments on visual–motor discordance was recently brought to my attention by Hiroshi Ono. Many experiments on the effects of displaced vision involved the use of only one eye during exposure. If the closed eye has a phoria, a not uncommon condition, it will turn inwards or outwards rather than remain converged on the visual target. It was reported in Section 7.6.3 that the position of a closed phoric eye affects the apparent headcentric position of a visual target. Furthermore, Ono has shown, in an as yet unpublished study, that such effects are modifiable through training of the same type that is involved in studies on the effects of visual displacement. Unless the effects of phoria are allowed for, they will therefore show up as a spurious adaptation to the optically-induced discordance.

Once the procedure to be employed in the exposure phase of an experiment has been selected, a decision must be made about the tests which are to be used before and after exposure. There are only a few general comments to be made about these tasks, because those used depend on the issue being studied. One obvious point is that pretests should resemble the post tests as closely as possible. A less obvious point, which has often been overlooked, is that tests should be neutral with respect to exposure conditions. By this I mean that if there are several exposure conditions in an experiment, the tests should not resemble one exposure condition more than they resemble other exposure conditions. If neutral tests cannot be found, the situation should be balanced by having several tests, each of which resembles one of the exposure conditions. For example, if one wishes to compare the effects of an exposure condition in which there are several visual targets with a condition in which there is only one target, pre- and postexposure tests in which there is one target and in which there are several targets should be given in association with each exposure condition.

I shall mention only one other point about testing procedures. Harris (1974) gave a warning about the interpretation of results from tests which require the subject to judge the straight ahead, either by pointing or by directing the head or the eyes straight ahead. He drew attention to the conflicting nature of conclusions in the literature which were based on this type of test, and pointed out that the criteria that a subject adopts in making a straight-ahead judgment may vary, in spite of attempts to explain what is required. One particularly troublesome feature of this type of measure is that people may judge the straight ahead not egocentrically, as required, but in terms of the walls of the room that they are in or, if the lights are out, in terms of where they think the walls are. A much safer task, which serves the same purpose as a straight-ahead judgment, is to ask subjects to point to some part of their body, such as a toe, which is known to be unaffected by the exposure to distorted vision.

The problem I have just mentioned is an example of the well-known problem of the demand characteristics of a task. Subjects are notoriously prone to adopt strategies and respond to cues other than those which the experimenter intends that he should adopt or respond to. There is no general remedy for this problem, except the experimenter's vigilance and imagination.

Most of the rest of this chapter will be devoted to the way in which human subjects adapt their visual–motor behaviour to an optical displacement of the visual scene. This is the most commonly studied discordance. Adaptation to rotated visual fields is discussed in Section 12.5. Adaptation to altered relationships between head, eye and image motions was discussed in Section 9.5.4.

12.4 Adaptation to displaced vision

12.4.1 Introduction

The recent revival of interest in the effects of visual distortion is due to Richard Held and his associates and to John Hay and Herbert Pick. In the early 1960's these investigators turned away from the complex procedure of exposing subjects to inverted or reversed vision and centred their attention on the more precise measurement of the effects of a simple lateral displacement of vision, which are manifest after only a few minutes of exposure.

Anyone with a wedge prism can experience these effects in the following simple way. A few numbers are marked on the edge of a piece of card which is then placed horizontally under the chin, as in Figure 12.8. With the prism before one eye and the other eye closed, the finger is directed towards a number on the far side of the card and allowed to come into view. This arm is then returned to the side of the body, after which the aiming movement is repeated several times to

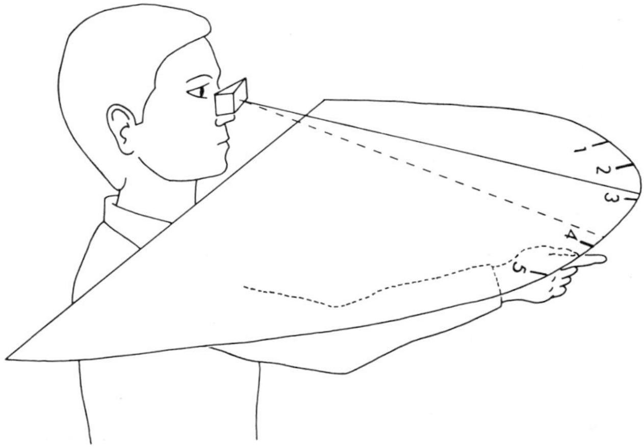

Figure 12.8 A simple apparatus for demonstrating adaptation of pointing to displaced vision

each of the numbers in random order. The error in pointing will be very evident for the first few trials but accuracy is soon restored. When the prism is removed, it will be found that the first few aims will be off target in the opposite direction to the error first experienced when the prism was in place. This demonstration involves a target-pointing task with terminal exposure.

Experiments of this kind have been directed towards two issues: the discovery of the nature of the changes underlying adaptation of perceptual judgments and motor performance and the elucidation of the conditions under which adaptation occurs. I shall deal with these two issues after I have described a model of the hand–eye system.

12.4.2 A model of the hand–eye coordination system

The hand–eye coordination system consists of the sensory, skeletal and motor components depicted in Figure 12.9(a). The variable elements in the system are called **articulations** (Howard, 1971a). Each of these articulations is free to vary in any plane, but I shall consider variations in the horizontal plane

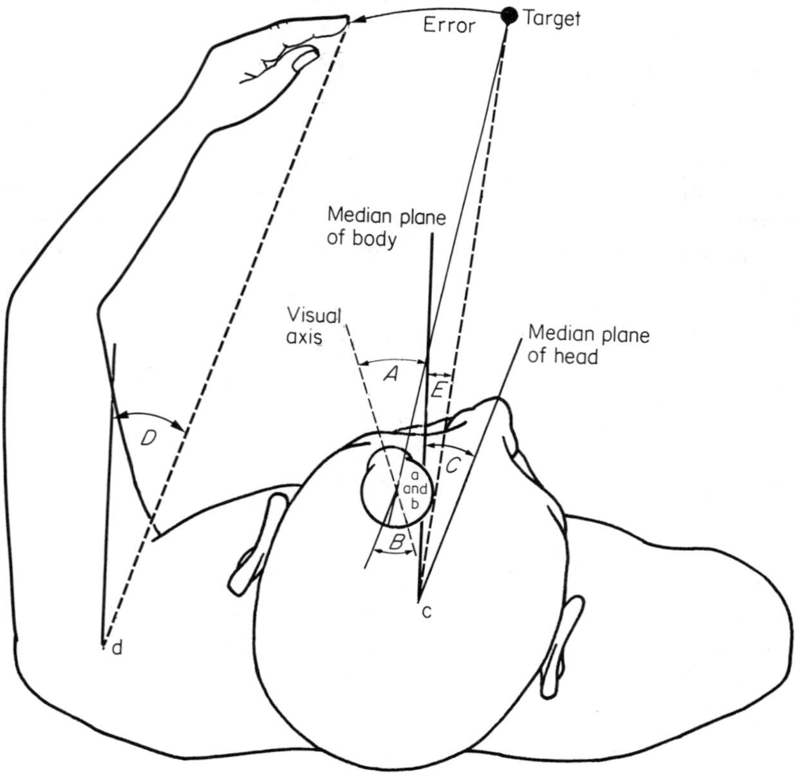

(a)

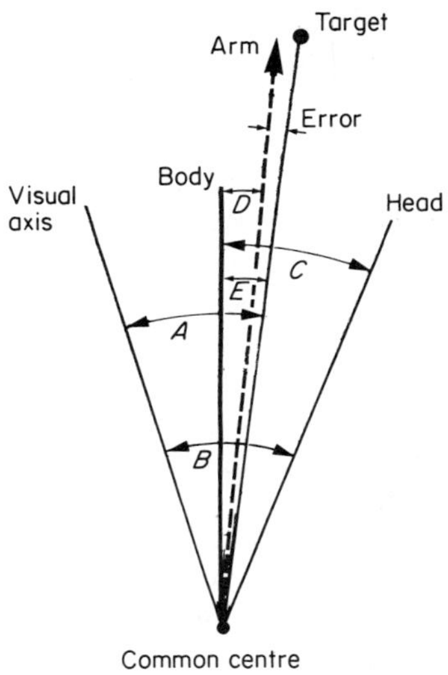

(b)

Figure 12.9 (a) The structural components of the hand–eye system showing the in-series arrangement of articulations (sense organs and joints). The centre of rotation of the eye and the nodal point of the eye (a and b) are assumed to be in the same place. The centre of rotation of the head and the arm are denoted by c and d. Angle A is the retinal eccentricity of the target, angle B is the eccentricity of the gaze, angle C specifies the direction of the head, angle D the direction of the arm, and angle E the direction of the target, all with respect to the median plane of the body. (b) The angular relationships of components depicted in (a) when the points of articulation are superimposed

only, with the observer upright, the target and hand at eye level, and the body fixed with reference to the ground. The articulations are: a, the nodal point of the eye, from which the eccentricity of the image of the target is measured; b, the centre of rotation of the eye, from which the direction of gaze is measured; c, the centre of the head, from which the direction of the head is measured; and d, the centre of the shoulder joint, from which the direction of the finger is measured. In Figure 12.9 the directions of the head (C) and the finger (D) are specified with respect to the median plane of the body, the direction of the eye (B) is specified with respect to the median plane of the head, E is the direction of the target with respect to the body, and A is the direction of the target with respect to the visual axis. If we assume that the four articulations are in vertical alignment, that is, that they are coaxial, as in Figure 12.9 (b), then $E = C - |A| + |B|$. Thus, by combining information about A, B, and C, a subject is able to estimate the direction of the target with respect to his body midline (E). By setting the direction of his arm (D) to equal E, he may then direct his arm towards the target. The angular error of his aim is given by $E - |D|$. If the error is displayed to the subject, the system of articulations forms an in-series, closed-loop control system. If the error is not displayed, it is an open-loop control system. Normally, the image of the target would fall on the fovea, but theoretically one must allow for the general case, which includes pointing to objects that are not viewed directly.

If a prism is placed before the eye, a fifth component is added to the system depicted in Figure 12.9(a). The pointing error is now the algebraic sum of the normal pointing error and the angular displacement of the prism. If the subject's task is to point his finger to a visual target, he can compensate for the error introduced by the prism only by adjusting the position of some part of the arm. If the task is to direct the gaze towards an unseen target which is felt by the hand, then the subject can compensate for a prism error only by adjusting the position of his eyes in his head or by adjusting the position of his head. Thus, the part of the system which must be physically adjusted is determined by the nature of the task. The theoretically important question is not which part of the system makes a physical adjustment, but for which part of the system is there a recalibration in neural coding. That is the question which I shall now consider.

Figure 12.10 is a flow chart for the stages of the neural processes which serve hand–eye coordination. This chart will provide a basis for distinguishing between theories of how the system adapts to displaced vision. The bold lines depict the structural components and articulations. The part of the body with respect to which spatial judgments are being made will be referred to as the **base**, and for purposes of exposition I shall assume that the torso is the base. An **articulated subsystem** may be defined as any in-series set of articulations which is used in making spatial judgments or spatially directed movements with respect to the base. Thus, the arm is an articulated subsystem which is involved in tasks like judging or setting the position of the finger relative to the body. Similarly, the retina, eye, and head together form an articulated subsystem which is involved in tasks like judging or setting the bodycentric direction of a visual target.

Sensory end organs are represented by small arrows and muscles by ovals. The oculocentric direction of the displaced target is signalled by the eccentricity of the retinal image. It is assumed that the direction of gaze is signalled by efference, and that is why no sensory end organ has been indicated at the extraocular muscles. The sense organs signalling position at each of the remaining articulations are assumed to be joint receptors. Sensory nerves lead from the sense organs to sensory spatial coders in the brain. These are depicted on the right of the diagram. The muscles at each articulation are controlled by motor nerves indicated on the left of the diagram. The motor neurones originate in motor spatial coders in the brain and constitute the final common path of the system. The sensory and motor spatial coders are represented by single elements in the chart, but are actually complex neural structures. Each sensory spatial coder transforms the input from the sense organs at one articulation into a spatial code. It is assumed that all judgments about the angular position of a given articulation rely on this code. Each motor spatial coder transforms a spatial code, derived from higher in the nervous system, into specific instructions to muscles to bring the joint to a particular angle. This transformation is straightforward in the case of eye movements; each eye position is associated with a unique and consistent

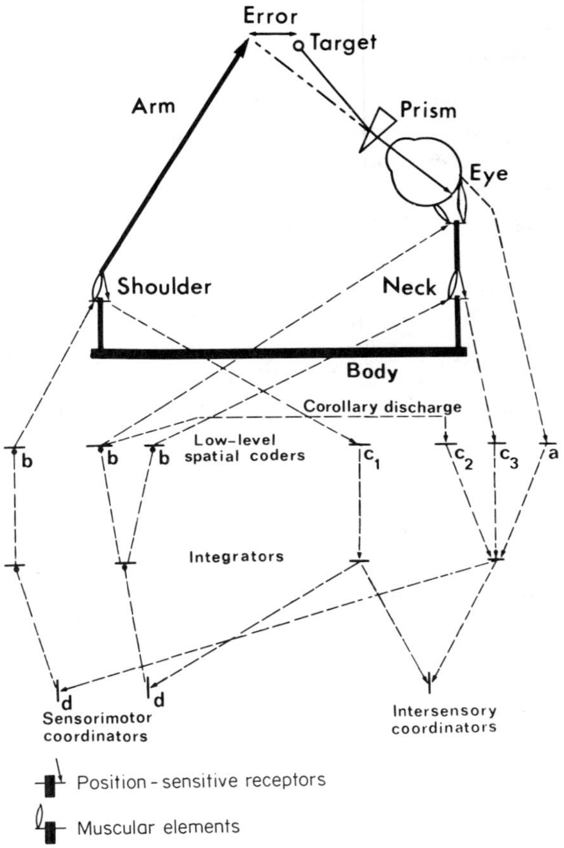

Figure 12.10 A flow chart for the eye–hand coordination system. The basic structural components of the eye–hand coordination system are shown in bold lines. Components are linked by articulations at which symbols defined in the diagram are used. Neural elements are drawn in dashed lines, junctions represent neural coding processes, small arrows depict inputs to sensory elements, and filled circles depict inputs to motor elements. The letters a, b, c, and d indicate possible sites of recalibration discussed in the text (Adapted from Howard, 1971a. Reproduced by permission of the British Council)

pattern of motor discharge, because the eye muscles act against a consistent load. In skeletal joints, however, the muscular effort required to reach or hold a given position depends upon the load against which the limb is working, which may vary; the motor spatial-coding elements of skeletal joints therefore require sensory feedback, so as to make the system load-compensated. Such feedback connections are not shown in Figure 12.10.

At the next level, the coded spatial information from separate articulations within a subsystem is integrated. On the sensory side, this integration is required for judgments about the position of the distal element of a multiply articulated system relative to the body. For instance, the integrator on the far right of the diagram receives a set of inputs, one of which codes the position of the retinal image, another the position of the eye in the socket, and a third the position of the head on the shoulders (note that the eye-in-socket code is in the form of a corollary discharge from the eye-muscle coding element). If it is assumed that the sensory coders code each input in terms of a common egocentre, then the algebraic sum of the three coded signals is all that is required to enable the subject to estimate the direction of the visual target with reference to the mid-body axis.

The integration of spatial information from the joint receptors of the arm is more complex because the joints are not even approximately coaxial. It is clearly necessary for the integrator to 'know' the length of each segment of the arm. Consider a two-segment arm acting in a plane, as in Figure 12.11(a). It is easily shown that the direction, θ, of the line joining the tip of the limb to the root of the limb with respect to the sagittal plane is given by:

$$\theta = \alpha \pm \tan^{-1} \frac{b \sin \beta}{a + b \cos \beta} \tag{1}$$

where α is the angular setting of the first limb segment with respect to the sagittal plane, β is the angular setting of the second limb segment with respect to the first, and a and b are the lengths of the first and second limb segments.

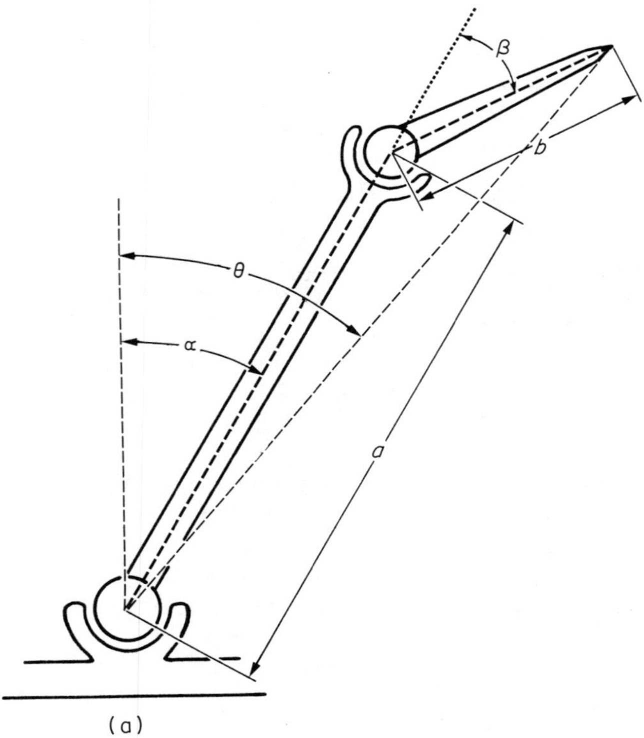

(a)

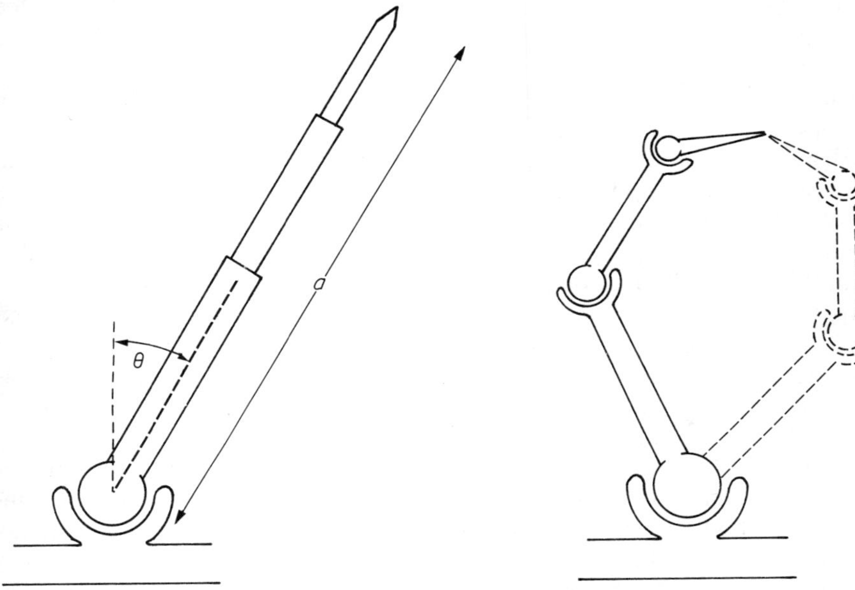

(b) (c)

Figure 12.11 (a) A schematic two-segment arm. (b) A schematic three-segment telescopic arm in which the position of the tip is unambiguously determined by angle α and the lengths of the arm segments. (c) A schematic three-segment arm showing how the same position of the tip may be achieved in more than one way

Formulae may be derived using Legrangian dynamics which incorporate the positional and velocity aspects of the movement of multisegmental limbs in three-dimensional space (see Saltzman, 1979, for a treatment of these methods). When a person judges the direction of the finger with reference to some point on the body he must integrate the angular settings and lengths of each arm segment in a way which is isomophic with an equation of the type just described. As the child grows older and the limbs grow, the system must update its 'knowledge' about the lengths of limb segments.

The problem of integrating spatial information from a limb is complicated further by the fact that a finger may be placed in the same position relative to the body with the limb in many different postures. Consider a simple 'limb' with a ball joint and three telescoping segments, as in Figure 12.11(b). The direction of the tip of such a limb is given by the angular setting of the joint (θ), and the distance of the tip from the base is given by the length of the limb (a). There is a one-to-one mapping of limb postures onto locations of the tip of the limb. But with a three-segment limb, as shown in Figure 12.11(c), there are many positions of the limb segment that place the tip in a given position. In other words, there is a many-to-one mapping of limb postures onto spatial locations of the tip of the limb. In such a limb, integration of spatial information requires the derivation of an equivalence function that defines this many-to-one mapping. For the two-segment coplanar limb this function is defined by equation (1). As long as the right-hand part of the equation is a constant, the tip of the limb remains in the same location.

On the motor side, the integrators enable the segments of a multijointed system to work as a team. In particular, the motor integrators enable a limb to achieve a given motor task in several ways and the gaze to be directed in the dark to a given location for different positions of the head. The sensory and motor integrators may be described as perceptual and motor spatial-constancy mechanisms, or as mechanisms for extracting and generating spatial invariants (Gibson, 1966). For a multisegmental limb moving in three-dimensional space, the task of specifying a particular set of efferent signals required for the execution of a desired movement is formidable. Pellionisz and Llinás (1980) have developed a model of the cerebellum which is designed to accomplish this task. I have described this model briefly at the end of Section 5.4.4, and readers are referred to the original paper for more details.

At the next level depicted in Figure 12.10, sensory and motor integrators for each articulated subsystem are coordinated. The coordination may be intersensory, that is, between two sensory integrators. For instance, sensory information from the eye–head system may be coordinated with sensory information from the arm, enabling the subject to estimate the position of his unseen finger in relation to a seen object. The coordination may also be between the sensory information from one articulated subsystem and the motor output to another subsystem. For example, the eye–hand coordinator maps sensory information from the arm onto the motor integrator of the eye–head system, so that a person is able to look at the position of his hidden hand. Finally, the coordination may be between the motor integrators of two subsystems; for instance, the movements of the two arms are often coordinated (see Kelso *et al.*, 1979).

Integrators deal with information from an articulated subsystem with respect to a base, whereas coordinators relate information from two or more articulated subsystems with respect to a base. It is assumed that information from each spatial coder, each integrator and each coordinator is available for making spatial judgments or executing motor control. The functioning of these coding elements may become elaborated or modified as a result of experience. Thus, compensation for visual displacement could in principle be achieved by a suitable recalibration of one or more sensory or motor spatial coders or by a recalibration of an integrator or of a coordinator. There is also the possibility that the system may be brought under the control of a higher-level non-specific controller, which one may equate with conscious control. I shall now consider theories and evidence bearing on this issue.

12.4.3 The nature of the changes underlying adaptation to displaced vision.

Nobody has made the unreasonable suggestion that signals generated by sense organs change during adaptation to displaced vision. Nor is it reasonable to suppose that there is any change in motor neurones or in neuromuscular junctions. The change is surely in the way signals are coded in the central nervous system, at one or more of the sites indicated by letters in Figure 12.10.

Recalibration of oculocentric spatial coding (site a). In the literature on adaptation to displaced vision there is constant mention of visual adaptation, but it is seldom made clear whether this refers to a recalibration of oculocentric spatial coding, which would involve a change in retinal local signs, or to a recalibration of headcentric visual coding, which would involve a change in the sense of position of the eyes in the head. The general opinion is that oculocentric recalibration does not occur. But I have argued elsewhere (Howard, 1971b) that this opinion is founded on an accidental feature of the experimental procedure generally adopted. In almost every experiment on prism adaptation the subject was allowed to fixate the visual target. A shift in oculocentric spatial coding under these circumstances requires the subject to behave as if the object he is fixated upon is not the object he is looking at—surely an unlikely state of affairs. No other articulation in the visual–motor system has such a salient landmark as the fovea, and it is no wonder that local sign is the least labile component when foveal fixation is allowed. Oculocentric shifts in spatial coding may well occur for objects imaged in the periphery of the retina. There are only two published experiments in which peripherally viewed targets have been used. Cohen (1966) asked the subject to watch his moving hand through a prism, with the gaze either centred on the hand or centred on a fixation point 20° on one or other side of the hand. Adaptation was measured by asking the subject to point to test targets with the unseen hand. After foveal exposure, adaptation was the same when the test target was imaged on the fovea as when it was imaged 20° to either side of the fovea. However, after exposure to a peripheral target, there was less adaptation with the foveal target than with the peripheral target. It looks as though the effects of adaptation are to some extent specific to retinal location for objects viewed peripherally during exposure, and this is evidence that there has been an oculocentric shift for such objects. However, Crawshaw and Craske (1974) pointed out that the optical displacement used by Cohen was not homogeneous and that he failed to control for the effects of asymmetrical eye posture. When the experiment was repeated with these factors corrected, no evidence of an oculocentric shift could be found. So, after all, it looks as though recalibration of oculocentric spatial coding does not occur after exposure to visual displacement.

There is a broader question of whether changes in the appearance of oculocentric space ever occur as a result of experience with visual–motor discordance of any type—not merely that caused by visual displacement. This question has a long history which is reviewed in Festinger *et al.* (1967). The best evidence that such changes may occur of which I am aware is from an experiment by Held and Rekosh (1963). In this experiment, subjects were asked to walk inside a large cylinder covered with a random pattern of dots, while wearing prisms which introduced visual displacement and curvature. The apparent curvature was not apparent in the array of dots, although it altered the flow pattern produced by the dots as the subjects walked about. To those subjects who had walked about actively inside the cylinder, straight lines seen subsequently appeared curved. It was argued that this purely oculocentric aftereffect had been produced by the discordance between active movement and the resulting retinal flow patterns. Subjects who inspected the same flow pattern while being passively moved about in the cylinder showed no aftereffect. Gyr *et al.* (1979) repeated the experiment but failed to find any effect of either active or passive exposure on the apparent straightness of lines. Miller and Festinger (1977) also conducted an experiment to see whether they could induce an oculocentric change in perceived straightness by exposure to visual-motor discordance. Subjects were required to scan their eyes along a curved line which changed its position as the eyes moved. Thus, subjects had to move their eyes in a path which differed from the visually apparent path. Subjects learned to adjust their eye movements, but this had no effect on the apparent curvature of lines seen subsequently. McLaughlin *et al.* (1968) trained subjects to make accurate saccades to a visual target which was suddenly displaced during the course of the saccade, but found that this had no effect on the perceived location of the peripheral targets towards which they had been making saccades. Oculocentric distances are judged in terms of retinal locus and not the efference involved in moving the eyes.

It would thus seem that there is no good evidence for believing that adjustments to sensorimotor conflict have any effect on oculocentric visual space. On the other hand, exposure to asymmetrical visual stimuli certainly produces oculocentric after effects. For instance, we saw in Section 4.6 that inspection of tilted lines causes straight lines to appear tilted the other way, or that inspection of curved lines causes straight lines to appear curved the other way. These normalization effects are never very large, and are best understood as attempts by the visual system to adjust the calibration of its internal metric in response to persisting stimulus asymmetries. Andrews (1964) has referred to such changes as 'error-correcting visual mechanisms'. However, the subjective metric of oculocentric visual space seems to be insulated from events outside vision.

Recalibration of motor codes (sites marked b) There is a general consensus that adaptation to displaced vision does not involve a recalibration of the efference coding system, that is, the system by which the decision to move a limb to a specific position is transformed into specific patterns of muscular innervation. But I wish to question this belief.

The sense of eye position is believed to be based on the corollary discharge, or efference copy, associated with motor outputs, because the eye has no joint receptors. The sense of position of a limb is based mainly on proprioceptors, but there is evidence that efference is probably also involved; for instance, it has been found that monkeys are able to make directed movements with hidden deafferented limbs, as long as they are not in unusual postures (Polit and Bizzi, 1979). If we assume that the efference copy is a precise copy of motor outflow, then the hypothesis that adaptation of the visual–motor system is based on a recalibration of the motor coding system is equivalent to stating that it is based on a recalibration of the corollary-discharge, or efference copy. I shall now consider the evidence for and against the view that efference may be recalibrated.

Prism adaptation has been found sometimes to involve a change in the sense of position of the eye (see following section). This is evidence that adaptation can involve a change in motor coding, because eye position is probably coded in terms of efference. The evidence upon which the contrary view is based is illustrated by the results of an experiment by Hardt *et al.* (1971). They found that training in which the subject watched his arm trace a rectangle transferred to a test in which the head was directed towards the stationary arm. This proves that adaptation transfers to a system that was not actively used during training but it does not prove, as Hardt *et al.* claimed, that recalibration does not involve motor coding. For one thing, there may still be a motor outflow to a limb which is held in a stationary position. But in any case, the pattern of motor outflow involved in tracing round a rectangle is not significantly affected if the rectangle is displaced to one side, so that one would not expect much motor recalibration in the training procedure which Hardt *et al.* used.

The training procedure most likely to lead to motor learning is one in which the subject is required to point ballistically towards visual targets, with terminal error feedback. Baily (1972a) is the only person to have considered the question in this way. He used two types of exposure. In the first exposure condition, which was designed to optimize motor adaptation, the subject pointed ballistically with terminal error feedback, and in the second condition, which was designed to optimize proprioceptive adaptation, the subject moved his unseen arm slowly from side to side until he was satisfied that it was under the target and was then shown his error. The testing procedures also involved either ballistic or slow movements of the trained and untrained arm to the apparent positions of several targets. It was found that the effects of ballistic exposure showed equally in both ballistic and slow-movement tests, but that the effects of slow-movement exposure were more evident in slow testing than in ballistic testing. Baily argued from these results, and from the patterns of intermanual and intertarget transfer, that there are two types of adaptation: an adaptation of the felt position of the arm and an adaptation of the programming of ballistic movements. This evidence would have been more convincing if the effects of each type of training had been largely confined to the equivalent type of test. Baily gave an *ad hoc* explanation of the high level of transfer from ballistic training to slow-movement testing. The experiment draws attention to the need for considering the type of training before deciding whether motor recalibration can occur.

A very convincing case for believing that motor learning may contribute to adaptation of the visual–motor system may be based on the results of experiments in which proprioception was absent during exposure to visual displacement. Taub and Goldberg (1974) deafferented monkeys by sectioning the sensory roots of the spinal cord. These monkeys could be trained to reach with the unseen arm towards visual targets and were able to adapt their pointing when viewing the targets through displacing prisms. Recalibration in this case must have involved the motor coding system because the sensory nerves of the neck were also severed. The really convincing and striking aspect of the results was that, whereas the adaptation of reaching was only 36% of the prism displacement in normal monkeys, it was 100% effective in deafferented monkeys. Thus, *animals adapt better without joint proprioception.* It looks as though the motor coding system is very easily recalibrated but that the joint senses are less easily recalibrated and normally prevent the motor recalibration from showing. In other words, the general opinion that the motor system does not adapt is probably wrong; recalibration is more likely to affect the motor coding system than it is to affect the sensory proprioceptive system.

The results of an experiment by Wallace and Garrett (1973) seem to contradict this conclusion. They anaesthetized the arm of the subject by hypnosis and asked him to observe it through a prism as he moved it to and fro. This procedure produced no evidence of adaptation in postexposure tests in which the subject pointed to his nose with either hand after the effects of the hypnosis had worn off. It was concluded that sensations from the arm are required for adaptation, a conclusion which apparently conflicts with the results obtained from deafferented monkeys. This conclusion is wrong. The posttests were done after the hypnosis had worn off, so that the absence of any adaptive shift in pointing probably meant that, in testing, the non-recalibrated joint sense dominated the recalibrated motor system. In any case, the subjects were unable to point to the nose when the arm was 'hypnotized', so that this procedure knocked out both the sensory system and the efference copy system, and the results prove nothing about the site of recalibration because there was no discordance during exposure.

I conclude that the presence of inputs from joint receptors does not enhance adaptation to a discordance as long as there are other sensory systems which can detect the discordance. When discordance is detected in some other way, the motor coding system shows itself to be very labile when inputs from joint receptors are blocked. On the other hand, if the detection of discordance depends on information from joint receptors, as it probably did in the rectangle-tracing procedure used by Hardt, Held and Steinbach, then removal of this input will prevent any adaptation from occurring. Furthermore, under these circumstances, any enhancement of kinaesthetic inputs should enhance the detection of discordance and hence lead to more adaptation. This probably explains why Kravitz and Wallach (1966) found that vibrating the arm seen through prisms produced more adaptation than was produced without the vibration (see also Mather and Lackner, 1975).

Thus, when we talk about the contribution of a sensory system to visual–motor adaptation we must distinguish between the role that the sensory system plays in detecting the discordance and the role it plays in being the vehicle for the recalibration. The best general assumptions are that *discordance may be detected by any pair of sensory systems under the right conditions*, and that *human beings may adapt any component of the visual–motor system* (except perhaps retinal local sign). Experimenters have been too ready to conclude that certain types of adaptation cannot occur on the basis of results from experiments involving a restricted range of stimuli and only one type of training.

Recalibration of the sense of eye position (site c_2). When exposure to displaced vision is confined to one eye, the degree of adaptation is the same whichever eye is open during testing. In other words, adaptation to displaced vision shows 100% interocular transfer (Hajos and Ritter, 1965; Pick *et al.*, 1966; Foley and Miyanshi, 1969). If interocular transfer were incomplete, one could argue that adaptation involves a change in the position sense of the exposed eye, given that oculocentric changes do not occur. But even if adaptation does involve a change in the felt position of the eye, the effects will almost certainly transfer fully to the other eye, because Hering's law of equal innervation (Section 5.3.5) tells us that the movements of the two eyes are closely linked. Thus, we need evidence of a different kind to establish that adaptation to displaced vision involves a change in the position sense of the eyes. There are three lines of evidence which bear on this issue, which I shall now describe.

Mann *et al.* (1979b) argued that people with alternating strabismus may have learned to dissociate the sense of position of the two eyes. In alternating strabismus, either eye can be used to fixate an object, but the two eyes cannot be simultaneously converged on the same point. They showed that people with alternating strabismus did not display interocular transfer of prism adaptation, whichever eye was used in training. This supports their suggestion that these people have learned to dissociate their eyes, although it is odd that they showed *no* interocular transfer, because one would have expected them to show a change in the felt position of the arm which would transfer from one eye to the other. Perhaps strabismics have a very labile sense of eye position, which they recalibrate in preference to other sensory components.

The second line of evidence bearing on the question of changes in the felt position of the eyes comes from experiments in which tests are used which are sensitive to changes in the sense of eye position. Hay and Pick (1966) used a variety of testing procedures at intervals during a period of exposure to displaced vision extending over several days. Some of these tests, such as pointing to a hidden sound source, were designed to indicate changes in the position sense of the arm, while others, such as looking at a hidden sound source, were designed to indicate changes in the position sense of the eyes. It was found that adaptation in the first few hours of exposure was largely confined to a change in the sense of arm position, but that, as exposure continued, adaptation of the sense of arm position decreased and was replaced by a change in the sense of eye position.

Finally, changes in the sense of eye position may be registered by objectively recording the position of the eyes when the subject is attempting to direct his gaze straight ahead or towards a hidden part of his body. This procedure was adopted by Kalil and Freedman (1966), Craske (1967), and McLaughlin and Webster (1967). They found that the objectively determined position of the eyes when the subject was attempting to look straight ahead in the dark was shifted after exposure to displacing prisms. Furthermore, the shifts in eye position were sufficient to account for the measured changes in pointing to visual targets with unseen hand. Craske showed, in a control condition, that the change in recorded eye position was not due merely to the asymmetrical posture of the eyes during the exposure phase (see also Craske and Crawshaw, 1975, 1978).

In a subsequent study Crawshaw and Craske (1976) established that the effects of this type of exposure are not mediated by a change in the sense of direction of the arm or of the head, and reaffirmed their belief that they are due to a change in the sense of eye position. They also found that the changed sense of eye position showed 100% interocular transfer, which supports the idea that the sense of position of one eye is closely linked to the sense of position of the other, at least in the adult.

Craske and Templeton (1968) found that when a person was placed in complete darkness after inspecting his own feet through a displacing prism for some time, the gaze began to execute a pendular oscillation in the plane of the prismatic displacement. This oscillation was found to persist until the lights were put on. Perhaps the eyes oscillated between the straight-ahead position, defined by the changed sense of eye position, and the straight ahead, defined by the normal balance of innervation in the agonistic and antagonistic muscles.

Ivo Kohler (1964) exposed subjects to a wedge prism which affected only the upper half of the visual field, leaving the lower half undistorted, and reported that adaptations to the various distortions produced by the prism were manifest only when the subjects directed their gaze upwards. These **gaze-contingent aftereffects** of spatial distortion were confirmed by Pick and Hay (1966). Pick, Hay and Martin (1969) measured the eye movements of the subjects after exposing them to split-field prisms and produced evidence which suggests that the gaze-contingent aftereffects are due to changes in the felt position of the eyes rather than to gaze-contingent oculocentric changes.

Several attempts have been made to produce opposite shifts in the position sense of the two eyes. Foley and Miyanshi (1969) had subjects walk about for 15 minutes with an 11° displacing prism, base down, over one eye and a similar prism, base up, over the other eye. Monocular tests of eye–hand coordination revealed that adaptation was to some extent contingent on the eye being tested. The effects produced in the eye viewing through the base-down prism were stronger than those produced in the eye viewing through the base-up prism. When both eyes were open but only one eye had a prism before it, no adaptation was produced in either eye, which shows that a veridical input dominates over a non-veridical input. However, the dominance of a base-down prism over a base-up prism and the dominance of a veridical input over a non-veridical input could in each case be reversed by changing the relative levels of illumination in the two eyes (Foley, 1970a, b). An even larger degree of eye-specific adaptation was obtained by Hajos (1968) when he placed a base-left prism before one eye and a base-right prism before the other.

Ebenholtz (1970b) pointed out that oppositely directed prisms induce binocular diplopia, which the subject would attempt to overcome by altering the vergence of the eyes: vertical vergence in the Foley and Miyanshi study and horizontal vergence in the Hajos study. Exposure to unusual conditions of vergence is known to produce aftereffects which show as errors of distance judgments based on vergence (Wallach and Frey, 1972; Paap and Ebenholtz, 1977), and it may have been these aftereffects which the prism studies of Foley and Hajos revealed. Craske and Crawshaw (1974) produced opposite changes in the sense of direction of the two eyes by allowing each eye in turn to view the legs through opposite prisms and found that visual estimates of distance were distorted in just the way one would expect from a recalibration of vergence (see Foley, 1974, for further discussion on this point). Prablanc *et al.* (1975b) exposed the left eye and the left arm to a base-right prism followed by exposure of the right eye and right arm to a base-left prism. Opposite adaptive shifts were induced in each arm irrespective of the eye used in testing for these adaptive shifts. This shows that arm-specific adaptation is more likely to occur than eye-specific adaptation when the system is presented with the option.

The evidence reviewed in this section supports the conclusion that visual–motor adaptation can sometimes involve a change in the sense of position of the eyes. Opposite effects in the two eyes are probably best interpreted as changes in the vergence system.

Recalibration of the sense of head position (site c_3). Several investigators have reported that, after exposure to displaced vision, subjects tend to rotate the head several degrees in the direction of the prism displacement when attempting to set the head in a straight-ahead position (I. Kohler, 1955; Harris, 1965b; Lackner, 1973). This presumably reflects a change in the apparent position of the straight-ahead posture of the head.

A change in the apparent position of the head should be reflected in a change in the apparent direction of sound sources relative to the body. Such changes have been reported by Freedman and Rekosh (1968).

Recalibration of the sense of arm position (site c_1) If one arm only is used in the exposure phase of an experiment on visual displacement, and the aftereffect is the same for the two arms, the adaptation is said to show **intermanual transfer**. If there is no intermanual transfer, the conclusion seems inescapable that the adaptive change is confined to the sensory or motor mechanisms which code the sense of position or movements of the trained arm. The occurrence of full intermanual transfer suggests, but does not prove, that the change occurs in a part of the control loop which is common to both arms, that is, the coding of head position or of eye position or perhaps conscious correction for the effects of the prism. This argument is not conclusive; a change in the felt position of one arm may transfer to the other arm because of some cross-linkage in the systems controlling the two arms—the sort of thing found in the eyes. Evidence for such a cross linkage between the arms comes from a study by Kelso *et al.* (1979) on the coordination of two-handed movements.

The evidence on the intermanual transfer of prism-adapted pointing is confused and often contradictory. However, there seems to be some agreement that there is more intermanual transfer when training involves terminal visual feedback than when it involves continuous viewing of the arm (M. M. Cohen, 1967; Uhlarik and Canon, 1971). We shall see in Section 12.4.4 that there is some evidence that adaptation is more likely to involve a part of the system which is not being attended to. Terminal visual feedback may lead the subject to attend to the felt position of the arm, thus making it less likely that any adaptation which occurs will involve a change in the felt position of the arm, and therefore increasing the probability that the adaptation will affect both arms equally. Continuous viewing of the arm, on the other hand, may lead the subject to disregard the felt position of the arm, thus increasing the probability of arm-specific adaptation.

Taub and Goldberg (1973) suggested that the difference between terminal and continuous exposure in the amount of intermanual transfer is due to the fact that pointing trials are given more frequently in the terminal exposure procedure than in the continuous exposure procedure, and produced evidence in favour of this view. However, Cohen (1973b) found that both the type of exposure and the frequency of trials contributed to the amount of intermanual transfer. The only explanation offered for why there is less intermanual transfer when trials are more frequent was that, when there is less time between trials, the subject does not need to depend on conscious recollection of the error and is therefore less likely to correct consciously for the effects of the prism.

Intermanual transfer has never been found to be complete and is usually less than 50% of the effect shown with the trained arm. Thus, insofar as intermanual transfer is incomplete, at least part of the adaptive change to displaced vision involves a change in the sense of position of the arm. This could be a change in the way information from joint receptors is coded or a change in the spatial coding based on efference.

The presence of intermanual transfer is not conclusive evidence that adaptive changes have occurred in the sense of position of the head or the eyes. However, we have seen that there is other evidence that, under certain circumstances, such adaptive changes do occur. The only type of adaptive change for which there is no evidence is a change in the coding of oculocentric direction.

This does not exhaust the question about the site of the recalibration underlying adaptation to displaced vision. We have yet to consider at what level in the hierarchy of processing depicted in Figure 12.10 the recalibration occurs.

The level of recalibration in the nervous system Harris (1965a) championed the view that adaptation to displaced vision involves a change in the felt position of the arm. Later, he admitted that this theory would not account for the fact that under certain circumstances there is some intermanual transfer, nor for the reported changes in the sense of direction of the eyes. Harris broadened his theory to include changes in the felt position of the head and of the eyes. He also realised that the felt position of a limb may depend both on proprioceptors and on efference. This theory has become known as the **felt-position theory** and, in terms of the flow chart in Figure 12.10, asserts that recalibration is at one or more of the sites marked 'c'. These sites indicate the simple sensory coding processes responsible for the sense of position of each of the mechanical joints in the system and the motor coding processes responsible for the efference copy. The theory excludes changes in oculocentric coding.

In contrast to Harris's general felt-position theory, Held has championed the view that adaptation involves a change at a higher level of processing—at a level where the felt position and potential movements of the arm (or other adapted organ) are assessed in relation to the sensed direction of an external target or of a target consisting of some other part of the body. This **sensorimotor theory**, like the felt-position theory, excludes changes in the oculocentric system. It also predicts, as does the felt-position theory, that adaptation transfers to other modalities, or shows in a test in which the trained arm points to the hand of the untrained arm. However, Held claimed that, unlike the felt-position hypothesis, the sensorimotor hypothesis predicts no effect of adaptation on a test in which the subject is asked to relocate his unseen arm into a position into which he was trained to put it before exposure to the prism. Held interpreted this task as a pure measure of a change in felt position, because the arm is not being directed to a target and is therefore not affected by a sensorimotor change. Using this and other tests, Hardt *et al.* (1971) obtained results which they claimed supported the sensorimotor hypothesis.

In terms of the scheme depicted in Figure 12.10, the adaptive changes suggested by Held are at the sites marked 'd', i.e. at the level of what I have called coordinators. Held suggested that what is adapted is a very general coordinator that controls directed movements of a particular limb with respect to all targets. This does not have the specificity of the coordinators depicted in Figure 12.10. However, these are meant to indicate only the logical status of coordinators in the total system. The scheme sets no limit on the types or complexity of coordinators.

Figure 12.10 brings out the main logical point underlying the debate between theorists like Harris, who emphasize changes at a low level in the spatial coding process, and theorists like Held, who emphasize changes at a higher level of processing. The important point is that, in a hierarchical sensory system, higher-level processors receive their information from the lower-level processors. Therefore, any change at a given level affects only those tasks that involve that level of processing and those levels above it.

To decide between these two theories, we need tasks that diagnose specific changes at the various levels of the system. Held proposed that his task, in which the arm is relocated to a remembered position, is a task for diagnosing low-level changes in the system. However, it is not clear just what sort of task this is, because it assumes that the system which remembers limb position is immune to the effects of training. In any case, Kennedy (1969) found that the relocated position of the arm was affected by training, so that the evidence resulting from the use of this test is equivocal.

Templeton *et al.* (1974) attempted to resolve this issue by applying, both during and after training, a series of tests which was designed to diagnose changes at each level of the system. A test of pointing at the unseen toe measured changes in the felt position of the arm, and a test of looking towards the unseen toe measured changes in the felt position of the neck or eye. If felt-position changes alone account for adaptation of pointing behaviour, then changes in these two subtests should add to the total adaptive shift, as measured by the error in pointing to visual targets with unseen hand. If the total adaptive shift exceeds the sum of the shifts on the subtests, this would be evidence of a shift in a higher part of the control hierarchy. The two felt-position shifts were found to add to the total shift in pointing at a visual target in trials given when the prism was at its maximum displacement of 15°. This additivity of component effects after training is complete confirms the results of a similar experiment conducted by Wilkinson (1971) and has received further recent confirmation (Wallace, 1977; Redding, 1978). However, Templeton *et al.* found that tests given during training, while the prism displacement was increasing, revealed that the shift in pointing to visual targets exceeded the sum of the felt-position shifts. This suggests that a higher-level system intervenes while learning is proceeding, but that the effects of training consolidate at lower levels of the system when adaptation is complete. This result was not in any obvious way due to conscious intervention, because the gradual shaping procedure was used and subjects were not aware that the visual targets had been displaced. Redding and Wallace (1978) found that when training involved several visual targets, component tasks summed to the total shift in pointing to visual targets, but that when training involved pointing to only one target, the sum of the component shifts exceeded the total shift. The conditions under which additivity of components does and does not hold have been further explored by Welch (1974), Redding and Wallace (1978), and Wallace and Redding (1979).

It would seem that changes in the felt position of parts of the system are mainly responsible for visual–motor adaptation, but that changes higher in the system may also occur. These experiments, like others, were conducted with certain arbitrary stimulus constraints and demand characteristics. It may be that the relative importance of felt-position changes and changes at higher levels will be reversed under other circumstances. No experiment in psychology reveals which mechanism *must* operate, but only which mechanism *may* operate under certain circumstances. The human central nervous system is very flexible, and works more like a programmable computer than like a machine with a single fixed way of doing things. Basic mechanisms of behaviour which experimenters claim to have 'discovered' often turn out to be the result of hidden assumptions, arbitrary stimulus configurations, or particular demand characteristics. Several arbitrary constraints of these kinds are evident in studies of prism adaptation. I have already mentioned how the procedural accident of allowing the subject to fixate the visual target invalidated the conclusion that adaptation is not in the oculocentric system. Another example of a hidden constraint is the fact that most investigators have used a training procedure in which the subject moves his arm rather than other parts of the body, and perhaps this procedural accident has also biased the results.

This thought prompted Howard *et al.* (1974) to compare the amount of intermanual transfer of prism adaptation produced by two types of training. The subject looked through a prism which displaced the apparent position of his hand, but not the apparent position of a light fixed in the median plane of his head (in front of his nose). During training, the subject was placed in the dark and repeatedly ran his finger along a horizontal track in a frontal plane until he judged it to be in front of his stationary nose, or repeatedly aimed his nose towards his stationary finger. In both cases, the subject was allowed to see his error at the end of each completed movement. It was found that there was less intermanual transfer, i.e. the adaptation was more specifically related to the trained arm, when the head was mobile than when the arm was mobile during training. Thus, the position sense of a stationary joint adapts more than that of a mobile joint. This result makes sense when considered in conjunction with the well-known fact that we are less certain about the felt position of a stationary limb than about the position of a moving limb, and the fact that adaptation tends to be associated with that part of the system about which there is most position uncertainty (Wilkinson, 1971). It is the 'weakest' part of the system that adapts.

The locus of attention has also been shown to be a factor determining the site of adaptation to displaced vision. In a study by Canon (1970), subjects were exposed to conflicting visual and auditory cues regarding the direction of a sound-emitting visible object. It was found that adaptation occurred in the modality which was not attended to during exposure to intersensory discordance. This finding has been confirmed in a study using a conflict between vision and proprioception (Kelso *et al.*, 1975). These results are compatible with the finding that adaptation is associated with that part of the system about which there is most uncertainty, because presumably there is more uncertainty about the position of something to which one is not attending than there is about the position of something to which one is attending.

Adaptation to prismatic displacement depends on some kind of learning process. Ever since the early studies of Pavlov it has been known that the effects produced by a given learning experience often become associated with the general surroundings in which the learning occurs. In a similar way it might be expected that the effects of prism adaptation may be evoked more readily when conditions resemble those of the exposure phase. There have been several reports that the effects of even short periods of exposure to displacing prisms persist for over 24 hours when the subject is returned to the same situation in which he was trained, even though his normal behaviour is unaffected (Klapp *et al.*, 1974; Lackner and Lobovitz, 1977; McGonigle and Flook, 1978). This behaviour suggests that two alternative calibrations of the visual–motor system may be held in storage at the same time, each evoked by the particular conditions in which the subject finds himself.

Thus, the question about the site of the adaptive shift underlying adaptation to displaced vision is one that does not have a single answer. All systems beyond the most peripheral processes in sense organs are probably capable of adapting to unusual circumstances, given the correct set of constraints and demands.

12.4.4 The conditions under which adaptation to displaced vision occurs

The second question that has engaged the attention of investigators in this field is that concerning the conditions under which adaptation occurs. Held (1961) proposed that active movement is necessary for visuomotor adaptation to displaced vision. This hypothesis has inspired most of the subsequent work in this area. The schematized process which Held proposed is shown in Figure 12.12. Except for the addition of the 'correlation storage' it is the same as the model which von Holst (1954) developed in connection with his work on visual–motor coordination in animals. In Held's words:

> 'The reafferent visual signal is compared (in the Comparator) with a signal selected from the Correlation Storage by the monitored efferent signal. The Correlation Storage acts as a kind of memory which retains traces of previous combinations of concurrent efferent and reafferent signals. The currently monitored efferent signal is presumed to select the trace combination containing the identical efferent part and to activate the reafferent trace combined with it. The resulting revived reafferent signal is sent to the Comparator for comparison with the

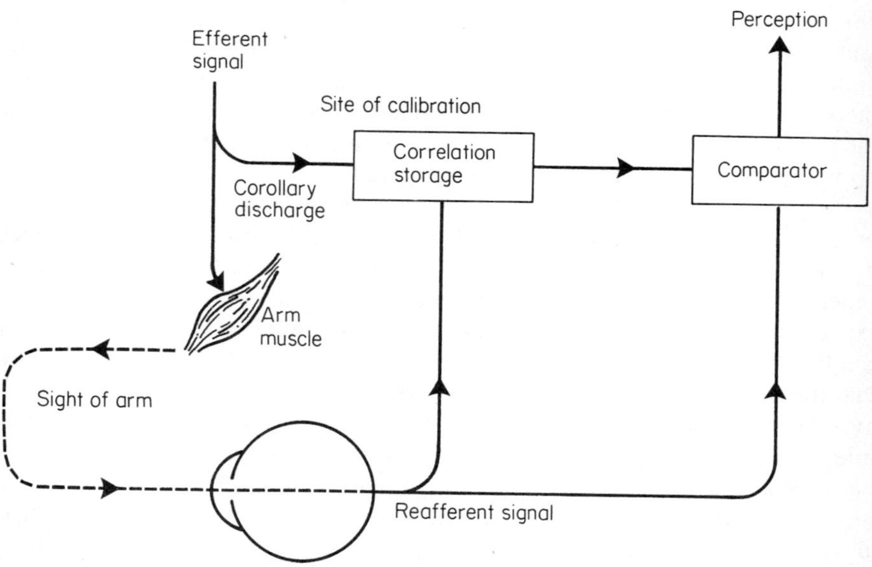

Figure 12.12 Schematized process assumed by Held to underlie the adaptation of the visuomotor system to discordant reafference (From Held, 1961. © 1961, The Williams & Wilkins Co., Baltimore)

current reafferent signal. The outcome of this comparison determines further performance.' (Held, 1961, p. 30)

The model predicts that a self-produced efferent signal is required for the build-up of stored reafferent traces upon which are based further modifications of the system. The evidence which Held and Hein (1958) produced to demonstrate the necessity of reafference is well known and may be summarized as follows. The subject's ability to aim at a visual target with an unseen hand was first established; the subject then viewed his hand for a few minutes through prisms which shifted its apparent position to one side. During this time the hand was waved from side to side, either actively by the subject or passively by the experimenter. Only after the active condition did subjects show evidence of having adapted to the prisms when they were tested in an aiming test.

Held and Hein concluded that self-produced movement coupled to visual reafferent stimulation is necessary for a change of visual–motor coordination. However, if Held and Hein's exposure procedure is examined it can be seen that Held's model is not appropriate to account for their conclusion. The motor outflow determined mainly the amplitude of the arm movement during the exposure condition, and the amplitude of movement was not affected by the prisms. The arm was, and appeared to be, moving through the same amplitude as the motor outflow commanded. In other words, the prisms did not significantly alter the correlation between corollary discharge and the visual reafference. The motor outflow required to hold the arm in a given position is very little affected by a change of 2.2 in which was the displacement produced by Held's prism, and in any case, motor innervation may not be the principal source of position sense in a limb (Howard and Templeton, 1966). The main discordance in Held's experiment was between the visual position of the arm and its position as detected by kinaesthetic receptors. But Held's model makes no mention of a correlation between vision and Kinaesthesis, and it is therefore inappropriate for this situation.

Had Held studied the effects of visual minification or magnification on visual–motor adaptation, he might have found some support for his model; at least it would have been the appropriate model to use. In any case, a model which accounts for only one selected set of experimental conditions cannot be proposed as a general principle.

It is surprising that there was no adaptation in Held's passive condition, for the discordance between vision and the joint sense should have been just as great in the passive condition as in the active condition. There are four possible reasons for this lack of passively induced adaptation. (1) The displacement produced by a prism is greater on the prism-base side than on the other side, so that the seen excursion of the hand may have been minified or magnified depending on its position in the field of view. Such an effect would disturb the correlation between corollary discharge and visual reafference and may therefore be expected to produce differential effects between active and passive movement. However, it is not clear in Held's account where the hand was in the visual field, and hence it is not possible to assess the effect of this factor, which in any case must have been small. (2) There was probably a difference in the degree of visual or motor asymmetry between the active and passive conditions. There is evidence that asymmetry alone can induce a change in visual–motor behaviour (see Section 7.6.4). (3) The test procedure used by Held and Hein involved active movement, and hence was more like the active training than the passive training procedure. This gives an unfair advantage to the active condition. It has been shown that a passive training procedure produces a change in a passive test (Pick and Hay, 1965). (4) Adaptation may have been greater in Held and Hein's active condition because subjects paid more attention to what they were doing when moving their own arms.

Held has produced other evidence for his reafference model, notably the now famous experiment in which it was found that only those kittens which had moved actively in relation to their visual surroundings developed visual–motor behaviour. However, these experiments were biased in favour of the active kitten, for the passively moved kitten was provided with virtually no information regarding the relationship between body moment and vision (see Section 11.5.1 and Howard and Templeton, 1966, p. 394).

Several investigators have demonstrated that self-produced movement is not necessary for visual–motor adaptation (Pick and Hay, 1965; Templeton *et al.*, 1966; Baily, 1972b). In these studies, the subject had to estimate when his passively moved, hidden finger was under a visual target seen through displacing prisms. He was shown his error after making his estimation. This discordance between passive arm-position sense and the sight of the arm generated some adaptation, as revealed in an active pointing task. Note that in this procedure the subject's attention is directed to the discordance, both in the active and in the passive exposure conditions. Singer and Day (1966), Fishkin (1969), and Mather and Lackner (1980b) found evidence of adaptation of pointing behaviour after passive training similar to that used by Held and Hein.

Further evidence against Held's interpretation of his data is provided by Moulden (1971), who found that adaptation occurred when the subject was allowed only momentary glimpses of his actively moved arm, which prevented him from seeing the arm in motion. According to Held's reafference theory, adaptation should not occur if the subject does not see his arm in motion.

Finally, evidence against Held's hypothesis comes from several studies in which it has been shown that adaptation of pointing to displaced visual targets occurs after exposure to discordant exafferent stimulation. For instance, Howard *et al.* (1965) obtained shifts in pointing to visual targets after subjects were exposed to the experience of a rod, seen through prisms, coming toward them and hitting them at an unexpected place on the forehead. Although Halper-Smith (1971) failed to replicate this effect, a similar effect was reported by Radeau and Bertelson (1969), who exposed passive subjects to a conflict between visual and auditory cues to the direction of an object and found evidence of an adaptive shift in a task of pointing to visual targets and in a task of pointing to an auditory target (see also Radeau, 1973; Radeau and Bertelson, 1974).

Held no longer insists that self-produced movement is necessary, only that this type of discordance is more effective in producing adaptation than other types of discordance. In this he may be right, but I think the whole issue should be placed in a broader theoretical framework.

Howard and Templeton (1966) argued that adaptation occurs in response to many forms of discordant information and that the important thing is that salient information regarding the discordance (not necessarily consciously perceived) should be available to the subject. The discordance is the mismatch between the unusual stimulus covariance function and the subject's covariance schema. According to this view, adaptation is enhanced by anything which increases the saliency and amount of information about the discordance. There are several factors which determine the effectiveness of information about discordance (Howard, 1978a), which I shall now list:

1. *The extent to which the subject is convinced that the discordant stimuli arise from the same object or part of the body*. In the typical prism adaptation study, this is the degree to which the subject relates his seen hand with his felt hand. Several studies have shown the importance of this factor. For instance, Welch (1972) showed that subjects who believed that the finger they saw was their own adapted more readily than subjects who did not believe this.

2. *The vividness of the display of the discordance.* For instance, Dewar (1971) found that a suddenly introduced visual displacement of the visual scene produced a greater amount of adaptation of pointing behaviour than that produced when the displacement was introduced gradually by the 'shaping' technique. It is not surprising that the amount of adaptation has been found to be greater with longer periods of exposure to the displacement and with greater amounts of prism displacement (Efstathiou, 1969). However, Dewar (1970) found that the amount of adaptation increased with increasing prism displacement only up to a displacement of 20°, after which it tended to decline. Presumably, displacements of this magnitude are too large for the system to cope with.

It has been shown that the opportunity for error correction is an important factor in prism adaptation (Webster, 1969; Welch and Abel, 1970). The error feedback need not be visual. For instance, Lackner (1974b) found that tactile feedback provided by contact between the hand and visual target is sufficient and Radeau (1974) that a buzzing sound emitted from the actual location of the visual target is sufficient. Uhlarik (1973) obtained adaptation by informing subjects verbally about the extent of their pointing error.

3. *The precision of the sense of position of the various joints involved in the task being studied.* There is abundant evidence that, in a training procedure in which arm proprioception is required for the detection of the discordance, anything which enhances the sense of position of the arm leads to a greater degree of adaptation. For instance, Kravitz and Wallach (1966) found evidence of adaptation in a pointing task after the subject inspected his immobile arm, the muscles of which were passively vibrated. The vibration presumably enhanced the sense of position of the arm. Melamed *et al.* (1973) found that adaptation

occurred in a passive-movement training condition when visible contours were provided on the background against which the arm was passively moved. These contours presumably enhanced the visual impression of the position of the arm. Thus, it may be concluded that adaptation of target-pointing behaviour is enhanced by anything which enhances the sense of position of the arm when the training procedure involves a discordance between the seen and felt position of the passive arm. The reason is that, under these circumstances, if the position of the arm were not sensed, the discordance would be undetected and adaptation would obviously not occur. I have already shown in Section 12.4.3 that, where the sense of arm position is *not* required for the detection of the discordance, a weakening of the sense of arm position may *increase* the amount of adaptation to visual displacement. Thus, the effect on adaptation of an increase or decrease of the sense of position of the arm, or of the sense of position of any other part of the body, depends on the type of training procedure adopted.

4. *The similarity between exposure and test procedures*. Wallace (1974) found that there was more adaptation when the target-pointing task used in the training procedure was the same as that used in the test procedure, compared with the case where the pointing task differed in the two procedures. Baily (1972a) found that the effect of training on pointing to targets with rapid or ballistic arm movements revealed itself with both rapid and slow test movements, but that the effect of training with slow arm movements revealed itself only with slow test movements.

5. *Differences in the way different people react to what is apparently the same situation*. Some subjects adapt more readily than others, and the site of adaptation varies among subjects, even after objectively similar conditions of exposure to prisms. For instance, some people show more adaptation of the sense of position of the eyes and others show more adaptation of the sense of position of the arm (Templeton *et al*., 1974). Warren and Platt (1974) found that subjects who were more precise at visually tracking a target and repositioning the gaze into a previously held position were more likely to show changes in the felt position of the arm after adapting to prisms, whereas subjects who were more precise at repositioning an arm in a previously held position were more likely to show changes in the felt position of the eyes. Thus, the system which is least precise in ordinary use is most likely to be recalibrated. This finding is consistent with the conclusion arrived at under (2) above.

The most reasonable general conclusion is that any consistent covariance between stimuli within modalities or between stimuli in different modalities or any covariance between stimuli and responses will be learned and stored as covariance schemata. Any change in the way stimuli, or stimuli and responses, covary will be detected because of the discordance with the stored schemata, and the schemata will be recalibrated accordingly.

12.5 Adaptation to visual tilt

An optical device which causes the visual scene to appear rotated through a certain angle about the visual axis disturbs the visual input in two distinct ways. It disturbs the retinal orientation of those objects, such as people, trees, and houses, which are normally seen in one orientation. Such objects are called **mono-oriented objects**. This type of disturbance is apparent as soon as familiar, mono-oriented objects are viewed through the optical device. Such objects appear tilted, or, if the rotation is through 180°, they appear inverted. The way people behave in the presence of tilted scenes was discussed in Section 10.3, and the perceptual effects of inverting familiar objects will be discussed in Section 14.4. The second type of disturbance caused by rotation of the visual scene is the discordance between the visually-detected orientation of the scene and its orientation as detected by the non-visual senses. This discordance affects unfamiliar objects as well as familiar objects, and mono-oriented objects as well as poly-oriented ones. In studying this discordance it is best to use simple stimuli which lack a recognizable top and bottom, so as to avoid complications due to things looking tilted or inverted. A person cannot, by mere visual inspection, tell whether a scene is optically tilted, and therefore discordant with respect to other senses, or whether it is actually tilted, and therefore not discordant. A discordance produced by optical rotation becomes apparent only when the subject either sees his own body or receives information about external objects from a sense organ not affected by the optical rotation. This section is concerned with how people adapt to this type of discordance, but attention will be confined to the discordance between vision and visually-directed movements of the hand.

The normal relationship between visually-directed movements and the orientation of visual objects may be disturbed either by viewing a visual scene through some optical device, such as a Dove prism, which rotates the visual input or by surgically rotating the eyeball. I shall first review experiments in which the eye was surgically rotated.

As part of a treatment for a detached retina, Barrios *et al.* (1959) severed each rectus muscle of one eye in several human patients, rotated the eyeball through 90°, and sutured each muscle back onto a stump of tendon which was 90° away from the muscle's normal insertion. The oblique muscle, which limited the selected rotation, was left in a severed condition. After recovering from the operation, the patients were allowed to use both eyes during the six-month period in which the eye was left in the rotated state. During the whole of this period the patients manifested the following behaviour when tested with only the rotated eye open. They reported that the visual scene appeared rotated 90°, their optokinetic nystagmus, pursuit eye movements, and visually-directed saccades were performed with a 90° difference with respect to the target motion, and visually-directed movements made with their unseen hand were also performed with a 90° difference with respect to the visual targets. The total absence of adaptation in these patients may have been due to suppression of vision in the rotated eye during their ordinary daily experience. The evidence I shall now review suggests that, had these patients been forced to use only their rotated eye, they would have adapted.

Blakemore *et al.* (1975) studied the effects of eye rotation in cats. In four kittens and one cat the muscles of one eye were severed and the eyeball rotated through various angles between 30 and 115°. The animals were tested several weeks later after the muscles had become reattached and eye movements had returned. All the animals showed visually mediated behaviour, but only the kitten with a 30° eye rotation was able to place its paw accurately on a prong, and none of them was able to avoid obstacles. This failure to adapt to the eye rotation may also have been due to the fact that the kittens were allowed the use of both eyes during the recovery period, and were therefore not forced to use their rotated eye. Gordon *et al.* (1979) detected severe visual-field deficits (amblyopia) in the rotated eyes of cats raised with one eye rotated and one eye normal, but no deficits in rotated eyes of cats raised with one eye rotated and the normal eye sutured. This shows that a rotated eye, when competing with a normal eye, becomes severely amblyopic. To overcome this problem, Blakemore *et al.* conducted a second experiment in which an animal with a 90° eye rotation was allowed to recover over a period of one year, during which time its normal eye was sutured. When tested with its rotated eye at the end of this period, this animal showed accurate visually-guided paw placement and obstacle avoidance. Similar results were obtained by Peck and Crewther (1975). Crewther *et al.* (1980) rotated one or both eyes of kittens at various angles at the time the eyes were beginning to open. Kittens were given visual experience each day. In the case of kittens with single eye rotation this experience was confined to the rotated eye. All the animals showed good visual–motor coordination when tested at four weeks of age.

Blakemore *et al.* investigated the retinotopic mapping of the rotated eye onto the visual cortex and found that it was unchanged except for a reduction in the number of binocularly driven cells. Yinon (1977) and Crewther *et al.* also found no change in retinotopic mapping in cats with one eye rotated 180°.

Cynader *et al.* (1978) found that the direction-of-motion specificity of cells in the superior colliculus became adapted in kittens reared with one eye rotated. Crewther *et al.* found some cells in cortical layer V which showed a significant shift in direction-of-motion preference in kittens in which the two eyes were up to 90° out of register. All these cortical cells were non-orientation specific and had properties which indicated that they projected to the superior colliculus. The superior colliculus is concerned with controlling saccades (see Section 5.5.3) and perhaps this type of adaptation is confined to the part of the visual system which serves to control saccadic eye movements.

The only evidence that the orientation preferences of cortical cells adapt to anomalous cyclocorrespondence of the two retinae comes from a study by Shinkman and Bruce (1977). They reared kittens with prisms before their eyes, which produced an opposed rotation of the scene through 8° in each eye. After this experience the interocular preferred orientations of cortical cells were matched to the rotated scenes. Blakemore *et al.* and Crewther *et al.* found no convincing evidence for this kind of adaptation, although they did find that the orientation tuning of cells in rotated eyes was broader than that of cells in normal kittens. But in these latter studies the eyes were rotated through larger angles than in the study by Shinkman and Bruce. Perhaps the capacity of binocular cortical cells to adjust their orientation preferences in the two eyes is limited to small angular changes.

A capacity to compensate behaviourally for eye rotation is apparently not present in all animal phyla. Sperry (1948) found no evidence of behavioural compensation to eye rotation in fish, and Sperry (1943), Stone (1944), and Skarf and Jacobson (1974) found no compensation to 180° eye rotation in amphibians. On the other hand, birds have been shown to be capable of adjusting to an optically induced disturbance of the correlation between eye movements and head movements (Rossi, 1968).

Mitchell *et al.* (1976) rotated one eye through 90° in each of two cats for a period of nine months, during which period the normal eye was kept closed. When tested with the rotated eye at the end of this period the animals showed appropriately directed head movements, had accurate paw placement, and avoided obstacles. The adequacy of the movements of the rotated eye was difficult to assess because the eye did not move far. The cats were then trained, with the normal eye, to distinguish between vertical and horizontal stripes on a Lashley jumping stand. When the normal eye of each cat was closed and the animals were tested with the rotated eye, they jumped to the same distal stimulus that they had been trained to jump to with the normal eye. Thus, for these cats, the vertical meridian in the normal eye was behaviourally equivalent to what should have been the horizontal meridian in the other eye. Electrophysiological investigations revealed that this adaptation was not due to changes in the preferred orientation of cells in the visual cortex (Blakemore *et al.*, 1975). Eye movements may have played a part, but this question has not been studied. A similar procedure was followed by Peck *et al.* (1979). They obtained results similar to those of Mitchell *et al.* with eye rotations of 90 and 120°, but kittens with eye rotations of 180°, although able to make accurate paw placements and avoid obstacles, were not able to learn to distinguish between patterns with the rotated eye. Perhaps the 180° rotation constricted blood circulation and impaired vision.

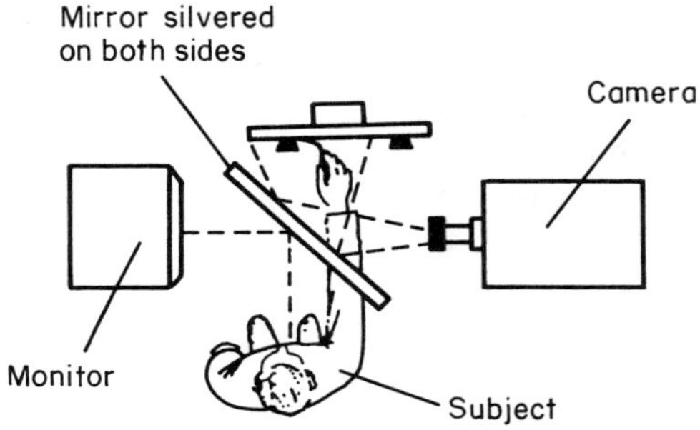

Figure 12.13 Arrangement for studying visual–motor coordination in the presence of a rotated image of the hand

The visual field may be rotated in human subjects either by placing Dove prisms in front of the eyes, in the way described in Section 12.2, or by having the subject view his own arm in a television monitor. A convenient arrangement of television camera, monitor, and two-sided mirror is shown in Figure 12.13. With this arrangement the subject sees an image of the scene and his own hand rotated in a frontal plane through any desired angle. Several studies have been done in which subjects moved about, either actively or passively, in a normal interior environment whilst wearing devices which rotated the scene (Ebenholtz, 1968, 1973; Quinlan, 1970). There is poor stimulus control with this type of procedure, so that the results are not easy to interpret.

McIntyre and Pick (1974) used the television monitor procedure to expose the subject to a 90° rotated image of his own arm as he moved it back and forth for 16 minutes along a track lying in a horizontal plane. In the pretest and posttest phases of the experiment the subject drew freehand along a line which he could see oriented at each of several angles, but with his hand hidden from view. The direction of lines drawn after exposure was displaced several degrees towards the direction predicted from adaptation to the discordance. There was some positive transfer of this adaptation to directions of hand movement other than that used during exposure to the discordance, although the amount of this transfer varied in a complex way as a function of the direction of the movement. When subjects were asked to draw a line in a given direction, rather than copy a line which was seen, they showed evidence of the same adaptation, but when they were asked to select visually a line in a given orientation, they showed no effects of adaptation. These tests demonstrated that the adaptation was in the motor–proprioceptive system and not in the visual system. With a 90° rotation one would not expect any visual tilt aftereffects or visual frame effects.

Prablanc *et al.* (1975a) did a similar experiment. In the exposure phase, the subject drew lines on a horizontal surface and saw only a luminous point moving on an oscilloscope. The direction of movement of the point was rotated 20° with respect to the direction of movement of the hand. Testing consisted of asking the subject to trace simple line figures with the unseen hand. Prablanc *et al.* also found that the effects of adaptation transferred to untrained directions of movement and that they showed in the task of moving the hand in a specified direction. Visual aftereffects were apparently not tested for.

The results of these two studies suggest that adaptation to a discordance between visual direction and the direction of hand movements in a given plane leads to a general recalibration of the motor-proprioceptive system underlying the control of arm movements in that plane.

Chapter 13

Distinguishing, identifying and describing the orientation of shapes

13.1 Introduction

Chapter 3 was about how people discriminate small differences in the orientation of very simple figures under the most favourable conditions. The present chapter is about how people distinguish between, identify and describe the orientations of visible objects under conditions that do not allow fine discriminations to be made.

When a person judges the orientation of an object, he relates the direction of some recognizable axis in the object, an **intrinsic axis**, to some axis of reference outside the object, an **extrinsic axis**. There are four types of intrinsic axes: (1) **axes of balance**, or axes about which an object will balance on a surface, (2) **axes of symmetry**, (3) **main-line axes**, or the main lines of an object, and (4) **polar axes**, or axes between pairs of significant landmarks in an object. There are three types of extrinsic axes: (1) **egocentric axes**, or axes fixed with reference to some part of the observer's body; a special case of an egocentric axis is a meridian of the eye, or an oculocentric axis, (2) **vertical and horizontal axes**, that is, axes defined with reference to gravity, and (3) a **visual framework**, or axes defined with reference to some other visible object or some feature of the visual surroundings such as the direction of a source of light. This then is the general stimulus domain with which we are concerned in this chapter.

Section 13.2 is about how each of the intrinsic axes enables people to judge the orientation of an isolated object. In Sections 13.3 and 13.4 I define the stimulus and task domains of the relative orientation of two shapes in the frontal plane, as a preliminary to Sections 13.5 and 13.6 in which empirical work on distinguishing between and identifying orientations is discussed.

13.2 Intrinsic axes for judging a shape's orientation

13.2.1 Axes of balance and symmetry and main-line axes

An **axis of balance** is a line through the centre of gravity of an object and which runs perpendicularly through a base. A base is the area formed by joining together points of contact which an object makes with a plane surface. Every object has at least one axis of balance (if this were not so, we could design a vehicle that would run indefinitely on a level surface without fuel). An unsupported object must have at least one of its axes of balance vertical if it is resting on a horizontal surface. As far as I know there are no studies of how people use axes of balance to judge an object's orientation.

An **axis of mirror-image symmetry** in a two-dimensional shape is a line which divides the shape into two identical mirror-image halves. A two-dimensional shape may have any number of axes of mirror symmetry: an arrow has one, a square has four, and a circle has an infinite number. Solid objects may have any number of **planes** of mirror symmetry, orthogonal sets of which intersect in axes of symmetry.

In objects in which mass is evenly distributed, all axes of symmetry are axes of balance. The position of an axis of mirror symmetry is easy to judge and therefore such an axis is a particularly salient axis of balance. It is difficult to find an object in an ordinary room that does not have at least one plane of symmetry and, apart from pieces of modern sculpture, I have never seen a symmetrical mono-oriented object which does not have one of its planes of symmetry vertical when in its upright position. It is not surprising that people put such heavy reliance on the axis of symmetry when judging whether an object is vertical.

The salience of symmetry for judgments of verticality was revealed in a study by Schaller and Harris (1975), in which children between six and 12 years of age were asked to rotate abstract objects until they were 'right-side up'. It was found that the axis of symmetry was almost always oriented vertically when the objects were in the 'right-side-up' position. The only exceptions were letter-like forms such as a shape that resembled a letter C.

Main-line axes are the predominant lines in an object or shape. These lines are often, but not always, parallel to axes of symmetry. For instance, a square looks like an upright diamond when it stands on one corner, and in this position the vertical axis of symmetry is not parallel to any of the main lines.

Axes of balance, axes of symmetry, and main-line axes may indicate whether an object is tilted, but they do not reveal whether it is right-way up or upside-down. For this judgment we need polar axes.

13.2.2 Polar axes

Many objects have features which distinguish one end from the other. For instance, a person has a head at one end and feet at the other. These are **polar features** and the line joining any related pair of polar features is a **polar axis**. For example, a car has three pairs of polar features, front and back, underside and roof, and driver's side and passenger's side. In symmetrical objects, polar axes lie in planes of symmetry. Most natural as well as most man-made objects have polar features that we recognize as top and bottom, although it is only man-made objects, such as shoes and letters of the alphabet, which have polar features that we use consistently for making distinctions between one side and the other.

The first question about polar features is non-behavioural, or ecological, namely, whether the polar features of different objects have anything in common which would serve as a general basis for judging uprightness, even in novel abstract shapes. The evidence suggests that this is so. For instance, Arnheim (1954, p. 17) reported that adults could judge the intended orientation of abstract paintings with significant consistency, and Chou (1935) found that Americans could judge when Chinese characters were abnormally oriented at above a chance level of performance. General polar features that people may use when judging the uprightness of objects include the following.

(a) *The position of the centre of gravity.* An object is more stable when its centre of gravity is low. Any visual feature that indicates the position of the centre of gravity may therefore influence the apparent uprightness of an object. Taper is obviously such a feature, and certainly more objects taper upwards than taper downwards. Goldstein and Andrews (1962) found that people most often judged abstract shapes as upright when the longest side was down.

(b) *Texture gradients.* An object in a frontal plane can normally extend only about two metres below eye level, which is probably not far enough to produce a noticeable perspective texture gradient. But many objects, like buildings and trees, extend well above eye level. This should tend to make a pattern with a real texture gradient look upright only when the smaller detail is at the top. Schaller and Harris (1975) found the opposite to be the case, but the real texture gradient that they used may not have been interpreted as a perspective gradient.

(c) *The position of dominant detail.* Ghent (1961) suggested that the eye naturally first fixates the top of a picture and that there is a preference for the centre of gaze to move towards centres of dominant detail. When the dominant detail is in the top of the picture, these two eye-movement tendencies are in harmony and the picture looks upright. When the dominant detail is in the bottom of the picture, the eye-movement tendencies are opposed and the picture looks upside-down. Ghent supported her suggestion that the eye scans from top to bottom by pointing out that all written languages are read from top to bottom, even though varying in other orientational features.

In order to give experimental support to her theory, Ghent presented children with 24 pairs of non-realistic shapes and asked them to choose the member of each pair which appeared upright. She found that if there was a focal point, such as a dot at one end of the shapes, then this end was chosen as the top. In another experiment Ghent *et al.* (1960) tested whether preferences still appear when eye movements are eliminated by exposing the shapes for 1/25 sec. The preferred orientations were unaffected by this procedure. The authors concluded that the postulated eye-movement tendencies must involve 'an internal motor process facilitating perceptual activity in a particular sequence'.

It would be more parsimonious to account for the fact that children prefer dominant detail in the top of a shape by reference to the many common objects which have dominant detail at the top. Children would generalize from their experience with the familiar shapes and apply the generalization to the meaningless shapes. However, without additional assumptions neither theory can account for Ghent's finding that these preferences decline and alter with age. Certain of the shapes which Ghent used resembled the letters V, Y, C, and U and others may have suggested particular objects to the children, so that Ghent's description of the shapes as 'non-realistic' does not seem justified. The changes in preferred orientation with age which she reported were almost certainly due to the greater familiarity of older children with letters. Antonovsky and Ghent (1964) found that other similar abstract shapes were consistently preferred in one orientation by children in Iran, showing that her original results with younger children were not peculiar to American children.

It was a feature of Ghent's procedure that the children were shown the shapes in only two orientations 180° apart. Harris and Schaller (1971) used a similar set of shapes seen against a circular background and allowed the children to rotate each of them to any angle. With many of the shapes, their results agreed with Ghent's results, but certain shapes, particularly letters, were often oriented at angles other than those that Ghent used. However, Harris and Schaller used only school children as subjects, so that their results should not be compared with those from Ghent's study on pre-school children, and these latter results are the ones that most probably reveal more basic factors determining orientation preference (Braine, née Ghent, 1973). The main problem with Ghent's notion of dominant detail is that in more complex shapes it is not always clear which detail is dominant.

13.3 The stimulus domain of relative orientation of shapes in a frontal plane

This section and the three following sections are about relative orientations of two-dimensional shapes in a plane. The stimulus domain of orientation in the frontal plane will first be described. The task domain will then be analysed and, finally, the literature will be reviewed according to whether the task is one of distinguishing between orientations, identifying orientations, or describing orientations.

Select any simple asymmetrical shape lying in a plane. Call this shape A. Select another shape, shape B, lying in the same plane. Shapes A and B may be identical, but for convenience I have chosen to let shape A be l and shape B be ɭ.

For each shape, specify the following:

1. An **intrinsic axis**; for l and ɭ, this is the axis passing through, and parallel to the long stroke of each shape.
2. A pair of **polar features** with which to define the uprightness of each shape with respect to the normally vertical meridian of the eye; l and ɭ are defined as upright when the long stroke is up and parrallel to the vertical meridan of the eye.
3. An **unreflected form**, as opposed to a reflected form produced by reflecting the shape about its intrinsic axis; for l and ɭ, this distinction is obvious.

For the two shapes, specify the following:

4. An **intershape axis**; in the case of l and ɭ, this is the axis passing through the midpoints of the long strokes of the two shapes.
5. A **standard orientation of the intershape axis**; for this purpose I have selected the orientation of the intershape axis which is parallel to the normally vertical meridian of the eye. The orientation of the intershape axis defines the **relative disposition** of the two shapes.

The stimulus domain of interest is the set of absolute and relative orientations of two shapes and the set of relative dispositions of two shapes in a frontal plane. I shall be concerned only with changes in orientation in steps of 90 or of 45°. The two shapes may be presented successively or simultaneously, in the same or in different locations. The observer may be upright, prone, tilted or inclined. Meridia parallel to the normally vertical and horizontal meridia of the retina will be referred to as vertical and horizontal meridia whether they are actually vertical and horizontal or not. When I refer to 'the plane', let it be understood that this is the plane in which the shapes lie, and which is a frontal plane with respect to an observer's head.

We need a system which allows us to describe, without ambiguity, all possible members of the stimulus domain. In this instance, such a system corresponds to what mathematicians call a **group**. Let us start with shape A, unreflected, upright and vertically above shape B, which is also unreflected and upright. The group needed is the set of all possible arrangements in the plane of shapes A and B formed by the following operations.

1. *The rotation of A* through steps of 90° in the plane, with respect to the intershape axis, with B remaining in line with the axis. This gives four orientations of shape A. The sign of rotation is arbitrarily designated positive if the small limb leads. A positive rotation produces an anticlockwise rotation of a shape, unless the shape is reflected, in which case it produces a clockwise rotation.
2. *The rotation of B* in a similar way for each of the four orientations of A, making 16 relative orientations.
3. *The reflection of A* about its own axis for each of the 16 relative orientations. We now have 32 relative orientations.
4. *The reflection of B* in a similar way for each of the 32 relative orientations, making 64 relative orientations. Note that in 16 of these, both A and B are reflected.
5. *The rotation of the intershape axis in 45° steps in the plane.* The sign of this rotation is immaterial. This is done for each of the 64 relative orientations, in each case keeping the orientations of A and B relative to the intershape axis constant. There are, thus, eight relative dispositions of A and B, unless A and B are identical, in which case there are only four distinct dispositions.

The shapes will be said to be disposed in vertical alignment when one is above the other, in lateral alignment when they are side by side, and in oblique alignment when the intershape axis is disposed at an angle of 45°. Each of the five operations defines a factorial subgroup of the total domain which we are considering. Subgroups are factorial when they are formed from independent operations and together form the complete group. These operations, or variables, therefore provide a convenient notation for relative orientation, because all possible absolute and relative orientations of A and B and all relative dispositions of A and B may be specified, together with rules for moving from any state to any other state. It is the experimenter's ideal descriptive system for orientation in this domain, or what may be called the **ideal perceiver** for orientation. Furthermore, this system enables one to construct any desired subgroups of the domain.

A subgroup of particular interest is the group formed by varying the orientation of A and B together, keeping the relative orientation constant at 90° and the intershape axis horizontal (Figure 13.1a). Analogous 90° orientation subgroups may be formed for other orientations of the intershape axis. Sets of analogous subgroups of this kind, which differ from each other in the value of some parameter, are known as **cosets**. Cosets have no elements in common, that is, they are mutually exclusive. If either A or B (but not both) is reflected, A and B will be said to be reflected with respect to each other. Note that when a shape is reflected, the clockwise rotation becomes positive. A subgroup may be formed as before, with A and B relatively reflected, by varying the orientation of A and B together, keeping the reflection of A to B constant and the intershape axis horizontal (Figure 13.1b). Again, cosets of this subgroup can be formed by

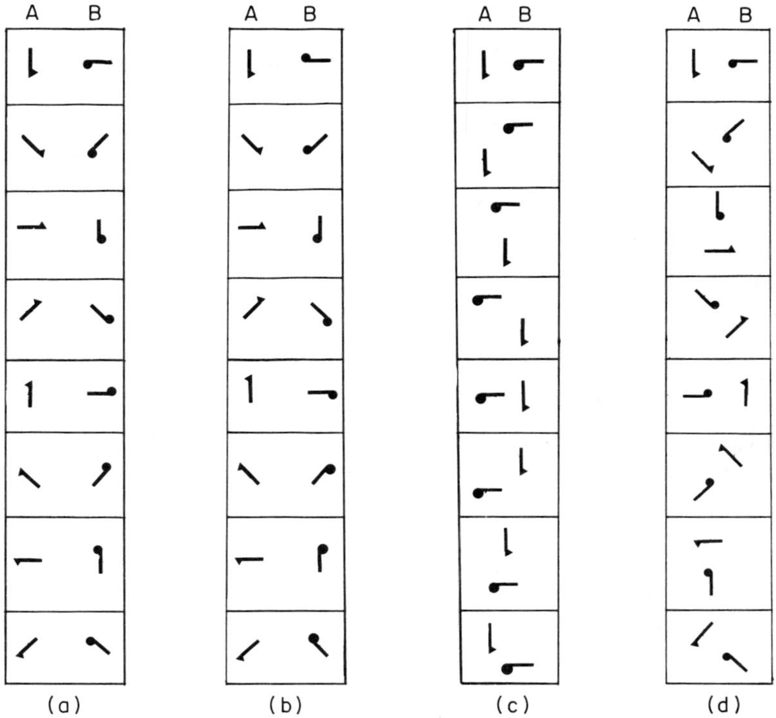

Figure 13.1 Subgroups of the domain of orientations of two objects A and B. (a) The subgroup of orientations of A and B with constant 90° relative orientation and constant orientation of the intershape axis. (b) Same as (a), with A and B relatively reflected. (c) Disposition of the intershape axis with constant 90° relative orientation and constant absolute orientation of A and B. The 'ferris-wheel' group. (d) Disposition of the intershape axis with constant 90° relative orientation and constant orientation relative of the intershape axis. The 'cartwheel' group

varying the orientation of the intershape axis. Other examples of subgroups are shown in Figure 13.1(c) and 13.1(d).

If two similar shapes, not relatively reflected, are placed at random within the same plane, it is always possible to rotate either of them about a suitably chosen axis, at right angles to the plane, so as to bring the shapes into coincident alignment. The geometrical procedure for determining the position of the centre of rotation in the plane, which brings the two shapes into coincidence, is indicated in Figure 13.2. Note that the required centre of rotation is easy to see in Figure 13.3(a), but is not so easy to see in Figure 13.3(b). On the other hand, if two shapes are reflected with respect to each other, then they can be brought into coincidence only by flipping one of them out of the plane, about the axis of symmetry which lies between them. If the shapes are reflected and rotated with respect to each other, there is no axis of symmetry between them and one shape will have to be both rotated and reflected to be brought into coincidence with the

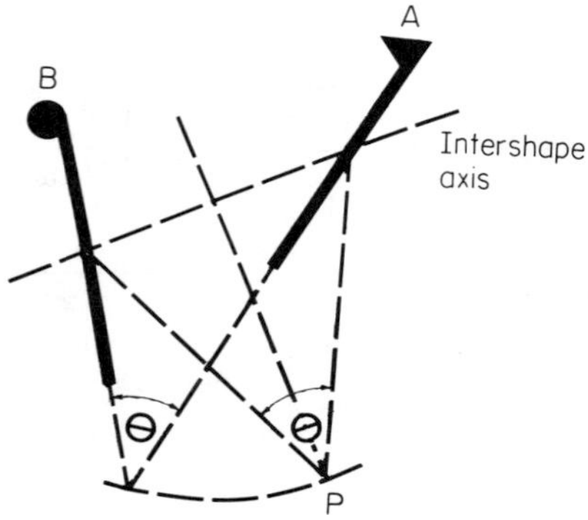

Figure 13.2 Showing how to construct the centre of the rotation in the plane, about which shape A may be rotated into coincidence with a similar shape B; this centre (P) lies on the perpendicular bisector of the intershape axis at a point which subtends angle θ to the midpoints of the two shapes. Angle θ is the angle of relative orientation of shape A with respect to Shape B

Figure 13.3 Showing how the centre of rotation about which one shape may be brought into coincidence with another may be (a) easy or (b) difficult to see

Figure 13.4 Illustrating how two shapes in mirror-image orientations may be (a) symmetrically or (b) non-symmetrically disposed

other. For instance, the two shapes in Figure 13.4(a) may be reflected into coincident alignment because they are reflected with respect to each other and symmetrically disposed, whereas the shapes in Figure 13.4(b) may not be reflected into coincidence without also rotating one of them, because they are not symmetrically disposed. This illustrates the point that shapes in orientations which are reflections of each other need not be symmetrically disposed, although shapes which are symmetrically disposed (in the sense of mirror symmetry) must be in orientations which are reflections of each other.

Reflections are intrinsically different from rotations in the plane. The reason is that reflections involve rotating the shape out of the plane about an axis in the plane, whereas rotations in the plane are about axes at right angles to the plane. In three-dimensional space, there are no distinctions between rotations about different axes. However, reflections of asymmetrical three-dimensional objects about a mirror *plane* are distinct from rotations. Reflections of such objects produce enantiomorphs, like the two hands or the objects in Figure 14.9. Reflections of three-dimensional objects in three-dimensional space cannot be achieved by rotation about any axis in space. If you doubt this, try bringing one hand into congruence with the other by changing the orientations of the hands. One can regard enantiomorphs as orientable into each other only by a rotation through four-dimensional 'space'.

A final attribute of the stimulus domain which must be considered is the nature of the surrounding field. This may be blank, either white or black, or it may consist of a square or circular frame round each or both of the shapes.

This completes the description of the stimulus domain of relative orientation in a frontal plane. If a perceiver is defined as a system which can describe members of a stimulus domain, then the ideal perceiver for the domain of relative orientation in a plane is one which knows the group structure which has just been outlined. This means that this group structure is the abstract theory of a perceiver for this domain (Howard, 1975). This ideal perceiver is the best basis for assessing the performance of an actual perceiver on tasks drawn from this domain. An experimenter who has incomplete knowledge of this domain is an imperfect perceiver of the domain and cannot possibly fully understand other perceivers. Two experimenters each with a different and incomplete knowledge of the stimulus domain cannot understand each other, and necessarily generate a confused literature.

13.4 The response domain for relative orientation in a frontal plane

We are not concerned in this chapter with the fine discrimination of differences in orientation. That is why only differences of 90 and 45° are considered and why the shapes being compared are spatially or temporally separated. There are three basic tasks to be considered.

13.4.1 The task of making distinctions between orientations

The first task is that of deciding whether or not two objects are in the same orientation. This task will be referred to as the task of **distinguishing between** the orientations of two objects. Consider the case where the subject is shown pairs of shapes and is asked to say 'same' or 'different' according to whether or not he judges the shapes in each pair to be oriented in a similar way. If he succeeds, he will have partitioned the set of stimulus pairs into a set in which the two shapes in each pair have the same orientation and a second set in which the shapes have different orientations. Next, consider the case where the subject is presented with a series of test shapes, one at a time, and for each one is asked to select a shape which has the same orientation from an array of reference shapes in various orientations. If he succeeds, he will have partitioned the stimulus set into distinct subsets. These will be **identity subsets** (within the domain of orientation) if all the stimuli in each subset are identical with respect to all dimensions of orientation, and they will be **equivalence subsets** if the stimuli in any subset are alike with respect to the dimensions that the subject is asked to attend to but not with respect to one or more dimensions of orientation which he is asked to ignore. For instance, it would be an equivalence set if all the shapes in the set were upright but some faced one way and some another.

The task of distinguishing between orientations may be generalized to include the case where the component stimuli being compared are themselves pairs of shapes. For instance, the subject may be asked to say whether two shapes have the same orientation with respect to each other as do two other shapes, irrespective of the orientation of the intershape axes. This task would, for instance, tap the subject's ability to distinguish a pair of shapes which are parallel to each other from a pair of shapes which are perpendicular, irrespective of whether the shapes in each pair lie above each other or side by side. To take another example, the subject may be asked to say whether the orientation of the intershape axis between one pair of shapes is the same as that of the intershape axis between another pair of shapes, irrespective of the orientation of the shapes. This task would tap the subject's ability to distinguish between an arrangement in which two shapes are above one another from one in which they are side by side, irrespective of the orientations of the component shapes.

In general, the task of distinguishing between orientations may be defined as *the partitioning of the stimulus domain into at least two identity or equivalence sets by a process of same – different matching.*

In a task of distinguishing between two things, the stimuli being compared are normally presented either together or in quick succession. However, it is still the same task, even with longer intervals between the stimuli, as long as the subject is not required to remember which of several past stimuli he must use for comparison. If he must do this, the task becomes one of identification.

13.4.2 The task of identifying orientations

The second basic task within the domain of orientation of shapes in the plane is that of responding in characteristic ways towards a shape in a particular orientation, or towards a pair of shapes which have a particular relative orientation or a particular orientation of the intershape axis. This task may be called *identification or recognition of orientations.*

The subject must not merely say whether the orientations of two stimuli are the same or different: he must respond in a distinct and consistent way to each of at least two different members of the stimulus set which he is shown.

One type of identification response consists of pairing a particular orientation with a particular stimulus of an entirely different kind. For instance, the subject may be asked to place each object with a particular orientation into a separate distinctly coloured box. But he must not be allowed to see the contents of the boxes because, if he were, he could perform the task by matching orientations. For the same reason, the 'labels' on the boxes must be neutral with respect to the orientation of the shapes, and the boxes must not be spatially arranged in a way that relates to the orientation of the stimuli.

Another type of identification response consists of pairing a particular orientation with a particular response that the subject executes. For instance, the subject may be asked to give a name to each orientation. But not all differential responses qualify as identifying responses. Consider the case where a person points in the direction of an arrow when it is in different orientations. It can be said that he has learned to differentially respond to different orientations, but really all we know is that he can match the orientation of his hand to that of the object: we do not know whether he can identify the orientations that he has matched. This type of differential responding is really an act of distinguishing between orientations. To prove that a person has identified an orientation, he must make differential responses which are not spatially related to the stimuli. For instance, if a person says 'left' when a pattern faces left and 'right' when it faces right, we have grounds for saying that he has identified the orientations of the pattern, because the words 'left' and 'right' are not intrinsically related to the stimuli (see Corballis and Beale, 1976, for an elaboration of this idea).

The task of saying whether or not a stimulus is in the same orientation as another stimulus presented at the same time is not an identification task. Even if the first stimulus is removed before the second stimulus is presented, the task of saying whether or not they have the same orientation would still not be one of identification, unless the subject has to remember which of several previous stimuli the present stimulus resembles. If there are several absent comparison stimuli it is an identification task, because the subject has to pair a particular orientation with a distinct non-spatial aspect of the comparison stimulus, namely, the identity of the comparison stimulus selected from memory. If a person merely says that he has seen a shape in this particular orientation before, this could be called an identification because he has attached a specific response to a set of orientations that are equivalent with respect to having been seen before, but the distinction between identifying and distinguishing between orientations becomes tenuous in this case.

The general feature of an identification is that each member of at least one proper (non-empty) identity or proper equivalence subset of the stimulus domain must evoke the same response, and each member of at least one other non-overlapping subset must evoke a different response. When a person says 'I can recognize two flowers, one of them is a rose and the other isn't', he is identifying the roses with respect to non-roses in the domain of flowers. If he says 'I don't know any flowers, but I can tell a flower when I see one', he is identifying flowers with respect to non-flowers in the domain of 'things'. The following profoundly important generalization emerges from these considerations.

A particular act of identification can be defined only with reference to the complete set of alternative identifications that the subject might make in a defined stimulus domain. Consider the case where the stimuli are a vertical line and a horizontal line, presented side by side, and the subject is asked to say 'vertical' to the vertical line and 'horizontal' to the other. If the subject succeeds, it is normally concluded that he has learned to identify vertical and horizontal lines. This is true insofar as he has attached a distinct response to each member of a particular stimulus set. But, because the stimulus set is so restricted, the performance tells us very little about the actual sets of things which the subject would regard as the same. For instance, we do not know whether a line tilted at 40° or another vertical shape would be accepted as vertical. Perhaps two dots, one above the other, would be acceptable, or even one dot near the top of the visual field. He may say 'vertical' to only a vertical line and yet call anything else horizontal, whatever it is. Consider another example. If a person says he identifies, or recognizes a car, this could mean that he recognizes it as (1) a means of transport and not as food, a person or a toy, (2) a car and not a bus, cycle or plane, (3) a Ford and not a Rover or Rolls Royce, (4) Bill Smith's Ford and not somebody else's Ford.

The point is that very little can be inferred about what orientation a subject has identified (included in his identity or equivalence set) unless he is tested with a sample of stimuli from the stimulus domain of relative orientation which is representative of the set about which the experimenter wishes to make conclusions. It may also be necessary to test the subject with stimuli from other domains, in order to be sure that it is orientations that have been identified and not merely some incidental features associated with particular orientations of particular shapes. Thus, when we speak about a subject identifying a feature of an object, we must, if we wish to be informative and precise, indicate the set of stimuli which the subject regards as equivalent in that act of identification. Such a set of stimuli may be referred to as the **domain of the identification**.

In nearly all the experiments which have been reported on simple identification of orientation, a shape was presented in two orientations, and the subject was asked to respond one way to the shape in one orientation and in another way to the shape in the other orientation. However, the task of identifying orientations may be generalized beyond this simple domain. For instance, the subject may be presented with different pairs of stimuli and asked to identify the relative orientation of the shapes within each pair, or the orientation of their intershape axis, or both. In other words, he may be asked to give differential responses to any distinct members of the stimulus domain of absolute orientations, relative orientations, or relative dispositions in the plane.

Logically, *an identification task is one that requires the subject to pair all members of at least one proper subset of a defined stimulus domain with the same identifying response, and members of another, non-overlapping proper subset with a different identifying response*. The net result of making identifications is that the stimulus set becomes partitioned into subsets by a process of response pairing. An **identifying response** is one that is not spatially related to the stimulus, and does not involve simply saying whether two stimuli are the same or different.

Note that when a person is asked to identify the orientation of an object, by making some distinctive response, he is not required to give a structural description, indeed he may not have any implicit or explicit notion of what orientation is. For instance, the subject might store an iconic image of each stimulus and attach each image to a distinctive response. He might even learn to give a distinctive name to each image without in any sense knowing that they differ in orientation. It has been argued that subjects do not behave this way and that implicit or explicit structural descriptions are normally involved in human perception (Pylyshyn, 1973). However, only by studying differential responses over cleverly designed sets of stimuli can one infer whether or not the subject is using structural descriptions.

The more interesting identification tasks are those that require the subject to identify as the same, orientations which are equivalent in some respect, but not identical. The rules defining the equivalence can take any form, but such rules can be classified according to the degree of order which they impose on the stimuli. The rule which imposes least order is that of **random assignment** of stimuli to different classes. The subject could be asked to associate a name with each set or class. He can succeed only by rote learning, as there are no rules to help him. A more highly ordered rule is that of **partial ordering**. For example, if a subject is asked to say whether a shape is more tilted or less tilted than a standard, he is using a partial ordering. The sets he is required to form are known as partially ordered sets, or **posets**, and the equivalence relationship is that of being 'more than' or 'less than'. Another assignment rule might be the **conjunction of features** from the domain of orientation. For instance, the subject may be asked to regard the pairs of Figure 13.5(a) as equivalent because they are symmetrical with short limbs near each other, and those of Figure 13.5(b) as equivalent because they are symmetrical with short limbs separated. These feature sets may be called **conjunctive**, and are a subset of the broader category of **taxonomic** feature sets.

The rules which produce the most highly ordered equivalence sets are those which generate the **subgroups** of the stimulus domain. For instance, the subject may be presented with the shape pairs in Figures 13.1(a) and 13.1(b) and be required to respond in one way if the shapes have a 90° relative orientation to

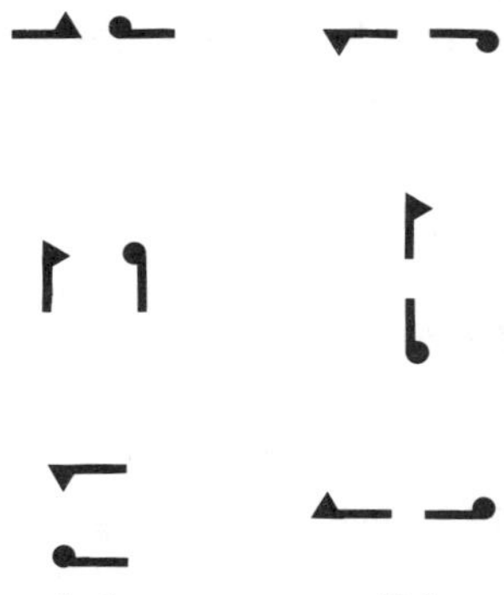

(a) (b)

Figure 13.5 Examples of conjunctive groups of relative orientation described in the text

each other and in another way if they are reflections of each other. The stimulus features which remain constant are known as **invariants**, and those invariants that form the basis for the subject's decision are known as **distinctive features**. Other variable or invariant features are **incidental features** with regard to that task.

Success in this type of task does not prove that the subject has implicitly or explicitly abstracted the invariant dimension from the incidental dimensions. He may simply have learned by rote to give the same response to each member of each subgroup, without in any sense realizing that they have anything in common. However, *if the subject correctly classifies stimuli which were not presented during training*, this provides evidence that the invariant dimensions have been abstracted. The subject may still not be conscious of the basis of his decision. For instance, people use perspective to judge relative distances although they may have no explicit knowledge of the principles of perspective.

13.4.3 The task of describing orientations

If a task is presented to the subject in the form of a description of the operations which he is required to execute, and if he successfully executes those instructions, then we are entitled to infer that the subject has a conscious descriptive knowledge of at least part of the stimulus domain. This may also be inferred if the subject can generate descriptions of stimuli in terms of defined operations within the stimulus domain, or if, on command, he can change the orientation of an object in a defined way or instruct someone else how to do this. Performance at this level is equivalent to what Piaget has called the stage of formal operations. I shall refer to it as the **descriptive level**. The descriptive performance of a subject can be assessed only by an experimenter with a knowledge of an ideal descriptive domain. Once we have evidence that a person can explicitly operate on the dimensions of a given domain, we can begin to ask questions about how quickly and accurately he carries out these operations. For studies along these lines see Section 14.3 and Olson (1970).

This abstract analysis has provided a basis for a discussion of the literature on the perception of relative orientations of shapes in the plane. First, work on the question of distinguishing between orientations will be reviewed, then work on the identification of orientations.

13.5 Distinguishing between orientations

13.5.1 Procedures

There are several ways to measure a person's performance on the task of distinguishing between orientations. The simplest way is to count the errors made when the subject is asked to say whether or not two shapes are in the same orientation. This procedure is more useful with young children than with adults, because adults make very few mistakes with simple stimuli. A more sensitive procedure with adults is to measure how long it takes them to make 'same' or 'different' decisions. One must use this method with caution, because the results may be contaminated by the fact that it may not take the same time to decide that two things are different as it takes to decide that they are the same.

In all these procedures the subject is essentially being asked to report on the similarity or dissimilarity of orientations; he must not be asked to name orientations, because that constitutes an identification.

13.5.2 The saliency of differences in orientation

When looking at a picturebook young children often seem indifferent as to whether the book is upside-down or not, and seem to recognize objects in the pictures even though they are upside-down by adult standards. One must not conclude too much from this behaviour. The book is usually lying horizontally, so that the pictures are not inverted to gravity; the relation of the contents of the picture to the baseline of the picture is not disturbed; and the objects are usually easy to recognize from features such as colour, texture, or local features that are not affected by changes in retinal orientation.

McGurk (1970) presented infants of between two and 26 weeks of age with two identical life-like models of a human face in different orientations and observed how long they spent looking at each. The infants showed no spontaneous preference for any particular orientation of the face. In another study, McGurk (1972b) asked three- and four-year-old children to judge the degree of similarity between a picture and copies of the same picture that varied in size, colour, and orientation. The children most often selected pictures which were matched in colour and size and usually ignored differences in orientation. Note that in both these studies the stimuli were presented in a horizontal plane, so that we cannot infer from the results that children ignore differences in orientation with respect to gravity. Apparently nobody has taken the trouble to find out, but I suspect that a three-year-old, or even a one-year-old, would spontaneously react if he woke in the morning to find his bedroom upside-down, or even if he saw a car upside-down in the street, and I suspect that in these cases orientation would be more salient than either colour or size.

A certain indifference to orientation is revealed in the way even older children and some adults perform when asked to copy a pattern which is tilted from the vertical in a frontal plane. There are test items of this kind in most of the better-known intelligence tests. One of the best known is the Kohs Block Test, in which subjects are required to copy a simple design using small blocks. Such tests are normally administered on a horizontal table surface, so that one cannot speak of orientation which respect to gravity. Young children tend to place their copy with its axis of symmetry, or one of its edges, in alignment with the normally vertical meridian of the eye whatever the orientation of the model, even though they are asked to make the copy exactly like the model. The same tendency has been found among adults in educationally primitive societies (Shapiro, 1960; Jahoda, 1976; Deregowski, 1972, 1977). This tendency to normalize the orientation of the copy has been claimed to be a diagnostic feature of certain neurological and psychiatric disorders (see Royer and Holland, 1975, for a recent review of this literature).

A closely related task is to ask subjects to draw a copy of a standard pattern when the standard pattern is in one of several orientations. Eldred (1973) used such a task and reported that children tend to normalize their drawings to an upright orientation. However, in this study, as in most other studies of this kind, the model and the copy were lying flat on a table-top, so that normalization was certainly not to the gravitational vertical. Furthermore, in this study, as in others, the patterns were presented on rectangular pieces of paper, so that normalization may not even have been to the normally vertical meridian of the eye, but rather to the edges of the paper. Gombrich (1969) has argued that the act of drawing involves first making an internal representation of the thing to be copied. In the context of the task we are considering, Gombrich would argue that the child simplifies this internal representation by first normalizing it with respect to the sides of the paper, or with respect to his own body if the paper is circular, and then draws the description which he has compiled and ignores the model.

If children do ignore differences in orientation under certain circumstances, that does not mean that they cannot discriminate differences in orientation, nor that they cannot distinguish between or even identify orientations. Whether or not children ignore differences in orientation when making same – different judgments may depend on how the question is asked. For instance, McGonigle and Chalmers (1980) found that children tended to ignore differences in orientation when given a non-specific instruction such as 'Pick me out one exactly the same as this one' but did not ignore orientation when the instructions specified the identity of the shapes as both 'standing up' or 'lying down'.

There is no more reason to doubt that young children can discriminate large differences in orientation than to doubt that they can discriminate large differences in location. If a child plays with a pair of scissors, he surely notices the changing orientation of the blades. The question is whether the child can decide whether or not two objects have the same orientation when they do not intersect or when they are presented one after the other. Under these circumstances, the task entails transferring the gaze or the attention from one object to the other and comparing the iconic image of one with the current image of the other, or perhaps forming an internal abstract representation, or description, of one and comparing it with a description of the other. That is why the term 'distinguishing' is more appropriate than the term 'discriminating' for tasks of this kind.

There are several recent studies bearing on this issue, but in most of them there is some uncertainty about which perceptual domain the subjects were using. McGurk (1970, 1972b) worked with infants between the ages of six and 52 weeks. The infants were repeatedly shown an object in one orientation until their interest flagged. They were then shown the same object in a new orientation whereupon their interest immediately revived. A similar experiment was conducted by Wiener and Kagan (1976). On the face of it, these studies indicate that six-week-old infants can distinguish between different orientations. However, the domain

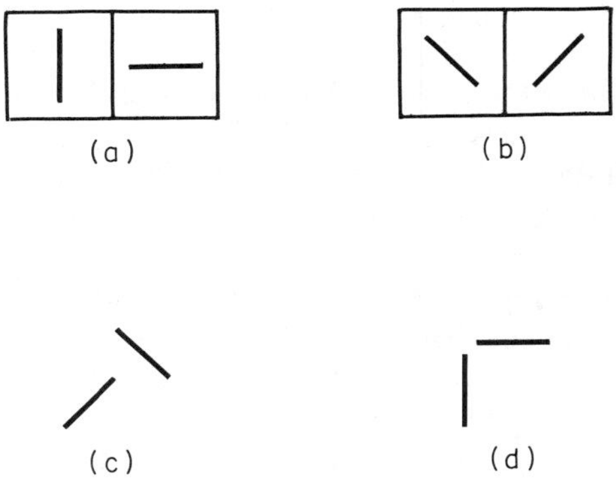

Figure 13.6 Shapes (a) and (b) were used by Over and Over (1967) and by Rudel and Teuber (1963) to investigate the relative difficulty of distinguishing between verticals and obliques. Shapes (c) and (d) should have been used also, to control for the relative dispositions of the shapes

of these distinctions may not be orientation differences. For instance, if the faces were always presented in the same location the infant may have responded to the change in location of the eyes or the change in position of some feature on the face relative to some feature in the background.

Over and Over (1967) presented four-year-olds with a simple black line lying flat on a table and asked them to indicate which of two other lines was in the same orientation. The lines were on square cards and presented as in Figures 13.6(a) and 13.6(b). Most of the children succeeded at this task. If the child checked in say the top left corner of each card, the correct matching card would be the one with something in that position, so that success on the task may not reflect an ability to distinguish between orientations; the children need not have realized that the other line was the same shape, although in this instance they probably did, because the shape was so simple.

Even though such experiments may not be conclusive, there is little reason to doubt that younger children can distinguish between differences in orientation when the shapes are not presented in the same location. But even if young children can tell when two objects do not have the same orientation, that is not to say that they can distinguish between all differences in orientation equally well. That is the question I shall now consider.

13.5.3 The distinguishability of mirror-image orientations

It has been claimed repeatedly that animals, children, and even adults find it particularly difficult to distinguish between the orientations of a mirror-image pair of shapes, that is, a pair in which the members are relatively reflected and symmetrically disposed. It must be stressed that we are talking about distinguishing between *orientations*, not between *shapes*. The shapes whose orientations are being compared need not be the same, as long as there is a criterion for deciding whether or not they have the same orientation (see Figure 13.12).

Let us now look at a typical study which was supposed to demonstrate that mirror-image orientations are more easily confused than are other orientations. Sekuler and Houlihan (1968) presented adults with each pair of patterns shown in Figure 13.7, displayed in a vertical plane. They were asked to press a switch one way or the other depending on whether they judged the two shapes in a pair to be in the same, or in different orientations. They found that it took longer for adults to decide whether mirror-image pairs (pairs d and g) were different than whether non-mirror-image pairs (pairs f and b) were different or whether identical pairs were the same. However, the decision that took the longest time was that between laterally arranged mirror-image pairs (pair g). This special difficulty with mirror-image pairs has been confirmed by Wolff (1971) and by Sekuler and Pierce (1973). Bornstein *et al.* (1978) habituated 4-month-old infants to a shape in one orientation and then showed them test shapes one at a time in other orientations. The time spent looking at test face profiles indicated that the infants distinguished between different faces but not between left and right profiles of the same face. But this does not tell us whether they would also have ignored other differences in the orientation of faces. Bornstein *et al.* also showed that differences between 180° rotations of a U-shaped figure were ignored by infants, especially when the orientations were in left-right symmetry, as opposed to up-down symmetry, whereas differences between 45 and 90° rotations were not ignored.

It has been concluded that mirror-image orientations are more difficult to distinguish than other relative orientations; but each of these studies had at least one of the following failings in design.

1. In the first place, the shapes used in these and all other studies on this question, when arranged as in pairs (d) and (g), may be regarded as 180° relative rotations rather than reflections. Therefore, it is equally proper to conclude that 180° rotations are more confusable than 90° rotations. The shapes should have been asymmetrical, as in Figure 13.8.
2. The mirror-image pairs are not in the same subgroup of the stimulus domain

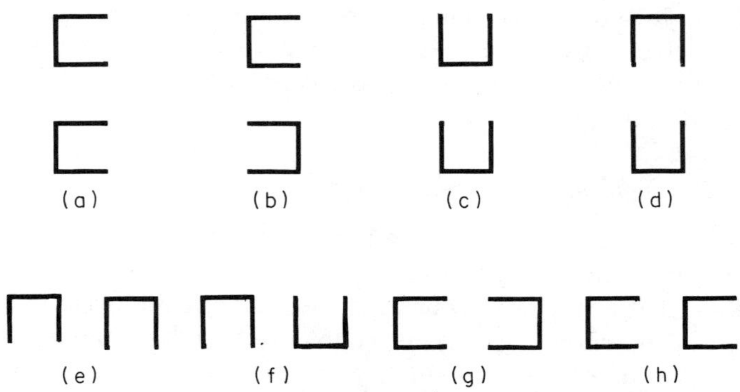

Figure 13.7 Shapes used by Sekuler and Houlihan (1968) (Reproduced by permission of Academic Press Inc.)

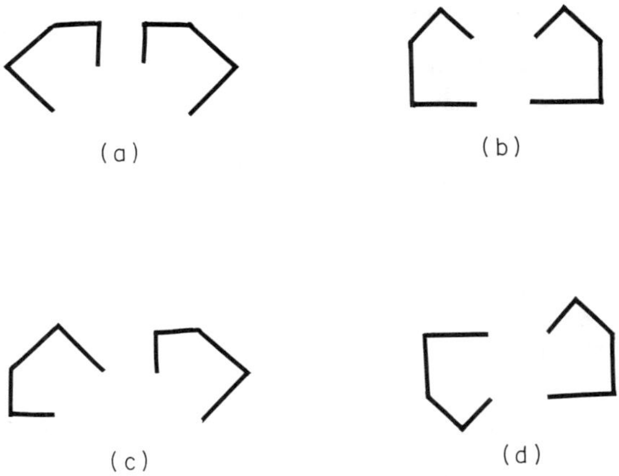

Figure 13.8 A sample of shapes that might be used to decide whether the property of being aligned with the intershape axis (b and d) effects the apparent similarity of the orientations of shapes more than the property of being symmetrically disposed (a and b). The shapes in (c) are neither aligned nor symmetrically disposed

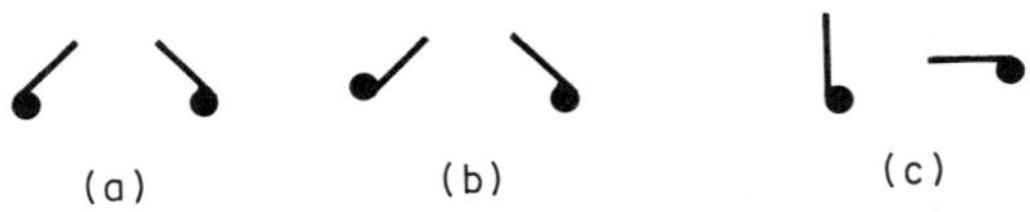

Figure 13.9 Illustrating how 90° relative orientation (a) may be more difficult to distinguish than mirror-image orientations (b and c)

as the non-mirror-image pairs. In other words, the two relative orientations differ in more than one respect. The shapes in pairs (d) and (g) of Figure 13.7 are aligned with the intershape axis whereas the shapes in pairs (b) and (f) are not. One might just as well have concluded from these studies that shapes in alignment are more easily confused than shapes which are not in alignment. The proper comparison is between pairs of shapes such as those shown in Figure 13.8.
3. In none of these studies was a distinction made between shapes which are relatively reflected but not symmetrically arranged and shapes which are reflections and symmetrically arranged, so that we do not know which of these factors is more confusing. This distinction is illustrated in Figure 13.4.
4. The sampling from the stimulus domain was very narrow and the results cannot be generalized beyond the U shape that these and most other investigators used.

A deficiency in any one of these points casts doubt on the validity of the conclusion that subjects are more likely to confuse reflections than other relative orientations. I suspect that there are many instances in which 90° relative orientations are more difficult to distinguish than mirror-image orientations. For example, I suspect that people may well confuse the pair of shapes in

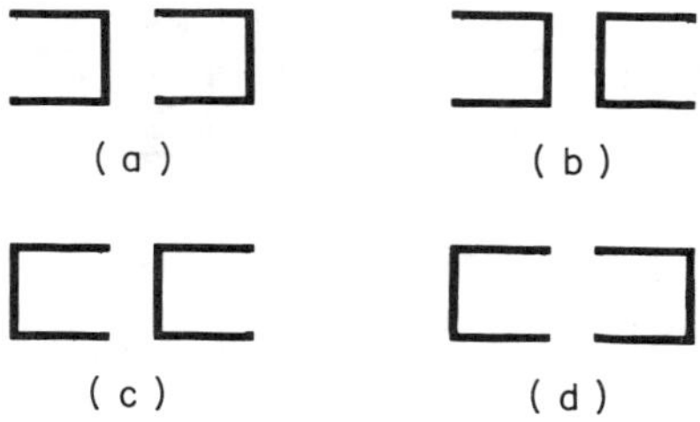

Figure 13.10 A sample of the shapes used by Staller and Sekuler (1977)

Figure 13.9(b) more than they confuse those in Figure 13.9(c), or even those in Figure 13.9(a).

5. There are problems associated with the categories 'same' and 'different' which were used in instructions to subjects or were required as responses from subjects. A study by Staller and Sekuler (1977) highlights this problem. One group of adults and pre-school children was asked to call the members of pairs (a) and (b) in Figure 13.10 the same and those of pairs (c) and (d) different. Another group was asked to call the members of pairs (c) and (d) the same because they both open inwards or outwards and to call those of pairs (a) and (b) different. The subjects in the first group responded more slowly on the mirror-image pairs, as one would expect, but the subjects in the other group were faster on the mirror-image pairs than on the identical pairs. This suggests that 'same' responses have a shorter reaction time than 'different' responses, and that this is part of the reason why people take longer to say that symmetrically arranged shapes are in different orientations (see Nickerson, 1972, for a review of this topic). That is not the whole story, because Staller and Sekuler also found that, especially among the children, those in the first group made more errors with the mirror-image shapes, whereas those in the second group made more errors with the identically oriented shapes. This suggests that children often say that members of mirror-image pairs are in the same orientation not because they confuse them with pairs whose members have the same orientation, but because, for them, mirror-image orientations are the same. Mirror-image shapes both face in or they both face out, and that is a similarity. On this view, young children do not confuse mirror-image orientations; they simply adopt an idiosyncratic criterion of 'same orientation'. This criterion of sameness should be evoked more readily with figures symmetrically oriented about a vertical axis than with those oriented about a horizontal axis, because most objects, including the human body, are symmetrical about a vertical axis; few are symmetrical about a horizontal axis. Staller and Sekuler found a pattern of errors that fitted this expectation. According to this view, the problem of why children call mirror-image orientations the same is only a problem when we suppose that children have the same criterion of 'sameness' that we have. This study too is based on an extremely narrow sample of the stimulus domain, so that the authors cannot be sure that mirror-image shapes are, in general, likely to be said to be in the same orientation.

13.5.4 The distinguishability of oblique orientations

It has often been claimed that it is more difficult to distinguish between two oppositely oriented oblique lines (at 45° on either side of the vertical) than between a vertical line and a horizontal line. It has also been claimed that obliques are more difficult to identify than are verticals and horizontals, but this is a different issue which is considered in Section 13.6.3. These two issues have been confused and the term 'discrimination' has been inappropriately applied to both.

Over and Over (1967) drew attention to this confusion in this context but persisted in using the term 'discrimination' for what is more appropriately called the task of distinguishing between orientations. In their experiment they used children between the ages of three and six years of age.Each child was shown an oblique line on a square card and asked to select a matching line from two oppositely inclined oblique lines, each on its own card, placed below the standard line. Using the same procedure, the child was also shown a vertical or horizontal line and asked to select a matching line from two lines, one vertical and the other horizontal. The children did not make many errors, but most of these were in the oblique 'discrimination' condition. The experiment succeeded in demonstrating that children can distinguish between oblique orientations even if they are unable to identify them. However, the conclusion that it is more difficult to distinguish between obliques than between verticals and horizontals was unwarranted for the following reasons.

1. The cards were placed on a table, so that the terms 'vertical' and 'horizontal' which the authors used did not refer to gravity, as one is entitled to suppose unless told otherwise.
2. The relative disposition of the comparison lines was not the same for obliques as it was for the vertical and horizontal lines. That is, the obliques were not disposed with respect to the intershape axis in the same way as were the vertical and horizontal lines. The vertical and horizontal lines formed a 'T' pattern, whereas the obliques formed an 'L' pattern (see Figures 13.6a and 13.6b). In other words, the pairs were not drawn from the same subgroup of the stimulus domain. This had the consequence that the obliques were symmetrically disposed whereas the other lines were not. Additional lines should have been arranged as in Figures 13.6(c) and 13.6(d).
3. The orientation of the intershape axis was not varied. The full set of dispositions that should have been used is shown in Figures 13.1(a) and 13.1(b).
4. Square frames should not have been used. They introduce a contaminating factor, namely, that the vertical and horizontal lines are parallel to the sides of the frames, whereas the obliques are not. The lines should have been presented within a large circular frame or as luminous lines in the dark.

Bryant (1969) claimed to have demonstrated the importance of the frame by showing that young children can correctly match obliques when they are presented on a diamond-shaped surround. However, the sides of the diamond which Bryant used were coloured differently so that the task became one of deciding which coloured line was parallel to the line to be matched. This gets us away from the intent of the original enquiries, which is to discover whether it is more difficult to distinguish between obliques than between other orientations. But in any case, the strategy that Bryant suggests, namely, that of distinguishing between obliques according to which of them is parallel to a given side of a diamond, will not work. It begs the very question that is being asked, because it requires the child to distinguish between the directions of the oblique sides of a diamond, and that is the problem the child is trying to solve. No wonder that, when Fellows and Brooks (1973) repeated Bryant's experiment with uncoloured diamonds, or with coloured diamonds when the colour was not brought to the attention of the child, they found that obliques were just as difficult to distinguish as were principal orientations, whatever the orientation of the frame. Bryant (1974, p. 72) reported another study on this issue which goes some way to meeting the objection raised against his first study. He demonstrated that a single fixed reference line which was parallel to one of the oblique lines but not to the other made it easier to distinguish between the oblique lines in a task in which the two obliques were presented successively along with the reference line. Bryant concluded that the young child uses a simple match–mismatch code to distinguish orientations. That is, the child distinguishes between one oblique and the other on the basis of which of them is parallel to a fixed feature of the visual scene. He argued that only older children develop the ability to judge the egocentric orientation of an object without the aid of a visual frame of reference. But there is a flaw in this theory, because it assumes that the child recognizes that there is a feature of the visual scene which remains in the same egocentric orientation, and this begs the question that is being asked. In Bryant's experiment the fixed feature was a bold red line which the children were *told* did not change its orientation, and the task was thus transformed into that of distinguishing between two parallel lines and two orthogonal lines. We do not know whether children naturally recognize the stability of such a reference line without being told. If they do, it demonstrates that they can at least appreciate the egocentric stability of an isolated oblique line, and if they don't, the reference line would not help them with other lines.

Bryant (1973) found that children have difficulty in distinguishing between obliques in a successive matching task, even when the lines are not tilted to the same extent, and concluded that the difficulty has nothing to do with symmetry. Corballis and Zalik (1977) pointed out that this does not prove that symmetry has no effect, only that it is not the only factor. In their own study they found that both similarity of slope and symmetry had an effect, although they did not vary the relative disposition of the lines (see also Harris *et al.*, 1974).

Proper controls for the relative dispositions of stimuli and for the orientation of the intershape axis were instituted by Olson and Hildyard (1977). One line was presented in various orientations in the centre of a circular display and a second line was presented either before or after the first, in one of several orientations and radial positions. Subjects took longer to decide about the similarity of orientation of the lines when the line which varied in position was oblique, no matter what the order or orientation of the other line. Conclusions would have been easier to draw if the lines had been presented simultaneously, or successively in the same location. However, it seems from this experiment that obliqueness introduces a special difficulty in the task of comparing orientations, apart from any effects due to relative disposition.

According to Piaget and Inhelder (1956) a child can draw a circle at three, a square at four, a triangle at five, and a diamond at seven years of age. Nevertheless, children can readily distinguish among these shapes at all these ages. Several factors have been suggested for this lag between perceiving and drawing (see Maccoby and Bee, 1965, for an insightful analysis of this issue). It is the difficulty children have in drawing the oblique sides of triangles and diamonds that is of interest here. It is clear that the main problem with this task has to do with that same contaminating factor, the shape of the paper upon which the shapes are drawn. Naeli and Harris (1976) showed that children of about four years of age drew a triangle on a triangular piece of paper just as easily as a square on a square piece of paper. These results were independently confirmed by Brittain (1976). This tendency in children to use the allocentric reference of the edges of the drawing paper or of nearby reference lines, rather than an egocentric reference, is well documented (see Howard and Templeton, 1966, p. 344, and Elliot and Connolly, 1978). Naeli and Harris also found that children were very inaccurate at placing a cut-out pattern in the same orientation as a model when the two patterns were placed on sheets of paper which were not the same shape as the patterns. This shows that the difficulties the children had with drawing shapes were not all due to motor difficulties associated with drawing. However, a square pattern on a circular paper was more accurately placed in the same orientation as a model, than was a diamond on a circular paper, showing that the orientation of a square is more accurately perceived than that of a diamond, quite apart from the effect of the background.

13.5.5 Theories about the distinguishability of relative orientations

I suggest that all theories that have been proposed to account for why certain orientations are more readily confused than others are special cases of a general principle. *The more alike two things are with respect to dimensions being judged, the more probable it is that they will be called the same.* When I refer to the 'dimension being judged' I include any dimension that the subject perceives as covarying with the dimension being judged. It remains to be decided in what ways certain pairs of orientations are more alike than other pairs. Again, I suggest a general principle. *The simpler the operation of bringing features of one shape into coincidence with corresponding features of the other shape, the more likely are the orientations of the two shapes to be confused.* I suggest that the simplicity of the operation of bringing features of shapes into coincidence is determined by the following factors.

1. If the shapes are in the same orientation they can be brought into complete coincidence by simply translating one with respect to the other, and this may be the mental operation which is applied when two shapes are said to be in the same orientation.
2. Even if two shapes are not in the same orientation, a person, and particularly a child, may be deceived into thinking that a translation brings two shapes into coincidence when, in fact, it does not. This is likely to occur when all the features of one shape may be joined to the corresponding features in the other shape by a set of parallel lines. When this is the case, as in Figure 13.9(a), it is possible to bring any pair of local features into coincidence by transferring the gaze from one shape to the other in the same direction. The pairs of features may not necessarily be matched for orientation, nor will the whole set of pairs of features necessarily be brought into coincidence simultaneously, but the subject may be deceived into accepting the coincidence of feature pairs as evidence that the shapes are in the same orientation. Even if shapes are in the same orientation, the test of coincident alignment of feature pairs by translation is probably done piecemeal, especially by a child. Only similar shapes of the same size, which have the same orientation, or which are disposed symmetrically with respect to each other, will have feature pairs joined in parallel. However, shapes in certain orientations, like those in Figure 13.9(b), which are not disposed symmetrically, may have feature pairs approximately in parallel, so that one would expect these shapes to be readily confused.
3. A person is even more likely to be deceived into thinking that two shapes may be brought into coincidence by translation if, in addition to feature pairs being connectable by parallel lines, the features themselves are parallel to each other. This is one aspect of a theory of orientation confusions proposed by Bryant (1974, chapter 5). He found that the pairs of shapes in Figure 13.11(a) were more readily confused than those in Figure 13.11 (b). Corresponding feature pairs are joined by parallel lines in both pairs of shapes, but corresponding features are also parallel to each other in the shapes in Figure 13.11(a). I am suggesting that Bryant's theory, while perfectly sound, is only one aspect of the broader theory which is expressed by the two general principles underlined above.
4. If corresponding features of two shapes are symmetrically disposed about an axis of symmetry between them, they will be easier to find, and hence easier to bring into coincidence, than if they are asymmetrically disposed. Only symmetrically disposed mirror-image shapes have this property, and on that account their orientations will be more readily confused. Note that the orientations of mirror-image shapes which are not symmetrically disposed, like those of Figure 13.9(c), are not likely to be confused under any circumstances.

Figure 13.11 Shapes used by Bryant (1974)

5. If corresponding features are symmetrically disposed about a vertical axis (for the upright observer) they will be easier to find and match than if they are symmetrical about an axis in any other orientation. People spot symmetries about a vertical axis most easily, probably because this is the commonest type of symmetry. This factor predicts that the shapes in Figure 13.7(d) are less easily confused than those in Figure 13.7(g).

According to a popular theory which has been proposed to account for confusions between left–right mirror-images such confusions have to do with the symmetry of the two cerebral hemispheres. Suffice it to say here that Sekuler and Pierce (1973) found that the reaction time for distinguishing mirror-image orientations was not affected by whether the two images were projected to the same hemisphere or to opposite hemispheres (see Corballis, 1976).

6. The orientations of shapes which can be rotated into coincidence through a small angle are more likely to be confused than the orientations of shapes which must be rotated through a larger angle before they come into coincidence.

7. The orientations of shapes which can be rotated into coincidence through a certain angle about an easily located centre of rotation (see Figure 13.3a) are more likely to be confused than are the orientations of shapes which can be rotated into coincidence through the same angle but about a centre which is more difficult to find (see Figure 13.3b).

8. The orientations of shapes in which at least some corresponding features are close together are more likely to be confused than are the orientations of shapes in which no corresponding features are close together. Barroso and Braine (1974) proposed this factor as a theory of why the orientations of symmetrically disposed shapes are confused. Their theory grew out of a study in which they asked children between the ages of three and five years to place a shape to the

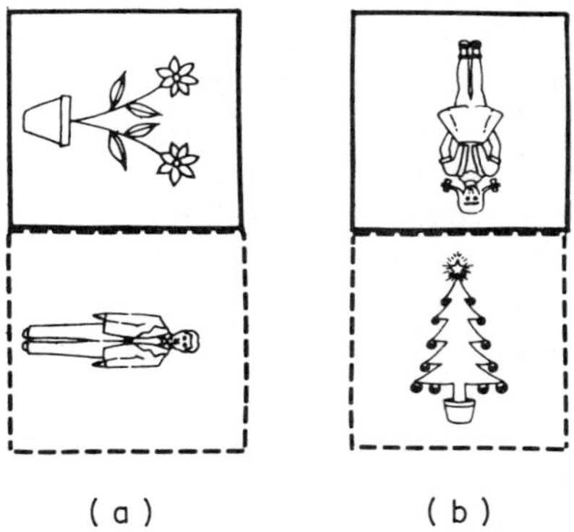

Figure 13.12 (a) Children tend to place a matching card (dashed) correctly with respect to a standard card when the shapes whose orientations are to be matched are parallel and not aligned. (b) Children often solved the above task incorrectly when the shapes are in alignment. In this case, they tend to place similar features close together (From Braine, 1978. Copyright, 1978, by the American Psychological Association. Reprinted by permission)

right of, or below a standard shape '. . . the same way as the one that is already there'. The shapes were all pictures of mono-oriented familiar objects, as in Figure 13.12, but the matching shape was not necessarily a picture of the same object as the standard. When the shapes were to be placed in axial alignment they were sometimes placed as if they were mirror-images, as in Figure 13.12(a). When the figures were to be placed parallel to each other, as in Figure 13.12(b), they were almost always correctly oriented. This pattern of results was unaffected by whether or not the figures were the same. Barroso and Braine suggested that, when children try to match the orientations of the two shapes, they place comparable parts close to each other. This strategy yields correct responses when the figures are parallel, but incorrect responses when the shapes are aligned. This may be the correct explanation for these particular orientations but it is only a special case of the general principle of similarity of orientations which I have proposed, and cannot account by itself for some of the other confusions that have been mentioned.

9. A final factor which may affect the operation of bringing two shapes into coincidence was suggested to me by Ulric Neisser. If a person walks halvway round an oblique rod, its spatial disposition relative to his body will be reversed. Thus, two rods tilted at equal and opposite angles are equivalent in the sense that they may be regarded as the same object viewed from opposite sides. Vertical and horizontal objects may be regarded as different views of the same object only by tilting the head, which is a less likely change in vantage point. Similarly, objects disposed as mirror-images may be regarded as the same object viewed from opposite sides, whereas objects disposed in other ways may not.

All these factors may be subsumed under the general principle that the simpler the operation of bringing corresponding features of two shapes into coincidence, the more likely it is that the orientations of the shapes will be confused. *In order to test the theory one would need an independent measure of the difficulty of bringing shapes in different orientations into coincident alignment, and then see whether, in each case, this predicts the error rates for distinguishing between orientations.*

All the above factors are **biasing factors**, that is, they affect the number of false 'same' responses without necessarily affecting the number of false 'different' responses. There are other factors, which may be referred to as **uncertainty factors**, which increase the general difficulty of a comparison task and thereby increase false 'same' and false 'different' responses equally. Uncertainty factors include the degree of spatial and temporal separation between compared shapes, the presence of distraction or masking stimuli, and fatigue. For instance, a decision about whether or not two shapes are in the same orientation is made more difficult when they are spatially separated than when they are presented on top of each other. It is difficult to imagine how a child could fail to distinguish between the orientations of the two oblique arms of a cross. The factor of spatial disposition has been confounded with that of simultaneous versus successive presentation. For instance, Fisher (1980) claimed that children can identify successively presented oblique lines more easily than lines presented simultaneously. But the successive lines were presented side by side and the simultaneous lines were presented in the same location, so that we cannot tell whether the difficulty with simultaneous lines was due to their simultaneity or to their separation in space.

13.6 Identifying orientations

13.6.1 Introduction

In Section 13.2.2 the studies of Ghent were described which showed that children from at least the age of three can make consistent decisions about when certain unfamiliar patterns are upright. More recently, McGurk (1972a) showed that three-year-olds have no difficulty indicating which of two pictures of familiar objects is upright. These studies demonstrate that children can identify at least two orientations of certain objects. The objects used in these studies had polar features that the subjects recognized as top and bottom before the experiments began. Such objects are not suitable for studying how people learn to identify particular orientations, which is the subject of interest in this section.

13.6.2 The identification of mirror-image orientations

Rudel and Teuber (1963) trained children to select either the open-upward or open-downward shape from each of the pairs (a) and (b) in Figure 13.13, and also to select the open-left or open-right shape from each of the pairs (c) and (d). The child was simply told that he was correct when he chose the orientation which the experimenter had designated 'correct'. Children between the ages of three and eight years of age were tested until they reached a certain criterion of success, or for 50 trials. All the children could select the 'correct' orientation from pairs (a) and (b), but most of the children of three, four and five years of age failed to select the 'correct' orientation from the mirror-image pairs (c) and (d). Rudel

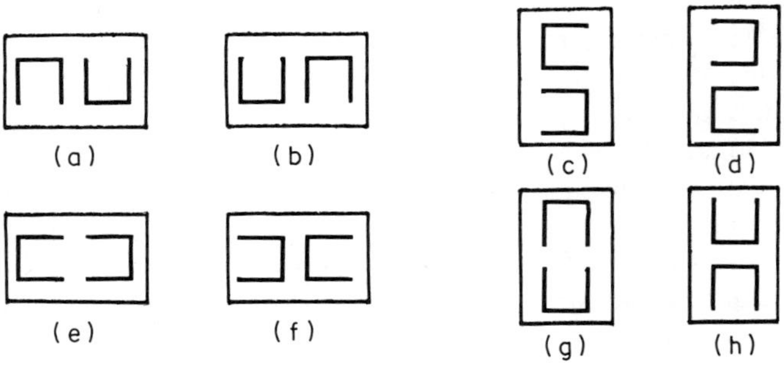

Figure 13.13 (a)–(d) Shapes used by Rudel and Teuber (1963) to study the ability of children to identify particular orientations. (e)–(h) Shapes added in a subsequent study by Huttenlocher (1967). (Copyright, 1967, by the American Psychological Association. Reprinted by permission)

and Teuber concluded that children '. . .readily discriminate right-side-up from upside-down U-figures, while a right–left reversal in the U-figure presents considerable difficulty.' This was not a 'discrimination' task, but in any case this conclusion is not justified by the evidence, for the following reasons.

1. The intershape axis was always horizontal, so that we do not know whether the difficulty with pairs (c) and (d) was to do with their left–right positions or with the fact that they were symmetrically disposed. Huttenlocher (1967) corrected for this by repeating the experiment with the addition of the pairs of shapes in Figures 13.13(e) – (h). She found that members of all mirror-image pairs were more difficult to identify than the members of 90° rotation pairs, although members of the left–right mirror-image pairs were still the most difficult to learn to identify (Huttenlocher still used the term 'discriminate'). There are now two conclusions: that members of mirror-image pairs are more difficult to identify than 90° rotation pairs, and that members of mirror-image left- and right-facing pairs are more difficult to identify than mirror-image up- and down-facing pairs. Neither of these conclusions is warranted by the results of Huttenlocher's experiment. The following critical issues apply to Rudel and Teuber's study and to Huttenlocher's study.
2. The shapes in themselves were symmetrical, so that the differences between mirror-image pairs may equally well be regarded as 180° rotations. We do not know which factor is causing the difficulty. The experiment should be repeated with asymmetrical objects.
3. The different pairs of shapes did not have the same relative dispositions with respect to the intershape axis. Pairs like those shown in Figure 13.8 should have been used.
4. The shapes were all symmetrical and also identical, so that we do not know whether it was the symmetry of each shape that caused the difficulty or the symmetry of the orientations of the two shapes in each pair.

A recent study by Farrell (1979) overcame these problems. He presented adults with a single arrow pointing left, right, up or down and timed the subjects as they either moved a lever in the same direction as the arrow or named the arrow's direction. The reaction times for the lever-moving response were the same for all directions of the arrow, but were 20 msec longer for naming the left and right arrows than for naming the up and down arrows. This confirms that the relative difficulty with left and right has to do with the task of identification and does not show in the task of spatially matching responses to stimuli. In control conditions, Farrell showed that the greater difficulty with left and right was not due to greater difficulty with the words 'left' and 'right' as opposed to the words 'up' and 'down'. He concluded that '. . .it is more difficult to discriminate left from right than up from down.' This is the wrong conclusion. In the first place, the task was that of identifying the orientations of the arrows, not discriminating between them. In the second place, the stimulus set consisted of arrows pointing left, right, up and down, so that the longer reaction time for left and right may have been due to the subjects taking longer to decide that the arrow was *either* a left or right arrow rather than an up or down arrow. Having made that decision, the subjects may then have taken no longer to decide between left and right than between up and down. The experiment should be repeated with the stimulus set properly confined to the pair 'left and right' in one set of trials and to the pair 'up and down' in another set of trials. Farrell also found that it took longer for subjects to decide whether a circle was either to the left or to the right of the fixation point than it took for them to decide that it was either above or below the fixation point. Similar results have been obtained by Maki *et al.* (1977) and by Maki *et al.* (1979).

The ability to identify the letters b and d, p and q, u and n, and m and w is essentially one of recognizing orientations within mirror-image sets. Children who are just learning to read have little difficulty copying these letters from a sample, or selecting one that matches a sample, but they often confuse them when required to draw one that has been named, or to name one when it is shown (Asso and Wyke, 1971). In other words, they can distinguish between these letters but have difficulty identifying them. Until the child is taught the letters of the alphabet he has not encountered any objects which change their identity when rotated or reflected, in fact letters are the only objects that I can think of that have this property. Thus, the child's indifference to the orientation of letters is very understandable. The members of mirror-image pairs, such as b and d and p and q, are particularly difficult to identify, and one of the reasons may be that the recognition of b and d and p and q required the child to make a left–right distinction, which is rendered more difficult because of the symmetry of the human body. However, this is not the whole story. Cairns and Steward (1970) asked children to place a matching letter in the empty box when shown displays like those in Figure 13.14. When shown letters such as H, M, and T, which are symmetrical about a vertical axis, the children tended to place them in mirror-image orientations when they were arranged one above the other, as in Figure 13.14(b). Letters such as B, C, and D, which are symmetrical about a horizontal axis, were often placed in mirror-image orientations when they were arranged side by side, as in Figure 13.14(a). The problem therefore seems to be with symmetry about any axis and not merely with symmetry about the body axis.

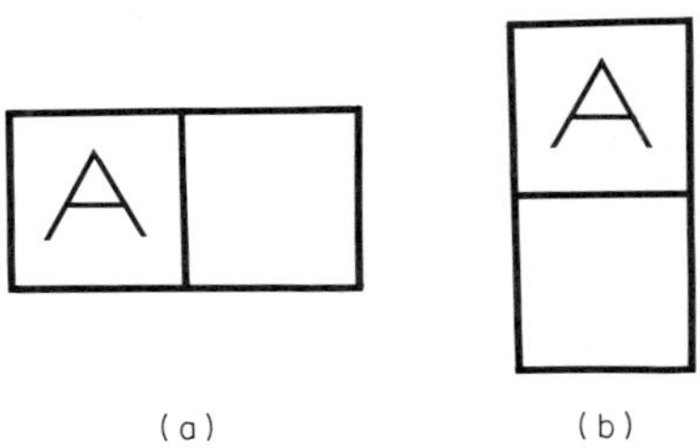

Figure 13.14 Displays used by Cairns and Steward (1970). Children were asked to place a matching letter in the empty box

It is not only young children who find it difficult to learn to attach names to orientations of simple shapes. Frith (1971) found that older children, who had learned to identify letters, had just as much difficulty as younger children when required to identify the orientations of novel simple shapes. Furthermore, the older children tended to copy a reversed letter as a normal letter even when instructed to copy the letter as they saw it. They copied their internal representation of the letter rather than the model provided.

There is a large but contradictory literature on the relationship between handedness and left–right confusions in reading and writing. For reviews of this literature see Flescher (1962), Lyle (1969), and Corballis and Beale (1976).

13.6.3 The identification of oblique orientations

Lashley (1938) was apparently the first to claim that animals (rats) learn to make differential responses to vertical and horizontal patterns more easily than to patterns which are in opposite oblique orientations. Since then this pattern of relative difficulties has been reported for many other species (see Appelle, 1972). Rudel and Teuber (1963) were the first to study this phenomenon in human subjects. They trained children to select the left or right oblique from pairs like those in Figure 13.6(b) and to select the vertical or horizontal line from pairs like those in Figure 13.6(a) by saying 'correct' when the child selected the designated line. All but one of the three-year-olds learned the task with the vertical and horizontal lines, but all 12 children of this age failed to learn, within 50 trials, with the oblique lines. Rudel and Teuber concluded: '. . .figures are virtually indiscriminable, if they are oriented obliquely, ...'. Again, they meant 'identifiable' not 'discriminable', as Over and Over (1967) pointed out. But in any case, all the criticisms of the studies on the distinguishability of oblique lines which were listed in Section 13.5.4 apply to this experiment also and, in varying degrees, to all other studies on identifying orientations that I have come across. The lines were on square cards and no controls using circular cards were included; the orientation of the intershape axis was not varied; the oblique lines were symmetrically disposed, and the others were not; the two sets of lines were not equally disposed about the intershape axis; and the lines were flat on a table. Of particular interest here is the fact that the oblique lines were symmetrically disposed whereas the vertical and horizontal lines were not. The controls shown in Figures 13.6(c) and (d) should have been applied *and nobody has yet applied them.* So we do not know whether the oblique lines were more difficult to identify because they oblique or because they were symmetrically disposed, and without this information, all theories about why children find obliques more difficult to identify than other orientations are premature.

Rudel and Teuber did not try to teach the young children how to identify oblique lines, so that we do not know whether their failure on this task was due to an absolute incapacity or a lack of attention to the correct procedure. Jeffrey (1966), using the same stimuli and testing procedure that Rudel and Teuber had used, instructed the children on how to solve the problem by placing arrowheads on the lines and by getting the children to point in the direction of the arrow. After this instruction, eight out of 17 four-year-olds showed improvement on the original task of identifying obliques. This experiment shows that, whatever causes the difficulty with oblique lines on cards certain four-year-olds require only experience to overcome it. But we do not know what maturational processes had been involved in bringing the children to the stage where experience could help them, nor do we know whether lack of maturation caused about half of the four-year-olds to fail even after instruction, and of course we still do not know the cause of their difficulty.

Goldstein and Wicklund (1973) trained children to copy patterns of pegs on a peg board. Only children over three years of age could learn to place pegs along a diagonal of the square board after having seen such a pattern made by the experimenter. It is not clear whether the difficulty experienced by the younger children was due to the square shape of the board, the arrangement of the holds in rows and columns, the oblique orientation of the rows of pegs in the frontal plane, or the symmetrical disposition of the diagonals of the board.

Attneave and Olson (1967) taught adults to attach the names Adam, Abner, Albert, and Andrew to left oblique, right oblique, vertical and horizontal lines respectively. They then showed that reaction times in naming the lines were longer for the oblique than for the main axis lines. In a similar way, Essock (1980) asked subjects to press one of several buttons according to the orientation in which a line was displayed. Reaction times for deciding that a line belonged to the set of vertical and horizontal lines were the same as those for deciding that the line belonged to the set of oblique lines at 45°. However, reaction times for deciding whether a line was a left oblique or a right oblique were longer than those required to decide whether it was vertical or horizontal. These experiments reveal a proper appreciation of the question of stimulus sets and the design overcomes problems associated with relative dispositions of lines, because only one line was presented at a time and in the same location. There is still the problem of knowing whether the longer reaction times for oblique lines was due to their obliqueness, as Essock concluded, or whether it was due to their symmetry about the body axis.

Attneave and Olson (1967) found that the pairing of names with oblique lines gave longer reaction times, even when they were viewed with the head tilted at 45° in the plane of the display. Attneave and Olson concluded that the gravitational orientations of the lines rather than their retinal orientations are what determine which are most difficult to identify, although in a later study it was found that this result could be modified by instructions (Attneave and Reid, 1968).

Most of these experiments involved simple meaningless shapes, and Mandler and Day (1975) wondered whether children and adults show the same difficulty when identifying members of mirror-image pairs of familiar shapes. All their subjects, from the age of five years to adult, made fewer errors when identifying the orientations of familiar shapes than when identifying the orientations of non-familiar shapes. It is a pity that children younger than five years of age were not tested, and also other relative orientations, because the stimuli and procedure that Mandler and Day used were better than those employed in other studies.

13.6.4 Theories about the identifiability of relative orientations

There is no need for a theory to account specifically for why members of mirror-image pairs or pairs of obliques are more difficult to identify than other relative orientations, because we are not sure of the facts. Nevertheless, I suggest that a simple extension of the two general principles which were presented in Section 13.5.5 can be used to predict the particular patterns of relative difficulties that have been reported, while not making claims which are more specific than is warranted.

The first principle is that *things which are more alike with respect to dimensions being responded to are more difficult to learn to identify (make differential responses to) than are things which are less alike*. If this principle is applied to the domain of relative orientations in the plane, it leads to the second principle. *The simpler the operation of bringing features of one shape into coincidence with corresponding features of the other shape, the more alike the orientations of the shapes will be judged to be, and the more difficult it will be to learn to identify those orientations.* The same factors listed in Section 13.5.5 apply to the task of identifying orientations.

But other principles may also be needed to account for relative difficulties of identifying orientations. I suggest that the error rates, and particularly the reaction times, in recognizing a given orientation depend on the size of the set from which the subject perceives the stimulus to be drawn. This set corresponds to what Garner (1962, 1974) has called the subject's **equivalence set**, and its size determines what he called **stimulus redundancy**; the smaller the subject's equivalence set, the larger is the stimulus redundancy, or put another way, the less information is conveyed to the subject by showing him a particular member of the set. Experiments have demonstrated that reaction times for identification of patterns depend on stimulus redundancy (see, for instance, Clement and Varnadoe, 1967). I further suggest that error rates and reaction times depend on the nature of the cognitive operations that the subject uses to 'define' the categories into which he classifies the stimuli. This corresponds to what Garner has called the **form of stimulus redundancy.**

I shall now apply these ideas to the task of identifying the orientations of lines. The orientations of oblique lines may take a long time to identify, because subjects perceive oblique lines as drawn from a set consisting of lines in all orientations except vertical and horizontal. Verticals and horizontals may be quickly identified because they are perceived as drawn from the much smaller set consisting of only verticals and horizontals. This would be an example of how reaction times vary because of a difference in subjective redundancy of the stimuli. The descriptive rules used to perceptually code oblique orientations are almost certainly more complex than the rules used to code verticals and horizontals. This is because people, at least adults, use vertical and horizontal as their basic descriptive categories for orientation and 'compose' descriptions for oblique orientations from these basic categories. This would be an example of how the form of stimulus redundancy may influence shape identification. For further discussion of the effects of stimulus structure on identification, see Royer (1966, 1971), Evans (1967), and Hock (1973).

13.6.5 Judged uprightness and extrinsic axes of reference

In the experiments reviewed up to now in this chapter, the subjects remained erect, so that the normally vertical meridian of the retina was in line with the direction of gravity. If a subject judges the orientation of a shape while his head is tilted, he may use either the direction of gravity or the normally vertical axis of the head as his frame of reference.

Thouless (1947) conducted an experiment to determine which axis of reference people use when judging the uprightness of a shape. The details of this experiment are described in Section 14.5.2. It need only be noted here that subjects were about equally divided between those who used an egocentric reference and those who used a gravitational reference when judging the uprightness of shapes viewed with head inverted. Some subjects were uncertain which answer to give. These differences in behaviour may reflect only the different ways in which people interpret the question. With prompting they may all be capable of basing their judgments on either criterion.

Braine and Knox (1975) presented children between three and nine years of age with pairs of pictures of familiar mono-oriented objects, such as a table, a horse, and a face. The subject viewed the pictures in a miniature room that could be upright or tilted through 90 or 180°. The subject's head was also either upright or tilted through 90 or 180°. The picture of one object was placed in line with the axis of the head, and the picture of another object was placed in line with either the direction of gravity or the walls of the miniature room. The children were asked to name the object which was upside-down. None of the subjects used the miniature room as their frame of reference in judging the uprightness of the pictures. This was probably because the room was a miniature room, and the children were allowed to see it being tilted. Furthermore, tilt illusions are not normally induced by visual frames tilted 90 or 180°. If the room had been larger and the angles of tilt smaller, the children would surely have been influenced by the visual frame in making judgments about uprightness (see Section 10.3.3). With 90° tilts of the head most subjects based their decision about uprightness on the orientation of the pictures with respect to gravity. This is probably related to the fact that people often tilt their heads up to 90°, and, when they do so, objects in their visual surroundings remain upright with respect to gravity , not with respect to the retina. The pictures which Braine used were all of familiar mono-oriented objects, so that the children would have no difficulty recognizing the intrinsic top in each of the pictures. With 180° tilts of the head most of the three-year-old children based their judgments on the orientation of the pictures relative to the head, whereas the older children were about equally divided between those who used an egocentric reference and those who used a gravitational reference. It was concluded that the younger children base their judgment about the uprightness of objects disoriented by 180° on the extent to which the objects look familiar. We shall see in Section 14.5 that familiar objects which are inverted on the retina but upright with respect to gravity look less familiar than pictures which are inverted with respect to gravity but upright on the retina. Older children and adults are also subject to this effect, but many of them are able to disregard the criterion of strangeness and concentrate on whether or not the shapes are inverted with respect to gravity. A similar experiment was performed by Gajzago and Day (1972) with very similar results. Further implications of these findings will be discussed in Section 14.5.

Chapter 14

Orientation and Shape Perception

14.1 Introduction

There are two questions to be considered in this chapter. The first is, how do people make allowances for differences in orientation when comparing unfamiliar shapes? The second question is, how do people recognize familiar shapes when the shapes are in an unfamiliar orientation?

Two objects have the same shape when they can be brought into spatial congruence by applying nothing more than a similarity transformation to one of them. A similarity transformation is any transformation that leaves angles and relative distances unchanged. Similarity transformations include the isometric transformations of translation and rotation along or about any of three orthogonal axes, as well as dilation and contraction. In the case of objects which are asymmetrical about three orthogonal planes, mirror reflection is a distinct similarity transformation. Such an object and its reflection are known as an **enantiomorphic pair** (see Figure 14.4). Enantiomorphs may, or may not, be regarded as the same shape, that is a matter of definition. If the definition of 'same shape' is extended to include cases when congruence is achieved by projective transformations, we can say that a flat object seen lying in perspective out of a frontal plane is the same shape as the same object seen in a frontal plane. However, this aspect of shape is not relevant to the subject of this book, so that for present purposes shape is defined as that set of static spatial features of an object which remains invariant under similarity transformations.

The ability to abstract invariant spatial features of objects confers great biological advantage on an animal because it allows the animal to recognize objects in different positions and orientations and at different distances. The problem of abstracting shape would be bypassed if the retinal image of each object were **normalized**. An image is normalized when it is transformed to a standard location, orientation, and size. When this is done, the task of deciding whether or not two objects have the same shape is reduced to the simple task of deciding whether their outlines are congruent. Normalization would also simplify the task of remembering the shape of an object so that it could be recognized on a subsequent occasion. With complete normalization each shape could be stored as a template consisting of the converged inputs from all those receptors which the outline of a given shape stimulates. At any time, the template with the greatest number of active units would indicate the shape of the stimulus. Animals, particularly more primitive animals, partially normalize their visual inputs by moving their eyes or heads so as to bring the image into a given location and orientation on the retina. However, size normalization of the retinal image is not usually achieved. Normalization of orientation is the easiest type of normalization to achieve because most objects in the visual world are mono-oriented, that is, they remain the same way up relative to gravity, and an animal can simultaneously normalize the images of all such objects by simply keeping the head, and hence the eyes, upright. This alone greatly reduces the complexity of the neural coding processes responsible for shape comparison and recognition. It does mean, though, that comparison and recognition will fail, or become slow or difficult, if objects are not in their normal orientations. This chapter is about such failures and difficulties. Let us take a look at some examples.

Consider the two shapes in Figure 14.8. It does not take long to see that they are not the same. However, it takes much longer to decide whether or not the shapes in each pair in Figure 14.6 are the same. The reason is that these shapes are in different orientations, and we find ourselves having to make allowance for this disorientation when comparing them. Section 14.3 is concerned with the question of how the shapes of two objects in different orientations are compared. In investigating this problem, it is better to use unfamiliar shapes, so that questions associated with the recognizability of the individual shapes do not intrude. Note that the question concerns a person's ability to tell whether or not two shapes are the same, not his ability to recognize either of the shapes, nor his ability to judge their orientations.

Consider Figure 14.1. The person in the picture is not easy to recognize until the picture is turned round. Look also at Figure 14.2, and notice that the perceived identity of the object changes when it is inverted. These demonstrations illustrate that the way certain familiar objects are stored in memory

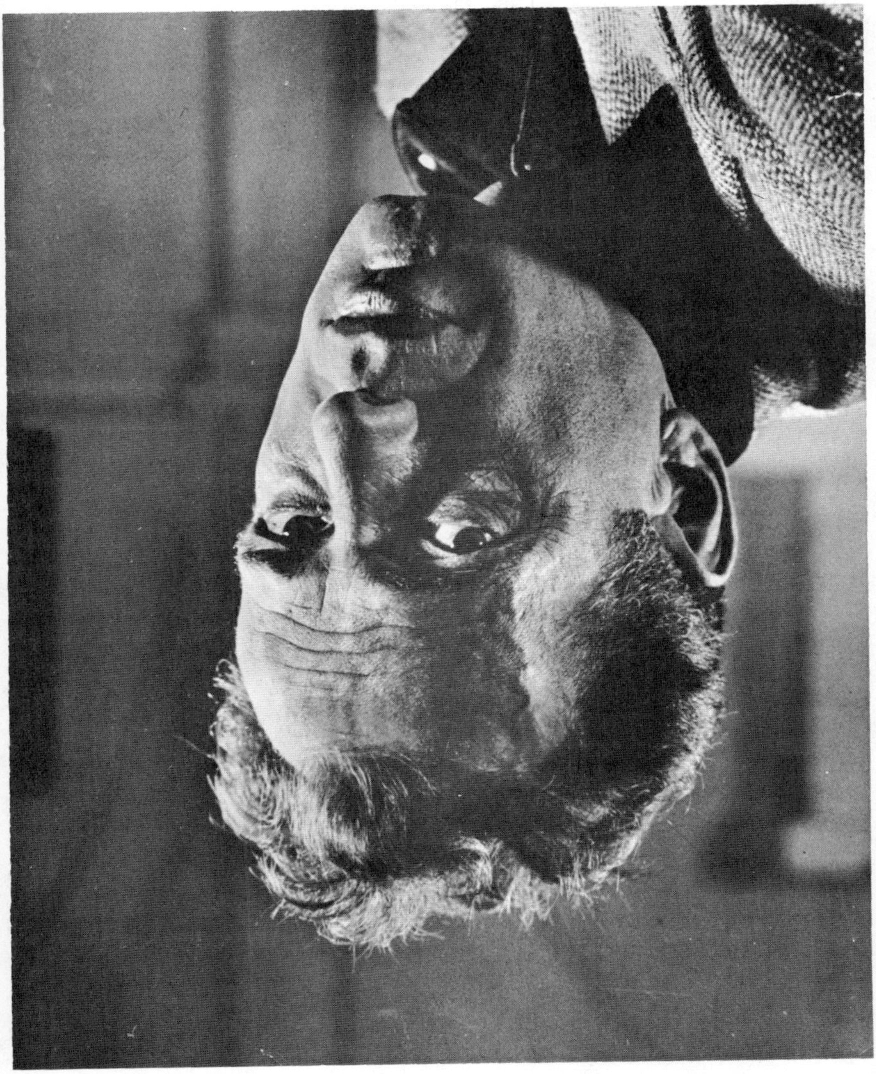

Figure 14.1 A well-known person is not so easily recognized when upside-down (Photograph by Philippe Halsman, *Life Magazine*. Copyright, 1944, Time Inc.)

is tied to the orientation in which we normally see those objects. I refer to this aspect of visual memory as **visual polarity**. The nature of the orientation-specific memory codes which are responsible for visual polarity is discussed in Section 14.4.

The mere fact that the ability to compare or recognize objects is disrupted by changes in their orientation while an observer remains erect does not tell us whether the difficulty is due to disorientation of objects relative to the retina, to

Figure 14.2 An ambiguous object which changes its appearance when turned round

gravity, or to other objects in view. This question is discussed in Section 14.5.

It has often been claimed that when optical devices which invert the visual scene are worn for some days the world no longer appears upside-down. In several textbooks it is stated that one part of the visual scene may geometrically right itself in experience while other parts remain apparently inverted. In Section 14.4.3, I shall examine the evidence upon which such statements are based and show that a less dramatic interpretation of it is possible.

Before any of these issues are discussed, we need to consider the stimulus domain with which we are dealing and the memory codes and schemata that people use in recognizing objects. The question of the way information is represented in memory arises for the first time in this chapter. The stimuli and tasks involved in previous chapters were so simple that the issue of memory could be ignored. The way visual features and knowledge about objects are represented in memory is crucial when considering shape perception.

14.2 Stimulus domains and schemata for shape perception

When a person attempts to recognize a particular object, he must recall certain of its features from memory. When comparing two objects, the relevant features of one of the objects need be committed to memory for only as long as it takes to inspect the second object, make necessary adjustments for any relative disorientation, and decide about the similarity of the two objects. Unnecessary complications will be avoided if we consider only rigid objects with constant physical properties. To understand how changing the orientation of an object may affect a person's memory of certain of its features, we need to know in general which features may be coded independently of its orientation and which may be coded only by including a specification of the object's orientation. But before discussing this central issue I must define a few terms.

A **visual feature** is any discriminable attribute which an experimenter uses to describe stimuli. Experimenters usually select simple features, like colour, length, and tilt, with which to work, but complex attributes, like triangularity or a facial expression, may also be regarded as visual features. An **element** is a localized feature of an object, like a part of an object or a coloured patch on its surface. A feature may be said to have **values**. Thus, the feature 'length' has various particular lengths and the feature 'smile' has various types of smile. The partitioning of the attributes of an object into features and values is entirely at the experimenter's discretion.

A **descriptive system** is a set of rules regarding the way the values of certain features vary over a set of stimulus objects, and which an experimenter uses to generate or describe members of that set. A **stimulus domain** is a set of stimulus objects each of which exemplifies a descriptive system. **Distinctive features** are those features whose varying values are used to define a given descriptive system. If two descriptive systems generate the same stimulus domain, they are **equivalent**. For example, the descriptive rule 'three-sided polygons' is equivalent to the rule 'three-cornered polygons', because they both generate the same stimulus domain of 'triangles'.

A **perceptual schema** is a piece of knowledge about a descriptive system which an experimenter infers is possessed by a subject in an experiment. This inference is based on the way the subject recognizes objects within that stimulus domain. An **orientation-invariant feature** is any feature which does not discriminably change when a stimulus object is rotated in an observer's frontal plane. An **orientation-free code** is any mnemonic representation of a feature which does not analogically embody or refer to the orientation of the stimulus object. An **orientation-specific code** is one which does embody or refer to the orientation of the object.

For present purposes, an initial distinction will be made between intrinsic features and extrinsic features. As we are dealing with only rigid objects with constant physical properties, **intrinsic features** may be defined as those features which remain objectively invariant as the object is rotated or moved about with respect to a reference system. Visual reference systems fall into two main classes: (a) egocentric reference systems, such as the normally-vertical retinal meridian; (b) the visual frame, consisting of other objects in view. **Extrinsic features** are those which are defined with respect to one or other of these reference systems. All intrinsic features of rigid objects are, by definition, orientation-invariant. What I want to show is that certain features are coded in orientation-free form, that others may or may not be coded in this way, and that others are necessarily coded in an orientation-specific form. I shall discuss intrinsic features first.

Non-spatial intrinsic features. All objects have certain intrinsic visual features which are non-spatial and invariant under all spatial transformations, even non-similarity transformations (see Figure 12.1). These features are colour, whiteness and opacity or their derivatives, such as albedo and glossiness. If we have a set of objects the members of which differ consistently in terms of such features, a person may recognize members of the set equally well in any orientation, provided he has learned these distinctive intrinsic features. However, the extraction of these features requires the subject to dissociate them from closely related features which are not necessarily orientation-invariant. For instance, the whiteness of a reflecting surface is invariant but its luminance depends on the strength and direction of the illumination. These two features cannot be distinguished unless due allowance is made for the strength and direction of incident illumination (Beck, 1965; Gilchrist, 1977).

Figure 14.3 illustrates how apparent spatial relationships may be affected by a change in the apparent direction of illumination. The assumption responsible for these effects is that illumination normally comes from a source above an object. Nevertheless, people usually recognize the colour and whiteness of objects in a way which is independent of changes in orientation and store this information in an orientation-free form.

Figure 14.3 Illustrating how an assumption about which way the light is coming affects the interpretation of the spatial disposition of parts of a scene. Turn the picture round to see the scene in its original orientation

Non-metric (topological) intrinsic features. Topological features are those which do not require the specification of metric attributes, and which are therefore invariant under similarity and non-similarity (rubber sheet) continuous transformations. Certain topological features do not require the specification of directional relationships within the object. For instance, the feature 'hole' is a topological feature that is based on the spatial property of inclusion of one contour in another, plus the property that it can be seen through. Neither of these spatial properties contains any reference to the relative directions of elements, and their appearance is not affected in any consistent way by changes in orientation. For these reasons, such features are easily coded in memory in an orientation-free form. Other examples of intrinsic topological features which are coded this way are 'connectedness' and 'number of sides'. There are other topological features which involve specific spatial dispositions of elements of an object. There features have to do with the way things are knotted and, although they could be coded as orientation-invariant features, may well be coded in memory in an orientation-specific form. Enantiomorphic knots, like those in Figure 14.4, cannot be identified unless referred to a fixed reference system.

Metric intrinsic features. Spatial features which are invariant only under rotation and translation will be referred to as metric features. These features are related to size and include texture density and spatial frequency. Like the non-directional topological features already mentioned, they are clearly orientation-

Figure 14.4 Two shapes which differ with respect to a topological feature involving the spatial disposition of elements

invariant and not easily confused with other features which are not orientation-invariant. Therefore, they are all easily represented in the visual system in an orientation-free form.

Directional intrinsic features. Features which have to do with the relative positions or directions of parts of an object will be referred to as directional features, but I shall also refer to them as **shape** or **pattern features**, because it is these features that define what is commonly referred to as the shape of an object or the pattern that an array of objects makes. They are invariant under all similarity transformations. The objective specification of intrinsic shape features may be precise, in terms of angular relationships and relative distances, or it may be more general, in terms of attributes like 'opposite', 'symmetrical' or 'obtuse'.

In geometry, there are two ways to describe patterns. One way is to use a Euclidean or other fixed coordinate system and specify all elements of the pattern in terms of their positions in relation to these coordinates. This is clearly an extrinsic, or orientation-specific way of coding patterns. On the other hand, one may use coordinate-free geometry, in which the pattern is specified in terms of its internal relationships only. This is an intrinsic, or orientation-free code. The nervous system is presented with the same choice. Wherever spatial relationships are represented in memory in an **analogical** fashion, that is, in a form which is directly isomorphic with the spatial relationships of the stimulus, then pattern features cannot be represented in an orientation-free form. That is not to say that disoriented objects cannot be distinguished or recognized by such a system, but in order to perform these tasks the system must possess some mechanism for rotating (normalizing) the stimulus, or the internal representation of the stimulus, unless it stores a distinct memory image for each orientation of the object.

On the other hand, if spatial relationships are represented in **propositional** form, the coding of pattern features may or may not be orientation-free. An orientation-free propositional code would contain such descriptors as 'opposite to', 'in line with', 'perpendicular', 'symmetrical', etc. An orientation-specific propositional code would contain such descriptors as 'above', 'to the left of', 'East of', etc. Human beings certainly possess both types of propositional code, both at the prelinguistic and linguistic levels. The initial input of images into the visual system is certainly not orientation-free, nor propositional, and one must therefore conclude that orientation-free forms of representation are derived by a process of abstraction from codes which are orientation-specific.

Certain objects contain features which **covary** in a consistent fashion. When this is the case, the value of one of a number of features is sufficient to specify the values of the rest, given that the subject knows about the covariance. For instance, the intrinsic pattern features of a triangle are completely specified by noting that the pattern has three straight sides of such and such a length. If the perceiver knows about triangles, he can infer that this object must also have three angles, is scalene or equilateral, etc. Memories for specific objects can be very efficiently coded by a system which embodies knowledge about covariance. Such knowledge structures have been called **frames** by Minsky (1975), but I see no reason to abandon the older term 'schema'. The important point about covarying features in the present context is that, even if some of the features are coded in an orientation-specific code, the observer may still be able to recognize the disoriented object on the basis of the values of those covarying features which are coded in an orientation-free form.

There is only one type of feature of shapes that cannot, even in principle, be specified in terms of orientation-free descriptions, and that is the feature that distinguishes between two-dimensional, mirror-image pairs or between three-dimensional enantiomorphs. This feature, although objectively orientation-invariant, must necessarily be coded with reference to extrinsic axes, which must themselves be asymmetrical. For example, we can recognize whether the thread of a bolt is a left-hand thread or a right-hand thread only by referring the object to some asymmetrical feature of our own body or visual memory or to some external reference. The reference frame may be dispensed with for the purpose of deciding whether two shapes are identical or enantiomorphs, because one object may be compared with the other. However, without a reference frame the enantiomorphs, although distinguished, cannot be named (identified).

Extrinsic visual features of on object may be classified in a similar way. Extrinsic features may be non-spatial, such as colour contrast between an object and its background. They may be non-directional topological features like 'contained in' or 'next to', or they may be non-directional metric features, like the size or texture of shadows cast on the surface of the object. Extrinsic features of these types, like their intrinsic counterparts, can easily be coded in an orientation-free form, except perhaps features having to do with illumination, such as shadows. Directional extrinsic features of an object may be coded in an orientation-free form *with respect to retinal coordinates* if these features relate to the orientation of the object *with respect to a visible framework*. But if the frame of reference is an internalized standard such as the normally-vertical retinal meridian, the features are, by definition, not coded in an orientation-free form.

Those features of objects which we call 'top' and 'bottom' have a special status. They are not unambiguously either intrinsic or extrinsic features. If we look at an unfamiliar shape, its top is that part which is gravitationally uppermost. In this case, we have an **extrinsic top**, because when the object is rotated, a different element becomes the top. If we look at a familiar mono-oriented object, the top is that element of the object which is normally up. In this case, we have an **intrinsic top**, because we call the same element top even when the object is rotated.

If certain intrinsic spatial features of an object are represented in memory in a particular orientation, with respect to retinal coordinates, then recognition of such features will be disrupted when the object is not in this orientation. There are three ways in which this disruption may be avoided. In the first place, there may be other features which are coded in an orientation-free form which allow the observer to infer the characteristics of the disrupted features. For instance, although the inverted picture of a friend may look strange, we may recognize the person because of the colour of the hair. Secondly, an observer may acquire orientation-specific memory codes of pattern features for more than one orientation of an object. For instance, a person may learn to read upside-down writing as well as erect writing. Thirdly, an observer may physically rotate the

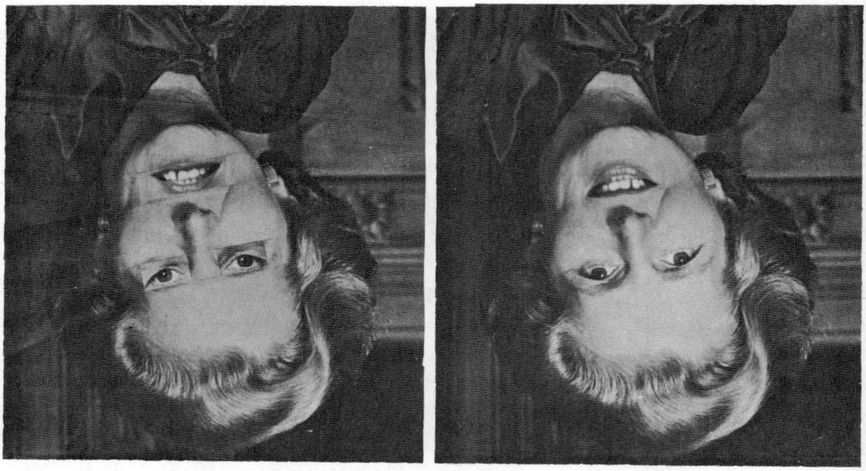

Figure 14.5 Showing how distortions in the features of a face are not evident when the face is upside-down (From Thompson, 1980, with apologies to Mrs Thatcher. Reproduced by permission of Pion Ltd.)

disoriented object into its normal orientation, or he may have the ability to rotate the internal representation of the object. This issue is discussed in Section 14.3. If a person normalizes an object or its image, he must have recognized that it was an object which needed normalizing and, if he is to avoid a trial-and-error procedure, he must recognize which way and by how much the object should be rotated. These issues have led to a spurious debate about whether recognition of the object precedes normalization or is achieved only after normalization. Spurious debates like this arise when recognition is regarded as a unitary act, rather than one which depends on the stimulus domain within which the observer is working. For most natural objects, there are features which are coded in orientation-free form, and, in any case, mono-oriented objects usually have an intrinsic top, which is recognizable even when inverted. These features may be recognized before the object is normalized, and indicate the normalization that is required. The orientation-specific features are recognized after normalization. For instance, when we see an upside-down picture of a face we rely on our orientation-free memory codes to inform us that this is a face, and we can see that it is inverted because its intrinsic top (hair) is down. But we cannot recognize the person (the orientation-specific features) until we turn the picture around. Distorted details of a face may not be apparent when a face is upside-down, even though they are very apparent when the face is erect (see Figure 14.5).

14.3 Distinguishing between unfamiliar shapes in different orientations

This section deals with the way people visually distinguish between a pair of unfamiliar shapes when they are placed in different orientations relative to each other in a frontal plane. In Howard and Templeton (1966, p. 320), we suggested, on the basis of the evidence then available, that in distinguishing between two unfamiliar shapes in different orientations:

> '. . .the observer first looks at one shape and forms a memory image of the shape which he then carries over as he attends to the second shape. If the two shapes are in a similar orientation, the outlines of the memory-image and the second shape will match, and minor differences will be easily detected. If the shapes are differently oriented, then the image and the second shape will not match unless some internal operation takes place equivalent to rotating the memory image.'

(a)
(b)
(c)

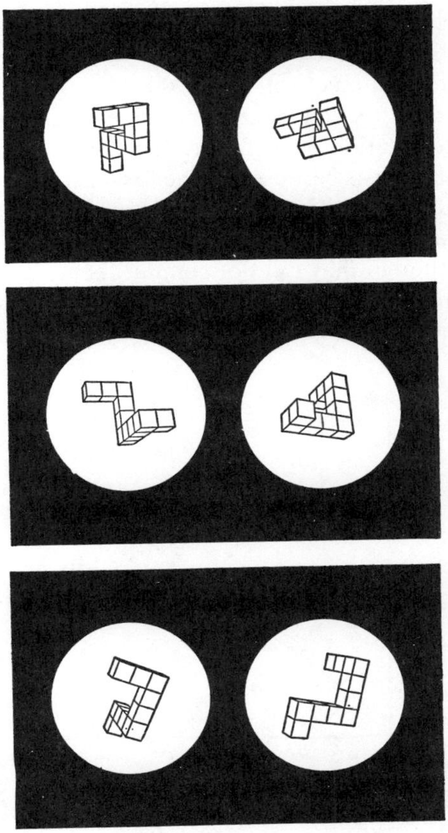

Figure 14.6 Examples of pairs of shapes used by Shepard and Metzler (1971). (a) A 'same' pair, differing by 80° rotation in the picture plane. (b) A 'same' pair, differing by 80° rotation in depth. (c) A 'different' pair, one of which must be both rotated and reflected to be brought into coincidence with the other. (Copyright, 1971, by the American Association for the Advancement of Science)

In 1971, Shepard and Metzler developed a procedure for measuring the rate at which subjects rotate mental images. They used pictures of the enantiomorphic pairs of objects shown in Figure 14.6. The members of each pair of identical objects differed in orientation by some multiple of 20°, either about an axis at right angles to the plane of the picture (see sample a) or about an axis in the plane of the picture (see sample b). The members of each non-identical pair were reflected about some plane in three-dimensional space and then, like the identical pairs, rotated relative to each other. The subject's task was to press a key as soon as he had decided whether the objects depicted in a given pair of pictures were identical or whether they were mirror-image reflections. The mean reaction time for individual subjects varied from about 1 sec when the objects were in the same orientation to between 4 and 6 sec when they were 180° apart, either in the plane of the picture or in three-dimensional picture space. But the most significant finding was that the function relating reaction time to angle of rotation was linear for all subjects, even though the slope and ordinate intercept of the function varied from subject to subject (see Figure 14.7). It was concluded that each subject rotated the mental image of one of the shapes at a constant rate unless he knew in advance the direction and axis of the required rotation, in which case the rate of mental rotation was faster. This experiment has caught the imagination of psychologists because it is the first clear case where an operation of such complexity on a mental image has been quantified. Shepard and Metzler concluded that a mental process occurs which is an analogue of the act of rotating a shape in space, although they did not commit themselves to a more detailed description of the mental process than that.

The success of the experiment was due largely to the clever use of enanti-

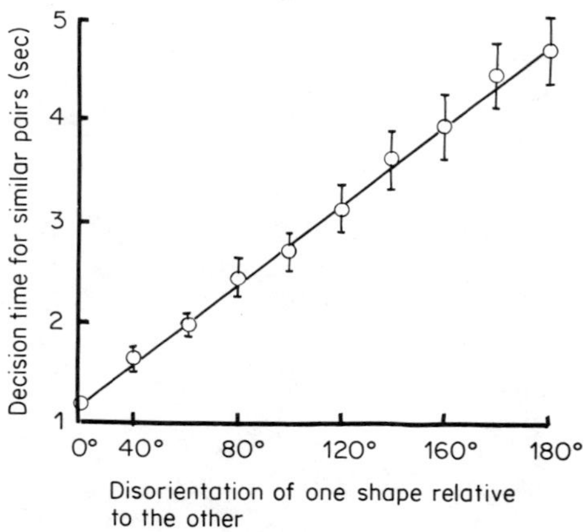

Figure 14.7 Time taken to decide that pairs of similar shapes like those in Figure 14.6(b) are the same, as a function of the orientation of one shape relative to the other. Bars are standard errors, based on $N = 8$ (From Shepard and Metzler, 1971. Copyright, 1971, by the American Association for the Advancement of Science)

omorphs. In previous studies of the effects of disorientation on shape recognition, the shapes being compared differed in local features, and the subject's task was to detect these differences. When this is done, the task may always be solved by noting the absence or presence of these critical local features, rather than by mentally rotating one shape with respect to the other. An *orientation-free* description of a shape is possible for all orientation-invariant properties of an object other than the enantiomorphic property. For instance, a triangle may be defined as a shape with three sides, and thus a triangle may be recognized as a triangle without the need to rotate it to a standard form. The distinction between enantiomorphs can be made only in terms of the relationship between each object and some *extrinsic* frame of reference, or by superimposing the two objects. If the two objects are in different orientations, this can be done by rotating the mental representation of one of them with respect to the frame of reference, or with respect to the other object if one of the objects is serving as a fixed reference. The use of enantiomorphs forces the subject to carry out some mental operation equivalent to such a rotation.

If we wish to guarantee that the subject has no other option than to mentally rotate the whole of the object, and not merely part of it, in order to solve the problem, something else is required. To guarantee this, the objects must be **irreducibly asymmetrical**. An object with irreducible asymmetry is one which cannot be simplified without destroying its asymmetry. For instance, the shape 'L' has irreducible asymmetry, and so has the pair of three-dimensional shapes shown in Figure 14.8. If a shape made from linear segments is used, which is more complex than these shapes, it is always possible for the subject to distinguish it from its mirror-image by paying attention to only part of the shape, and that is the only part that he needs to mentally rotate. The objects used by Metzler and Shepard were almost irreducibly asymmetrical, so that their subjects were virtually forced to mentally rotate one of them in an attempt to bring them into coincidence before deciding whether or not they were mirror-images. There is only one other procedure for deciding whether or not two objects are mirror-images. For example, with the shape 'L', a person may ask himself whether a rotation of each shape, with short limb leading, gives a rotation of the same sign (clockwise or anticlockwise). If the rotations have the same sign, the shapes are the same, otherwise they are mirror-images. He need not rotate either mental image far to make this decision. In a similar way a person can make a decision about the enantiomorphs shown in Figure 14.9. Any experienced mechanical

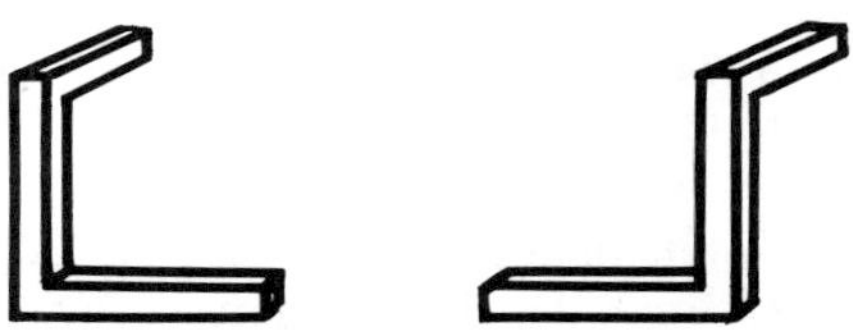

Figure 14.8 A three-dimensional shape with irreducible asymmetry, and its enantiomorph

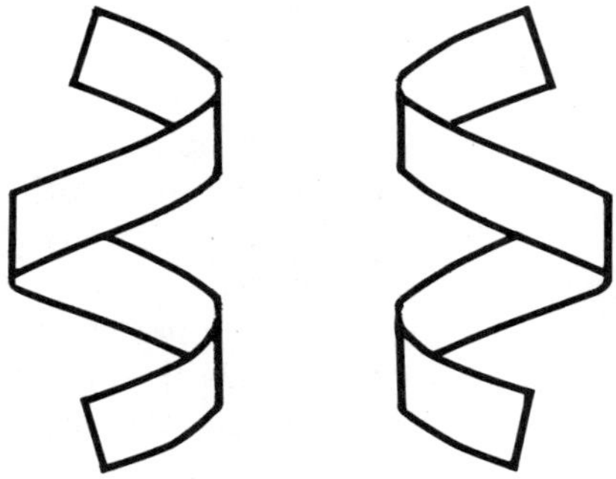

Figure 14.9 Enantiomorphic spirals

engineer has no difficulty deciding whether a spiral (thread) is a right-hand spiral or a left-hand spiral, whichever way round it is. Actually, this strategy does not really exempt the subject from the need to carry out at least a partial mental rotation. When he applies this type of test he rotates a frame of reference with respect to each object in turn. For instance, when a person decides whether a bolt has a left-hand or right-hand thread, he imagines looking out along the bolt (or holding it) and then deciding whether he must turn clockwise or anticlockwise in order to follow the spiral. Engineers do this very quickly because they are used to putting bolts in from any direction. An engineer has a ready-made set of frames of reference, one for each direction of space. These frames of reference are probably linked to various postures of the hand. It is unlikely that subjects in Metzler and Shepard's experiment resorted to this strategy.

Distinguishing between two objects by mental rotation is a method which we are rarely forced to use in ordinary life, because distinctions between objects may usually be based on orientation-free descriptions. These descriptions let us down only when the shapes to be distinguished differ in subtle ways, as for instance when we try to recognize the expression on an upside-down face. But mental rotation does not help in such cases either, because it is not easy to mentally rotate the image of a complex object.

The linear function relating reaction time to rotation angle was also found in the task of deciding whether two hands are the same or opposite in laterality (Cooper and Shepard, 1975) and in that of deciding whether an alphanumeric character is in its normal or mirror-image version (Cooper and Shepard, 1973). Mental rotation times have been found to be shorter for familiar shapes than for unfamiliar shapes (Hock and Ross, 1975; Hock and Tromley, 1978). A similar dependence of reaction time on rotation angle has been found for blind people making judgments by touch (Marmor and Zaback, 1976; Carpenter and Eisenberg, 1978), which shows that visual imagery is not necessarily involved. Men have been found to rotate images faster than women (Tapley and Bryden, 1977; Petrusic *et al.*, 1978), and older children have been found to be faster than younger children (Kail *et al.*, 1980).

Cooper and Shepard (1973) found that the reaction time for deciding whether an alphanumeric character was a normal or mirror-image version was independent of its orientation when the subject was told in advance what the character was and in what orientation it was to be presented. This suggests that subjects mentally prepared for the rotated character by imagining the normal version rotated into the appropriate orientation before the stimulus arrived. However, they needed time to prepare themselves in this way, and it was found that the time they needed depended in a linear fashion on the rotation angle of the character, so that here again was evidence that a mental image is rotated at a constant speed (see also Cooper, 1975).

Roldan and Phillips (1980) have questioned the generality of this finding. Using a similar paradigm they also found that reaction time increased monotonically with increasing separation between a prior image of an alphanumeric character and a test character. However, when the prior image was inverted, reaction times did not continue to increase as the test character approached the upright position. The familiarity of the upright orientation of letters must interact with the rotation process.

In another experiment, Cooper (1976) asked subjects to imagine that a designated two-dimensional shape was constantly rotating within a circular field. The mental rotation rate was determined for each subject by a method like that described above. The experimenter was thus able to present the same shape or its mirror-image briefly to the subject in an orientation which was congruent with his mental image at the time of presentation. It was found that the reaction time for deciding whether the briefly exposed shape was the same or the reflection of the mental image was short and independent of the orientation of the briefly-exposed shape as long as the shape coincided with the mental image at the time of presentation.

All these results seem to confirm that people operate on mental images in a way which is analogous to rotating a shape at a constant angular velocity through each intermediate angle. However, given that the task involving enantiomorphs requires something like mental rotation, this result is not very surprising, and one might ask what other theory of the mental processes involved there could be. A theory which was claimed to be different has been proposed by Pylyshyn (1973). In this theory it is suggested that shapes are stored mentally in propositional form, or as 'structured descriptions', rather than as analogue representations, or images. I do not intend to debate the issue of image *versus* propositional models of mental representations, which, according to some writers, is a pseudo-issue (Kosslyn and Pomerantz, 1977). The important point in the present context is that, in the propositional mechanism, orientation predicates are reassigned to shape elements in discrete rather than continuous intervals.

An empirical test to decide between these two theories was suggested by Cooper and Podgorny (1976) and by Pylyshyn (1977). It was suggested that the propositional model predicts that the time required to mentally rotate a structured description should increase with the complexity of a visual object. This, they claimed, is a consequence of the fact that, as complexity increases, more time will be needed to reassign orientation predicates to the various structural components, or elements, of the representation. Cooper (1975) found that the time to decide whether two random shapes were identical when one was rotated with respect to the other was not affected by the complexity of the shapes. However, the shapes which Cooper used were far from being irreducibly asymmetrical, so that subjects could have decided about the symmetry of each shape by attending to only one part of it, and that may have been why increasing complexity had no effect. Cooper and Podgorny (1976) attempted to eliminate this possibility by using sets of shapes like those shown in Figure 14.10. In each set there were several forms of the standard shape in which details varied. It was hoped that this would prevent subjects relying on a local feature for their judgments about symmetry. They found that mental rotation time was still unaffected by complexity. It seems to me that, with the shapes used by Cooper and Podgorny, a reasonably accurate performance could still be achieved on the basis of one or two local features. Better shapes for this purpose were used by Hochberg and Gellman (1977), as shown in Figure 14.11. With these shapes, it is somewhat more difficult to base a decision on local features, and subjects took much longer to distinguish between complex shapes and their rotated and reflected versions than between simple shapes and their rotated and reflected versions.

I think it is obvious by introspection that in looking at these complex shapes

Figure 14.10 One of the sets of shapes used by Cooper and Podgorny (1976) to investigate the effect of shape complexity on rate of mental rotation (Copyright, 1976, by the American Psychological Association. Reprinted by permission)

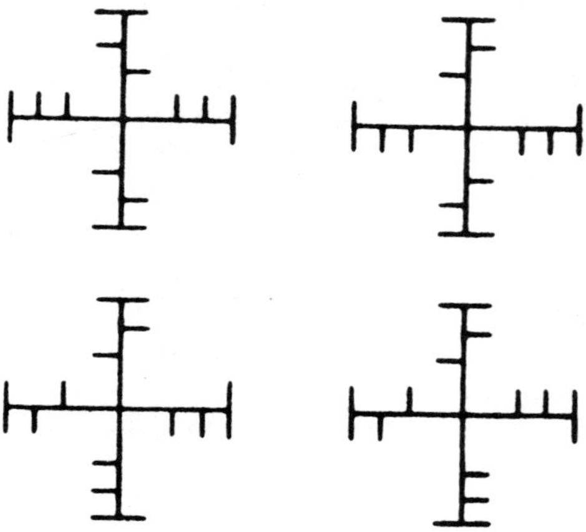

Figure 14.11 Some of the shapes used by Hochberg and Gellman (1977) to investigate the effect of shape complexity on rate of mental-image rotation (Reproduced by permission of Psychonomic Society, Inc)

one makes several glances from one to the other and several mental rotations of parts before deciding whether or not they are mirror-images. It may well be that the rotation time for each component rotation is a linear function of the rotation angle. Others, including Pylyshyn (1977) and Pringle and Cowan (1978), have also found that it takes longer to distinguish between complex rotated shapes than simple ones, and I think that the reason in both cases is that the subject could not store the complex shapes in iconic memory long enough to rotate them in one piece.

I do not think the shape complexity issue has anything to do with the form of internal representations: it has simply to do with the capacity of iconic memory. Those who believe in mental rotation must postulate a limit to the complexity of the image that can be stored and rotated as a unit. Simple shapes are seemingly rotated mentally in a continuous fashion after one glance, and this supports the analogical account of the way rotation is mentally represented. However, there is no reason why a continuous rotation operator should not exist in a system which also uses structural descriptions, and indeed Bobrow (1975) and Marr and Nishihara (1976) have described such models. In any case, perception involves a variety of mechanisms at many levels in the nervous system (see Howard, 1975; Palmer, 1977), and perhaps the form of representation is analogical at some levels and descriptive at others. In many tasks, a person is able to decide which form of mental representation is appropriate.

14.4 Visual polarity

14.4.1 Introduction

Objects like trees, people, and houses are normally seen in one orientation with respect to gravity and with respect to the normally-vertical retinal meridia. Such objects have an intrinsic top and bottom and will be referred to as **mono-oriented** objects, to distinguish them from **poly-oriented** objects, such as scissors, tooth-brushes, and tennis rackets, which have no preferred orientation. Asymmetrical mono-oriented objects also have distinguishable left and right sides although this aspect of visual polarity is not perceptually compelling. In Section 13.2, I discussed the possibility that certain mono-oriented objects share certain general features, such as the way they taper, which we associate with their upright orientation. Some of these general polarized features of the visual world may determine innate responses. For instance, fish swim with their backs to the light, even when lacking utricles, and many vertebrates, including infant humans, have a reflex tendency to adopt a standing posture with respect to a visual surface (see Section 10.5). However, the perceived uprightness of most objects must surely depend on learning, because the upright, as opposed to the upside-down position of many objects is purely a matter of convention.

I shall refer to the consistent orientation of most features of the visual world as **stimulus polarity** and to the perceptual effects of the polarity as **visual polarity**. For instance, a mono-oriented object, such as a face, is more easily recognized when upright than when upside-down. I discussed in Section 4.4 how visual acuity and sensitivity are better for stimuli which lie on the normally-vertical or horizontal meridia of the retina, rather than on oblique meridia. Those were **anisotropies**, that is, differences *between* retinal meridia. Visual polarity has to do with differences in visual perception due to the way a stimulus lies on a particular meridian. Note that I do not refer here to a particular *retinal* meridian, because whether visual polarity is dependent on the orientation of stimuli with respect to a retinal meridian, a head meridian or a gravitationally-defined meridian is the question which will be discussed in Section 14.5.

In this section I shall assume that the subject remains in an upright posture, and consider the effect that inverting mono-oriented objects has on his ability to recognize them. An understanding of how object recognition is disturbed by inversion, like all other problems in perception, depends on a specification of the stimulus domain and task domain with which one is working. For example, there is little point in stating that inverted faces are difficult to recognize unless the stimulus domain within which the subject is operating is specified. The statement could mean that the subject has difficulty recognizing the face as a face and not some other object, a man's face and not a woman's face, a Caucasian face and not a Mongolian face, a particular person's face and not somebody else's, or a smiling face and not a frowning face. Each of these distinctions requires the subject to attend to particular *distinctive features* and to operate with the rules governing the way these features vary over the set of faces within that stimulus domain. An individual face is an example of a potentially infinite number of sets of faces: thus, a face may at the same time be a man's face, a Caucasian face, Bill Jones' face, and a smiling face. Each descriptive domain is associated with a schema, or knowledge structure, within the observer. It is not easy to know which descriptive domain a subject is using, especially with complex stimuli like faces, but without such information, theorizing becomes meaningless.

Learning must surely account for most, if not all of the disruptive effects of inversion on shape recognition. I shall now consider what type of learning this might be.

14.4.2 Developmental aspects of visual polarity

Anyone who has dealt with young children will have noticed that they are often indifferent to the orientation in which a picture is placed. They make no attempt to right a picture if it is upside-down, and seem to be able to recognize objects in the picture however it is oriented. The same indifference to the orientation of pictures has been noted in primitive peoples (Wyndham, 1936). I think children behave this way because, when they look at pictures, they are interested only in recognizing the gross features of the objects depicted, and this ability is not seriously impaired by inversion. They may also be unaware of the physical operations required to bring the picture into an upright position. It does not mean that children cannot distinguish between different orientations of a picture, as I have shown in Section 13.5.2. But even though children may be able to distinguish one orientation from another, they must learn to make differential responses or explicit reference to the orientation of an object in space, and Piaget and Inhelder (1956) claim that these skills do not develop before the age of nine years. But even after these skills have developed, shape recognition may still depend on orientation. Thus, the face in Figure 14.1 is difficult to recognize even though one is aware that it is upside-down.

I suggest that the perceptual schemata which the infant develops are rather crude at first and rely on orientation-invariant intrinsic features. For instance, a child may code a face in terms of its colour, size, ovoid shape, and the presence of eyes and mouth. The child may also code intrinsic spatial features, such as the fact that the eyes and nose are on opposite sides of the mouth. Such orientation-invariant features are readily coded in an orientation-free form. At this stage, the child does not make fine distinctions between objects and is not too perturbed if things are upside-down. As the child gets older and adopts a consistent upright posture, it is repeatedly exposed to mono-oriented objects and begins to code intrinsic directional features in an orientation-specific form. For instance, a face may now be represented in memory as an object with eyes above the mouth, either with respect to a retinal or a gravitational frame of reference. At the same time, the repertoire of schemata is enlarged and the child begins to construct schemata which allow him to make more subtle distinctions between objects. Many of the refined schemata will embody intrinsic features which are coded in an orientation-specific form, and so the child will begin to be disturbed by inverted objects. There will also be a development of the ability to code the intrinsic directional features of objects in an orientation-free, propositional form. I have already argued that this ability represents a degree of abstraction from analogical, and therefore orientation-specific, coding processes in the peripheral stages of the visual process. So here we have two opposed developmental trends, an increasing exposure to mono-oriented objects on the one hand, which fosters orientation-specific coding, and, on the other hand, an increasing ability to form orientation-free descriptive codes. One trend predicts an increasing degree of disruption with inversion and the other predicts a decreasing degree of disruption. Which trend is dominant at a given age is likely to depend on the types of objects used and the fineness of the recognition that is required. No wonder that the literature which was reviewed in Howard and Templeton (1966) was contradictory. Some investigators concluded that young infants are less affected than older children by inversion and other investigators concluded the opposite. I shall not review that literature again, but will look briefly at some recent studies on children's recognition of faces and print.

Human faces are the most complex class of objects that human beings normally recognize. A face illustrates beautifully how an object may be responded to as belonging to any one of many descriptive systems. Yin (1969) allowed adults to inspect sets of pictures of unfamiliar faces, houses, aeroplanes, and caricatures. Some sets were erect and some were upside-down. The subjects were then required to indicate which pictures they had seen before, when each of the pictures was paired, either erect or inverted, with an erect or inverted picture not previously shown. He found that all these mono-oriented objects were more difficult to recognize when upside-down, but that recognition of the faces was much more adversely affected than the recognition of the other objects. Yin concluded that there is some factor peculiar to faces which makes them difficult to recognize when upside-down. But the faces were also more difficult to recognize when erect, which suggests that the variations among the faces were more difficult to distinguish. The faces used by Yin were the same sex and had the same hair colour and general shape, which means that differences between them were probably rather subtle differences in the relative proportions and distances between the elements of the faces. These are just the features that are likely to be coded in orientation-specific form. Perhaps the distinctive features of the other objects included more non-directional features. The fact that faces differ in subtle ways having to do with relative proportions and distances makes them particularly useful for studying developmental aspects of visual polarity.

Fagan (1972) presented infants with a familiar and a novel face, either both upright or both upside-down with respect to the child's head. Recognition of the familiar face was inferred from the infant's differential visual fixation of the two faces. In a recent application of this procedure, Fagan and Shepherd (1979) showed that at 16 weeks a child's ability to recognize a familiar face is disrupted by inversion or 90° tilt of the face. Brooks and Goldstein (1963) and Goldstein (1975) showed inverted photographs of classmates to school children and students, between the ages of three and 20 years, and asked them to name the person in each photograph. When the data from these two studies were combined, they revealed that the ability to recognize upside-down faces increased with age up to the age of about 12 or 13, after which it declined again. It was established in each case that the subjects could recognize all the faces when they were upright. These results suggest that three-year-old children code faces in an orientation-specific way. Their improvement at recognizing upside-down faces could have been due to an increasing ability to mentally rotate their orientation-specific representations, but it could also have been due to the development of the ability to abstract orientation-free descriptions. The decline in ability after the age of 13 was most probably due to an increased reliance on orientation-specific codes, unless it was due to decreased exposure to the faces of their fellow students.

The ability to recognize upside-down print has also been used to study the development of visual polarity. The distinctions between letters such as p and q, d and b, u and n, and m and w are extrinsic and directional, and are therefore necessarily coded in an orientation-specific code. Such letters change their identity when inverted. Other letters, like C, I, O, and X, have no distinct top and bottom. The remaining letters of the alphabet are mono-oriented and possess a distinct top and bottom. The shapes of words also have distinct polar features which are mono-oriented. In addition, many letters and most words have a left–right polarity. For this reason it is important to invert print without also reversing it when studying the effects of inversion.

In several experiments, the time taken to recognize erect and inverted letters or words has been measured (for example, Clay, 1970; Staller and Sekuler, 1975). In each case, the ability to read both erect and inverted print improved with age.

However, with older children and better readers, the difference in reading speed between erect and inverted print was much greater than it was for younger children or slower readers. In other words, better readers increase their ability to read erect print at a greater rate than their ability to read inverted print. There is no indication in these studies of an absolute decline with age in the ability to read upside-down print. However, in none of these studies was testing done beyond the age of 10, so that if a decline in the ability to read upside-down print occurs later, as did the ability to recognize upside-down faces in Goldstein's experiment, then these experiments would not have detected it.

Several experiments have been done to determine whether it is easier to recognize rotated print, inverted print or reflected print. Kolers (1968) and Kolers and Perkins (1969) measured the speed at which adult subjects could read letters and text which had been rotated 180° in the plane of the paper, inverted, or left–right reversed. Samples of these three types of transformation are shown in Figure 14.12. It was found that rotated print was easiest to read, inverted print next, and mirror-reversed print most difficult to read. This same order of difficulty for the three types of transformation was also found by Taylor (1972), Navon (1978), and Corballis, Zbrodoff, Shetzer and Butler (1978).

Before it can be concluded that mental rotations about the visual axis are easier to perform than mental inversions (rotations about a transverse axis) or mental reflections (rotations about a vertical axis), it is necessary to show that the greater difficulties caused by inversion of print are not due to the specific characteristics of print. For instance, we see rotated print more frequently than we see inverted or reflected print. But the particular difficulty with mirror-reversed print may also have to do with the fact that there are fewer differences between a mirror-reversed word and a normal word than there are between a rotated or inverted word and a normal word. In other words, letters and words are more prominently polarized up and down than they are from left to right. The effect of this may be that subjects can more easily recognize the type of transformation that has been imposed when print is rotated or inverted than they can when it has been reversed. Recognition of the type of transformation must clearly precede any correction that the subject mentally imposes on the internal representation of the words or the processes by which they are recognized. In objects like faces, which are more or less bilaterally symmetrical, reversal does not affect recognition as much as either rotation or inversion. But this is because we discount the small left–right differences between faces. We cannot discount left–right differences between letters and words.

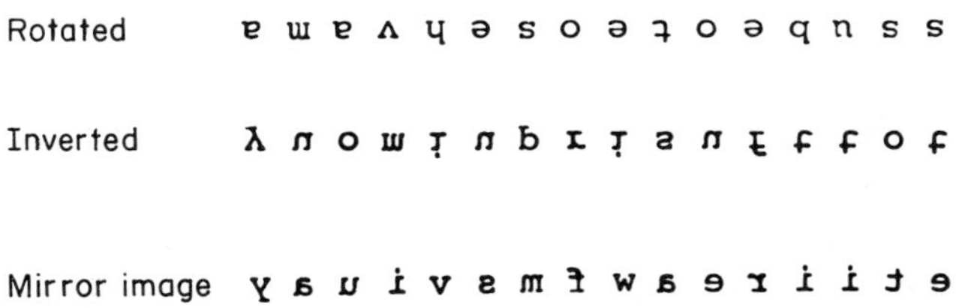

Figure 14.12 Examples of transformed letters used by Kolers and Perkins (1969)

In order to prove that mental operations equivalent to rotation about the visual axis are easier than those equivalent to inversion (rotation about a transverse axis) or reflection (rotation about a vertical axis), one would have to use unfamiliar shapes or objects which are no more prominently polarized one way than they are the other. The only experiment which involved such stimuli is the one by Shepard and Metzler (1971), described in Section 14.3. They found that subjects could 'rotate' mental images about each of the three axes of space with equal speed. This suggests that the differences found for print are peculiar to letters and words, although not only English letters and words, because Kolers found the same order of difficulty among readers of Hebrew.

14.4.3 Adapting to disturbed visual polarity

In the seventeenth cenury, Kepler was the first to demonstrate clearly that the retinal image is reversed and inverted with respect to the object being viewed. Since that time, philosophers and scientists have debated the question of how it is that the visual word is not experienced as inverted. The related question of why it is not experienced as reversed from left to right has been strangely ignored. The history of this debate is recounted in Walls (1951b) and Polyak (1957).

Stratton (1897a, b) is credited with being the first to conduct experiments on this issue. For several days he wore lenses which inverted and reversed the visual scene, and kept a record of his experiences. His aim was to decide whether a person wearing such lenses ever comes to experience the visual world right-way up. This is an ambiguous question, and attempts to answer it have led to controversy and confusion. Stratton's daily accounts of his experiences were as clear as one might expect, given the complexity of the situation. However, people have expected to find in Stratton's account a clear answer to an ambiguous question. Many writers (for example, Carr, 1935) concluded that Stratton came to experience an upright world, while others (for example, Woodworth, 1934; Ewert, 1937) concluded that he did not. Snyder and Pronko (1952) repeated Stratton's experiment but gave no conclusive answer to the question.

The question is ambiguous because it can refer to any one of three distinct criteria of uprightness. In the first place, there is the criterion of visual–motor coordination. According to this criterion, a person experiences an upright visual world if he successfully executes skills such as reaching for things, avoiding obstacles, riding a bicycle, etc. (see Section 12.5). In the second place, there is the criterion of intersensory coordination. According to this criterion, a person experiences an upright world if the seen direction of an object coincides with the direction of the object judged by other senses. Of particular concern in the context of inverted vision is the coincidence between the seen and felt position of the body of the person wearing the inverting lenses. Questions associated with this criterion are discussed in Chapter 11 and Section 12.5. Finally, there is the criterion of the experienced polarity of the visual world, which is the criterion which will be discussed in this section. Visual polarity has the following characteristics:

(a) It applies only to objects which have a distinguishable top and bottom.
(b) It applies only to objects which are normally experienced in one orientation.
(c) It can be applied to objects which are visually inspected without the intervention of other senses.

Note that the criteria of visual–motor and intersensory coordination apply to all objects, whether or not they have a distinguishable top and bottom and whether or not they are normally seen in one orientation. Furthermore, these two criteria do not apply to passive and purely visual inspection of a scene. The three characteristics listed above specify the meaning of polarity but do not specify what measurements can be taken to indicate the extent to which a person's visual polairty is disturbed. Several such measurements are listed below:

(a) The speed and accuracy of recognition of mono-oriented objects. Faces and print are particularly useful stimuli for this purpose, as we have already seen in Section 14.4.2.
(b) The speed and accuracy of identification of the intrinsic tops and bottoms of mono-oriented objects.
(c) The speed and accuracy of recognition of polarized movements, such as the recognition that smoke is rising or that an object is falling.
(d) The first figure that a subject recognizes when presented with ambiguous figures like those shown in Figure 14.2.

In each case the response measure must be the speed or accuracy of recognition of some feature of the visual stimulus. The response must not involve pointing or any other movement which is spatially related to the stimulus. For instance, if a subject were asked to point to the bottom of an object or to point in the direction in which he anticipates an object will fall, he may succeed by pointing objectively downwards, whichever way the object appears visually. Our experience of mono-oriented objects has led to the development of such words as 'upright' and 'upside-down', but these words are ambiguous. For instance, if a person places his head upside-down and looks at a page of print which is upright to gravity, he may say that the print appears upside-down because it is difficult to read, or he may say that the print is erect because it is erect with respect to gravity. For this reason, it is pointless to ask a subject whether the visual scene in an experiment appears upside-down. This question is ambiguous for another reason. This ambiguity has to do with whether the subject answers purely in terms of his present impressions while wearing inverting lenses or whether he compares what he is experiencing with a recollection of how things appeared before the inverting lenses were worn. This point may be illustrated by the answer made by one of Snyder and Pronko's subjects when he was asked while wearing inverting lenses, whether things looked upside-down. The subject replied, 'I wish you hadn't asked me. Things were all right until you popped the question at me. Now when I recall how they *did* look, *before* I put on these lenses, I must answer that they do look upside-down *now*. But until the moment that you asked me, I was absolutely unaware of it and hadn't given a thought to the question of whether things were right-side-up or upside-down.' (Snyder and Pronko, 1952, p. 113). Even if this source of ambiguity is avoided, there are other ambiguities in the general question, 'Do things appear upside-down?'. A change in one aspect of visual polarity may occur without a change in some other aspect. For instance, a person may learn to anticipate correctly the direction in which things fall in an inverted scene, but still be unable to read upside-down print. In other words, visual polarity is not a unitary aspect of perception but a composite of distinct recognition skills, each associated with distinct polarized features of particular mono-oriented objects. I shall have more to say about this source of ambiguity when discussing the results of experiments on inverted vision.

Measuring the speed and accuracy with which a subject recognizes specified polarized features of mono-oriented objects avoids the difficulties and ambiguities associated with motor and verbal responses.

Most, if not all behaviour associated with the orientation of mono-oriented objects is learned, so that it is reasonable to suppose that people can eventually learn to recognize such objects when they are inverted just as quickly and accurately as when they are upright. But before this issue is discussed, one more source of confusion must be explained.

Many writers on this subject have assumed that if a person comes to experience the visual world as upright while wearing inverting lenses, this implies that the mental image of the world has righted itself in a geometrical sense. For instance, Taylor (1962) maintained, on the basis of reports published by Kohler (1953), that when inverting spectacles have been worn for some time, objects which have been handled are experienced as erect but objects which have not been handled are experienced as inverted. Taylor gave the example of a person in a room who may, he claimed, experience the contents of the room as erect but the scene outside the window as inverted, in the manner depicted in Figure 14.13. But this view confuses geometry with visual polarity. Visual polarity has to do with how we recognize polarized features of mono-oriented objects—not with the geometry of the disposition of those objects in the experiential visual field. The distinction between objects which have lost their upside-down appearance and those that have not is not a distinction having to do with apparent relative orientations. The differences in visual polarity between things inside and outside a room are not revealed in the way the subject draws his visual experiences. Such differences are revealed only by the tests listed above. The subject will recognize objects in the room quickly, correctly identify their tops and bottoms, and predict the direction in which they fall. Objects outside the window will tend to be recognized more slowly and less accurately. These things cannot be depicted in a drawing, because there are no geometrical criteria involved.

There is no empirical justification for the belief that the upside-down scene rights itself in a geometrical sense, or swings round in experience, so as to look as it does normally. No subject in an experiment has reported such an experience, in fact they have usually stressed that such a thing does not happen. For instance, one of Kohler's subjects (Kottenhoff, 1957a) reported that things sometimes looked upright, but this did not mean that things appeared to turn round. He said, 'I see always the same but the interpretation is different'. Kohler himself

Figure 14.13 A representation of the type of visual experience which, according to certain textbooks, a person wearing inverting spectacles may have

(1953), when describing the left–right appearance of things seen through reversing lenses, wrote, ' . . . but this isn't a sudden reversal; it remains the same picture experienced differently'. Kohler experienced the inscriptions on the number plates of cars as reversed even though the cars had come to be experienced as moving along the correct side of the road. Kohler commented that this experience seems to be geometrically paradoxical but, nevertheless, wrote, ' . . . the purely pictorial impression remains the same'. The experiences are not geometrically paradoxical; it is simply a question of different polarity habits holding simultaneously for different objects in the field of view. Erisman and Kohler (1953) produced a film in which they optically reversed one part of a visual scene with respect to another part of the scene in an effort to depict photographically the paradoxical experiences of subjects wearing inverting lenses. This film is very misleading, as Erisman and Kohler must have realized. In the film, a subject wearing inverting lenses was presented with two faces, one erect and the other inverted. At first the subject reported that the erect face was inverted and the inverted face erect; but when the truly erect face began to smoke, the direction of the smoke was incompatible with its inverted appearance, and the subject reported that both faces now appeared erect, although the erect face retained some inverted features. For instance, the hair had the appearance of a beard. In the film, this experience is depicted as a geometrical righting of the face that had been experienced as inverted. Taylor (1962, p. 206) was probably not the only person to misinterpret this report. He concluded that, 'The erect face appeared to have acquired a beard, as if the crown of the head had failed to jump with the rest and now occupied the same space as the chin'. Taylor insisted that Kohler's subject experienced a change in the geometrical dispositions of parts of the scene.

A simple illustration of how mistaken it is to interpret changes in visual polarity as geometrical changes consists of imagining what would happen if a person were allowed to view only upside-down print. At first, reading would be difficult and the material would look strange. The person might attempt to invert his mental image of words he is reading in order to decipher them. But after a while, he would learn to recognize the words in their new orientation and his reading speed would gradually return to normal. There would come a time when he would no longer regard the inverted print as upside-down; it would lose its strange upside-down appearance. However, if he recalled how normal print appears, he would realize that inverted print had not geometrically righted itself in his experience so as to look pictorially like normal print. It is possible that, after long experience with inverted print, normal print would begin to look strange and be difficult to read. It would occasion no surprise to discover after such training that inverted and erect print seen side by side would both look 'upright' in the sense of being familiar and readable, and yet differently oriented in a geometrical sense. This illustration should make it clear how two things can appear to be erect and yet, at the same time, appear to be in different orientations.

Other than the verbal reports of subjects who wore inverting or reversing lenses, there are no data on how, and under what cirmcumstances, people adapt to disturbances of visual polarity. Kohler contended that, 'Only through manipulation of objects does the simultaneously-seen world obtain its directions; . . . he who wants to *see* correctly must first be able to manipulate correctly.' (Kohler, 1953, trans. Gleitman, p. 19). Taylor (1962) has proposed a similar theory. It is not clear what counts as manipulation, but presumably eye movements are not included. In any case, the experiments necessary to prove this point of view as far as visual polarity is concerned have not been done. Such experiments would be most easily performed by showing subjects physically inverted objects for so many hours a day, under controlled conditions, rather than by having subjects wear inverting lenses in everday environments. For instance, one could measure how rapidly people learn to read inverted print. People obviously can learn to read inverted print, and this alone is sufficient to show that manipulation in a narrow sense is not required for such learning. See

Section 12.3.4 for futher discussion of the relation between action and perception.

14.5 Shape perception and extrinsic axes

14.5.1 Introduction

Those axes of reference external to an object by means of which its orientation may be judged are **extrinsic axes**. There are three types of extrinsic axes, as follows:

(a) *The egocentric axes.* Egocentric axes may be oculocentric, headcentric or bodycentric. However, I shall assume in what follows that the orientation of the stimulus with respect to the normally vertical meridian of the retina is the only important egocentric variable. I shall assume that the results of experiments in which the subject tilted his head relative to his torso would have been the same had the head and torso been tilted together. I shall also ignore complications due to countertorsion of the eyes.
(b) *The direction of gravity,* as sensed by the utricles and by proprioception and touch.
(c) *The visual frame.* This is usually the square or rectangular card upon which the stimulus is mounted. If this factor is not being investigated, circular cards are used or the stimuli are viewed against a dark background or an illuminated *Ganzfeld.*

In Sections 14.3 and 14.4 it was shown that a person's ability to distinguish between or recognize shapes may be disrupted by changing the relative or absolute orientation of the shapes. However, in the experiments reviewed in those sections, the observer and the visual frame were kept in an upright position. Thus, the shapes were disoriented in the same way with respect to all three extrinsic axes, and it was not possible to tell from the results of such experiments what effects would have been produced by disorienting shapes relative to each type of extrinsic axis separately.

Since the literature on this whole question was reviewed in Howard and Templeton (1966) and more recently in a book entitled *Orientation and Form* by Rock (1973), there is no point in reviewing it again here, and very little has appeared on this issue since 1973. I shall merely outline what I consider to be the main conclusions to be drawn from the literature. Before doing that, I must point out that there are three distinct questions that have often been confused.

The first question has to do with whether the ability to recognize objects possessing a distinct polar axis is disrupted more by disorientation with respect to the observer than by disorientation with respect to gravity or the visual frame. The second question has to do with how the ability to recognize objects lacking a distinct polar axis is disrupted by their disorientation with respect to each of these extrinsic axes. The third question concerns whether a person is more likely to call a mono-oriented object erect when it is egocentrically erect but inverted with respect to gravity than when it is erect with respect to gravity but not egocentrically erect. The first two of these questions involve judgments about shape and will be discussed next. The third question, which involves judgments about orientation, has already been discussed in Section 13.6.5. It will be mentioned again here only to make clear that it is distinct from the question of shape recognition.

14.5.2 Recognition of disoriented shapes which have polar features

Familiar mono-oriented objects have an intrinsic top and bottom (polar axis) which we can usually recognize, even when the object is upside-down, either egocentrically or with respect to gravity. Certain unfamiliar shapes also possess an easily recognized polar axis. For instance, an axis of bilateral symmetry in a shape like that in Figure 14.14 is an easily recognized polar axis. We can readily recognize when such a shape is tilted. However, such shapes are not mono-oriented, because there is no ready-made criterion for deciding when they are erect as opposed to upside-down. I shall deal with mono-oriented shapes first.

Certain facts about the relative effects of disorienting a mono-oriented object with respect to the retina and gravity may be established from some very simple observations with a page of print. You can tell straight away that inverted print is upside-down, but it is nevertheless difficult to read. Now look at inverted print with the head inverted. Again, you can tell that the print is inverted relative to gravity, but the print is now easy to read. Finally, look at the page of print with inverted head, but with the print upright with respect to gravity. The print is now difficult to read. Clearly, disorienting print relative to the retina disrupts our ability to read it more than does disorienting it relative to gravity.

With the head erect, note that print does not become difficult to read until the page is tilted by more than 90°. We clearly tolerate this amount of retinal disorientation in reading. Similar simple observations may be carried out with a photograph of a familiar person, such as that shown in Figure 14.1. Note that

Figure 14.14 An example of an unfamiliar shape which possesses an easily recognized polar axis

with disoriented familiar mono-oriented objects, like print or a face, there is no noticeable disruption of the ability to recognize the intrinsic top and bottom (polar axis) of the object. Furthermore, the orientation of the polar axis with respect to retinal, gravitational or frame axes is easily assessed.

Other simple observations may be made using the ambiguous picture shown in Figure 14.2. Thouless (1947) found that, with head inverted, most of his subjects, whether adults or children, when presented with an ambiguous picture of this kind, reported that they saw the alternative that was egocentrically erect rather than the alternative which was erect with respect to gravity. Rock (1956) did a similar experiment, using the shapes shown in Figure 14.15. He found that most people reported seeing the shape that was gravitationally upright. However, the alternatives in these pictures are only 90° apart, so that the degree of retinal disorientation never exceeded 90°, which is probably not enough to bias subjects towards selecting the retinally erect alternative. Without a strong retinal bias they selected the alternative in which the intrinsic top was uppermost with respect to gravity. The reason for this is, I think, that we often tilt our heads by up to 90°, and when we do so we expect objects to remain upright to gravity.

Thouless went on to ask his subjects whether the face which they had seen when observing the ambiguous pattern with inverted head was erect or upside-down. There was considerable disagreement among subjects, and several did not know how to answer this question. It is, of course, an ambiguous question, as I have already explained.

There can be little doubt that a 180° disorientation with respect to the retina disrupts the ability to recognize mono-oriented objects more than does a 180° disorientation with respect to gravity. However, gravitational disorientation of print with no retinal disorientation apparently has some adverse effect on reading speed, because Rock (1973, p. 100) found that the time needed for recognition of inverted letters by an upright observer is slightly greater than the time needed for recognition of upright letters by an inverted observer.

Casual observation does not reveal any serious disruption of our ability to recognize mono-oriented objects when such objects are rotated through 90° with respect to either the retina or gravity. Nevertheless, Rock and Heimer (1957)

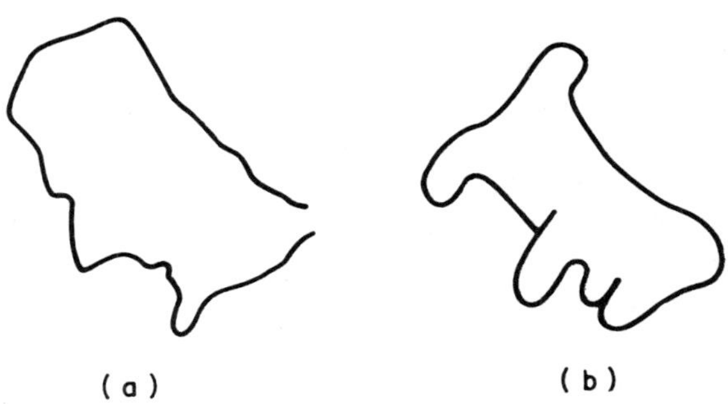

Figure 14.15 Ambiguous shapes used by Rock (1956). (a) is a face or a map and (b) is a dog or a chef

were able to reveal detrimental effects of a 90° disorientation when the task was made difficult by fragmenting the pictures of mono-oriented objects. A 90° disorientation of fragmented pictures with respect to gravity (with erect egocentric orientation) adversely affected the ability of subjects to recognize the pictures. However, this adverse effect was eradicated when the subjects were told beforehand what the orientation of the picture would be. This suggests that the effect of gravitational disorientation was due to a 'set' to look in the wrong location for the intrinsic top of the tilted object. Once the subjects knew where to look, this difficulty was overcome, and gravitational disorientation had no other detrimental effect.

A 90° retinal disorientation of the fragmented pictures (with gravitational orientation normal) also disrupted the ability of subjects to recognize the pictures. The effects of retinal disorientation persisted when subjects were forewarned about the disorientation. The detrimental effects of retinal (egocentric) disorientation are therefore not caused by a set to look in the wrong location for the top of the picture, but must be due to a less trivial cause, having to do with the coding in memory of certain features of the pictures in a form which is tied to their retinal orientation.

Braine (1965) was able to reveal a detrimental effect of a 90° retinal disorientation by exposing pictures of familiar mono-oriented objects for only 0.03 sec, and she also found that a forewarning about the orientation of the pictures did not abolish this detrimental effect.

Under ordinary conditions of viewing the ability to recognize mono-oriented objects is not noticeably affected by a 90° disorientation with respect to egocentric or gravitational axes. We tolerate disorientations of up to about 90° because we often tilt our heads this far or observe objects which are tilted by this amount. Furthermore, most mono-oriented objects have an easily recognized intrinsic top, so that little time is lost in finding the top even when it is not in its usual position on the retina or in relation to gravity. Once the top is found, the orientation of the object is discerned and allowed for in the final stages of the recognition process, in which the detailed structure of the stimulus is recognized. The retinal orientation of the stimulus is the crucial factor for this stage of the recognition process, but a 90° disorientation is not noticeably detrimental.

Corballis and Roldan (1975) asked subjects to judge whether dot patterns were or were not symmetrical about a line when the line was in various orientations. When subjects held their heads erect, it was found that they mentally rotated the patterns until the line was subjectively vertical before deciding about the symmetry of the pattern. When subjects performed the same task with their heads tilted 45°, it was found that they mentally rotated the line into coincidence with the normally vertical meridian of the retina rather than into coincidence with the gravitational vertical. This suggests that the recognition of visual symmetry, like the recognition of print and faces, is affected more by retinal disorientation than by disorientation with respect to gravity. However, to be fully convincing, the experiment should be repeated with head tilts of 90 and 180°. Corballis *et al.* (1976) found that when subjects were asked to judge whether rotated alphanumeric characters were normal or mirror-reversed with heads tilted 60°, they mentally rotated them to the gravitational upright, rather than the retinally upright position. This result suggests, in contrast to the evidence already cited, that the recognition of print is more disrupted by gravitational than by retinal disorientation. However, the subjects tilted their heads by only 60°, so that the degree of retinal disorientation was well within the range where reading is little, if at all, affected by retinal disorientation. If the experiment were repeated with head tilts of over 90°, I suspect that the results would be very different.

14.5.3 *Recognition of disoriented shapes lacking polar features*

Rock (1956) carried out the most extensive experiments on the ability of people to recognize shapes which lack a polar axis when the shapes are disoriented with respect to the retina, gravity, and the visual frame. In one series of experiments the subject's task was to select from a briefly exposed circular array of six shapes that shape which most resembled a training shape shown earlier. Two of the six shapes (the critical shapes) were identical to the training shape. One of the critical shapes had the same objective orientation as the training shape and the other was rotated through 45°. There were three such displays, each mounted on a square card seen in the context of a room (see Figure 14.16). In one experiment the subject inspected the training shape with head erect and was then shown the circular array while his head was tilted 45°. The effect of this was that one of the critical shapes was in the same orientation as the training shape with respect to the retina, and the other critical shape was in the same orientation as the training shape with respect to gravity and the visual frame. Most subjects selected the

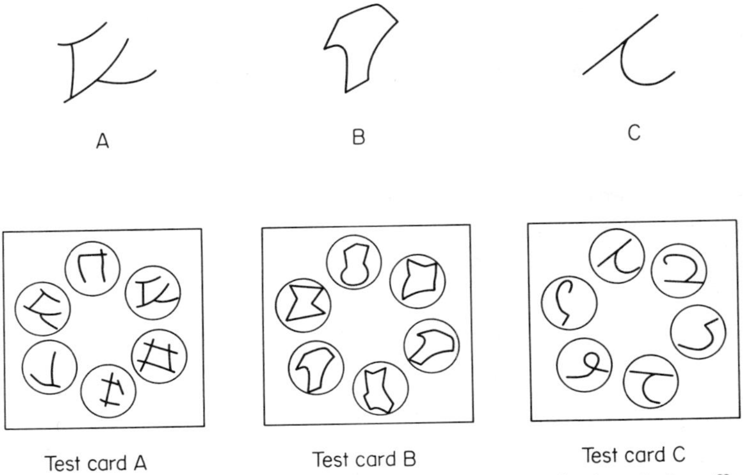

Figure 14.16 Nonsense shapes used by Rock (1956) to study the relative effects of egocentric and gravitational disorientation on shape recognition (Copyright, 1957, by the Board of Trustees of the University of Illinois. Reprinted with the permission of the University of Illinois Press)

gravitationally erect shape as the one resembling the training shape. When the head was left upright and a Dove prism was used to tilt the visual frame through 45° while the comparison shapes were inspected, the critical shape which was 'erect' with respect to the frame was more often chosen than the shape 'erect' on the retina. When the frame was eliminated, and the head tilted during the presentation of the comparison shapes, the critical shape which was 'erect' with respect to gravity was chosen more often than the shape 'erect' on the retina. Thus, for 45° of disorientation, the ability to recognize abstract shapes was found to depend more on a consistent orientation of the shape to gravity and to the visual frame than on a consistent orientation of the shape on the retina. The reason is probably that, in the case of shapes lacking an intrinsic top, people tend to look towards the gravitationally uppermost point or at the top of the visual frame for those features of the shape which they saw at the top of the training shape. This is a rational strategy, because things usually stay in the same orientation with respect to gravity and the visual frame when the head is tilted. In any case, a 45° retinal disorientation does not disrupt shape recognition appreciably, so that people have no difficulty in recognizing the critical shape which is tilted on the retina once they have found its 'top', or some other corresponding feature. In another condition of the experiment, the two critical shapes differed in orientation by 180° and the subject inverted his head during the presentation of the comparison shapes. The retinally 'erect' critical shape was now selected as frequently as the gravitationally 'erect' shape. In a further experiment (Rock and Heimer, 1957) the shapes were placed flat on the floor and the subjects looked down at them, thus eliminating the gravitational reference. The retinally upright critical shape was selected more often than the shape remaining in line with the subject's torso. It was suggested that, under these circumstances, the head axis assumes the function of the gravitational reference.

Corballis, Nagourney, Shetzer and Stefanatos (1978) asked subjects to decide whether abstract shapes, like the one shown in Figure 14.17, were in a form designated as standard or in a form designated as mirror-reversed. The shapes were presented one at a time in various orientations, while the subjects held their heads tilted at an angle of 60°. It was found that subjects mentally rotated the shape to near the gravitational upright, rather than to the retinal upright position, before making their decision about its orientation. Subjects behaved

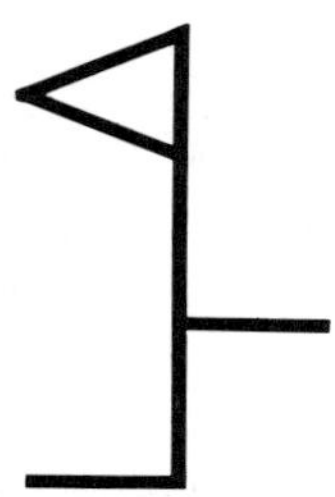

Figure 14.17 An example of the shapes used by Corballis, Nagourney, Shetzer and Stefanatos (1978)

this way even when the frame containing the shape was also tilted in line with the tilted head, although they could mentally rotate the shapes to the retinal upright under these circumstances when specifically instructed to do so. That subjects used the gravitational upright when not prompted was probably due to the fact that people often tilt their heads 60°, and, when they do so, objects which they are viewing, normally remain upright with respect to gravity. A 60° tilt of the retinal image does not seriously disrupt the ability to recognize gross differences between patterns. I suspect that the result of such an experiment would be different if the head were inverted. The results are certainly different when the visual task is more demanding, as the experiment of Corballis and Roldan which was described at the end of the last section showed.

In summary, it may be stated that the ability to recognize familiar mono-oriented objects is disrupted more by a disorientation of 180° with respect to the retina than by a similar disorientation with respect to gravity or the visual frame. The intrinsic top of such an object is easy to find, even when it is not in its expected position. A disorientation of 90° or less with respect to extrinsic axes has very little effect on the ability to recognize mono-oriented objects under ordinary conditions. Meaningless, asymmetrical shapes lack an intrinsic polar axis. If we are asked to recognize such an object after tilting our heads, we assume that the object remains upright to gravity, and look for matching features within a gravitationally-defined system of reference. We are therefore quicker at finding a shape which matches the training shape gravitationally than we are at finding a shape which matches it retinally. This is especially true for head tilts of less than 90°, because we often tilt our heads by this amount while observing a scene which remains gravitationally upright. When the head is upside-down, the disruptive effect due to the subject's set to use a gravitational reference in searching for corresponding features of an inverted shape is about equal to the disruption produced by viewing a shape inverted with respect to the retina.

References

Abadi, R. V. 1974. 'The aftereffect of early anomalous visual inputs on orientation selectivity.' *Perception*, **3**, 141–50.

Abadi, R. V. 1980. 'Pattern contrast thresholds in latent nystagmus.' *Acta Ophthal.*, **58**, 210–18.

Abadi, R. V., and King-Smith, P. E. 1979. 'Congenital nystagmus modifies orientational detection.' *Vision Res.*, **19**, 1409–11.

Abend, W. K. 1977. 'Functional organization of the superior vestibular nucleus of the squirrel monkey.' *Brain Res.*, **132**, 65–84.

Abrahams, V. C., and Rose, P. K. 1975a. 'Projections of extraocular, neck muscle, and retinal afferents to superior colliculus in the cat. Their connection to cells of origin of tectospinal track.' *J. Neurophysiol.*, **38**, 10–18.

Abrahams, V. C., and Rose, P. K. 1975b. 'Velocity and displacement characteristics of passive movements which initiate unit discharges in the superior colliculus.' *J. Physiol.*, **246**, 101 P.

Adams, H. F. 1912. 'The autokinetic sensations.' *Psychol. Rev. Monogr. Suppl.*, **14**, 1–45.

Adamük, E. 1870. 'Über die Innervation der Augenbewegungen.' *Zbl. med. Wiss.*, **8**, 65–7.

Adler, F. H., and Fliegelman, M. 1934. 'Influence of fixation on the visual acuity.' *A. M A. Arch. Opthal.*, **12**, 475–83.

Albrecht, D. G., De Valois, R. L., and Thorell, L. G. 1980. 'Visual cortical neurones: are bars or gratings the optimal stimulus.' *Science*, **207**, 88–90.

Albus, K. 1975a. 'A quantitative study of the projection area of the central and paracentral visual field in area 17 of the cat. I. The precision of the topography.' *Exp. Brain Res.*, **24**, 159–79.

Albus, K. 1975b. 'A quantitative study of the projection area of the central and the paracentral visual field in area 17 of the cat. II. The spatial organization of the orientation domain. *Exp. Brain Res.*, **24**, 181–202.

Aldridge, J. W., Anderson, R. J., and Murphy, J. T. 1980. 'The role of basal ganglia in controlling a movement initiated by a visually presented cue.' *Brain Res.*, **192**, 3–16.

Allen, M. J. 1954. 'The dependence of cyclophoria on convergence, elevation and the system of axes.' *Amer. J. Optom.*, **31**, 297–307.

Allen, M. J., and Carter, J. H. 1967. 'The torsional component of the near reflex.' *Amer. J. Optom.*, **44**, 343–9.

Allum, J. H. J, and Graf, W. 1977. 'Time constants of vestibular nuclei neurons in the goldfish: a model with ocular proprioception.' *Biol. Cybern.*, **28**, 95–9.

Allum, J. H. J., Tole, J. R., and Weiss, A. D. 1975. 'MITNYS-II—A digital program for on-line analysis of nystagmus.' *IEEE Trans. biomed. Engin.*, **BME–22**, 196–202.

Amblard, B., and Cremieux, J. 1976. 'Rôle de l'information visuelle du mouvement dans le maintien de l'équilibre.' *Aggressologie*, **17c**, 25–37.

Anderson, J. H., and Precht, W. 1979. Otolith responses of extraocular muscles during sinusoidal roll rotations.' *Brain Res.*, **160**, 150–4.

Anderson, K. V., and Symmes, D. 1969. The superior colliculus and higher visual functions in monkeys.' *Brain Res.*, **13**, 37–52.

Anderson, M. E., Yoshida, M., and Wilson, V. J. 1971. 'Influence of the superior colliculus on cat neck motoneurons.' *J. Neurophysiol.*, **34**, 898–907.

Andrews, D. P. 1964. 'Error-correcting perceptual mechanisms.' *Quart. J. exp. Psychol.*, **16**, 104–15.

Andrews, D. P. 1967a. 'Perception of contour orientation in the central fovea. Part I: Short lines.' *Vision Res.*, **7**, 975–97.

Andrews, D. P. 1967b. 'Perception of contour orientation in the central fovea. Part II: Spatial integration.' *Vision Res.*, **7**, 999–1013.

Andrews, D. P., Butcher, A. K., and Buckley, B. R. 1973. 'Acuties for spatial arrangement in line figures: human and ideal observers compared.' *Vision Res.*, **13**, 599–620.

Angel, R. W., and Garland, H. 1972. 'Transfer of information from manual to oculomotor control system.' *J. exp. Psychol.*, **96**, 92–6.

Angel, R. W., Hollander, M., and Wesley M. 1973. 'Hand–eye coordination: The role of motor memory.' *Percept. Psychophys.*, **14**, 506–10.

Annis, R. C., and Frost, B. 1973. 'Human visual ecology and orientation anisotropies in acuity.' *Science*, **182**, 729–31.

Anstis, S. M. 1974. 'Chart demonstrating variations in acuity with retinal position.' *Vision Res.*, **14**, 589–92.

Anstis, S. M., Mayhew, J. W., and Morley, T. 1969. 'The perception of where a face or television "portrait" is looking.' *Amer. J. Psychol.*, **82**, 474–89.

Antonovsky, H. F., and Ghent, L. 1964. 'Cross-cultural consistency of children's preferences for the orientation of figures.' *Amer. J. Psychol.*, **77**, 295–7.

Appelle, S. 1972. 'Perception and discrimination as a function of stimulus orientation: The "oblique effect" in man and animals.' *Psychol. Bull.*, **78**, 266–78.

Appelle, S. 1973. 'Orientation of test-object affects visibility throughout visual field.' *Amer. J. Optom.*, **50**, 940–3.

Applle, S. 1976. 'Figure embeddedness depends on contour orientation.' *Percept. Psychophys.*, **19**, 109–12.

Arndts, F. 1924. 'Zur Frage nach der Lagewahrnehmung dienenden Sinnesfunktionen.' *Pflüg. Arch. ges. Physiol.*, **204**, 539–40.

Arnheim, R. 1954. *Art and Visual Perception*. London: Faber and Faber.

Arnulf, A., and Dupuy, O. 1960. 'La transmission des contrastes par le système optique de l'oeil et les seuils des contrastes rétiniens.' *Compt. Rendu Acad. Sci., Paris*, **250**, 2727–59.

Aronson, E., and Rosenbloom. S. 1971. 'Space perception in early infancy: Perception within a common auditory–visual space.' *Science*, **172**, 1161–3.

Asch, S. E., and Witkin, H. A. 1948. 'Studies in space orientation: II. Perception of the upright with displaced visual fields and with body tilted.' *J. exp. Psychol.*, **38**, 455–77.

Aschan, G. 1956. 'Nystagmography. Recording of nystagmus in clinical neuro-otological examinations.' *Acta oto-laryng., Suppl.*, **129.**

Aschoff, J. C., and Cohen, B. 1971. 'Changes in saccadic eye movements produced by cerebellar cortical lesions.' *Exp. Neurol.*, **32**, 123–33.

Aslin, R. N., and Salapatek, P. 1975. 'Saccadic localization of visual targets by the very young human infant.' *Percept. Psychophys.*, **17**, 293–302.

Asso, D., and Wyke, M. 1971. 'Discrimination of spatially confusable letters by young children.' *J. exp. child Psychol.*, **11**, 11–20.

Astruc, J. 1971. 'Corticofugal projection of area 8 (frontal eye field) in Malaca Maulatta.' *Brain Res.*, **33**, 241–56.

Atkin, A. 1969. 'Shifting fixation to another pursuit target: selective and anticipatory control of ocular pursuit initiation.' *Exp. Neurol.*, **23**, 157–73.

Atkinson, J. 1972. 'The effect of size, retinal locus and orientation on the visibility of a single afterimage.' *Percept. Psychophys.*, **12**, 213–7.

Atkinson, J. 1973. 'Properties of human visual orientation detectors: A new approach using patterned afterimages.' *J. exp. Psychol.*, **98**, 55–63.

Atkinson, J., Braddick, O., and French, J. 1980. 'Infant astigmatism: its disappearance with age.' *Vision Res.*, **20**, 891–5.

Atkinson, J., and Campbell, F. W. 1974. 'The effects of phase on the perception of compound gratings.' *Vision Res.*, **14**, 159–62.

Attneave, F., and Olson, R. K. 1967. 'Discriminability of stimuli varying in physical and retinal orientation.' *J. exp. Psychol.*, **74**, 149–57.

Attneave, F., and Reid, K. W. 1968. 'Voluntary control of frame of reference and slope equivalence under head rotation.' *J. exp. Psychol.*, **78**, 153–9.

Aubert, H. 1961. 'Eine scheinbare bedeutende Drehung von Objekten bei Neigung des Kopfes nach rechts oder links.' *Virchows Arch.*, **20**, 381–93.

Aubert, H. 1886. 'Die Bewegungsempfindungen., *Pflüg. Arch. ges. Physiol.*, **39**, 347–70'

Aubert, H. 1887. 'Die Bewegungsempfindungen.' *Pflüg. Arch. ges. Physiol.*, **40**, 459–80.

Auerbach, C., and Sperling, P. 1974. 'A common auditory–visual space: Evidence for its reality.' *Percept. Psychophys.*, **16**, 129–35.

Aulhorn, E., and Harms, H. 1972. 'Visual perimetry.' In D. Jameson and L. M. Hurvich (Eds.), *Handbook of Sensory Physiology*, Vol. VII/4. Berlin: Springer-Verlag, pp. 102–45.

Austin, M., Singer, G., and Day, R. H. 1969. 'Visual orientation illusion following judgments with a tilted visual field.' *Nature*, **221**, 583–4.

Azzena, G. B. 1969. 'Role of the spinal cord in compensating the effects of hemi-labyrinthectomy.' *Arch. Ital. Biol.*, **107**, 43–53.

Bach-y-Rita, P. 1971. 'Neurophysiology of eye movements.' In P. Bach-y-Rita, C. C. Collins and J. E. Hyde (Eds.), *The Control of Eye Movements*. New York: Academic Press, pp. 7–46.

Bach-y-Rita, P. 1972. 'Extraocular muscle inhibitory reflex during active contraction.' *Arch. Ital. Biol.*, **110**, 1–15.

Bach-y-Rita, P. 1975. 'Structural functional correlations in eye muscle fibres. Eye muscle proprioception.' In G. Lennerstrand and P. Bach-y-Rita (Eds.). *Basic Mechanisms of Ocular Motility and Their Clinical Implications*. New York: Pergamon, pp. 91–108.

Bach-y-Rita, P., Collins, C. C., and Hyde, J. E. 1971. *The Control of Eye Movements*. New York: Academic Press.

Bahill, A. T., Adler, B., and Stark, L. 1975 'Most naturally occurring human saccades have magnitudes of 15 degrees or less.' *Invest. Ophthal.*, **14**, 468–9.

Bahill, A. T., Ciuffreda, K. J., Kenyon, R., and Stark, L. 1976. 'Dynamic and static violations of Hering's Law of Equal Innervation.' *Amer. J. Optom. & physiol. Optics*, **53**, 786–96.

Bahill, A. T., Clark, M. R., and Stark, L. 1975a. 'The main sequence, a tool for studying human eye movements.' *Math. Biosci.*, **24**, 191–204.

Bahill, A. T., Clark, M. R., and Stark, L. 1975b. 'Dynamic overshoot in saccadic eye movements is caused by neurological control signal reversals.' *Exp. Neurol.*, **48**, 107–22.

Bahill, A. T., Hsu, F. K., and Stark, L. 1978. 'Glissadic overshoots are due to pulse width errors.' *Arch. Neurol.*, **35**, 138–42.

Bahill, A. T., and Stark. L. 1975. 'Overlapping saccades and glissades are produced by fatigue in the saccadic eye movement system.' *Exp. Neurol.*, **48**, 95–106.

Bahill, A. T., and Stark, L. 1977. 'Oblique saccadic movements: Independence of horizontal and vertical channels.' *A. M. A. Arch. Ophthal.*, **95**, 1258–61.

Baily, J. S. 1972a. 'Adaptation to prisms: Do proprioceptive changes mediate adapted behaviour with ballistic arm movements?' *Quart. J. exp. Psychol.*, **24**, 8–20.

Baily, J. S. 1972b. 'Arm–body adaptation with passive arm movements.' *Percept. Psychophys.*, **12**, 39–44.

Baizer, J. S., Robinson, D. L., and Dow, B. M. 1977. 'Visual responses of area 18 neurones in awake behaving monkey.' *J. Neurophysiol.*, **40**, 1024–37.

Baker, J., Gibson, A., Glickstein, M., and Stein, J. 1976. 'Visual cells in the pontine nuclei of the cat.' *J. Physiol.*, **255**, 415–33.

Baker, K. D., and Sullivan, G. D. 1981. 'Multiple band-pass filters in image-processing.' *IEEE*, in press.

Baker, R. 1977. 'The nucleus prepositus hypoglossi.' In B. A. Brooks and F. J. Bajandas (Eds.), *Eye Movements*. New York: Plenum, pp. 145–78.

Baker, R., and Berthoz, A. 1974. 'Organization of vestibular nystagmus in oblique oculomotor system.' *J. Neurophysiol.*, **37**, 195–217.

Baker, R., and Berthoz, A. 1975. 'Is the prepositus hypoglossi nucleus the source of another vestibulo-ocular pathway?' *Brain Res.*, **86**, 121–7.

Baker, R., and Berthoz, A. 1977. *Control of Gaze by Brain Stem Neurons*. New York: Elsevier.

Baker, R., Berthoz, A., and Delgado-Garcia, J. 1977. 'Monosynaptic excitation of trochlear motoneurons following electrical stimulation of the prepositus hypoglossi nucleus.' *Brain Res.*, **121**, 157–61.

Baker, R., Gresty, M., and Berthoz, A. 1976. 'Neuronal activity in the prepositus hypoglossi nucleus correlated with vertical and horizontal eye movement in the cat.' *Brain Res.*, **101**, 366–71.

Baker, R., and Highstein, S. M. 1975. 'Physiologic identification of interneurones and motoneurones is the abducens nucleus.' *Brain Res.*, **91**, 292–8.

Baker, R., Precht, W., and Berthoz, A. 1973. 'Synaptic connections to trochlear motoneurones determined by individual vestibular nerve branch stimulation in the cat.' *Brain Res.*, **64**, 402–6.

Baker, R., Precht, W., and Llinás, R. 1972a. 'Mossy and climbing fiber projections of extraocular muscle afferents to the cerebellum.' *Brain Res.*, **38**, 440–5.

Baker, R., Precht, W., and Llinás, R. 1972b. 'Cerebellar modulatory action on the vestibulo-trocheal pathway in the cat.' *Exp. Brain Res.*, **15**, 364–85.

Ball, W. A., and Tronick, E. 1971. Infant responses to impending collision: Optical and real.' *Science*, **171**, 818–20.

Balliet, R., and Nakayama, K. 1978a. 'Training of voluntary torsion.' *Invest. Ophthal.*, **17**, 303–14.

Balliet, R., and Nakayama, K. 1978b. 'Egocentric orientation is influenced by trained voluntary cyclorotary movements.' *Nature, New Biol.*, **275**, 214–6.

Baloh, R. W., Konrad, H. R., and Honrubia, V. 1975. 'Vestibulo-ocular function in patients with cerebellar atrophy.' *Neurology*, **25**, 160–8.

Baloh, R. W., Yee, R. D., and Honrubia, V. 1980. 'Optokinetic asymmetry in patients with maldeveloped foveas.' *Brain Res.*, **186**, 211–16.

Banks, M. S., and Stolarz, S. J. 1975. 'The effect of head tilt on meridional differences in acuity: Implications for orientation constancy.' *Percept. Psychophys.*, **17**, 17–22.

Bárány, R. 1908. 'Die Untersuchungen der optischen und vestibulären reflektorischen Augenbewegungen in einem Falle von einseitigen Blicklähmung.' *Mschr. Ohrenheilk*, **42**, 109–13.

Barbeito, R., and Ono, H. 1979. 'Four methods of locating the egocentre: A comparison of their predictive validities and reliabilities.' *Behav. res. methods Instrument.*, **11**, 31–6.

Barber, H. O., and Wright, G. 1973. 'Positional nystagmus in normals.' *Adv. Oto-Rhino-Laryng.*, **19**, 276–85.

Barlow, H. B. 1952. 'Eye movements during fixation. '*J. Physiol.*, **116**, 290–306.

Barlow, H. B. 1961. 'Possible principles underlying the transformations of sensory messages.' In W. A. Rosenblith (Ed.), *Sensory Communication*. Cambridge, Mass.: MIT Press, pp. 217–34.

Barlow, H. B., Blakemore, C., and Pettigrew, J. D. 1976, 'The neural mechanism of binocular depth discrimination.' *J. Physiol.*, **193**, 327–42.

Barlow, H. B., Fitzhugh, R., and Kuffler, S. W. 1957. 'Change of organization in the receptive fields of the cat's retina during dark adaptation.' *J. Physiol.*, **137**, 338–54.

Barlow, J. S., and Cigánek, L. 1969. 'Lambda responses in relation to visual evoked responses in man.' *E. E. G. clin. Neurophysiol.*, **26**, 183–92.

Barmack, N. H. 1970. 'Modification of eye movements by instantaneous changes in the velocity of visual targets.' *Vision Res.*, **10**, 1431–41.

Barmack, N. H., Bell, C. C., and Rence, B. G. 1971. 'Tension and rate of tension development during isometric responses of extraocular mucle.' *J. Neurophysiol.*, **34**, 1072–9.

Barnes, G. R. 1979. 'Vestibulo-ocular function during coordinated head and eye movements to acquire visual targets.' *J. Physiol.*, **287**, 127–47.

Barnes, G. R. 1980. 'Vestibular control of oculomotor and postural mechanisms. *Clin. Phys. Physiol. Meas.*, **1**, 3–40.

Barnes, G. R., Benson, A. J., and Prior, A. R. J. 1978. 'Visual vestibular interaction in the control of eye movement.' *Aviat. space environ. Med.*, **49**, 557–64.

Barnes, G. R., and Forbat, L. N. 1979. 'Cervical and vestibular afferent control of oculomotor responses in man.' *Acta oto-laryng.*, **88**, 79–87.

Barnes, G. R., and Gresty, M. A. 1973. 'Characteristics of eye movements to targets of short duration.' *Aerosp. Med.*, **44**, 1236–40.

Barr, C. C., Schultheis, L. W., and Robinson, D. A. 1976. 'Voluntary non-visual control of the human vestibulo-ocular reflex.' *Acta oto-laryng.*, **81**, 365–75.

Barrios, R. R., Recalde, E. M., and Mendilaharzu, C. 1959. 'Surgical rotation of the eyeball.' *Brit. J. Ophthal.*, **43**, 584–90.

Barroso, F., and Braine, L. G. 1974. 'Mirror-image errors without mirror-image stimuli.' *J. exp. child Psychol.*, **18**, 213–25.

Barten, S., Birns, B., and Ronch, J. 1971. 'Individual differences in the visual pursuit behavior of neonates.' *Child Devel.*, **42**, 313–9.

Bartz, A. E. 1966. 'Eye and head movements in peripheral vision: Nature of compensatory eye movements.' *Science*, **152**, 1644–5.

Batini, C., and Horcholle-Bossavit, G. 1977. 'Interaction entre activation visuelle et activation proprioceptive au niveau des neurones du colliculus superieur.' *Compt. Rendu Acad. Sci., Paris*, **285**, 1491–3.

Bauer, J. A., Owens, D. A., Thomas, J., and Held, R. 1979. 'Monkeys show an oblique effect.' *Perception*, **8**, 247–53.

Bauer, J. A., Woods, G., and Held, R. 1969. 'A device for recording of positioning responses in two dimensions.' *Behav. res. Methods & Instrum.*, **1**, 157–9.

Bauer, R., Dow, B. M., and Vautin, R. G. 1980. 'Laminar distribution of preferred orientations in foveal striate cortex of the monkey.' *Exp. Brain Res.*, **41**, 54–9.

Bauermeister, M. 1964. 'Effect of body tilt on apparent verticality, apparent body position, and their relation.' *J. exp. Psychol.*, **67**, 142–7.

Beaubaton, D., Trouch, E., and Amato, G. 1980. 'Dentate and parallel control of a goal-directed movement in monkeys.' In G. E. Stelmach and J. Requin (Eds.), *Tutorials in Motor Behaviour.* New York: North-Holland Publishing Co., pp. 315–27.

Beck, J. 1965. 'Apparent spatial location and the perception of lightness.' *J. exp. Psychol.*, **69**, 170–9.

Becker, W. 1972. 'The control of eye movements in the saccadic system.' In J. Dichgans and E. Bizzi (Eds.), *Cerebral Control of Eye Movements.* New York: Karger, pp. 308–16.

Becker, W. 1976. 'Do correction saccades depend exclusively on retinal feedback? A note on the possible role of non-retinal feedback. *Vision Res.*, **16**, 425–7.

Becker, W., and Fuchs, A. F. 1969. 'Further properties of the human saccadic system. Eye movements and correction saccades with and without visual fixation points.' *Vision Res.*, **9**, 1247–59.

Becker, W., and Jürgens, R. 1975. 'Saccadic reactions to double step stimuli: evidence for model feedback and continuous information uptake.' In G. Lennerstrand and P. Bach-y-Rita (Eds.), *Basic Mechanisms of Ocular Motility and their Clinical Implications.*

Oxford: Pergamon, pp. 519–24.
Becker, W., and Klein, H. M. 1973. 'Accuracy of saccadic eye movements and maintenance of eccentric eye positions in the dark.' *Vision Res.*, **13**, 1021–34.
Beggs, W. D. A., and Howarth, C. I. 1972. 'The movement of the hand towards a target.' *Quart. J. exp. Psychol.*, **24**, 448–53.
Beh, H. C., and Wenderoth, P. M. 1972. 'The effect of variation of frame shape on the angular function of the rod-and-frame illusion.' *Percept. Psychophys.*, **11**, 35–8.
Beh, H., Wenderoth, P. M., and Purcell, A. T. 1971. 'The angular function of a rod-and-frame illusion.' *Percept. Psychophys.*, **9**, 353–5.
Békésy, G. von. 1953. 'Shearing microphonics produced by vibrations near the inner and outer hair cells.' *J. acoust. Soc. Amer.*, **25**, 786–90.
Bell, C., Sierra, G., Buendia, N., and Segundo, J. P. 1964. 'Sensory properties of neurones in the mesencephalic reticular formation.' *J. Neurophysiol.*, **27**, 961–87.
Bender, M. B. 1980. 'Brain control of conjugate horizontal and vertical eye movements: a survey of the structural and functional correlates.' *Brain*, **103**, 23–70.
Bender, M. B., and Krieger, H. P. 1951. 'Visual function in perimetrically blind fields.' *Arch. Neurol. Psychiat.*, **65**, 72–8.
Bender, M. B., and Shanzer, S. 1964. 'Oculomotor pathways defined by electrical stimulation and lesions in the brainstem of monkey.' In M. Bender (Ed.), *The Oculomotor System*. New York: Harper and Row, pp. 81–140.
Bender, M. B., and Teuber, H. L. 1946. 'Phenomena of fluctuation, extinction and completion in visual perception.' *Arch. Neurol. Psychiat.*, **55**, 627–58.
Benevento, L. A., Creutzfeldt, O. D., and Kuhn, H. 1972. 'Significance of intracortical inhibition.' *Nature, New Biol.*, **238**, 124–6.
Benevento, L. A., and Fallon, J. H. 1975. 'The ascending projections of the superior colliculus in the rhesus monkey.' *J. comp. Neurol.*, **160**, 339–62.
Bengi, H., and Thomas, J. G. 1968. 'Fixation tremor in relation to eyeball-muscle mechanics.' *Nature*, **217**, 773–4.
Bennett, M. J., Smith, E. L., Harwerth, R. S., and Crawford, M. L. J. 1980. 'Ocular dominance, eye alignment and visual acuity in kittens reared with an optically induced squint.' *Brain Res.*, **193**, 33–45.
Benson, A. J. 1968. 'Lateral eye movements produced by a rotating linear acceleration vector.' *J. Physiol.*, **197**, 85–6 P.
Benson, A. J. 1970. 'Interactions between semicircular canals and gravireceptors.' In D. E. Busby (Ed.), *Recent Advances in Aerospace Medicine*. Dordrecht, Holland: Reidel, pp. 249–61.
Benson, A. J. 1974. 'Modification of the response to angular accelerations by linear accelerations.' In H. H. Kornhuber (Ed.), *Handbook of Sensory Physiology*, Vol. VI/2. New York: Springer, pp. 281–320.
Benson, A. J., and Barnes, G. R. 1973. 'Responses to rotating linear acceleration vectors considered in relation to a model of the otolith organs.' In *5th Symposium on the Role of the Vestibular Organs in Space Exploration*. Washington, D. C.: NASA SP-314, 221–236.
Benson, A. J., and Barnes, G. R. 1978. 'Vision during angular oscillation; The dynamic interaction of visual and vestibular mechanisms.' *Aviat. space environ. Med.*, **49**, 340–5.
Benson, A. J., and Bodin, M. A. 1966a. 'Interaction of linear and angular acceleration on vestibular receptors in man.' *Aerosp. Med.*, **37**, 144–54.
Benson, A. J., and Bodin, M. A. 1966b. 'Comparison of the effect of the direction of the gravitational acceleration on post-rotational response in yaw, pitch and roll.' *Aerosp. Med.*, **37**, 889–97.
Benson, A. J., and Guedry, F. E. 1971. 'Comparison of tracking-task performance and nystagmus during sinusoidal oscillation in yaw and pitch.' *Aerosp. Med.*, **42**, 593–601.
Benson, A. J., Guedry, F. E., and Melvill Jones, G. 1970. 'Responses of semicircular canal dependent units in vestibular nuclei to rotation of a linear acceleration vector without angular acceleration.' *J. Physiol.*, **210**, 475–94.

Bergen, J. R., Wilson, H. R., and Cowan, J. D. 1979. 'Further evidence for four mechanisms mediating vision at threshold: Sensitivities to complex gratings and aperiodic stimuli.' *J. opt. Soc. Amer.*, **69**, 1580–7.

Berkley, M. 1978. 'Vision: Geniculocortical system.' In R. B. Masterton (Ed.), *Handbook of Behavioral Neurobiology*, Vol. I. *Sensory Integration*. New York: Plenum, pp. 165–208.

Berkley, M. A., Kitterle, F., and Watkins, D. W. 1975. 'Grating visibility as a function of orientation and retinal eccentricity.' *Vision Res.*, **15**, 239–44.

Berman, N., and Cynader, M. 1976. 'Early versus late visual cortex lesions: effects on receptive fields in cat superior colliculus.' *Exp. Brain Res.*, **25**, 131–8.

Berman, N., Payne, B. R., Garcia-Kennedy, R., and Murphy, E. H. 1981. 'Orientation anistropy in cat visual cortex.' *Invest. Ophthal., ARVO Abstr.*, **20**, 147.

Berthoz, A., Baker, R., and Goldberg, A. 1974. 'Neuronal activity underlying vestibular nystagmus in the oblique oculomotor system of the cat.' *Brain Res.*, **71**, 233–8.

Berthoz, A., Jeannerod, M., Vital-Durand, F., and Oliveras, J. L. 1975. 'Development of vestibulo-ocular responses in visually deprived kittens.' *Exp. Brain Res.*, **23**, 425–42.

Berthoz, A., Pavard, B., and Young, L. R. 1975. 'Perception of linear horizontal self-motion induced by peripheral vision (linear vection): basic characteristics and visual–vestibular interactions.' *Exp. Brain Res.*, **23**, 471–89.

Beverley, K. I., and Regan, D. 1973. 'Evidence for the existence of neural mechanisms selectively sensitive to the direction of movement in space.' *J. Physiol.*, **235**, 17–29.

Beverley, K. I., and Regan, D. 1979. 'Visual perception of changing size: the effect of object size.' *Vision Res.*, **19**, 1093–104.

Beyerstein, B. L., and Freeman, R. D. 1976. 'Increment sensitivity in humans with abnormal visual experience.' *J. Physiol.*, **260**, 497–514.

Beyerstein, B. L., and Freeman, R. D. 1977. 'Lateral spatial interaction in humans with abnormal visual experience.' *Vision Res.*, **17**, 1029–36.

Bilodeau, E. A. 1966. *Acquisition of Skill*. New York: Academic Press.

Bischof, N., and Kramer, E. 1968. 'Untersuchungen and Uberlegungen zur Richtungswahrnehmung bei willkurlichen sakkadischen Augenbewegungen.' *Psychol. Forsch.*, **32**, 185–218.

Bischof, N., and Scheerer, E. 1970. 'Systemanalyse der optisch-vestibulären Interaktion bei der Wahrnehmung der Vertikalen.' *Psychol. Forsch.*, **34**, 99–181.

Bishop, P. O., Coombs, J. S., and Henry, G. H. 1971a. 'Responses to visual contours: spatio-temporal aspects of excitation in the receptive fields of simple striate neurones.' *J. Physiol.*, **219**, 625–57.

Bishop, P. O., Coombs, J. S., and Henry, G. H. 1971b. 'Interaction effects of visual contours in the discharge frequency of simple striate neurones.' *J. Physiol.*, **219**, 659–87.

Bishop, P. O., Dreher, B., and Henry, G. H. 1973. 'Simple striate cells: A comparison of response to stationary and moving stimuli.' *J. Physiol.*, **227**, 15–17 P.

Bisti, S., Clement, R., Maffei, L., and Mecacci, L. 1977. 'Spatial frequency and orientation tuning curves of visual neurones in the cat: Effects of mean luminance.' *Exp. Brain Res.*, **17**, 335–45.

Bisti, S., and Maffei, L. 1979. 'Binocular interaction in bilateral strabismic kittens.' *Brain Res.*, **170**, 359–61.

Bisti, S., Maffei, L., and Piccolino, M. 1974. 'Visuovestibular interactions in the cat superior colliculus.' *J. Neurophysiol.*, **37**, 146–55.

Bizzi, E., Kalil, R. E., and Morasso, P. 1972. 'Two modes of active eye–head coordination.' *Brain Res.*, **40**, 45–8.

Bizzi, E., Kalil. R. E., and Tagliasco, V. 1971. 'Eye–head coordination in monkeys: Evidence for centrally patterned organization.' *Science*, **173**, 452–4.

Bizzi, E., and Schiller, P. H. 1970. 'Single unit activity in the frontal eye fields of unanaesthetized monkeys during eye and head movement.' *Exp. Brain. Res.*, **10**, 151–8.

Blackwell, H. R. 1952. 'Studies of psychophysical methods for measuring thresholds.' *J. opt. Soc. Amer.*, **42**, 606–16.

Blair, C. J., Goldberg, M. F., and Noorden, G. K. von. 1967. 'Voluntary nystagmus. Electro-oculographic findings in four cases.' *A. M. A. Arch. Ophthal.*, **77**, 349–54.

Blair, S., and Gavin, M. 1979a. 'Response of the vestibulo-ocular reflex to differing programs of acceleration.' *Invest. Ophthal.*, **18**, 1086–90.

Blair, S., and Gayin, M. 1979b. 'Modification of the macaque's vestibulo-ocular reflex after ablation of the cerebellar vermis.' *Acta oto-laryng.*, **88**, 235–43.

Blakemore, C. 1970. 'The representation of three-dimensional visual space in the cat's striate cortex.' *J. Physiol.*, **209**, 155–78.

Blakemore, C. 1974. 'Development of the mammalian visual system.' *Brit. med. Bull.*, **30**, 152–7.

Blakemore, C. 1976. 'Conditions required for the maintenance of binocularity in the kitten's visual cortex.' *J. Physiol.*, **261**, 423–6.

Blakemore, C., and Campbell, F. W. 1969. 'On the existence of nurones in the human visual system selectively sensitive to the orientation and size of retinal images.' *J. Physiol.*, **203**, 237–60.

Blakemore, C., and Cooper, G. F. 1970. 'Development of the brain depends on the visual environment.' *Nature*, **228**, 477–8.

Blakemore, C., and Mitchell, D. E. 1973. 'Environmental modification of the visual cortex and the neural basis of learning and memory.' *Nature, New Biol.*, **24**, 467–8.

Blakemore, C., Muncey, J. P., and Ridley, R. M. 1973. 'Stimulus specificity in the human visual system.' *Vision Res.*, **13**, 1915–31.

Blakemore, C., and Nachmias, J. 1971. 'The orientation specificity of two visual aftereffects.' *J. Physiol.*, **213**, 157–74.

Blakemore, C., and Sutton, P. 1969. 'Size adaptation: a new aftereffect.' *Science*, **166**, 245–7.

Blakemore, C., and Tobin, E. A. 1972. 'Lateral inhibition between orientation detectors in the cat's visual cortex.' *Exp. Brain Res.*, **15**, 439–40.

Blakemore, C., and van Sluyters, R. C. 1975. 'Innate and environmental factors in the development of the kitten's visual cortex.' *J. Physiol.*, **248**, 663–716.

Blakemore, C., van Sluyters, R. C., Peck, C. K., and Hein, A. 1975. 'Development of cat visual cortex following rotation of one eye.' *Nature, New Biol.*, **257**, 584–7.

Blanks, R. H. I., Curthoys, I. S., and Markham, C. H. 1975. 'Planar relationships of semicircular canals in man.' *Acta oto-laryng.*, **80**, 185–96.

Blanks, R. H. I., Volkind, R., Precht, W., and Baker, R. 1977. 'Responses of cat prepositus hypoglossi neurons to horizontal angular acceleration.' *Neurosci.*, **2**, 391–403.

Blasdel, G. G., Mitchell, D. E., Muir, D. W., and Pettigrew, J. D. 1977. 'A physiological and behavioural study in cats of the effect of early visual experience with contours of a single orientation.' *J. Physiol.*, **265**, 615–36.

Blodi, F. C., and Van Allen, M. W. 1957. 'Electromyography of the extraocular muscles in fusional movements.' *Amer. J. Ophthal.*, **44**, 136–44.

Bobrow, D. G. 1975. 'Dimensions of representation.' In D. G. Bobrow and A. M. Collins (Eds.), *Representation and Understanding: Studies in Cognitive Science*. New York: Academic Press.

Boeder, P. 1961. 'The co-operation of extraocular muscles.' *Amer. J. Ophthal.*, **51**, 469–81.

Boeder, P. 1962. 'Co-operative action of extraocular muscles.' *Brit. J. Ophthal.*, **46**, 397–403.

Boltz, R. L., Harwerth, R. S., and Smith, E. L. 1979. 'Orientation anisotropy of visual stimuli in rhesus monkey: A behavioral study.' *Science*, **205**, 511–13.

Boothe, R. G., Williams, R. A., Kiorpes, L., and Teller, D. J. 1980. 'Development of contrast sensitivity in infant Macaca nemestrina monkeys.' *Science*, **208**, 1290–1.

Bornstein, M. H., Gross, C. G., and Wolf, J. Z. 1978. 'Perceptual similarity of mirror images in infancy.' *Cognition*, **6**, 89–116.

Borresen, C. R. 1973. 'Autokinetic movement as a function of the implied movement of target shape.' *J. exp. Psychol.*, **97**, 89–92.

Bosco, C., and Komi, P. V. 1979. 'Potentiation of the mechanical behaviour of the human sceletal muscle through prestretching.' *Acta physiol.*, **106**, 467–72.

Bouma, H., and Andriessen, J. J. 1968. 'Perceived orientation of isolated line segments.' *Vision Res.*, **8**, 493–507.

Bouma, H., and Andriessen, J. J. 1970. 'Induced changes in the perceived orientation of line segments.' *Vision Res.*, **10**, 333–49.

Bouman, M. A. 1953. 'Visual thresholds for line shaped targets.' *J. opt. Soc. Amer.*, **43**, 209–11.

Bower, T. G. R. 1974. *Development in Infancy*, San Francisco: Freeman.

Bower, T. G. R., Broughton, J. M., and Moore, M. K. 1970. Demonstrations of intention in the reaching behavior of neonate humans.' *Nature*, **228**, 679–81.

Bower, T. G. R., and Wishart, J. G. 1972. 'The effects of motor skill on object permanence.' *Cognition*, **1**, 165–72.

Boyce, P. R. 1967. 'Monocular fixation in human eye movement.' *Proc. R. Soc. B*, **167**, 293–315.

Boynton, R. M., and Onley, J. W. 1962. 'A critique of the special status assigned by Brindley to "psychophysical linking hypotheses" of "class A".' *Vision Res.*, **2**, 383–90.

Braddick, O. J. 1969. 'Eye movements and visual direction.' Report of work carried out at Brown University.

Braine, L. G. 1965. 'Disorientations of forms: An examination of Rock's theory.' *Psychonom. Sci.*, **3**, 541–2.

Braine, L. G. 1973. 'Perceiving and copying the orientation of geometric shapes.' *J. Res. Devel. Education*, **6**, 44–55.

Braine, L. G. 1978. 'A new slant on orientation perception.' *Amer. Psychologist*, **3**, 10–22.

Braine, L. G., and Knox, C. 1975. 'Children's orientation judgments: Retinally or environmentally determined?' *Percept. Psychophys.*, **17**, 473–9.

Brandt, T., Büchele, W., Arnold, F., Fallert, M., and Wassermeyer, B. 1977. 'Arthrokinetic nystagmus and ego-motion sensation.' *Exp. Brain Res.*, **30**, 331–8.

Brandt, T., Dichgans, J., and Büchele, W. 1974. 'Motion habituation: Inverted self-motion perception and optokinetic after-nystagmus.' *Exp. Brain Res.*, **21**, 337–52.

Brandt, T., Dichgans, J., and Koenig, E. 1973. 'Differential effects of central versus peripheral vision on egocentric and exocentric motion perception.' *Exp. Brain Res.*, **16**, 476–91.

Brandt, T., Wenzel, D., and Dichgans, J. 1976. 'Die Enwicklung der visuellen Stabilisation des aufrechten Standes beim Kind: Ein Reifezeichen in der Kinderneurologie.' *Arch. Psychiat. Nervenkr.*, **223**, 1–13.

Brandt, T., Wist, E. R., and Dichgans, J. 1975. 'Foreground and background in dynamic spatial orientation.' *Percept. Psychophys.*, **17**, 497–503.

Brazelton, T. B., Scholl, M. L., and Robey, J. S. 1966. 'Visual response in the newborn.' *Pediatrics*, **37**, 284–90.

Brecher, G. A. 1934. 'Die optokinetische Auslösung von Augenrollung und rotatorischem Nystagmus.' *Pflüg. Arch. ges Physiol.*, **234**, 13–28.

Brecher, G. A., Brecher, M. H., Kommerell, G., Sauter, F. A., and Sellerbeck, J. 1972. 'Relation of optical and labyrinthean orientation.' *Optica acta*, **19**, 467–71.

Breinin, G. M. 1955. 'The nature of vergence revealed by electromyography.' *A. M. A. Arch. Ophthal.*, **54**, 407–12.

Breitmeyer, B. G. 1978. 'Disinhibition in metacontrast masking of vernier acuity targets: Sustained channels inhibit transient channels.' *Vision Res.*, **18**, 1401–5.

Breitmeyer, B. G., and Ganz, L. 1976. 'Implications of sustained and transient channels for theories of visual pattern masking, saccadic suppression and information processing.' *Psychol. Rev.*, **83**, 1–36.

Bridgeman, B. 1972. 'Visual receptive fields sensitive to absolute and relative motion during tracking.' *Science*, **178**, 1106–8.
Bridgeman, B., Hendry, D., and Stark, L. 1975. 'Failure to detect displacement of the visual world during saccadic eye movements.' *Vision Res.*, **15**, 719–22.
Bridgeman, B., Lewis, S., Heit, G., and Nagle, M. 1979. 'Relation between cognitive and motor-oriented systems of visual position perception.' *J. exp. Psychol.: Hum. Percept. Perform.*, **5**, 692–700.
Bridgeman, B., and Palca, J. 1980. 'The role of microsaccades in high acuity observational tasks.' *Vision Res.*, **20**, 813–17.
Bridgeman, B., and Stark, L. 1979. 'Omnidirectional increase in threshold for image shifts during saccadic eye movements.' *Percept. Psychophys.*, **25**, 241–3.
Bridges, C. C., and Bitterman, M. E. 1954. 'The measurement of autokinetic movement.' *Amer. J. Psychol.*, **67**, 525–9.
Brindley, G. S. 1970. *Physiology of the Retina and Visual Pathway*. London: Arnold.
Brindley, G. S. 1973. 'Sensory effects of electrical stimulation of the visual and paravisual cortex in man.' In R. Jung (Ed.), *Handbook of Sensory Physiology*, Vol. VII/3B. New York: Springer, pp. 583–94.
Brindley, G. S., Gauthier-Smith, P. C., and Lewin, W. 1969. 'Cortical blindness and the functions of the non-geniculate fibres of the optic tracts.' *J. Neurol. Neurosurg. Psychiat.*, **32**, 259–65.
Brindley, G. S., Goodwin, G. M., Kulikowski, J. J., and Leighton, D. 1976. 'Stability of vision with a paralyzed eye'. *J. Physiol.*, **258**, 65–6 P.
Brindley, G. S., and Lewin, W. S. 1968. 'The sensations produced by electrical stimulation of the visual cortex.' *J. Physiol.*, **196**, 479–93.
Brindley, G. S., and Merton, P. A. 1960. 'The absence of position sense in the human eye.' *J. Physiol.*, **153**, 127–30.
Brinkman, J., and Kuypers, H. G. J. M. 1973. 'Cerebral control of contralateral and ipsilateral arm, hand and finger movements in the split-brain Rhesus monkey.' *Brain*. **96**, 653–74.
Brittain, W. L. 1976. 'The effect of background shape on the ability of children to copy geometric forms.' *Child Devel.*, **47**, 1179–81.
Brodal, A. 1974. 'Anatomy of the vestibular nuclei and their connections.' In H. H. Kornhuber (Ed.), *Handbook of Sensory Physiology*, Vol. VI/1. New York: Springer, pp. 239–352.
Brodal, P. 1972a. 'The corticopontine projection from the visual cortex of the cat: I. The total projection and the projection from area 17.' *Brain Res.*, **39**, 297–317.
Brodal, P. 1972b. 'The corticopontine projection from the visual cortex of the cat: II. The projection from areas 18 and 19.' *Brain Res.*, **39**, 319–35.
Brodkey, J., and Stark, L. 1968. 'New direct evidence against intermittency or sampling in human smooth pursuit eye movements.' *Nature*, **218**, 273–5.
Brooks, B. A., and Bajandas, F. J. 1977. *Eye Movements*. New York: Plenum.
Brooks, B. A., Impelman, D. M., and Lum, J. T. 1980. 'Influence of background luminance on visual sensitivity during saccadic eye movements.' *Exp. Brain Res.*, **40**, 322–29.
Brooks, B. A., Yates, J. T., and Coleman, R. D. 1980. 'Perception of images moving at saccadic velocities during saccades and during fixation.' *Exp. Brain Res.*, **40**, 71–8.
Brooks, D. C. 1968. 'Waves associated with eye movement in the awake and sleeping cat.' *E. E. G. clin. Neurophysiol.*, **24**, 532–41.
Brooks, R. M., and Goldstein, A. G. 1963. 'Recognition by children of inverted photographs of faces.' *Child Devel.*, **34**, 1033–40.
Brosgole, L. 1968. 'The autokinesis of an after-image.' *Psychonom. Sci.*, **12**, 233–4.
Brosgole, L., and Cristal, R. M. 1967. 'Vertically induced autokinesis.' *Psychonom. Sci.*, **7**, 337–8.
Brown, A. E., and Hopkins, H. K. 1967. 'Interaction of the auditory and visual sensory

modalities.' *J. opt. Soc. Amer.*, **41**, 1–6.

Brown, D. L., and Salinger, W. L. 1975. 'Loss of X-cells in lateral geniculate nucleus with monocular paralysis: Neural plasticity in the adult cat.' *Science*, **189**, 1011–2.

Brown, J. H. 1966. 'Magnitude estimation of angular velocity during passive rotation.' *J. exp. Psychol.*, **72**, 169–72.

Brown, J. H. 1968. 'Cross-modal estimation of angular velocity.' *Percept. Psychophys.*, **3**, 115–7.

Brown, J. H., and Wolfe, J. W. 1969. 'Adaptation to prolonged constant angular acceleration.' *Ata oto-larying.*, **67**, 389–98.

Brown, J. L. 1961. 'Orientation to the vertical during water immersion.' *Aerosp. Med.*, **32**, 209–17.

Bruell, J. H., and Albee, G. W. 1955. 'Notes towards a motor theory of visual egocentric localization.' *Psychol. Rev.*, **62**, 391–9.

Bruner, J. S. 1973. 'Organization of early skilled action.' *Child Devel.*, **44**, 1–11.

Bruner, J. S., and Koslowski, B. 1972. 'Visually preadapted constituents of manipulating action.' *Perception*, **1**, 3–14.

Bryant, P. E. 1969. 'Perception and memory of the orientation of visually presented lines by children.' *Nature*, **224**, 1331–2.

Bryant, P. E. 1973. 'Discrimination of mirror images by young children.' *J. comp. physiol. Psychol.*, **82**, 415–25.

Bryant, P. E. 1974. *Perception and Understanding in Young Children*. London: Methuen.

Buettner, U. W., Büttner, U., and Henn, V. 1978. 'Transfer characteristics of neurons in vestibular nuclei of the alert monkey.' *J. Neurophysiol.*, **41**, 1614–28.

Buisseret, P., Gary-Bobo, E., and Imbert, M. 1978. 'Ocular motility and recovery of orientational properties of visual cortical neurones in dark-reared kittens.' *Nature, New Biol.*, **272**, 816–7.

Buisseret, P., and Imbert, M. 1976. Visual cortical cells: Their developmental properties in normal and dark reared kittens.' *J. Physiol.*, **255**, 511–25.

Buisseret, P., and Maffei, L. 1977. 'Extraocular proprioceptive projections to the visual cortex.' *Exp. Brain Res.*, **28**, 421–5.

Buizza, A., Léger, A., Droulez, J., Berthoz, A., and Schmid, R. 1980. 'Influence of otolithic stimulation by horizontal linear acceleration on optokinetic nystagmus and visual motion perception.' *Exp. Brain Res.*, **39**, 165–76.

Bullier, J., and Henry, G. H. 1979a. 'Ordinal position of neurones in cat striate cortex.' *J. Neurphysiol.*, **42**, 1251–63.

Bullier, J., and Henry, G. H. 1979b. 'Laminar distribution of first-order neurones and afferent terminals in cat striate cortex.' *J. Neurophysiol.*, **42**, 1271–81.

Bullier, J., and Henry, G. H. 1979c. 'Neural path taken by afferent streams in striate cortex of the cat.' *J. Neurophysiol.*, **42**, 1264–70.

Bullier, J., and Norton, T. T. 1979a. 'X and Y relay cells in cat lateral geniculate nucleus: Quantitative analysis of receptive-field properties and classificaion.' *J. Neurphysiol.*, **42**, 244–73.

Bullier, J., and Norton, T. T. 1979b. 'Comparison of receptive field properties of X and Y ganglion cells with X and Y lateral geniculate cells in the cat.' *J. Neurphysiol.*, **42**, 274–81.

Bunt, A. H., and Minckler, D. S. 1977. 'Foveal sparing. New anatomical evidence for bilateral representation of central retina.' *A.M.A. Arch. Ophthal.*, **95**, 1445–7.

Burbeck, C. A. 1981. 'Criterion effects in pattern and flicker experiments.' *Invest. Ophthal.*, **20**, ARVO Abstr., 178.

Burian, H. M., and Noorden, G. K. von. 1974. *Binocular Vision and Ocular Motility*. St. Louis: Mosby.

Burns, B. Delisle, Gassanov, U., and Webb, A. C. 1972. 'Responses of neurones in the cat's visual cerebral cortex to relative movement of patterns.' *J. Physiol.*, **226**, 113–51.

Burns, B. Delisle, and Pritchard, R. 1971. 'Geometrical illusions and the response of neurones in the cat visual cortex to angle patterns.' *J. Physiol.*, **213**, 599–616.

Burr, D. C. 1980. 'Sensitivity to spatial phase.' *Vision Res.*, **20**, 391–6.

Butter, C. M., Weinstein, C., Bender, D. B., and Gross, C. G. 1978. 'Localization and detection of visual stimuli following superior colliculus lesions in rhesus monkeys.' *Brain Res.*, **156**, 33–49.

Butterworth, G., and Castillo, M. 1976. 'Coordination of auditory and visual space in newborn human infants.' *Perception*, **5**, 155–60.

Butterworth, G., and Hicks, L. 1977. 'Visual proprioception and postural stability in infancy. A developmental study.' *Perception*, **6**, 255–62.

Büttner, U., Büttner-Ennever, J. A., and Henn, V. 1977. 'Vertical eye movement related unit activity in the rostral mesenchephalic reticular formation of the alert monkey.' *Brain Res.*, **130**, 239–52.

Büttner, U., Hepp. K., and Henn, V. 1977. 'Neurons in the rostral mesencephalic and paramedian pontine reticular formation generating fast eye movements.' In R. Baker and A. Berthoz (Eds.), *Control of Gaze by Brain Stem Neurones*. New York: Elsevier, pp. 309–18.

Büttner, U., and Lang, W. 1979. 'The vestibulocortical pathway: neurophysiological and anatomical studies in the monkey.' In R. Granit and O. Pompeiano (Eds.), *Reflex Control of Posture and Movement*. New York: Elsevier, pp. 581–8.

Büttner-Ennever, J. A., and Henn, V. 1976. 'An autoradiographic study of the pathways from the pontine reticular formation involved in horizontal eye movement.' *Brain Res.*, **108**, 155–64.

Buys, E., and Rijlant, P. 1939. 'Le seuil d'excitation (accélération angulaire) des canaux semi-circulaires.' *Arch. int. Physiol.*, **49**, 101–12.

Cairns, N. U., and Steward, M. S. 1970. 'Young children's orientation of letters as a function of axis of symmetry and stimulus alignment.' *Child Devel.*, **41**, 993–1002.

Camarda, R., Grupp, L. A., Pisa, M., and Rizzolatti, G. 1974. 'Interaction of visual stimuli in cat superior colliculus: spatial and temporal factors.' *Brain Res.*, **66**, 358–9.

Camisa, J. M., Blake, R., and Lema, S. 1977. 'The effects of temporal modulation on the oblique effect in humans.' *Perception*, **6**, 165–71.

Campbell, F. W., Cooper, G. F., and Enroth-Cugell, C. 1969. 'The spatial selectivity of the visual cells of the cat.' *J. Physiol.*, **203**, 223–35.

Campbell, F. W., and Corfield, R. 1977. 'Methods for demonstrating what one can and cannot see during a saccade.' *J. Physiol.*, **271**, 3P.

Campbell, F. W., and Green, D. G. 1965. 'Optical and retinal factors affecting visual resolution. *J. Physiol.*, **181**, 576–93.

Campbell, F. W., and Gubisch, R. W. 1966. 'Optical quality of the human eye.' *J. Physiol.*, **186**, 558–78.

Campbell, F. W., Howell, E. R., and Johnstone, J. R. 1978. 'A comparison of threshold and suprathreshold appearance of gratings with components in the low and high spatial frequency range.' *J. Physiol.*, **284**, 193–201.

Campbell, F. W., and Kulikowski, J. J. 1966. 'Orientational selectivity of the human visual system.' *J. Physiol.*, **187**, 437–45.

Campbell, F. W., Kulikowski, J. J., and Levinson, J. 1966. 'The effect of orientation on the visual resolution of gratings.' *J. Physiol.*, **187**, 427–36.

Campbell, F. W., and Maffei, L. 1970. 'Electrophysiological evidence for the existence of orientation and size detectors in the human visual system.' *J. Physiol.*, **207**, 635–52.

Campbell, F. W., and Maffei, L. 1971. 'The tilt after-effect: A fresh look.' *Vision Res.*, **11**, 833–40.

Campbell, F. W., and Robson, J. G. 1968. 'Application of Fourier analysis to the visibility of gratings.' *J. Physiol.*, **197**, 551–66.

Campos, E. C., Bedell, H. E., Enoch, J. M., and Fitzgerald, C. R. 1978. 'Retinal receptive field-like properties and Stiles–Crawford effect in a patient with a traumatic choroidal rupture.' *Doc. Opthalmologica*, **45**, 381–95.

Canon, L. K. 1970. 'Intermodality inconsistency of input and directed attention as determinants of the nature of adaptation.' *J. exp. Psychol.*, **84**, 141–7.

Carlson, C. R., Cohen, R. W., and Gorog, I. 1977. 'Visual processing of simple two-dimensional sine-wave luminance gratings.' *Vision Res.*, **17**, 351–8.

Carpenter, M. B. 1971. 'Central oculomotor pathways.' In P. Bach-y-Rita, C. C. Collins and J. E. Hyde (Eds.), *The Control of Eye Movements*. New York: Academic Press, pp. 67–103.

Carpenter, P. A., and Eisenberg, P. 1978. 'Mental rotation and the frame of reference in blind and sighted individuals.' *Percept. Psychophys.*, **23**, 117–24.

Carpenter, R. H. S. 1972. 'Cerebellectomy and the transfer function of the vestibulo-ocular reflex in the decerebrate cat.' *Proc. roy. Soc. B*, **181**, 353–74.

Carpenter, R. H. S. 1977. *Movements of the Eyes*. New York: Academic Press.

Carpenter, R. H. S., and Blakemore, C. 1973. 'Interactions between orientations in human vision.' *Exp. Brain Res.*, **18**, 287–303.

Carr, H. A. 1910. 'The autokinetic sensation.' *Psychol. Rev.*, **17**, 42–75.

Carr, H. A. 1935. *An Introduction to Space Perception*. New York: Longmans, Green.

Cawthorne, T. E., Dix, M. R., Hallpike, C. S., and Hood, J. D. 1956. 'Vestibular function.' *Brit. med. Bull.*, **12**, 131–42.

Cawthorne, T. E., Fitzgerald, G., and Hallpike, C. S. 1942. 'Studies in human vestibular function: II. Observations on the directional preponderance of caloric nystagmus ('Nystagmusbereitschaft') resulting from unilateral labyrinthectomy.' *Brain*, **65**, 138–80.

Cazin, L., Precht, W., and Lannou, J. 1980a. 'Optokinetic responses of vestibular nucleus neurons in the rat.' *Pflüg. Arch. ges. Physiol.*, **384**, 31–8.

Cazin, L., Precht, W., and Lannou, J. 1980b. 'Firing characteristics of neurons mediating optokinetic responses to rat's vestibular neurons.' *Pflüg. Arch. ges. Physiol.*, **386**, 221–30.

Chalupa, L. M., Macadar, A. W., and Lindsley, D. B. 1975. 'Response plasticity of lateral geniculate neurons during and after pairing of auditory and visual stimuli.' *Science*, **190**, 290–2.

Cheng, M., and Outerbridge, J. S. 1974a. 'Inter-saccadic interval analysis of vestibular nystagmus.' *Acta oto-laryng.*, **77**, 348–53.

Cheng, M., and Outerbridge, J. S. 1974b. 'Inter-saccadic interval analysis of optokinetic nystagmus.' *Vision Res.*, **14**, 1053–8.

Cheng, M., and Outerbridge, J. S. 1975. Optokinetic nystagmus during selective retinal stimulation.' *Exp. Brain Res.*, **23**, 129–39.

Chiang, C. 1968. 'A new theory to explain geometrical illusions produced by crossing lines.' *Percept. Psychophys.*, **3**, 174–6.

Chou, S. K. 1935. 'Reading and legibility of Chinese characters: IV. An analysis of judgments of positions of Chinese characters by American subjects.' *J. exp. Psychol.*, **18**, 318–47.

Christian, P. 1940. 'Experimentelle Beiträge zur intermodalen vestibulo-optischen Wechselbeziehung der Sinnesorgane,' *Pflüg. Arch. ges. Physiol.*, **243**, 370–87.

Churchill, A. V. 1969. 'Effect of head movement in visual–kinaesthetic localization.' *Percept. mot. Skills*, **28**, 785–6.

Clark, B. 1967. 'Thresholds for the perception of angular acceleration in man.' *Aerosp. Med.*, **38**, 443–50.

Clark, B., and Graybiel, A. 1951. 'Visual perception of the horizontal following exposure to radial acceleration on a centrifuge.' *J. comp. physiol. Psychol.*, **44**, 525–34.

Clark, B., and Graybiel, A. 1963. 'Perception of the postural vertical in normals and subjects with labyrinthine defects.' *J. exp. Psychol.*, **65**, 490–4.

Clark, B., and Graybiel, A. 1966a. 'Factors contributing to the delay in the perception of the oculogravic illusion.' *Amer. J. Psychol.*, **79**, 377–88.

Clark, B., and Graybiel, A. 1966b. 'Perception of the visual horizontal in normal and labyrinthine defective observers during prolonged rotation.' *Amer. J. Psychol.*, **79**, 608–12.

Clark, B., and Graybiel, A. 1967. 'Egocentric localization of the visual horizontal in normal and labyrinthine-defective observers as a function of head and body tilt.' *Percept. Psychophys.*, **2**, 609–11.
Clark, B., and Stewart, J. D. 1962. 'Perception of angular acceleration about the yaw axis of a flight simulator.' *Aerosp. Med.*, **33**, 1426–32.
Clark, B., and Stewart, J. D. 1968a. 'Comparison of sensitivity for the perception of bodily rotation and the oculogyral illusion.' *Percept. Psychophys.*, **3**, 253–6.
Clark, B., and Stewart, J. D. 1968b. 'Comparison of three methods to determine thresholds for perception of angular acceleration.' *Amer. J. Psychol.*, **81**, 207–16.
Clark, B., and Stewart, J. D. 1968c. 'Magnitude estimates of rotational velocity during and following prolonged increasing, constant, and zero angular acceleration.' *J. exp. Psychol.*, **78**, 329–39.
Clark, B., and Stewart, J. D. 1969. 'Effects of angular accelaration on man: Thresholds for the perception of rotation and the oculogyral illusion.' *Aerosp. Med.*, **40**, 952–6.
Clark, B., and Stewart, J. D. 1972. 'The power law for the perception of rotation by airline pilots.' *Percept. Psychophys.*, **11**, 433–6.
Clay, M. M. 1970. 'An increasing effect of disorientation of print: A developmental study.' *J. exp. Child Psychol.*, **9**, 297–306.
Cleland, B. G., Dubin, M. W., and Levick, W. R. 1971. 'Sustained and transient neurons in the cat's retina and lateral geniculate nucleus.' *J. Physiol.*, **217**, 473–96.
Cleland, B. G., and Enroth-Cugell, C. 1968. 'Quantitative aspects of sensitivity and summation in the cat retina.' *J. Physiol.*, **198**, 17–38.
Cleland, B. G., Harding, T. H., and Tulunay-Keesey, U. 1979. 'Visual resolution and receptive field-size: Examination of two kinds of cat retinal ganglion cell.' *Science*, **205**, 1015–7.
Clement, D. E., and Varnadoe, K. W. 1967. 'Pattern uncertainty and the discrimination of visual patterns.' *Percept. Psychophys.*, **2**, 427–31.
Clément, G., Courjon, J. H., Jeannerod, M., and Schmid, R. 1981. 'Unidirectional habituation of vestibulo-ocular responses by repeated rotational or optokinetic stimulations in the cat.' *Exp. Brain Res.*, **42**, 34–42.
Cogan, A. I. 1979. 'The relationship between the apparent vertical and the vertical horopter.' *Vision Res.*, **19**, 655–65.
Cogan, D. C. 1974. 'Paralysis of down-gaze.' *A.M.A. Arch. Ophthal.*, **91**, 192–9.
Cohen, B. 1974. 'The vestibulo-ocular reflex arc.' In H. H. Kornhuber (Ed.), *Handbook of Sensory Physiology*, Vol. VI/1. New York: Springer, pp. 477–540.
Cohen, B., and Feldman, M. 1971. 'Potential changes associated with rapid eye movement in the calcarine cortex.' *Exp. Neurol.*, **31**, 100–13.
Cohen, B., and Komatsuzaki, A. 1972. 'Eye movements induced by stimulation of the pontine reticular formation: evidence for integration in oculomotor pathways.' *Exp. Neurol.*, **36**, 101–17.
Cohen, B., Matsuo, V., and Raphan, T. 1977. 'Quantitative analysis of the velocity characteristics of optokinetic nystagmus and optokinetic after-nystagmus.' *J. Physiol.*, **270**, 321–44.
Cohen, B., Suzuki, J., Shanzer, S., and Bender, M. B. 1964. 'Semicircular canal control of eye movements.' In M. B. Bender (Ed.), *The Oculomotor System*. New York: Harper and Row, Ch. 6.
Cohen, B., Uemura, T., and Takemori, S. 1973. 'Effects of labyrinthectomy on optokinetic nystagmus (OKN) and optokinetic after-nystagmus (OKAN).' *Equil. Res.*, **3**, 80–93.
Cohen, H. B. 1966. 'Some critical factors in prism adaptation.' *Amer. J. Psychol.*, **79**, 285–90.
Cohen, M. M. 1967. 'Continuous versus terminal visual feedback in prism aftereffects.' *Percept. mot. Skills*, **24**, 1295–1302.
Cohen, M. M. 1973a. 'Elevator illusion: Influences of otolith organ activity and neck

proprioception.' *Percept. Psychophys.*, **14**, 401–6.

Cohen, M. M. 1973b. 'Visual feedback, distribution of practice, and intermanual transfer of prism aftereffects.' *Percept. mot. Skills*, **37**, 599–609.

Collewijn, H 1972. 'An analog model of the rabbit's optokinetic system.' *Brain Res.*, **36**, 71–88.

Collewijn, H. 1975a. 'Oculomotor areas in the rabbit's midbrain and pretectum.' *J. Neurophysiol.*, **6**, 3–22.

Collewijn, H. 1975b. 'Direction-selective units in the rabbit's nucleus of the optic tract.' *Brain Res.*, **100**, 489–508.

Collewijn, H. 1976. Impairment of optokinetic (after-) nystagmus by labyrinthectomy in the rabbit.' *Exp. Neurol.*, **52**, 146–56.

Collewijn, H. 1977. 'Optokinetic and vestibulo-ocular reflexes in dark-reared rabbits.' *Exp. Brain Res.*, **27**, 287–300.

Collewijn, H., Winterson B. J., and van der Steen, J., 1980. 'Post-rotatory nystagmus and optokinetic after-nystagmus in the rabbit; linear rather than exponential decay.' *Exp. Brain Res.*, **40**, 330–45.

Collin, N. G., and Cowey, A. 1980. 'The effect of frontal eye-fields and superior colliculi on visual stability and movement discrimination in Rhesus monkeys.' *Exp. Brain Res.*, **40**, 251–60.

Collins, C. C. 1975. 'The human oculomotor control system.' In G. Lennerstrand and P. Bach-y-Rita (Eds.), *Basic Mechanisms of Ocular Motility and Their Clinical Implications*. New York: Pergamon, pp. 145–80.

Collins, C. C., Carlson, M. R., Scott, A. B., and Jampolsky, A. 1981. 'Extraocular muscle forces in normal human subjects.' *Invest. Ophthal.*, **20**, 652–664.

Collins, C. C., O'Meara, D., and Scott, A. B. 1975. 'Muscle tension during unrestrained human eye movements.' *J. Physiol.*, **245**, 351–69.

Collins, J. K., and Singer, G. 1968. 'Interaction between sensory spatial aftereffects and persistence of response following behavioural compensation.' *J. exp. Psychol.*, **77**, 301–7.

Collins, W. E. 1965. 'Subjective responses and nystagmus following repeated unilateral caloric stimulation.' *Ann. Otol Rhinol. Laryngol.*, **74**, 1034–54.

Collins, W. E. 1966. 'Vestibular responses from figure skaters.' *Aerosp. Med.*, **37**, 1098–104.

Collins, W. E. 1968a. 'Special effects of brief periods of visual fixation on nystagmus and sensations of turning.' *Aerosp. Med.*, **39**, 257–66.

Collins, W. E. 1968b. 'Coriolis vestibular stimulation and the influence of different visual surrounds.' *Aerosp. Med.*, **39**, 125–30.

Collins, W. E. 1974a. 'Arousal and vestibular habituation.' In H. H. Kornhuber (Ed.), *Handbook of Sensory Physiology*, Vol. VI/2. New York: Springer, pp. 361–8.

Collins, W. E. 1974b. 'Habituation of vestibular responses with and without visual stimulation.' In H. H. Kornhuber (Ed.), *Handbook of Sensory Physiology*, Vol. VI/2. New York: Springer, pp. 369–88.

Collins, W. E., Crampton, G. H., and Posner, J. B. 1961. 'Effects of mental activity on vestibular nystagmus and the electroencephalogram.' *Nature*, **190**, 194–5.

Colonnier, M. 1964. 'The tangential organization of the visual cortex.' *J. Anat.*, **98**, 327–44.

Coltheart, M. 1971. 'Visual feature-analyzers and aftereffects of tilt and curvature.' *Psychol. Rev.*, **78**, 114–21.

Coltheart, M., and Cooper, C. M. 1972. 'The retinal reference of the tilt aftereffect.' *Percept. Psychophys.*, **11**, 321–4.

Condry, S. M., Haltom, M., and Neisser, U. 1977. 'Infant sensitivity to audio-visual discrepancy: A failure to replicate.' *Bull. Psychonom. Soc.*, **9**, 431–2.

Cooper, L. A. 1975. 'Mental rotation of random two-dimensional shapes.' *Cognit. Psychol.*, **7**, 20–43.

Cooper, L. A. 1976. 'Demonstration of a mental analog of an external rotation.' *Percept. Psychophys.*, **19**, 296–302.
Cooper, L. A., and Podgorny, P. 1976. 'Mental transformations and visual comparison processes: Effects of complexity and similarity.' *J. exp. Psychol.: Hum. Percept. Perform.*, **2**, 503–14.
Cooper, L. A., and Shepard, R. N. 1973. 'The time required to prepare for a rotated stimulus.' *Mem. Cognit.*, **1**, 246–50.
Cooper, L. A., and Shepard, R. N. 1975. 'Mental transformations in the identification of left and right hands.' *J. exp. Psychol: Hum. Percept. perform.*, **104**, 48–56.
Corah, N. L. 1965. 'Effects of the visual field upon perception of change in spatial orientation.' *J. exp. Psychol.*, **70**, 598–601.
Corballis, M. C., and Beale, I. L. 1976. *The Psychology of Left and Right.* New York: Wiley.
Corballis, M. C., Nagourney, B. A., Shetzer, L. I., and Stefanatos, G. 1978. 'Mental rotation under head tilt: Factors influencing the location of the subjective reference frame.' *Percept. Psychophys.*, **24**, 263–73.
Corballis, M. C., and Roldan, C. E. 1975. 'Detection of symmetry as a function of angular orientation.' *J. exp. Psychol.: Hum. Percep. Perform.*, **104**, 221–30.
Corballis, M. C., and Zalik, M. C. 1977. 'Why do children confuse mirror-image obliques?' *J. exp. Child Psychol.*, **24**, 516–23.
Corballis, M. C., Zbrodoff, J., and Roldan, C. E. 1976. 'What's up in mental rotation?' *Percept. Psychophys.*, **19**, 525–30.
Corballis, M. C., Zbrodoff, N. J., Shetzer, L. I., and Butler, P. B. 1978. 'Decisions about identity and orientation of rotated letters and digits.' *Mem. Cognit.*, **6**, 98–107.
Coren, S. 1969. 'The influence of optical aberrations on the magnitude of the Poggendorff illusion.' *Percept.* Psychophys., **6**, 185–6.
Coren, S., and Festinger, L. 1967. 'An alternative view of the "Gibson normalization effect".' *Percept. Psychophys.*, **2**, 621–6.
Coren, S., and Hoenig, P. 1972. 'Effect of non-target stimuli upon length of voluntary saccades.' *Percept. mot. Skills*, **34**, 499–508.
Corfield, R., Fosdick, J. P., and Campbell, F. W. 1978. 'Grey-out elimination: The roles of spatial waveform, frequency and phase.' *Vision Res.*, **18**, 1305–11.
Cornsweet, T. 1956. 'Determinants of stimuli for involuntary drifts and saccadic eye movements.' *J. opt. Soc. Amer.*, **46**, 987–93.
Cornsweet, T. N. 1970. *Visual Perception.* New York: Academic Press
Correia, M. J., Hixson, W. C., and Niven, J. I. 1968. 'On predictive equations for subjective judgments of vertical and horizon in a force field.' *Acta oto-laryng., Suppl. 230*, 3–20.
Correia, M. J., and Money, K. E. 1970. 'The effect of blockage of all six semicircular canal ducts on nystagmus produced by dynamic linear acceleration in the cat.' *Acta oto-laryng.*, **69**, 7–16.
Corteen, R. S. 1970. 'The autokinetic effect and signal detection.' *J. gen. Psychol.*, **82**, 77–80.
Corwin, T. R., Moskowitz-Cook, A., and Green, M. A. 1977. 'The oblique effect in a vernier acuity situation.' *Percept. Psychophys.*, **21**, 445–9.
Courjon, J. H., Jeannerod, M., Ossuzio, I., and Schmid, R. 1977. 'The role of vision in compensation of vestibulo-ocular reflex after hemilabyrinthectomy in the cat.' *Exp. Brain Res.*, **28**, 235–48.
Cowey, A. 1964. 'Projection of the retina on to striate and prestriate cortex in the squirrel monkey.' *J. Neurophysiol.*, **27**, 366–93.
Cowey, A., 1979. 'Cortical maps and visual perception: The Grindley Memorial Lecture.' *Quart. J. exp. Psychol.*, **31**, 1–17.
Cowey, A., and Perry, V. H. 1980. 'The projection of the fovea to the superior colliculus in Rhesus monkeys.' *Neurosci.*, **5**, 53–62.

Cowey, A., and Rolls, E. T. 1974. 'Human cortical magnification factor and its relation to visual acuity.' *Exp. Brain Res.*, **21**, 447–54.
Craske, B. 1967. 'Adaptation to prisms: Change in internally registered eye-position.' *Brit. J. Psychol.*, **58**, 329–35.
Craske, B., and Crawshaw, M. 1974. 'Adaptive changes of opposite sign in the oculomotor systems of the two eyes.' *Quart. J. exp. Psychol.*, **26**, 106–13.
Craske, B., and Crawshaw, M. 1975. 'Oculomotor adaptation to prisms is not simply a muscle potentiation effect.' *Percept. Psychophys.*, **18**, 105–6.
Craske, B., and Crawshaw, M. 1978. 'Spatial discordance is a sufficient condition for oculomotor adaptation to prisms: Eye muscle potentiation need not be a factor.' *Percept. Psychophys.*, **23**, 75–9.
Craske, B., Crawshow, M., and Heron, P. 1975. 'Disturbance of the oculomotor system due to lateral fixation.' *Quart. J. exp. Psychol.*, **27**, 459–65.
Craske, B., and Templeton, W. B. 1968. 'Prolonged oscillation of the eyes induced by conflicting position input.' *J. exp. Psychol.*, **76**, 387–93.
Crassini, B., and Broerse, J. 1980. 'Auditory–visual integration in neonates: A signal detection analysis.' *J. exp. Child Psychol.*, **29**, 144–55.
Crawford, B. H. 1940. 'The effect of field size and pattern on the change of visual sensitivity with time.' *Proc. roy. Soc., B.* **129**, 94–106.
Crawford, M. L. J., and Noorden, G. K. von. 1980. 'Optically induced concomitant strabismus in monkeys.' *Invest. Ophthal.*, **19**, 1105–9.
Crawshaw, M., and Craske, B. 1974. 'No retinal component in prism adaptation.' *Acta psychol.*, **38**, 421–3.
Crawshaw, M., and Craske, B. 1976. 'Oculomotor adaptation to prisms: complete transfer between eyes.' *Brit. J. Psychol.*, **67**, 475–8.
Creutzfeldt, O. D. 1977. 'Generality of the functional structure of the neocortex.' *Naturwissenschaften*, **64**, 507–17.
Creutzfeldt, O. D., and Heggelund, P. 1975. 'Neural plasticity in visual cortex of adult cats after exposure to visual patterns.' *Science*, 188, 1025–7.
Creutzfeldt, O. D., Kuhnt, U., and Benevento, L. A. 1974. 'An intracellular analysis of visual cortical neurones to moving stimuli: responses in a co-operative neuronal network.' *Exp. Brain Res.*, **21**, 251–74.
Creutzfeldt, O. D., Sakmann, B., Scheich, H., and Korn, A. 1970. 'Sensitivity distribution and spatial summation within receptive field center of retinal on-center ganglion cells and transfer function of the retina.' *J. Neurophysiol.*, **33**, 654–71.
Crewther, S. G., Crewther, D. P., Peck, C. K., and Pettigrew, J. D. 1980. 'Visual cortical effects of rearing cats with monocular or binocular deprivation.' *J. Neurophysiol.*, **44**, 97–118.
Critchley, M. 1969. *The parietal lobes*. New York: Hafner.
Crommelinck, M., Guitton, D., and Roucoux, A. 1977. 'Retinotopic versus spatial coding of saccades: Clues obtained by stimulating deep layers of cat's superior colliculus.' In R. Baker and A. Berthoz (Eds.), *Control of Gaze by Brain Stem Neurones*. New York: Elsevier, pp. 425–36.
Crone, R. A. 1975. 'Optokinetically induced eye torsion.' *v. Graefes Arch. Ophthal.*, **196**, 1–7.
Crone, R. A., and Verduyn Lunel, H. F. E. 1969. 'Autokinesis and the perception of movement: The physiology of eccentric fixation.' *Vision Res.*, **9**, 89–101.
Cunitz, R. J., and Steinman, R. M. 1969. 'Comparison of saccadic eye movements during fixation and reading.' *Vision Res.*, **9**, 683–93.
Cynader, M., and Berman, N. 1972. 'Receptive-field organization of monkey superior colliculus.' *J. Neurophysiol.*, **35**, 187–201.
Cynader, M., Blakemore, C., and Van Sluyters, R. C. 1978. 'Congruent binocular preferred directions in the superior colliculus of kittens reared with one eye rotated.' *Soc. Neurosci. Abstr.*, **4**, 633.
Cynader, M., and Harris, L. 1980. Eye movement in strabismic cats.' *Nature*, **286**, 64–5.

Cynader, M., and Regan, D. 1978. 'Neurons in cat parastriate cortex selective to the direction of motion in three-dimensional space.' *J. Physiol.*, **274**, 549–69.

Damasio, A. R., and Benton, A. L. 1979. 'Impairment of hand movements under visual guidance.' *Neurol.*, **29**, 170–8.

Daniel, P. M., and Whitteridge, D. 1961. 'The representation of the visual field on the cerebral cortex in monkeys.' *J. Physiol.*, **159**, 203–21.

Daniels, J. D., Norman, J. L., and Pettigrew, J. D. 1977. 'Biases for oriented moving bars in lateral geniculate nucleus of normal and stripe-reared cats.' *Exp. Brain Res.*, **29**, 155–72.

Darian-Smith, I., Johnson, K. O., and Goodwin, A. W. 1979. 'Posterior parietal cortex: relations of unit activity to sensorimotor function.' *Ann Rev. Physiol.*, **41**, 141–57.

Darlot, C., Lopezbarneo, J., and Tracey, D. 1981. 'Asymmetries of vertical vestibular nystagmus in the cat.' *Exp. Brain Res.*, **41**, 420–26.

Daugman, J. G. 1980. 'Two-dimensional spectral analysis of cortical receptive field profiles.' *Vision Res.*, **20**, 847–56.

Daunton, N., and Thomsen, D. 1979. 'Visual modulation of otolith-dependent units in cat vestibular nuclei.' *Exp. Brain Res.*, **37**, 173–6.

Davidoff, J. B. 1974. 'An observation concerning the preferred perception of the visual horizontal and vertical.' *Perception*, **3**, 47–8.

Davson, H. (Ed.) 1969. *The Eye, Vol. 3.* New York: Academic Press.

Davson, H. 1978. *The Eye, Vol. 4.* New York: Academic Press.

Day, R. H. 1962. 'Excitatory and inhibitory processes as the basis of contour shift and negative aftereffect.' *Psychologica*, **5**, 185–93.

Day, R. H., and Singer, G. 1964. 'Spatial aftereffects within and between kinesthesis and vision.' *J. exp. Psychol.*, **68**, 337–43.

Day, R. H., and Wade, N. J. 1966. 'Visual spatial aftereffect from prolonged head-tilt.' *Science*, **154**, 1201–2.

Day, R. H., and Wade, N. J. 1968. 'Involvement of neck proprioceptive system in visual after-effect from prolonged head tilt.' *Quart. J. exp. Psychol.*, **20**, 290–3.

Day, R. H., and Wade, N. J. 1969. 'The reference for visual normalization.' *Amer. J. Psychol.*, **82**, 191–7.

Dealy, R. S., and Tolhurst, D. J. 1974. 'Is spatial adaptation an after-effect of prolonged inhibition?' *J. Physiol.*, **241**, 261–70.

Deckert, G. H. 1964. 'Pursuit eye-movements in the absence of a moving visual stimulus.' *Science*, **143**, 1192–3.

de Monasterio, F. M. 1978a. 'Properties of concentrically organized X and Y ganglion cells of macaque retina.' *J. Neurophysiol.*, **41**, 1394–1417.

de Monasterio, F. M. 1978b. 'Properties of ganglion cells with atypical receptive-field organization in retina of macaques.' *J. Neurophysiol.*, **41**, 1435–49.

Denney, D., and Adorjani, C. 1972. 'Orientation specificity of visual cortical neurons after head tilt.' *Exp. Brain Res.*, **14**, 312–7.

Denny-Brown, D., and Chambers, R. A. 1958. 'The parietal lobe and behavior.' *Res. Publ. Assoc. Res. nerv. ment. Dis.*, **36**, 35–117.

Denny-Brown, D., and Chambers, R. A. 1976. 'Physiological aspects of visual perception–I. Functional aspects of visual cortex.' *Arch. Neurol.*, **33**, 219–27.

Deregowski, J. B. 1972. 'Reproduction of orientation of Kohs-type figures: A cross-cultural study.' *Brit. J. Psychol.*, **63**, 283–96.

Deregowski, J. B. 1977. 'A study of orientation errors in response to Kohs-type figures.' *Internat. J. Psychol.*, **12**, 183–91.

Derrington, A. M., and Fuchs, A. F. 1979. 'Spatial and temporal properties of X and Y cells in the cat lateral geniculate nucleus.' *J. Physiol.*, **293**, 347–64.

DeSisto, M. J., and McLaughlin, S. C. 1968. 'Change in Judgment of direction of gaze after autokinetic perception.' *Psychonom. Sci.*, **10**, 221–2.

De Valois, K. K. 1977a. 'Spatial frequency adaptation can enhance contrast sensitivity.'

Vision Res., **17**, 1057–65.

De Valois, K. K. 1977b. 'Independence of black and white: phase-specific adaptation.' *Vision Res.*, **17**, 209–15.

De Valois, K. K., De Valois, R. L., and Yund, E. W. 1979. 'Responses of striate cortex cells to grating and checkerboard patterns.' *J. Physiol.*, **291**, 483–505.

De Valois, R. L., Albrecht, D. G., and Thorell, L. G. 1978. Cited in De Valois, K. K. 1978. 'Interactions among spatial frequency channels.' In S. J. Cool and E. L. Smith (Eds.), *Frontiers in Visual Science*. Berlin: Springer, pp. 277–85.

de Vries, H. 1950. 'The mechanics of the labyrinth otoliths.' *Acta oto-laryng.*, **38**, 262–73.

Dewar, R. 1970. 'Adaptation to displaced vision: Amount of optical displacement and practice.' *Percept. Psychophys.*, **8**, 313–6.

Dewar, R. 1971. 'Adaptation to displaced vision: Variations on the "prismatic shaping" technique.' *Percept. Psychophys.*, **9**, 155–7.

Dichgans, J. 1974. 'Spinal afferents to the oculomotor system.' In G. Lennerstrand and P. Bach-y-Rita (Eds.), *Basic Mechanisms of Ocular Motility and Their Clinical Implications*. New York: Pergamon, pp. 299–320.

Dichgans, J. 1977. 'Optokinetic nystagmus as dependent on the retinal periphery via the vestibular nucleus.' In R. Baker and A. Berthoz (Eds.), *Control of Gaze by Brain Stem Neurons*. New York: Elsevier, pp. 261–8.

Dichgans, J., and Bizzi, E. (Eds.), 1972. *Cerebral Control of Eye Movements and Motion Perception*. Basel: Karger.

Dichgans, J., Bizzi, E., Morasso, P., and Tagliasco, V. 1973. 'Mechanisms underlying recovery of eye–head coordination following bilateral labyrinthectomy in monkeys.' *Exp. Brain Res.*, **18**, 548–62.

Dichgans, J., Bizzi, E., Morasso, P., and Tagliasco, V. 1974. 'The role of vestibular and neck afferents during eye–head coordination in the monkey.' *Brain Res.*, **71**, 225–32.

Dichgans, J., and Brandt. T. 1973. 'Optokinetic motion sickness and pseudo-Coriolis effects induced by moving visual stimuli.' *Acta oto-laryng.*, **76**, 339–48.

Dichgans, J., and Brandt, T. 1978. 'Visual–vestibular interaction: Effects on self-motion perception and postural control.' In *Handbook of Sensory Physiology*, Vol. VIII. New York: Springer, pp. 755–804.

Dichgans, J., Diener, H. C., and Brandt. T. 1974. 'Optokinetic–graviceptive interaction in different head positions.' *Acta oto-laryng.*, **78**, 391–8.

Dichgans, J., Held, R., Young, L. R., and Brandt, T. 1972. 'Moving visual scenes influence the apparent direction of gravity.' *Science*, **178**, 1217–9.

Dichgans, J., Körner, F., and Voigt, K. 1969. 'Verleichende Skalierung des afferenten und efferenten Bewegungssehens beim Menschen: Lineare Funktionen mit verschiedener Anstiegssteilheit.' *Psych. Forsch.*, **32**, 277–95.

Dichgans. J., Mauritz. K. H., Allum, J. H. J., and Brandt. T. 1976. 'Postural sway in normals and atactic patients: Analysis of the stabilizing and destabilizing effects of vision.' *Aggressologie*, **77c**, 15–24.

Dichgans, J., Reuturn, G. M. von, and Römmelt, U. 1978. 'Impaired suppression of vestibular nystagmus by fixation in cerebellar and noncerebellar patients.' *Arch. Psychiat. Nervenkr.*, **226**, 183–99.

Dichgans, J., Wist, E., Diener, H. C., and Brandt. T. 1975. 'The Aubert–Fleischl phenomenon: A temporal frequency effect on perceived velocity in afferent motion perception,' *Exp. Brain Res.*, **23**, 529–33.

Diener, H. C., Wist, E. R., Dichgans, J., and Brandt. T. 1976. 'The spatial frequency effect on perceived velocity.' *Vision Res.*, **16**, 169–76.

Dietzel, H. 1924. 'Untersuchungen über die optische Lokalisation der Mediane.' *Z. Biol.*, **80**, 289–316.

Di Franco, D., Muir, D. W., and Dodwell, P. C. 1978. 'Reaching in very young infants.' *Perception*, **7**, 385–92.

Dineen, J., and Keating, E. G. 1981. 'The primate visual system after bilateral removal of striate cortex–survival of complex pattern vision.' *Exp. Brain Res.*, **41**, 338–45.

Dix, M. R., and Hood, J. D. 1969. 'Observations upon the nervous mechanism of vestibular habituation.' *Acta oto-laryng.*, **67**, 310–8.

Dixon, N. F., and Dixon, P. M. 1966. '"Sloping water" and related framework illusions: Some informal observations.' *Quart. J. exp. Psychol.*, **18**, 369–70.

Dobelle, W. H., and Mladejovsky, M. G. 1974. 'Phosphenes produced by electrical stimulation of human occipital cortex and their application to the development of prothesis for the blind.' *J. Physiol.*, **243**, 553–76.

Dobson, V., and Teller, D. Y. 1978. 'Visual acuity in human infants: a critical review and comparison of behavioral and electrophysiological studies.' *Vision Res.*, **18**, 1469–83.

Dockstader, S. L. 1971. 'Comparison of cupulometric and psychophysical thresholds for perception of rotation and the oculogyral illusion.' *Percept. Psychophys.*, **9**, 299–302.

Dodwell, P. C., Muir, D., and Di Franco, D. 1976. 'Responses of infants to visually presented objects.' *Science*, **194**, 209–11.

Dohlman, G. 1935. 'Towards a method for quantitative measurement of the functional capacity of the vestibular apparatus.' *Acta oto-laryng.*, **23**, 50–62.

Donaldson, I. M. L., and Long, A. C. 1977. 'Suppression of visual responses in the cat superior colliculus following stretch of extraocular muscles.' *J. Physiol.*, **272**, 94–95 P.

Donaldson, I. M. L., and Long, A. C. 1980. 'Interactions between extraocular proprioceptive and visual signals in the superior colliculus of the cat.' *J. Physiol.*, **298**, 85–110.

Donaldson, I. M. L., and Nash, J. R. G. 1975. 'Variability of the relative preference for stimulus orientation and direction of movement in some units of the cat visual cortex.' *J. Physiol.*, **245**, 305–24.

Donders, F. C. 1848. 'Beitrag zur Lehre von den Bewegungen des menschilichen Auges.' *Holländ Beitr. Anat. Physiol., Wiss.*, **1**, 104–384.

Doty, R. L. 1969. 'Effect of duration of stimulus presentation on the angular acceleration threshold.' *J. exp. Psychol.*, **80**, 317–21.

Doty, R. W. 1973. 'Ablation of visual areas in the central nervous system.' In R. Jung (Ed.), *Handbook of Sensory Physiology*, Vol. VII/3B. New York: Springer, pp. 483–542.

Dowd, P. J. 1974. 'Sleep deprivation effects on the vestibular habituation process.' *J. appl. Psychol.*, **59**, 748–52.

Dowd, P. J., and Cramer, R. L. 1967. 'Habituation transference in Coriolis acceleration.' *Aerosp. Med.*, **38**, 1103–7.

Dowling, J. E., and Boycott, B. B. 1966. 'Organization of the primate retina: electronmicroscopy.' *Proc. roy Soc., B*, **166**, 80–111.

Downer, J. L. de C. 1959. 'Changes in visually guided behaviour following midsagittal division of the optic chiasma and corpus callosum in monkeys.' *Brain*, **82**, 251–9.

Drasdo, N. 1977. 'The neural representation of visual space.' *Nature*, **266**, 554–6.

Dreher, B. 1972. 'Hypercomplex cells in the cat's striate cortex.' *Invest. Ophthal.*, **11**, 355–6.

Dreher, B., Fukada, Y., and Rodieck, R. W. 1976. 'Identification, classification and anatomical segregation of cells with X-like and Y-like properties in the lateral geniculate nucleus of old-world primates.' *J. Physiol.*, **258**, 433–52.

Dubois, M. F. W., and Collewijin, H. 1979. 'Optokinetic reactions in man elicited by localized retinal motion stimuli.' *Vision Res.*, **19**, 1105–15.

Dubrovsky, B. O., and Barbas, H. 1977. 'Frontal projections to dorsal neck and extraocular muscles.' *Exp. Neurol.*, **55**, 680–93.

Duensing, F., and Schaefer, K. P. 1958. 'Die Aktivität einzelner Neurone im Bereich der Vestibulariskerne bei Horizontalbeschleunigungen unter besonderer Berücksichtigung des vestibulären Nystagmus.' *Arch. Psychiat. Nervenks.*, **198**, 225–52.

Dufossé, M., Ito, M., Jastreboff, P. J., and Miyashita, Y. 1978. 'A neuronal correlate in rabbit's cerebellum to adaptive modification of the vestibulo-ocular reflex.' *Brain Res.*, **150**, 611–6.

Duncker, K. 1929. 'Ueber induzierte Bewegung.' *Psychol. Forsch*, **22**, 180–259.
Dunkeld, J., and Bower, T. G. R. 1980. 'Infant response to impending collision.' *Perception*, **9**, 549–54.
Easter, S. S. 1973. 'A comment on the "glissade".' *Vision Res.*, **13**, 881–92.
Easton, T. A. 1972. 'Patterned inhibition from single eye muscle stretch in the cat.' *Exp. Neurol.*, **34**, 497–510.
Ebenholtz, S. M. 1968. 'Readaptation and decay after exposure to optical tilt.' *J. exp. Psychol.*, **78**, 350–1.
Ebenholtz, S. M. 1970a. 'Perception of the vertical with body tilt in the median plane.' *J. exp. Psychol.*, **83**, 1–6.
Ebenholtz, S. M. 1970b. 'On the relation between interocular transfer of adaptation and Hering's law of equal innervation.' *Psychol. Rev.*, **77**, 343–7.
Ebenholtz, S. M. 1972. 'The constancy of object orientations: Effects of target inclination.' *Psychol. Forsch.*, **35**, 178–86.
Ebenholtz, S. M. 1973. 'Optimal input rates for tilt adaptation'. *Amer. J. Psychol.*, **86**, 193–200.
Ebenholtz, S. M., 1974. 'The possible role of eye-muscle potentiation in several forms of prism adaptation.' *Perception*, **3**, 477–85.
Ebenholtz, S. M. 1976. 'Additivity of aftereffects of maintained head and eye rotations: An alternative to recalibration.' *Percept. Psychophys.*, **19**, 113–6.
Ebenholtz, S. M., 1977. 'Determinants of the rod and frame effect: The role of retinal size.' *Percept. Psychophys.*, **22**, 531–8.
Ebenholtz, S. M., and Benzschawel, T. L. 1977. 'The rod and frame effect and induced head tilt as a function of observation distance.' *Percept. Psychophys.*, **22**, 491–6.
Ebenholtz, S. M., and Paap. K. R. 1973. 'The constancy of object orientation: Compensation for ocular rotation.' *Percept. Psychophys.*, **14**, 458–70.
Ebenholtz, S. M., and Shebilske, W. 1975. 'The doll reflex: Ocular counterrolling with head–body tilt in the median plane.' *Vision Res.*, **15**, 713–7.
Eccles, J. C., Ito, M., and Szentágothai, J. 1967. *The Cerebellum as a Neuronal Machine.* New York: Springer.
Eckmiller, R. 1974. 'Hysteresis in the static characteristics of eye position coded neurons in the alert monkey.' *Pflüg. Arch. ges. Physiol.*, **350**, 249–58.
Eckmiller, R. 1975. 'Differences in the activity of eye-position coded neurons in the alert monkey during fixation and tracking movements.' In G. Lennerstrand and P. Bach-y-Rita (Eds.), *Basic Mechanisms of Ocular Motility and Their Clinical Implication.* New York: Pergamon, pp. 447–52.
Eckmiller, R., and Mackeben, M. 1978. 'How does a monkey keep a slowly moving light spot on both foveas?' *Invest. Ophthal.*, **17**, *Suppl.*, *ARVO Proc.*, 271.
Eckmiller, R., and Mackeben, M. 1980. 'Pre-motor unit activity in the monkey brain stem correlated with eye velocity during pursuit.' *Brain Res.*, **184**, 210–4.
Edginton, E. S. 1953. 'Kinaesthetically guided movements of head and arm.' *J. Psychol.*, **36**, 51–57.
Edwards, W. 1954. 'Two- and three-dimensional autokinetic movement as a function of size and brightness of stimuli.' *J. exp. Psychol.*, **48**, 391–8.
Efstathiou, A. 1969. 'Effects of exposure time and magnitude of prism transform on eye–hand coordination.' *J. exp. Psychol.*, **81**, 235–40.
Eldred, C. A. 1973. 'Judgments of right side up and figure rotation by young children.' *Child Devel.*, **44**, 395–9.
Elliott. J. M., and Connolly, K. J. 1978. 'The effect of visual frame of reference on a judgement of plane stimulus orientation by children.' *Perception*, **7**, 139–49.
Ellis, S. R. 1975. 'Fragmentation of fixated line stimuli as a function of gravitational orientation.' *Percept. Psychophys.*, **17**, 601–6.
Elsner, A. 1978. 'Hue difference contours can be used in processing orientation information.' *Percept. Psychophys.*, **24**, 451–6.

Elsner, W. 1971. 'Power laws for the perception of rotation and the oculogyral illusion.' *Percept. Psychophys.*, **9**, 418–20.

Emerson, P., Wenderoth, P., Curthoys, I., and Edmonds, I. 1975. 'Measuring perceived orientation.' *Vision Res.*, **15**, 1031–3.

Emerson, R. C., and Gerstein, G. L. 1977. 'Simple striate neurons in the cat. II. Mechanisms underlying directional asymmetry and directional selectivity.' *J. Neurophysiol.*, **40**, 136–73.

Emsley, H. H. 1946. *Visual Optics*, 4th Ed. London: Hatton Press.

Engström, H. 1961. 'The innervation of the vestibular sensory cells.' *Acta oto-Laryng.*, **163**, 30–41.

Enoch, J. M., Birch, D. G., and Birch, E. E. 1979. 'Monocular light exclusion for a period of days reduces directional sensitivity of the human retina.' *Science*, **206**, 705–7.

Enoksson, P. 1963. 'Binocular rivalry and monocular dominance studied with optokinetic nystagmus.' *Acta ophthal., Kbh.*, **41**, 544–63.

Enright, J. T. 1980. 'Ocular translation and cyclotorsion due to changes in fixation distance.' *Vision Res.*, **20**, 595–601.

Enroth-Cugell, C., and Robson, J. G. 1966. 'The contrast sensitivity of retinal ganglion cells of the cat.' *J. Physiol.*, **187**, 517–52.

Enroth-Cugell, C., and Shapley, R. M. 1973. 'Adaptation and dynamics of cat retinal ganglion cells.' *J. Physiol.*, **233**, 271–309.

Erickson, R. P. 1968. 'Stimulus coding in topographic and non-topographic afferent modalities: on the significance of the activity of individual sensory neurons.' *Psychol. Rev.*, **75**, 447–65.

Erisman, T., and Kohler, I. 1953. 'Upright vision through inverting spectacles.' Penn State College: Psychol. Cinema. Reg. No. 2070.

Essock, E. A. 1980. 'The oblique effect of stimulus identification considered with respect to two classes of oblique effects.' *Perception*, **9**, 37–46.

Estévez, O., and Cavonius, C. R. 1976. 'Low-frequency attenuation in the detection of gratings: sorting out the artefacts. *Vision Res.*, **16**, 497–500.

Evans, S. H. 1967. 'Redundancy as a variable in pattern perception.' *Psychol. Bull.*, **67**, 104–13.

Evinger, C., and Fuchs, A. F. 1978. 'Saccadic, smooth pursuit, and optokinetic eye movements of the trained cat. *J. Physiol.*, **285**, 209–29.

Ewert, P. H. 1937. 'Factors in space localization during inverted vision: II. An explanation of interference and adaptation.' *Psychol. Rev.*, **44**, 105–16.

Eysel, U. T., Gonzalez-Aguilar, F., and Mayer, U. 1980. 'A functional sign of reorganization in the visual system of adult cats: lateral geniculate neurons with displaced receptive fields after lesions of the nasal retina.' *Brain Res.*, **181**, 285–300.

Fagan, J. F, 1972, 'Infant recognition memory for faces.' *J. exp. Child Psychol.*, **14**, 453–75.

Fagan, J. F., and Shepherd, P. A. 1979. 'Infants' perception of face orientation.' *Infant Behav. Devel.*, **2**, 227–33.

Farrell, W. S. 1979. 'Coding left and right.' *J. exp. Psychol: Hum. Percept. Perform.*, **5**, 42–51.

Faugier-Grimaud, S., Frenois, C., and Stein, D. G. 1978. 'Effects of posterior parietal lesions on visually guided behavior in monkeys.' *Neuropsychologia*, **16**, 151–68.

Favilla, M., Ghelarducci, B., Hill, C. D., and Spyer, K. M. 1980. 'Vestibular inputs to the fastigial nucleus; evidence of convergence of macular and ampular inputs.' *Pflüg. Arch. Europ. J. Physiol.*, **384**, 193–206.

Feinberg, T. E., Pasik, T., and Pasik, P. 1978. 'Extrageniculostriate vision in the monkey. VI. Visually guided accurate reaching behaviour.' *Brain Res.*, **152**, 422–8.

Feinstein, R., and Williams, W. J. 1972. 'Interactions of the horizontal and vertical human oculomotor systems: The saccadic system.' *Vision Res.*, **12**, 33–44.

Feldman, M., and Cohen, B. 1968. 'Electrical activity in the lateral geniculate body of the alert monkey associated with eye movements.' *J. Neurophysiol.*, **31**, 455–66.

Fellows, B. J., and Brooks, B. 1973. 'An investigation of the role of matching and mismatching frameworks upon the discrimination of differently oriented line stimuli in young children.' *J. child Psychol. Psychiat.*, **14**, 293–9.

Fernández, C., and Goldberg, J. M. 1971. 'Physiology of peripheral neurons innervating semicircular canals of the squirrel monkey. II. Response to sinusoidal stimulation and dynamics of peripheral vestibular system.' *J. Neurophysiol.*, **34**, 661–75.

Fernández, C., and Goldberg, J. M. 1976a. 'Physiology of peripheral neurons innervating otolith organs of the squirrel monkey. I. Response to static tilts and to long-duration centrifugal force.' *J. Neurophysiol.*, **39**, 970–84.

Fernández, C., and Goldberg, J. M. 1976b. 'Physiology of peripheral neurons innervating otolith organs of the squirrel monkey. II. Directional selectivity and force – response relations.' *J. Neurophysiol.*, **39**, 985–95.

Fernández, C., and Goldberg, J. M. 1976c. 'Physiology of peripheral neurons innervating otolith organs of the squirrel monkey. III. Response dynamics.' *J. Neurophysiol.*, **39**, 996–1008.

Ferrell, W. R. 1980. 'The adequacy of stretch receptors in the cat knee joint for signalling joint angle throughout a full range of movement.' *J. Physiol.*, **299**, 85–100.

Ferrier, D. 1874. 'The localization of function in the brain.' *Proc. roy. Soc., B*, **22**, 229–32.

Festinger, L., and Holtzman, J. D. 1978. 'Retinal image smear as a source of information about magnitude of eye movement.' *J. exp. Psychol.: Hum. Percept. Perform.*, **4**, 573–85.

Festinger, L., Ono, H., Burnham, C. A., and Bamber, D. 1967. 'Efference and the conscious experience of perception.' *J. exp. Psychol.*, **74**, 1–36.

Field, J. 1977. 'Coordination of vision and prehension in young infants'. *Child Devel.*, **48**, 97–103.

Filehne, W. 1922. 'Über das optische Wahrnehmen von Bewegungen.' *Z. Sinnesphysiol.*, **53**, 134–44.

Findlay, J. M. 1974. 'Direction perception and human fixation eye movements.' *Vision Res.*, **14**, 703–11.

Findlay, J. M. 1980a. 'The Visual stimulus for saccadic eye movements in human observers.' *Perception*, **9**, 7–21.

Findlay, J. M. 1980b. 'Local and global influences on saccadic eye movements.' In D. F. Fisher, R. A. Monty and J. W. Senders (Eds.), *The Last Whole Earth Eye Movement Conference*. New York: Erlbaum.

Findlay, J. M., and Parker, D. M. 1972. 'An investigation of visual orientation constancy using orientation-specific properties of acuity and adaptation.' *Perception*, **1**, 305–13.

Fine, B. J., and Danforth, A. V. 1975. 'Field-dependence, extraversion, and perception of the vertical: empirical and theoretical perspectives of the rod-and-frame test.' *Percept. mot. Skills*, **40**, 683–93.

Finke, R. A., and Held, R. 1978. 'State reversals of optically induced tilt and torsional eye movements.' *Percept. Psychophys.*, **23**, 337–40.

Finlay, B. L., Schiller, P. H., and Volman, S. F. 1976. 'Meridional differences in orientation sensitivity in monkey striate cortex.' *Brain Res.*, **105**, 350–2.

Fiorentini, A., Ghez, C., and Maffei, L. 1972. 'Physiological correlates of adaptation to a rotated visual field.' *J. Physiol.*, **227**, 313–22.

Fischer, B. 1973. 'Overlap of receptive field centers and representation of the visual field in the cat's optic tract.' *Vision Res.*, **13**, 2113–20.

Fischer, M. H. 1927. 'Messende Untersuchungen über die Gegenrollung der Augen und die Lokalisation der schienbaren Vertikalen bei seitlicher Neigung (des Kopfes, des Stammes und des Gesamtkorpers). I. Neigungen biz zu 40°.' *v. Graefe's Arch. Ophthal.*, **118**, 633–80.

Fischer, M. H. 1930. 'Messende Untersuchungen über die Gegenrollung der Augen und die Lokalisation der scheinbaren Vertikalen bei seitlicher Neigung des Gesamtkörpers bis zu 360°: II Mitteilung. Untersuchungen an Normalen. *v. Graefe's Arch. Ophthal.*, **123**, 476–508.

Fischer, M. H., and Kornmüller, A. E. 1930. 'Optokinetisch ausgelöste Bewegungswahrnehmung und optokinetischer Nystagmus.' *J. Psychol. Neurol. (Lpz)*, **41**, 273–308.

Fisher, C. B. 1980. 'Children's memory for line orientation: A reexamination of the oblique effect.' *J. exp. Child Psychol.*, **29**, 446–59.

Fisher, G. H. 1960. Intersensory localization in three modalities.' *Bull. Brit. Psychol. Soc.*, **41**, 24–5.

Fisher, G. H. 1962a. 'Intersensory localization.' *Ph.D. Thesis*, Univ. Hull, England.

Fisher, G. H. 1962b. 'Resolution of spatial conflict.' *Bull. Brit. Psychol. Soc.*, **46**, 3A.

Fisher, G. H. 1969. 'An experimental study of angular subtension.' *Quart. J. exp. Psychol.*, **21**, 356–66.

Fisher, R. A. 1966. *The Design of Experiments*, 8th Edn. London: Oliver & Boyd.

Fishkin, S. M. 1969. 'Passive vs. active exposure and other variables related to the occurrence of hand adaptation to lateral displacement.' *Percept. mot. Skills*, **29**, 291–7.

Fishman, M. C., and Michael, C. R. 1973. 'Integration of auditory information in the cat's visual cortex.' *Vision Res.*, **13**, 1415–9.

Fitts, P. M. 1954. 'The information capacity of the human motor system in controlling the amplitude of movement.' *J. exp. Psychol.*, **47**, 381–91.

Fitts, P. M., and Peterson, J. R. 1964. 'Information capacity of discrete motor responses.' *J. exp. Psychol.*, **67**, 103–12.

Fitts, P. M., and Posner, M. I. 1967. *Human Performance*. Belmont, Calif.: Brooke-Cole.

Fitzgerald, G., and Hallpike, C. S. 1942. 'Studies in human vestibular function: I. Observations on the directional preponderance ('Nystagmusberietschaft') of caloric nystagmus resulting from cerebral lesions.' *Brain*, **65**, 115–37.

Flandrin, J-M., and Jeannerod, M. 1977. 'Developmental constraints of motion detection mechanisms in the kitten.' *Perception*, **6**, 513–27.

Flandrin, J-M., Kennedy, H., and Amblard, B. 1976. 'Effects of stroboscopic rearing on the binocularity and directionality of cat's superior colliculus neurons.' *Brain Res.*, **101**, 576–81.

Fleischl, E. von. 1882. 'Physiologisch-optische Notizen. 2. Mitteilung.' *S.-B. Akad. Wiss. Wien.*, **86**, 8–25.

Flescher, I. 1962. 'Ocular–manual laterality and perceptual rotation of literal symbols.' *Genet. Psychol. Monogr.*, **66**, 3–48.

Flock, Å. 1964. 'Structure of the macula utriculi with special reference to directional interplay of sensory responses as revealed by morphological polarization.' *J. Cell Biol.*, **22**, 413–31.

Flock, Å. 1965. 'Electron microscopic and electrophysiological studies on the lateral line canal organ.' *Acta oto-laryng.*, **Suppl. 199**, 1–90.

Flom, M. C., Heath, G. G., and Takahashi, E. 1963. 'Contour interaction and visual resolution: contralateral effects.' *Science*, **142**, 979–80.

Flom, M. C., Weymouth, F. W., and Kahneman, D. 1963. 'Visual resolution and contour interaction.' *J. opt. Soc. Amer.*, **53**, 1026–32.

Flood, D. G., and Coleman. P. D. 1979. 'Demonstration of orientation columns with [^{14}C]2-deoxyglucose in the cat reared in a striped environment.' *Brain Res.*, **173**, 538–42.

Fluur, E. 1970. 'The interaction between the utricle and the saccule.' *Acta oto-laryng.*, **69**, 17–24.

Fluur, E. 1973a. 'Interaction between the utricles and the horizontal semicircular canals IV. Tilting of human patients with acute unilateral vestibular neuritis.' *Acta oto-laryng.*, **76**, 349–52.

Fluur, E. 1973b. 'The reason why the patient with acute labyrinthine destruction lies on his sound ear.' *Adv. Oto-Rhino-Laryng.*, **35**, 253–7.

Fluur, E. 1974a. 'Positional and positioning nystagmus as a result of utriculocupular integration.' *Acta oto-laryng.*, **78**, 19–27.

Fluur, E. 1974b. 'Investigation of positional and positioning nystagmus by using a new type of electrically-driven rotation table.' *Acta oto-laryng.*, **78**, 385–90.

Fluur, E., and Mellström, A. 1970a. 'Ultricular stimulation and oculomotor reactions.' *Laryngoscope*, **80**, 1701–12.

Fluur, E., and Mellström, A. 1970b. 'Saccular stimulation and oculomotor reactions.' *Laryngoscope*, **80**, 1713–21.

Fluur, E., and Mellström, A. 1971a. 'Vestibular nystagmus—a differential reaction.' *Acta oto-laryng.*, **71**, 299–302.

Fluur, E., and Mellström, A. 1971b. 'The otolith organs and their influence on oculomotor movements.' *Exp. Neurol.*, **30**, 139–47.

Fluur, E., and Mendel, L. 1966. 'Relation between strength of stimulus and duration of latency time in vestibular rotatory nystagmus.' *Acta oto-laryng.*, **61**, 463–74.

Fluur, E., and Mendel, L. 1970. 'Crescendo time for per-rotatorily elicited nystagmus.' *Acta oto-laryng.*, **69**, 239–46.

Fluur, E., and Siegborn, J. 1973a. 'Interaction between the utricles and the horizontal semicircular canals. I. Unilateral selective sectioning of the horizontal ampullar nerve followed by tilting around the longitudinal axis.' *Acta oto-laryng.*, **75**, 17–20.

Fluur, E., and Siegborn, J. 1973b. 'Interaction between the utricles and the horizontal semicircular canals. II. Unilateral selective section of the horizontal ampullar and the utricular nerve followed by tilting around the longitudinal axis.' *Acta oto-laryng.*, **75**, 393–5.

Fluur, E., and Siegborn, J. 1973c. 'Interaction between the utricles and the horizontal semicircular canals. III. Sectioning of the horizontal ampullar nerve on one side and of the utricular nerve on the other, following by tilting around the longitudinal axis. *Acta oto-laryng.*, **75**, 485–8.

Fluur, E., and Siegborn, J. 1973d. 'The otolith organs and the nystagmus problem.' *Acta oto-laryng.*, **76**, 438–42.

Fluur, E., and Siegborn, J. 1974a. 'Interaction between the utricles and the vertical semicircular canals. V. Bilateral selective sectioning of anterior or posterior ampullar nerves followed by tilting around their bitemporal or longitudinal axis. *Acta oto-laryng.*, **77**, 80–5.

Fluur, E., and Siegborn, J. 1974b. 'Interaction between the utricles and the vertical semicircular canals. VI. Unilateral selective sectioning of the horizontal and vertical ampullar nerves, followed by tilting around the horizontal axis. *Acta oto-laryng.*, **77**, 167–70.

Fogelgren, L. A., and Shebilske, W. L. 1979. 'Central visual learning and illusory visual direction after backward head tilts.' *Percept. Psychophys.*, **25**, 519–23.

Foley, J. E. 1970a. 'Prism adaptation with opposed base orientation: The weighting of direction information from the two eyes.' *Percept. Psychophys.*, **8**, 23–5.

Foley, J. E. 1970b. 'Prism adaptation with simultaneous receipt of normal input.' *Percept. Psychophys.*, **8**, 393–5.

Foley, J. E. 1974. 'Factors governing interocular transfer of prism adaptation.' *Pyschol. Rev.*, **81**, 183–6.

Foley, J. E., and Miyanshi, K. 1969. 'Interocular effects in prism adaptation.' *Science*, **165**, 311–2.

Foley, J. M. 1975. 'Error in visually directed manual pointing.' *Percept. Psychophys.*, **17**, 69–74.

Foley, J. M., and Held, R. 1972. 'Visually directed pointing as a function of target distance, direction, and available cues.' *Percept. Psychophys.*, **12**, 263–8.

Foley, P. J. 1962. 'Stimulus orientation and retinal summation.' *J. opt. Soc. Amer.*, **52**, 474–5.

Ford, A. S. 1910. 'The pendular whiplash illusion.' *Psychol. Rev.*, **17**, 192–204.

Fox, R., Lehmkuhle, S., and Leguire, L. E. 1978. 'Stereoscopic contours induce

optokinectic nystagmus.' *Vision Res.*, **18**, 1189–92.
Fraenkel, G. S., and Gunn, D. L. 1940. *The Orientation of Animals*. Oxford: Clarendon.
Fraser, J. 1908. 'A new illusion of direction.' *Brit. J. Psychol.*, **2**, 307–20.
Fredrickson, J. M., and Schwarz, D. W. F. 1977. 'Vestibulo-cortical projection.' In R. F. Naunton (Ed.), *The Vestibular System*. New York: Academic Press, pp. 203–10.
Freedman, D. G. 1974. *Human Infancy: An Evolutionary Perspective*. Hillside, N. J.: Erlbaum.
Freedman, S. J., and Rekosh, J. H. 1968. 'The functional integrity of spatial behavior.' In S. J. Freedman (Ed.), *The Neuropsychology of Spatially Oriented Behavior*. Homewood, Ill.: Dorsey, pp. 153–62.
Freeman J. A., and Nicholson, C. N. 1970. 'Space–time transformation in the frog cerebellum through an intrinsic tapped delay line.' *Nature*, **226**, 640–2.
Freeman, R. D., and Pettigrew, J. D. 1973. 'Alteration of visual cortex from environmental asymmetries.' *Nature*, **246**, 359–60.
Freeman, R. D., and Thibos, L. N. 1973. 'Electrophysiological evidence that abnormal early visual experience can modify the human brain.' *Science*, **180**, 876–8.
Freeman, R. D., and Thibos, L. N. 1975a. 'Contrast sensitivity in humans with abnormal visual experience.' *J. Physiol.*, **247**, 687–710.
Freeman, R. D., and Thibos, L. N. 1975b. 'Visual evoked responses in humans with abnormal visual experience.' *J. Physiol.*, **247**, 711–24.
Frégnac, Y., and Imbert, M. 1978. 'Early development of visual cortical cells in normal and dark-reared kittens: relationship between orientation, selectivity and ocular dominance.' *J. Physiol.*, **278**, 27–44.
Frégnac, Y., Trotter, E., Bienenstock, P., Buisseret, E. Gary-Bobo, E., and Imbert, M. 1981. 'Effect of neonatal unilateral enucleation on the development of orientation selectivity in the primary visual cortex of normally and dark-reared kittens.' *Exp. Brain Res.*, **42**, 453–466.
Freund, H. J. 1973. 'Neuronal mechanisms of the lateral geniculate body.' In R. Jung (Ed.), *Handbook of Sensory Physiology*, Vol. VII/3B. New York: Springer, pp. 177–246.
Friedmann, G. 1971. 'The influence of unilateral labyrinthectomy on orientation in space.' *Acta oto-laryng.*, **71**, 289–98.
Fries, W., Albus, K., and Creutzfeldt, O. D. 1977. 'Effects of interacting visual patterns on single cell responses in cat's striate cortex.' *Vision Res.*, **17**, 1001–8.
Frith, U. 1971. 'Why do children reverse letters?' *Brit. J. Psychol.*, **62**, 459–68.
Frome, F. S., Levinson, J. Z., Danielson, J. T., and Clavadetscher, J. E. 1979. 'Shifts in perception of size after adaptation to gratings.' *Science*, **206**, 1327–29.
Frost, B. J., and Kaminer, J. J. 1975. 'The orientation anistropy and orientation constancy: A visual evoked potential study.' *Perception*, **4**, 51–8.
Frost, D., and Pöppel, E. 1976. 'Different programming modes of human saccadic eye movements as a function of stimulus eccentricity: Indications of a functional subdivision of the visual field.' *Biol. Cybernetics*, **23**, 39–48.
Fry, G. A. 1950. 'Visual perception of space.' *Amer. J. Optom.*, **27**, 531–53.
Fry, G. A., Treleavan, C. L., and Baxter, R. C. 1945. 'Specification of the direction of regard.' *Amer. J. Optom.*, **22**, 351–60.
Fry, G. A., Treleavan, C. L., Walsh, R., Higgins, E. L., and Radde, C. A. 1947. 'Definition and measurement of torsion.' *Amer. J. Optom.*, **24**, 329–34.
Fuchs, A. F. 1967. 'Saccadic and smooth pursuit eye movements in the monkey.' *J. Physiol.*, **191**, 609–31.
Fuchs, A. F., and Kimm, J. 1975. 'Unit activity in vestibular nucleus of the alert monkey during horizontal angular acceleration and eye movement.' *J. Neurophysiol.*, **38**, 1140–61.
Fuchs, A. F., and Kornhuber, H. H. 1969. Extraocular muscle afferents to the cerebellum of the cat.' *J. Physiol.*, **200**, 713–22.
Fuchs, A. F., and Luschei, E. S. 1970. 'Firing patterns of abducens neurons of alert

monkeys in relationship to horizontal eye movement.' *J. Neurophysiol.*, **33**, 382–92.

Fukuda, Y., and Stone, J. 1974. 'Retinal distribution and central projections of Y-X- and W-cells of the cat's retina.' *J. Neurophysiol.*, **37**, 749–72.

Fuller, J. H. 1980. 'The dynamic neck-eye reflex in mammals.' *Exp. Brain Res.*, **41**, 29–35.

Funaishi, S. 1926. 'Weiteres über das Zentrum des Sehrichtungen.' *V. Graefe's Arch. Ophthal.*, **117**, 296–303.

Funaishi, S. 1927. 'Uber die falsche Lichtlokalisation bei geschlossenen Lidern sowie uber das subjektive Zyklopenauge.' *V. Graefe's Arch. Ophthal.*, **119**, 227–34.

Furchner, C. S., and Ginsburg, A. P. 1978. 'Monocular rivalry of a complex waveform.' *Vision Res.*, **18**, 1641–8.

Furchner, C. S., Thomas, J. P., and Campbell, F. W. 1977. 'Detection and discrimination of simple and complex patterns at low spatial frequencies.' *Vision Res.*, **17**, 827–36.

Gaarder, K. 1966. 'Fine eye movements during inattention.' *Nature*, **209**, 83–4.

Gaarder, K. 1967. 'Mechanisms in fixation saccadic eye movements.' *Brit. J. physiol. Opt.*, **24**, 28–44.

Gabor, D. 1946. Theory of communication. *J. I.E.E. (London)*, **93,** 429–57.

Gacek, R. R. 1974. 'Morphological aspects of the efferent vestibular system.' In H. H. Kornhuber (Ed.), *Handbook of Sensory Physiology*, Vol. VI/1. New York: Springer, pp. 213–20.

Gacek, R. R. 1975. 'The innervation of the vestibular labyrinth.' In R. F. Naunton (Ed.), *The Vestibular System*. New York: Academic Press, pp. 21–30.

Gajzago, C., and Day, R. H. 1972. 'Uprightness constancy with head inversion in young children and adults.' *Child Psychol.*, **14**, 43–52.

Ganz, L. 1966. 'Mechanism of the figural aftereffects.' *Psychol. Rev.*, **73**, 128–50.

Gardner, E. P., and Fuchs, A. F. 1975. 'Single-unit responses to natural vestibular stimuli and eye movements in deep cerebellar nuclei of the alert monkey.' *J. Neurophysiol.*, **38**, 627–49.

Gardner, R., and Weitzman, E. D. 1967. 'Examination for optokinetic nystagmus in sleep and waking.' *A. M. A. Arch. Neurol.*, **16**, 415–20.

Garey, L. J., and Blakemore, C. 1977. 'Monocular deprivation: Morphological effects on different classes of neurons in the lateral geniculate nucleus.' *Science*, **195**, 414–6.

Garey, L. J., and Powell, T. P. S. 1971. 'An experimental study of the termination of the lateral geniculo-cortical pathway in the cat and monkey.' *Proc. roy. Soc. B,* **179,** 41–63.

Garner, W. R. 1962. *Uncertainty and Structure as Psychological Concepts*. New York: Wiley.

Garner, W. R. 1974. *The Processing of Information and Structure*. New York: Wiley.

Garten, S. 1920. 'Über die Grundlagen unserer Orientierung im Raum.' *Abh. sächs Ges. (Akad.) Wiss.*, **36**, 431–510.

Gattass, R., Oswaldo-Cruz, E., and Sousa, A. P. B. 1979. 'Visual receptive fields of units in the pulvinar of cebus monkey.' *Brain Res.*, **160**, 413–30.

Gattass, R., Sousa, A. P. B., and Oswaldo-Cruz, E. 1978. 'Single unit response type in the pulvinar of the cebus monkey to multisensory stimulation.' *Brain Res.*, **158**, 75–88.

Gauthier, G. M., and Hofferer, J.-M. 1976. 'Eye tracking of self-moved targets in the absence of vision.' *Exp. Brain Res.*, **26**, 121–39.

Gauthier, G. M., and Robinson, D. A. 1975. 'Adaptation of the human vestibular – ocular reflex to magnifying lenses.' *Brain Res.*, **92**, 331–5.

Gaze, R. M. 1970. *The Formation of Nerve Connections*. London: Academic Press.

Gazzaniga, M. S. 1969, 'Cross cuing mechanisms and ipsilateral eye–hand control in split-brain monkeys.' *Exp. Neurol.*, **23**, 11–17.

Gazzaniga, M. S., Bogen, J. E., and Sperry, R. W. 1967. 'Dyspraxia following division of the cerebral commissures.' *Arch. Neurol.*, **16**, 606–12.

Georgeson, M. A. 1973. 'Spatial frequency selectivity of a visual tilt illusion.' *Nature*, **245**, 43–5.

Georgeson, M. A., 1976a. 'Antagonism between channels for pattern and movement in human vision.' *Nature*, **259**, 413–5.

Georgeson, M. A. 1976b. 'Psychophysical hallucinations of orientation and spatial frequency.' *Perception*, **5**, 99–111.

Georgeson, M. A., and Sullivan, G. D. 1975. 'Contrast constancy: deblurring in human vision by spatial frequency channels.' *J. Physiol.*, **252**, 627–56.

Gernandt, B. E. 1974. 'Vestibulo-spinal mechanisms.' In H. H. Kornhuber (Ed.), *Handbook of Sensory Physiology*, Vol. VI/1. New York: Springer, pp. 541–64.

Ghent, L. 1961. 'Form and its orientation: A child's-eye view.' *Amer. J. Psychol.*, **74**, 177–90.

Ghent, L., Bernstein, L., and Goldweber, A. M. 1960. 'Preferences for orientation of form under varying conditions.' *Percept. mot. Skills*, **11**, 46.

Gibson, J. J. 1933. 'Adaptation, after-effect and contrast in the perception of curved lines.' *J. exp. Psychol.*, **16**, 1–31.

Gibson, J. J. 1937a. 'Adaptation, after-effect and contrast in the perception of tilted lines: II. Simultaneous contrast and the areal restriction of the after-effect.' *J. exp. Psychol.*, **20**, 553–69.

Gibson, J. J. 1937b. 'Adaptation with negative after-effects.' *Psychol. Rev.*, **44**, 222–44.

Gibson, J. J. 1950. *The Perception of the Visual World.* Boston: Houghton Mifflin.

Gibson, J. J. 1958. 'Visually controlled locomotion and visual orientation in animals.' *Brit. J. Psychol.*, **49**, 182–94.

Gibson, J. J. 1966. *The Sense Considered as Perceptual Systems.* Boston: Houghton Mifflin.

Gibson, J. J. 1968. 'What gives rise to the perception of motion?' *Psychol. Rev.*, **75**, 335–64.

Gibson, J. J., Olum, P., and Rosenblatt, F. 1955. 'Parallax and perspective during aircraft landings.' *Amer. J. Psychol.*, **68**, 373–85.

Gibson, J. J., and Radner, M. 1937. 'Adaptation, after-effect, and contrast in the perception of tilted lines. I. Quantitative studies.' *J. exp. Psychol.*, **20**, 453–67.

Gilbert, C. D. 1977. 'Laminar differences in receptive field properties of cells in cat primary visual cortex.' *J. Physiol.*, **268**, 391–420.

Gilbert, C. D., and Wiesel, T. N. 1979. 'Morphology and intracortical projections of functionally characterized neurones in the cat visual cortex.' *Nature*, **280**, 120–5.

Gilbert, D. S. 1969. 'Monocular estimates of distance and direction with stabilized and non-stabilized retinal images.' *Vision Res.*, **9**, 103–15.

Gilbert, D. S., and Fender, D. H. 1969. 'Contrast thresholds measured with stablized and non-stabilized sine-wave gratings.' *Optica Acta*, **16**, 191–204.

Gilchrist, A. L. 1977. 'Perceived lightness depends on perceived spatial arrangement.' *Science*, **195**, 185–7.

Gilinsky, A. S. 1968. 'Orientation-specific effects of patterns of adapting light on visual acuity.' *J. opt. Soc. Amer.*, **58**, 13–8.

Gilinsky, A. S., and Doherty, R. S. 1969. 'Interocular transfer of orientational effects.' *Science*, **164**, 454–5.

Gilinsky, A. S., and Mayo, T. 1971. 'Inhibitory effects of orientational adaptation.' *J. opt. Soc. Amer.*, **61**, 1710–4.

Gilson, R. D., Stockwell, C. W., and Guedry, F. E. 1973. Nystagmus responses during triangular velocity about the y- and z-axes.' *Acta oto-laryng.*, **75**, 21–6.

Ginsburg, A. P., Cannon, M. W., and Nelson, M. A. 1980. 'Suprathreshold processing of complex visual stimuli: evidence for linearity in contrast perception.' *Science*, **208**, 619–21.

Glenny, G., and Heywood, S. 1979. 'Hans Gertz revisited: the different effects of invisibility and darkness on pursuit eye movements.' *Perception*, **8**, 31–6.

Gogel, W. C. 1977. 'Independence of motion induction in separated portions of the visual field.' *Bull. Phychonom. Soc.*, **10**, 408–10.

Gogel, W. C., and Newton, R. E. 1975. 'Depth adjacency and the rod-and-frame illusion.' *Percept. Psychophys.*, **18**, 163–71.

Goldberg, J. M., and Fernández, C. 1971a. 'Physiology of peripheral neurons innervating semicircular canals of the squirrel monkey. I. Resting discharge and response to constant angular accelerations.' *J. Neurophysiol.*, **34**, 635–60.

Goldberg, J. M., and Fernández, C. 1971b. 'Physiology of peripheral neurons innervating semicircular canals of the squirrel monkey. III. Variations among units in their discharge properties.' *J. Neurophysiol.*, **34**, 676–84.

Goldberg, J. M., and Fernández, C. 1975. 'Responses of peripheral vestibular neurons to angular and linear accelerations in the squirrel monkey.' *Acta oto-laryng.*, **80**, 101–10.

Goldberg, J. M., and Fernández, C. 1977. 'Conduction times and background discharge of vestibular afferents.' *Brain Res.*, **122**, 545–50.

Goldberg, J. M., and Fernández, C. 1980. 'Efferent vestibular system in the squirrel monkey: anatomical location and influence on afferent activity.' *J. Neurophysiol.*, **43**, 986–1025.

Goldberg, M. E., and Robinson, D. L. 1978. 'Visual system: Superior colliculus.' In R. B. Masterton (Ed.), *Handbook of Behavioral Neurobiology*, Vol. I. *Sensory Integration.* New York: Plenum, pp. 119–64.

Goldberg, M. E., and Wurtz, R. H. 1972a. 'Activity of superior colliculus in behaving monkey. I. Visual receptive fields of single neurons.' *J. Neurphysiol.*, **35**, 542–51.

Goldberg, M. E., and Wurtz, R. H. 1972b. 'Activity of superior colliculus in behaving monkey. II. Effect of attention on neuronal responses.' *J. Neurophysiol.*, **35**, 560–74.

Goldstein, A. G. 1975. 'Recognition of inverted photographs of faces by children and adults.' *J. gen. Psychol.*, **127**, 109–23.

Goldstein, A. G., and Andrews, J. 'Perceptual uprightness and complexity of random shapes.' *Amer. J. Psychol.*, **75**, 667–69.

Goldstein, D. M., and Wicklund, D. A. 1973. 'The acquisition of the diagonal concept.' *Child Devel.*, **44**, 210–13.

Gombrich, E. H. 1969. *Art and Illusion.* Princeton, N. J.: Princeton Univ. Press.

Gonshor, A., and Melvill Jones, G., 1973. Changes of human vestibulo-ocular response induced by vision-reversal during head rotation. *J. Physiol.*, **234**, 102–103P.

Gonshor, A., & Melvill Jones, G. 1976a. 'Extreme vestibulo-ocular adaptation induced by prolonged optical reversal of vision.' *J. Physiol.*, **256**, 381–414.

Gonshor, A., and Melvill Jones, G. 1976b. 'Short-term adaptive changes in the human vestibulo-ocular reflex.' *J. Physiol.*, **256**, 361–79.

Gonshor, A., and Melvill Jones, G. 1980. 'Postural adaptation to prolonged optical reversal of vision in man.' *Brain Res.*, **192**, 239–48.

Goodale, M. A. 1973. 'Cortico-tectal and intertectal modulation of visual responses in the rat's superior colliculus.' *Exp. Brain Res.*, **17**, 75–86.

Goodale, M. A., Foreman, N. P., and Milner, A. D. 1977. 'Visual orientation in the rat: a dissociation of deficits following cortical and collicular lesions.' *Exp. Brain Res.*, **30**, 1–13.

Goodenough, D. R., Oltman, P. K., Sigman, E., Rosso, J., and Mertz, H. 1979. 'Orientation contrast effects in the rod-and-frame test.' *Percept. Psychophys.*, **25**, 419–24.

Goodenough, D. R., Sigman, E., Oltman, P., Rosso, J., and Mertz, H. 1979. 'Eye torsion in response to a tilted visual stimulus.' *Vision Res.*, **19**, 1177–80.

Goodwin, A. W., and Fender, D. H. 1973a. 'Recognition of component differences in two-dimensional oculomotor tracking tasks.' *Vision Res.*, **13**, 1905–13.

Goodwin, A. W., and Fender, D. H. 1973b. 'The interaction between horizontal and vertical eye-rotations in tracking tasks.' *Vision Res.*, **13**, 1701–12.

Gordon, B. 1973. 'Receptive fields in deep layers of cat superior colliculus.' *J. Neurophysiol.*, **36**, 157–78.

Gordon, B., Moran, J., and Presson, J. 1979. 'Visual fields of cats reared with one eye intorted.' *Brain Res.*, **174**, 167–71.

Gordon, B., Presson, J., Packwood, J., and Scheer, R. 1979. 'Alteration of cortical orientation selectivity: Importance of asymmetric input. *Science*, **204**, 1109–11.

Göthlin, G. F. 1946. 'Entoptic analysis of vestibular nystagmus.' *Acta oto-laryng.*, **34**, 230–45.

Gouras, P. 1972. 'Colour opponency from fovea to striate cortex.' *Invest. Ophthal.*, **11**, 427–34.

Graham, N. 1977. 'Visual detection of aperiodic spatial stimuli by probability summation among narrowband channels.' *Vision Res.*, **17**, 637–52.

Graham, N. 1980. 'Spatial-frequency channels in human vision: detecting edges without edge detectors.' In C. S. Harris (Ed.), *Visual Coding and Adaptability*. Hillside, N. J.: Erlbaum, pp. 215–62.

Graham, N., and Nachmias, J. 1971. 'Detection of grating patterns containing two spatial frequencies: A comparison of single-channel and multiple-channel models.' *Vision Res.*, **11**, 251–9.

Granit, R. 1971. 'The probable role of muscle spindles and tendon organs in eye movement control.' In P. Bach-y-Rita, C. C. Collins and J. E. Hyde (Eds.), *The Control of Eye Movements*. New York: Academic Press, pp. 3–6.

Grantyn, A., Grantyn, R., and Ronbiné, K.-P. 1977. 'Neuronal organization of the tecto-oculomotor pathways.' In R. Baker and A. Berthoz (Eds.), *Control of Gaze by Brain Stem Neurons*. New York: Elsevier, pp. 197–207.

Grantyn, A., Grantyn, R., Robiné, K.-P., and Berthoz, A. 1979. 'Electroanatomy of tectal efferent connections related to eye movements in the horizontal plane.' *Exp. Brain Res.*, **37**, 149–72.

Graybiel, A. 1952. 'The oculogravic illusion.' *A. M. A. Arch. Ophthal.*, **48**, 605–15.

Graybiel, A., and Clark, B. 1945. 'The autokinetic illusion and its significance in night flying.' *J. aviat. Med.*, **16**, 111–51.

Graybiel, A., and Hupp, D. I. 1946. 'The oculo-gyral illusion, a form of apparent motion which may be observed following stimulation of the semicircular canals.' *J. aviat. Med.*, **17**, 3–27.

Graybiel, A., Johnson, W. H., Money, K. E., Malcolm, R. E., and Jennings, G. L. 1979. 'Oculogravic illusion in response to straight-ahead acceleration of a CF-104 aircraft.' *Aviat. space, environ, Med.*, **50**, 383–6.

Graybiel, A., and Kellogg, R. S. 1967. 'The inversion illusion in parabolic flight: Its probable dependence on otolith function.' *Aerosp. Med.*, **38**, 1099–102.

Graybiel, A., Kerr, W. A., and Bartley, S. H. 1948. 'Stimulus thresholds of the semicircular canals as a function of angular acceleration.' *Amer. J. Psychol.*, **61**, 21–36.

Graybiel, A., Miller, E. F., Billingham, J., Waite, R., Berry, C. A., and Deitlein, L. F. 1967. 'Vestibular experiments in Gemini flights V and VII'. *Aerosp. Med.*, **38**, 360–70.

Graybiel, A., Miller, E. F., Newsom, B. D., and Kennedy, R. S. 1968. 'The effect of water immersion on perception of the oculogravic illusion in normal and labyrinthine-defective subjects.' *Acta oto-laryng.*, **65**, 599–610.

Graybiel, A. M. 1974. 'Visuo-cerebellar and cerebello-visual connections involving the ventral lateral geniculate nucleus.' *Exp. Brain Res.*, **20**, 303–6.

Graybiel, A. M. 1976. 'Evidence for banding of the cat's ipsilateral retinotectal connections.' *Exp. Brain Res.*, **114**, 318–27.

Green, D. G. 1970. 'Regional variations in the visual acuity for interference fringes on the retina.' *J. Physiol.*, **207**, 351–6.

Green, M. 1980. 'Orientation-specific adaptation: effects of checkerboards on the detectability of gratings.' *Perception*, **9**, 369–78.

Gregory, R. L. 1958. 'Eye movements and the stability of the visual world.' *Nature*, **182**, 1214–6.

Gregory, R. L. 1963. 'Distorsions of visual space as inappropriate constancy scaling.' *Nature*, **199**, 678–80.

Gregory, R. L. 1965. 'Inappropriate constancy explanation of spatial distortions.' *Nature*, **207**, 891–3.

Gregory, R. L., and Zangwill, O. L. 1963. 'The origin of the autokinetic effect.' *Quart. J. exp. Psychol.*, **15**, 252–61.

Gresty, M. A. 1974. 'Coordinations of head and eye movements to fixate continuous and intermittent targets.' *Vision Res.*, **14**, 395–403.

Gresty, M., and Baker, R. 1976. 'Neurons with visual receptive field, eye movement and neck displacement sensitivity within and around the nucleus prepositus hypoglossi in the alert cat.' *Exp. Brain Res.*, **24**, 429–33.

Gresty, M., and Leech, J. 1977. 'Coordination of the head and eyes in pursuit of predictable and random target motion.' *Aviat. space. environ. Med.*, **48**, 741–4.

Gresty, M., Trinder, E., and Leech, J. 1976. 'Perception of everyday visual environments during saccadic eye movements.' *Aviat. space environ. Med.*, **47**, 991–2.

Groen, J. J. 1957. 'Cupulometry.' *Laryngoscope*, **67**, 894–910.

Groen, J. J. 1960. 'Problems of the semicircular canal from a mechanico-physiological point of view.' *Acta oto-laryng., Suppl.*, **163**, 59–67.

Groen, J. J. 1961. 'The problems of the spinning top applied to the semi-circular canals.' *Confin. neurol., Basel*, **21**, 454–5.

Groen, J. J., and Jongkees, L. B. W. 1948. 'The threshold of angular acceleration perception.' *J. Physiol.*, **107**, 1–7.

Groen, J. J., Lowenstein, O., and Vendrik, A. J.H. 1952. 'The mechanical analysis of the responses from the end organs of the horizontal semicircular canal in the isolated elasmobranch labyrinth.' *J. Physiol.*, **117**, 329–46.

Gross, C. G. 1973. 'Visual functions of inferotemporal cortex.' In R. Jung (Ed.), *Handbook of Sensory Physiology*, Vol. VII/3B. New York: Springer, pp. 451–82.

Grüttner, R. 1939. 'Experimentale Untersuchungen über den optokinetischen Nystagmus.' *Z. Sinnesphysiol.*, **68**, 1–48.

Guedry, F. E. 1964. 'Visual control of habituation to complex vestibular stimulation in man.' *Acta oto-laryng.*, **58**, 377–89.

Guedry, F. E. 1965. 'Orientation of the rotation-axis relative to gravity.' *Acta oto-laryng.*, **60**, 30–48.

Guedry, F. E. 1974. 'Psychophysics of vestibular sensation.' In H. H. Kornhuber (Ed.), *Handbook of Sensory Physiology*, Vol. VI/2. New York: Springer, pp. 3–154.

Guedry, F. E., and Benson, A. J. 1971. 'Nystagmus and visual performance during sinusoidal stimulation of the vertical semicircular canals.' *Nav. Aerosp. Med. Res. Lab. Report*, 1131, Pensacola, Florida.

Guedry, F. E., Collins, W. E., and Sheffey, P. L. 1961. 'Perceptual and oculomotor reactions to interacting visual and vestibular stimulation.' *Percept. mot. Skills*, **12**, 307–24.

Guedry, F. E., Lentz, J. M., and Jell, R. M. 1979. 'Visual – vestibular interactions: I. Influence of peripheral vision on suppression of the vestibulo-ocular reflex and visual acuity.' *Aviat. space environ. Med.*, **50**, 205–11.

Guedry, F. E., Stockwell, C. W., and Gilson, R. D. 1971. 'Comparison of subjective responses to semicircular canal stimulation produced by rotation about different axes.' *Acta oto-laryng.*, **72**, 101–6.

Guedry, F. E., Stockwell, C. W., Norman, J. W., and Owens, G. G. 1971. 'Use of triangular waveforms of angular velocity in the study of vestibular function.' *Acta oto-laryng.*, **71**, 439–48.

Guilford, J. P., and Dallenbach, K. M. 1928. 'A study of the autokinetic sensation.' *Amer. J. Psychol.*, **40**, 83–91.

Guitton, D., Crommelinck, M., and Roucoux, A. 1980. 'Stimulation of the superior colliculus in the alert cat. I. Eye movements and neck EMG activity evoked when the head is restrained.' *Exp. Brain Res.*, **39**, 63–74.

Guitton, D., and Mandl, G. 1980a. 'A comparison between saccades and quick phases of vestibular nystagmus in the cat.' *Vision Res.*, **20**, 865–73.

Guitton, D., and Mandl, G. 1980b. 'Oblique saccades of the cat: A comparison between the durations of horizontal and vertical components.' *Vision Res.*, **20**, 875–81.

Gundry, A. J. 1978. 'Experiments on the detection of roll motion.' *Aviat. space environ, Med.*, **49**, 657–64.

Gwiazda, J., Brill, S., Mohindra, I., and Held, R. 1980. 'Infant visual acuity and its meridional variation.' *Vision Res.*, **18**, 1557–64.

Gyr, J., Willey, R., and Henry, A. 1979. 'Motor-sensory feedback and geometry of visual space: an attempted replication.' *Behav. Brain Sci.*, **2**, 59–94.

Haaxma, R., and Kuypers, H. G. J. M. 1975. 'Intrahemispheric cortical connexions and visual guidance of hand and finger movements in the rhesus monkey. *Brain*, **98**, 239–60.

Haciska, D. T. 1973. 'The influence of drugs on caloric-induced nystagmus.' *Acta oto-laryngol.*, **75**, 477–84.

Haddad, G. M., Demer, J. L., and Robinson, D. A. 1980. 'The effect of lesions of the dorsal cap of the inferior olive on the vestibulo-ocular and optokinetic systems of the cat.' *Brain Res.*, **185**, 265–75.

Haddad, G. M., Friendlich, A. R., and Robinson, D. A. 1977. 'Compensation of nystagmus after VIIIth nerve lesions in vestibulo-cerebellectomized cats.' *Brain Res.*, **135**, 192–6.

Haddad, G. M., and Steinman, R. M. 1973. 'The smallest voluntary saccade: implications for fixation.' *Vision Res.*, **13**, 1075–6.

Haegerstrom-Portnoy, G., and Brown, B. 1979. 'Contrast effects on smooth-pursuit eye movement velocity.' *Vision Res.*, **19**, 169–74.

Hajos, A. 1968. 'Sensumotorische Koordinationsprozesse bei Richtungs localization.' *Zeit. exp. angew. Psychol.*, **15**, 435–61.

Hajos, A., and Ritter, M. 1965. 'Experiments to the problem of interocular transfer.' *Acta Psychol.*, **24**, 81–90.

Hallett, P. E. 1963. 'Spatial summation.' *Vision Res.*, **3**, 9–24.

Hallett, P. E., and Lightstone, A. D. 1976a. 'Saccadic eye movements towards stimuli triggered by prior saccades.' *Vision Res.*, **16**, 99–106.

Hallett, P. E., and Lightstone, A. D. 1976b. 'Saccadic eye movements to flashed targets.' *Vision Res.*, **16**, 107–14.

Halliday, A. M., and Michael, W. F. 1970. 'Changes in pattern-evoked responses in man assiociated with the vertical and horizontal meridians of the visual field.' *J. Physiol.*, **208**, 499–513.

Hallpike, C. S. 1975. 'Directional preponderance 1942–1974: A review.' *Acta oto-laryng.*, **79**, 409–18.

Halper-Smith, C. 1971. Unpublished Doctoral Diss. Stanford University.

Hamasaki, D. I., and Cohen, H. I. 1977. 'Differential response of X and Y retinal ganglion cells to moving stimuli results from a difference in the surround mechanism.' *Brain Res.*, **122**, 157–61.

Hamasaki, D. I., and Sutija, V. G. 1979. 'Classification of cat retinal ganglion cells into X- and Y-cells with a contrast reversal stimulus.' *Exp. Brain Res.*, **35**, 25–36.

Hamilton, C. R. 1967. 'Effects of brain bisection on eye–hand coordination in monkeys wearing prisms.' *J. comp. physiol. Psychol.*, **64**, 434–43.

Hammond, P. 1975. 'Receptive field mechanisms of sustained and transient retinal ganglion cells in the cat.' *Exp. Brain Res.*, **23**, 113–28.

Hammond, P., and Andrews, D. P. 1978a, 'Orientation tuning of cells in areas 17 and 18 of the cat's visual cortex. *Exp. Brain Res.*, **31**, 341–52.

Hammond, P., and Andrews, D. P. 1978b. 'Collinearity tolerance of cells in areas 17 and 18 of the cat's visual cortex: relative sensitivity to straight lines and chevrons.' *Exp. Brain Res.*, **31**, 329–40.

Hammond, P., Andrews, D. P., and James, C. R. 1975. 'Invariance of orientational and directional tuning in visual cortical cells of the adult cat.' *Brain Res.*, **96**, 56–9.

Hammond, P., and MacKay, D. M. 1977. 'Differential responsiveness of simple and complex cells in cat striate cortex to visual texture.' *Exp. Brain Res.*, **30**, 275–96.

Hannen, R. A., Kabrisky, M., Replogle, C. R., Hartzler, V. L., and Roccaforte, P. A. 1966. 'Experimental determination of a portion of the human vestibular system response through measurement of eyeball counter-roll.' *IEEE Trans. biomed. Engin.*, **13**, 65–70.

Hansen R. M. 1979. 'Spatial localization during pursuit eye movements.' *Vision Res.*, **19**, 1213–21.

Hardt, M. E., Held, R., and Steinbach, M. J. 1971. 'Adaptation to displaced vision: A change in the central control of sensorimotor coordination.' *J. exp. Psychol.*, **89**, 229–39.

Hardy, M. 1934. 'Observations on the innervation of the macula saculi in man.' *Anat. Rec.*, **59**, 403–78.

Harrington, T. L., Harrington, M. K., Wilkins, C. A., and Koh, Y. O. 1980. 'Visual orientation by motion-produced blur patterns: detection of divergence.' *Percept. Psychophys.*, **28**, 293–305.

Harris, C. S. 1965a. 'Perceptual adaptation to inverted, reversed and displaced vision.' *Psychol. Rev.*, **72**, 419–44.

Harris, C. S. 1965b. 'Comments on change in straight ahead during adaptation to prism.' *Psychonom. Sci.*, **2**, 285–6.

Harris, C. S. 1974. 'Beware of the straight-ahead shift—a nonperceptual change in experiments on adaptation to displaced vision.' *Perception*, **3**, 461–76.

Harris, L., and Schaller, M. J. 1971. 'Form and its orientation: Re-examination of a child's eye view.' *Amer. J. Phychol.*, **84**, 218–34.

Harris, L. R. 1980. 'The superior colliculus and movements of the head and eyes in cats.' *J. Physiol.*, **300**, 367–92.

Harris, L. R., Blakemore, C., and Donaghy, M. 1980. 'Integration of visual and auditory space in the mammalian superior colliculus.' *Nature*, **288**, 56–9.

Harris, L. R., and Cynader, M. 1981. 'Modification of gain and balance of the vestibulo-ocular reflex: Effects of dark-rearing and surgical section of crossed visual pathways.' In W. Becker and A. F. Fuchs, (Eds.), *Progress in Oculomotor Research.* New York: Elsevier, in press.

Harris, L. R., Leporé, F., Guillemot, J. P., and Cynader, M. 1980. 'Abolition of optokinetic nystagmus in the cat.' *Science*, **210**, 91–2.

Harris, P., and MacFarlane, A. 1974. 'The growth of the effective visual field from birth to seven weeks.' *J. exp. Child Psychol.*, **18**, 340–8.

Harris, P. L., Letendre, J. B., and Bishop, A. 1974. 'The young child's discrimination of obliques.' *Perception*, **3**, 261–5.

Hartline, H. K. 1940. 'The receptive fields of optic nerve fibres.' *Amer. J. Physiol.*, **130**, 690–9.

Harwerth, R. S., and Levi, D. L. 1978. 'Raction time as a measure of suprathreshold grating detection.' *Vision Res.*, **18**, 1579–86.

Hay, J. C. 1968. 'Visual adaptation to an altered correlation between eye movement and head movement.' *Science*, **160**, 429–30.

Hay, J. C., and Goldsmith, W. M. 1973. 'Space–time adaptation of visual position constancy.' *J. exp. Psychol.*, **99**, 1–9.

Hay, J. C., and Pick, H. L. 1966. 'Visual and proprioceptive adaptation to optical displacement of the visual stimulus.' *J. exp. Psychol.*, **71**, 150–8.

Hay, J. C., Pick, H. L., and Ideda, K. 1965. 'Visual capture produced by prism spectacles.' *Psychonom. Sci.*, **2**, 215–6.

Hay, J. C., Pick, H. L., and Roser, E. 1963. 'Adaptation to chromatic aberration by the human visual system.' *Science*, **141**, 167–9.

Hay, J. C., and Sawyer, S. 1969. 'Position constancy and binocular convergence.' *Percept. Psychophys.*, **5**, 310–2.

Hay, L. 1978. 'Accuracy of children on an open-loop pointing task.' *Percept. mot. Skills*, **47**, 1079–82.

Hay, L. 1979, 'Spatial–temporal analysis of movements in children: motor programs versus feedback in the development of reaching.' *J. Motor Behav.*, **11**, 189–200.

Hayashi, Y. 1969. 'Recurrent collateral inhibition of visual cortical cells projecting to the superior colliculus in cats.' *Vision Res.*, **9**, 1367–80.

Hecht, S., and Mintz, E. U. 1939. 'The visibility of single lines at various illuminations and the retinal basis of visual resolution.' *J. gen. Physiol.*, **22**, 593–612.

Heeley, D. W. 1979. 'A perceived spatial frequency shift at orientations orthogonal to adapting gratings.' *Vision Res.*, **19**, 1229–36.

Heggelund, P. 1981a. 'Receptive field organization of simple cells in cat striate cortex.' *Exp. Brain Res.*, **42**, 89–98.

Heggelund, P. 1981b. 'Receptive field organization of complex cells in cat striate cortex.' *Exp. Brain Res.*, **42**, 99–107.

Heggelund, P., and Albus, K. 1978a. 'Orientation selectivity of single cells in striate cortex of cat: The shape of orientation tuning curves.' *Vision Res.*, **18**, 1067–71.

Heggelund, P., and Albus, K. 1978b. 'Response variability and orientation discrimination of single cells in striate cortex of cat.' *Exp. Brain Res.*, **32**, 197–211.

Hein, A., and Diamond, R. M. 1971. 'Independence of the cat's scotopic and photopic systems in acquiring control of visually guided behavior.' *J. comp. physiol. psychol.*, **76**, 31–8.

Hein, A., Diamond, R., and De Martino, R. 1981. 'Input from eye muscle is essential to developing visual control of behavior.' *Invest. Ophthal.*, **20**, ARVO Abstr., 55.

Hein, A., Gower, E. C., and Diamond, R. M. 1970. 'Exposure requirements for developing the triggered component of the visual-placing response.' *J. comp. physiol. Psychol.*, **73**, 188–92.

Hein, A., and Held, R. 1967. 'Dissociation of the visual placing response into elicited and guided components.' *Science*, **158**, 390–2.

Hein, A., Held, R., and Gower, E. C. 1970. 'Development and segmentation of visually controlled movement by selective exposure during rearing.' *J. comp. physiol. Psychol.*, **73**, 181–7.

Hein, A., Vital-Durand, F., Salinger, W., and Diamond, R. 1979. 'Movements initiate visual–motor development in the cat.' *Science*, **204**, 1321–2.

Held, R. 1961. 'Exposure history as a factor in maintaining stability of perception and coordination.' *J. nerv. ment. Dis.*, **132**, 26–32.

Held, R., and Bauer, J. A. 1967. 'Visually guided reaching in infant monkeys after restricted rearing.' *Science*, **155**, 718–20.

Held, R., Dichgans, J., and Bauer, J. 1975. 'Characteristics of moving visual areas influencing spatial orientation.' *Vision Res.*, **15**, 357–65.

Held, R., Gwiazda, J., Brill, S., Mohindra, I., and Wolfe, J. 1979. 'Infant visual acuity is underestimated because near threshold gratings are not preferentially fixated.' *Vision Res.*, **19**, 1377–9.

Held, R., and Hein, A. 1958. 'Adaptation of disarranged hand–eye coordination contingent upon re-afferent stimulation.' *Percept. mot. Skills*, **8**, 87–90.

Held, R., and Hein, A. 1963. 'Movement-produced stimulation in the development of visually guided behavior.' *J. comp. physiol. Psychol.*, **56**, 872–6.

Held, R., and Rekosh, J. 1963. 'Motor-sensory feedback and geometry of visual space.' *Science*, **141**, 722–3.

Helmholtz, H. von. 1962. *Treatise on Physiological Optics*. New York: Dover.

Henn, V., and Cohen, B. 1973. 'Quantitative analysis of activity in eye muscle motoneurons during saccadic eye movements and positions of fixation.' *J. Neurophysiol.*, **36**, 115–26.

Henn, V., and Cohen, B. 1976. 'Coding of information about rapid eye movements in the pontine reticular formation of alert monkeys.' *Brain Res.*, **108**, 307–25.

Henn, V., Cohen, B., and Young, L. R. 1980. 'Visual-vestibular interaction in motion perception and the generation of nystagmus.' *Neurosciences Res. Prog. Bull.*, **18**, 459–651.

Henn, V., Young, L. R., and Finley, C. 1974. 'Vestibular nucleus units in alert monkeys are also influenced by moving visual fields.' *Brain Res.*, **71**, 144–9.

Hennebert, P. E. 1960. 'Nystagmus audiocinétique.' *J. aud. Res.*, **1**, 84–7.

Henriksson., N. G., Novotny, M., and Tjernström, Ö. 1974. 'Eye movements as a function of active head turnings.' *Acta oto-laryng.*, **77**, 86–91.

Henry, G. H. 1977. 'Receptive field classes of cells in the striate cortex of the cat.' *Brain Res.*, **133**, 1–28.

Henry,G. H., Bishop, P. O., and Dreher, B. 1974. 'Orientation, axis and direction as stimulus parameters for striate cells.' *Vision Res.*, **14**, 767–77.

Henry, G. H., Dreher, B., and Bishop, P. O. 1974. 'Orientation specificity of cells in cat striate cortex.' *J. Neurophysiol.*, **37**, 1394–409.

Henson, D. B. 1978. 'Corrective saccades: Effects of altering visual feedback.' *Vision Res.*, **18**, 63–7.

Henson, D. B. 1979. 'Investigation into corrective saccadic eye movements for refixation amplitudes of 10 degrees and below.' *Vision Res.*, **19**, 57–61.

Hering, E. 1942 (1868). *Spatial Sense and Movements of the Eye*. Baltimore: Amer. Acad. Science.

Hess,W. R., Bürgi, S., and Bucher, V. 1946. 'Motorische Function des Tectal und Tegmentalgebietes.' *Mschr. Psychiat. Neurol.*, **112**, 1–52.

Heywood, S. 1973. 'Asymmetries related to cerebral dominance in returning the eyes to specified target positions in the dark.' *Vision Res.*, **13**, 81–94.

Heywood, S., and Chessell, K. 1977. 'Expanding angles? Systematic distortions of space involving angular figures.' *Perception*, **6**, 571–82.

Heywood, S., and Churcher, J. 1971. 'Eye movements and the afterimage. I. Tracking the afterimage.' *Vision Res.*, **11**, 1163–8.

Heywood, S., and Churcher, J. 1972. 'Eye movements and the after-image. II. The effect of foveal and non-foveal after-images on saccadic behaviour.' *Vision Res.*, **12**, 1033–43.

Heywood, S., and Churcher, J. 1980. 'Structure of the visual array and saccadic latency: implications for oculomotor control.' *Quart. J. exp. Psychol.*, **32**, 335–41.

Heywood, S., and Churcher, J. 1981. 'Saccades to step-ramp stimuli.' *Vision Res.*, **21**, 479–90.

Heywood, S., and Ratcliff, G. 1975. 'Long-term oculomotor consequences of unilateral colliculectomy in man.' In G. Lennerstand and P. Bach-y-Rita (Eds.), *Basic Mechanisms of Ocular Motility*. Oxford: Pergamon, pp. 562–4.

Hick, W. E. 1952. 'On the rate of gain of information.' *Quart. J. exp. Psychol.*, **4**, 11–26.

Higgins, D. C., and Daroff, R. B. 1966. 'Overshoot and oscillation in ocular dysmetria.' *A. M. A. Arch. Ophthal.*, **75**, 742–5.

Highstein, S. M. 1972. 'Electrophysiological investigation of the organization of the vestibulo-ocular pathways in rabbit.' In J. Dichgans and E. Bizzi (Eds.), *Cerebral Control of Eye Movements and Motion Perception.* New York: Karger, pp. 89–98.

Hikosaka, O., and Kawakami, T. 1977. 'Inhibitory reticular neurons related to the quick phase of vestibular nystagmus—their location and projection.' *Exp. Brain Res.*, **27**, 377–96.

Hikosaka, O., and Maeda, M. 1973. 'Cervical effects on abducens motoneurons and their interaction with vestibulo-ocular reflex.' *Exp. Brain Res.*, **18**, 512–30.

Hill, A. L., 1972. 'Direction constancy.' *Percept. Psychophys.*, **11**, 175–8.

Hilz, R., and Cavonius, C. R. 1974. 'Functional organization of the peripheral retina: sensitivity to periodic stimuli.' *Vision Res.*, **14**, 1333–7.

Hines, M. 1976. 'Line spread function variation near the fovea.' *Vision Res.*, **16**, 567–72.

Hirsch, H. V. B., and Spinelli, D. N. 1970. 'Visual experience modifies distribution of horizontally and vertically oriented receptive fields in cats.' *Science,* **168**, 869–71.

Hirsch, H. V. B., and Spinelli, D. N. 1971. 'Modification of the distribution of receptive field orientation in cats by selective visual exposure during development.' *Exp. Brain Res.*,**13**, 509–27.

Hirsch, J., Schneider, B., and Vitiello, M. 1974. 'The effects of adaptation to square-wave gratings as a function of grating orientation.' *Percept. Psychophys.*, **15**, 475–8.

Hixson, W. C., Niven, J. I., and Correia, M. J. 1966. 'Kinematics nomenclature for physiological accelerations: With special reference to vestibular applications.' Monograph 14. Nav. Aero. Med. Instit., Pensacola, Florida.

Hochberg, J., and Gellman, L. 1977. 'The effect of landmark features on mental rotation times.' *Mem. Cognit.*, **5**, 23–6.

Hochstein, S., and Shapley, R. M. 1976a. 'Quantitative analysis of retinal ganglion cell classifications.' *J. Physiol.*, **262**, 237–64.

Hochstein, S., and Shapley, R. M. 1976b. 'Linear and nonlinear spatial subunits in Y cat retinal ganglion cells.' *J. Physiol.*, **262**, 265–84.

Hock, H. S. 1973. 'The effects of stimulus structure and familiarity on same–different comparisons.' *Percept. Psychophys.*, **14**, 413–20

Hock, H. S., and Ross, K. 1975. 'The effect of familiarity on rotational transformation.' *Percept. Psychophys.*, **18**, 15–20.

Hock, H. S., and Tromley, C. L. 1978. 'Mental rotation and perceptual uprightness.' *Percept. Psychophys.*, **24**, 529–33.

Hoekstra, J., van der Goot, D. P. J., van den Brink, G., and Bilsen, F. A. 1974. 'The influence of the number of cycles upon the visual contrast threshold for spatial since wave patterns.' *Vision Res.*, **14**, 365–8.

Hoffmann, K. P. 1973. 'Conduction velocity in pathways from retina to superior colliculus in the cat: A correlation with receptive-field properties.' *J. Neurophysiol.*, **36**, 409–24.

Hoffmann, K. P., Morrone, C. M., and Reuter, J. H. 1980. 'A comparison of the responses of single cells in the LGN and visual cortex to bar and noise stimuli in the cat.' *Vision Res.*, **20**, 771–8.

Hoffmann, K. P., and Schoppmann, A. 1975. 'Retinal input to direction selective cells in the nucleus opticus of the cat.' *Brain Res.*, **99**, 359–67.

Hoffmann, K. P., and Sherman, S. M. 1974. 'Effects of early monocular deprivation on visual input to cat superior colliculus.' *J. Neurophysiol.*, **37**, 1276–86.

Hoffmann, K. P., and Sherman, S. M. 1975. 'Effects of early binocular deprivation on visual input to cat superior colliculus.' *J. Neurophysiol.*, **38**, 1049–59.

Hofsten, C. von, and Lindhagen, K. 1979. 'Observations on the development of reaching for moving objects.' *J. exp. Child Psychol.*, **28**, 158–73.

Hökendorf, H., Bles, W., and Brandt, T. 1977. 'Direction specific reduction of circularvection latencies as a measure of latent labyrinthine vertigo.' *Pflüg. Arch. ges. Physiol. Suppl.*, **368**, R45.

Holding, D. H. 1965. *Principles of Training*. London: Pergamon.

Holding. D. H. 1968. 'Accuracy of delayed aiming responses.' *Psychonom. Sci.*, **12**, 125–6.

Holly, F. 1975. 'Saccadic presentation of a moving target.' *Vision Res.*, **15**, 331–5.

Holmes,G. 1918. 'Disturbances of vision by cerebral lesions.' *Brit. J. Ophthal.*, **2**, 353–84.

Holst, E. von. 1954. 'Relations between the central nervous system and the peripheral organs.' *Brit. J. anim. Behav.*, **2**, 89–94.

Holst, E. von, and Mittelstaedt, H. 1950. 'Das Reafferenzprinzip.' *Naturwissenschaften*, **37**, 464–76.

Honeyman, W. M., Cowper, M. C., and Rose, E. W. 1946. 'The autokinetic illusion.' Flying Personnel Research Committee, Rep. 664, Air Ministry, London.

Honisett, J., and Oldfield, R. C. 1961. 'Movement and distortion in visual patterns during prolonged fixation.' *Scand. J. Psychol.*, **2**, 49–55.

Honrubia, V., Downey, W. L., Mitchell, D. P., and Ward, P. H. 1968. 'Experimental studies on optokinetic nystagmus: II. Normal humans.' *Acta oto-laryng.*, **65**, 441–8.

Hood, J. D. 1967. 'Observations upon the neurological mechanism of optokinetic nystagmus with especial reference to the contribution of peripheral vision.' *Acta oto-laryng.*, **63**, 208–15.

Hood, J. D. (Ed.) 1978. *Vestibular Mechanisms in Health and Disease*. New York: Academic Press.

Hood, J. D., and Leech, J. 1974. 'The significance of peripheral vision in the perception of movement.' *Acta oto-laryng.*, **77**, 72–9.

Horn, G., Stechler, G., and Hill, R. M. 1972. 'Receptive fields of units in the visual cortex of the cat in the presence and absence of bodily tilt.' *Exp. Brain Res.*, **15**, 113–31.

Hotopf, W. H. N., and Robertson, S. H. 1975. 'The regression to right angles tendency, lateral inhibition and the transversals in the Zöllner and Poggendorff illusions.' *Percept. Psychophys.*, **18**, 453–9.

Hou, R. L., and Fender, D. H. 1979. 'Processing of direction and magnitude by the saccadic eye-movement system.' *Vision Res.*, **19**, 1421–6.

Houk, J. 1967. 'A viscoelastic interaction which produces one component of adaptation in responses of Golgi tendon organs.' *J Neurophysiol.*, **30**, 1482–93.

Howard, I. P. 1961. 'An investigation of a satiation process in the reversible perspective of revolving skeletal shapes.' *Quart. J. exp. Psych.*, **13**, 19–31.

Howard, I. P. 1968. 'Displacing the optical array.' In S. J. Freedman (Ed.), *The Neuropsychology of Spatially Oriented Behaviour*. Homewood, Ill.: Dorsey, pp. 19–36.

Howard, I. P. 1971a. 'Perceptual learning and adaptation.' *Brit. med. Bull.*, **27**, 248–52.

Howard, I. P. 1971b. 'The adaptability of the visual–motor system.' In K. J. Konnolly (Ed.), *Mechanisms of Motor Skill Development*. London: Academic Press, pp. 337–52.

Howard, I. P. 1975. 'Proposals for the study of adaptation to anomalous schemata.' *Perception*, **3**, 497–513.

Howard, I. P. 1978a. 'The effects of exposure to spatially distorted sensory inputs.' In R. D. Walk and H. L. Pick (Eds.), *Perception and Experience*. New York: Plenum, pp. 269–306.

Howard, I. P. 1978b. 'Recognition and knowledge of the water-level principle.' *Perception*, **7**, 151–60.

Howard, I. P., and Anstis, T. 1974. 'Muscular and joint-receptor components in postural persistence.' *J. exp. Psychol.*, **103**, 167–70.

Howard, I. P., Anstis, T., and Lucia, H. C. 1974. 'The relative lability of mobile and stationary components in a visual–motor adaptation task.' *Quart. J. exp. Psychol.*, **26**, 293–300.

Howard, I. P., Craske, B., and Templeton, W. B. 1965. 'Visuo-motor adaptation to discordant exafferent stimulation.' *J. exp. Psychol.*, **70**, 189–91.

Howard, I. P., and Templeton, W. B. 1963. 'A critical note on the use of the human centrifuge.' *Amer. J. Psychol.*, **76**, 150–2.

Howard, I. P., and Templeton, W. B. 1964a. 'The effect of steady fixation on the judgment of relative depth.' *Quart. J. exp. Psychol.*, **16**, 193–203.

Howard, I. P., and Templeton, W. B. 1964b. 'Visually-induced eye torsion and tilt adaptation.' *Vision Res.*, **4**, 433–7.

Howard, I. P., and Templeton, W. B. 1966. *Human Spatial Orientation*. London: Wiley.

Howard, R. B. 1971. 'Neurophysiological models of figural aftereffects and visual illusions.' *Psychonom. Monogr. Suppl.*, **4**, 57–72.

Howell, E. R., and Hess, R. F. 1978. 'The functional area for summation to threshold for sinusoidal gratings,' *Vision Res.*, **18**, 369–74.

Hubel, D. H., LeVay, S., and Wiesel, T. N. 1975. 'Mode of termination of retinotectal fibres in macaque monkey: An autoradiographic study.' *Brain Res.*, **96**, 25–40.

Hubel, D. H., and Wiesel, T. N. 1959. 'Receptive fields of single neurones in the cat's striate cortex.' *J. Physiol.*, **148**, 574–91.

Hubel, D. H., and Wiesel, T. N. 1962. 'Receptive fields, binocular interaction, and functional architecture in the cat's visual cortex.' *J. Physiol.*, **160**, 106–54.

Hubel, D. H., and Wiesel, T. N. 1963. 'Receptive fields of cells in striate cortex of very young, visually inexperienced kittens.' *J. Neurophysiol.*, **26**, 994–1002.

Hubel, D. H., and Wiesel, T. N. 1965. 'Receptive fields and functional architecture in two nonstriate visual areas (18 and 19) of the cat.' *J. Neurophysiol.*, **28**, 229–89.

Hubel, D. H., and Wiesel, T. N. 1967. 'Cortical and callosal connections concerned with the vertical meridian of visual fields in the cat.' *J. Neurophysiol.*, **30**, 1561–73.

Hubel, D. H., and Wiesel, T. N. 1968. 'Receptive fields and functional architecture of monkey striate cortex.' *J. Physiol.*, **195**, 215–43.

Hubel, D. H., and Wiesel, T. N. 1969. 'Anatomical demonstration of columns in the monkey striate cortex.' *Nature*, **221**, 747–50.

Hubel, D. H., and Wiesel. T. N. 1972. 'Laminar and columnar distribution of geniculo-cortical fibres in the macaque monkey.' *J. comp. Neurol.*, **146**, 421–50.

Hubel, D. H., and Wiesel, T. N. 1974a. 'Sequence regularity and geometry of orientation columns in the monkey striate cortex.' *J. comp. Neurol.*, **158**, 267–94.

Hubel, D. H., and Wiesel, T. N. 1974b. 'Uniformity of monkey striate cortex: a parallel relationship between field size, scatter, and magnification factor.' *J. comp. Neurol.*, **158**, 295–306.

Hubel, D. H., and Wiesel, T. N. 1977. 'Functional architecture of macaque monkey visual cortex.' *Proc. roy. Soc., B.*, **198**, 1–59.

Hubel, D. H., Wiesel, T. N., and LeVay, S. 1977. 'Plasticity of ocular dominance columns in monkey striate cortex.' *Phil. Trans. roy. Soc., B.*, **278**, 377–409.

Hubel, D. H., Weisel, T. N., and Stryker, M. P. 1977. 'Orientation columns in macaque monkey visual cortex demonstrated by the 2 deoxyglucose autoradiographic technique.' *Nature*, **269**, 328–9.

Hughes, J. R. 1958. 'Post-tetanic potentiation.' *Physiol. Rev.*, **38**, 91–113.

Hughes, P. C., Brecher, G. A., and Fishkin, S. M. 1972. 'Effects of rotating backgrounds upon the perception of verticality.' *Percept. Psychophys.*, **11**, 135–8.

Hulls, J. R., and Stockwell, C. W. 1974. 'Suppression of optokinetic nystagmus.' *Minerva otorinolaringologica*, **24**, 261–5.

Humphrey, N. K. 1974. 'Vision in a monkey without striate cortex: a case study.' *Perception*, **3**, 241–55.

Humphrey, N. K., and Weiskrantz, L. 1967. 'Vision in monkeys after removal of the striate cortex.' *Nature*, **215**, 595–7.

Huttenlocher, J. 1967. 'Discrimination of figure orientation: Effects of relative position.' *J. comp. Physiol.*, **63**, 359–61.

Hutton, R.S., Smith, J. L., and Eldred, E. 1973. 'Postcontraction sensory discharge from muscle and its source.' *J. Neurophysiol.*, **36**, 1090–103.

Ibbotson, A., and Bryant, P. E. 1976. 'The perpendicular error and the vertical effect in children's drawing.' *Perception*, **5**, 319–26.

Igarashi, M., Takahashi, M., Reschke, M. F., and Wright, W. K. 1977. 'Effect of otolith end organ ablation on pendular rotation nystagmus in squirrel monkey.' *Arch. Oto-Rhino-Laryng.*, **217**, 183–8.

Ikeda, H., and Obonai, T. 1953. 'The quantitative analysis of figural aftereffects.' *Jap. J. Psychol.*, **24**, 84–5.

Ikeda, H., and Wright, M. J. 1972a. 'The outer disinhibitory surround of the retinal ganglion cell receptive field.' *J. Physiol.*, **266**, 511–44.

Ikeda, H., and Wright, M. J. 1972b. 'Receptive field organization of sustained and transient retinal ganglion cells which subserve different functional roles.' *J. Physiol.*, **227**, 769–800.

Ikeda, H., and Wright, M. J. 1974. 'Evidence for "sustained" and "transient" neurones in the cat's visual cortex.' *Vision Res.*, **14**, 133–6.

Ikeda, H., and Wright, M. J. 1975a. 'Spatial and temporal properties of "sustained" and "transient" neurones in area 17 of the cat's visual cortex.' *Exp. Brain Res.*, **22**, 363–83.

Ikeda, H., and Wright, M. J. 1975b. 'Retinotopic distribution, visual latency and orientation tuning of "sustained" and "transient" cortical neurones in Area 17 of the cat.' *Exp. Brain Res.*, **22**, 385–98.

Ingling, C. R., and Drum, B. A. 1973. 'Retinal receptive fields: Correlations between psychophysics and electrophysiology.' *Vision Res.*, **13**, 1151–63.

Ito, M. 1972. 'Neural design of the cerebellar motor control system.' *Brain Res.*, **40**, 81–4.

Ito, M. 1976. 'Cerebellar learning control of vestibulo-ocular mechanisms.' In T. Desiraju (Ed.), *Mechanisms in Transmission of Signals for Conscious Behavior*. Amsterdam:

Elsevier, pp. 1–21.
Ito, M., Nisimaru, N., and Yamamoto, M. 1976a. 'Inhibitory interaction between the vestibulo-ocular reflexes arising from semicircular canals of rabbits.' *Exp. Brain Res.*, **26**, 89–103.
Ito, M., Nisimaru, N., and Yamamoto, M. 1976b. 'Postsynaptic inhibition of oculomotor neurons involved in vestibulo-ocular reflexes arising from semicircular canals of rabbits.' *Exp. Brain Res.*, **24**, 273–83.
Ito, M., Nisimaru, N., and Yamamoto, M. 1976c. 'Pathways for the vestibulo-ocular reflex excitation arising from semicircular canals of rabbits.' *Exp. Brain Res.*, **24**, 257–71.
Iurato, S. 1967. 'Light microscope features.' In S. Iurato *et al.* (Eds.), *Submicroscopic Structure of the Inner Ear.* Oxford: Pergamon, pp. 18–37.
Jack, C. E., and Thurlow, W. R. 1973. 'Effects of degree of visual association and angle of displacement on the "ventriloquism" effect.' *Percept. mot. Skills*, **37**, 967–79.
Jackson, C. V. 1953. 'Visual factors in auditory localization.' *Quart. J. exp. Psychol.*, **5**, 52–65.
Jacobs, L., Anderson, P. J., and Bender, M. B. 1973. 'The lesions producing paralysis of downward but not upward gaze.' *Arch. Neurol.*, **28**, 319–23.
Jacobs, R. J. 1979. 'Visual resolution and contour interaction in the fovea and periphery.' *Vision Res.*, **19**, 1187–95.
Jacobson, S. G., Eames, R. A., and McDonald, W. I. 1979. 'Optic nerve fibre lesions in adults cats: pattern of recovery of spatial vision.' *Exp. Brain Res.*, **36**, 491–508.
Jäger, J., and Henn, V. 1981. 'Habituation of the vestibulo-ocular reflex (VOR) in the monkey during sinusoidal rotation in the dark.' *Exp. Brain Res.*, **41**, 108–14.
Jahoda, G. 1976. 'Reproduction of Kohs-type figures by Ghanaian children: Orientation errors revisited.' *Brit. J. Psychol.*, **67**, 203–11.
Jakiela, H. G., Enroth-Cugell, C., and Shapley, R. M. 1976. 'Adaptation and dynamics in X-cells and Y-cells of the cat's retina.' *Exp. Brain Res.*, **24**, 335–42.
Jampel, R. S. 1966. 'The action of the superior oblique muscle.' *A. M. A. Arch. Ophthal.*, **75**, 535–44.
Janeke, J. B., Jongkees, L. B. W., and Oosteveld, W. J. 1970. 'Relationship between otoliths and nystagmus.' *Acta oto-laryng.*, **69**, 1–6.
Jassik-Gerschenfeld, D. 1965. 'Somesthetic and visual responses of superior colliculus neurones.' *Nature*, **208**, 898–900.
Jastrow, J. 1892. 'On the judgment of angles and the position of lines.' *Amer. J. Psychol.*, **5**, 214–48.
Jeannerod, M., and Chouvet, G. 1973. 'Saccadic displacements of the retinal image: Effects on the visual system in the cat.' *Vision Res.*, **13**, 161–9.
Jeannerod, M., and Putkonen, P. T. S. 1971. 'Lateral geniculate activity eye movements: saccade-locked changes in dark and light.' *Exp. Brain Res.*, **13**, 533–46.
Jeannerod, M., and Sakai, K. 1970. 'Occipital and geniculate potentials related to eye movements in the unanesthetized cat.' *Brain Res.*, **19**, 361–77.
Jeffrey, W. E. 1966. 'Discrimination of oblique lines by children.' *J. comp. physiol. Psychol.*, **62**, 154–6.
Jennings, J. A. M., and Charman, W. N. 1981. 'Off-axis image quality in the human eye.' *Vision Res.*, **21**. 445–55.
Johansson, G. 1973. 'Visual perception of biological motion and a model for its analysis.' *Percept. Psychophys.*, **14**, 201–11.
Johansson, G. 1974. 'Vector analysis in visual perception of rolling motion. A quantitative approach.' *Psychol. Forsch.*, **36**, 311–19.
Johansson, G. 1977. 'Studies on visual perception of locomotion.' *Perception*, **6**, 365–76.
Johnston, I. R., White, G. R., and Cumming, R. W. 1973. 'The role of optical expansion patterns in locomotor control.' *Amer. J. Psychol.*, **86**, 311–24.
Jones, R. M., and Tulunay-Keesey, U. 1975. 'Local retinal adaptation and spatial

frequency channels.' *Vision Res.*, **15**, 1239–44.

Jones, R. M., and Tulunay-Keesey, U. 1980. 'Phase selectivity of spatial frequency channels.' *J. opt. Soc. Amer.*, **70**, 66–70.

Jongkees, L. B. W. 1969. 'Cervical vertigo.' *Laryngoscope*, **79**, 1473–83.

Jongkees, L. B. W. 1975. 'On the physiology and examination of the vestibular labyrinths.' In R. F. Naunton (Ed.), *The Vestibular System*. New York: Academic Press, pp. 227–47.

Jordan, S. 1968. 'Autokinesis and felt eye-position.' *Amer. J. Psychol.*, **81**, 497–512.

Jordan, S. 1970. 'Ocular pursuit movement as a function of visual and proprioceptive stimulation.' *Vision Res.*, **10**, 775–80.

Judge, S. J. 1973. 'Temporal interaction between human voluntary saccades and rapid phases of optokinetic nystagmus.' *Exp. Brain Res.*, **18**, 114–8.

Judge, S. J., Wurtz, R. H., and Richmond, B. J. 1980. 'Vision during saccadic eye movements. I. Visual interactions in striate cortex'. *J. Neurophysiol.*, **43**, 1133–55.

Julesz, B. 1971. *Foundations of Cyclopean Perception*. Chicago: University of Chicago Press.

Jung, R., Kornhuber, H. H., and Da Fonseca, J. S. 1963. 'Multisensory convergence on cortical neurons: Neuronal effects of visual, acoustic and vestibular stimuli in the superior convolutions of the cat's cortex.' In G. Moruzzi, A. Fessard and H. H. Jasper (Eds.), *Progress in Brain Research*. Amsterdam: Elsevier, pp. 207–40.

Jürgens, R., and Becker, W. 1975. 'Is there a linear addition of saccades and pursuit movements?' In G. Lennerstrand and P. Bach-y-Rita (Eds.), *Basic Mechanisms of Ocular Motility and Their Clinical Implications*. New York: Pergamon, pp. 525–9.

Kail, R., Pellegrino, J., and Carter, P. 1980. 'Developmental changes in mental rotation.' *J. exp. Child Psychol.*, **29**, 102–16.

Kalil, R. F., and Freedman, S. J. 1966. 'Persistence of ocular rotation following compensation for displaced vision.' *Percept. mot. Skills*, **22**, 135–9.

Karwoski, T. F., Redner, H., and Wood, W. O. 1948. 'Autokinetic movement of large stimuli.' *J. gen. Psychol.*, **39**, 29–37.

Kasahara, M., and Uchino, Y. 1974. 'Bilateral semicircular canal inputs to neurons in cat vestibular nuclei.' *Exp. Brain Res.*, **20**, 285–96.

Kasai, T., and Zee, D. S. 1978, 'Eye–head coordination in labyrinthine-defective human begins.' *Brain Res.*, **144**, 123–42.

Kase, M., Miller, D. C., and Noda, H. 1980. 'Discharges of Purkinje cells and mossy fibres in the cerebellar vermis of the monkey during saccadic eye movements and fixation.' *J. Physiol.*, **300**, 539–55.

Kase, M., Noda, H., Suzuki, D. A., and Miller, D. C. 1979. 'Target velocity signals of visual tracking in vermal Purkinje cells of the monkey.' *Science*, **205**, 717–20.

Kato, H., Bishop, P. O., and Orban, G. A. 1978. 'Hypercomplex and simple/complex cell classification in cat.' *J. Neurophysiol.*, **41**, 1071–95.

Katz, M. S., Metlay, W., and Cirincione, P. A. 1965. 'Effects of stimulus and field size on the accuracy of orientation in the homogeneous environment.' *Percept. mot. Skills*, **20**, 167–72.

Kaufman, L., and Richards, W. 1969. 'Spontaneous fixation tendencies for visual forms.' *Percept. Psychophys.*, **5**, 85–8.

Kaufmann, E. L., Reese, E. P., Volkmann, J., and Rogers, S. 1949. In E. P. Reese, (Ed.), Summary Report No. 5DC 131–1–5. Psychophysical Res. Unit, Mount Holyoke College, 1953.

Keating, E. G. 1974. 'Impaired orientation after primate tectal lesions.' *Brain Res.*, **67**, 538–41.

Keating, E. G. 1980. 'Residual spatial vision in the monkey after removal of striate and preoccipital cortex.' *Brain Res.*, **187**, 271–90.

Keck, M. J., Palella, T. D., and Pantle, A. 1976. 'Motion aftereffect as a function of the contrast of sinusoidal gratings.' *Vision Res.*, **16**, 187–91.

Keene, G. C. 1963. 'The effect of response codes on the accuracy of making absolute judgments of linear inclination.' *J. gen. Psychol.*, **69**, 37–50.

Keller, E. L. 1974. 'Participation of medial pontine reticular formation in eye movement generation in monkey.' *J. Neurophysiol.*, **37**, 316–32.

Keller, E. L. 1976. 'Behaviour of horizontal semicircular canal afferents in alert monkey during vestibular and optokinetic stimulation.' *Exp. Brain Res.*, **24**, 459–71.

Keller, E. L. 1977. 'The role of the brain-stem reticular formation in eye movement control.' In B. A. Brooks and F. J. Bajandas (Eds.), *Eye Movements*. New York: Plenum, pp. 105–26.

Keller, E. L. 1978. 'Gain of the vestibulo-ocular reflex in monkey at high rotational frequencies.' *Vision Res.*, **18**, 311–5.

Keller, E. L., and Daniels, P. D. 1975. 'Oculomotor related interaction of vestibular and visual stimulation in vestibular nucleus cells in alert monkey.' *Exp. Neurol.*, **46**, 187–98.

Keller, E. L., and Kamath, B. Y. 1975. 'Characteristics of head rotation and eye movement related neurons in alert monkey vestibular nucleus.' *Brain Res.*, **100**, 182–7.

Keller, E. L., and Precht, W. 1979a. 'Adaptive modification of central vestibular neurons in response to visual stimulation through reversing prisms.' *J. Neurophysiol.*, **42**, 896–911.

Keller, E. L., and Precht, W. 1979b. 'Visual–vestibular responses in vestibular nuclear neurons in the intact and cerebellectomized, alert cat.' *Neurosci.*, **4**, 1599–1613.

Keller, E. L., and Robinson, D. A. 1971. 'Absence of a stretch reflex in extraocular muscles of the monkey.' *J. Neurophysiol.*, **34**, 908–19.

Keller, W., and Kinchla, R. A. 1968. 'Visual movement discrimination.' *Percept. Psychophys.*, **3**, 233–6.

Kellogg, R. S., and Graybiel, A. 1967. 'Lack of response to thermal stimulation of the semicircular canals in the weightless phase of parabolic flight.' *Aerosp. Med.*, **38**, 487–90.

Kelly, D. H. 1974. 'Spatio-temporal frequency-characteristics of color vision mechanisms.' *J. opt. Soc. Amer.*, **64**, 983–90.

Kelly, D. H. 1975a. 'Spatial frequency selectivity in the retina.' *Vision Res.*, **15**, 665–72.

Kelly, D. H. 1975b. 'No oblique effect in chromatic pathways.' *J. opt. Soc. Amer.*, **65**, 1512–4.

Kelly, D. H. 1976. 'Pattern detection and the two-dimensional Fourier transform: Flickering checkerboards and chromatic mechanism.' *Vision Res.*, **16**, 277–87.

Kelly, D. H. 1977. 'Visual contrast sensitivity.' *Optica Acta*, **24**, 107–29.

Kelly, D. H., and Magnuski, H. S. 1975. 'Pattern detection and the two-dimensional Fourier transform: Circular targets.' *Vision Res.*, **15**, 911–5.

Kelso, J. A. S., Cook, E., Olson, M. E., and Epstein, W. 1975. 'Allocation of attention and the locus of adaptation to displaced vision.' *J. exp. Psychol: Hum. Percept. Perform.*, **104**, 237–45.

Kelso, J. A. S., Southard, D. L., and Goodman, D. 1979. 'On the coordination of two-handed movements.' *J. exp. Psychol: Human Percept. Perform.*, **5**, 229–38.

Kennedy, H., and Orban, G. A. 1979. 'preferences for horizontal or vertical orientation in cat visual cortex neurones.' *J. Physiol.*, **296**, 61–2P.

Kennedy, J. M. 1969. 'Prismatic displacement and the remembered location of targets.' *Percept. Psychophys.*, **5**, 218–20.

Kenyon, R. V., Ciuffreda, K. J., and Stark, L. 1978. 'Binocular eye movements during accommodative vergence.' *Vision Res.*, **18**, 545–55.

Kertesz, A. E. 1980. 'Human fusional vergence. Proc.' Eye Movement Cont. (OMS 80), Calif. Inst. Technol., Pasadena.

Kertesz, A. E., and Jones, R. W. 1969. 'The effect of angular velocity of stimulus on human torsional eye movements.' *Vision Res.*, **9**, 995–8.

Kertesz, A. E., and Jones, R. W. 1970. 'Human cyclofusional response.' *Vision Res.*, **10**, 891–6.

Kertesz, A. E., and Sullivan, M. J. 1978. 'The effect of stimulus size on human cyclofusional response.' *Vision Res.*, **18**, 567–71.

Kiff, R. D., and Lepard, C. 1966. 'Visual response of premature infants.' *A. M. A.* Arch. *Ophthal.*, **75**, 631–3.

Kinchla, R. A. 1976. 'A psychophysical model of visual-movement perception.' In R. Monty and J. Senders (Eds.), *Eye Movements and Psychological Processes.* New York: Wiley, pp. 263–76.

Kinchla, R. A., and Allen, L. G. 1969. 'A theory of visual movement perception.' *Psychol. Rev.*, **76**, 537–58.

Kinchla, R. A., and Smyzer, F. 1967. 'A diffusion model of perceptual memory.' *Percept. Psychophys.*, **2**, 219–29.

King, W. M., and Fuchs, A. F. 1979. 'Reticular control of vertical saccadic eye movements by mesencephalic burst neurons.' *J. Neurophysiol.*, **42**, 861–76.

King-Smith, P. E., and Kulikowski, J. J. 1973. 'Lateral interaction in the detection of composite patterns'. *J. Physiol.*, **234**, 5–6P.

King-Smith, P. E., and Kulikowski, J. J. 1975a. 'The detection of gratings by independent activation of line detectors.' *J. Physiol.*, **247**, 237–71.

King-Smith, P. E., and Kulikowski, J. J. 1975b. 'Pattern and flicker detection analysed by subthreshold summation.' *J. Physiol.*, **249**, 519–48.

King-Smith, P. E., and Kulikowski, J. J. 1980. 'Pattern and movement detection in a patient lacking sustained vision.' *J. Physiol.*, **300**, 60P.

Kinney, J. S., and Luria, S. M. 1970. 'Conflicting visual and tactual–kinaesthetic stimulation.' *Percept. psychophys.*, **8**, 189–92.

Kirkham, W. R., Collins, W. E., Grape, P. M., Simpson, J. M., and Wallace, T. F. 1978. 'Spatial disorientation in general aviation accidents.' *Aviat. space environ. Med.*, **49**, 1080–6.

Kitahara, M., and Uno, R. 1967. 'Equilibrium and vertigo in a tilting environment.' *Ann. Otol.*, **76**, 166–78.

Kitterle, F. L. 1973. 'The possible locus of lightness contrast.' *Percept. Psychophys.*, **14**, 585–9.

Klapp, S. T., Nordell, S. A., Hoekenga, K. C., and Patton, C. B. 1974. 'Long-lasting aftereffect of brief prism exposure.' *Percept. Psychophys.*, **15**, 399–400.

Kleinschmidt, H. J., and Collewijn, H. 1975. 'A search for habituation of vestibulo-ocular reactions to rotatory and linear sinusoidal accelerations in the rabbit.' *Exp. Neurol.*, **47**, 257–67.

Klinke, R., and Galley, N. 1974. 'Efferent innervation of vestibular and auditory receptors.' *Physiol. Rev.*, **54**, 316–57.

Klüver, H. 1941. 'Visual functions after removal of the occipital lobes.' *J. Psychol.*, **11**, 23–45.

Klüver, H. 1942. 'Functional significance of the geniculostriate system.' Cold Springs Harb. Symp. quant. Biol., **7**, 253–99.

Koenderink, J. J., and van Doorn, A. J. 1980. 'Spatial summation for complex bar patterns.' *Vision Res.*, **20**, 169–76.

Koenig, E., Allum, J. H. J., and Dichgans, J. 1978. 'Visual–vestibular interaction upon nystagmus slow phase velocity in man.' *Acta oto-laryng.*, **85**, 397–410.

Kohler, I. 1953. 'Umgewöhnung in Wahrnehmungsbereich.' *Pyramide*, **5**, 92–5; **6**, 109–13.

Kohler, I. 1955. 'Experiments with prolonged optical distortions.' *Acta Psychol.*, **11**, 176–8.

Kohler, I. 1964. 'The formation and transformation of the perceptual world.' *Psychol. Issues*, **3**, Monogr. No. 12.

Köhler, W., and Wallach, H. 1944. 'Figural-aftereffects: An investigation of visual processes.' *Proc. Amer. Philos. Soc.*, **88**, 269–357.

Kolata, G. B. 1974. 'Developmental neurobiology: Specificity in the visual system.' *Science*, **186**, 1013–14.

Kolers, P. A. 1968. 'The recognition of geometrically transformed text.' *Percept. Psychophys.*, **3**, 57–64.

Kolers, P. A. 1972. *Aspects of Motion Perception.* Oxford: Pergamon.

Kolers, P. A., and Perkins, D. N. 1969. 'Orientation of letters and their speed of recognition.' *Percept. Psychophys.*, **5**, 275–80.

Kommerell, G., and Klein, U. 1971. 'Uber die visuelle Regelung der Okulomotorik: die optomotorische Wirkung exzentrischer Nachbilder.' *Vision Res.*, **11**, 905–20.

Komoda, M. K., Festinger, L., Phillips, L. J., Duckman, R. H., and Young, R. A. 1973. 'Some observations concerning saccadic eye movements.' *Vision Res.*, **13**, 1009–20.

Körner, F., and Schiller, P. H. 1972. 'The optokinetic response under open and closed loop conditions in the monkey.' *Exp. Brain Res.*, **14**, 318–30.

Kornheiser, A. S. 1976. 'Adaptation to laterally displaced vision: A review.' *Psychol. Bull.*, **83**, 783–816.

Kornhuber, H. H. 1974a. 'Nystagmus and related phenomena in man: An outline of otoneurology.' In H. H. Kornhuber (Ed.), *Handbook of Sensory Physiology*, Vol. VI/2. New York: Springer, pp. 193–232.

Kornhuber, H. H. 1974b. 'The vestibular system and the general motor system.' In H. H. Kornhuber (Ed.), *Handbook of Sensory Physiology*, Vol. VI/2. New York: Springer, pp. 581–620.

Kornmuller, A. E. 1931. 'Eine experimentalle Anästhesie de äusseren Augenmuskeln am menschen und ihre Auswirkungen.' *J. Psychol. Neurol.*, **41**, 354–66.

Kosslyn, S. M., and Pomerantz, J. R. 1977. 'Imagery, propositions, and the form of internal representations.' *Cognit. Psychol.*, **9**, 52–76.

Kossut, M., Michalski, A., and Zernicki, B. 1978. 'The ocular following reflex in cats deprived of pattern vision from birth.' *Brain Res.*, **141**, 77–88.

Kottenhoff, H. 1957a. 'Situational and personal influences on space perception with experimental spectacles. I. Prolonged experiments with inverting glasses.' *Acta psychol.*, **13**, 79–97.

Kottenhoff, H. 1957b. 'Situational and personal influences on space perception with experimental spectacles. II. Semi-prolonged tests with inverting glasses.' *Acta Psychol.*, **13**, 151–61.

Kowler, E., and Steinman, R. M. 1979a. 'The effect of expectations on slow oculomotor control—I. Periodic target steps.' *Vision Res.*, **19**, 619–32.

Kowler, E., and Steinman, R. M. 1979b. 'The effect of expectation on slow oculomotor control—II. Single target displacements.' *Vision Res.*, **19**, 633–46.

Kratz, K. E., Mangel, S. C., Lehmkuhle, S., and Sherman, S. M. 1979. 'Retinal X- and Y-cells in monocularly lid-sutured cats: Normality of spatial and temporal properties.' *Brain Res.*, **172**, 545–51.

Krauskopf, J. 1962. 'Light distribution in human retinal images.' *J. opt. Soc. Amer.*, **52**, 1046–50.

Krauskopf, J., Cornsweet, T. N., and Riggs, L. A. 1960. 'Analysis of eye movements during monocular and binocular fixation.' *J. opt. Soc. Amer.*, **50**, 572–8.

Kravitz, J. H., and Wallach, H. 1966. 'Adaptation to displaced vision contingent upon vibrating stimulation.' *Psychonom. Sci.*, **6**, 465–6.

Krejčová, H., Highstein, S., and Cohen, B. 1971. 'Labyrinthine and extra-labyrinthine effects on ocular counterrolling.' *Acta oto-laryng.*, **72**, 165–71.

Kremenitzer, J. P., Vaughan, H. G., Kurtzberg, D., and Dowling, K. 1979. 'Smooth-pursuit eye movements in the newborn infant.' *Child Devel.*, **50**, 442–8.

Kroon, J. N., Rijsdijk, J. P., and van der Wildt, G. J. 1980. 'Peripheral contrast sensitivity for sine-wave gratings and single periods. *Vision Res.*, **20**, 243–52.

Kroon, J. N., and van der Wildt, G. J. 1980. 'Spatial frequency tuning studies: weighting as a prerequisite for describing psychometric curves by probability summation.' *Vision Res.*, **20**, 253–63.

Kubin, L., Magherini, P. C., Manzoni, D., and Pompeiano, O. 1980. 'Responses of lateral reticular neurons to sinusoidal stimulation of labyrinth receptors in decerebrate cats.' *J. Neurophysiol.*, **44**, 922–36.

Kuffler, S. W. 1953. 'Discharge patterns and functional organization of mammalian retina.' *J. Neurophysiol.*, **16**, 37–68.

Kulikowski, J. J. 1969. Reported in Ikeda and Wright (1972).

Kulikowski, J. J. 1972. 'Orientation selectivity of human binocular and monocular vision revealed by simultaneous and successive masking.' *J. Physiol.*, **226**, 67–8P.

Kulikowski, J. J. 1980. 'Processing of patterns by simple cells in the cat visual cortex.' *Neurosci. Letters*, Suppl. 53.

Kulikowski. J. J., Abadi, R., and King-Smith, P. E. 1973. 'Orientation selectivity of grating and line detectors in human vision.' *Vision Res.*, **12**, 1479–86.

Kulikowski, J. J., Bishop, P. O., and Kato, H. 1979. 'Sustained and transient responses by cat striate cells to stationary flashing light and dark bars.' *Brain Res.*, **170**, 362–7.

Kulikowski, J. J., and King-Smith, P. E. 1973. 'Spatial arrangement of line, edge and grating detectors revealed by subthreshold summation.' *Vision Res.*, **13**, 1455–75.

Kulikowski, J. J., and Leighton, D. A. 1977. 'Lack of a centrifugal component in saccadic suppression.' *J. Physiol.*, **265**, 28–9P.

Kulikowski, J. J., and MacCana, F. B. 1980. 'Is there antagonism between pattern and movement detection?' *J. Physiol.*, **298**, 22P.

Kulikowski, J. J., and Tolhurst, D. J. 1973. 'Psychophysical evidence for sustained and transient detectors in the human visual system.' *J. Physiol.*, **232**, 149–62.

Kurtz, D., and Butter, C. M. 1980. 'Impairments in visual discrimination performance and gaze shifts in monkeys with superior colliculuar lesions.' *Brain Res.*, **196**, 109–24.

Kuypers, H. G. J. M., Szwarcbart, M. K., Mishkin, M., and Rosvold. H. E. 1965. 'Occipitotemporal corticocortical connections in the rhesus monkey.' *Exp. Neurol.*, **11**, 245–62.

Lackner, J. R. 1973. 'The role of posture in adaptation to visual rearrangement.' *Neuropsychologia*, **11**, 33–44.

Lackner, J. R. 1974a. 'Proprioceptive facilitation of open-loop visual pointing.' *Percept. mot. Skills*, **39**, 263–5.

Lackner, J. R. 1974b. 'Adaptation to displaced vision: Role of proprioception.' *Percept. mot. Skills*, **38**, 1251–6.

Lackner, J. R. 1975. 'Pursuit eye movements elicited by muscle afferent information.' *Neurosci. Letters*, **1**, 25–8.

Lackner, J. R. 1977. 'Induction of illusory self-rotation and nystagmus by a rotating sound-field.' *Aviat. space environ. Med.*, **48**, 129–31.

Lackner, J. R., and Evanoff, J. N. 1977. 'Smooth pursuit eye movements elicited by somatosensory stimulation.' *Neurosci. Letters*, **4**, 43–48.

Lackner, J. R., and Levine, M. S. 1978. 'Visual direction depends on the operation of spatial constancy mechanisms: The oculobrachial illusion.' *Neurosci. Letters*, **7**, 207–12.

Lackner, J. R., and Lobovitz, D. 1977. 'Adaptation to displaced vision: Evidence for prolonged aftereffects.' *Quart. J. exp. Psychol.*, **29**, 65–9.

Lackner, J. R., and Zabkar, J. J. 1977. 'Proprioceptive information about target location suppresses autokinesis.' *Vision Res.*, **17**, 1225–9.

Lacour, M., Roll, J. P., and Appaix, M. 1976. 'Modifications and development of spinal reflexes in the alert baboon (Papio papio) following a unilateral neurotomy.' *Brain Res.*, **113**, 255–69.

Lamontagne, C. 1973. 'A new experimental paradigm for the investigation of the secondary system of human visual motion perception.' *Perception*, **2**, 167–80.

LaMotte, R. H., and Acuna, C., 1978. 'Defects in accuracy of reaching after removal of posterior parietal cortex in monkeys.' *Brain Res.*, **139**, 309–26.

Lanman, J., Bizzi, E., and Allum, J. 1978. 'The coordination of eye and head movements during smooth pursuit.' *Brain Res.*, **153**, 39–54.

Lashley, K. S. 1938. 'The mechanisms of vision: XV. Preliminary studies of the rat's capacity for detail vision.' *J. gen. Psychol.*, **18**, 123–93.

Laties, A. M., and Sprague, J. M. 1966. 'The projection of optic fibres to the visual centers in the cat. *J. comp Neurol.*, **127**, 35–70.

Latto, R., and Cowey, A. 1971a. 'Visual field defects after frontal eye-field lesion in monkeys.' *Brain Res.*, **30**, 1–24.

Latto, R., and Cowey, A. 1971b. 'Fixation changes after frontal eye-field lesions in monkeys.' *Brain Res.*, **30**, 25–36.

Lechner-Steinleitner, S. 1978. 'Interaction of labyrinthine and somatoreceptor inputs as determinants of the subjective vertical.' *Psychol. Res.*, **40**, 65–76.

Lechner-Steinleitner, S., and Schöne, H. 1980. 'The subjective vertical under "dry" and "wet" conditions at clockwise and anticlockwise changed positions and the effect of a parallel background field.' *Psychol. Res.*, **41**, 305–18.

Lee, B. B., Albus, K., Heggelund, P., Hume, M. J., and Creutzfeldt, O. D. 1977. 'The depth distribution of optimal stimulus orientations for neurones in cat area 17.' *Exp. Brain Res.*, **27**, 301–14.

Lee. B. B., Cleland, B. G., and Creutzfeldt, O. D. 1977. 'The retinal input to cells in area 17 of the cat's cortex.' *Exp. Brain Res.*, **30**, 527–38.

Lee, D. N., and Aronson, E. 1974. 'Visual proprioceptive control of standing in human infants.' *Percept. Psychophys.*, **15**, 529–32.

Leehey, S. C., Moskowitz-Cook, A., Brill, S., and Held, R. 1975. 'Orientational anisotropy in infant vision.' *Science*, **190**, 900–1.

Legge, D. 1970. 'Visual–proprioceptive correspondence in the para-median plane.' *J. mot. Behav.*, **11**, 149–62.

Legge, G. E. 1976. 'Adaptation to a spatial impulse: implications for Fourier transform models of visual processing.' *Vision Res.*, **16**, 1407–18.

Legge, G. E. 1978. 'Sustained and transient mechanisms in human vision: Temporal and spatial properties.' *Vision Res.*, **18**, 69–81.

Lehmkuhle, S., Kratz, K. E., Mangel, S. C., and Sherman, S. M. 1980. 'Spatial and temporal sensitivity of X- and Y-cells in dorsal lateral geniculate nucleus of the cat.' *J. Neurophysiol.*, **43**, 520–41.

Leibowitz, H. W., Johnson, C. A., and Isabelle, E. 1972. 'Peripheral motion detection and refractive error.' *Science*, **177**, 1207–8.

Leibowitz, H. W., Rodemer, C. S., and Dichgans, J. 1979. 'The independence of dynamic spatial orientation from luminance and refractive error.' *Percept. Psychophys.*, **25**, 75–9.

Leinonen, L. M., and Hyvärinen, J. 1980. 'Parietal association cortex of the monkey as revealed by cellular recordings.' In G. E. Stelmach and J. Requin (Eds.), *Tutorials in Motor Behavior*. New York: North-Holland Publ. Co., pp. 117–27.

Lennerstrand, G., and Bach-y-Rita, P. 1974. 'Spindle responses in pig eye muscles.' *Acta physiol.*, **90**, 795–7.

Lennerstrand, G., and Bach-y-Rita, P. (Eds.) 1975. *Basic Mechanisms of Ocular Motility and Their Clinical Implications*. New York: Pergamon.

Lennie, P. 1971. 'Distortions of perceived orientation.' *Nature New Biology*, **233**, 155–6.

Lennie, P. 1974. 'Head orientation and meridional variations in acuity.' *Vision Res.*, **14**, 107–11.

Lennie, P. 1980a. 'Parallel visual pathways: a review.' *Vision Res.*, **20**, 561–94.

Lennie, P. 1980b. 'Perceptual signs of parallel pathways.' *Phil. Trans. roy. Soc.*, **B 290**, 23–37.

Lennie, P., and MacLeod, D. I. A. 1973. 'Background configuration and rod threshold.' *J. Physiol.*, **233**, 143–56.

Lennie, P., and Sidwell, A. 1978. 'Saccadic eye movements and visual stability.' *Nature*, **275**, 766–8.

Lester, G. 1968. 'The rod-and-frame test: some comments on methodology.' *Percept. mot. Skills*, **26**, 1307–14.

Lester, G., and Morant, R. B. 1970. 'Apparent sound displacement during vestibular stimulation.' *Amer. J. Psychol.*, **83**, 554–66.

Lestienne, F., Soechting, J., and Berthoz, A. 1977. 'Postural readjustments induced by linear motion of visual scenes.' *Exp. Brain Res.*, **28**, 363–84.

Le Vay, S., and Ferster, D. 1977. 'Relay cell classes in the lateral geniculate nucleus of the cat and the effect of visual deprivation.' *J. comp. Neurol.*, **172**, 563–84.

Le Vay, S., and Gilbert, C. D. 1976. 'Laminar patterns of geniculocortical projection in the cat.' *Brain Res.*, **113**, 1–19.

Leventhal, A. G. 1979. 'Evidence that the different classes of relay cells of the cat's lateral geniculate nucleus terminate in different layers of the striate cortex.' *Exp. Brain Res.*, **37**, 349–72.

Leventhal, A. G., and Hirsch, H. V. B. 1975. 'Cortical effect of early selective exposure to diagonal lines.' *Science*, **190**, 902–4.

Levick, W. R. 1972. 'Receptive fields of retinal ganglion cells.' In M. G. F. Fuortes (Ed.), *Handbook of Sensory Physiology*, Vol. VII/2. New York: Springer, pp. 530–66.

Levick, W. R., and Cleland, B. G. 1974. 'Receptive fields of cat retinal ganglion cells having slowly conducting axons.' *Brain Res.*, **74**, 156–60.

Levick, W. R., Oyster, C. W., and Takahashi, E. 1969. 'Rabbit lateral geniculate nucleus: sharpener of directional information.' *Science*, **165**, 712–4.

Levick, W. R., and Thibos, L. N. 1980. 'Orientation bias of cat retinal ganglion cells.' *Nature*, **286**, 389–90.

Levy, J. 1972. 'Autokinetic illusion. A systematic review of theories, measures and independent variables.' *Psychol. Bull.*, **78**, 457–74.

Levy, J. 1973. 'Autokinesis direction during and after eye turn.' *Percept. Psychophys.*, **13**, 337–42.

Lévy-Schoen, A., and Blanc-Garin, J. 1974. 'On oculomotor programming and perception.' *Brain Res.*, **71**, 443–50.

Linsenmeier, R. A., and Jakiela, H. G. 1979. 'Non-linear spatial summation in cat retinal ganglion cells at different background levels.' *Exp. Brain Res.*, **36**, 301–9.

Lisberger, S. G., and Fuchs, A. F. 1974 'Response of flocculus Purkinje cells to adequate vestibular stimulation in alert monkey: fixation vs. compensatory eye movements.' *Brain Res.*, **69**, 347–53.

Lisberger, S. G., and Fuchs, A. F. 1978a. 'Role of primate flocculus during rapid behavioral modification of vestibuloocular reflex. I. Purkinje cell activity during visually guided horizontal smooth-pursuit eye movements and passive head rotation.' *J. Neurophysiol.*, **41**, 733–63.

Lisberger, S. G., and Fuchs, A. F. 1978b. 'Role of primate flocculus during rapid behavioral modification of vestibuloocular reflex. II. Mossy fiber firing patterns during horizontal head rotation and eye movement.' *J. Neurophysiol.*, **41**, 764–77.

Lisberger, S. G., Fuchs, A. F., King, W. M., and Evinger, L. C. 1975. 'Effect of mean reaction time of saccadic responses to two-step stimuli with horizontal and vertical components. '*Vision Res.*, **15**, 1021–26.

Lisberger, S. G., and Miles, F. A. 1980. 'Role of primate medial vestibular nucleus in long-term adaptive plasticity of vestibuloocular reflex.' *J. Neurophysiol.*, **43**, 1725–45.

Lishman, J. R., and Lee, D. N. 1973. 'The autonomy of visual kinaesthesis.' *Perception*, **2**, 287–94.

Llewellyn, K. R. 1971. 'Visual guidance of locomotion.' *J. exp. Psychol.*, **91**, 245–61.

Louie, A. W., and Kimm, J. 1976. 'The response of 8th nerve fibers to horizontal sinusoidal oscillation in the alert monkey.' *Exp. Brain Res.*, **24**, 447–57.

Lovegrove, W. J. 1976. 'Inhibition in simultaneous and successive contour interaction in human vision.' *Vision Res.*, **16**, 1519–21.

Lovegrove, W. J., and Over, R. 1973. 'Colour selectivity in orientation masking and aftereffect.' *Vision Res.*, **13**, 895–902.

Lowenstein, O. 1975. 'The Peripheral neuron.' In R. F. Naunton (Ed.), *The Vestibular System.* New York: Academic Press, pp. 99–108.

Lowenstein, O., and Wersäll, J. 1959. 'A functional interpretation of the electron-microscopic structure of the sensory hairs in the cristae of the elasmobranch, *Raja clavata*, in terms of directional sensitivity.' *Nature*, **184**, 1807–8.

Luchins, A. S. 1954a. 'The autokinetic effect in central and peripheral vision.' *J. gen. Psychol.*, **50**, 39–44.

Luchins, A. S. 1954b. 'The autokinetic effect and gradients of illumination of the visual field.' *J. gen. Psychol.*, **50**, 29–37.

Ludvigh, E., and McKinnon, P., 1967. 'The effect of orientation on the three-dot alignment test.' *Amer. J. Ophthal.*, **64**, 261–5.

Lund, J. S., Downer, J. L. de C., and Lumley, J. S. P. 1970. 'Visual control of limb movement after section of optic chiasm and corpus callosum in the monkey.' *Cortex*, **6**, 323–46.

Luneburg, R. K. 1948. *Mathematical Analysis of Binocular Vision.* Princeton, NJ: Princeton University Press.

Luschei, E. S., and Fuchs, A. F. 1972. 'Activity of brain stem neurons during eye movements of alert monkeys.' *J. Neurophysiol.*, **35**, 445–61.

Luster, T. 1972. 'Vestibular adaptation in man: Effects of increased acceleration during different phases of adaptation.' *J. exp. Psychol.*, **96**, 263–72.

Lyle, J. G. 1969. 'Reading retardation and reversal tendency: a factorial study.' *Child Devel.*, **40**, 833–43.

Lynch, J. C. 1980. 'The functional organization of posterior parietal association cortex.' *Behav. Brain Sciences*, **3**, 485–534.

Lynch, J. C., Mountcastle, V. B., Talbot, W. H., and Yin, T. C. T. 1977. 'Parietal lobe mechanisms for directed visual attention.' *J. Neurophysiol.*, **40**, 362–89.

Maccoby, E. E., and Bee, H. L. 1965. 'Some speculations concerning the lag between perceiving and performing.' *Child Devel.*, **36**, 367–78.

MacDougall, R. 1903. 'The subjective horizon.' *Psychol. Rev. Monogr.*, Suppl. 4.,

Mach, E. 1875. *Grundlinien der Lehre von der Bewegungsempfindungen.* Leipzig: Engelmann.

Mack, A. 1970. 'An investigation of the relationship between eye and retinal image movement in the perception of movement.' *Percept. Psychophys.*, **8**, 291–8.

Mack, A., and Bachant, J. 1969. 'Perceived movement of the afterimage during eye movements.' *Percept. Psychophys.*, **6**, 379–84.

Mack, A., Fendrich, R., and Pleune, J. 1978. 'Adaptation to an altered relation between retinal image displacements and saccadic eye movements.' *Vision Res.*, **18**, 1321–7.

Mack, A., Fendrich, R., and Pleune, J. 1979. 'Smooth pursuit eye movements: Is perceived motion necessary?' *Science*, **203**, 1361–3.

Mack, A., and Herman, E. 1973. 'Position constancy during pursuit eye movement: An investigation of the Filehne illusion.' *Quart. J. exp. Psychol.*, **25**, 71–84.

Mack, A., and Herman. E. 1978. 'The loss of position constancy during pursuit eye movements.' *Vision Res.*, **18**, 55–62.

MacKay, D. M. 1970a. 'Elevation of visual threshold by displacement of retinal image.' *Nature*, **225**, 90–2.

MacKay, D. M. 1970b. 'Mislocation of test flashes during saccadic image displacements.' *Nature*, **227**, 731–2.

MacKay, D. M. 1970c. 'Interocular transfer of suppressive effects of retinal image displacement.' *Nature*, **225**, 872–3.

MacKay, D. M. 1972. 'Voluntary eye movements as questions.' In J. Dichgans and E.

Bizzi (Eds.), *Cerebral Control of Eye Movements and Motion Perception.* Basel: Karger, pp. 369–76.

MacKay, D. M. 1973. 'Visual stability and voluntary eye movements.' In R. Jung (Ed.), *Handbook of Sensory Physiology*, Vol. VII/3A. New York: Springer, pp. 307–32.

Mackeben, M., Haegerstrom-Portnoy, G., and Brown, B. 1980. 'Correctional movements during foveal pursuit in man.' *Proc. OMS 80 Conf.*, Pasadena, Calif.

Maclean, I. E., and Stacey, B. G. 1971. 'Judgment of angle size.' *Percept. Psychophys.*, **9**, 499–504.

Maekawa, K., and Simpson, J. I. 1973. 'Climbing fiber responses evoked in the vestibulocerebellum of rabbit from visual system.' *J. Neurophysiol.*, **36**, 649–66.

Maffei, L., and Campbell, F. W. 1970. 'Neurophysiological localization of the vertical and horizontal visual coordinates in man.' *Science*, **167**, 386–7.

Magnin, M., Jeannerod, M., and Putkonen, P. T. S. 1974. 'Vestibular and saccadic influence on dorsal and ventral nuclei of the lateral geniculate body.' *Exp. Brain Res.*, **21**, 1–18.

Magnussen, S., and Kurtenbach, W. 1979. 'A test for contrast-polarity selectivity in the tilt aftereffect.' *Perception*, **8**, 523–8.

Magnussen, S., and Kurtenbach, W. 1980. 'Linear summation of tilt illusion and tilt aftereffect.' *Vision Res.*, **20**, 39–42.

Magoun, H. W., and Ranson, S. W. 1935. 'The central path of the light reflex: A study of the effect of lesions.' *A. M. A. Arch. Ophthal.*, **13**, 791–811.

Maki, R. H., Grandy, C. A., and Hauge, G., 1979. 'Why is telling right from left more difficult than telling above from below?. *J. exp. Psychol: Human Percept. Perform.*, **5**, 52–67.

Maki, R. H., Maki, W. S., and Marsh, L. G. 1977. 'Processing locational and orientational information.' *Mem. Cognit.*, **5**, 602–12.

Malcolm, R. 1974. 'A mechanism by which the hair cells of the inner ear transduce mechanical energy into a modulated train of action potential.' *J. gen. Physiol.*, **63**, 757–72.

Malcolm, R., and Melvill Jones, G. 1970. 'A quantitative study of vestibular adaptation in humans.' *Acta oto-laryng.*, **70**, 126–35.

Malcolm, R., and Melvill Jones, G. 1974. 'Erroneous perception of vertical motion by humans seated in the upright position.' *Acta oto-laryng.*, **77**, 274–83.

Mandl, G., and Melvill Jones, G. 1979. 'Rapid visual–vestibular interaction during visual tracking in strobe light.' *Brain Res.*, **165**, 133–8.

Mandler, J. M., and Day, J. 1975. 'Memory for orientation of forms as a function of their meaningfulness and complexity.' *J. exp. child Psychol.*, **20**, 430–43.

Mann, C. W., Berthelot-Berry, N. H., and Dauterive, H. J. 1949. 'The perception of the vertical: I. Visual and non-labyrinthine cues.' *J. exp. Psychol.*, **39**, 538–47.

Mann, V. A., Hein, A., and Diamond, R. 1979a. 'Localization of targets by strabismic subjects: Contrasting patterns in constant and alternating suppressors.' *Percept. Psychophys.*, **25**, 29–34.

Mann, V. A., Hein, A., and Diamond, R. 1979b. 'Patterns of interocular transfer of visuomotor coordination reveal differences in the representation of visual space.' *Percept. Psychophys.*, **25**, 35–41.

Manni, E., Palmieri, G., Marini, R., and Pettorossi, V. E. 1975. 'New observations on the representation of the eye muscle proprioception within the descending trigeminal nucleus.' *Experientia*, **31**, 944–5.

Mansfield, R. J. W. 1974. 'Neural basis of orientation perception in primate vision.' *Science*, **186**, 1133–4.

Mansfield, R. J. W., and Ronner, S. F. 1978. 'Orientation anistropy in monkey visual cortex.' *Brain Res.*, **149**, 229–34.

Marek, P., and Markel, E. 1971. 'Phasic stretch reflex in the extraocular muscles of *encéphale isolé* cats.' *Acta physiol. Acad. Scien. Hung.*, **40**, 359–65.

Marini, G., Provini, L., and Rosina, A. 1976. 'Gravity responses of Purkinje cells in the nodulus.' *Exp. Brain Res.*, **24**, 311–23.

Marks, M. R. 1949. 'Some phenomena attendant on long fixation.' *Amer. J. Psychol.*, **62**, 392–98.

Marmor, G. S., and Zaback, L. A. 1976. 'Mental rotation by the blind: Does mental rotation depend on visual imagery?' *J. exp. Psychol: Hum. Percept. Perform.*, **2**, 515–21.

Marr, D. A. 1969. 'A theory of cerebellar cortex.' *J. Physiol.*, **202**, 437–70.

Marr, D., and Nishihara, H. K. 1976. 'Representation and recognition of the spatial organization of three-dimensional shapes.' A. I. Memo 377, Artificial Intelligence Lab. Mass. Inst. Technol.

Marrocco, R. T. 1978. 'Saccades induced by stimulation of the frontal eye fields: interaction with voluntary and reflexive eye movements.' *Brain Res.*, **146**, 23–34.

Marrocco, R. T., and Li, R. H. 1977. 'Monkey superior colliculus: Properties of single cells and their afferent inputs.' *J. Neurophysiol.*, **40**, 844–60.

Marshall, W. H., and Talbot, S. A. 1942. 'Recent evidence for neural mechanisms in vision leading to a general theory of sensory acuity.' *Biol. Symp.*, **7**, 117–64.

Mateeff, S. 1978, 'Saccadic eye movements and localization of visual stimuli.' *Percept. Psychophys.*, **24**, 215–24.

Mather, J. A., and Lackner, J. R. 1975. 'Adaptation to visual rearrangement elicited by tonic vibration reflexes.' *Exp. Brain Res.*, **24**, 103–5.

Mather, J. A., and Lackner, J. R. 1980a. 'Visual tracking of active and passive movements of the hand.' *Quart. J. exp. Psychol.*, **32**, 307–16.

Mather, J. A., and Lackner, J. R. 1980b. 'Adaptation to visual displacement with active and passive limb movements: effect of movement frequency and predictability of movement.' *Quart. J. exp. Psychol.*, **32**, 317–24.

Matin, E. 1974a. 'Saccadic suppression: A review and an analysis.' *Psychol. Bull.*, **81**, 899–917.

Matin, E. 1974b. 'Light adaptation and the dynamics of induced tilt.' *Vision Res.*, **14**, 255–65.

Matin, E., and Drivas, A. 1979. 'Acuity for orientation measured with a sequential recognition task and signal detection methods.' *Percept. Psychophys.*, **25**, 161–8.

Matin, L. 1976a. 'A possible hybrid mechanism for modification of visual direction associated with eye movements: The paralysed-eye experiment reconsidered.' *Perception*, **5**, 233–9.

Matin, L. 1976b. 'Saccades and extraretinal signal for visual direction.' In R. A. Monty and J. W. Senders (Eds.), *Eye Movements and Psychological Processes*. New Jersey: Erlbaum, pp. 205–20.

Matin, L., and Kibler, G. 1966. 'Acuity of visual perception of direction in the dark for various positions of the eye in the orbit.' *Percept. mot. Skills*, **22**, 407–20.

Matin, L., and MacKinnon, G. E. 1964. 'Autokinetic movement: selective manipulation of directional components by image stabilization.' *Science*, **143**, 147–8.

Matin, L., Matin, E., and Pearce, D. G. 1969. 'Visual perception of direction when voluntary saccades occur. I. Relation of visual direction of a fixed target extinguished before a saccade to a flash presented during a saccade.' *Percept. Psychophys.*, **5**, 65–80.

Matin, L., Matin, E., and Pearce, D. G., 1970. Eye movements in dark during the attempt to maintain a prior fixation position. *Vision Res.*, **10**, 837–58.

Matin, L., Matin, E., and Pola, J. 1970. 'Visual perception of direction when voluntary saccades occur: II. Relation of visual direction of a fixation target extinguished before a saccade to a subsequent test flash presented before the saccade.' *Percept. Psychophys.*, **8**, 9–14.

Matin, L., and Pearce, D. G. 1965. 'Visual perception of direction for stimuli flashed during voluntary saccadic eye movements.' *Science*, **148**, 1485–8.

Matin, L., Pearce, D., Matin, E., and Kibler, G. 1966. 'Visual perception of direction in the dark. Roles of local sign, eye movement, and ocular proprioception.' *Vision Res.*, **6**,

453–69.

Matin, L., Picoult, E., Stevens, J. K., Edwards, M., Young, D., and MacArthur, R. 1981. 'Oculoparalytic illusion: visual-field dependent spatial mislocalization by humans with experimentally paralysed extraocular muscles.' *Science*. in press.

Matsuo, V., Cohen, B., Raphan, T., de Jong, V., and Henn, V. 1979. 'Asymmetric velocity storage for upward and downward nystagmus.' *Brain Res.*, **176**, 159–64.

Matthews, P. B. C. 1972. *Mammalian Muscle Receptors and Their Central Action*. London: Arnold.

Maturana, H. R., Lettvin, J. Y., McCulloch, W. S., and Pitts, W. H. 1960. 'Anatomy and physiology of vision in the frog (Rana Pipiens).' *J. gen. Physiol.*, **43**, Suppl. No. 2, 129–76.

Mauritz, K. H., Dichgans, J., and Hufschmidt, A. 1977. 'The angle of visual roll motion determines displacement of subjective visual vertical.' *Percept. Psychophys.*, **22**, 557–62.

Mayer, D. L., and Dobson, V. 1980. 'Assessment of vision in young children: a new operant approach yields estimates of acuity.' *Invest. Ophthal.*, **19**, 566–70.

Mayer, M. J. 1977. 'Development of anisotropy in late childhood,' *Vision Res.*, **17**, 703–10.

Mayne, R. 1950. 'The dynamic characteristics of the semicircular canals.' *J. comp. physiol. psychol.*, **43**, 309–19.

Mayne, R. 1952. 'The audiogyral illusion and the mechanism of spatial representation.' *Bull. Math. Biophysics*, **14**, 27–34.

Mays,L. E., and Sparks, D. L. 1980a. 'Saccades are spatially, not retinotopically coded.' *Science*, **208**, 1163–4.

Mays, L. E., and Sparks, D. L. 1980b. 'Dissociation of visual and saccade-related responses in superior colliculus neurones.' *J. Neurophysiol.*, **43**, 207–32.

McCann, J. J., and Hall, J. A. 1980. 'Effects of average-luminance surrounds on the visibility of sine-wave gratings.' *J. opt. Soc. Amer.*, **70**, 212–9.

McCann, J. J., Savoy, R. L., and Hall, J. A. 1978. 'Visibility of low-frequency sine-wave targets: Dependence on number of cycles and surround parameters.' *Vision Res.*, **18**. 891–4.

McDonnell, P. M. 1975. 'The development of visually guided reaching.' *Percept. Psychophys.*, **18**, 181–5.

McDonnell, P. M., and Abraham, W. C. 1979. 'Adaptation to displacing prisms in human infants.' *Perception*, **8**, 175–85.

McDonnell, P. M., and Duffett, J. 1972. 'Vision and touch: A reconsideration of conflict between the two senses.' *Canad. J. Psychol.*, **26**, 171–80.

McGinnis, J. M. 1930. 'Eye movements and optic nystagmus in early infancy.' *Genet. Psychol. Monogr.*, **8**, 321–430.

McGonigle, B. O., and Chalmers, M. 1980. 'The selective impact of verbally communicated criteria in orientation judgements of young children.' *Perception*, **9**, 345–51.

McGonigle, B. O., and Flook, J. 1978. 'Long-term retention of single and multistate prismatic adaptation by humans.' *Nature*, **272**, 364–6.

McGraw, M. 1943. *Neuromuscular Maturation of the Human Infant*. New York: Columbia University Press.

McGurk, H. 1970. 'The role of object orientation in infant perception.' *J. exp. Child Psychol.*, **9**, 363–73.

McGurk, H. 1972a. 'Infant discrimination of orientation.' *J. exp. Child Psychol.*, **14**, 151–64.

McGurk, H. 1972b. 'The salience of orientation in young children's perception of form.' *Child Devel.*, **43**, 1047–52.

McGurk, H., and Lewis, M. 1974. 'Space perception in early infancy: Perception within a common auditory–visual space?' *Science*, **186**, 649–50.

McGurk, H., Turnure, C., and Creighton, S. J. 1977. 'Auditory–visual coordination in neonates.' *Child Devel.*, **48**, 138–43.

McIlwain, J. T. 1975. 'Visual receptive fields and their images in superior colliculus of the cat.' *J. Neurophysiol.*, **38**, 219–30.

McIlwain, J. T. 1976. 'Large receptive fields and spatial transformations in the visual system.' In R. Porter (Ed.), *International Review of Physiology*. Baltimore: University Park Press, pp. 223–48.

McIlwain, J. T. 1978. 'Properties of cells projecting rostrally from the superficial layers of the cat's superior colliculus.' *Brain Res.*, **143**, 445–58.

McIntyre, A. K. 1939. 'The quick component of nystagmus.' *J. Physiol.*, **97**, 8–16.

McIntyre, C. W., and Pick, H. L. 1974. 'Transfer of adaptation to rotation of the visual field.' *J. exp. Psychol.*, **101**, 782–91.

McKee, S. P., and Westheimer, G. 1978. 'Improvement in vernier acuity with practice. *Percept. Psychophys.*, **24**, 258–62.

McLaren, J. W., and Hillman, D. E. 1979. 'Displacement of the semicircular canal cupula during sinusoidal rotation.' *Neurosci.*, **4**, 2001–8.

McLaughlin, S. C. 1967. 'Parametric adjustment in saccadic eye movements.' *Percept. Psychophys.*, **2**, 359–62.

McLaughlin, S. C., Kelly, M. J., Anderson, R. E., and Wenz, T. G. 1968. 'Localization of a peripheral target during parametric adjustment of saccadic eye movements.' *Percept. Psychophys.*, **4**, 45–8.

McLaughlin, S. C., Rifkin, K. I., and Webster, R. G. 1966. 'Oculomotor adaptation to wedge prisms with no part of the body seen.' *Percept. Psychophys.*, **1**, 452–8.

McLaughlin, S. C., and Webster, R. G. 1967. 'Changes in straight-ahead eye position during adaptation to wedge prisms.' *Percept. Psychophys.*, **2**, 37–44.

Medawar, P. B. 1967. *The Art of the Soluble*. London: Methuen.

Megaw, E. D., and Armstrong, W. 1973. 'Individual and simultaneous tracking of a step input by horizontal saccadic eye movement and manual control systems.' *J. exp. Psychol.*, **100**, 18–28.

Meiry, J. L. 1971. 'Vestibular and proprioceptive stabilization of eye movements.' In P. Bach-y-Rita, C. C. Collins and J. E. Hyde (Eds.), *The Control of Eye Movements*. New York: Academic Press, pp. 483–96.

Melamed, L. E., Halay, M., and Gildow, J. W. 1973. 'Effect of external target presence on visual adaptation with active and passive movement.' *J. exp. Psychol.*, **98**, 125–30.

Melvill Jones, G. 1964. 'Predominance of anti-compensatory oculomotor response during rapid head rotation.' *Aerosp. Med.*, **35**, 965–8.

Melvill Jones, G. 1970. 'Origin significance and amelioration of coriolis illusions from the semicircular canals: a non-mathematical appraisal.' *Aerosp. Med.*, **41**, 483–90.

Melvill Jones, G. 1976. 'The vestibular system for eye movement control.' In R. A. Monty and J. W. Senders (Eds.), *Eye Movements and Psychological Processes*. Hillside, N. J.: Erlbaum, pp. 3–18.

Melvill Jones, G., Barry, W., and Kowalsky, N. 1964. 'Dynamics of the semicircular canals compared in yaw, pitch and roll.' *Aerosp. Med.*, **35**, 984–9.

Melvill Jones, G., and Davies, P. 1976. 'Adaptation of cat vestibulo-ocular reflex to 200 days of optically reversed vision.' *Brain Res.*, **103**, 551–4.

Melvill Jones, G., Downing, D., and Rolf, R. 1973. 'Human subjective and reflex responses to sinusoidal vertical acceleration.' Ames Prog. Rep., NASA. Cited in Melvill Jones (1976), p. 15.

Melvill Jones, G., and Mandl, G. 1979. 'Effects of strobe light on adaptation of vestibulo-ocular reflex (VOR) to vision reversal.' *Brain Res.*, **164**, 300–3.

Melvill Jones, G., and Milsum, J. H. 1970. 'Characteristics of neural transmission from the semicircular canal to the vestibular nuclei of cats.' *J. Physiol.*, **209**, 295–316.

Melvill Jones, G., and Milsum, J. H. 1971. 'Frequency-response analysis of central vestibular unit activity resulting from rotational stimulation of the semicircular canals.' *J. Physiol.*, **219**, 191–215.

Melvill Jones, G., Rolph, R., and Downing, G. H. 1980. 'Comparison of human subjective and oculomotor responses to sinusoidal vertical linear acceleration.' *Acta*

oto-laryng., **90**, 431–40.
Melvill Jones, G., and Spells, K. E. 1963. 'A theoretical and comparative study of the functional dependence of the semicircular canal upon its physical dimensions.' *Proc. roy. Soc. B.*, **157**, 403–19.
Melvill Jones, G., and Sugie, N. 1972. 'Vestibulo-ocular responses in man during sleep.' *Electroenceph. clin. Neurophysiol.*, **32**, 43–53.
Melvill Jones, G., and Young, L. R. 1978. 'Subjective detection of vertical acceleration: A velocity-dependent response.' *Acta oto-laryng.*, **85**, 45–53.
Merker, B. H., and Held, R. 1981. 'Eye torsion and the apparent horizon under head tilt and visual field rotation.' *Vision Res.*, **21**, 543–7.
Merton, P. A. 1961. 'The accuracy of directing the eyes and the hand in the dark.' *J. Physiol.*, **156**, 555–77.
Metz, H. S., and Scott, A. B. 1970. 'Innervational plasticity of the oculomotor system.' *A.M.A. Arch. Ophthal.*, **84**, 86–91.
Metzler, J., and Spinelli, D. N. 1979. 'Tilt-constant receptive fields in the kitten visual cortex.' *Brain Res.*, **163**, 344–8.
Michael, C. R. 1966. 'Receptive fields of directionally selective units in the optic nerve of the ground squirrel.' *Science*, **152**, 1092–5.
Michael, C. R. 1970. 'Integration of retinal and cortical information in the superior colliculus of the ground squirrel.' *Brain Behav. Evol.*, **3**, 205–9.
Michael, C. R. 1972. Visual Receptive fields of single neurons in superior colliculus of the ground squirrel.' *J. Neurophysiol.*, **35**, 815–32.
Michael, J. A., and Melvill Jones, G. 1966. 'Dependence of visual tracking capability upon stimulus predictability.' *Vision Res.*, **6**, 707–16.
Miles, F. A. 1974. 'Single unit firing patterns in the vestibular nuclei related to voluntary eye movements and passive body rotation in conscious monkeys.' *Brain Res.*, **71**, 215–24.
Miles, F. A. 1977. 'The primate flocculus and eye–head coordination.' In B. A. Brooks and F. J. Bajandas (Eds.), *Eye Movements.* New York: Plenum, pp. 75–92.
Miles, F. A., and Braitman, D. J. 1980. 'Long-term adaptive changes in primate vestibuloocular reflex. II. Electrophysiological observations on semicircular canal primary afferents.' *J. Neurophysiol.*, **43**, 1426–36.
Miles, F. A., Braitman, D. J., and Dow, B. M. 1980. 'Long-term adaptive changes in primate vestibuloocular reflex. IV. Electrophysiological observations in flocculus of adapted monkeys.' *J. Neurophysiol.*, **43**, 1477–93.
Miles, F. A., and Eighmy, B. B. 1980. 'Long-term adaptive changes in primate vestibuloocular reflex. I. Behavioral observations.' *J. Neurophysiol.*,**43**, 1406–25.
Miles, F. A., and Fuller, J. H. 1974. 'Adaptive plasticity in the vestibuloocular responses of the rhesus monkey.' *Brain Res.*, **80**, 512–6.
Miles, F. A., and Fuller, J. H. 1975. 'Visual tracking and the primate flocculus.' *Science*, **189**, 1000–2.
Miles, F. A., Fuller, J. H., Braitman, D. J., and Dow, B. M. 1980. 'Long-term adaptive changes in primate vestibuloocular reflex. III. Electrophysiological observations in flocculus of normal monkey.' *J. Neurophysiol.*, **43**, 1437–76.
Miller, E. A. 1972. 'Interaction of vision and touch in conflict and nonconflict form perception tasks.' *J. exp. Psychol.*, **96**, 114–23.
Miller, E. F., Fregly, A. R., and Graybiel, A. 1968. 'Visual horizontal-perception in relation to otolith-function.' *Amer. J. Psychol.*, **81**, 488–96.
Miller, E. F., Fregly, A. R., van den Brink, G., and Graybiel, A. 1965. 'Visual localization of the horizontal as a function of body tilt up to $\pm 90°$ from gravitational vertical.' NSAM-942. NASA Order No. R-47. Nav. Sch. Aviat. Med., Pensacola, Florida.
Miller, E. F., and Graybiel, A. 1963. 'Rotary autokinesis and displacement of the visual horizontal associated with head (body) position.' USN Sch. Aviat. Med. NASA & Res. Rep. Proj. MROOS. 13-6001, Subtask 1, Rep. No. 77.
Miller, E. F., and Graybiel, A. 1966. 'Magnitude of gravitoinertial force, an independent

variable in egocentric visual localization of the horizontal.' *J. exp. Phychol.*, **71**, 425–60.

Miller, E. F., and Graybiel, A. 1971. 'Effect of gravitoinertial force on ocular counterrolling.' *J. appl. Physiol.*, **31**, 697–700.

Miller, E. F., and Graybiel, A. 1974. 'Human ocular counterrolling measured during eight hours of sustained body tilt.' *Minerva otorinolaringologica*, **24**, 247–52.

Miller, E. F., and Graybiel, A. 1975. 'Thresholds for the perception of angular acceleration as indicated by the oculogyral illusion.' *Percept. Psychophys.*, **17**, 329–32.

Miller, J. M. 1977. 'Information available to the perceptual and oculomotor systems regarding saccadic and pursuit movements.' Ph. D. Diss., New School for Social Research.

Miller, J. M. 1980. 'Information used by the perceptual and oculomotor systems regarding the amplitude of saccadic and pursuit eye movements.' *Vision Res.* **20**, 59–68.

Miller, J. M., Anstis, T., and Templeton, W. B. 1979. 'Parametric adjustment of saccadic asymmetries and transfer.' *Invest. Ophthal.*, **18**, ARVO Suppl., 101.

Miller, J. M., and Festinger, L. 1977. 'Impact of oculomotor retraining on the visual perception of curvature.' *J. exp. Psychol: Hum. Percept. Perform.*, **3**, 187–206.

Miller, J. M., Ono, H., and Steinbach, M. J. 1980. 'Additivity of fusional vergence and pursuit eye movements.' *Vision Res.*, **20**, 43–8.

Miller, J. W., and Hall, R. J. 1963. 'The problem of motion perception and orientation in the Ganzfeld.' *Aerosp. Med.*, **34**, 337–41.

Miller, L. K. 1969. 'Eye latency as a function of age, stimulus uncertainty, and position in the visual field.' *Percept. mot. Skills*, **28**, 631–6.

Millodot, M. 1972. 'Variation of visual acuity in the central region of the retina.' *Brit. J. physiol. Opt.*, **27**, 24–8.

Milsum, J. H., and Melvill Jones, G. 1969. 'Dynamic asymmetry in neural components of the vestibular system.' *Ann. New York Acad. Sci.*, **156**, 851–71.

Minsky, M. 1975. 'A framework for representing knowledge.' In P. H. Winston (Ed.), *The Psychology of Computer Vision.* New York: McGraw-Hill, pp. 211–77.

Minsky, M., and Papert, S. 1969. *Perceptrons.* Cambridge, Mass.: MIT Press.

Mishkin, S., and Melvill Jones, G. 1966 'Predominant direction of gaze during slow head rotation.' *Aerosp. Med.*, **37**, 897–900.

Mitchell, D. E., and Blakemore, C. 1972. 'The site of orientational constancy.' *Perception*, **1**, 315–20.

Mitchell, D. E., Freeman, R. D., Millodot, M., and Haegerstrom, G. 1973. 'Meridional amylopia: Evidence for modification of the human visual system by early visual experience.' *Vision Res.*, **13**, 535–58.

Mitchell, D. E., Freeman, R. D., and Westheimer, G. 1967. 'Effect of orientation on the modulation sensitivity for interference fringes on the retina.' *J. opt. Soc. Amer.*, **57**, 246–9.

Mitchell, D. E., Giffin, F., Muir, D., Blakemore, C., and Van Sluyters, R. C. 1976. 'Behavioural compensation for cats after early rotation of one eye.' *Exp. Brain Res.*, **25**, 109–13.

Mitchell, D. E., and Muir, D. W. 1976. 'Does the tilt aftereffect occur in the oblique meridian?' *Vision Res.*, **16**, 609–13.

Mitchell, D. E., and Ware, C. 1974. 'Interocular transfer of a visual after-effect in normal and stereoblind humans.' *J. Physiol.*, **236**, 707–21.

Mitchell, D. E., and Wilkinson, F. 1974. 'The effect of early astigmatism on the visual resolution of gratings.' *J. Physiol.*, **243**, 739–56.

Mitrani, L., Mateef, St., and Yakimoff, N. 1971. 'Is saccadic suppression really saccadic?' *Vision Res.*, **11**, 1157–61.

Mitrani, L., Yakimoff, N., and Mateef, St. 1973. 'Saccadic suppression in the presence of a structured background.' *Vision Res.*, **13**, 517–21.

Mitson, L., Ono, H., and Barbeito, R. 1976. 'Three methods of measuring the location of the egocentre: Their reliability, comparative locations and intercorrelations.' *Canad. J. Psychol.*, **30**, 1–8.

Mize, R. R., and Murphy, E. H. 1973. 'Selective visual experience fails to modify receptive field properties of rabbit striate cortex neurons.' *Science*, **180**, 320–1.

Mohler, C. W., Goldberg, M. E., and Wurtz, R. H. 1973. 'Visual receptive fields of frontal eye field neurons.' *Brain Res.*, **61**, 385–9.

Mohler, C. W., and Wurtz, R. H. 1976. 'Organization of monkey superior colliculus: intermediate layer cells discharging before eye movements.' *J. Neurophysiol.*, **39**, 722–44.

Mohler, C. W., and Wurtz, R. H. 1977. 'Role of striate cortex and superior colliculus in visual guidance of saccadic eye movements in monkey.' *J. Neurophysiol.*, **40**, 74–94.

Money, K. E., Bonen, L., Beatty, J. D., Kuehn, L. A., Sokoloff, M., and Weaver, R. S. 1971. 'Physical properties of fluids and structures of vestibular apparatus of the pigeon.' *Amer. J. Physiol.*, **220**, 140–7.

Money, K. E., Johnson, W. H., and Corlett, B. M. A. 1965. 'Role of semicircular canals in positional alcohol nystagmus.' *Amer. J. Physiol.*, **208**, 1065–70.

Money, K. E., and Myles, W. S. 1974. 'Heavy water nystagmus and effects of alcohol.' *Nature*, **247**, 404–5.

Monty, R. A., and Senders, J. W. 1976. *Eye Movements and Psychological Processes.* Hillside, N. J.: Erlbaum.

Morant, R. B., and Aronoff, J. 1966. 'Starting position, adaptation, and visual framework as influencing the perception of verticality.' *J. exp. Psychol.*, **71**, 684–6.

Morant, R. B., and Beller, H. K. 1965. 'Adaptation to prismatically rotated visual fields.' *Science*, **148**, 530–1.

Morant, R. B., and Mikaelian, H. H. 1960. 'Inter-field tilt after-effects.' *Percept. mot. Skills*, **10**, 95–8.

Morasso, P., Bizzi, E., and Dichgans, J. 1973. 'Adjustment of saccade characteristics during head movements.' *Exp. Brain Res.*, **16**, 492–500.

Morgan, C. L. 1978. 'Constancy of egocentric visual direction.' *Percept. Psychophys.*, **23**, 61–8.

Morgan, M. J., and Turnbull, D. F. 1978. 'Smooth eye tracing and the perception of motion in the absence of real movement.' *Vision Res.*, **18**, 1053–9.

Morrell, F. 1972. 'Visual system's view of acoustic space.' *Nature*, **238**, 44–6.

Moulden, B. 1971 'Adaptation to displaced vision: Reafference is a special case of the cue-discrepancy hypothesis.' *Quart. J. exp. Psychol.*, **23**, 113–7.

Mountcastle, V. B., Lynch, J. C., Geogopoulos, A., Sakata, H., and Acuna, C. 1975. 'Posterior parietal association cortex of the monkey: Command functions for operations within extrapersonal space.' *J. Neurophysiol.*, **38**, 871–908.

Movshon, J. A., and Blakemore, C. 1973. 'Orientation specificity and spatial selectivity in human vision.' *Perception*, **2**, 53–60.

Movshon, J. A., and Lennie, P. 1979. 'Pattern-selective adaptation in visual cortical neurones.' *Nature, New Biol.*, **278**, 850–2.

Movshon, J. A., Thompson, I. D., and Tolhurst, D. J. 1978a. 'Spatial summation in the receptive fields of simple cells in the cat's striate cortex.' *J. Physiol.*, **283**, 53–78.

Movshon, J. A., Thompson, I. D., and Tolhurst, D. J. 1978b. 'Receptive field organization of complex cells in the cat's striate cortex.' *J. Physiol.*, **283**, 79–100.

Movshon, J. A., Thompson, I. D., and Tolhurst, D. J. 1978c. 'Spatial and temporal contrast sensitivity of neurones in areas 17 and 18 of the cat's visual cortex.' *J, Physiol.*, **283**, 101–20.

Mower, G., Gibson, A. Robinson, F., Stein, J., and Glickstein, M. 1980. 'Visual pontocerebellar projections in the cat.' *J. Neurophysiol.*, **43**, 355–66.

Muir, D., and Field, J. 1979. 'Newborn infants orient to sounds.' *Child Devel.*, **50**, 431–6.

Muir, D. W., and Mitchell, D. E. 1973. 'Visual resolution and experience: Acuity deficits in cats following early selective visual deprivation.' *Science*, **180**, 420–2.

Muir, D. W., and Mitchell, D. E. 1975. 'Behavioral deficits in cats following early selected visual exposure to contours of a single orientation.' *Brain Res.*, **85**, 459–77.

Muir, D. W., and Over, R. 1970. 'Tilt aftereffects in central and peripheral vision.' *J. exp. Psychol.*, **85**, 165–70.

Müller, G. E. 1916. 'Über das Aubertsche Phänomenon.' *Z. Psychol. Physiol. Sinnesorg.*, **49**, 109–246.

Muller, P. E., Sidorsky, R. C., Slivinske, A. J., Alluisi, E. A., and Fitts, P. M. 1955. 'The symbolic coding of information on cathode ray tubes and similar displays.' U.S.A.F. WADC. Tech. Rep. No. 55–375.

Mulligan, R. M., and Shaw, M. L. 1980. 'Multimodal signal detection: independent decision vs. integration.' *Percept. Psychophys.*, **28**, 471–8.

Muratore, R., and Zee, D. S. 1979. 'Pursuit after-nystagmus.' *Vision Res.*, **19**, 1057–9.

Murphy, B. J., Kowler, E., and Steinman, R. M. 1975. 'Slow oculomotor control in the presence of moving backgrounds.' *Vision Res.*, **15**, 1263–8.

Myers, R. E., Sperry, R. W., and McCurdy, N. M. 1962. 'Neural mechanisms in visual guidance of limb movement.' *Arch. Neurol.*, **7**, 195–202.

Nachmias, J. 1959. 'Two-dimensional motion of the retinal image during monocular fixation.' *J. opt. Soc. Amer.*, **49**, 901–8.

Nachmias, J., and Weber, A. 1975. 'Discrimination of simple and complex gratings.' *Vision Res.*, **15**, 217–23.

Naeli, H., and Harris, P. L. 1976. 'Orientation of the diamond and the square.' *Perception*, **5**, 73–7.

Nagle, M., Bridgeman, B., and Stark, L. 1980. 'Voluntary nystagmus, saccadic suppression, and stabilization of the visual world.' *Vision Res.*, **20**, 717–23.

Nakao, S., Curthoys, I. S., and Markham, C. H. 1980. 'Eye movement related neurones in the cat pontine reticular formation; projection to the flocculus.' *Brain Res.*, **183**, 291–300.

Nakayama, K. 1975. 'Coordination of extraocular muscles.' In G. Lennerstrand and P. Bach-y-Rita (Eds.), *Basic Mechanisms of Ocular Motility and Their Clinical Implications*. New York: Pergamon, pp. 193–208.

Nakayama, K. 1978. 'A new method of determining the primary position of the eye using Listing's law.' *Amer. J. Optom.*, **55**, 331–6.

Nakayama, K., Appelman, J., and Tyler, C. W. 1981. 'Human binocular correspondence in the vertical dimension: an ecological tilt?' *Science*, in press.

Nakayama, K., and Balliet, R. 1977. 'Listing's Law, eye position sense, and perception of the vertical.' *Vision Res.*, **17**, 453–7.

Nakayama, K., and Loomis, J. M. 1974. 'Optical velocity patterns, velocity-sensitive neurons, and space perception. A hypothesis.' *Perception*, **3**, 63–80.

Nakayama, K., and Tyler, C. W. 1976. 'Local versus global factors in the perception of orientation.' *J. opt. Soc. Amer.*, **66**, 1090.

Nashner, L., and Berthoz, A. 1978. 'Visual contribution to rapid responses during postural control.' *Brain Res.*, **150**, 403–7.

Naunton, R. F. (Ed.) 1975. *The Vestibular System*. New York: Academic Press.

Navon, D. 1978. 'Perception of misoriented words and letter strings.' *Canad. J. Psychol.*, **32**, 129–40.

Nelson, J. R., and Cope, D. 1971. 'The otoliths and the ocular countertorsion reflex.' *Arch. oto-laryng.*, **94**, 40–50.

Nemet, P., and Ron, S. 1977. 'Cerebellar role in smooth pursuit movement.' *Doc. Ophthal.*, **43**, 101–8.

Nickerson, R. S. 1972. 'Binary-classification reaction time: A review of some studies of human information-processing capabilities.' *Psychonom. Monog. Suppl.*, **4**, 275–317.

Niven, J. I., Hixson, W. C., and Correia, M. J. 1966. 'Elicitation of horizontal nystagmus by periodic linear acceleration.' *Acta oto-laryng.*, **62**, 429–41.

Noda, H., and Adey, W. R. 1974. 'Retinal ganglion cells of the cat transfer information on saccadic eye movement and quick target motion.' *Brain Res.*, **70**, 340–5.
Noda, H., Freeman, R. B., and Creutzfeldt, O. D. 1972. 'Neuronal correlates of eye movements in the visual cortex of the cat.' *Science*, **175**, 661–3.
Noda, H., and Suzuki, D. A. 1970a. 'The role of the flocculus of the monkey in saccadic eye movements.' *J. Physiol.*, **294**, 317–35.
Noda, H., and Suzuki, D. A. 1979b. 'The role of the flocculus of the monkey in fixation and smooth pursuit eye movements.' *J. Physiol.*, **294**, 335–48.
Noda, H., and Suzuki, D. A. 1979c. 'Processing of eye movement signals in the flocculus of the monkey.' *J. Physiol.*, **294**, 349–64.
Noorden, G. K. von, and Mackensen, G. 1962. 'Pursuit movements of normal and amblyopic eyes. An electro-ophthalmographic study. II. Pursuit movements in amblyopic patients.' *Amer. J. Ophthal.*, **53**, 477–87.
Norcia, A. M., Yonas, A., and Peck, J. 1980. 'Saccadic eye movement latencies as a function of target directional uncertainty.' *Proc. Oculomotor Symp.*, Caltech., Pasadena.
Norton, T. T. 1974. 'Receptive-field properties of superior colliculus cells and development of visual behavior in kittens.' *J. Neurophysiol.*, **37**, 674–89.
Nyberg-Hansen R. 1975. 'Anatomical aspects of the functional organization of the vestibulo-spinal pathways.' In R. F. Naunton (Ed.), *The Vestibular System.* New York: Academic Press, pp. 71–96.
Nyborg, H. 1971. 'Tactile stimulation and perception of the vertical. I. Effects of diffuse vs. specific tactile stimulation.' *Scand. J. Psychol.*, **12**, 1–13.
Nyborg, H. 1974. 'A method for analysing performance in the rod-and-frame test. I.' *Scand. J. Psychol.*, **15**, 119–23.
Nyborg, H. 1980. 'Psychological differentiation in school children. Maturation, cognition, and personality development.' *Psychological Reports Aarhus*, Vol. 5, No. 2. University of Aarhus, Denmark.
O'Brien, B. 1946. 'A theory of the Stiles–Crawford effect.' *J. opt. Soc. Amer.*, **36**, 506–9.
Ochs, A. L. 1979. 'Is Fourier analysis performed by the visual system or by the visual investigator.' *J. opt. Soc. Amer.*, **69**, 95–8.
Ogilvie, J. C., and Taylor, M. M. 1958. 'Effect of orientation on the visibility of fine wires.' *J. opt. Soc. Amer.*, **48**, 628–9.
Ogle, K. N. 1964. *Researches in Binocular Vision.* New York: Hafner.
Olson, C. B., and Swett, C. P. 1971. 'Effect of prior activity on properties of different types of motor units.' *J. Neurophysiol.*, **34**, 1–16.
Olson, C. R., and Freeman, R. D. 1980. 'Rescaling of the retinal map of visual space during growth of the kitten's eye.' *Brain Res.*, **186**, 55–66.
Olson, D. R. 1970. *Cognitive Development. The Acquisition of Diagonality.* New York: Academic Press.
Olson, D. R., and Hildyard, A. 1977. 'On the mental representation of oblique orientation.' *Canad. J. Psychol.*, **31**, 3–13.
Oman, C. M., Bock, O. L., and Huang, J. K. 1980. 'Visually-induced self-motion sensation adapts rapidly to left–right visual reversal.' *Science*, **209**, 706–8.
Oman, C. M. and Young, L. R. 1972. 'The physiological range of pressure difference and cupula deflection in the human semicircular canal.' *Acta oto-laryng.*, **74**, 324–31.
Onley, J. W., and Volkmann, J. 1958. 'The visual perception of perpendicularity.' *Amer. J. Psychol.*, **71**, 504–516.
Ono, H. 1979. 'Axiomatic summary and deductions from Hering's principles of visual direction.' *Percept. Psychophys.*, **25**, 473–77.
Ono, H. 1980. 'Hering's law of equal innervation and vergence eye movement.' *Amer. J. Optom. physiol Optics,* **57**, 578–85.
Ono, H., and Angus, R. 1974. 'Adaptation to sensory–motor conflict produced by the

visual direction of the hand specified from the cyclopean eye.' *J. exp. Psychol.*, **103**, 1–9.

Ono, H., Angus, R., and Gregor, P. 1977. 'Binocular single vision achieved by fusion and suppression.' *Percept. Psychophys.*, **21**, 513–21.

Ono, H., and Gonda, G. 1978. 'Apparent movement, eye movements and phoria when two eyes alternate in viewing a stimulus.' *Perception*, **7**, 75–83

Ono, H., Wilkinson, A., Muter, P., and Mitson, L. 1972. 'Apparent movement and change in perceived location of a stimulus produced by a change in accommodative vergence.' *Percept. Psychophys.*, **12**, 187–92.

Oosterveld, W. J., and van der Laarse, W. D. 1969. 'Effect of gravity on vestibular nystagmus.' *Aerosp. Med.*, **40**, 382–5.

Optican, L. M., and Robinson, D. A. 1980. 'Cerebellar-dependent adaptive control of primate saccadic system.' *J. Neurophysiol.*, **44**, 1058–76.

Orban, G., Duysens, J., and Callens, M. 1973. 'Movement perception during voluntary saccadic eye movements.' *Vision Res.*, **13**, 1343–53.

Orban, G., and Vandenbussche, E. 1979. 'Behavioral evidence for the oblique effect in the cat.' *J. Physiol.*, **295**, 15–16P.

Ornitz, E. M., Atwell, C. W., Walter, D. O., Hartmann, E. E., and Kaplan, A. R. 1979. 'The maturation of vestibular nystagmus in infancy and childhood.' *Acta oto-laryng.*, **88**, 244–56.

Osterhammel, P., Terkildsen, K., and Zilstorff, K. 1968a. 'Vestibular responses following high velocity rotation.' *Acta oto-laryng.*, **66**, 145–50.

Osterhammel, P., Terkildsen, K., and Zilstorff, K. 1968b. 'Vestibular habituation in ballet dancers.' *Acta oto-laryng.*, **66**, 221–8.

O'Toole, B. I. 1979. 'Exposure-time and spatial-frequency effects in the tilt illusion.' *Perception*, **8**, 557–64.

O'Toole, B. I., and Wenderoth, P. 1977. 'The tilt illusion: Repulsion and attraction effects in the oblique meridian.' *Vision Res.*, **17**, 367–74.

Outerbridge, J. S., and Melvill Jones, G. 1971. 'Reflex vestibular control of head movement in man.' *Aerosp. Med.*, **42**, 935–40.

Over, R. 1966. 'An experimentally induced conflict between vision and proprioception.' *Brit. J. Psychol.*, **57**, 335–41.

Over, R., and Broerse, J. 1972. 'Lack of directional specificity in tilt aftereffect induced with moving contours.' *Psychonom. Sci.*, **28**, 235–6.

Over, R., Broerse, J., and Crassini. B. 1972. 'Orientation illusion and masking in central and peripheral vision.' *J. exp. Psychol.*, **96**, 25–31.

Over, R., and Over J. 1967. 'Detection and recognition of mirror-image obliques by young children.' *J. comp. physiol. Psychol.*, **64**, 467–70.

Oyama, T. 1953. 'Experimental studies of figural aftereffects: I. Temporal factors.' *Jap. J. Psychol.*, **23**, 239–54.

Oyama, T. 1975. 'Determinants of the Zöllner illusion.' *Psychol. Res.*, **37**, 261–80.

Oyster, C. W., Takahashi, E., and Collewijn, H. 1972. 'Direction-selective retinal ganglion cells and control of optokinetic nystagmus in the rabbit.' *Vision Res.*, **12**, 183–93.

Paap, K. R., and Ebenholtz, S. M. 1976. 'Perceptual consequences of potentiation in the extraocular muscles: An alternative explanation for adaptation to wedge prisms.' *J. exp. Psychol.*, **2**, 457–68.

Paap, K. R., and Ebenholtz, S. M. 1977. 'Concomitant direction and distance aftereffect of sustained convergence: A muscle potentiation explanation for eye-specific adaptation.' *Percept. Psychophys.*, **21**, 307–14.

Padden, D. A. 1959. 'Ability of deaf swimmers to orient themselves when submerged in water.' *Res. Quart.*, **30**, 214–26.

Paillard, J., Brouchon, M. 1968. 'Active and passive movements in the calibration of position sense.' In S. J. Freedman (Ed.), *The Neuropsychology of Spatially Oriented Behavior*. Homewood, Ill.: Dorsey, pp. 37–56.

Palmer, S. E. 1977. 'Hierarchical structure in perceptual representation.' *Cognitive*

Psychol., **9**, 441–74.
Park, J. N. 1969. 'Displacement of apparent straight ahead as an after effect of deviation of the eyes from normal position.' *Percept. mot. Skills*, **28**, 591–7.
Park, R. S., and Park, G. E. 1933. 'The center of ocular rotation in the horizontal plane.' *Amer. J. Physiol.*, **104**, 545–52.
Parker, D. E., Gulledge, W. L., Tubbs, R. L., and Littlefield, V. M. 1978. 'A temporary threshold shift for self-motion detection following sustained oscillating acceleration'. *Percept. Psychophys.*, **23**, 461–7.
Parker, D. M. 1972. 'Contrast and size variables and the tilt aftereffect.' *Quart. J. exp. Psychol.*, **24**, 1–7.
Parsons, R. D. 1970. 'Magnitude estimates of the oculogyral illusion during and following angular acceleration.' *J. exp. Psychol.*, **84**, 230–8.
Pasik, P., and Pasik, T. 1964. 'Oculomotor functions in monkeys with lesions of the cerebrum and the superior colliculus.' In M. B. Bender (Ed.), *The Oculomotor System.* New York: Harper & Row, pp. 40–80.
Pasik, P., Pasik, T., and Bender, M. B. 1969. 'The pretectal syndrome in monkeys: I. Disturbances of gaze and body posture.' *Brain*, **92**, 521–34.
Pasik, P., Pasik, T., and Kreiger, H. P. 1959. 'Effects of cerebral lesions upon optokinetic nystagmus in monkeys.' *J. Neurophysiol.*, **22**, 297–304.
Pasik, P., Pasik. T., and Schilder, P. 1969. 'Extrageniculostriate vision in the monkey: discrimination of luminous flux-equated figures.' *Exp. Neurol.*, **24**, 421–37.
Pasik, P., Pasik, T., Valciukas, J. A., and Bender, M. B. 1971. 'Vertical optokinetic nystagmus in the split-brain monkey.' *Exp. Neurol.*, **30**, 162–71.
Pasik, T., and Pasik, P. 1971. 'The visual world of monkeys deprived of striate cortex: Effective stimulus parameters and the importance of the accessory optic system.' *Vision Res.* **Suppl. 3**, 419–35.
Patel, H. H., and Sillito, A. M. 1978. 'Inhibition and velocity tuning in the cat visual cortex.' *J. Physiol.*, **284**, 113P.
Payne, B. R., Berman, N., and Murphy, E. H. 1981. 'Preferred directions of cells are clustered in cat visual cortex.' *Invest. Ophthal., ARVO Abstr.*, **20**, 148.
Pearce, D. G., and Matin, L. 1966. 'The measurement of autokinetic speed.' *Canad. J. Psychol.*, **20**, 160–72.
Peck, C. K., and Crewther, S. G. 1975. 'Perceptual effects of surgical rotation of the eye in kittens.' *Brain Res.*, **99**, 213–9.
Peck, C. K., Crewther, S. G., Barber, G., and Johannsen, C. J. 1979. 'Pattern discrimination and visuomotor behavior following rotation of one or both eyes in kittens and in adult cats.' *Exp. Brain Res.*, **34**, 401–18.
Pellionisz, A., and Llinás, R. 1977. 'A computer model of cerebellar Purkinje cells.' *Neurosci.*, **2**, 37–48.
Pellionisz, A., and Llinás, R. 1979. 'Brain modelling by tensor network theory and computer simulation. The cerebellum: distributed processor for predictive coordination.' *Neurosci.*, **4**, 323–48.
Pellionisz, A., and Llinás, R. 1981. 'Tensorial approach to the geometry of brain function: cerebellar coordination via metric tensor.' *Neurosci.*, **5**, in press.
Perenin, M. T. 1978. 'Visual function within the hemianopic field following early cerebral hemidecortication in man—II. Pattern discrimination.' *Neuropsychologia*, **16**, 697–708.
Perenin, M. T., and Jeannerod, M. 1975. 'Residual vision in cortically blind hemifields.' *Neuropsychologia*, **13**, 1–7.
Perenin, M. T., and Jeannerod, M. 1978. 'Visual function within the hemianopic field following early hemidecortication in man—I. Spatial localization.' *Neuropsychologia*, **16**, 1–13.
Perenin, M. T., Jeannerod, M., and Prablanc, C. 1977. 'Spatial localization with paralyzed eye muscles.' *Ophthalmologica*, **175**, 206–14.
Perkel, D. H., and Bullock, T. H. 1968. 'Neural coding.' *Neurosciences Res. Prog. Bull.*, **6**, 221–348.

Petrov, A. P., and Zenkin, G. M. 1973. 'Torsional eye movements and constancy of the visual field.' *Vision Res.*, **13**, 2465–77.

Petrusic, W. A., Varro, L., and Jamieson, D. G. 1978. 'Mental rotation validation of two spatial ability tests.' *Psychol. Res.*, **40**, 139–48.

Pettigrew, J. D. 1973. 'Binocular neurones which signal change of disparity in area 18 of cat visual cortex.' *Nature, New Biol.*, **241**, 123–4.

Pettigrew, J. D. 1974. 'The effect of visual experience on the development of stimulus specificity by kitten cortical neurones.' *J. Physiol.*, **237**. 49–74.

Pettigrew, J. D., and Freeman, R. D. 1973. 'Visual experience without lines: Effect on developing cortical neurons.' *Science*, **182**, 599–600.

Pettigrew, J. D., and Garey, L. J. 1974. 'Selective modification of single neuron properties in the visual cortex of kittens.' *Brain Res.*, **66**, 160–4.

Pettigrew, J. D., Nikara, T., and Bishop, P. O. 1968. 'Responses to moving slits by single units in cat striate cortex.' *Exp. Brain Res.*, **6**, 373–90.

Pettigrew, J. D., Olson, C., and Hirsch, H. V. B. 1973. 'Cortical effect of selective visual experience: degeneration or reorganization?' *Brain Res.*, **51**, 345–51.

Pfaltz, C. R., and Ohtsuka, Y. 1975. 'The influence of optokinetic training upon vestibular habituation.' *Acta oto-laryng.*, **79**, 253–8.

Piaget, J. 1954. *The Origins of Intelligence.* New York: Basic Books.

Piaget, J., and Inhelder, B. 1956. *The Child's Conception of Space.* London: Routledge & Kegan Paul.

Pick, H. L., and Hay, J. C. 1965. 'A passive test of the Held reafference hypothesis.' *Percept. mot. Skills*, **20**, 1070–2.

Pick, H. L., and Hay, J. C. 1966. 'Gaze-contingent adaptation to prismatic spectacles.' *Amer. J. Psychol.*, **79**, 443–50.

Pick, H. L., Hay, J. C., and Martin, R. 1969. 'Adaptation to split-field wedge prism spectacles.' *J. exp. Psychol.*, **80**, 125–32.

Pick, H. L., Hay, J. C., and Willoughby, R. H. 1966. 'Interocular transfer of adaptation to prismatic distortion.' *Percept. mot. Skills*, **23**, 131–5.

Pick, H. L., Pick, A. D., and Klein, R. 1967. 'Perceptual integration in children.' In L. P. Lipsitt and C. C. Spiker (Eds.), *Advances in Child Development and Behavior*, Vol. 3. New York: Academic Press. pp. 192–224.

Pick, H. L., Warren, D. H., and Hay, J. C. 1969. 'Sensory conflicts in judgments of spatial direction.' *Percept. Psychophys.*, **6**, 203–5.

Piggins, D. J. 1965. 'Autokinesis with an afterimage.' *Amer. J. Psychol.*, **78**, 509–10.

Poggio, G. F., Baker, F. H., Mansfield, R. J. W., Sillito, A., and Grigg, P. 1975. 'Spatial and chromatic properties of neurons subserving foveal and parafoveal vision in Rhesus monkey.' *Brain Res.*, **100**, 25–59.

Pola, J., and Robinson, D. A. 1978. 'Oculomotor signals in medial longitudinal fasciculus of the monkey.' *J. Neurophysiol.*, **41**, 245–59.

Pola, J., and Wyatt, H. J. 1980. 'Target position and velocity: the stimulus for smooth pursuit eye movements.' *Vision Res.*, **20**, 523–34.

Polit, A., and Bizzi, E. 1979. 'Characteristics of motor programs underlying arm movements in monkeys.' *J. Neurophysiol.*, **42**, 183–94.

Pollen, D. A., and Ronner, S. F. 1981. 'Phase relationships between adjacent simple cells in the visual cortex of the cat.' *Invest. Ophthal*, **20**, ARVO Abstr., 148.

Polyak, S. 1957. *The Vertebrate Visual System.* Chicago: University of Chicago Press.

Pompeiano, O. 1974. 'Cerebello-vestibular interrelations.' In H. H. Kornhuber (Ed.), *Handbook of Sensory Physiology*, Vol. VI/1. New York: Springer, pp. 417–76.

Pompeiano, O. 1975. 'Vestibulo-spinal relationships.' In R. F. Naunton (Ed.), *The Vestibular System.* New York: Academic Press, pp 147–84.

Pöppel, E. 1973. 'Comment on "Visual system's view of acoustic space,".' *Nature, New Biol.*, **243**, 231.

Pöppel, E., Held, R., and Frost, D. 1973. 'Residual visual function after brain wounds involving the central visual pathways in man.' *Nature, New Biol.*, **243**, 295–6.

Posner, M. I. 1967. 'Chatacteristics of visual and kinesthetic memory codes.' *J. exp. Psychol.*, **75**, 103–7.
Prablanc, C., Echallier, J. F., Komilis, E., and Jeannerod, M. 1979. 'Optimal response of eye and hand motor systems in pointing at a visual target.' *Biol. Cybernetics*, **35**, 113–24.
Prablanc, C., and Jeannerod, M. 1975. 'Corrective saccades: Dependence on retinal reafferent signals.' *Vision Res.*, **15**, 465–9.
Prablanc, C., Massé, D., and Echallier, J. F. 1978. 'Error-correcting mechanisms in large saccades.' *Vision Res.*, **18**, 557–60.
Prablanc, C., Tzavaras, A., and Jeannerod, M. 1975a. 'Adaptation of hand tracking to rotated visual coordinates.' *Percept. Psychophys.*, **17**, 325–8.
Prablanc, C., Tzavaras, A., and Jeannerod, M. 1975b. 'Adaptation of the two arms to opposite prism displacements.' *Quart. J. exp. Psychol.*, **27**, 667–71.
Precht, W., Shimazu, H., and Markham, C. H. 1966. 'A mechanism of central compensation of vestibular function following hemilabyrinthectomy.' *J. Neurophysiol.*, **29**, 996–1010.
Precht, W., and Strata, P. 1980. 'On the pathways mediating optokinetic responses in vestibular nuclear neurons.' *Neuroscience*, **5**, 777–87.
Prechtl, H. F. R. 1958. 'The directed headturning response and allied movements of the human baby.' *Behaviour,* **13**, 212–42.
Prentice, W. C. H., and Beardslee, D. C. 1950. 'Visual "normalization" near the vertical and horizontal.' *J. exp. Psychol.*, **40**, 355–64.
Pringle, R., and Cowan, T. M. 1978. 'Mental rotation of possible and impossible four-cornered toruses.' *Percept. Psychophys.*, **24**, 84–92.
Puckett, J. De W., and Steinman, R. M. 1969. 'Tracking eye movements with and without saccadic correction.' *Vision Res.*, **9**, 695–703.
Purcell, T., Wenderoth, P., and Moore, D. 1978. 'The angular function of orientation illusions induced by projected images of tilted real object scenes.' *Perception*, **7**, 229–38.
Putkonen, P. T. S., Courjon, J. H., and Jeannerod, M. 1977. 'Compensation of postural effects of hemilabyrinthectomy in the cat. A sensory substitution process?' *Exp. Brain Res.*, **28**, 249–57.
Pylyshyn, Z. W. 1973. 'What the mind's eye tells the mind's brain.' *Psychol. Bull.*, **80**, 1–24.
Pylyshyn, Z. W. 1977. 'Factors influencing the apparent rate of mental rotation.' Research Bull., 412, Psychology Department, University of Western Ontario.
Quereau, J. 1954. 'Some aspects of torsion.' *A. M. A. Arch. Ophthal.*, **51,** 783–8.
Quick, R. F. 1974. 'A vector-magnitude model of contrast detection.' *Kybernetik*, **16**, 65–7.
Quick, R. F., and Reichert, T. A. 1975. 'Spatial-frequency selectivity in contrast detection.' *Vision Res.*, **15**, 637–43.
Quinlan, D. 1970. 'Effects of sight of the body and active locomotion in perceptual adaptation.' *J. exp. Psychol.*, **86**, 91–6.
Radeau, M. 1973. 'The locus of adaptation to auditory–visual conflict.' *Perception,* **2**, 327–32.
Radeau, M. 1974. 'Adaptation au déplacement prismatique sur la base d'une discordance entre la vision et l'audition.' *Année Psychol.*, **74**, 23–33.
Radeau, M., and Bertelson, P. 1969. 'Adaptation a un déplacement prismatique sur la base de stimulations exafferente en conflit.' *Psychol. Belgica.*, **9**, 133–40.
Radeau, M., and Bertelson, P. 1974. 'The after-effects of ventriloquism.' *Quart. J. exp. Psychol.*, **26**, 63–71.
Radeau, M., and Bertelson, P. 1976. 'The effect of a textured visual field on modality dominance in a ventriloquism situation.' *Percept. Psychophys.*, **20**, 227–35.
Radeau, M., and Bertelson, P. 1977. 'Adaptation to auditory–visual discordance and ventriloquism in semirealistic situations.' *Percept. Psychophys.*, **22**, 137–46.
Radeau, M., and Bertelson, P. 1978. 'Cognitive factors and adaptation to auditory–visual discordance.' *Percept. Psychophys.*, **23**, 341–3.

Rademaker, G. G. J., and ter Braak, J. W. G. 1948. 'On the central mechanism of some optic reactions.' *Brain*, **71**, 48–76.

Ransom-Hogg, A., and Spillmann, L. 1980. 'Perceptive field size in fovea and periphery of the light- and dark-adapted retina.' *Vision Res.*, **20**, 221–8.

Raphan, T., and Cohen, B. 1978. 'Brainstem mechanisms for rapid and slow eye movements.' *Ann. Rev. Physiol.*, **40**, 527–52.

Raphan, T., Cohen, B., and Matsuo, V. 1977. 'A velocity-storage mechanism responsible for optokinetic nystagmus (OKN), optokinetic after-nystagmus (OKAN) and vestibular nystagmus,' In R. Baker and A. Berthoz (Eds.), *Control of Gaze by Brain Stem Neurons*. New York: Elsevier, pp. 37–48.

Raphan, T., Matsuo, V., and Cohen, B. 1979. 'Velocity storage in the vestibulo-ocular reflex arc (VOR).' *Exp. Brain Res.*, **35**, 229–48.

Rashbass, C. 1961. 'The relationship between saccadic and smooth tracking eye movements.' *J. Physiol.*, **159**, 326–38.

Ratcliff, G., Ridley, R. M., and Ettlinger, G. 1977. 'Spatial disorientation in the monkey.' *Cortex*, **13**, 62–5.

Ratliff, F. 1965. *Mach Bands: Quantitative Studies on Neural Networks in the Retina*. San Francisco: Holden Day.

Raybourn, M. S., and Keller, E. L. 1977. 'Colliculoretinal organization in primate oculomotor system.' *J. Neurophysiol.*, **40**, 861–78.

Rayner, K. 1978. 'Eye movements in reading and information processing.' *Psychol. Bull.*, **85**, 618–60.

Reason, J. T., and Brand, J. J. 1975. *Motion Sickness*. New York: Academic Press.

Reason, J. T., and Graybiel, A. 1971. 'The effect of varying the time interval between equal and opposite Coriolis accelerations.' *Brit. J. Psychol.*, **62**, 165–73.

Rebelsky, F. 1964. 'Adult perception of the horizontal.' *Percept. mot. Skills*, **19**, 371–4.

Redding, G. M. 1978. 'Additivity in adaptation to optical tilt.' *J. exp. Psychol. Hum. Percept. Perform.*, **4**, 178–90.

Redding, G. M., and Wallace, B. 1978. 'Sources of "overadditivity" in prism adaptation.' *Percept. Psychophys.*, **24**, 58–62.

Reese, E. P. 1953. Summary Report No. SDC 131–1–5. Psychophysical Res. Unit, Mount Holyoke College, Mass.

Regan, D., and Beverley, K. I. 1973. 'Electrophysiological evidence for existence of neurones sensitive to direction of depth movement.' *Nature*, **246**, 504–6.

Regan, D., and Beverley, K. I. 1978. 'Illusory motion in depth: aftereffect of adaptation to changing size.' *Vision Res.*, **18**, 209–12.

Regan, D., and Beverley, K. I. 1979. 'Visually guided locomotion: Psychophysical evidence for a neural mechanism sensitive to flow patterns.' *Science*, **205**, 311–3.

Regan, D., Beverley, K. I., and Cynader, M. 1979. 'Stereoscopic subsystems for position in depth and for motion in depth.' *Proc. roy. Soc., B*, **204**, 485–501.

Regan, D., and Cynader, M. 1979. 'Neurons in cat parastriate cortex selectively sensitive to changing size: nonlinear interactions between responses to two edges.' *Vision Res.*, **19**, 699–711.

Reinhart, R. J., and Zuber, B. L. 1971. 'Abducens nerve signals controlling saccadic eye movements in the cat.' *Brain Res.*, **34**, 331–44.

Reisine, H., and Highstein, S. M. 1979. 'The ascending tract of Deiters' conveys a head velocity signal to medial rectus motorneurons.' *Brain Res.*, **170**, 172–6.

Rentschler, I., and Fiorentini, A. 1974. 'Meridional anisotropy of psychophysical spatial interactions.' *Vision Res.*, **14**, 1467–73.

Rentschler, I., and Hilz, R. 1976. 'Evidence for disinhibition in line detectors.' *Vision Res.*, **16**, 1299–302.

Richmond, B. J., and Wurtz, R. H. 1980. 'Vision during saccadic eye movements. II. A corollary discharge to monkey superior colliculus.' *J. Neurophysiol.*, **43**, 1156–67.

Riddoch, G. 1917. 'Dissociation of visual perceptions due to occipital injuries, with especial reference to appreciation of movement.' *Brain*, **40**, 15–57.

Rijsdijk, J. P., Kroon, J. N., and van der Wildt, G. J. 1980. 'Contrast sensitivity as a function of position on the retina.' *Vision Res*, **20**, 235–41.

Ritchie, L. 1976. 'Effects of cerebral lesions on saccadic eye movements.' *J. Neurophysiol.*, **39**, 1246–56.

Rizzolatti, G., Camarda, R., Grupp, L. A., and Pisa, M. 1973. 'Inhibition of visual responses of single units in the cat superior colliculus by the introduction of a second visual stimulus.' *Brain Res.*, **61**, 390–4.

Roberts, T. D. M. 1978. *Neurophysiology of Postural Mechanisms*, 2nd ed. London: Butterworth.

Robertson, T. W., Christen, W. G., Winters, R. W., and Cohen, H. I. 1979. 'A comparison of on-inhibition and off-excitation measures of the surround response mechanism in cat retinal ganglion cells.' *Brain Res.*, **160**, 509–13.

Robertson, T. W., Winters, R. W., Christen, W. G., and Cohen, H. I. 1978. 'The effect of adapting target location on the gain of the surround mechanism in cat retinal ganglion cells.' *Brain Res.*, **156**, 360–3.

Robinson, D. A. 1963. 'A method of measuring eye movement using a scleral search coil in a magnetic field.' *IEEE Trans. Biomed. Electronics*, 10, 137–145.

Robinson, D. A. 1964. 'The mechanics of human saccadic eye movement.' *J. Physiol.*, **174**, 245–64.

Robinson, D. A. 1965. 'The mechanics of human smooth pursuit eye movement.' *J. Physiol.*, **180**, 569–91.

Robinson, D. A. 1968. 'Eye movement control in primates.' *Science*, **161**, 1219–24.

Robinson, D. A. 1970. 'Oculomotor unit behavior in the monkey.' *J. Neurophysiol.*, **33**, 393–404.

Robinson, D. A. 1972. 'Eye movements evoked by collicular stimulation in the alert monkey.' *Vision Res.*, **12**, 1795–808.

Robinson, D. A. 1974. 'The effect of cerebellectomy on the cat's vestibulo-ocular integrator.' *Brain Res.*, **71**, 195–207.

Robinson, D. A. 1975. 'A quantitative analysis of extraocular muscle cooperation and squint.' *Invest. Ophthal.*, **14**, 801–25.

Robinson, D. A. 1976. 'Adaptive gain control of vestibulo-ocular reflex by the cerebellum.' *J. Neurophysiol.*, **39**, 954–69.

Robinson, D. A. 1977. 'Linear addition of optokinetic and vestibular signals in the vestibular nucleus.' *Exp. Brain Res.*, **30**, 447–50.

Robinson, D. A. 1981. 'The use of control systems analysis in the neurophysiology of eye movements.' *Ann. Rev. Neurosci.*, **4**, 463–503.

Robinson, D. A., and Fuchs, A. F. 1969. 'Eye movements evoked by stimulation of frontal eye fields.' *J. Neurophysiol.*, **32**, 637–48.

Robinson, D. L., Goldberg, M. E., and Stanton, G. B. 1978. 'Parietal association cortex in the primate: sensory mechanisms and behavioral modifications.' *J. Neurophysiol.*, **41**, 910–32.

Robinson, D. L., and Jarvis, C. D. 1974. 'Superior colliculus neurones studied during head and eye movements of the behaving monkey.' *J. Neurophysiol.*, **37**, 533–40.

Robinson, D. L., and Wurtz, R. H. 1976. 'Use of an extraretinal signal by monkey superior colliculus neuron to distinguish real from self-induced stimulus movement.' *J. Neurophysiol.*, **39**, 852–70.

Robinson, J. O. 1972. *The Psychology of Visual Illusions*. London: Hutchinson.

Robinson, J. S., and Fish, S. E. 1974. 'A cat's form-experienced but visual-motor deprived eye lacks focal vision.' *Devel. Psychobiol.*, **7**, 331–42.

Robson, J. G. 1966. 'Spatial and temporal contrast sensitivity functions of the visual system.' *J. opt. Soc. Amer.*, **56**, 1141–2.

Rock I. 1954. 'The perception of the egocentric orientation of a line.' *J. exp. Psychol.*, **48**, 367–74.

Rock, I. 1956 'The orientation of forms on the retina and in the environment.' *Amer J. Psychol.*, **69**, 513–28.
Rock, I. 1966. *The Nature of Perceptual Adaptation.* New York: Basic Books.
Rock, I. 1968. 'The basis of position-constancy during passive movement.' *Amer. J. Psychol.*, **81**, 262–5.
Rock, I. 1973. *Orientation and Form.* New York: Academic Press.
Rock, I., and Heimer, W. 1957. 'The effect of retinal and phenomenal orientation on the perception of form.' *Amer. J. Psychol.*, **70**, 493–511.
Rock, I. and Victor, J. 1964. 'Vision and touch: an experimentally created conflict between the senses.' *Science*, **143**, 594–6.
Rodieck, R. W. 1965. 'Quantitative analysis of cat retinal ganglion cell response to visual stimuli.' *Vision Res.*, **5**, 583–601
Rodieck, R. W., and Dreher, B. 1979. 'Visual suppression from nondominant eye in the lateral geniculate nucleus: A comparison of cat and monkey.' *Exp. Brain Res.*, **35**, 465–77.
Roelofs, C. O. 1935. 'Optische Lokalisation.' *Arch. Augenheilk*, **109**, 395–415.
Roelofs, C. O. 1959. 'Considerations on the visual egocentre.' *Acta Psychol.*, **16**, 226–34.
Roland, P. E., Skinhøj, E., Lassen, N. A., and Larsen, B. 1980. 'Different cortical areas in man in organization of voluntary movements in extra-personal space.' *J. Neurophysiol.*, **43**, 137–50.
Roldan, C. E., and Phillips, W. A. 1980. 'Functional differences between upright and rotated images.' *Quart. J. exp. Psychol.*, **32**, 397–412.
Rolls, E. T., and Cowey, A. 1970. 'Topography of the retina and striate cortex and its relationship to visual acuity in rhesus and squirrel monkeys.' *Exp. Brain Res.*, **10**, 298–310.
Ron, S., and Nemet, P. 1977. 'The cerebellum involvement in the generation of saccades.' *Doc. Ophthal.*, **43**, 109–14.
Ron, S., and Robinson, D. A. 1973. 'Eye movements evoked by cerebellar stimulation in the alert monkey.' *J. Neurophysiol.*, **36**, 1004–22.
Ron, S., Robinson, D. A., and Skavenski, A. A. 1972. 'Saccades and the quick phase of nystagmus.' *Vision Res.*, **12**, 2015–22.
Ronchi, L., and Fidanzati, G. 1972. 'Changes of psychophysical organization across the light-adapted retina.' *J. opt. Soc. Amer.*, **62**, 912–5.
Rondot, P., Decondo, J., and Ribadeau Dumas, J. L. 1977. 'Visuomotor ataxia.' *Brain*, **100**, 355–77.
Rose, D. 1977. 'Response of single units in cat visual cortex to moving bars of light as a function of bar length.' *J. Physiol.*, **271**, 1–23.
Rose, D. 1979. 'Mechanisms underlying the receptive field properties of neurons in cat visual cortex.' *Vision Res.*, **19**, 533–44.
Rose, D., and Blakemore, C. 1974. 'An analysis of orientation selectivity in the cat's visual cortex.' *Exp. Brain Res.*, **20**, 1–17.
Rose, P. K., and Abrahams, V. C. 1975. 'The effect of passive eye movement on unit discharge in the superior colliculus of the cat.' *Brain Res.*, **97**, 95–106.
Rosenhall, U. 1972. 'Vestibular macular mapping in man.' *Ann. Otol. Rhinol. Laryngol.*, **81**, 339–51.
Ross, H. E. 1974. *Behaviour and Perception in Strange Environments.* London: Allen & Unwin.
Ross, H. E., Crickmar, S. D., Sills, N. V., and Owen E. P. 1969. 'Orientation to the vertical in free divers.' *Aerosp. Med.*, **40**, 728–32.
Ross, J., and Johnstone, J. R. 1980. 'Phase and detection of compound gratings.' *Vision Res.*, **20**, 189–92.
Rossi, P. J. 1968. 'Adaptation and negative aftereffect to lateral opitical displacement in newly hatched chicks.' *Science*, **160**, 430–2.
Roucoux, A., and Crommelinck, M. 1976. 'Eye movements evoked by superior colliculus

stimulation in the alert cat.' *Brain Res.*, **106**, 349–63.

Roucoux, A., and Crommelinck, M. 1977. 'Influence of supranuclear structures on brain stem neurons B: Collicular, thalmic and cortical influence on brain stem neurons.' In R. Baker and A. Berthoz (Eds.), *Control of Gaze by Brain Stem Neurons.* New York: Elsevier, pp. 483–88.

Rovamo, J., and Virsu., V. 1979. 'An estimation and application of the human cortical magnification factor.' *Exp. Brain Res.*, **37**, 495–510.

Royce, J. R., Carran, A. B., Aftanas, M., Lehman, R. S., and Blumenthal, A. 1966. 'Autokinetic phenomenon: A critical review.' *Psychol. Bull.*, **65**, 243–60.

Royce, J. R., Stayton, W. R., and Kinkade, R. C. 1962. 'Experimental reduction of autokinetic movements.' *Amer. J. Psychol.*, **75**, 221–31.

Royer, F. L. 1966. 'Figural goodness and internal structure in perceptual discrimination.' *Percept. Psychophys.*, **1**, 311–4.

Royer, F. L. 1971. 'Spatial orientation and figural information in free recall of visual figures.' *J. exp. Psychol.*, **91**, 326–32.

Royer, F. L., and Holland, T. R. 1975. 'Rotational transformation of visual figures as a clinical phenomenon.' *Psychol. Bull.*, **82**, 843–68.

Rudel, R. G., and Teuber, H. 1963. 'Discrimination of direction of line in children.' *J. comp. physiol. Psychol.*, **56**, 892–8.

Ruff, H. A., and Halton, A. 1978. 'Is there directed reaching in the human neonate?' *Devel. Psychol.*, **14**, 425–6.

Ryu, J. H., and McCabe, B. F. 1976. 'Central vestibular compensation.' *Arch. Otolaryng.*, **102**, 71–6.

Sabin, T. D., and Poche, J. A. 1969. 'Pure torsional nystagmus as a consequence of head trauma.' *J. Neurol. Neurosurg. Psychiat.*, **32**, 265–7.

Sachs, M. B., Nachmias, J., and Robson, J. G. 1971. 'Spatial frequency channels in human vision.' *J. opt. Soc. Amer.*, **61**, 1176–86.

Sakata, H., Shibutani, H., and Kawano, K. 1980. 'Spatial properties of visual fixation neurons in posterior parietal association cortex of the monkey,' *J. Neurophysiol.*, **43**, 1654–72.

Saltman, D. C., and Garner, W. R. 1950. 'Accuracy of visual estimation of azimuth position.' *J. Psychol.*, **29**, 453–69.

Saltzman, E. 1979. 'Levels of sensorimotor representation.' *J. math. Psychol.*, **20**, 91–163.

Sanders, M. D., Warrington, E. K., Marshall, J., and Wieskrantz, L. 1974. '"Blindsight": vision in a field defect.' *The Lancet*, April, 707–9.

Sandström, C. I. 1951, *Orientation in the Present Space.* Uppsala: Almquist and Wiksell.

Sandström, C. I., and Lundberg, I. 1956. 'A genetic approach to sex differences in localization.' *Acta Psychol.*, **12**, 247–53.

Sansbury, R. V., Skavenski, A. A., Haddad, G. M., and Steinman, R. M. 1973. 'Normal fixation of eccentric targets.' *J. opt. Soc. Amer.*, **63**, 612–4.

Saslow, M. G. 1967a. 'Effects of components of displacement-step stimuli upon latency for saccadic eye movements.' *J. opt. Soc. Amer.*, **57**, 1024–9.

Saslow, M. G. 1967b. 'Latency for saccadic eye movement.' *J. opt. Soc. Amer.*, **57**, 1030–3.

Savoy, R. L., and McCann, J. J. 1975. 'Visibility of low-frequency sine-wave targets: dependence on number of cycles.' *J. opt. Soc. Amer.*, **65**, 343–50.

Schaefer, K.-P. 1972. 'Neuronal elements of the orienting response.' In J. Dichgans and E. Bizzi (Eds.), *Cerebral Control of Eye Movements and Motion Perception.* Basel: Karger, pp. 139–48.

Schaefer, K.-P., and Meyer, D. L. 1974. 'Compensation of vestibular lesions.' In H. H. Kornhuber (Ed.), *Handbook of Sensory Physiology*, Vol. VII/2. Berlin: Springer, pp. 463–90.

Schaller, M. J., and Harris, L. J. 1975. '"Upright" orientations of forms change with subject age and with features of form.' *Percept. Psychophys.*, **17**, 179–88.

Schiff, W. 1965. 'The perception of impending collision: A study of visually directed avoidant behavior.' *Psychol. Monogr.*, **79**, No. 604.

Schiff, W., Caviness, J. A., and Gibson, J. J. 1962. 'Persistent fear responses in rhesus monkeys to the optical stimulus of "looming".' *Science,* **136**, 982–3.

Schilder, P., and Wechler, D. 1936. 'The illusion of the oblique intercept.' *J. exp. Psychol.*, **19**, 747–57.

Schiller, P. H. 1970. 'The discharge characteristics of single units in the oculomotor and abducens nuclei of the unanesthetized monkey.' *Exp. Brain Res.*, **10**, 347–62.

Schiller, P. H. 1977. 'The effect of superior colliculus ablation on saccades elicited by cortical stimulation.' *Brain Res.*, **122**, 154–6.

Schiller, P. H., Finlay, B. L., and Volman, S. F. 1976a. 'Quantitative studies of single cells in monkey striate cortex. II. Orientation specificity and ocular dominance.' *J. Neurophysiol.*, **39**, 1320–33.

Schiller, P. H., Finlay, B. L., and Volman, S. F. 1976b. 'Quantitative studies of single cell properties in the monkey striate cortex. III. Spatial frequency.' *J. Neurophysiol.*, **39**. 1334–51.

Schiller, P. H., and Koerner, F. 1971. 'Discharge characteristics of single units in superior colliculus of the alert Rhesus monkey.' *J. Neurophysiol.*, **34**, 920–36.

Schiller, P. H., and Malpeli, J. G. 1977. 'Properties and tectal projections of monkey retinal ganglion cells.' *J. Neurophysiol.*, **40**, 428–45.

Schiller, P. H. and Malpeli, J. G. 1978. 'Functional specificity of lateral geniculate nucleus laminae of the rhesus monkey.' *J. Neurophysiol.*, **41**, 788–97.

Schiller, P. H., Malpeli, J. G., and Schein, S. J. 1979. 'Composition of geniculostriate input to superior colliculus of the rhesus monkey.' *J. Neurophysiol.*, **42**, 1124–33.

Schiller, P. H., and Stryker, M. P. 1972. 'Single-unit recording and stimulation in superior colliculus of the alert rhesus monkey.' *J. Neurophysiol.*, **35**, 915–24.

Schiller, P. H., Stryker, M. P., Cynader, M., and Berman, N. 1974. 'Response characteristics of single cells in the monkey superior colliculus following ablation or cooling of visual cortex.' *J. Neurophysiol.*, **37**, 181–94.

Schiller, P. H., True, S. D., and Conway, J. L. 1979. 'Effects of frontal eye field and superior colliculus ablations on eye movements.' *Science*, **206**, 590–2. (See also *J. Neurophysiol.*, **44**, 1175–90.)

Schmidt, M. J., Cosgrove, M. P., and Brown, D. R. 1972. 'Stabilized images: Functional relationships among populations of orientation-specific mechanisms in the human visual system.' *Percept. Psychophys.*, **11**, 389–92.

Schneider, C. W. 1966. 'Monocular and binocular perception of verticality and the relationship of ocular dominance.' *Amer. J. Psychol.*, **79**, 632–6.

Schneider,, G. E. 1969. 'Two visual system.' *Science*, **163**, 895–902.

Scholtes, A. M. W., and Bouman, M. A. 1977. 'Psychophysical experiments on spatial summation at threshold level of human peripheral retina.' *Vision Res.*, **17**, 867–73.

Schöne, H. 1962. 'Über den Einfluss der Schwerkraft auf die Augenrollung und auf die Wahrnehmung der Lage im Raum.' *Z. vergl. Physiol.*, **46**, 57–87.

Schöne, H. 1964. 'On the role of gravity in human spatial orientation.' *Aerosp. Med.*, **35**, 764–72.

Schöne, H. 1980. *Orientierung im Raum.* Stuttgart: Wissenschaftliche Verlagsgesellschaft.

Schöne, H., and Lechner-Steinleitner, S. 1978. 'The effect of preceding tilt on the perceived vertical.' *Acta oto-laryng.*, **85**, 68–73.

Schöne, H., and Mortag, H. G. 1968. 'variation of the subjective vertical on the parallel swing at different body positions.' *Psychol. Forsch.*, **32**, 124–34.

Schöne, H., Parker, D. E., and Mortag, H. G. 1967. 'Subjective vertical as a function of body position and gravity magnitude.' *Naturwissenschaften*, **54**, 288.

Schöne, H., and Udo de Haes, H. 1968. 'perception of gravity—vertical as a function of head and trunk position.' *Z. vergl. Physiol.*, **60**, 440–4.

Schöne, H., and Wade, N. J. 1971. 'The influence of force magnitude on the perception of body position: II. Effect of body posture.' *Brit. J. Psychol.*, **62**, 347–52.

Schor, C. M., and Levi, D. M. 1980. 'Disturbances of small-field horizontal and vertical optokinetic nystagmus in amblyopia.' *Invest. Ophthal.*, **19**, 668–83.

Schroeder, D. J. 1971. 'Influence of alcohol on vestibular responses to angular accelerations.' *Aerosp. Med.*, **42**, 959–70.

Schuck, J. R., Cross, H. A., and Mills, D. H. 1970. 'A signal detection analysis of the rod and frame test.' *Percept. Psychophys.*, **7**, 276–80.

Schwartz, E. L. 1980. 'Computational anatomy and functional architecture of striate cortex: a spatial mapping approach to perceptual coding.' *Vision Res.*, **20**, 645–70.

Schwartzkroin, P. A. 1972. 'The effect of body tilt on the directionality of units in cat visual cortex.' *Exp. Neurol.*, **36**, 498–506.

Schwarz, D. W. F., and Tomlinson, R. D. 1977. 'Neuronal responses to eye muscle stretch in cerebellar lobule VI of the cat.' *Exp. Brain Res.*, **27**, 101–11.

Schwindt, P. C., Richter, A., and Precht, W. 1973."Short latency utricular and canal input to ipsilateral abducens motoneurons.' *Brain Res.*, **60**, 259–62.

Scobey, R. P., and Gabor, A. J. 1979. 'The orientation tuning curve: variability of preferred orientation due to methodology.' *Vision Res.*, **19**, 1389–94.

Scott, A. B. 1967. 'Extraocular muscles and head tilting. *A.M.A. Arch. Ophthal.*, **78**, 397–9.

Scott, D. F., Groethuysen, U. C., and Bickford, R. G. 1967. 'Lambda responses in the human electroencephalogram.' *Neurology*, **17**, 770–8.

Sekuler, R., and Bauer, J. A. 1966. 'Adaptation to prismatic displacements: Hand position and target location.' *J. exp. Psychol.*, **72**, 207–12.

Sekuler, R., and Houlihan, K. 1968. 'Discrimination of mirror-images: Choice time analysis of human adult performance.' *Quart. J. exp. Psychol.*, **20**, 204–7.

Sekuler, R., Hutman, L. P., and Owsley, C. J. 1980. 'Human aging and spatial vision.' *Science*, **209**, 1255–6.

Sekuler, R., and Littlejohn, J. 1974. 'Tilt aftereffect following very brief exposures.' *Vision Res.*, **14**, 151–2.

Sekuler, R., and Pierce, S. 1973. 'Perception of stimulus direction: Hemispheric homology and laterality.' *Amer. J. Psychol.*, **86**, 679–95.

Selfridge, O. G. 1956. 'Pattern recoginition and learning.' In *Proceedings of the Third London Symposium of Information Theory*. New York: Academic Press, p. 345.

Senders, J. W., Fisher, D. F., and Monty, R. A. (Eds.) 1978. *Eye Movements and the Higher Psychological Functions*. Hillsdale, N. J.: Erlbaum.

Shapiro, M. B. 1960. 'Rotation of drawings by illiterate Africans.' *J. soc. Psychol.*, **52**, 17–30.

Shapley, R, and So, Y. T. 1980. 'Is there an effect of monocular deprivation on the proportions of X and Y cells in the cat lateral geniculate nucleus? *Exp. Brain Res.*, **39**, 41–8.

Shapley, R., and Tolhurst, D. J. 1973. 'Edge detectors in human vision.' *J. Physiol.*, **229**, 165–83.

Sharp, R. H., and Whiting, H. T. A. 1974. 'Exposure and occluded duration effects in a ball-catching skill.' *J. mot. Behav.*, **6**, 139–47.

Sharpe, C. R., and Tolhurst, D. J. 1973a. 'The effects of temporal modulation on the orientation channels of the human visual system. *Perception*, **2**, 23–9.

Sharpe, C. R., and Tolhurst, D. J. 1973b. 'Orientation and spatial frequency channels in peripheral vision.' *Vision Res.*, **13**, 2103–12.

Sharpe, J. A., Lo, A. W., and Rabinovitch, H. E. 1979. 'Control of the saccadic and smooth pursuit systems after cerebral hemidecortication.' *Brain*, **102**, 387–403.

Shebilske, W. L. 1976. 'Extraretinal information in corrective saccades and inflow vs. outflow theories of visual direction constancy.' *Vision Res.*, **16**, 621–8.

Shebilske, W. L., and Fogelgren, L. 1977. 'Eye-position aftereffects of backward head tilt manifested by illusory visual direction.' *Percept. Psychophys.*, **21**, 77–82.

Shebilske, W. L., and Karmiohl, C. M. 1979. 'Illusory visual direction during and after backward head tilts.' *Percept. Psychophys.*, **24**, 543–5.

Shebilske, W. L., and Nice, D. S. 1976. 'Optical insignificance of the nose and the Pinocchio effect in free-scan visual straight-ahead judgments.' *Percept. Psychophys.*, **20**, 17–20.

Sheedy, J. E., and Fry, G. A. 1979. 'The perceived direction of the binocular image.' *Vision Res.*, **19**, 201–11.

Shepard, R. N., and Metzler, J. 1971. 'Mental rotation of three-dimensional objects.' *Science*, **171**, 701–3.

Sherk, H., and Stryker, M. P. 1976. 'Quantitative study of cortical orientation selectivity in visually inexperienced kitten.' *J. Neurophysiol.*, **39**, 63–9.

Sherman, S. M. 1974. 'Visual fields of cats with cortical and tectal lesions.' *Science*, **185**, 355–7.

Shinkman, P. G., and Bruce, C. J. 1977. 'Binocular differences in cortical receptive fields of kittens after rotationally disparate binocular experience.' *Science*, **197**, 285–7.

Siebeck, R. 1954. 'Wahrnehmungsstörung und Störungswahrnehmung bei Augenmuskellähmungen.' *v. Graefes Arch. Ophthal.*, **155**, 26–34.

Sigman, E., Goodenough, D. R., and Flannagan M. 1979. 'Instructions, illusory self-tilt and the rod-and-frame test.' *Quart. J. exp. Psychol.*, **31**, 155–65.

Sillito, A. M. 1975. 'The contributions of inhibitory mechanisms to the receptive field properties of neurones in the striate cortex of the cat.' *J. Physiol.*, **250**, 305–29.

Sillito, A. M. 1977. 'The spatial extent of excitatory and inhibitory zones in the receptive field of superficial layer hypercomplex cells.' *J. Physiol.*, **273**, 791–803.

Sillito, A. M., Kemp, J. A., Milson, J. A., and Berardi, N. 1980. 'A re-evaluation of the mechanisms underlying simple cell orientation selectivity.' *Brain Res.*, **194**, 517–20.

Sillito, A. M., and Versiani, V. 1977. 'The contribution of excitatory and inhibitory inputs to the length preference of hypercomplex cells in Layers II and III of the cat's striate cortex.' *J. Physiol.*, **273**, 775–90.

Silverman, M. S., Tootell, R. B., and De Valois, R. L. 1980. 'Deoxyglucose mapping of orientation and spatial frequency in cat visual cortex.' *Invest. Ophthal. ARVO Abstr.*, **19**, 225.

Simpson, J. I., Precht, W., and Llinás, R. 1974. 'Sensory separation in climbing and mossy fiber inputs to cat vestibulocerebellum.' *Pflüg. Arch. ges. Physiol.*, **351**, 183–95.

Singer, G., and Day, R. H. 1966. 'Spatial adaptation and aftereffect with optically transformed vision: Effects of active and passive responding and the relationship between test and exposure responses.' *J. exp. Psychol.*, **71**, 725–31.

Singer, G., Purcell, A. T., and Austin, M. 1970. 'The effect of structure and degree of tilt on the tilted room illusion.' *Percept. Psychophys.*, **7**, 250–2.

Singer, W. 1977. 'Control of thalamic transmission by corticofugal and ascending reticular pathways in the visual system.' *Physiol Rev.*, **57**, 386–420.

Singer, W., and Bedworth, N. 1973. 'Inhibitory interaction between X- and Y-units in the cat LGN.' *Brain Res.*, **49**, 291–307.

Singer, W., Freeman, B., and Rauschecker, J. 1981. 'Restriction of visual experiences to a single orientation affects the organization of orientation columns in cat visual cortex—a study with deoxyglucose.' *Exp. Brain Res.*, **41**, 199–215.

Singer, W., Tretter, F., and Cynader, M. 1975. 'Organization of striate cortex: a correlation of receptive field properties with afferent and efferent connections.' *J. Physiol.*, **38**, 1080–98.

Sireteanu, R., and Singer, W. 1980. 'The "vertical effect" in human squint amblyopia.' *Exp. Brain Res.*, **40**, 354–7.

Skarf, B., and Jacobson, M. 1974. 'Development of binocularly driven single units in frogs raised with asymmetrical visual stimulation.' *Exp. Neurol.*, **42**, 669–86.

Skavenski, A. A. 1971. 'Extraretinal correction and memory for target position.' *Vision Res.*, **11**, 743–6.

Skavenski, A. A. 1972. 'Inflow as a source of extraretinal eye position information.' *Vision Res.*, **12**, 221–9.

Skavenski, A. A., Haddad, G., and Steinman, R. M. 1972. 'The extraretinal signal for the visual perception of direction.' *Percept. Psychophys.*, **11**, 287–90.

Skavenski, A. A., and Robinson, D. A. 1973. 'Role of abducens neurons in the vestibulo-ocular reflex.' *J. Neurophysiol.*, **36**, 724–38.

Skavenski, A. A., and Steinman, R. M. 1970. 'Control of eye position in the dark.' *Vision Res.*, **10**, 193–203.

Skowbo, D., Timney, B. N., Gentry, T. A., and Morant, R. B. 1975. 'McCollough effects: experimental findings and theoretical accounts.' *Psychol. Bull.*, **82**, 497–510.

Smedslund, J. 1963. 'The effect of observation on children's representation of the spatial orientation of a water surface.' *J. genet, Psychol.*, **102**, 195–201.

Smiles, K. A., Hite, D., Hyams, V. J., and Junker, A. M. 1975. 'Effect of labyrinthectomy on the dynamic vestibulo-ocular counterroll reflex in the Rhesus monkey.' *Aviat. space environ. Med.*, **46**, 1017–22.

Smith, A. T., and Over, R. 1976. 'Color-selective tilt aftereffects with subjective contours.' *Percept. Psychophys.*, **20**, 305–8.

Smith, A. T., and Over, R. 1977. 'Orientation masking and the tilt-illusion with subjective contours.' *Perception*, **6**, 441–7.

Smith, K. U. 1937. 'The postoperative effects of removal of the striate cortex upon certain unlearned visually controlled reactions in the cat.' *J. genet. Psychol.*, **50**, 137–56.

Smith, K. U., and Bojar, S. 1938. 'The nature of optokinetic reactions in mammals and their significance in the experimental analysis of the neural mechanisms of visual functions.' *Psychol. Bull.*, **35**, 193–219.

Smith, S. L. 1962. 'Angular estimation.' *J. appl. Psychol.*, **46**, 240–6.

Smothergill, D. W. 1973. 'Accuracy and variability in the localization of spatial targets at three age levels.' *Devel. Psychol.*, **8**, 62–6.

Snell, A. C. 1939. 'The optokinetoscope.' *Trans. Amer. Acad. Ophthal. Oto-laryng.*, **44**, 396.

Snyder, A. W., and Srinivasan, M. V. 1979. 'Human psychophysics: Functional interpretation for contrast sensitivity versus spatial frequency curve.' *Biol. Cybernetics*, **32**, 9–17.

Snyder, F. W., and Pronko, N. H. 1952. *Vision with Spatial Inversion*. Wichita, Kansas: McCormick-Armstrong.

Sokolovski, A. 1966. 'The influence of mental activity and visual fixation upon caloric-induced nystagmus in normal subjects.' *Acta oto-laryng.*, **61**, 209–20.

Solomon, A. D. 1947. 'Visual field factors in the perception of direction.' *Amer. J. Psychol.*, **60**, 68–88.

Sparks, D. L. 1978. 'Functional properties of neurones in the monkey superior colliculus: Coupling of neuronal activity and saccade onset.' *Brain Res.*, **156**, 1–16.

Sparks, D. L., and Pollack, J. G. 1977. 'The neural control of saccadic eye movements: The role of the superior colliculus.' In B. A. Brooks and F. J. Bajandas (Eds.), *Eye Movements*. New York: Plenum, pp. 179–220.

Spelke, E. 1976. 'Infants' intermodal perception of events.' *Cognit. Psychol.*, **8**, 553–60.

Spelke, E. 1979. 'Perceiving bimodally specified events in infancy.' *Devel. Psychol.*, **15**, 626–36.

Sperling, G., and Speelman, R. A. 1966. 'Visual spatial localization during object motion, apparent object motion and image motion produced by eye movements.' *J. opt. Soc. Amer.*, **55**, 1576–7.

Sperry, R. W. 1943. 'Effect of 180 degree rotation of the retinal field on visuomotor coordination.' *J. exp. Zool.*, **92**, 263–79.

Sperry, R. W. 1948. 'Patterning of central synapses in regeneration of the otpic nerve in Teleosts.' *Physiol. Zool.*, **21**, 351–61.

Spiegel, E. A., and Scala, N. P. 1943. 'Response of the labyrinthine apparatus to electrical stimulation.' *Arch. Otolaryng.*, **38**, 131–8.

Spigel, I. M. 1963. 'Autokinetic movement of an intermittent lumittent luminance.' *Psychol. Rec.*, **13**, 149–53.

Spinelli, D. N., and Jensen, F. E. 1979. 'Plasticity: The mirror of experience.' *Science*, **203**, 75–8.

Spinelli, D. W., Starr, A., and Barrett, T. W. 1968. 'Auditory specificity in unit recordings from cat's visual cortex.' *Exp. Neurol.*, **22**, 75–84.

Spoendlin, H. H. 1965. 'Ultrastructural studies of the labyrinth in squirrel monkeys.' In *The Role of the Vestibular Organs in the Exploration of Space*. NASA SP-77. Washington, D. C.: National Aeronautics and Space Administration, pp. 7–22.

Sprague, J. M. 1966. 'Interaction of cortex and superior colliculus in mediation of visually guided behavior in the cat.' *Science*, **153**, 1544–7.

Sprague, J. M., Levy, J., DiBerardino, A., and Berlucchi, G. 1977. 'Visual cortical area mediating form discrimination in the cat.' *J. comp. Physiol.*, **172**, 441–88.

Spyer, K. M., Ghelarducci, B., and Pomeiano, O. 1974. 'Gravity responses of neurons in main reticular formation.' *J. Neurophysiol.*, **37**, 705–20.

Staller, J., and Sekuler, R. 1975. 'Children read normal and reversed letters: A simple test of reading skill.' *Quart. J. exp. Psychol.*, **27**, 539–50.

Staller, J., and Sekuler, R. 1977. 'Stimulus and response factors in mirror-image discrimination.' *Percept. Psychophys.*, **22**, 592–8.

Stark, L., Kong, R., Schwartz, S., and Hendry, D. 1976. 'Saccadic suppression of image displacement.' *Vision Res.*, **16**, 1185–7.

St. Cyr, G. J., and Fender, D. H. 1969a. 'The interplay of drifts and flicks in binocular fixation.' *Vision Res.*, **9**, 245–65.

St. Cyr, G. J., and Fender. D. H. 1969b. 'Nonlinearities of the human oculo-motor system: gain.' *Vision Res.*, **9**, 1235–46.

St. Cyr, G. J., and Fender, D. H. 1969c. 'Nonlinearities of the human oculo-motor system: time delays.' *Vision Res.*, **9**, 1491–503.

Stein, B. E., and Arigbede, M. O. 1972. 'Unimodal and multimodal response properties of neurons in the cat's superior colliculus.' *Exp. Neurol.*, **36**, 179–96.

Stein, B. E., Goldberg, S. J., and Clamann H. P. 1976. 'The control of eye movements by the superior colliculus in the alert cat.' *Brain Res.*, **118**, 469–74.

Stein, B. E., and Magalhães-Castro, B. 1975. 'Effects of neonatal cortical lesions upon the cat superior colliculus.' *Brain Res.*, **83**, 480–5.

Stein, B. E., and Magalhães-Castro, B., and Kruger, L. 1976. 'Relationship between visual and tactile representations in cat superior colliculus.' *J. Neurophysiol.*, **39**, 401–19.

Stein, R. B., Pearson, K. G., Smith, R. S., and Redford, J. B. (Eds.) 1973. *Control of Posture and Locomotion.* New York: Plenum Press.

Steinbach, M. J. 1969. 'Eye tracking of self-moved targets: The role of efference.' *J. exp. Psychol.*, **82**, 366–76.

Steinbach, M. J. 1970. 'Aligning the eye to the actively or passively positioned hand.' *Percept. Psychophys.*, **8**, 287–8.

Steinbach, M. J. 1976. 'Pursuing the perceptual rather than the retinal stimulus.' *Vision Res.*, **16**, 1371–6.

Steinbach, M. J., and Anstis, S. M. 1977. 'Smoothly pursuing moving cyclopean gratings.' *J. opt. Soc. Amer.*, **67**, 1415.

Steinbach, M. J., and Held, R. 1968. 'Eye tracking of observer-generated target movements.' *Science*, **161**, 187–8.

Steinbach, M. J., and Money, K. E. 1973. 'Eye movements of the owl.' *Vision Res.*, **13**, 889–91.

Steinbach, M. J., and Pearce, D. G. 1972. 'Release of pursuit eye movements using after-images.' *Vision Res.*, **12**, 1307–11.

Steinhausen, W. 1931. 'Über den Nachweis der Bewegung der Cupola in der intakten Bogengansampulle des Labyrinths bei der natürlichen rotatorischen und calorischen

Reizung.' *Pflüg. Arch. ges. Physiol.*, **228**, 322–8.
Steinman, R. M., and Collewijn, H. 1980. 'Binocular retinal image motion during active head rotation.' *Vision Res.*, **20**, 415–30.
Steinman, R. M., Cunitz, R. J., Timberlake, G. T., and Herman, M. 1967. 'Voluntary control of microsaccades during maintained monocular fixation.' *Science*, **155**, 1577–9.
Steinman, R. M., Haddad, G. M., Skavenski, A. A., and Wyman, D. 1973. 'Miniature eye movement.' *Science*, **181**, 810–9.
Steinman, R. M., Skavenski, A. A., and Sansbury, R. V. 1969. 'Effect of lens accommodation on holding the eye in place without saccades.' *Vision Res.*, **9**, 629–31.
Stern, L. D., and Emelity, D. 1978. 'Evidence for frames of reference based on pursuit eye movements.' *Percept. Psychophys.*, **24**, 521–8.
Stevens, J. K., Emerson, R. C., Gerstein, G. L., Kallos, T., Neufeld, G. R., Nichols, C. W., and Rosenquist, A. C. 1976. 'Paralysis of the awake human: Visual perceptions.' *Vision Res.*, **16**, 93–8.
Stiles, W. S., and Crawford, B. H. 1933. 'The luminous efficiency of rays entering the eye pupil at different points.' *Proc. roy. Soc., B*, **112**, 428–50.
Stockwell C. W., Gilson, R. D., and Guedry, F. E. 1973. 'Adaptation of horizontal semicircular canal responses.' *Acta oto-laryng.*, **75**, 471–6.
Stockwell, C. W., and Guedry, F. E. 1970. 'The effect of semicircular canal stimulation during tilting on the subsequent perception of the visual vertical.' *Acta oto-laryng.*, **70**, 170–5.
Stone, J. 1965. 'A quantitative analysis of the distribution of ganglion cells in the cat's retina.' *J. comp. Neurol.*, **124**, 337–52.
Stone, J., and Dreher, B. 1973. 'Projection of X- and Y-cells of the cat's lateral geniculate nucleus to areas 17 and 18 of visual cortex.' *J. Neurophysiol.*, **36**, 551–67.
Stone, J., Dreher, B., and Leventhal, A. 1979. 'Hierarchical and parallel mechanisms in the organization of visual cortex.' *Brain Res. Rev.*, **1**, 345–94.
Stone, J., and Hoffmann, K.-P. 1972. 'Very slow conducting ganglion cells in the cat's retina: A major, new functional type?' *Brain Res.*, **43**, 610–6.
Stone, L. S. 1944. 'Functional polarization in retinal development and its reestablishment in regenerating retinae of rotated grafted eyes.' *Proc. Soc. exp. Biol.*, **57**, 13–14.
Stoper, A. E. 1967. 'Vision during pursuit movement: the role of oculomotor information.' Ph.D. Diss., Brandeis University.
Stoper, A. E. 1973. 'Apparent motion of stimuli presented stroboscopically during pursuit movement of the eye.' *Percept. Psychophys.*, **13**, 301–11.
Straschill, M., and Rieger, P. 1973. 'Eye movements evoked by focal stimulation of the cat's superior colliculus.' *Brain Res.*, **59**, 211–27.
Straschill, M., and Schick, F. 1974. 'Neuronal activity during eye movements in a visual association area of cat cerebral cortex.' *Exp. Brain Res.*, **19**, 467–77.
Straschill, M., and Schick, F. 1977. 'Discharges of superior colliculus neurons during head and eye movements of the alert cat.' *Exp. Brain Res.*, **27**, 131–41.
Stratton, G. M. 1897a. 'Upright vision and the retinal image.' *Psychol. Rev.*, **4**, 182–7.
Stratton, G. M. 1897b. 'Vision without inversion of the retinal image.' *Psychol. Rev.*, **4**, 341–60, 463–81.
Stromeyer, C. F., and Klein, S. 1975. 'Evidence against narrow-band spatial frequency channels in human vision: the detectability of frequency modulated gratings.' *Vision Res.*, **15**, 899–910.
Stromeyer, C. F., Kronauer, R. E., Madsen, J. C., and Cohen, M. A. 1980. 'Spatial adaptation of short-wavelength pathways in humans.' *Science*, **207**, 555–7.
Stryker, M P., and Schiller, P. H. 1975. 'Eye and head movements evoked by electrical stimulation of monkey superior colliculus.' *Exp. Brain Res.*, **23**, 103–12.
Stryker, M P., and Sherk, H. 1975. 'Modification of cortical orientation selectivity in the cat by restricted visual experience: A re-examination.' *Science*, **190**, 904–6.
Stryker, M. P., Sherk, H., Leventhal, A. G., and Hirsch, H. V. B. 1978. 'Physiological

consequences for the cat's visual cortex of effectively restricting early visual experience with oriented contours.' *J. Neurophysiol.*, **41**, 896–909.

Sugarman, R. C., and Cohen, W. 1968. 'Perceived target displacement as a function of field movement and asymmetry.' *Percept. Psychophys.*, **3**, 169–73.

Sullivan, G. D., Georgeson, M. A., and Oatley, K. 1972. 'Channels for spatial frequency selection and the detection of single bars by the human visual system.' *Vision Res.*, **12**, 383–94.

Sullivan, G. D., Oatley, K., and Sutherland, N. S. 1972. 'Vernier acuity as affected by target length and separation.' *Percept. Psychophys.*, **12**, 438–44.

Suzuki, J. I., Tokumasu, K., and Goto, K. 1969. 'Eye movements from single utricular nerve stimulation in the cat.' *Acta oto-laryng.*, **68**, 350–62.

Switkes, E., Mayer, M. J., and Sloan, J. A. 1978. 'Spatial frequency analysis of the visual environment: Anistropy and the carpentered environment hypothesis.' *Vision Res.*, **18**, 1393–9.

Syka, J., Popelář, J., and Božkov., V. 1979. 'Responses of neurons in the superior colliculus of the cat to stationary and moving visual stimuli.' *Vision Res.*, **19**, 213–9.

Syka, J., and Radil-Weiss, T. 1971. 'Electrical stimulation of the tectum in freely moving cats.' *Brain Res.*, **28**, 567–72.

Szentágothai, J. 1950. 'The elementary vestibulo-ocular reflex arc.' *J. Neurophysiol.*, **13**, 395–407.

Szentágothai, J. 1973. 'Neuronal and synaptic architecture of the lateral geniculate nucleus.' In R. Jung (Ed.), *Handbook of Sensory Physiology*, Vol. VII/3B. New York: Springer-Verlag, pp. 141–76.

Takemori, S., and Cohen, B. 1974. 'Loss of visual suppression of vestibular nystagmus after flocculus lesions.' *Brain Res.*, **72**, 213–24.

Takemori, S., and Suzuki, J. I. 1971. 'Eye deviations from neck torsion in humans.' *Ann. Otol. Rhinol. Laryngol.*, **80**, 439–44.

Talbot, S. A., and Marshall, W. H. 1941. 'Physiological studies in neural mechanism of visual localization and discrimination.' *Amer. J. Ophthal.*, **24**, 1255–64.

Tamler, E., Jampolsky, A., and Marg, E. 1958. 'An electromyographic study of asymmetric convergence.' *Amer. J. Ophthal.*, **46**, 174–82.

Tapley, S. M., and Bryden, M. P. 1977. 'An investigation of sex differences in spatial ability: Mental rotation of three-dimensional objects.' *Canad. J. Psychol.*, **31**, 122–30.

Tarlov, E. 1975. 'Synopsis of current knowledge about ascending projections from the vestibular nuclei.' In R. F. Naunton (Ed.), *The Vestibular System.* New York: Academic Press, pp. 55–70.

Tastevin, J. 1937. 'En partant de l'experience d'Aristotle.' *L'Encephale*, **1**, 57–84, 140–58.

Taub, E., and Goldberg, I. A. 1973. 'Prism adaptation: Control of intermanual transfer by distribution of practice.' *Science*, **180**, 755–6.

Taub, E., and Goldberg, I. A. 1974. 'Use of sensory recombination and somatosensory deafferentation techniques in the investigation of sensory–motor integration.' *Perception*, **3**, 393–408.

Tauber, E. S., and Atkin, A. 1968. 'Optomotor responses to monocular stimulation: relation to visual system organization.' *Science*, **160**, 1365–7.

Täumer, R. 1975. 'Three relative mechanisms of the saccadic system in response to a double jump.' In G. Lennerstrand and P. Bach-y-Rita (Eds.), *Basic Mechanisms of Ocular Motility and Their Clinical Implications.* New York: Pergamon, pp. 515–8.

Taylor, J. G. 1962. *The Behavioural Basis of Perception.* New Haven: Yale University Press.

Taylor, M. M. 1963. 'Visual discrimination and orientation.' *J. opt. Soc. Amer.*, **53**, 763–5.

Taylor, R. L. 1972. 'Reading spatially transformed digits.' *J. exp. Psychol.*, **96**, 396–9.

Teixeira, R. A., and Lackner, J. R. 1979. 'Optokinetic motion sickness: attenuation of visually-induced apparent self-rotation by passive head movements.' *Aviat. space environ. Med.*, **50**, 264–6.

Teller, D. Y., Morse, R., Borton, R., and Regal, D. 1974. 'Visual acuity for vertical and diagonal gratings in human infants.' *Vision Res.*, **14**, 1433–9.
Templeton, W. B. 1972. 'Visual tilt normalization: The method of kinaesthetic matching.' *Percept. Psychophys.*, **12**, 422–4.
Templeton, W. B. 1973. 'The role of gravitational cues in the judgment of visual orientation.' *Percept. Psychophys.*, **14**, 451–7.
Templeton, W. B., and Anstis, T. 1975. 'Egocentric localization of a brief flash presented before, during or after a saccadic eye movement.' Unpublished Report, Psychology Dept., York University, Toronto.
Templeton, W. B., Howard, I. P., and Easting, G. 1965. 'Satiation and the tilt after-effect.' *Amer. J. Psychol.*, **78**, 656–9.
Templeton, W. B., Howard, I. P., and Lowman, A. E. 1966. 'Passively generated adaptation to prismatic distortion.' *Percept. mot. Skills*, **22**, 140–2.
Templeton, W. B., Howard, I. P., Wilkinson, D. A. 1974. 'Additivity of components of prismatic adaptation.' *Percept. Psychophys.*, **15**, 249–57.
ter Braak, J. W. G. 1972. 'Ambivalent optokinetic stimulation and motion detection.' In J. Dichgans and E. Bizzi (Eds.), *Cerebral Control of Eye Movements.* New York: Karger, pp. 308–16.
ter Braak, J. W. G., and Buis, C. 1970. 'Optokinetic nystagmus and attention.' *Internat. J. Neurol.*, **8**, 34–42.
ter Braak, J. W. G., Schenk, V. W. D., and Van Vliet, A. G. M. 1971. 'Visual reactions in a case of long-lasting cortical blindness.' *J. Neurol. Neurosurg. Psychiat.*, **34**, 140–7.
ter Braak, J. W. G., and Van Vliet, A. G. M. 1963. 'Subcortical optokinetic nystagmus in the monkey.' *Psychiat. Neurol. Neurochir. (Amst.)*, **66**, 277–83.
Thomas, H., Jamison, W., and Hummel, D. D. 1973. 'Observation is insufficient for discovering that the surface of still water is invariantly horizontal.' *Science*, **181**, 173–4.
Thomas, J. P., and Gille, J. 1979. 'Bandwidths of orientation channels in human vision.' *J. opt. soc. Amer.*, **69**, 652–60.
Thomas, J. P., and Shimamura, K. K. 1975. 'Inhibitory interaction between visual pathways tuned to different orientations.' *Vision Res.*, **15**, 1373–80.
Thompson, J. G., and Schiffman, H. R. 1974. 'The influence of figure size and orientation on the magnitude of the horizontal-vertical illusion.' *Acta Psychol.*, **38**, 413–20.
Thompson, P. 1980. 'Margaret Thatcher: a new illusion.' *Perception*, **9**, 483–4.
Thouless, R. H. 1947. 'The experience of "upright" and "upside-down" in looking at pictures.' In *Miscellanea Psychologia Albert Michotte.* Louvain: University of Louvain Press, pp. 130–7.
Thurlow, W. R., and Jack, C. E. 1973. 'Certain determinants of the "ventriloquism effect".' *Percept. mot. Skills*, **36**, 1171–84.
Thurlow, W. R., and Rosenthal, T. M. 1976. 'Further study of existence regions for the "ventriloquism effect".' *J. Amer. audiol. Soc.*, **1**, 280–6.
Tibbling, L. 1969. 'The rotatory nystagmus response in children.' *Acta oto-laryng.*, **68**, 459–67.
Timberlake, G. T., Wyman, D., Skavenski, A. A., and Steinman, R. M. 1972. 'The oculomotor error signal in the fovea.' *Vision Res.*, **12**, 1059–64.
Timney, B. N., and Muir, D. W. 1976. 'Orientation anisotropy: Incidence and magnitude in Caucasian and Chinese subjects.' *Science*, **193**, 699–700.
Toates, F. M. 1975. *Control Theory in Biology and Experimental Psychology.* London: Hutchinson.
Tokunaga, O. 1977. 'The influence of linear acceleration on optokinetic nystagmus in human subjects.' *Acta oto-laryng.*, **84**, 338–43.
Tolhurst, D. J. 1972. 'On the possible existence of edge detector neurones in the human visual system.' *Vision Res.*, **12**, 797–804.
Tolhurst, D. J. 1973. 'Separate channels for the analysis of the shape and the movement of a moving visual stimulus.' *J. Physiol.*, **231**, 385–402.
Tolhurst, D. J. 1975a. 'Reaction times in the detection of gratings by human observers: A

probabalistic mechanism.' *Vision Res.*, **15**, 1143–9.

Tolhurst, D. J. 1975b. 'Sustained and transient channels in human vision.' *Vision Res.*, **15**, 1151–5.

Tolhurst, D. J. 1977. 'Colour-coding properties of sustained and transient channels in human vision.' *Nature, New Biol.*, **266**, 266–8.

Tolhurst, D. J., and Barfield, L. P. 1978. 'Interactions between spatial frequency channels.' *Vision Res.*, **18**, 951–8.

Tolhurst, D. J., and Movshon, J. A. 1975. 'Spatial and temporal contrast sensitivity of striate cortical neurones.' *Nature*, **257**, 674–5.

Tolhurst, D. J., Movshon, J. A., and Thompson, I. D. 1981. 'The dependence of response amplitude and variance of cat visual cortical neurones on stimulus contrast.' *Exp. Brain Res.*, **41**, 414–19.

Tolhurst, D. J., and Thompson, P. G., 1975 'Orientation illusions and after-effects: inhibition between channels.' *Vision Res.*, **15**, 967–72.

Tomlinson, R. D., Saunders, G. E., and Schwarz, D. W. F. 1980. 'Analysis of human vestibulo-ocular reflex during active head movements.' *Acta oto-laryng.*, **90**, 184–90.

Tomlinson, R. D., Schwarz, D. W. F., and Fredrickson, J. M. 1978. 'Cerebellar and brainstem responses to eye muscle stretch in the cat.' In J. D. Hood (Ed.), *Vestibular Mechanisms in Health and Disease*, New York: Academic Press, pp. 45–51.

Toyama, K., Kimura, M., Tanaka, K., and Shiida, T. 1976. 'Neuronal connections and receptive field organization of area 19 cells of the cat.' *Exp. Brain Res.*, Suppl. l, 370–3.

Toyama, K., Maekawa, K., and Takeda, T. 1977, 'Convergence of retinal inputs onto cortical cells. I. A study of the cells monosynaptically excited from the lateral geniculate body.' *Brain Res.*, **137**, 207–20.

Travis, R. C., and Dodge, R. 1928. 'Experimental analysis of the sensorimotor consequences of passive oscillation—rotary and rectilinear.' *Psychol. Monogr.*, **38**, 1–96.

Treisman, M. 1977. 'Motion sickness: An evolutionary hypothesis.' *Science*, **197**, 493–5.

Tretter, F., Cynader, M., and Singer, W. 1975. 'Modification of direction selectivity of neurons in the visual cortex of kittens.' *Brain Res.*, **84**, 143–9.

Trincker, D. E. W. 1962. 'The transformation of mechanical stimulus into nervous excitation by the labyrinthine receptors.' In *Biological Receptor Mechanisms, Symp. Soc. exp. Biol.*, No. 16. Cambridge: Cambridge University Press.

Tsumoto, T., Creutzfeldt, O. D., and Legéndy, C. R. 1978. Functional organization of the corticofugal system from visual cortex to lateral geniculate nucleus in the cat (with an appendix on geniculocortical monosynaptic connections). *Exp. Brain Res.*, **32**, 345–64.

Tsumoto, T., Eckhart, W., and Creutzfeldt, O. D. 1979. 'Modification of orientation sensitivity of cat visual cortex neurons by removal of GABA-mediated inhibition.' *Exp. Brain Res.*, **34**, 351–63.

Tulunay Keesey, U. 1960. 'Effects of involuntary eye movements on visual acuity.' *J. opt. Soc. Amer.*, **50**, 769–74.

Tulunay Keesey, U. 1972, 'Flicker and pattern detection: a comparison of threshold.' *J. opt. Soc. Amer.*, **62**, 446–8.

Tumarkin, I. A. 1937. 'Some observations on the function of the labyrinth.' *Proc. roy. Soc. Med. otol. Sec.*, **30**, 599–610.

Turkewitz, G., Birch, H. G., Moreau, T., Levy, L., and Cornwell, A. C. 1966. 'Effect of intensity of auditory stimulation on directional eye movements in the human neonate.' *Animal Behav.*, **14**, 93–101.

Twitchell, T. E. 1970. 'Reflex mechanisms and the development of prehension.' In K. Connolly (Ed.), *Mechanisms of Motor Skill Development*. London: Academic Press, pp. 25–37.

Tyler, C. W. 1975. 'Stereoscopic tilt and size aftereffects.' *Perception*, **4**, 187–92.

Tyler, C. W., and Chang, J. J. 1977. 'Visual echoes: The perception of repetition in quasi-random patterns.' *Vision Res.*, **17**, 109–16.

Tyler, C. W., and Mitchell, D. E. 1977. 'Orientation differences for perception of sinusoidal line stimuli.' *Vision Res.*, **17**. 83–8.

Tyler, C. W., and Scott, A. B. 1979. 'Binocular vision.' In R. E. Records (Ed.), *Physiology of the Human Eye and Visual System.* New York: Harper and Row, pp. 643–74.

Uchino, Y., Suzuki, S., and Watanabe, S. 1980. 'Vertical semicircular canal inputs to cat extraocular motoneurons.' *Exp. Brain Res.*, **41**, 45–53.

Udo de Haes, H. A. 1970. 'Stability of apparent vertical and ocular countertorsion as a function of lateral tilt.' *Percept. Psychophys.*, **8**, 137–42.

Udo de Haes, H. A., and Schöne, H. 1970. 'Interaction between statolith organs and semicircular canals on apparent vertical and nystagmus. Investigations on the effectiveness of the statolith organs.' *Acta oto-laryng.*, **69**, 25–31.

Uemura, T., Arai, Y., and Shimazaki, C. 1980. 'Eye-head coordination during lateral gaze in normal subjects.' *Acta oto-laryng.*, **90**, 191–8.

Uemura, T., and Cohen B. 1973. 'Effects of vestibular nuclei lesions on vestibulo-ocular reflexes and posture in monkeys.' *Acta oto-laryng., Suppl.*, **315**, 1–71.

Uhlarik, J. J. 1973. 'Role of cognitive factors on adaptation to prismatic displacement.' *J. exp. Psychol.*, **98**, 223–32.

Uhlarik, J. J., and Canon, L. K. 1971. 'Influence of concurrent and terminal exposure conditions on the nature of perceptual adaptation.' *J. exp. Psychol.*, **91**, 233–9.

Updyke, B. V. 1974. 'Characteristics of unit responses in superior colliculus of the Cebus monkey.' *J. Neurophysiol.*, **37**, 896.

Vaegan, 1976. 'The position of random autokinetic movement and the physiological position of rest are frequently stable and identical.' *Percept. Psychophys.*, **19**, 240–5.

Van der wildt, G. J., Keemink, C. J., and van den Brink, G. 1976. 'Gradient detection and contrast transfer by the human eye.' *Vision Res.*, **16**, 1047–55.

Van Hof-Van Duin, J. 1976. 'Early and permanent effects of monocular deprivation on pattern discrimination and visuomotor behavior in cats.' *Brain Res.*, **111**, 261–76.

Van Meeteren, A. 1978. 'On the detective quantum efficiency of the human eye.' *Vision Res.*, **18**, 257–67.

Van Nes, F. L., Koenderink, J. J., Nas. H., and Bouman, M. A. 1967. 'Spatiotemporal modulation transfer in the human eye.' *J. opt. Soc. Amer.*, **57**, 1082–8.

Van Sluyters, R. C., and Blakemore, C. 1973. 'Experimental creation of unusual neuronal properties in visual cortex of kitten.' *Nature, New Biol.*, **246**, 506–8.

Vassilev, A., and Penchev, A. 1976. 'Spatial and temporal summation in the perception of lines.' *Vision Res.*, **16**, 1329–35.

Vidyasagar, T. R. 1979. 'Differential susceptibility of optic nerve components to a neurotoxin.' *Neurosci. Letters*, Suppl. 3, S296.

Virsu, V., and Taskinen, H. 1975. 'Central inhibition interactions in human beings.' *Exp. Brain Res.*, **23**, 65–74.

Vital-Durand, F., and Jeannerod, M. 1974a. 'Role of visual experience in the development of optokinetic response in kittens.' *Exp. Brain Res.*, **20**, 297–302.

Vital-Durand, F., and Jeannerod, M. 1974b. 'Maturation of the optokinetic responses: genetic and environmental factors.' *Brain Res.*, **71**, 249–57.

Viviani, P., and Berthoz, A. 1977. 'Voluntary deceleration and perceptual activity during oblique saccades.' In R. Baker and A. Berthoz (Eds.), *Control of Gaze by Brain Stem Neurons.* New York: Elsvier, pp. 23–8.

Viviani, P., Berthoz, A., and Tracey, D. 1977. 'The curvature of oblique saccades.' *Vision Res.*, **17**, 661–4.

Volkmann, F. C., and Pufall, P. B. 1972. 'Adjustment of visual tilt as a function of age.' *Percept. Psychophys.*, **11**, 187–92.

Volkmann, F. C., Riggs, L. A., White, K. D., and Moore, R. K. 1978. 'Contrast sensitivity during saccadic eye movements.' *Vision Res.*, **18**, 1193–9.

von Grünau, M. W. 1978. 'Interaction between sustained and transient channels: Form inhibits motion in the human visual system.' *Vision Res.*, **18**, 197–201.

Wade, N. J. 1968. 'Visual orientation during and after lateral head, body, and trunk tilt.' *Percept. psychophys.*, **3**, 215–9.

Wade, N. J. 1969a. 'Visual orientation as a function of head tilt.' *Percept. mot. Skills*, **29**, 573–4.

Wade, N. J. 1969b. 'The effect of nonocular and binocular observation on visual orientation during head tilt.' *Amer. J. Psychol.*, **82**, 384–8.

Wade, N. J. 1969c. 'The effect of stimulus line variations on visual orientation with head upright and tilted.' *Austral. J. Psychol.*, **21**, 177–85.

Wade, N. J. 1970a. 'Effect of instructions on visual orientation.' *J. exp. Psychol.*, **83**, 331–2.

Wade, N. J. 1970b. 'The effect of different psychophysical methods on visual orientation during tilt.' *Psychon. Sci.*, **19**, 201–3.

Wade, N. J. 1970c. 'Visual orientation during lateral head tilt when upright and supine.' *Quart. J. exp. Psychol.*, **22**, 420–2.

Wade, N. J. 1970d. 'Effect of prolonged tilt on visual orientation.' *Quart. J. exp. Psychol.*, **22**, 423–39.

Wade, N. J. 1972. 'Effect of forward head inclination on visual orientation during lateral body tilt.' *J. exp. Psychol.*, **96**, 203–5.

Wade, N. J. 1973a. 'Orientation and spatial frequency effects on linear afterimages.' *Percept. Psychophys.*, **13**, 446–50.

Wade, N. J. 1973b. 'The effect of water immersion on perception of the visual vertical.' *Brit. J. Psychol.*, **64**, 351–61.

Wade, N. J. 1980. 'The influence of colour and contour rivalry on the magnitude of the tilt illusion.' *Vision Res.*, **20**, 229–33.

Wade, N. J., and Day, R. H. 1968a. 'Development and dissipation of a visual spatial aftereffect from prolonged head tilt.' *J. exp. Psychol.*, **76**, 439–43.

Wade, N. J., and Day, R. H. 1968b. 'Apparent head position as a basis for a visual aftereffect of prolonged head tilt.' *Percept. Psychophys.*, **3**, 324–6.

Wade, N. J., and Day, R. H. 1973. 'A correction to "The retinal reference of the tilt aftereffect" by Coltheart and Cooper.' *Percept. Psychophys.*, **13**, 319–20.

Wade, N. J., and Schöne, H. 1971. 'The influence of force magnitude on the perception of body position: I. Effect of head posture.' *Brit. J. Psychol.*, **62**, 157–63.

Waespe, W., and Henn, V. 1977a. 'Neuronal activity in the vestibular nuclei of the alert monkey during vestibular and optokinetic stimulation.' *Exp. Brain Res.*, **27**, 523–38.

Waespe, W., and Henn, V. 1977b. 'Vestibular nuclei activity during optokinetic after-nystagmus (OKAN) in the alert monkey.' *Exp. Brain Res.*, **30**, 323–30.

Waespe, W., and Henn, V. 1978a. 'Conflicting visual–vestibular stimulation and vestibular nucleus activity in alert monkeys.' *Exp. Brain Res.*, **33**, 203–12.

Waespe, W., and Henn, V. 1978b. 'Reciprocal changes in primary and secondary optokinetic after-nystagmus (OKAN) produced by repetitive optokinetic stimulation in the monkey.' *Arch. Psychiat. Nervenkr.*, **225**, 23–30.

Waespe, W., and Henn, V. 1979. 'The velocity response of vestibular nucleus neurones during vestibular, visual, and combined angular acceleration.' *Exp. Brain Res.*, **37**, 337–47.

Waespe, W., Henn, V., and Miles, T. S. 1977. 'Activity in the vestibular nuclei of the alert monkey during spontaneous eye movements and vestibular or optokinetic stimulation.' In R. Baker and A. Berthoz (Eds.), *Control of Gaze by Brain Stem Neurons*. New York: Elsevier, pp. 269–78.

Waespe, W., Huber, T., and Henn, V. 1978. 'Dynamic changes of optokinetic after-nystagmus (OKAN) caused by brief visual fixation periods in monkey and in man.' *Arch. Psychiat. Nervenkr.*, **226**, 1–19.

Walberg, F. 1975. 'The vestibular nuclei and their connections with the eighth nerve and the cerebellum.' In R. F. Naunton (Ed.), *The Vestibular System*. New York: Academic Press, pp. 31–54.

Walk, R. D., and Bond, E. K. 1971. 'The development of visually guided reaching in monkeys reared without sight of the hands.' *Psychonom. Sci.*, **23**, 115–6.

Wallace, B. 1974. 'Target and arm observation effects on adaptation to prism displacement.' *Percept. Psychophys.*, **15**, 145–8.

Wallace, B. 1977. 'Stability of Wilkinson's linear model of prism adaptation over time for various targets.' *Perception*, **6**, 145–51.

Wallace, B., and Garrett, J. B. 1973. 'Reduced felt arm sensation effects on visual adaptation.' *Percept. Psychophys.*, **14**, 597–600.

Wallace, B., and Redding, G. M. 1979. 'Additivity in prism adaptation as manifested in intermanual and interocular transfer.' *Percept. Psychophys.*, **25**, 133–6.

Wallace, G. K. 1969, 'The critical distance of interaction in the Zöllner illusion.' *Percept. Psychophys.*, **5**, 261–4.

Wallace, G. K., and Crampin, D. J. 1969. 'The effect of background density on the Zöllner illusion.' *Vision Res.*, **9**, 167–77.

Wallace, G. K., and Moulden, B. 1973. 'The effect of body tilt on the Zöllner illusion.' *Quart. J. exp. Psychol.*, **25**, 10–21.

Wallace, M., Singer, G., Mottram. J., and Purcell, T. 1975. 'Effects of exposure to a tilted room on kinesthetic postexposure responses.' *Perception*, **4**, 447–51.

Wallach, H., and Bacon, J. 1976. 'The constancy of the orientation of the visual field.' *Percept. Psychophys.*, **19**, 492–8.

Wallach, H., and Frey, K. J. 1969. 'Adaptation in the constancy of visual direction measured by a one-trial method.' *Percept. Psychophys.*, **5**, 249–56.

Wallach, H., and Frey, K. J. 1972. 'Adaptation in distance perception based on oculomotor cues.' *Percept. Psychophys.*, **11**, 77–83.

Wallach, H., Frey, K. J., and Romney, G. 1969. 'Adaptation to field displacement during head movement unrelated to the constancy of visual direction.' *Percept. Psychophys.*, **5**, 253–6.

Wallach, H., and Lewis, C. 1966. 'The effect of abnormal displacement of the retinal image during eye movements.' *Percept. Psychophys.*, **1**, 25–9.

Walls, G. L. 1951a. 'A theory of ocular dominance.' *A. M. A. Arch. Ophthal.*, **45**, 387–412.

Walls, G. L. 1951b. 'The problem of visual direction. II. The history to 1900.' *Amer. J. Optom.*, **28**, 55–83.

Walls, G. L. 1962. 'The evolutionary history of eye movements.' *Vision Res.*, **2**, 69–80.

Walsh, E. G. 1961. 'Role of the vestibular apparatus in the perception of motion on a parallel swing.' *J. Physiol.*, **155**, 506–13.

Walsh, E. G. 1962 'The perception of rhythmically repeated linear motion in the horizontal plane.' *Brit. J. Psychol.*, **53**, 439–45.

Walsh, E. G. 1964. 'The perception of rhythmically repeated linear motion in the vertical plane.' *Quart. J. exp. Physiol.*, **49**, 58–65.

Ward, F. 1976. 'Pursuit eye movements and visual localization.' In R. A. Monty and J. W. Senders (Eds.), *Eye Movements and Psychological Processes.* New York: Wiley, pp. 289–98.

Ward, R., and Morgan, M. J. 1978. 'Perceptual effect of pursuit eye movements in the absence of a target.' *Nature, New Biol.*, **274**, 158–9.

Ware, C., and Mitchell, D. E. 1974a. 'The spatial selectivity of the tilt aftereffect.' *Vision Res.*, **14**, 735–7.

Ware, C., and Mitchell, D. E. 1974b. 'On interocular transfer of various aftereffects in normal and stereoblind observers.' *Vision Res.*, **14**, 731–4.

Warren, D. H. 1979. 'Spatial localization under conflict conditions: is there a single explanation?' *Perception*, **8**, 323–37.

Warren, D. H. 1980. 'Response factors in intermodality localization under conflict conditions.' *Percept. Psychophys.*, **27**, 28–32.

Warren, D. H., and Pick, H. L. 1970. 'Intermodality relations in localization in blind and sighted people.' *Percept. Psychophys.*, **8**, 430–2.

Warren, D. H., and Platt, B. B. 1974. 'The subject: a neglected factor in recombination research.' *Perception*, **3**, 421–38.
Warren, D. H., and Schmitt, T. L. 1978. 'On the plasticity of visual–proprioceptive bias effects.' *J. exp. Psychol: Hum. Percept. Perform.*, **4**, 302–10.
Warren, R. 1976. 'The perception of egomotion.' *J. exp. Psychol: Hum. Percept. Perform.*, **2**, 448–56.
Weale, R. A. 1978. 'Experiments on the Zöllner and related optical illusions.' *Vision Res.*, **18**, 203–8.
Weber, R. B., and Daroff, R. B. 1972. 'Corrective movements following refixation saccades: Type and control system analysis.' *Vision Res.*, **12**, 467–75.
Webster, R. G. 1969. 'The relationship between cognitive, motor–kinesthetic, and oculomotor adaptation.' *Percept. Psychophys.*, **6**, 33–8.
Weene, P., and Held, R. 1966. 'Changes in perceived size of angle as a function of orientation in the frontal plane.' *J. exp. Psychol.*, **71**, 55–9.
Weintraub, D. J., and Krantz, D. H. 1971. 'The Poggendorff illusion: Amputations, rotations, and other perturbations.' *Percept. Psychophys.*, **10**, 257–64.
Weintraub, D. J., and Virsu, V. 1972. 'Estimating the vertex of converging lines: Angle misperception?' *Percept. Psychophys.*, **11**, 277–83.
Weisfeld, G. E. 1972. 'Parametric adjustment to a shifting target alternating with saccades to a stationary reference point.' *Psychonom. Sci.*, **28**, 72–4.
Weiskrantz, L. 1963. 'Contour discrimination in a young monkey with striate cortex ablation.' *Neuropsychologia*, **1**, 145–64.
Weiskrantz, L., Warrington, E. K., Sanders, M. D., and Marshall, J. 1974. 'Visual capacity in hemianopic field following a restricted occipital ablation.' *Brain*, **97**, 709–28.
Weiss, A. D., and Tole, J. R, 1973. 'Effect of galvanic vestibular stimulation on rotation testing.' *Adv. Oto-Rhino-Laryng.*, **19**, 311–7.
Weisstein, N., Harris, C. S., Berman, K., Tangney, J., and Williams, A. 1977. 'Contrast reduction by small localized stimuli: Extensive spatial spread of above-threshold orientation-selective masking.' *Vision Res.*, **17**, 341–50.
Weitzman, D. O., Smith, J. M., and Karasik, R. 1973. 'Signal detection analysis of meridional variation to vertical and horizontal gratings.' *Vision Res.*, **13**, 2177–81.
Welch, R. B. 1972. 'The effect of experienced limb identity upon adaptation to simulated displacement of the visual field.' *Percept. Psychophys.*, **12**, 453–6.
Welch, R. B. 1974. 'Speculations on a model of prism adaptation.' *Perception*, **3**, 451–60.
Welch, R. B. 1978. *Perceptual Modification*. New York: Academic Press.
Welch, R. B., and Abel, M. R. 1970. 'The generality of the "target-pointing effect" in prism adaptation.' *Psychonom. Sci.* **20**, 226–7.
Welch, R. B., and Warren, D. H. 1980. 'Immediate perceptual response to intersensory discrepancy.' *Psychol. Bull.*, **88**, 638–67.
Welch, R. B., Widawski, M. H., Harrington, J., and Warren, D. H. 1979. 'An examination of the relationship between visual capture and prism adaptation.' *Percept. Psychophys.*, **25**, 126–32.
Welford, A. T. 1968. *Fundamentals of Skill*. London: Methuen.
Weller, R. E., Kaas, J. H., and Wetzel, A. B. 1979. 'Evidence for the loss of X-cells after long-term ablation of visual cortex in monkeys.' *Brain Res.*, **160**, 134–8.
Wells, W. C. 1792. *An Essay upon single vision with two eyes; together with experiments and observations on several other subjects in optics*. London: T. Cadell.
Wenderoth, P. 1974. 'The distinction between the rod-and-frame illusion and the rod-and-frame test.' *Perception*, **3**, 205–12.
Wenderoth, P., and Beh, H. 1977. 'Component analysis of orientation illusions.' *Perception*, **6**, 57–75.
Wenderoth, P., Beh, H., and White, D. 1978. 'Perceptual distortion of an oblique line in the presence of an abutting vertical line.' *Vision Res.*, **18**, 923–30.

Wenderoth, P., Parkinson, A., and White, D. 1979. 'A comparison of visual tilt illusions measured by the techniques of vertical setting, parallel matching and dot alignment.' *Perception*, **8**, 47–57.

Wenderoth, P., and Tyler, T. 1979. 'The role of apparent motion cues in orientation masking.' *Percept. Psychophys.*, **25**, 413–8.

Wersäll, J. 1956. 'Studies on the structure and innervation of the sensory opithelium of the critae ampullaris in the guinea pig.' *Acta oto-laryng, Suppl.*, 126, 1–85.

Wersäll, J., and Bagger-Sjöbäck, D. 1974. 'Morphology of the vestibular sense organs.' In H. H. Kornhuber (Ed.), *Handbook of Sensory Physiology*, Vol. VI/1. New York: Springer, pp. 123–70.

Wertheimer, M. 1912. 'Experimentelle Studien über das Sehen von Bewegung.' *z. Psychol. Physiol. Sinnesorg.*, **61**, 161–265.

Wertheimer, M. 1961. 'Psychomotor coordination of auditory and visual space at birth.' *Science*, **134**, 1692.

Westheimer, G. 1954a. 'Mechanism of saccadic eye movements.' *A.M.A. Arch. Ophthal.*, **52**, 710–24.

Westheimer, G. 1954b. 'Eye movement responses to a horizontally moving visual stimulus.' *A.M.A. Arch. Ophthal.*, **52**, 932–41.

Westheimer, G. 1960. 'Modulation thresholds for sinusoidal light distributions on the retina.' *J. Physiol.*, **152**, 67–74.

Westheimer, G. 1965. 'Spatial interaction in the human retina during scotopic vision.' *J. Physiol.*, **181**, 881–94.

Westheimer, G. 1967. 'Spatial interaction in human cone vision.' *J. Physiol.*, **190**. 139–54.

Westheimer, G. 1972. 'Visual acuity and spatial modulation thresholds.' In D. Jameson and L. M. Hurvish (Eds.), *Handbook of Sensory Physiology, Vol. VII/4*. New York: Springer, pp. 170–87.

Westheimer, G. 1976. 'Diffraction theory and visual hyperacuity.' *Amer. J. Optom. physiol Optics*, **53**, 262–364.

Westheimer, G. 1978. 'Spatial phase sensitivity for sinusoidal grating targets.' *Vision Res.*, **18**, 1073–4.

Westheimer, G., and Blair, S. M. 1974. 'Functional organization of primate oculomotor system revealed by cerebellectomy.' *Exp. Brain Res.*, **21**, 463–72.

Westheimer, G., and Campbell, F. W. 1962. 'Light distribution in the image formed by the living human eye.' *J. opt. Soc. Amer.*, **52**, 1040–5.

Westheimer, G., and Conover, D. W. 1954. 'Smooth eye movements in the absence of a moving visual stimulus.' *J. exp. Psychol.*, **47**, 283–4.

Westheimer, G., and McKee, S. P. 1973. 'Failure of Donder's Law during smooth pursuit eye movements.' *Vision Res.*, **13**, 2145–53.

Westheimer, G., and McKee, S. P. 1975. 'Visual acuity in the presence of retinal-image motion.' *J. opt. Soc. Amer.*, **65**, 847–50.

Westheimer, G., and McKee, S. P. 1977a. 'Integration regions for visual hyperacuity.' *Vision Res.*, **17**, 89–93.

Westheimer, G., and McKee, S. P. 1977b. 'Spatial configurations for visual hyperacuity.' *Vision Res.*, **17**, 941–7.

Westheimer, G., Shimamura, K., and McKee, S. P. 1976. 'Interference with line-orientation sensitivity.' *J. opt. Soc. Amer.*, **66**, 332–8.

Westheimer, G., and Wiley, R. W. 1970. 'Distance effects in human scotopic retinal interaction.' *J. Physiol.*, **206**, 129–43.

Wetherick, N. E. 1977. 'The significance of the nose for certain phenomena of visual perception.' *Nature, New Biol.*, **266**, 442–3.

Weymouth, F. W., Anderson, E. E., and Averill, H. L. 1923. 'Retinal mean local sign: a new view of the relation of the retinal mosaic to visual perception.' *Amer. J. Physiol.*, **63**, 410–1.

Wheeless, L. L., Boynton, R. M., and Cohen, G. H. 1966. 'Eye-movement responses to step and pulse-step stimuli.' *J. opt. Soc. Amer.*, **56**, 956–60.

Whipple, W. R., and Wallach, H. 1978. 'Direction-Specific motion thresholds for abnormal image shifts during saccadic eye movement.' *Percept. Psychophys.*, **24**, 349–55.

White, B. L. 1970. 'Experience and the development of motor mechanisms in infancy.' In K. Connolly (Ed.), *Mechanisms of Motor Skill Development.* London: Academic Press, pp. 95–113.

White, B. L., Castle, P., and Held, R. 1964. 'Observations on the development of visually-directed reaching.' *Child Devel.*, **35**, 349–63.

White, C. T., Eason, R. G., and Bartlett, N. R. 1962. 'Latency and duration of eye movements in the horizontal plane.' *J. opt. Soc. Amer.*, **52**, 210–3.

White, K. D., Post, R. B., and Leibowitz, H. W. 1980. 'Saccadic eye movements and body sway.' *Science*, **208**, 621–3.

White, K. G. 1975. 'Orientation effects on contour interaction in the Zöllner illusion.' *Percept. Psychophys.*, **17**, 387–92.

Whiting, H. T. A. 1969. *Acquiring Ball Skill.* London: Bell.

Wickelgren, B. G. 1971. 'Superior colliculus: Some receptive field properties of bimodally responsive cells.' *Science*, **173**, 69–72.

Wickelgren, B. G., and Sterling, P. 1969. 'Influence of visual cortex on receptive fields in the superior colliculus of the cat.' *J. Neurophysiol.*, **32**, 16–22.

Wiener, K., and Kagan, J. 1976. 'Infants' reaction to changes in orientation of figure and frame.' *Perception.* **5**, 25–8.

Wiesel, T. N. 1960. 'Receptive fields of ganglion cells in the cat's retina.' *J. Physiol.*, **153**, 583–94.

Wiesel, T. N., and Hubel, D. H. 1965 'Extent of recovery from the effects of visual deprivation in kittens.' *J. Neurophysiol.*, **28**, 1060–72.

Wiesel, T. N., and Hubel, D. H. 1974. 'Ordered arrangement of orientation columns in monkeys lacking visual experience.' *J. comp. Neurol.*, **158**, 307–18.

Wildt, G. J., van der, Keemink, C. J., and van den Brink, G. 1976. 'Gradient detection and contrast transfer by the human eye.' *Vision Res.*, **16**, 1047–54.

Wilkinson, D. A. 1971. 'The visual–motor control loop: a linear system?' *J. exp. Psychol.*, **89**, 250–7.

Wilkinson, I. M. S., Kime, R., and Purnell, M. 1974. 'Alcohol and human eye movement.' *Brain*, **97**, 785–92.

Willey, R., Gyr, J. W., and Henry, A. 1978. 'Changes in the perception of spatial location: A test of potentiation vs. recalibration theory.' *Percept. Psychophys.*, **24**, 356–60.

Williams, D., and Gassel, M. M. 1962. 'Visual functions in patients with homonymous hemianopia. Part I: The visual fields.' *Brain,* **85**, 175–250.

Williams, R. A., and Fender, D. H. 1979. 'Velocity precision in smooth pursuit eye movements.' *Vision Res.*, **19**, 343–8.

Willott, J. F. 1973. 'Perceptual judgments with discrepant information from audition and proprioception.' *Percept. Psychophys.*, **14**, 577–80.

Wilson, H. R. 1978a. 'Quantitative prediction of line spread function measurements: Implications for channel bandwidths.' *Vision Res.*, **18**, 493–6.

Wilson, H. R. 1978b. 'Quantitative characterization of two types of line spread function near the fovea.' *Vision Res.*, **18**, 971–81.

Wilson, H. R., and Bergen, J. R. 1979. 'A four mechanism model for threshold spatial vision.' *Vision Res.*, **19**, 19–32.

Wilson, H. R., and Giese, S. C. 1977. 'Threshold visibility of frequency gradient patterns.' *Vision Res.*, **17**, 1177–90.

Wilson, H. R., Phillips, G., Rentschler, I., and Hilz, R. 1979. 'Spatial probability summation and disinhibition in psychophysically measured line-spread functions.' *Vision Res.*, **19**, 593–8.

Wilson, M. E., and Toyne, M. J. 1970. 'Retino-tectal and cortico-tectal projections in Macaca Mulatta.' *Brain Res.*, **24**, 395–406.

Wilson, P. D., and Stone, J. 1975. 'Evidence of W-cell input to the cat's visual cortex via the C laminae of the lateral geniculate nucleus.' *Brain Res.*, **92**, 472–8.

Wilson, V. J., Gacek, R. R., Maeda, M., and Uchino, Y. 1977. 'Saccular and utricular input to cat neck motoneurons.' *J. Neurophysiol.*, **40**, 63–73.

Wilson, V. J., Gacek, R. R., Uchino, Y., and Susswein, A. J. 1978. 'Properties of central vestibular neurons fired by stimulation of the saccular nerve.' *Brain Res.*, **143**, 251–61.

Wilson, V. J., and Melvill Jones, G. 1979. *Mammalian Vestibular Physiology*. New York: Plenum.

Wilson, V. J., Peterson, B. W., Fukushima, K., Hirai, N., and Uchino, Y. 1979. 'Analysis of vestibulocollic reflexes by sinusoidal polarization of vestibular afferent fibers.' *J. Neurophysiol.*, **42**, 331–46.

Winston, P. H. (Ed.) 1975. *The Psychology of Computer Vision*. New York: McGraw-Hill.

Winterson, B. J., and Collewijn, H. 1976. 'Microsaccades during finely guided visuomotor tasks.' *Vision Res.*, **16**, 1387–90.

Winterson, B. J., and Steinman, R. M. 1978. 'The effect of luminance on human smooth pursuit of perifoveal and foveal targets.' *Vision Res.*, **18**, 1165–72.

Wishart, J. G., Bower, T. G. R., and Dunkeld, J. 1978. 'Reaching in the dark.' *Perception*, **7**, 507–12.

Wist, E. R., Diener, H. C., Dichgans, J., and Brandt, T. 1975. 'Perceived distance and the perceived speed of self-motion: Linear vs. angular velocity?' *Percept. Psychophys.*, **17**, 549–54.

Witkin, H. A., and Asch, S. E. 1948. 'Studies in space orientation: IV. Further experiments on perception of the upright with displaced visual fields.' *J. exp. Psychol.*, **38**, 762–82.

Witkin, H. A., Oltman, P. K., Cox, P. W., Ehrichman, E., Hamm, R. M., and Ringler, R. W. 1973. 'Field-dependence and psychological differentiation: A bibliography through 1972 with index.' Princeton, N. J.: Educational Testing Service Res. Bull., pp. 73–62.

Wolf, J. W. 1969. 'Mesodiencephalic and cerebellar influcences on optokinetic and vestibular nystagmus.' *Exp. Neurol.*, **25**, 24–34.

Wolfe, J. W., and Cramer, R. L. 1970. 'Illusions of pitch induced by centrifugal acceleration.' *Aerosp. Med.*, **41**, 1136–9.

Wolff,P. 1971. 'Mirror-image confusability in adults.' *J. exp. Psychol.*, **91**, 268–72.

Wong, S. C. P., and Frost, B. J. 1978. 'Subjective motion and acceleration induced by the movement of the observer's entire visual field.' *Percept. Psychophys.*, **24**, 115–20.

Wong-Riley, M. 1979. 'Columnar cortico-cortical interconnections within the visual system of the squirrel and Macaque monkeys.' *Brain Res.*, **162**, 201–17.

Wood, C. C., Spear, P. D., and Braun, J. J. 1973. 'Direction-specific deficits in horizontal optokinetic nystagmus following removal of visual cortex in the cat.' *Brain Res.*, **60**, 231–7.

Wood, R. W. 1895. 'The "haunted-swing" illusion.' *Psychol. Rev.*, **2**, 277–8.

Woodworth, R. S. 1934. *Psychology*. New York: Holt.

Worchel, S., and Burnham, C. A. 1967. 'Reduction of autokinesis with information about the registration of eye position.' *Amer. J. Psychol.*, **80**, 434–7.

Wright, M. J., and Ikeda, H. 1974. 'Processing of spatial and temporal information in the visual system.' In F. O. Schmitt and F. G. Warden (Eds.), *The Neurosciences: Third Study Program*. Cambridge, Mass.: MIT Press, pp. 115–22.

Wurtz, R. H. 1969a. 'Response of striate cortex neurons to stimuli during rapid eye movements in the monkey.' *J. Neurophysiol.*, **32**, 975–86.

Wurtz, R. H. 1969b. 'Comparison of effects of eye movements and stimulus movements on striate cortex neurons of the monkey.' *J. Neurophysiol.*, **32**, 987–94.

Wurtz, R. H. 1976. 'Extraretinal influences on the primate visual system.' In R. A. Monty and J. W. Senders (Eds.), *Eye Movements and Psychological Processes*. New York: Wiley, pp. 231–44.

Wurtz, R. H., and Goldberg, M. E. 1971. 'Superior colliculus cell responses related to eye movements in awake monkeys.' *Science*, **171**, 82–4.

Wurtz, R. H., and Goldberg, M. E. 1972a. 'Activity of superior colliculus in behaving monkey. III. Cells discharging before eye movements.' *J. Neurophysiol.*, **35**, 578–86.

Wurtz, R. H., and Goldberg, M. E. 1972b. 'Activity of superior colliculus in behaving monkey. IV. Effects of lesions on eye movements.' *J. Neurophysiol.*, **35**, 587–96.

Wurtz, R. H., and Mohler, C. W. 1974. 'Selection of visual targets for the initiation of saccadic eye movements.' *Brain Res.*, **71**, 209–14.

Wurtz,R. H., and Mohler, C. W. 1976. 'Organization of monkey superior colliculus: Enhanced visual response of superficial layer cells.' *J. Neurophysiol.*, **39**, 745–65.

Wurtz, R. H., Richmond, B. J., and Judge, S. J. 1980. 'Vision during saccadic eye movements. III. Visual interactions in monkey superior colliculus.' *J. Neurophysiol.*, **43**, 1168–81.

Wyatt, H. J., and Daw, N. W. 1975. 'Directionally sensitive ganglion cells in the rabbit retina: specificity for stimulus direction, size and speed.' *J. Neurophysiol.*, **38**, 613–26.

Wyatt, H. J., and Pola, J. 1979. 'The role of perceived motion in smooth pursuit eye movements.' *Vision Res.*, **19**, 613–8.

Wyman, D., and Steinman, R. M. 1973a. 'Small step tracking: Implications for the oculomotor "dead zone".' *Vision Res.*, **13**, 2165–72.

Wyman, D., and Steinman, R. M. 1973b. 'Latency characteristics of small saccades.' *Vision Res.*, **13**, 2173–5.

Wyndham, R. 1936. *The Gentle Savage*. New York: Morrow.

Yarbus, A. L. 1967. *Eye Movements and Vision* (trans. L. A. Riggs). New York: Plenum.

Yasui, S., and Young, L. R. 1975. 'Perceived visual motion as effective stimulus to pursuit eye movement system.' *Science*, **190**, 906–8.

Yin, R. K. 1969. 'Looking at upside-down faces.' *J. exp. Psychol.*, **81**, 141–5.

Yin, T. C. T., and Mountcastle, V. B. 1977. 'Visual input to the visuomotor mechanism of the monkey's parietal lobe.' *Science,* **197**, 1381–3.

Yinon, U, 1977. 'Inverted vision surgically induced in experienced cats: Physiology of the primary cortex.' *Exp. Brain Res.*, **28**, 141–51.

Yonas, A., Bechtold, A. G., Frankel, D., Gordon, F. R., McRoberts, G., Norcia, A., and Sternfels, S. 1977. 'Development of sensitivity to information for impending collision.' *Percept. Psychophys.*, **21**, 97–104.

Young, L. R. 1977. 'Pursuit eye movement—what is being pursued?' In R. Baker and A. Berthoz (Eds.), *Control of Gaze by Brain Stem Neurons*. New York: Elsevier, pp. 29–36.

Young, L. R., Dichgans, J., Murphy, R., and Brandt, T. 1973. 'Interaction of optokinetic and vestibular stimuli in motion perception.' *Acta oto-laryng.*, **76**, 24–31.

Young, L. R., and Henn, V. S. 1974. 'Selective habituation of vestibular nystagmus by visual stimulation.' *Acta oto-laryng.*, **77**, 159–66.

Young, L. R., and Meiry, J. L. 1968. 'A revised dynamic otolith model.' *Aersop. Med.*, **39**, 606–8.

Young, L. R., and Oman, C. M. 1969. 'Model for vestibular adaptation to horizontal rotation.' *Aerosp. Med.*, **40**, 1076–80.

Young, L. R., and Oman, C. M. 1970. 'Modeling adaptation in human semicircular canal response to rotation.' *Trans. New York Acad. Sci.*, **32**, 489–94.

Young, L. R., Oman, C. M., and Dichgans, J. M. 1975. 'Influence of head orientation on visually induced pitch and roll sensation.' *Aviat. space environ. Med.*, **46**, 264–9.

Young, L. R., and Sheena, D. 1975. 'Survey of eye movement recording methods.' *Behav. res. Meth. Instrument.*, **7**, 397–429.

Young, L. R., and Stark, L. 1963. 'Variable feedback experiments testing a sampled data model for eye tracking movements.' *IEEE Trans. Hum. Fac. Electron.*, **4**, 38–50.

Zee, D. S., Optican, L. M., Cook, J. D., Robinson, D. A., and Engel, W. K. 1976. 'Slow saccades in spinocerebeller degeneration.' *Arch. Neurol.*, **33**, 243–51.

Zee, D. S., Yee, R. D., and Robinson, D. A. 1976. 'Optokinetic responses in labyrinthine-defective human beings.' *Brain Res.*, **113**, 423–8.
Zeki, S. M. 1974a. 'Cells responding to changing image size and disparity in the cortex of the rhesus monkey.' *J. Physiol.*, **242**, 827–41.
Zeki, S. M. 1974b. 'Functional organization of a visual area in the posterior bank of the superior sulcus of the rhesus monkey.' *J. Physiol.*, **236**, 549–73.
Zeki, S. M. 1978. 'Uniformity and diversity of structure and function in rhesus monkey prestriate visual cortex.' *J. Physiol.*, **277**, 273–90.
Zihl, J., and von Cramon, D. 1979. 'Collicular function in human vision.' *Exp. Brain Res.*, **35**, 419–24.
Zuber, B. L., and Stark, L. 1965. 'Microsaccades and the velocity–amplitude relationships for saccadic eye movements.' *Science*, **150**, 1459–60.

Author Index

Subject Index

Numbers in **bold** type refer to pages where terms are defined.

ff following a page number indicates that the discussion continues on subsequent pages.